Springer Proceedings in Physics

Volume 250

Indexed by Scopus

The series Springer Proceedings in Physics, founded in 1984, is devoted to timely reports of state-of-the-art developments in physics and related sciences. Typically based on material presented at conferences, workshops and similar scientific meetings, volumes published in this series will constitute a comprehensive up-to-date source of reference on a field or subfield of relevance in contemporary physics. Proposals must include the following:

- name, place and date of the scientific meeting
- a link to the committees (local organization, international advisors etc.)
- scientific description of the meeting
- list of invited/plenary speakers
- an estimate of the planned proceedings book parameters (number of pages/articles, requested number of bulk copies, submission deadline).

More information about this series at http://www.springer.com/series/361

Domenico Elia · Giuseppe E. Bruno ·
Pietro Colangelo · Leonardo Cosmai
Editors

The XVIII International Conference on Strangeness in Quark Matter (SQM 2019)

Springer

Editors
Domenico Elia
Dipartimento Interateneo di Fisica
Istituto Nazionale di Fisica
Nucleare (INFN)—Sezione di Bari
Bari, Italy

Giuseppe E. Bruno
Dipartimento Interateneo di Fisica
Istituto Nazionale di Fisica
Nucleare (INFN) and Politecnico di Bari
Bari, Italy

Pietro Colangelo
Dipartimento Interateneo di Fisica
Istituto Nazionale di Fisica
Nucleare (INFN)—Sezione di Bari
Bari, Italy

Leonardo Cosmai
Dipartimento Interateneo di Fisica
Istituto Nazionale di Fisica
Nucleare (INFN)—Sezione di Bari
Bari, Italy

ISSN 0930-8989 ISSN 1867-4941 (electronic)
Springer Proceedings in Physics
ISBN 978-3-030-53447-9 ISBN 978-3-030-53448-6 (eBook)
https://doi.org/10.1007/978-3-030-53448-6

This Springer imprint is published by the registered company Springer Nature Switzerland AG
The registered company address is: Gewerbestrasse 11, 6330 Cham, Switzerland

More than 270 participants attended the SQM 2019 conference at Villa Romanazzi, Bari (Image credit: Domenico Elia)

Members of the SQM 2019 Organizing Committee at the conference social evening (Image credit: Domenico Elia)

Organizing Institutions

INFN (Istituto Nazionale di Fisica Nucleare)
Physics Department, University and Politecnico, Bari, Italy

Organizing Committee

International Advisory Committee

Joerg Aichelin, SUBATECH Nantes, France
Yasuyuki Akiba, RIKEN-RBRC, Japan
Federico Antinori, INFN Padova, Italy and CERN Geneva, Switzerland
Francesco Becattini, University and INFN Firenze, Italy
Marcus Bleicher, FIAS Frankfurt, Germany
Peter Braun-Munzinger, EMMI/GSI and University of Heidelberg, Germany
Helen Caines, Yale University, USA
Jean Cleymans, University of Cape Town, South Africa
Andrea Dainese, INFN Padova, Italy
Laura Fabbietti, Technical University Munich, Germany
Marek Gaździcki, Kielce University, Poland and University of Frankfurt, Germany
Paolo Giubellino, INFN Torino, Italy and GSI/FAIR Darmstadt, Germany
Boris Hippolyte, University of Strasbourg, France
Alexander Kalweit, CERN Geneva, Switzerland
Maria Paola Lombardo, INFN Firenze, Italy
Christina Markert, University of Texas at Austin, USA
Andre Mischke†, Utrecht University, Netherlands
Badangadas Mohanty, NISER, India
Berndt Müller, BNL and Duke University, USA
Grazyna Odyniec, LBNL, USA
Johann Rafelski, University of Arizona, USA

Alexander Sorin, JINR Dubna, Russia
George Stephans, MIT Cambridge, USA
Horst Stöcker, FIAS/Goethe University Frankfurt and GSI Darmstadt, Germany
Joachim Stroth, University of Frankfurt, Germany
Orlando Villalobos Baillie, University of Birmingham, UK
Fuqiang Wang, Purdue University, USA
Nu Xu, CCNU, China and LBNL, USA
In-Kwon Yoo, Pusan National University, South Korea
Pengfei Zhuang, Tsinghua University, China

Conference Chair

Domenico Elia, INFN Bari, Italy

Organizing Committee

Giuseppe Eugenio Bruno, Politecnico and INFN Bari
Fabio Colamaria, INFN Bari
Pietro Colangelo, INFN Bari
Domenico Colella, INFN Bari
Leonardo Cosmai, INFN Bari
Mauro de Palma, University and INFN Bari
Domenico Di Bari, University and INFN Bari
Domenico Elia, INFN Bari, chair
Rosa Anna Fini, INFN Bari
Vito Manzari, INFN Bari
Annalisa Mastroserio, University Foggia and INFN Bari
Marianna Mazzilli, University and INFN Bari
Eugenio Nappi, INFN Bari
Alessandra Pastore, INFN Bari
Alexis Pompili, University and INFN Bari
Giacomo Volpe, University and INFN Bari

Secretariat and Support

Anna Catalano, University Bari
Crescenza D'Alba, INFN Bari
Giacinto Donvito, INFN Bari
Antonio Franco, INFN Bari

Luigia Lamarca, University Bari
Loredana Napolitano, University Bari
Stefano Nicotri, INFN Bari
Cosimo Pastore, INFN Bari
Giuseppina Salente, INFN Padova
Antonio Silvestri, INFN Bari
Gioacchino Vino, INFN Bari

Preface

The XVIII International Conference on Strangeness in Quark Matter (SQM 2019) was held from 10 to 15 June 2019 in Bari, Italy. Hosted by the INFN (the Italian National Institute for Nuclear and Particle Physics), in collaboration with the Physics Department of the Bari "Aldo Moro" University and Polytechnic University, the conference attracted more than 270 participants from 32 countries, including a large number of graduate students and young scientists. The SQM series focuses on new experimental and theoretical developments on the role of strange and heavy-flavour quarks in high-energy heavy-ion collisions and in astrophysical phenomena. The main scientific topics addressed at SQM 2019 were the following: strangeness and heavy-quark production in nuclear collisions and hadronic interactions, hadron resonances in the strongly coupled partonic and hadronic medium, bulk matter phenomena associated with strange and heavy quarks, QCD phase structure, collectivity in small systems, strangeness in astrophysics, open questions and new developments.

The scientific programme consisted of 50 invited plenary talks, 76 contributed parallel talks and a quite rich poster session with more than 60 contributions. A state-of-the-art session opened the conference, with a tribute to the late Roy Glauber entitled "The Glauber model in high-energy nucleus-nucleus collisions" by Reinhard Stock (Goethe University Frankfurt), followed by two overview talks to set the scene on the theory and experiment sides, respectively. The first 2-day plenary sessions were dedicated to highlights from theory and experiments: they included reports on results from low- and high-energy collisions, as well as on hyperon interaction in lattice QCD and thermal model. Representatives from all major collaborations at CERN's LHC and SPS, Brookhaven's RHIC, the Heavy Ion Synchrotron SIS at the GSI Darmstadt and the NICA project at the JINR Dubna made special efforts to release new results at SQM 2019: thanks to the excellent performance of these accelerator facilities, a wealth of new data on the production of strangeness and heavy-flavour quarks in nuclear collisions have become available. The conference was organized in further plenary sessions dedicated to the main scientific topics, two half-day parallel session afternoons, a poster session evening and a final session on "Future experiments and physics perspectives" on the last morning before the "Summary and closing".

Among the highlights presented at the conference, identified particle yield measurements were shown to be progressing towards determining where phenomena such as strangeness enhancement are localised in phase space. This is currently addressed using complementary methods such as transverse multiplicity estimators and two-particle correlations: such new approaches promise to constrain theoretical models in unprecedented ways. Collective behaviour in small systems was also a highlighted and much discussed topic, with new results from PHENIX showing that p-Au, d-Au and 3He-Au exhibit elliptic flow coefficients consistent with expectations regarding their initial collision geometry. This remarkable finding indicates that even in these systems the observed v2 coefficient is not due to jet-like phenomena but is rather a true manifestation of initial conditions translating into momentum anisotropies, a hallmark feature of the classic flow paradigm. Further results from ALICE, CMS and STAR complete the picture and consistently corroborate the presence of elliptic flow in small systems. An increasing interest in transverse-momentum differential baryon-to-meson ratios in the heavy-flavour sector was also evident. Recent results from pp and Pb-Pb collisions from both ALICE and CMS suggest that the same dynamics observed in the ratio Λ/K^0_S may be present in the Λc/D despite the fact that strange and charm quarks are thought to be created in different stages of system evolution. Further studies and potential future measurements may still be needed to fully conclude on the similarities in these baryon-to-meson ratios.

A promising new perspective at the LHC is to use high-energy pp and p-Pb collisions as factories of identified hadrons created by a source of finite radius and then measure the ensuing interactions between these hadrons using femtoscopy. This technique has allowed the ALICE Collaboration to study interactions that were so far not measured at all and probe, for instance, the p-Ξ and p-Ω interaction potentials. These results provide fundamental constraints to the QCD community and are significant in the context of the astrophysics: such potentials indeed serve as input when modelling neutron stars under the hypothesis that hyperons are present in this extreme state of matter. New results on the onset of deconfinement were shown by the NA61/SHINE Collaboration, in particular, with the measurement of a charged kaon-to-pion ratio in Ar-Sc intermediate compared to results in pp/Be-Be and Pb-Pb, possibly suggestive of a change in the production mechanism that can involve a phase transition. Recent results on strangeness production at low energy from HADES and BM@N were also very welcome and enriched the discussion at SQM 2019.

Presentations at the final session on Saturday morning showed good prospects in the field for future measurements with FAIR at GSI Darmstadt, NICA at JINR Dubna, Heavy Ions at J-PARC Tokai and at CERN with the currently ongoing upgrades, opportunities for HL-LHC and next-generation experiments (HL/HE-LHC and FCC). Perspectives for QCD measurements at future Electron Ion Collider facilities were also presented. On the theory side, new developments and strong research efforts are currently taking place towards a better understanding of strangeness production and open heavy-flavour dynamics in heavy-ion collisions. Small system scan with more information about heavy flavour, crosstalk on the Equation of state (EOS) between lattice QCD, Heavy ion collisions (HIC), neutron stars, and global polarization in HIC are also topics of currently large interest.

Two young scientist prizes, sponsored by the NuPECC, were awarded to the best experimental and theory posters, respectively, Bong-Hwi Lim (Pusan National University, Korea) and Olga Soloveva (Goethe University Frankfurt, Germany). A special award dedicated to the memory of a friend and colleague has been established for this conference series: the inaugural "Andre Mischke Award" for the young scientist with the best experimental parallel talk at SQM 2019 was assigned to Erin Frances Gauger (University of Texas at Austin, United States).

The scientific and organizational success of SQM 2019 was the result of the work of many people. We would like to thank the colleagues of the International Advisory Committee for their valuable help and guidance in shaping the scientific programme of the conference. We are also warmly grateful to the members of the Local Organizing Committee, the team of secretaries, the technical and administrative support staff and the student helpers: with them we've been sharing a fantastic human and professional adventure over a year or so, in a common effort to make sure that all aspects of the meeting were handled smoothly and efficiently. Last but not least, we express our gratitude to the Italian and international organizations that provided financial contributions to the conference, allowing us to support the participation of more than 50 young scientists to SQM 2019.

More information about the conference, including all oral and poster presentations, is available on the conference website at https://sqm2019.ba.infn.it/. The next International Conference on Strangeness in Quark Matter (SQM 2021) will take place in Busan, Korea, in May 2021.

Bari, Italy

Domenico Elia
Giuseppe Eugenio Bruno
Pietro Colangelo
Leonardo Cosmai
Editors

Contents

Part I
Opening

Chapter 1
ROY GLAUBER: In Memoriam the Glauber Model in High Energy Nucleus-Nucleus Collisions

Reinhard Stock

Abstract This article is devoted to the memory of Roy Glauber. His multi-diffractive model of nucleus-nucleus collisions has served a wide variety of applications in our field. After a sketch of the model we illustrate a number of physics observables where, in particular, the construction of the initialization phase of a A+A collision rests on Monte Carlo simulations of the participant-spectator geometry of the incident configuration. This results, chiefly, in the determination of the nucleon participant , and the binary collision number, as well as in a specification of the spatial expansion modes of the source that lead to the observables of directed flow. Finally, one can attempt to pin down the colour glass initial state, in comparison to Glauber initialization.

At the opening of this Conference let us commemorate Roy Glauber who passed away last December. From among the many themes of his theoretical life-work it is the "Glauber-Model" of multiple hadronic scattering phenomena which has become basic to the understanding of nucleus-nucleus collisions at high energy. From Bevalac time until today we employ the model to define observables of increasing complexity: initially it served, in particular, to quantify the concept of "central" collisions and centrality classes but it then provided for the geometrical basis of the family of flow-observables from which we have concluded on the bulk properties of the QGP liquid. And, finally, it helps to quantify the basics of analysis of individual events, thus opening up the wide field of observables related to event-by-event fluctuations of the initial geometry. Thus, Roy Glaubers work has become a "household article" in essentially all bulk QCD matter questions where we talk about Glauber-initialization vs. colour glass condensate initialization, to give just one example. A few of these aspects will be recollected below. Let us begin, however, with a short look at Roy Glaubers trajectory, in life and science.

Roy Glauber was born Sept. 1, 1925 in New York. He attended the famous Bronx High School of Science, as a kind of *Wunderkind* with exceptional science talent,

R. Stock (✉)
FIAS and Institut für Kernphysik, Johann Wolfgang Goethe-Universität,
Max-von-Laue-Str. 1, 60438 Frankfurt am Main, Germany
e-mail: stock@ikf.uni-frankfurt.de

D. Elia et al. (eds.), *The XVIII International Conference on Strangeness in Quark Matter (SQM 2019)*, Springer Proceedings in Physics 250,
https://doi.org/10.1007/978-3-030-53448-6_1

much reminding of Richard Feynmans famous performance in his "teenage years". At age 16 he entered undergraduate study at Harvard, and in 1943 he was drafted into the Manhattan Project, the youngest scientist there, involved in critcal mass computations. Returning to Harvard he got his Ph.D. in 1949 with Julian Schwinger (Nobel laureate for the formulation of QED, later on). A period of *Wanderjahre* then sees him at Princeton, Zuerich (with W.Pauli) and Caltech (where he substituted for Feynman for a year). In 1976 Harvard again, he became Full Professor there, from now on his permanent academic home but interspersed with many foreign visiting positions including CERN. His principle research topic was Quantum Electrodynamics in the interaction of light with matter, addressed to Quantum Optics. He also first tackled the problems of multiple hadronic collisions at high energy, where he developed the foundations of the so-called Optical Model for diffractive forward scattering. The application to high energy proton-deuteron scattering then resulted in the formulation of the Glauber Model to which we shall turn below because it became extraordinarily useful later on, in the environment of nucleus-nucleus collision at relativistic energy where the concept of participant and spectator nucleons, seen along a straight trajectory in an *optical eikonal* approximation, proved highly useful. Figure 1.1 shows a photo of Roy Glauber as we all remember him, taken at

Fig. 1.1 Roy Glauber at the Nobel Prize ceremony 2005

the occasion of his Nobel Prize ceremony in 2005. What a remarkable picture! A deep, and passionate academic and research life trajectory culminates, at age 80, in this radiant and encouraging image of one of the Noblemen of science. We sadly commemorate, today, to his passing away on December 26, 2018. His ideas live onward in our research.

1.1 The Glauber Model

This is not the place for a detailed review of the Glauber Model of diffractive multi-hadronic scattering [1] but let us briefly sketch the main points of the optical *eikonal* approximation, the basis of the model. At very high energy the nucleon scattering from a nucleus will stay essentially undeflected because of the far excess of longitudinal over transverse momentum. Thus one might approximate its trajectory by a light ray, the nucleon summing up all the successive phase shifts received at the scattering centers inside the target. What emerges is a diffractive shadow image in the transverse plane: the target is *X-rayed*. Thus the name eikonal approximation, from greek eikona = image. This model neglects the fact that the participating nucleons are, in reality, quantum mechanical objects, but it allows to implement all the geometric aspects of the multiple scattering process, i.e. impact parameter of the projectile, and participant density distributions. We show in Fig. 1.2 the result by Franco and Glauber [2] for proton-deuteron scattering at 1 and 2 GeV incident energy. It illustrates the concept of participant and spectator nucleons, invented in this context. The steep forward peak reflects single nucleon-nucleon scattering which is by far the most frequent sub-process owing to the very dilute nucleon density distribution of the deuteron.

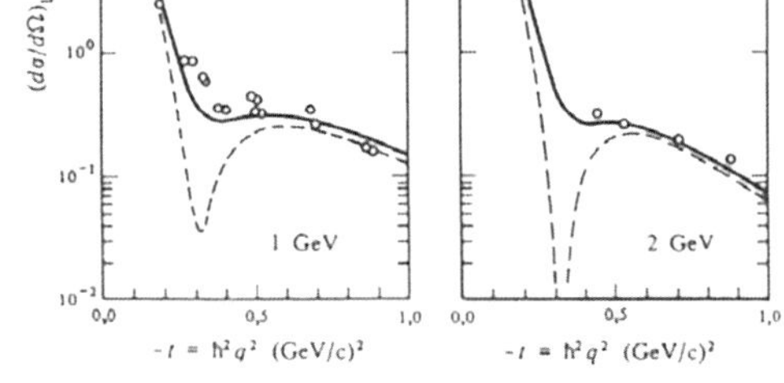

Fig. 1.2 Elastic proton-deuteron scattering with Glauber Model fits. From [2]

1.2 Participant, Spectator and Collision Number

The assumption of independent straight line trajectories of the constituent nucleons of target and projectile nuclei in a nucleus-nucleus collision makes it possible to simply count the number of participating nucleons from target and projectile and, moreover, the number of binary collisions occurring at the microscopic level, during interpenetration. In a Monte Carlo realization one has to dial an instantaneous, *frozen* position map of the nucleons of target and projectile, based e.g. on a Woods-Saxon density distribution. The nucleon trajectories are given finite transverse extension according to the know (or assumed) nucleon-nucleon cross section. The entire geometrical calculation takes place at a chosen impact parameter (see [3] for a comprehensive review). We illustrate this process in Fig. 1.3 where a snapshot of a Au+Au collision is shown. At an impact parameter 6 fm the instantaneous nucleon density distributions are seen approaching each other along the beam axis [3]. One can now read off the distributions of to-be participants (deep colours), and count the number of binary encounters of each participant, thus the total binary collision number. At first sight this picture, and procedure, looks utterly unrealistic. If the microscopic nature of the overall collision process is seen as independent and *successive* N-N scatterings there is hardly a justification to consider a participant nucleon to be still intact exerting its initial cross section, after, say, 6 successive encounters. The misconception is in the word successive. At the ultra-relativistic energies considered here the Lorentz factor exceeds thousand, and thus all aspects of both longitudinal and transverse motion of extended objects, such as nucleon form factors, must occur essentially simultaneously. There are no sequential instances of time resolvable. This point would clearly deserve more discussion! However, the most elementary aspects of the picture, the number of to-be involved or not involved nucleons, as a function of target/projectile nuclear size, and of impact parameter, remain well defined. And, please note, that this is all we need in the analysis of $A + B$ collisions, as the following examples will illustrate. We do not perform a full Glauber model analysis of the collisions

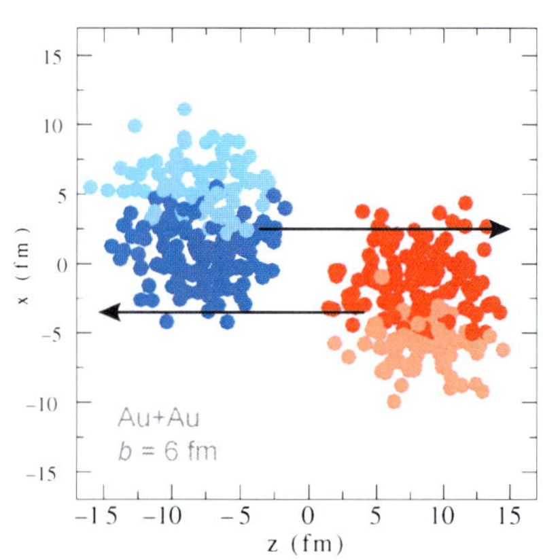

Fig. 1.3 Initial state of a simulated Au+Au collision at impact parameter 6fm, specifying participants (dark colour) and spectators (light colour). From [3]

but need to pin down all geometrical influences, including, more recently, also the event-by-event fluctuations in the geometrical positions of the impinging nucleons. Now to a bit of history.

1.3 Bevalac Physics: Carved—Out Fireballs, Rows on Rows

In the mid-70ties the Berkeley LBL Bevatron provided beams of ^{20}Ne at energies up to 2.1 GeV per projectile nucleon. The Fireball Model [4] was addressed to the proton p_t distributions of minimum bias Ne+U collisions. It combined the Glauber-type geometrical abrasion model [5] of Swiatecki and collaborators, employed to fix the interaction volume and the effective participant center of mass velocity, with a Hagedorn-inspired statistical thermal emission model. Note that this approach came to stay until today, where we initialize the A+B collision with a Glauber calculation and then interpolate the resulting momentum space distribution by the energy-momentum tensor from which a hydrodynamic evolution originates. Figure 1.4 shows a sketch of the abrasion ("clean cut") model applied to the Ne+U collision at a peripheral impact parameter [4].

The assumption of thermal equilibrium met with serious critique (not for the first time: Hagedorns entire life was accompanied by remarks like Feynmans about the "nonsense to smash delicate swiss watches against the wall and study the debris"). This motivated Huefner and Knoll [6] for their *rows on rows* model, a strict and complete Glauber model calculation of Ne on U collisions that we illustrate in Fig. 1.4. In fact they showed that the equilibrium assumption was a good approximation in collisions of heavy nuclei, as a consequence of multiple scattering at the microscopic level.

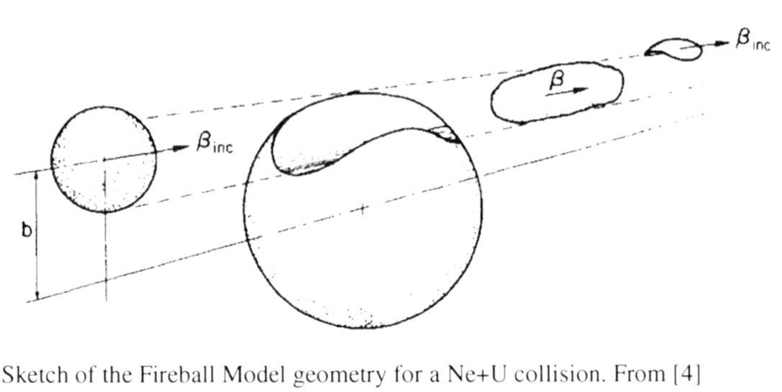

Fig. 1.4 Sketch of the Fireball Model geometry for a Ne+U collision. From [4]

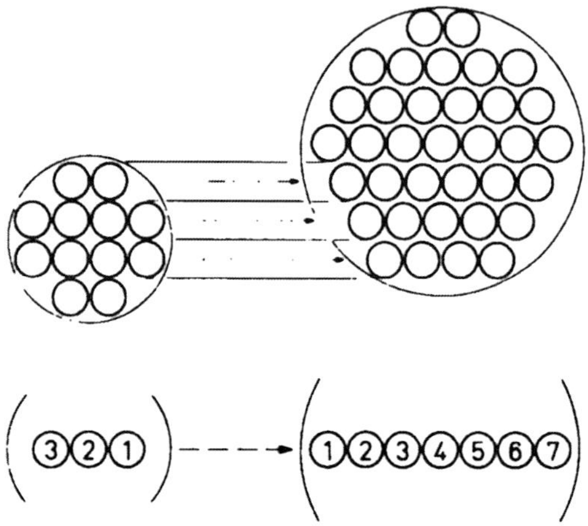

Fig. 1.5 Rows of nucleons along straight line Glauber trajectories for the combination of a light projectile and heavy target. From [6]

1.4 Asymmetric Collision Systems

Figure 1.5 illustrates a notorious difficulty in dealing with asymmetric collision systems, such as the Bevalac Ne+U collision. The effective center of mass varies drastically with impact parameter! In Fig. 1.5 a typical central Glauber-row of nucleons is shown, a 3 on 7 collision. Toward surface reactions one finally approaches a 1 on 1 case! The shifting effective center of mass is beautifully illustrated in Fig. 1.6 by the d+Au data from PHOBOS [7], the only RHIC experiment with a wide rapidity acceptance. One sees a systematic shift of the charged particle rapidity distribution with centrality (impact parameter), from symmetry about mid-rapidity for surface reactions (the 1 on 1 case), to a downward peak shift of about 2.5 units for the most central selection. This has important consequences [8], most often overlooked, for the analysis of p+A and d+A data from experiments with a narrow rapidity acceptance such as STAR, PHENIX and ALICE. Note that the center of mass for hard collisions (jet production) will always stay fixed at mid-rapidity: parton-parton scattering. Thus in central collisions the bulk soft production medium—oftentimes called the co-traveling plasma—moves with considerable longitudinal velocity opposed to the leading parton. Moreover, if the acceptance is placed symmetric to the N-N center of mass rapidity, i.e. $y = 0$ in collider experiments, the effective center of mass rapidity of the soft bulk production falls far outside the acceptance. This explains the high priority placed from Bevalac time onward on equal mass target and projectile collisions.

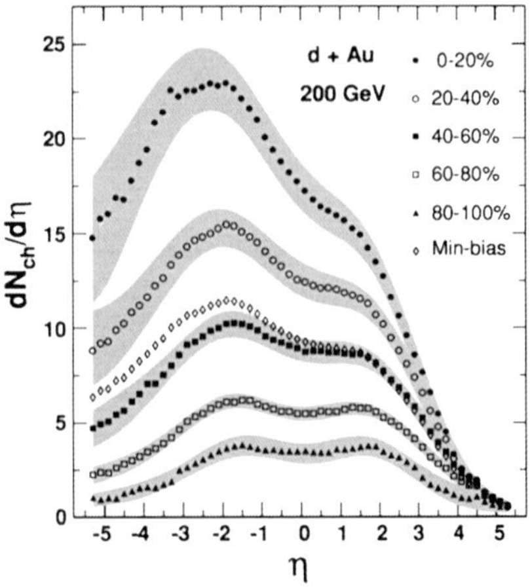

Fig. 1.6 The shift with centrality of the charged particle pseudorapidity distribution in d+Au collisions at 200 GeV, measured by PHOBOS at RHIC [7]

1.5 The Glauber - Relation Between $\langle N_{charge} \rangle$, $\langle N_{part} \rangle$ and Impact Parameter $\langle b \rangle$

From a minimum bias experiment with sufficient event statistics one obtains the distribution of the charged particle multiplicity cross section, or of some related quantity. This is sketched in Fig. 1.7 which illustrates the further steps in the Glauber-type analysis [3]. Integrating the cross section downward from maximum multiplicity one defines the successive multiplicity classes 0–5%, 5%–10% etc. A Glauber calculation then establishes the connection between impact parameter b and average corresponding participant number $\langle N_{part} \rangle$. Note that the latter varies from event to event due to nucleon position fluctuations in the impinging nuclei, so even a sharp b yields a broad Npart distribution, with mean $\langle N_{part} \rangle$. The converse is equally true. Finally, one integrates the resulting $\langle N_{part} \rangle$ distribution downward from its maximum into corresponding percentile classes, and associates this with the N_{ch} percentile classes. A typical final statement arises: in the 5%–10% class the average N_{ch} is about 1200, $\langle N_{part} \rangle$ is about 300 and the mean corresponding impact parameter is about 4 fm. This then sets the stage for a representation of other experimental results, such as e.g. strangeness per participant pair versus centrality. Furthermore, it provides the geometrical input for all model calculations, devoted to soft and hard production. The Nuclear Modification Factor R_{AA} makes further use of this calculation because a mean number of binary collisions is also associated with each of the percentile bins in Fig. 1.7. One can then consider R_{AA}, the ratio of a certain hard production cross section (photons, high p_t hadrons) observed in an $A + A$ collision, to the corresponding elementary proton-proton minimum bias collision cross section as multiplied by the appropriate number of collisions. The case $R_{AA} = 1$ stands for *nothing but trivial collision number scaling*. Figure 1.8 illustrates the modification factors for central Au+Au collisions at top RHIC energy obtained by PHENIX [9]. The photons stay close to unity, they do not significantly interact with the medium,

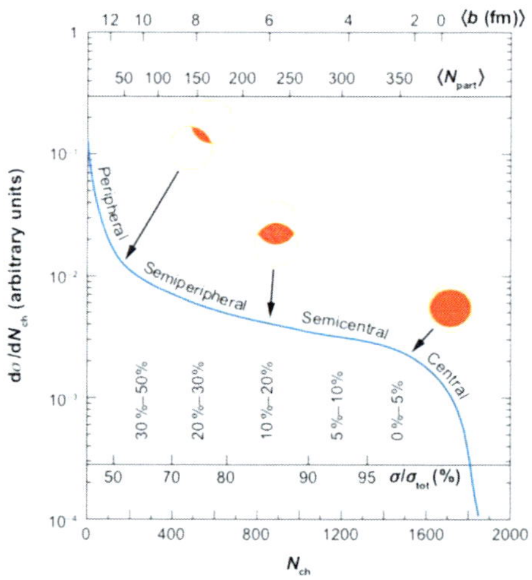

Fig. 1.7 Definition of centrality windows in the charged particle multiplicity distribution in Au+Au collisions, exhibiting the correspondence of impact parameter, participant nucleon number and charged particle multiplicity. From [3]

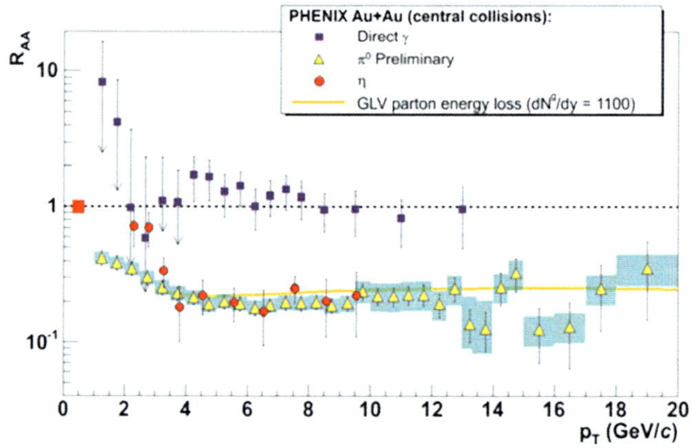

Fig. 1.8 The nuclear modification factor dependence on p_t for pions, η and direct photons, measured by PHENIX [9], compared with the parton energy loss model [10]

whereas the pions and η suffer a drastic suppression, they lose momentum in the "cotraveling" QCD plasma. Diagnosis of its properties thus becomes possible, as illustrated here by comparison to the GLV parton energy loss model [10]. Other models feature the QCD transport coefficient $\hat{q} = \langle q_t^2 \rangle / \lambda$ as the fit variable.

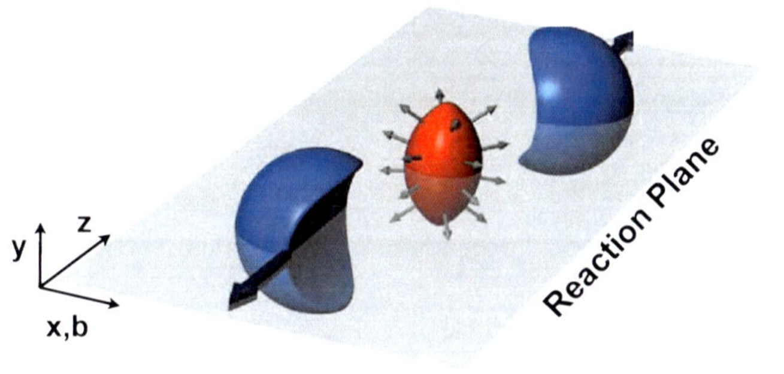

Fig. 1.9 Sketch of the initialization of elliptic flow, exhibiting the emerging primordial participant fireball

1.6 Elliptic Flow Initialization

Further toward new observables that are related to participant-spectator geometrical construction of an initial state in an A+A collision, let us turn to the famous image shown in Fig. 1.9. It shows the geometrical constellation that leads to the initialization of elliptic flow. The non-sperical *fireball* interaction volume of the participants, carved out in a semi-peripheral collision, is depicted here after about a few fm/c of elapsed time, as we can infer from the receding projectile remnants (note that Lorentz contraction is ignored here). That time interval was sufficient to describe the collision volume in a hydrodynamic view, exhibiting the pressure gradients governing the onset of its expansion. Clearly, the gradients are higher along the (x, z) reaction plane than perpendicular to it. This will create an unisotropic development of expansion flow, as attested by the eventually observed azimuthal emission distribution of all hadronization products. The analysis requires an event-by-event localization of the reaction plane, a technique first developed in the Bevalac studies of directed sideward flow, by the Plastic Ball Collaboration [11]. Its resolution rests on the high multiplicity of emitted hadrons, at unprecedented hight at top RHIC and LHC energies. This is, perhaps, the best-studied observable of the field, both experimentally and theoretically. The key question: to what extent does the hydrodynamic expansion evolution preserve the primordial eccentricity? This analysis allows for a determination (or estimate) of the dissipative deviations from ideal fluid evolution, "blurring" the initial image of the fireball source. Viscous damping would be the main influence, as controlled by the specific shear viscosity *eta*/s, and, in fact, we now have stringent estimates of this parameter which reflects another bulk property of the QCD plasma, in addition to the QCD transport coefficient addressed in the previous paragraph. Looking again at Fig. 1.9, this must be the most striking, and successful application of Glaubers eikonal approximation. The "shutter speed" involved here must be of unit fm/c dimension! Not often appreciated, the possibility to draw such a high time resolution diagram is owed to the extreme energies reached at RHIC and LHC where the interpenetration time of the interacting nuclei is shortened, by the high Lorentz factor, to fractions of a fm/c. Thus the primordial reaction volume is created almost instantaneously, giving rise to a well-defined initialization period, as implied in the above consideration. This period is, therefore, also accessible in principle to theoretical models, e.g. Glauber initialization vs. colour glass initial state: a remaining topic of research. The instantaneous positions of the nucleons inside the average Woods-Saxon density distribution vary from event to event, as we mentioned when looking at the relation between impact parameter and participant number. That has invited a study of the higher moments of the initial excentricity, up to n=6. The simultaneous hydrodynamic fit offers a far higher sensitivity to the specific shear viscosity. Thus we conclude that, with the help of the Glauber model, relativistic heavy ion collision studies have, in fact, succeeded to clarify the properties of the QCD plasma.

References

1. R.J. Glauber, High Energy Coll. Theory, Lectures in Theor. Phys. Interscience NY (1959)
2. V. Franco, R.J. Glauber, Effect of quadrupole deformation on high-energy scattering by deuterons. Phys. Rev. Lett. **22**, 370 (1969). https://doi.org/10.1103/PhysRevLett.22.370
3. M.L. Miller, K. Reygers, S.J. Sanders, P. Steinberg, Glauber modeling in high energy nuclear collisions. Ann. Rev. Nucl. Part. Sci. **57**, 205 (2007). https://doi.org/10.1146/annurev.nucl.57.090506.123020
4. G.D. Westfall, J. Gosset, P.J. Johansen, A.M. Poskanzer, W.G. Meyer, H.H. Gutbrod, A. Sandoval, R. Stock, Nuclear Fireball Model for Proton Inclusive Spectra from Relativistic Heavy Ion Collisions. Phys. Rev. Lett. **37**, 1202 (1976). https://doi.org/10.1103/PhysRevLett.37.1202
5. Swiatecki W J, et al.,Abrasion And Ablation Of Heavy Ions, unpublished, LBL-2908, see appendix of Gosset, J,et al., 1977, Phys. Rev. C16 629
6. J. Hufner, J. Knoll, Rows on Rows: A Theory for Collisions Between Heavy Ions at High-Energy. Nucl. Phys. A **290**, 460 (1977). https://doi.org/10.1016/0375-9474(77)90448-1
7. B.B. Back, [PHOBOS Collaboration], et al., Scaling of charged particle production in d + Au collisions at s(NN)**(1/2) = 200-GeV. Phys. Rev. C **72**, 031901 (2005). https://doi.org/10.1103/PhysRevC.72.031901
8. Steinberg, P, Inclusive pseudorapidity distributions in p(d) + A collisions modeled with shifted rapidity distributions, [nucl-ex/0703002 [NUCL-EX]]
9. Y. Akiba, [PHENIX Collaboration], Probing the properties of dense partonic matter at RHIC. Nucl. Phys. A **774**, 403 (2006). https://doi.org/10.1016/j.nuclphysa.2006.06.060
10. I. Vitev, M. Gyulassy, High p_T tomography of d + Au and Au+Au at SPS, RHIC, and LHC. Phys. Rev. Lett. **89**, 252301 (2002). https://doi.org/10.1103/PhysRevLett.89.252301
11. H.A. Gustafsson et al., Collective Flow Observed in Relativistic Nuclear Collisions. Phys. Rev. Lett. **52**, 1590 (1985). https://doi.org/10.1103/PhysRevLett.52.1590

Part II
Highlights from Theory and Experiments

Chapter 2
Recent Results from HADES

Manuel Lorenz

Abstract HADES measured Au+Au collisions at $\sqrt{s_{NN}} = 2.4$ GeV, which probe very similar temperature and density regions during their time evolution as neutron star merger. Measurements of light nuclei, which contribute to the bulk of the created matter at this energy, as well as rare and penetrating probes like dilepton radiation and strangeness production are presented. In addition, with the start of the FAIR Phase-0 program, new high-statistic data of Ag+Ag collisions at $\sqrt{s_{NN}} = 2.55$ GeV have been collected in March 2019.

2.1 Merging Neutron Stars in the Laboratory: Au+Au Collisions at $\sqrt{s_{NN}} = 2.4$ GeV

The discovery of gravitational wave signals from a neutron star merger event by the Laser Interferometer Gravitational Wave Observatory (LIGO) [1], marks the start of the gravitational-wave and true multi-messenger astrophysics era in which the signal transmitted by compact stellar objects are expected to deliver tight constraints on the equation of state of matter [2]. The next step is to explore the microscopic composition of neutron star mergers.

Heavy-ion collisions in the few GeV energy regime offer the unique opportunity to study the microscopic substance of neutron star mergers in the laboratory as both probe very similar temperature and density regions during their time evolution. This is visualized in Fig. 2.1, where the simulated density evolution in coordinate space of two merging neutron stars [3] (upper part) is confronted to the conditions realized in Au+Au collisions at $\sqrt{s_{NN}} = 2.4$ GeV (lower part), modelled with the microscopic transport code UrQMD [4]. The temperature in a neutron star merger is expected

---

Manuel Lorenz for the HADES Collaboration.

M. Lorenz (✉)
Institut für Kernphysik, Goethe-Universität, 60438 Frankfurt, Germany
e-mail: Lorenz@Physik.uni-frankfurt.de

D. Elia et al. (eds.), *The XVIII International Conference on Strangeness in Quark Matter (SQM 2019)*, Springer Proceedings in Physics 250,
https://doi.org/10.1007/978-3-030-53448-6_2

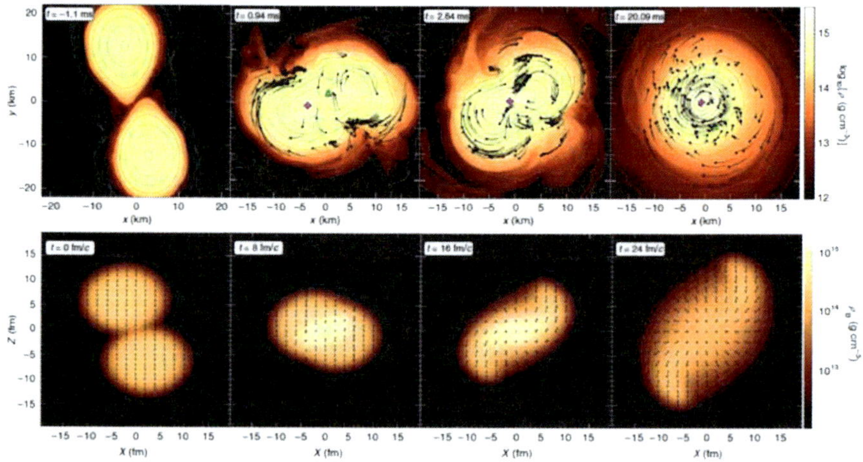

Fig. 2.1 Density evolution in coordinate space of two merging neutron stars [3] (upper part) compared to those of a Au+Au collisions at $\sqrt{s_{NN}} = 2.4$ GeV (lower part), modelled with the microscopic transport code UrQMD [4]. More details can be found in the text and in [6]

to reach 50–80 MeV [3, 5], while the one in Au+Au collisions is determined to be around 70 MeV, based on a measurement of dilepton radiation by HADES [6].

2.2 HADES

HADES, displayed in Fig. 2.2 in an expanded view, is a fixed-target setup served by the SIS18 accelerator at GSI, Darmstadt. It comprises a 6-coil toroidal magnet centered around the beam axis and six identical detection sections located between the coils, covering almost the full azimuthal angle. Each sector is equipped with a Ring-Imaging Cherenkov (RICH) detector followed by low-mass Mini-Drift Chambers (MDCs), two in front of and two behind the magnetic field, as well as a scintillator hodoscope (TOF) and a resistive plate chamber (RPC) at the end of the system. The RICH detector is used mainly for $e^{\pm}$ identification, the MDCs are the main tracking detectors, while the TOF and RPC are used for time-of-flight measurements in combination with a diamond start detector located in front of a 15-folded segmented gold target. The setup is completed by an electromagnetic calorimeter and a forward hodoscope used for event plane and centrality determination. A detailed description of the HADES detector is given in [7].

With the RICH as dedicated detector, HADES is predestinated for $e^{\pm}$, but the time resolution of the RPC and TOF detectors allow also for clear separation of different charged hadrons. In addition, the precision of the track reconstruction allows for the reconstruction of secondary vertices due to weak decay topologies. Both weak decay

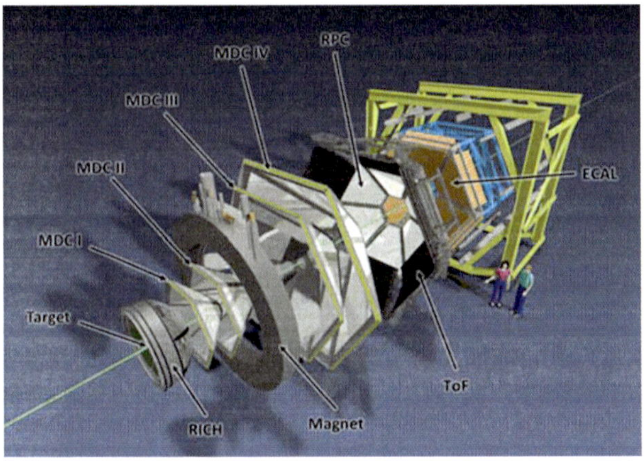

Fig. 2.2 Expanded view of the HADES sub-detectors. Not shown is the forward hodoscope, which is located 7 m down-stream from the target

topology reconstruction and $e^{\pm}$ detection are enforced by the use of specially trained artificial neural networks [8, 9]. The HADES heavy-ion program is supplemented by investigations on more elementary collisions systems using pion and proton induced reactions on nuclei [10, 11].

2.3 Bulk Properties and Light Nuclei

In Au+Au collisions at $\sqrt{s_{NN}} = 2.4$ GeV a baryon dominated rather long-living system is created, which is already visible from the correlations in the β versus momentum $\times$ polarity plane, which is used for particle identification, displayed in Fig. 2.3. The hadrons observed in central events show a clear hierarchy in yields, with about 100 free protons, 10 π, 10^{-2} K^+ and 10^{-4} K^- per event. Around 50 additional protons are bound in light nuclei, which makes them, besides free protons and neutrons, the most abundant hadrons. Combined with the intriguing results on light nuclei production at the LHC [12, 13], this offers the unique possibility to study light nuclei formation in two extreme regimes, one where they are an extremely rare probe and one where they contribute to the bulk of the created matter. Further details on light nuclei measurements with HADES can be found in the following contribution to this conference [14].

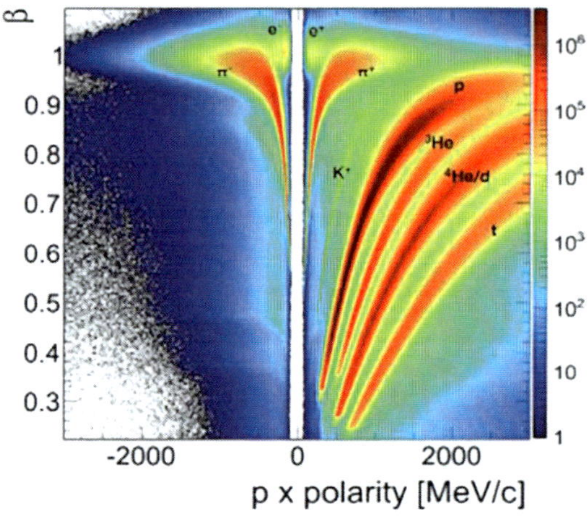

Fig. 2.3 Correlation between β and momentum × polarity. The baryon dominance over anti-baryons is clearly visible. Also clearly visible, are the high abundances of light nuclei

2.4 Flavour Production

Strangeness is of particular interest as its presence and interaction with the surrounding nucleons influences the EOS itself. It has been realized that inside neutron stars the appearance of hyperons is possible via the weak interaction and substantially softens the EOS and hence influences the stability of such objects [15–18]. Strangeness is the only exotic quark flavour which is produced in the HADES energy regime. Already the lightest strange hadrons are produced below their free NN-threshold, so the required energy must be supplied by the collision system. Hence, one expects their abundances to rise as a function of the geometrical overlap of the nuclei, which is a proxy of the number of nucleons taking part in the collision. Indeed, the abundances of hadrons carrying strangeness per mean number of participants, Mult/$\langle A_{part}\rangle$, show a stronger rise than linear with $\langle A_{part}\rangle$, as shown in Fig. 2.4 [19]. If one assumes energy accumulation in sequential nucleon-nucleon collisions, one expects to observe significantly different slopes, due to the clear hierarchy in the production thresholds, $\approx$ -150 MeV for K^+, K^0, Λ (NN$\rightarrow$NΛK) and $\approx$ -450 MeV, $\approx$ -490 MeV for the K^- (NN$\rightarrow$NNK^+K^-) and the ϕ meson (NN$\rightarrow$NNϕ). Yet, a global fit of the function Mult $\propto \langle A_{part}\rangle^{\alpha}$ to all strange hadron yields returns a satisfactory value of χ^2/NDF = 0.59, with $\alpha = 1.45 \pm 0.06$. This points to a more entangled system than assumed in the past, as the total amount of produced strangeness increases with the number of participants and seems to be only later redistributed to the finally observed hadron species, once the system freezes out. Details on flow anisotropies of kaons and the search for Λ polarization with HADES can be found in the following contributions [20, 21] to this conference.

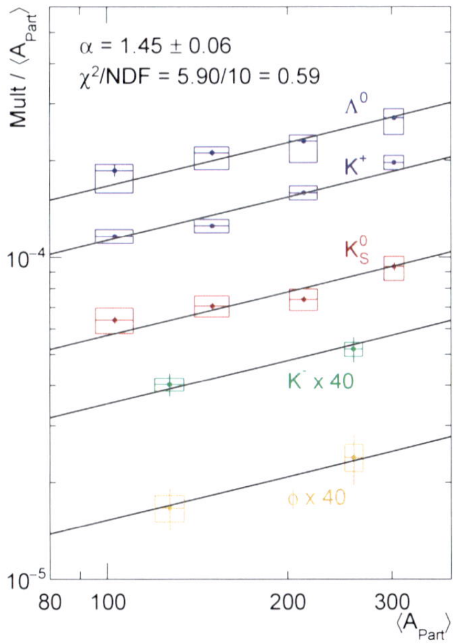

Fig. 2.4 Multiplicities per mean number of participants Mult/$\langle A_{part} \rangle$ as a function of $\langle A_{part} \rangle$. All hadron yields are fitted simultaneously with a function of the form Mult $\propto \langle A_{part} \rangle^{\alpha}$ with the result: $\alpha = 1.45 \pm 0.06$. More details can be found in the text and in [19]

2.5 FAIR Phase-0

As part of the Phase-0 of the FAIR project, HADES recorded in a very smooth 4 weeks measurement campaign in March 2019, 15×10^9 events of central Ag+Ag collisions at $\sqrt{s_{NN}} = 2.55$ GeV, meeting exactly the performance goals of the beam-time proposal. FAIR Phase-0 marks with upgrades of the existing GSI accelerators UNILAC, SIS18, ESR and CRYRING [22], as well as the use of CBM detector technology, e.g. the upgraded HADES-RICH [23] the start of the FAIR project. Those data will be used to further scrutinize the previously discussed observation with higher accuracy.

Acknowledgments SIP JUC Cracow, Cracow (Poland), National Science Center, 2016/23/P/ST2/040 POLONEZ, 2017/25/N/ST2/00580, 2017/26/M/ST2/00600; TU Darmstadt, Darmstadt (Germany), VH-NG-823, DFG GRK 2128, DFG CRC-TR 211, BMBF: 05P18RDFC1; Goethe-University, Frankfurt (Germany) and TU Darmstadt, Darmstadt (Germany), ExtreMe Matter Institute EMMI at GSI Darmstadt; TU Mnchen, Garching (Germany), MLL Mnchen, DFG EClust 153, GSI TMLRG131 6F, BmBF 05P15WOFCA, SFB 1258, DFG FAB898/2-2; NRNU MEPhI Moscow, Moscow (Russia), in framework of Russian Academic Excellence Project 02.a03.21. 0005, Ministry of Science and Education of the Russian Federation 3.3380.2017/ 4.6; JLU Giessen, Giessen (Germany), BMBF:05P12RGGHM; IPN Orsay, Orsay Cedex (France), CNRS/IN2P3; NPI CAS, Rez, Rez (Czech Republic), MSMT LM2015049, OP VVV CZ.02.1.01/0.0/0.0/16 013/0001677, LTT17003.

References

1. B. P. Abbott et al. [LIGO Scientific and Virgo Collab.], Phys. Rev. Lett. **116** (2016) no.6, 061102
2. J.S. Read, C. Markakis, M. Shibata, K. Uryu, J.D.E. Creighton, J.L. Friedman, Phys. Rev. D **79**, 124033 (2009)
3. E.R. Most, L.J. Papenfort, V. Dexheimer, M. Hanauske, S. Schramm, H. Stocker, L. Rezzolla, Phys. Rev. Lett. **122** 6, 061101 (2019)
4. S.A. Bass et al., Prog. Part. Nucl. Phys. **41**, 225 (1998)
5. A. Bauswein, S. Goriely, H.-T. Janka, Astrophys. J. **773**, 78 (2013)
6. J. Adamczewski-Musch et al. [HADES Collab.], Nature Phys. **15** 10, 1040 (2019)
7. G. Agakishiev, et al. [HADES Collab.], Eur. Phys. J. A **41**, (2009) 243
8. S. Harabasz, PhD Thesis, Tech. Univ. Darmstadt and Jagiellonian Univ. Cracow (2017)
9. S. Spies, Master Thesis, Goethe-University Frankfurt (2018)
10. B. Ramstein [HADES Collab.], AIP Conf. Proc. **1735** (2016) 080001
11. P. Salabura, et al. [HADES Collab.], Nucl. Phys. News **25** (2015) 2, 22
12. P. Braun-Munzinger, B. Dönigus, Nucl. Phys. A **987**, 144 (2019)
13. A. Andronic, P. Braun-Munzinger, K. Redlich and J. Stachel, Nature **561** (2018) no.7723, 321
14. M. Szala *et al.* [HADES Collab.], these proceedings
15. V.R. Pandharipande, Nucl. Phys. A **178**, 123 (1971)
16. H.A. Bethe, M.B. Johnson, Nucl. Phys. A **230**, 1 (1974)
17. N.K. Glendenning, S.A. Moszkowski, Phys. Rev. Lett. **67**, 2414 (1991)
18. S. Balberg, A. Gal, Nucl. Phys. A **625**, 435 (1997)
19. J. Adamczewski-Musch *et al.* [HADES Collab.], Phys. Lett. B **793** (2019) 457
20. F. Kornas *et al.* [HADES Collab.], these proceedings
21. L. Chlad *et al.* [HADES Collab.], these proceedings
22. M. Bai *et al.*, https://doi.org/10.18429/JACoW-IPAC2018-THYGBF3.
23. C. Pauly *et al.* [TRB Collab.], Nucl. Instrum. Meth. A **876** (2017) 164

Chapter 3
BM@N First Results

Mikhail Kapishin

Abstract The BM@N (Baryonic Matter at Nuclotron) is the first experiment undertaken at the accelerator complex of NICA-Nuclotron. The BM@N scientific program comprises studies of dense nuclear matter in heavy ion beams of the intermediate energy range between the SYS-18 and NICA/FAIR facilities. The first experimental runs were performed with the deuteron and carbon beams of kinetic energy per nucleon ranging from 3.5 to 4.5 GeV with fixed targets. The extended configuration of the BM@N set-up was realized in recent runs with the argon and krypton beams. First physics results are presented on Λ hyperon production in carbon-nucleus interactions at the beam kinetic energy of 4.0 AGeV.

3.1 Nuclotron Heavy-Ion Physics Program

Collisions of relativistic heavy ions provide a unique opportunity to study nuclear matter at extreme densities and temperatures. In such collisions, nuclear matter is heated and compressed for a very short amount of time (few fm/c). At the Nuclotron with beam kinetic energy per nucleon ranging from 1 to 4.5 GeV, baryons form the majority of the products in a nucleus-nucleus collision, in contrast to collisions that occur at higher energies at the RHIC or SPS accelerators. According to the Quark Gluon String Model (QGSM) calculations presented in [1], at the Nuclotron energies the nucleon density in a fireball created by two colliding gold nuclei is 3–4 times higher than the saturation density. At such densities, nucleons start to overlap. At the Nuclotron, the experimental research is focused on studies of hadrons with strangeness produced in the collision and not present in the initial state of two colliding nuclei, unlike the nucleons consisting of light (u and d) quarks. The measured yields of light and strange mesons, hyperons, and antihyperons are shown in

Mikhail Kapishin for the BM@N Collaboration.

M. Kapishin (✉)
Joint Institute for Nuclear Research, Dubna, Joliot Curie 6, 141980 Dubna, Russia
e-mail: kapishin@jinr.ru

D. Elia et al. (eds.), *The XVIII International Conference on Strangeness in Quark Matter (SQM 2019)*, Springer Proceedings in Physics 250,
https://doi.org/10.1007/978-3-030-53448-6_3

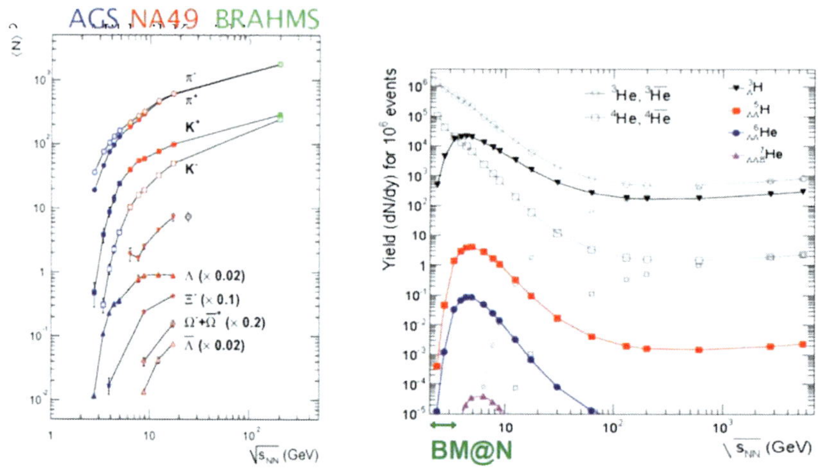

Fig. 3.1 Left plot: Yields of mesons and (anti-)hyperons measured in different experiments as a function of the energy per nucleon-nucleon collision in c.m.s. for Au+Au and Pb+Pb collisions [2]. The Nuclotron beam energy range corresponds to $\sqrt{s_{NN}} = 2.3-3.5$ GeV. Right plot: Yields of hyper-nuclei predicted by the thermal model in [3] as a function of the nucleon-nucleon collision energy in c.m.s. for Au+Au collisions. Predictions for the yields of ^{3}He and ^{4}He nuclei are presented for comparison. The Nuclotron BM@N energy range is specified

Fig. 3.1 (left) as a function of the energy per nucleon-nucleon collision in the center of mass system (c.m.s) in Au+Au / Pb+Pb collisions. The energy range of heavy ion beams at the Nuclotron corresponds to $\sqrt{s_{NN}} = 2.3-3.5$ GeV, these energies are high enough to study strange mesons and multi-strange hyperons produced in nucleus-nucleus collisions close to the kinematic threshold. Heavy ion collisions are an abundant source of strange hadrons that can bind with nucleons and initiates formation of a variety of light hypernuclei [4, 5]. Studies of hypernucleus production processes are expected to provide insight into the properties of hyperon-nucleon and hyperon-hyperon interactions. In Fig. 3.1 (right) the hypernucleus yields per event are presented as a function of the nucleon-nucleon collision energy in c.m.s in Au + Au collisions as predicted by the thermal model [3]. The maximum probability of hyper-nucleus production is predicted for energies of $\sqrt{s_{NN}} = 4-5$ GeV, that is close to the Nuclotron energy range. In general, the research program of the BM@N experiment aims to study heavy ion collisions at the Nuclotron including the following topics: exploration of the reaction dynamics and the equation of state (EoS) of nuclear matter, study of in-medium properties of hadrons, production of (multi)strange hyperons in the vicinity of the threshold and searches for hypernuclei [6–8].

3.2 BM@N Detector for Studies of Baryonic Matter at the Nuclotron

BM@N (Baryonic Matter at Nuclotron) is the first experiment operational at the Nuclotron/ NICA accelerating complex. The purpose of the BM@N experiment is to study relativistic heavy ion beam interactions with fixed targets [7]. The Nuclotron will provide the experiment with beams of a variety of particles, from protons to gold ions, with a kinetic energy ranging from 1 to 6 GeV/nucleon. The maximum kinetic energy of ions with the charge to atomic weight ratio of 0.5 is 6 GeV/nucleon. The maximum kinetic energy of gold ions with Z/A of 0.4 is 4.5 GeV/nucleon, while the maximum kinetic energy for protons is 13 GeV. Recently the BM@N experiment collected data with beams of carbon, argon, and krypton ions. The planned intensity of the gold ion beam at BM@N is 10^6 ions/s. The acceleration of the gold ion beam is planned in 2021, after the Nuclotron upgrade. In Fig. 3.2 the interaction rates are presented for different experiments with heavy ion collisions at different energies per nucleon-nucleon collision in the center of mass system. The beam energy of the BM@N experiment is in the intermediate range between experiments at the SIS-18 and NICA/FAIR facilities and partially overlaps the energy range of the HADES experiment. The acquisition rate of non-peripheral collisions, i.e., central or intermediate interactions is expected to range from 20 to 50 kHz at the second stage of the BM@N experiment in 2022 and later. The interaction rate is limited by the capacity of the data acquisition system and readout electronics.

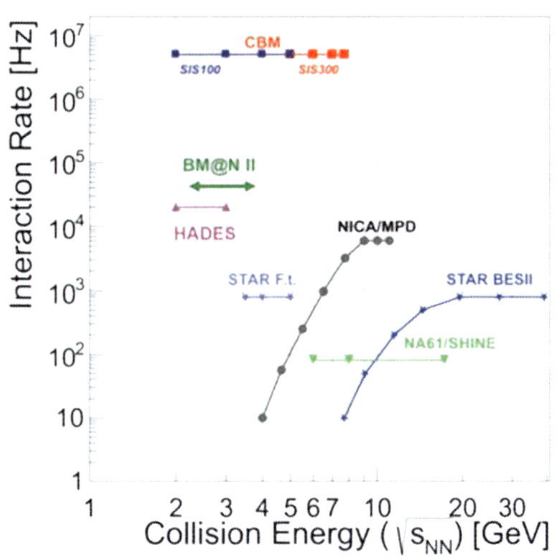

Fig. 3.2 Interaction rate and energy per nucleon-nucleon collision in c.m.s. in experiments with heavy ions [9]. The range for BM@N is superimposed

3.3 First Results of the BM@N Runs

The technical runs with the BM@N detector were performed in the deuteron beam in December 2016 and in the carbon beam in March 2017. The kinetic energy was 4 GeV/nucleon for the deuteron beam and was varied from 3.5 to 4.5 GeV/nucleon for the carbon beam. The starting configuration of the central tracker was based on a forward silicon strip detector and a set of GEM detectors [10]. The experimental data from the central tracker, outer drift chambers, time-of-flight detectors, zero degree calorimeter and trigger detectors were read out using the integrated data acquisition system. The extended configuration of the BM@N set-up was realized in the recent runs with the argon and krypton beams performed in March 2018. The set-up comprised GEM detectors with the size of 163 x 45 cm^2, forward silicon strip detectors, full time-of-flight system, extended trigger system, hadron and electro-magnetic calorimeters. The collected data were used to check efficiencies of sub-detectors and develop algorithms for the event reconstruction and analysis. In particular, experimental data of minimum bias interactions of the beam with different targets were analyzed with the aim to reconstruct tracks, primary and secondary vertices using the central tracking detectors [11–13]. The track reconstruction method was based on the so-called 'cellular automaton' approach [14]. Since the GEM tracker configuration was tuned to measure relatively high-momentum beam particles, the geometrical acceptance for relatively soft decay products of strange V0 particles was rather low.

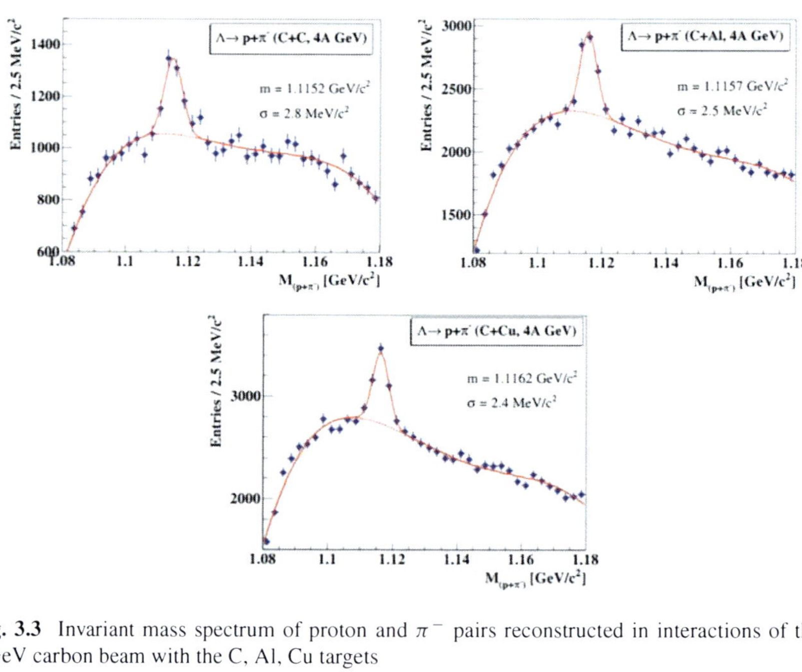

Fig. 3.3 Invariant mass spectrum of proton and π^- pairs reconstructed in interactions of the 4 AGeV carbon beam with the C, Al, Cu targets

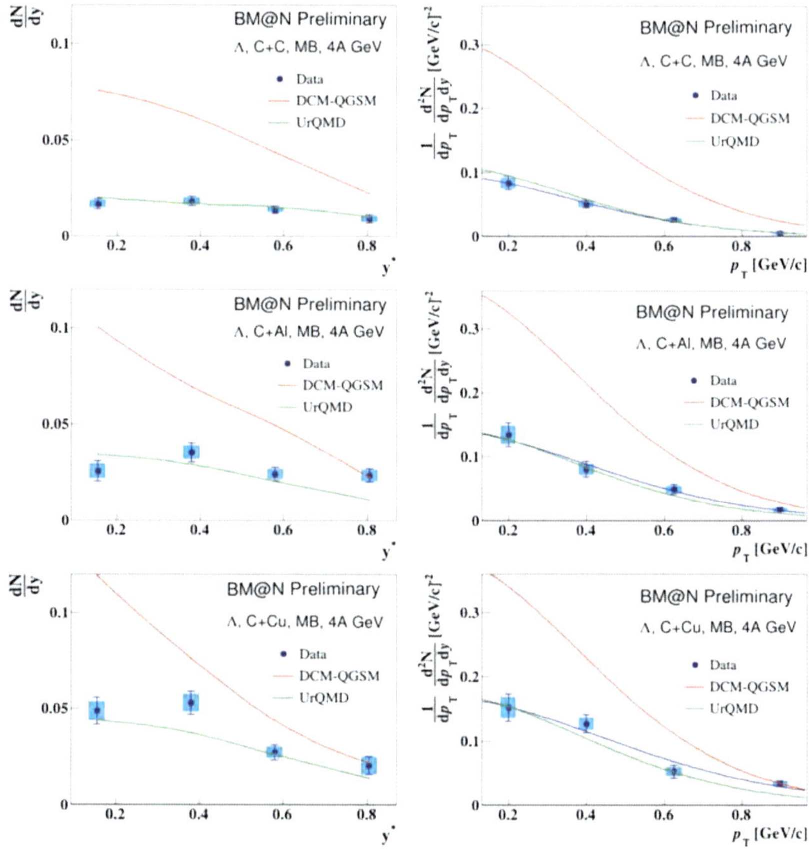

Fig. 3.4 Yields of Λ hyperons in minimum bias C+C, C+Al, C+Cu interactions vs rapidity y^* in c.m.s. (left plots) an transverse momentu p_T (right plots). The predictions of the DCM-QGSM and UrQMD models are shown as lines

The Monte Carlo simulation showed that only ~4% of Λ hyperons and ~0.8% of K_s^0 could be reconstructed. Λ hyperons were reconstructed using their decay mode into p, π^- pairs [15]. Since particle identification at this stage of the analysis was not used, all positive tracks were considered as protons and all negative as π^-. The invariant mass distributions of p and π^- are shown in Fig. 3.3 for reconstructed interactions of the carbon beam with the C, Al, Cu targets. The background under the signal will be reduced by introducing additional silicon tracking detectors to improve the primary and decay vertex resolution.

The yields of Λ hyperons in minimum bias interactions of the 4 AGeV carbon beam with the C, Al, Cu targets are measured in the kinematic range on the Λ transverse momentum of $0.1 < p_T < 1.05$ GeV/c and the Λ rapidity in c.m.s. of $0.03 < y^* < 0.93$. The y^* and p_T spectra of the Λ hyperon yields corrected for the detector acceptance and efficiency are presented in Fig. 3.4. The predictions

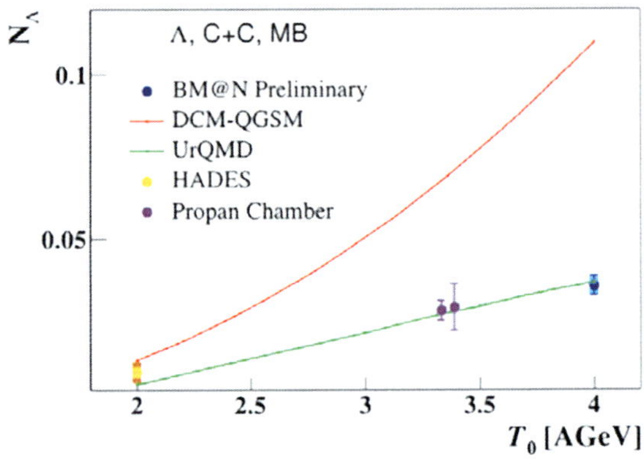

Fig. 3.5 Energy dependence of Λ hyperon yields in C+C interactions measured in different experiments. The BM@N result is compared with the data taken from [17–19]. The predictions of the DCM-QGSM and UrQMD models are shown

of the DCM-QGSM [16] and URQMD models are shown for comparison. The measured p_T-differential yields of the Λ hyperon are parameterized by the form: $1/p_T d^2N/dp_T dy \propto exp(-(m_T - m_\Lambda)/T)$, where $m_T = \sqrt(m_\Lambda^2 + p_T^2)$ is the transverse mass, and the inverse slope parameter (temperature) T is a free parameter of the fit. The value of T extracted from the fit of the p_T spectra is about 113 MeV for C+C interactions rising up to 170 MeV for C+Cu interactions. The fit results are consistent within the uncertainties with the model predictions. The measured yields of the Λ hyperons in minimum bias C+C interactions are extrapolated into the full kinematical range using averaged predictions of the DCM-QGSM and URQMD models and compared in Fig. 3.5 with the results of other experiments [17–19].

Acknowledgments This work is supported by the Russian Foundation for Basic Reasearch (RFBR) under grant No. 18-02-40036 mega.

References

1. B. Friman, W. Nörenberg, V.D. Toneev, Eur. Phys. J. A **3**, (1998)
2. C. Blume, J. Phys. G **31**, S57 (2005)
3. A. Andronic et al., Phys. Lett. B **695**, 203 (2011)
4. J. Steinheimer et al., Progress in Particle and Nuclear Physics **62**, 313317 (2009)
5. J. Steinheimer, K. Gudima, A. Botvina, I. Mishustin, M. Bleicher, H. Stocker, Phys. Lett. B **714**, 85 (2012)
6. NICA White Paper. http://theor0.jinr.ru/twiki-cgi/view/NICA/NICAWhitePaper
7. BMN Conceptual Design Report, http://nica.jinr.ru/files/BM@N/BMN_CDR.pdf
8. D. Suvarieva et al., J. Phys. Conf. Ser. 668 (2016) no.1, 012121
9. V. Friese, Strangeness Prospects with the CBM Experiment, talk at the SQM-2015 Conference, Dubna, Russia, July 6–11, 2015
10. D. Baranov et al., JINST 12 (2017) no.06, C06041
11. M. Kapishin (for the BMN Collaboration), Eur.Phys.J. A52 (2016) no.8, 213
12. M. Kapishin (for the BMN Collaboration), Phys.Atom.Nucl. 80 (2017) no.10, 1613-1619, Yad.Fiz. 7 (2016) no.6, 543-550
13. M. Kapishin (for the BMN Collaboration), Nucl.Phys. A982 (2019) 967-970

14. V. Akishina and I. Kisel, J. Phys.: Conf. Ser. 599, 012024 (2015), I. Kisel, Nucl. Instrum. Meth. A **566**, 85 (2006)
15. D. Baranov et al., Phys. Part. Nucl. Lett. **15**(2), 148–156 (2018)
16. N. Amelin, K. Gudima, V. Toneev, Sov. J. Nucl. Phys. **51**, 1093 (1990)
17. S. Arakelian et al., P1-83-354, JINR, Dubna
18. D. Armutlijsky et al., P1-85-220, JINR, Dubna
19. K. Kanaki, PhD Study of hyperon production in C+C collisions at 2A GeV beam energy with the HADES spectrometer, 2007

Chapter 4
Hyperons in Thermal QCD from the Lattice

Gert Aarts, Chris Allton, Davide de Boni, Jonas Glesaaen, Simon Hands, Benjamin Jäger, and Jon-Ivar Skullerud

Abstract We study the spectrum of light baryons and hyperons as a function of temperature using lattice gauge theory methods. We find that masses of positive parity states are temperature independent, within errors, in the hadronic phase. The negative parity states decrease in mass as the temperature increases. Above the deconfining temperature, lattice correlators and spectral functions show a degeneracy between parity sectors, i.e. parity doubling. We apply our findings to an in-medium Hadron Resonance Gas model. The techniques used in this study include direct analysis of the hadronic correlation functions, conventional fitting procedures, and the Maximum Entropy Method.

4.1 Introduction

Symmetries play a crucial role in the Standard Model and are especially significant in the QCD transition from the confining, chirally broken, hadronic phase to the deconfined, chirally symmetric, plasma phase. Chiral symmetry restoration in the meson sector at finite temperature has been studied extensively, but there have been very few studies in baryons [1–3].

In the baryonic case, a combination of unbroken chiral symmetry and parity leads to parity doubling, i.e. a degeneracy between positive and negative parity states. Hence it is expected that in the chirally symmetric phase, baryonic channels related

G. Aarts · C. Allton (✉) · D. de Boni · J. Glesaaen · S. Hands
Department of Physics, College of Science, Swansea University, Swansea SA2 8PP, Swansea, UK
e-mail: c.allton@swansea.ac.uk

B. Jäger
IAS, Department of Mathematics and Computer Science, University of Southern Denmark, 5230 Odense M, Odense, UK

J.-I. Skullerud
Department of Theoretical Physics, National University of Ireland Maynooth, Maynooth, County Kildare, Ireland

D. Elia et al. (eds.), *The XVIII International Conference on Strangeness in Quark Matter (SQM 2019)*, Springer Proceedings in Physics 250,
https://doi.org/10.1007/978-3-030-53448-6_4

by parity will display approximate degeneracy. This work tests this conjecture by studying octet and decuplet light baryons and hyperons as the temperature varies using our FASTSUM Collaboration's anisotropic, 2+1 flavour lattices [4].

We find evidence of parity doubling above the deconfining temperature, T_c,[1] with this degeneracy being most pronounced for the baryons with the smallest strange quark content. In the hadronic phase we find that the negative parity masses decrease as $T \to T_c$ while the corresponding positive parity masses remain constant within the uncertainty. We use these temperature-dependent masses to define an in-medium Hadron Resonance Gas, deriving results for susceptibilities and partial pressures.

The work presented here is detailed more fully in [5–7].

4.2 Parity in Baryons

We use the standard interpolation operator for a nucleon

$$O_N(\mathbf{x}, \tau) = \epsilon_{abc} u_a(\mathbf{x}, \tau) \left[u_b^T(\mathbf{x}, \tau) \mathcal{C} \gamma_5 d_c(\mathbf{x}, \tau) \right],$$

where u, d are the quark fields, a, b, c are colour indices, other indices are suppressed and $\mathcal{C}$ denotes the charge conjugation matrix. Similar operators are used for octet and decuplet cases, see [6]. Parity operators for the positive and negative parity channels are defined as $O_{N_\pm}(\mathbf{x}, \tau) = P_\pm O_N(\mathbf{x}, \tau)$, where $P_\pm = \frac{1}{2}(1 \pm \gamma_4)$. We study the usual Euclidean correlators of these operators, summed over the Dirac indices and projected to zero momentum,

$$G_\pm(\tau) = \int d^3x \left\langle O_{N_\pm}(\mathbf{x}, \tau) \overline{O}_{N_\pm}(\mathbf{0}, 0) \right\rangle.$$

From the properties of Euclidean time reflection, it follows that the forward (backward) propagation of $G_+(\tau)$ corresponds to the positive (negative) parity channel (see e.g. [5]). Hence both parities are obtained from one correlator.

4.3 Lattice Parameters

Our FASTSUM collaboration specialises in using a fixed-scale approach on anisotropic lattices where the temporal lattice spacing a_τ is smaller than the spatial one a_s. We use 2+1 quark flavours, where the strange quark mass has its physical value, but the two lightest flavours are heavier than in nature resulting in a pion mass of $M_\pi = 392(4)$ MeV. The inverse temporal lattice spacing is $a_\tau^{-1} = 5.63(4)$GeV with

[1] T_c is not uniquely defined. Here it's defined from the renormalised Polyakov loop [4]. * Invited plenary talk presented at the Strangeness in Quark Matter Conference (SQM 2019), Bari, Italy, 10–15 June 2019.

Table 4.1 Lattice parameters and temperatures studied. The ensemble at the lowest temperature was provided by the HadSpec Collaboration [8]

N_τ	128	40	36	32	28	24	20	16
T/T_c	0.24	0.76	0.84	0.95	1.09	1.27	1.52	1.90
T [MeV]	44	141	156	176	201	235	281	352

$a_s/a_\tau \approx 3.5$, and our spatial lattice volume is 24^3. We use a variety of temporal extents, N_τ, with the corresponding temperatures $T = 1/(a_\tau N_\tau)$ shown in Table 4.1, spanning both phases.

4.4 Results

The nucleon correlation functions are shown in Fig. 4.1 for both parity states. Other channels have similar behaviour. The backward movers (from the second half of the temporal range) have been reflected in τ to enable direct comparison with their positive counterparts.

We define the ratio $R(\tau) = [G_+(\tau) - G_+(1/T - \tau)]/[G_+(\tau) + G_+(1/T - \tau)]$ which measures the time reflection (a)symmetry of $G_+(\tau)$ and therefore the presence (absence) of parity doubling. $R(\tau) \sim 1$ signifies non-degenerate parity states, and $R(\tau) \sim 0$ parity doubling. It is convenient to average $R(\tau)$ over τ obtaining R, a quasi order parameter, and this is plotted in Fig. 4.2 (left). R clearly approaches zero around T_c indicating parity doubling. This effect is strongest for baryons with

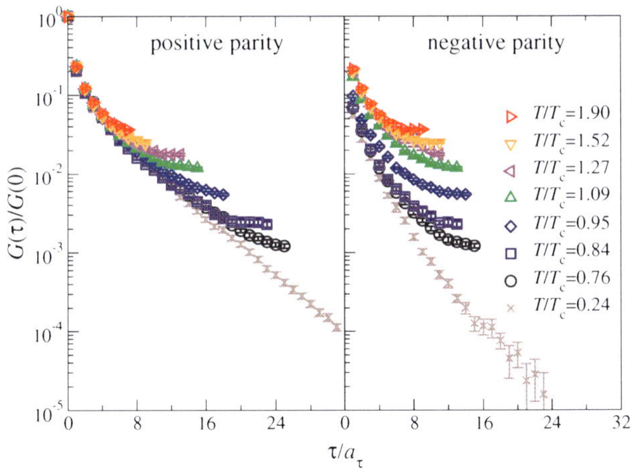

Fig. 4.1 Temperature dependent correlators for the +ve and −ve parity nucleon channels

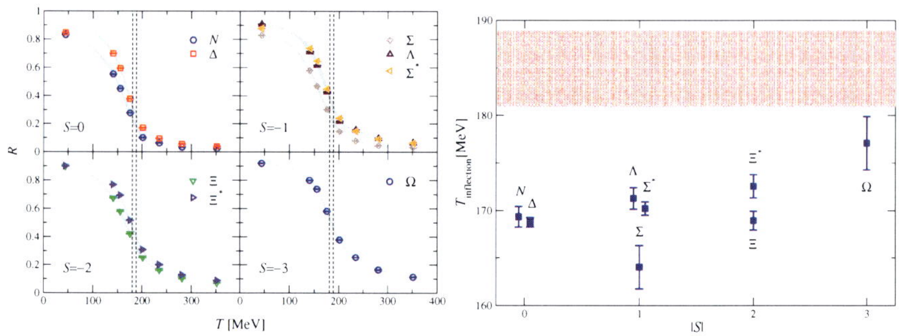

Fig. 4.2 **(Left)** The ratio R for various strangeness sectors. **(Right)** Inflection point temperature of R with T_c from the Polyakov loop shown in the band [4]

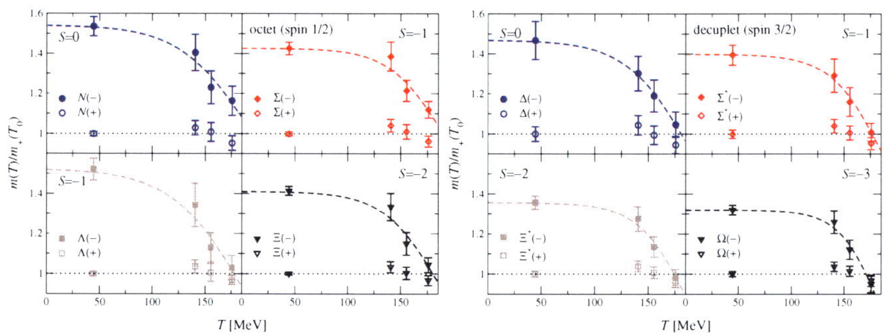

Fig. 4.3 Temperature dependent masses of the octet and decuplet baryons for both parity sectors

the smallest strangeness content, as expected due to the residual chiral symmetry breaking effects from the strange quark.

We obtain a measure of the transition temperature for each channel using the point of inflection of R and plot these in Fig. 4.2 (right) together with the T_c value obtained from the Polyakov loop [4].

The ground state masses are extracted from the correlators in the hadronic phase using conventional exponential fits with the results displayed in Fig. 4.3. We see that the $T = 0$ masses are heavier than the experimental values—this reflects the non-physical value of the two lightest dynamical quarks in our simulation. The positive parity states' masses appear to be T-independent, whereas the negative states' masses decrease as $T \to T_c$.

By using these temperature-dependent masses in the Hadron Resonance Gas (HRG) model, we obtain "in-medium HRG" predictions for the partial pressures and susceptibility $\chi_{BS} = \frac{1}{VT}\langle BS \rangle$, where B and S are the baryon and strangeness

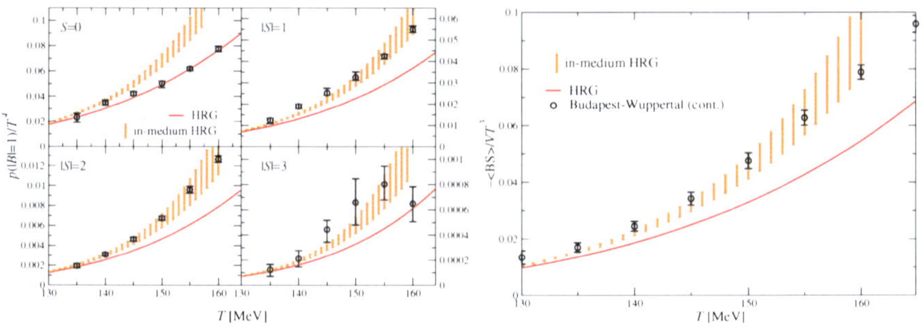

Fig. 4.4 In-medium Hadron Resonance Gas results. (Left) partial pressures for each strangeness sector, and (Right) the susceptibility $\langle BS \rangle$

number, see Fig. 4.4. For χ_{BS} and the strange partial pressure sectors, this in-medium HRG gives better agreement with independently obtained lattice results [10].

To interpret hadronic correlation functions in the plasma phase, where exponential fits no longer work and we assume there are no bound states, we introduce spectral functions [6],

$$G_{\pm}(\tau) = \int_{-\infty}^{\infty} \frac{d\omega}{2\pi}\, K(\tau,\omega)\, \rho_{\pm}(\omega), \qquad \text{where} \quad K(\tau,\omega) = \frac{e^{-\omega\tau}}{1+e^{-\omega/T}}.$$

We use the Maximum Entropy Method to solve the above inverse problem for $\rho(\omega)$, noting that the positive (negative) parity states appear for $\omega > 0$ ($\omega < 0$). Spectral function results for a selection of channels are shown in Fig. 4.5 for a representative

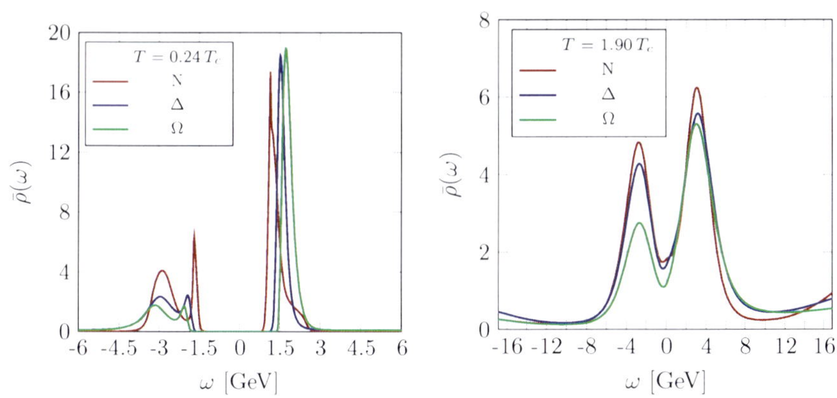

Fig. 4.5 Spectral function for indicative temperatures in the (left) hadronic and (right) plasma phases for three baryon channels. The details of our Maximum Entropy Method procedure, including choices of default model are detailed in [6]

temperature in each phase. We see clear ground states in both parity sectors in the hadronic phase, and signs of parity doubling in the plasma phase, particularly for the non-strange baryons.

4.5 Conclusion

This work uses lattice simulations of 2+1 flavour QCD to analyse the parity states in the baryonic spectrum as the temperature is varied from essentially zero to around 350 MeV. Our FASTUM collaboration's lattices are anisotropic which increases the sampling rate in the temporal direction, thereby enhancing the accuracy of our results. We employ a variety of approaches in this work: using the correlation functions directly, conventional fits to exponentials, and extracting the spectral functions using the Maximum Entropy Method.

In the hadronic phase, these methods indicate that the positive parity ground state masses are temperature independent (within errors) and that the negative parity states' masses decrease as T increases until becoming essentially degenerate with its positive parity partner at or close to T_c. This "parity-doubling" is particularly evident for baryons with the smallest strange content, and this is to be expected due to the strange quark's chiral symmetry breaking effects. This pattern is observed for both octet and decuplet states.

We plan to use both lighter quarks and lattices with finer (temporal) spacings to test the systematics of our approach in our next generation ensembles.

Acknowledgments We acknowledge PRACE for awarding us access to Marconi at CINECA, Italy. This work used the STFC DiRAC Blue Gene Q system at the University of Edinburgh, U.K. We have been supported by the STFC grant ST/P00055X/1, and the Swansea Academy for Advanced Computing.

References

1. C.E. DeTar, J.B. Kogut, Phys. Rev. Lett. **59**, 399 (1987)
2. Phys. Rev. D 36 (1987) 2828
3. I. Pushkina et al., QCD-TARO Collaboration. Phys. Lett. B **609**, 265 (2005). [hep-lat/0410017]
4. S. Datta, S. Gupta, M. Padmanath, J. Maiti, N. Mathur, JHEP **1302**, 145 (2013). [arXiv:1212.2927 [hep-lat]]
5. G. Aarts, C. Allton, A. Amato, P. Giudice, S. Hands and J. I. Skullerud, JHEP 1502 (2015) 186 http://arxiv.org/abs/1412.6411 [hep-lat]]. doi: https://doi.org/10.1007/JHEP02(2015)186
6. G. Aarts, C. Allton, S. Hands, B. Jger, C. Praki and J. I. Skullerud, Phys. Rev. D 92 (2015) no.1, 014503 http://arxiv.org/abs/1502.03603 [hep-lat]]
7. G. Aarts, C. Allton, D. De Boni, S. Hands, B. Jger, C. Praki, J.I. Skullerud, JHEP **1706**, 034 (2017). http://arxiv.org/abs/1703.09246 [hep-lat]]
8. G. Aarts, C. Allton, D. De Boni and B. Jger, Phys. Rev. D 99 (2019) no.7, 074503 http://arxiv.org/abs/1812.07393 [hep-lat]]

9. R.G. Edwards, B. Joo, H.W. Lin, Phys. Rev. D **78**, 054501 (2008). 10.1103/PhysRevD.78.054501. http://arxiv.org/abs/0803.3960 [hep-lat]]
10. M. Tanabashi et al. [Particle Data Group], Phys. Rev. D 98 (2018) no.3, 030001

Chapter 5
Recent Results from NA61/SHINE

Szymon Pulawski

Abstract The research programme of the NA61 collaboration covers a wide range of hadronic physics in the CERN SPS energy range, encompassesing measurements of hadron-hadron, hadron-nucleus as well as nucleus-nucleus collisions. The latter are analysed to better understand the properties of hot and dense nuclear matter. In this paper recent results of particle production properties as well event by event fluctuations in proton-proton, Be+Be and Ar+Sc interactions at beam energies of 19*A*/20*A*, 30*A*, 40*A*, 75*A*/80*A* and 158*A* GeV/c are presented.

5.1 The NA61/SHINE Facility

The NA61/SHINE detector [1] is a large acceptance hadron spectrometer with excellent capabilities in charged particle momentum measurements and identification by a set of eight Time Projection Chambers as well as Time-of-Flight detectors. The high resolution forward calorimeter, the Projectile Spectator Detector (PSD), measures energy flow around the beam direction, which in nucleus-nucleus reactions is primarily a measure of the number of projectile spectator (non-interacted) nucleons and is thus related to the violence (centrality) of the collision. A set of beam detectors identifies beam particles and measures precisely their trajectories.

NA61/SHINE performed a two-dimensional scan in collision energy (13*A*-150*A* GeV/c and system size (p+p, Be+Be, Ar+Sc, Xe+La, Pb+Pb) to study the phase diagram of strongly interacting matter. The main goals of NA61/SHINE are the search for the critical point and a study of the onset of deconfinemnet.

Szymon Pulawski for the NA61/SHINE Collaboration.

S. Pulawski (✉)
Institute of Physics, University of Silesia, Katowice, Poland
e-mail: szymon.pulawski@us.edu.pl

D. Elia et al. (eds.), *The XVIII International Conference on Strangeness in Quark Matter (SQM 2019)*, Springer Proceedings in Physics 250,
https://doi.org/10.1007/978-3-030-53448-6_5

5.2 Study of the Onset of Deconfinement

5.2.1 Particle Production Properties

The Statistical Model of the Early Stage (SMES) [2] predicts a 1st order phase transition from the QGP to a hadron matter phase between top AGS and top SPS energies. In the transition region constant temperature and pressure in the mixed phase and an increase of the number of internal degrees of freedom is expected.

A plateau ("step") in the energy dependence of the inverse slope parameter T was observed by the NA49 experiment in Pb+Pb collisions for m_T spectra of $K^{\pm}$. It was expected for the onset of deconfinement due to the presence of a mixed phase of hadron gas (HRG) and quark-gluon plasma (QGP). In p+p interactions at SPS energies the inverse slope parameter T of m_T spectra shows qualitatively similar enery dependence as in central Pb+Pb collisions ("step") and such a behaviour seems to emerge also in Be+Be reactions, as visible in Fig. 5.1. The values of the T parameter in Be+Be collisions are slightly above those in p+p interactions. The T parameter in Ar+Sc reactions is found between those in p+p/Be+Be and Pb+Pb collisions.

Finally, rapid changes of the ratios K^+/π^+ at mid-rapidity and $\langle K^+\rangle/\langle\pi^+\rangle$ as function of collision energy ("horn") were observed in Pb+Pb collisions by the NA49 experiment. These were predicted by the SMES model as a signature of the onset of deconfinement. These two ratios together with new NA61/SHINE results from Be+Be and Ar+Sc collisions are shown in Fig. 5.2. A plateau like structure is visible in p+p interactions. The ratio K^+/π^+ at mid-rapidity as well as the ratio of total yields from Be+Be collisions is close to the p+p measurements. For the five analysed energies of Ar+Sc collisions, the ratio K^+/π^+ at mid-rapidity and $\langle K^+\rangle/\langle\pi^+\rangle$ are higher than in p+p collisions but show a qualitatively similar energy dependence—no horn structure visible.

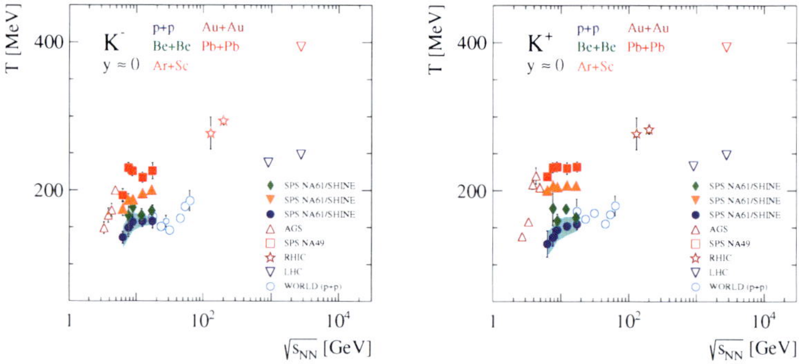

Fig. 5.1 Inverse slope parameter T of m_T spectra of $K^{\pm}$ as function of collision energy. Most results are shown with statistical uncertainties only. For the p+p data the shaded band indicates systematic uncertainties

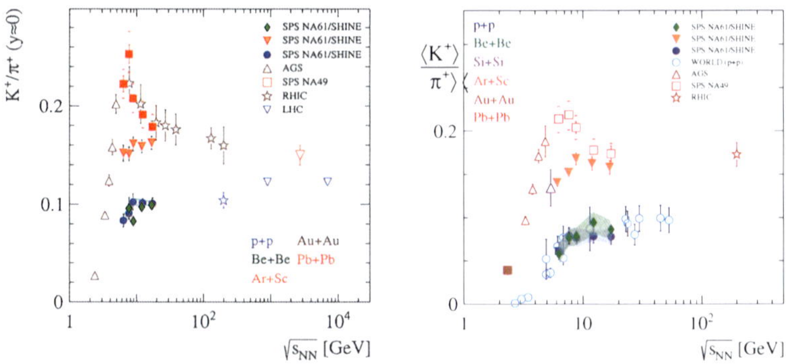

Fig. 5.2 Ratio of yields K^+/π^+ at mid-rapidity and the ratio of total yields $\langle K^+\rangle / \langle \pi^+\rangle$ produced in p+p, Be+Be and Pb+Pb collisions as function of collision energy

5.2.2 *Flow*

Directed flow v_1 was considered to be sensitive to the first order phase transition (strong softening of the Equation of State) [3–5]. The expected effect is a non-monotonic behaviour (change from positive to negative and again to positive values) of proton dv_1/dy as a function of beam energy. This effect is usually referred to as collapse of proton flow. The NA49 experiment measured anti-flow of protons at mid-rapidity [6]. A negative value of dv_1/dy was observed in peripheral Pb+Pb collisions at 40A GeV/c beam momentum ($\sqrt{s_{NN}} = 8.8$ GeV).

In 2018 the NA61/SHINE experiment reported the first results on anisotropic flow, measured in centrality selected Pb+Pb collisions at 30A GeV/c beam momentum ($\sqrt{s_{NN}} = 7.6$ GeV). According to the *horn* structure in the energy dependence of the K^+/π^+ ratio in Pb+Pb collisions, this is the energy of the onset of deconfinement. Therefore, studying the centrality dependence of flow at this specific energy may allow to better understand the properties of the onset of deconfinement.

The NA61/SHINE fixed target setup allows tracking and particle identification over a wide rapidity range. Flow coefficients were measured relative to the spectator plane estimated with the Projectile Spectator Detector (PSD), which is unique for NA61. Preliminary results on the centrality dependence of dv_1/dy at mid-rapidity, measured in Pb+Pb collisions at 30A GeV/c, are presented in Fig. 5.3 (*left*). One sees that the slope of pion v_1 is always negative. In contrast, the slope of proton v_1 changes sign for centrality of about 50%. Recently, preliminary results of directed flow for Pb+Pb collisions at 13A were released [7]. Proton directed flow as function of rapidity is shown in Fig. 5.3 (*right*). The results do not show evidence for the collapse of proton directed flow in Pb+Pb interactions at 13A GeV/c.

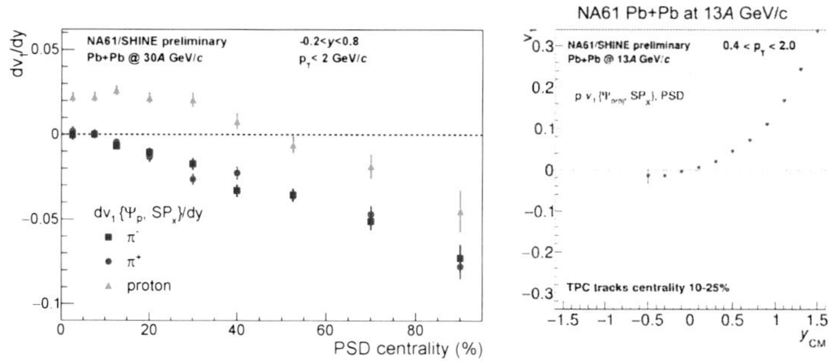

Fig. 5.3 Preliminary results on centrality dependence of dv_1/dy at mid-rapidity measured in Pb+Pb collisions at 30*A* GeV/c (*left*)and v_1 as the function of rapidity measured in Pb+Pb at 13*A* GeV/c (*right*)

5.3 Search for the Critical Point

An intermittency signal in protons was predicted close to the critical point. This is expected to become manifest in local power-law fluctuations of the baryon density which can be searched for by studying the scaling behaviour of second factorial moments $F_2(M)$ with the cell size or, equivalently, with the number of cells in (p_x, p_y) space of protons at mid-rapidity (see [8–10]). The transverse momentum space is partitioned into $M \times M$ equal-size bins, and the proton distribution is quantified by multiplicities in individual bins. The second order factorial moment in transverse momentum space is expressed as:

$$F_2(M) = \frac{\langle \frac{1}{M^2} \sum_{m=1}^{M^2} n_m(n_m - 1) \rangle}{\langle \frac{1}{M^2} \sum_{m=1}^{M^2} n_m \rangle^2}, \quad (5.1)$$

where M^2 is the number of bins (M bins in p_x and M bins in p_y) and n_m is the number of protons in the m-th bin. Combinatorial background subtracted (by mixed events) second factorial moments, $\Delta F_2(M)$, should scale according to a power-law (for $M \gg 1$):

$$\Delta F_2(M) \sim (M^2)^{\phi_2} \quad (5.2)$$

In the recent analysis of NA61/SHINE the intermittency effects were studied in central Be+Be and centrality selected Ar+Sc collisions at 150*A* GeV/c. The dE/dx method was used for the identification of protons. Centrality was determined from the energy deposited in the PSD detector. For Ar+Sc collisions protons were selected with at least 90% purity. Figure 5.4 shows preliminary results on $F_2(M)$ of mid-rapidity protons produced in 5–10% and 10–15% central Ar+Sc collisions at 150*A* GeV/c.

The result of $F_2(M^2)$seen in Ar+Sc collisions are higher in data than in mixed events. A detailed investigation of the significance of this result is in progress.

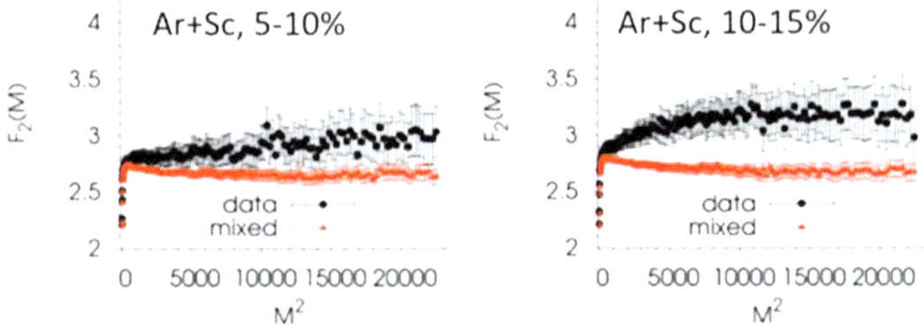

Fig. 5.4 Preliminary results on $F_2(M)$ of mid-rapidity protons measured in 5–10% (*left*) and 10–15% (*right*) central Ar+Sc collisions at 150*A* GeV/c

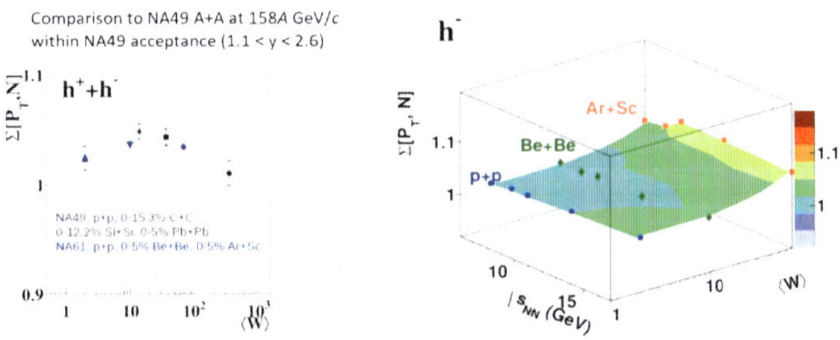

Fig. 5.5 $\Sigma[P_T, N]$ for all charged hadrons ($h^+ + h^-$) from the NA61/SHINE and NA49 experiments as function of system size at 150*A*/158*A* GeV/*c* (*left*) and $\Sigma[P_T, N]$ for negatively charged hadrons in inelastic p+p (blue squares), 0–5% Be+Be (green diamonds), and 0-5% Ar+Sc (orange squares) collisions obtained by NA61/SHINE. for NA61/SHINE only statistical uncertainties are shown. All NA61/SHINE results are preliminary

A critical point is also expected to lead to enhanced fluctuations of multiplicity and transverse momentum. For their study NA61/SHINE uses the *strongly intensive* measures $\Delta[P_T, N]$ and $\Sigma[P_T, N]$, see [11]. In the Wounded Nucleon Model (WNM) they depend neither on the number of wounded nucleons (W) nor on fluctuations of W. In the Grand Canonical Ensemble they do not depend on volume and volume fluctuations. Moreover, $\Delta[P_T, N]$ and $\Sigma[P_T, N]$ have two reference values, namely they are equal to zero in case of no fluctuations and one in case of independent particle production.

The system size dependence of $\Sigma[P_T, N]$ at 150*A*/158*A* GeV/*c* from the NA61/SHINE and NA49 experiments as function of system size (wounded nucleons) is shown in Fig. 5.5 (*left*). NA49 and NA61/SHINE measurements show consistent trends. Finally NA61/SHINE results for the NA61/SHINE acceptance for p+p, Be+Be and Ar+Sc collisions are presented in Fig. 5.5 (*right*). So far there are no prominent structures observed which could be related to a critical point.

5.4 Strangeness Production in P+p Interactions at 158 GeV/c

5.4.1 Ξ Production

Hyperons are excellent probes of the dynamics of proton-proton interactions as constituent strange quarks are not present in the initial state of this process. Therefore hyperon production has been studied in a long series of experiments in elementary hadron+hadron interactions. However, the experimental situation in this field remains inconclusive.

New data from p+p collisions on Ξ^- and $\overline{\Xi}^+$ hyperon production are presented. The event sample consists of 53 million registered interaction trigger events obtained at 158 GeV/*c* beam momentum corresponding to $\sqrt{s_{NN}}$ = 17.3 GeV/*c*. The results refer to primary Ξ^- and $\overline{\Xi}^+$ produced in strong and electromagnetic processes and are corrected for detector geometrical acceptance and reconstruction efficiency.

To find the Ξ candidates, all Λ candidates are combined with pion tracks of appropriate charge (daughter track). A fitting procedure is applied, using as parameters the decay position of the V^0 candidate, the momenta of both the V^0 decay tracks, the momentum of the daughter track, and finally the z position of the Ξ decay point. The x and y position of the Ξ decay position are not subject to the minimization, as they are determined from the parameters using momentum conservation. This procedure yields the decay position and the momentum of the Ξ candidate.

Preliminary results derived from two dimensional spectra (y vs p_T) are presented as transverse momentum distributions in bins of rapidity in Fig. 5.6. Statistical uncertainties are shown as vertical bars and preliminary estimates of systematic uncertainty are indicated by shaded bands. The blue lines show results of exponential fits to the measurements binned in m_T.

The obtained p_T spectra were used to calculate the rapidity spectrum of Ξ^- and $\overline{\Xi}^+$ production as the sum of measured points and extrapolation to the unmeasured region of p_T. The result is displayed in Fig. 5.7. Vertical bars show statistical, the shaded band systematic uncertainties. The rapidity distribution was fitted by a Gaussian function for extrapolation into the unmeasured regions. Based on summing the data points and the extrapolation of the fitted function resulted in the mean multiplicity $\langle \overline{\Xi}^+ \rangle = 0.00079 \pm 0.00002 \pm 0.00010$ and the mean multiplicity of $\langle \Xi^- \rangle = 0.0033$ $\pm$ 0.0001 $\pm$ 0.0006.

Additionally, the ratio of rapidity spectra $\overline{\Xi}^+ / \Xi^-$ was calculated and compared with UrQMD and EPOS1.99 model predictions in Fig. 5.8. UrQMD fails to describe $\overline{\Xi}^+ / \Xi^-$, which is a known problem of string models. EPOS1.99 describes the rapidity distributions of $\overline{\Xi}^+$, Ξ^- and their ratio, but not the shape of the transverse momentum spectra.

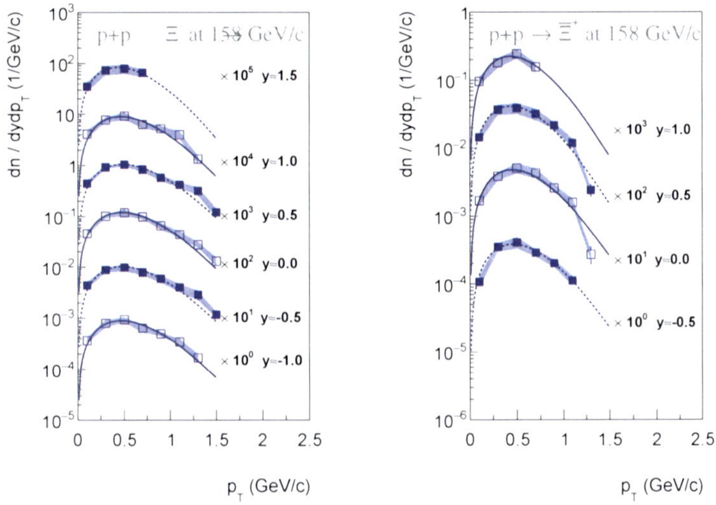

Fig. 5.6 Preliminary results on transverse momentum spectra of Ξ^- (*left*) and $\overline{\Xi}^+$ (*right*) hyperons produced in inelastic p+p interactions at 158 GeV/*c* in consecutive rapidity bins. Results are scaled for better separation, shaded bands indicate systematic uncertainty

Fig. 5.7 Preliminary results on rapidity spectrum of $\overline{\Xi}^+$ (red) and Ξ^- (blue) hyperons produced in inelastic p+p interactions at 158 GeV/*c* fitted by a Gaussian function (lines)

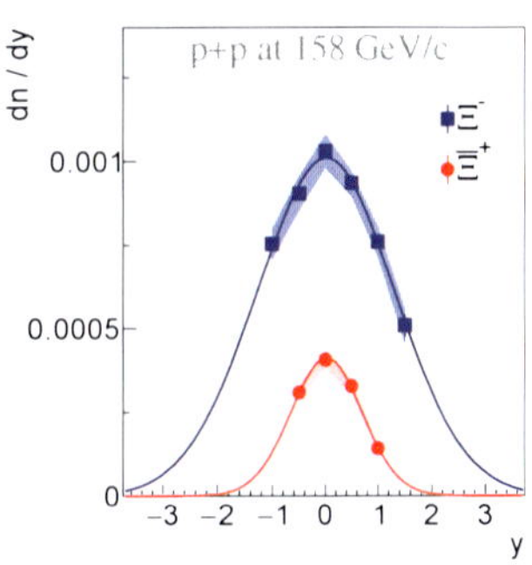

Fig. 5.8 Ratio of rapidity spectra $\overline{\Xi}^+ / \Xi^-$ compared with UrQMD (magenta solid line) and EPOS1.99 (blue solid line) models predictions

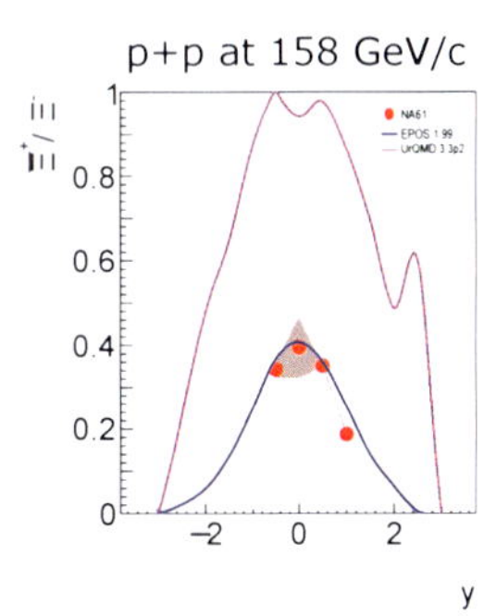

5.4.2 Search for Pentaquark Candidates

The NA49 Collaboration published evidence for the existence of a narrow $\Xi^- \pi^-$ baryon resonance with mass of 1.862 ± 0.002 GeV/c^2 and width below the detector resolution [12] in 2004. The significance was estimated to be 4.0σ. This state was a candidate for the hypothetical exotic $\Xi^{--}_{\frac{3}{2}}$ baryon with S=-2, I=$\frac{3}{2}$ and a quark content of $(dsds\bar{u})$. At the same mass a peak was observed in the $\Xi^- \pi^+$ spectrum which is a candidate for the $\Xi^0_{\frac{3}{2}}$ member of this isospin quartet with a quark content of $(dsus\bar{d})$. The corresponding antibaryon spectra also showed enhancements at the same invariant mass.

Recently a similar analysis was performed by NA61/SHINE based on an order of magnitude higher statistics. The first step in the analysis was the search for Λ candidates, which were then combined with the π^- to form the Ξ^- candidates. Next the $\Xi^{--}_{\frac{3}{2}}$ $\left(\Xi^0_{\frac{3}{2}}\right)$ were searched for in the $\Xi^- \pi^-$ ($\Xi^- \pi^+$) invariant mass spectrum, where the π^- (π^+) are primary vertex tracks. An analogous procedure was followed for the antiparticles.

To search for the exotic $\Xi^{--}_{\frac{3}{2}}$ state the selected Ξ^- candidates were combined with primary π^- tracks. The resulting $\Xi^- \pi^-$ invariant mass spectrum is shown in Fig. 5.9 (*top, left*). The shaded histogram is the mixed-event background, obtained by combining the Ξ^- and π^- from different events and normalising to the number of real combinations. The complete set of invariant mass distributions measured by NA61/SHINE ($\Xi^- \pi^-$, $\Xi^- \pi^+$, $\overline{\Xi}^+ \pi^-$, $\overline{\Xi}^+ \pi^+$) is shown in Fig. 5.9 (*left*). In addition to the described cuts, a lower cut of 3 GeV/c was imposed on the π^+ momenta to minimize the large proton contamination. Blue histograms show normalised mixed-event backgrounds. One sees that data overlap with the mixed-event backgrounds in the mass window 1.848 – 1.870 GeV/c^2 where the NA49 signal was observed; see Fig. 5.9 (*right*). Finally a narrow peak of $\Xi(1530)^0$ is observed in the invariant mass of $\overline{\Xi}^+ \pi^-$. The yield of observed $\Xi(1530)^0$ scales with the number of events compared to NA49 results.

In summary, this NA61/SHINE analysis of p+p interactions, with $\approx$ 10 times higher statistics, disproves the NA49 indication of the production of $\Xi^{--}_{\frac{3}{2}}$, $\Xi^0_{\frac{3}{2}}$ and their antiparticles. All four invariant mass distributions shown in Fig. 5.9 do not show significant signals in the mass window for which NA49 previously reported pentaquark candidates.

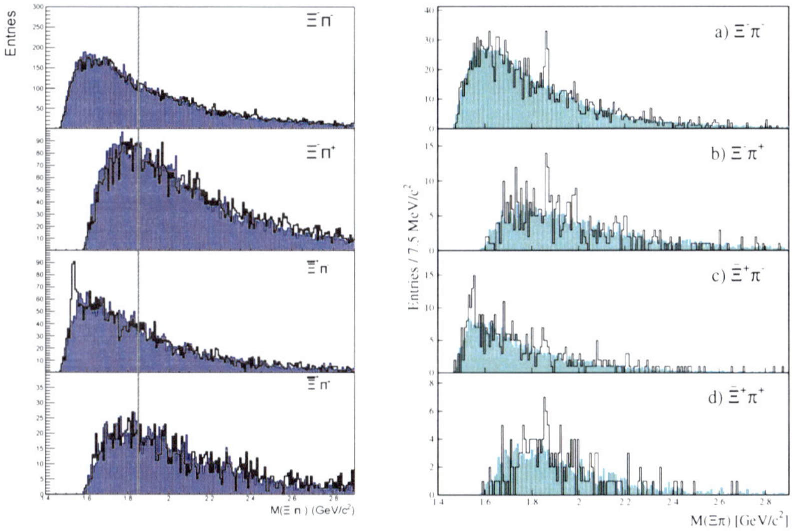

Fig. 5.9 *Left*: NA61/SHINE invariant mass spectra after selection cuts for $\Xi^- \pi^-$, $\Xi^- \pi^+$, $\overline{\Xi}^+ \pi^-$ (note that the $\Xi(1530)^0$ state is also visible) and $\overline{\Xi}^+ \pi^+$. The shaded histograms are the normalised mixed-event backgrounds. Gray line point to place where indication of the signal was observed by the NA49 Collaboration. NA61/SHINE results were obtained for 53M interaction trigger events (33M after all event cuts). *Right*: Similar plots but obtained by NA49 [12]. In NA49 6.5M recorded events (3.75M after event cuts) were used

Acknowledgments This work was supported by the Polish Ministry of Science and Higher Education (grants 667/N-CERN/2010/0, NN 202 48 4339 and NN 202 23 1837), the National Science Centre Poland 2015/18/M/ST2/00125.

References

1. N. Abgrall et al., [NA61/SHINE Collab.] *JINST* **9** (2014) P06005, arXiv:1401.4699 [physics.ins-det]
2. M. Gazdzicki and M. I. Gorenstein *Acta Phys.Polon.* **B30** (1999) 2705, arXiv:hep-ph/9803462 [hep-ph]
3. L. P. Csernai and D. Rohrich *Phys. Lett.* **B458** (1999) 454, arXiv:nucl-th/9908034 [nucl-th]
4. H. Stoecker *Nucl. Phys.* **A750** (2005) 121–147,arXiv:nucl-th/0406018 [nucl-th]
5. J. Brachmann, S. Soff, A. Dumitru, H. Stoecker, J.A. Maruhn, W. Greiner, L.V. Bravina, and D. H. Rischke Phys. Rev. C **61**, 024909 (2000). arXiv:nucl-th/9908010 [nucl-th]
6. C. Alt *et al.*, [NA49 Collab.] *Phys. Rev.* **C68** (2003) 034903, arXiv:nucl-ex/0303001
7. V. Klochkov and I. Selyuzhenkov, [NA61/SHINE Collab.] *Nucl. Phys.* **A982** (2019) 439–442, arXiv:1810.07579
8. A. Bialas and R. B. Peschanski *Nucl. Phys.* **B273** (1986) 703–718
9. L. Turko *Phys. Lett.* **B227** (1989) 149–152
10. F. K. Diakonos, N. G. Antoniou, and G. Mavromanolakis *PoS* **CPOD2006** (2006) 010
11. A. Aduszkiewicz *et al.*, [NA61/SHINE Collab.] arXiv:1510.00163
12. C. Alt *et al.*, [NA49 Collab.] *Phys. Rev. Lett.* **92** (2004) 042003, arXiv:hep-ex/0310014

Chapter 6
Recent Heavy-Flavor Results from STAR

Guannan Xie

Abstract In these proceedings, we report on the production of various open heavy-flavor hadrons and quarkonia in Au+Au collisions at $\sqrt{s_{\mathrm{NN}}} = 200$ GeV from the STAR experiment.

6.1 Introduction

Due to the intrinsic large mass (charm and bottom), measurements of heavy-flavor production (open heavy-flavor hadrons and quarkonia) are an important tool for studying the properties of the Quark-Gluon Plasma (QGP) formed in relativistic heavy-ion collisions. The modification of their distributions in transverse momentum (p_T) due to energy loss and in azimuth due to anisotropic flows is sensitive to heavy-quark dynamics in the partonic QGP phase [1, 2].

In these proceedings we present measurements of the D^0 nuclear modification factors and elliptic flow in Au+Au collisions from STAR, and compare to similar measurements for light-flavor hadrons. The $\Lambda_c^{\pm}$ and $D_s^{\pm}$ production are presented to study the coalescence mechanism for charm-quark hadronization. The measurements of open bottom production through the reconstruction of their displaced decay daughters (B $\rightarrow J/\psi$, D^0, e) are performed to test the mass dependence of parton-medium interactions in the QGP. The strong J/ψ suppression in heavy-ion collisions has a complicated interpretation as not only color-screening, but also the cold nuclear matter (CNM) effects and the regeneration mechanism play a role. Υ measurements are a cleaner probe of the color-screening effect at RHIC energies and the suppres-

---

Guannan Xie for the STAR Collaboration.

G. Xie (✉)
Lawrence Berkeley National Laboratory, Berkeley, CA 94706, USA
e-mail: xieguannanpp@gmail.com

D. Elia et al. (eds.), *The XVIII International Conference on Strangeness in Quark Matter (SQM 2019)*, Springer Proceedings in Physics 250,
https://doi.org/10.1007/978-3-030-53448-6_6

sion pattern of different bottomonium states will help to constrain the temperature of the medium. The J/ψ measurements as well as the Υ ones are also presented in these proceedings.

6.2 Nuclear Modification Factors for D^0

Figure 6.1 left panel shows the D^0 R_{AA}, which is the yield ratio between Au+Au and p+p scaled by the number of binary collisions [3]. From low to intermediate p_T region, the D^0 R_{AA} shows a characteristic structure which is qualitatively consistent with the expectation from model predictions in which charm quarks gain sizable collective motion during the medium evolution. In order to take advantage of the precision of the Au+Au spectra and avoid the large uncertainties from the p+p baseline, we construct the R_{CP} which is the yield ratio between central and peripheral Au+Au collisions. The right panel shows the D^0 R_{CP} for different centralities as a function of p_T with the 40–60% centrality spectrum as the reference. The measured D^0 R_{CP} in central 0–10% collisions shows a significant suppression at $p_T > 5\,\mathrm{GeV}/c$. The suppression level is similar to that of light-flavor hadrons and strange mesons and the suppression gradually decreases from central to mid-central and peripheral collisions, similarly as R_{AA}. The D^0 R_{CP} for $p_T < 4\,\mathrm{GeV}/c$ does not show a modification with centrality, in contrast to light-flavor hadrons. Calculations from the Duke group and the Linearized Boltzmann Transport (LBT) model match the data well [5, 6], while the improved precision of the new measurements is expected to further help constrain the theoretical model calculations.

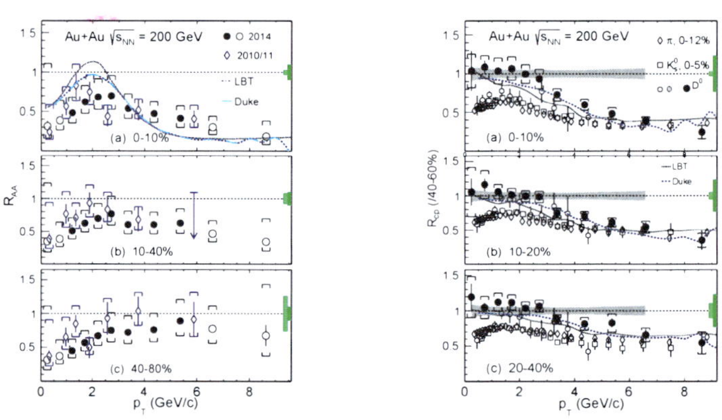

Fig. 6.1 (Left) D^0 R_{AA} in Au+Au collisions at $\sqrt{s_{\mathrm{NN}}}$ = 200 GeV for different centrality bins. (Right) D^0 R_{CP} with the 40–60% spectrum as the reference

6.3 D_s/D^0, Λ_c/D^0 Yield Ratios

Figure 6.2 left panel shows the Λ_c/D^0 yield ratio as a function of p_T for the 10–80% centrality class. The values show a significant enhancement compared to the calculations from PYTHIA. The model calculations which include coalescence hadronization of charm quarks can qualitatively reproduce the p_T dependence [7–9]. However, one needs measurements at low p_T to further differentiate between different models. The middle panel shows the measured Λ_c/D^0 ratio as a function of N_{part} in $3 < p_T < 6\,\text{GeV}/c$. There is a clear increasing trend towards more central collisions while the value in the peripheral collisions is comparable with the measurement in p+p collisions at $\sqrt{s_{\text{NN}}} = 7$ TeV from ALICE [10]. The right panel shows the D_s/D^0 ratio for two centrality classes. There is a strong enhancement compared to the PYTHIA fragmentation with no significant centrality dependence [11].

Besides the D^0, D_s and $\Lambda_c^{\pm}$, STAR also has performed measurements of $D^{\pm}$ in Au+Au collisions at $\sqrt{s_{\text{NN}}} = 200$ GeV. With these various charmed hadron measurements, the total charm quark cross section per binary nucleon-nucleon collision was obtained as listed in Table 6.1. The total $c\bar{c}$ cross section per binary nucleon-nucleon collision in Au+Au collisions is consistent with that in p+p within uncertainties. However, as demonstrated by the Λ_c/D^0 and D_s/D^0 yield ratios, the charm hadrochemistry is modified in heavy-ion collisions compared to p+p collisions.

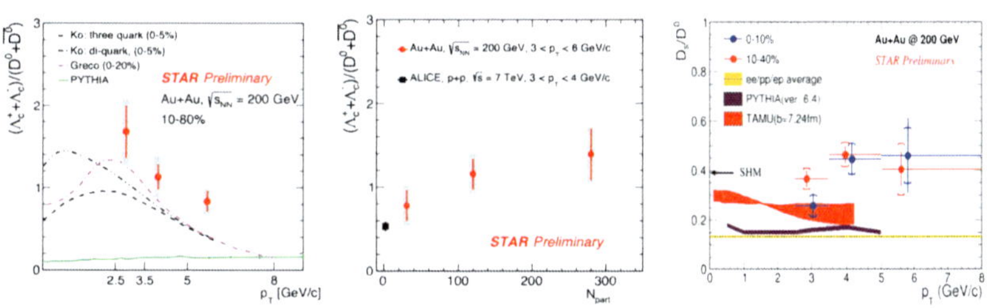

Fig. 6.2 (Left) Λ_c/D^0 ratio as a function of p_T for the 10–80% centrality class. (Middle) Λ_c/D^0 ratio as a function of N_{part} in $3 < p_T < 6\,\text{GeV}/c$. (Right) D_s/D^0 ratio as a function of p_T for the 0–10% and 10–40% centralities

Table 6.1 Total charm cross-section per binary nucleon-nucleon collision at midrapidity in Au+Au and p+p collisions at 200 GeV

Charm Hadron		Cross Section $d\sigma/dy(\mu b)$
Au+Au	D^0	41 ± 1 (stat) ± 5 (sys)
	D^+	18 ± 1 (stat) ± 3 (sys)
(10–40%)	D_s^+	15 ± 1 (stat) ± 5 (sys)
	Λ_c^+	78 ± 13 (stat) ± 28 (sys)
	total $c\bar{c}$	152 ± 13 (stat) ± 29 (sys)
p+p	total $c\bar{c}$	130 ± 30 (stat) ± 26 (sys)

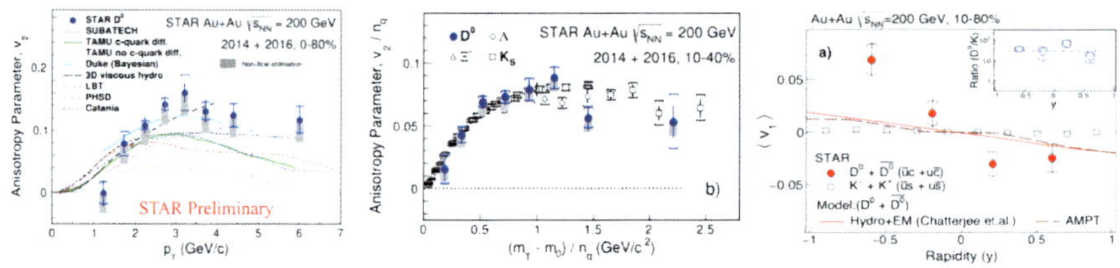

Fig. 6.3 D^0 elliptic flow v_2 vs p_T (Left) and the test of NCQ scaling (Middle) for D^0 and light-flavor hadrons. (Right) D^0 directed flow v_1 vs rapidity

6.4 D^0 Elliptic Flow (v_2) and Directed Flow (v_1)

Figure 6.3 left panel shows STAR results showing a large non-zero v_2 for D^0 mesons [12], which clearly follows the Number of Constituent Quarks (NCQ) scaling similarly as light-flavor hadrons below p_T of 1 GeV/c as shown in the middle panel. This suggests that charm quarks gain significant flow through interactions with the medium. The v_2 is compared to various model calculations and in particular the 3D viscous hydrodynamic model calculation can reproduce the results for $p_T < 4$ GeV/c. The other transport models with charm quark diffusion in the medium are consistent with the data when incorporating a diffusion coefficient ($2\pi T D_s$) in the range of $2 \sim 5$ around T_c [13].

The D-meson directed flow v_1 is expected to be sensitive to the initial tilt of the bulk medium, while the difference between D^0 and $\overline{D^0}$ is suggested to be sensitive to the initial electromagnetic field. Figure 6.3 right panel shows the first observation of a non-zero D-meson v_1 slope which is much larger than that of kaons. The v_1 values measured separately for D^0 and $\overline{D^0}$ are consistent within uncertainties. Future measurements with improved precision are needed to investigate the potential influence of the electromagnetic field on the v_1 values [14].

6.5 Measurements of R_{AA} for B-decayed J/ψ, D^0 and e

The STAR Heavy Flavor Tracker (HFT) provides the capability of using the impact parameter method to distinguish the daughter particles from decays of bottom hadrons. Figure 6.4 shows the R_{AA} of $B \rightarrow J/\psi$, D^0 and e. Strong suppressions for $B \rightarrow J/\psi$ and $B \rightarrow D^0$ at high p_T are observed. The production of $B \rightarrow e$ is less suppressed than that of $D \rightarrow e$ with a significance level of about 2σ, which is consistent with the expectation of mass hierarchy of parton energy loss [15].

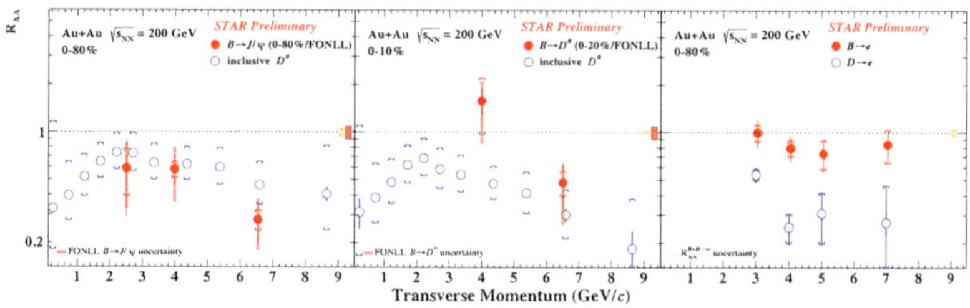

Fig. 6.4 R_{AA} of different daughter particles from decays of B-hadrons including $B \rightarrow J/\psi$, $B \rightarrow D^0$ and $B \rightarrow e$

6.6 Measurements of J/ψ Productions in Au+Au Collisions

Figure 6.5 shows the J/ψ R_{AA} reconstructed through the di-muon channel using the Muon Telescope Detector (MTD) as a function of p_T in Au+Au collisions [16]. As can be seen the J/ψ production is suppressed across the whole p_T range. The suppression at low p_T is likely due to the combination of the cold nuclear matter effects, the regeneration and the dissociation in the QGP. With increasing p_T the CNM effects are expected to diminish. The relative contribution from the b-hadron decays increases with p_T, and the suppression level of J/ψ originating from these decays is expected to be smaller than that of the prompt J/ψ. The centrality dependence of the J/ψ suppression is shown in the right panel. The R_{AA} decreases from peripheral to central collisions. Comparing the Au+Au measurements at $\sqrt{s_{\rm NN}} = 200$ GeV to the Pb+Pb measurements at $\sqrt{s_{\rm NN}}$ = 2.76 TeV from the LHC [17, 18], the STAR result shows more suppression in central and semi-central collisions, which is likely due to a smaller contribution from regeneration caused by the lower charm production cross-section at the RHIC energy. Models taking into account dissociation and regeneration can reasonably describe the data [19–21].

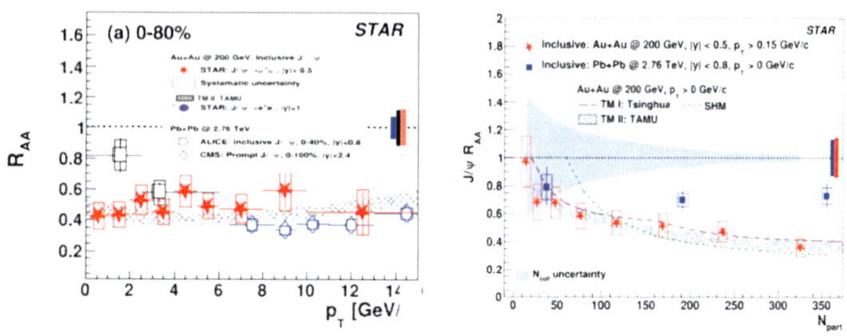

Fig. 6.5 (Left) J/ψ R_{AA} as a function of p_T in Au+Au collisions at $\sqrt{s_{\rm NN}}$ = 200 GeV. (Right) J/ψ R_{AA} as a function of N_{part} in Au+Au collisions, compared to that in Pb+Pb collisions at $\sqrt{s_{\rm NN}}$ = 2.76 TeV and model calculations

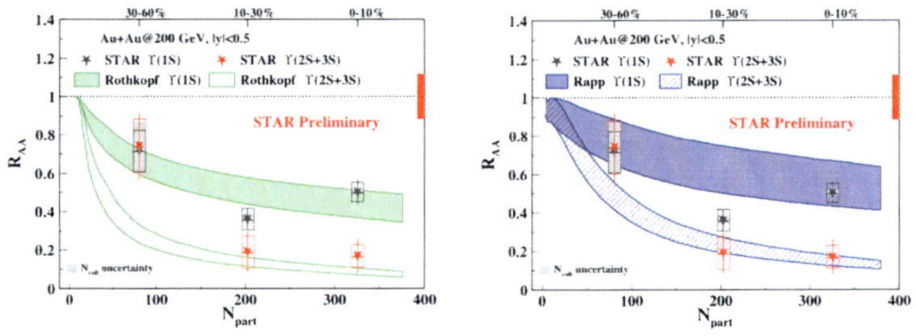

Fig. 6.6 $\Upsilon(1S)$ and $\Upsilon(2S+3S)$ R_{AA} as a function of N_{part} in Au+Au collisions, compared to two model calculations

6.7 Measurements of Υ Productions in Au+Au Collisions

Figure 6.6 shows the $\Upsilon(1S)$ and $\Upsilon(2S+3S)$ R_{AA} as a function of N_{part} in Au+Au collisions from the combined dielectron and dimuon results. The R_{AA} shows a decreasing trend from peripheral to central collisions for both Υ R_{AA}, while the $\Upsilon(2S+3S)$ are more suppressed than $\Upsilon(1S)$ in the most central collisions. This is consistent with the "sequential melting" expectation. The data are also compared with two model calculations. In the Rothkopf model [22] the Υ behavior in the QGP medium is described using a complex potential from lattice QCD calculations and there are no CNM or regeneration effects. While in the Rapp model [23], both CNM and regeneration effects are taken into account. These two models can describe well the measurements for the $\Upsilon(1S)$ and $\Upsilon(2S+3S)$ in mid-central and central collisions.

6.8 Summary

We have presented the recent measurements of various open heavy-flavor hadrons in Au+Au collisions at $\sqrt{s_{\rm NN}} = 200$ GeV utilizing the HFT at STAR. We have also reported on the measurements of the J/ψ and Υ production in Au+Au collisions at $\sqrt{s_{\rm NN}} = 200$ GeV enabled by the MTD.

References

1. G. Moore, D. Teaney, Phys. Rev. C **71**, 064904 (2005)
2. M. Cacciari et al., Phys. Rev. Lett. **95**, 122001 (2005)
3. J. Adam et al., STAR Collaboration. Phys. Rev. C **99**, 034908 (2019)
4. L. Adamczyk et al., STAR Collaboration. Phys. Rev. D **86**, 072013 (2012)
5. Y. Xu et al., Phys. Rev. C **97**, 014907 (2018)
6. S. Cao et al., Phys. Rev. C **94**, 014909 (2016)
7. I. Kuznetsova et al., The European Physical Journal C **51**, 113133 (2007)

8. S.H. Lee et al., Phys. Rev. Lett. **100**, 222301 (2008)
9. S. Ghosh et al., Phys. Rev. D **90**(5), 2–7 (2014)
10. S. Acharya *et al.* (ALICE Collaboration) JHEP **04** 108 (2018)
11. L. Zhou (STAR Collaboration), Nuclear Physics A **967**, 620-623 (2017)
12. L. Adamczyk et al., STAR Collaboration. Phys. Rev. Lett. **118**, 212301 (2017)
13. X. Dong, V. Greco, Progress in Particle and Nuclear Physics **104**, 97 (2019)
14. J. Adam *et al.* (STAR Collaboration), arXiv:1905.02052
15. S. Zhang. (STAR Collaboration), Int. J. Mod. Phys. Conf. Ser. 2018.46
16. J. Adam et al., STAR Collaboration. Phys. Lett. B **797**, 134917 (2019)
17. B. Abelev et al., ALICE Collaboration. Phys. Lett. B **734**, 314 (2014)
18. S. Chatrchyan et al., CMS Collaboration. JHEP **05**, 063 (2012)
19. K. Zhou et al., Phys. Rev. C **89**, 054911 (2014)
20. X. Zhao, R. Rapp, Phys. Rev. C **82**, 064905 (2010)
21. X. Zhao, R. Rapp, Nuclear Physics A **859**, 114–125 (2011)
22. B. Krouppa et al., Phys. Rev. D **97**, 016017 (2018)
23. X. Du et al., Phys. Rev. C **96**, 054901 (2017)

Chapter 7
Recent Results from ATLAS: Onia, Heavy-Flavor, and More

Martin Spousta

Abstract In this short report we review the recent results obtained using the ATLAS experiment at the LHC on quarkonia and heavy flavor measurements in heavy-ion collisions.

In this short report we review the recent results obtained using the ATLAS experiment [1] at the LHC on quarkonia and heavy flavor measurements in heavy-ion collisions. The text is divided into two sections. The first section briefly discusses the final state effects in the production of quarkonia and heavy flavor and it describes the measurements of suppression and flow of quarkonia and heavy-flavor in Pb+Pb collisions. The second section discusses the initial state effects and describes the measurements done in p+Pb collisions.

7.1 Suppression and Flow of Quarkonia and Heavy-Flavor in Pb+Pb Collisions

It was predicted a long time ago that the Debye screening of the quark colour charge in a hot quark-gluon plasma (QGP) would lead to a dissociation of quarkonium bound state in the medium, when the Debye length becomes smaller than the quarkonium binding radius [2]. While the strong suppression of quarkonia in nucleus-nucleus collisions is firmly established experimental fact, its interpretation remains open. Besides initial state effects and the interaction with co-moving hadrons, which are discussed in the context of p+Pb measurements in Sect. 7.2, it is also the parton energy loss which was recently suggested as an alternative physics mechanism for the

Martin Spousta on behalf of the ATLAS Collaboration.

M. Spousta (✉)
Faculty of Mathematics and Physics, Charles University V Holešovičkách 2, Prague 180 00, Czech Republic
e-mail: martin.spousta@mff.cuni.cz

D. Elia et al. (eds.), *The XVIII International Conference on Strangeness in Quark Matter (SQM 2019)*, Springer Proceedings in Physics 250,
https://doi.org/10.1007/978-3-030-53448-6_7

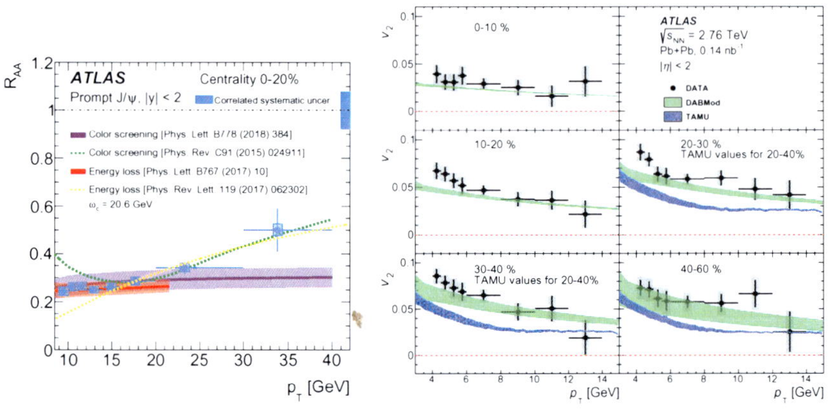

Fig. 7.1 Left: Comparison of $R_{\rm AA}$ for prompt J/ψ production with different theoretical models. The statistical uncertainty of each point is indicated by a narrow error bar. The error box plotted with each point represents the uncorrelated systematic uncertainty, while the shaded error box at $R_{\rm AA} = 1$ represents correlated scale uncertainties. Plot taken from [5]. Right: Comparison of the Pb+Pb heavy-flavor muon v2 with calculations from the TAMU [6] and DABMod [7] models. Each panel represents a different centrality interval. For the 20−30% and 30−40% centrality intervals, the plotted TAMU values correspond to the 20−40% centrality interval. For the data, the error bars and shaded bands represent statistical and total uncertainties, respectively. For the model calculations, the bands represent theoretical systematic uncertainties. Plot taken from [8]

observed quarkonia suppression [3, 4]. To help address the origin of the quarkonium suppression, ATLAS published a measurement [5] of prompt charmonia, originating from the formation of $c\bar{c}$ bound state which follows immediately after the hard process, and non-prompt charmonia, originating from b-hadron decays. In that measurement, prompt and non-prompt per-event yields, non-prompt fraction and nuclear modification factors, $R_{\rm AA}$, of the J/ψ and $\psi(2S)$ is performed. The measurement is performed in the dimuon decay channel in Pb+Pb collisions at $\sqrt{s_{NN}} = 5.02$ TeV, for quarkonia with transverse momentum $9 < p_{\rm T} < 40$ GeV and rapidity $|y| < 2.0$.

Strong suppression of prompt and non-prompt J/ψ and $\psi(2S)$ mesons is observed in Pb+Pb data. In the 0−10%, the $R_{\rm AA}$ of prompt J/ψ is approximately 0.2 for $p_{\rm T} = 10$ GeV and it grows and achieves a value of 0.3 at $p_{\rm T} \approx 25$ GeV. The $R_{\rm AA}$ of non-prompt J/ψ in 0−10% is approximately 0.3 and remains independent of $p_{\rm T}$ within the full $p_{\rm T}$ range of this measurement. The dependence of the $R_{\rm AA}$ on centrality is approximately the same for prompt and non-prompt J/ψ. The ratio of $\psi(2S)$ to J/ψ meson production is measured for both the prompt and non-prompt mesons, and as a function of centrality. Values consistent with unity are measured for the non-prompt mesons, while the values observed for the prompt mesons are below unity. The left panel of Fig. 7.1 shows a comparison of the J/ψ $R_{\rm AA}$ measured in 0 − 20% central collisions with theoretical calculations.

ATLAS also measured the elliptic flow coefficient (v_2) of prompt and non-prompt J/ψ in Pb+Pb collisions [9]. The measurement is done using similar techniques and kinematic cuts and using the same dataset as used for the measurement of the $R_{\rm AA}$

of J/ψ and $\psi(2S)$. In that measurement, the v_2 coefficient is evaluated relative to the event plane and the results are presented as a function of p_T, y and centrality. It is found that both prompt and non-prompt J/ψ mesons have non-zero elliptic flow. Prompt J/ψ v_2 is found to decrease as a function of p_T, while the non-prompt J/ψ is found to be, with limited statistical significance, consistent with a flat behaviour over the studied kinematic region. There is no observed dependence on y or centrality.

The suppression and the azimuthal anisotropy was measured also for muons from heavy-flavor decays in 2.76 TeV Pb+Pb collisions [8]. The measurement is performed over the muon transverse momentum range $4 < p_T < 14$ GeV. Backgrounds arising from in-flight pion and kaon decays, hadronic showers, and mis-reconstructed muons is statistically removed using a template-fitting procedure. The resulting R_{AA} is observed to be independent of p_T within uncertainties. For the 10% most central Pb+Pb events, the measured R_{AA} is approximately 0.35 indicating a clear suppression of production of heavy-flavor muons in Pb+Pb collisions.

The azimuthal modulation of the heavy-flavor muon yields is measured and the associated Fourier coefficients v_n for $n = 2, 3$ and 4 are given as a function of p_T and centrality. Significant v_2 values up to about 0.08 are observed at $p_T = 4$ GeV. In the 10–20%, 20–30%, and 30–40% intervals, the v_2 decreases with p_T but is still significant at 10 GeV. At fixed p_T, the v_2 values show a systematic variation with centrality which is typical of elliptic-flow measurements. For most centrality intervals, v_3 also decreases with increasing p_T over the 4–8 GeV p_T range. At a given p_T and centrality, the v_3 values are smaller than the v_2 values by a factor of 2–4. The v_3 values show a much weaker centrality dependence than the v_2. The right panel of Fig. 7.1 shows a comparison of heavy-flavor muon v_2 in five centrality bins with theory calculations.

7.2 Quarkonia and Heavy Flavor in p+Pb Collisions

In order to understand quarkonium suppression in Pb+Pb collisions it is necessary to disentangle effects due to interaction between quarkonium and the QGP medium from those that can be ascribed to initial state effects, such as: modifications of the nuclear parton distribution functions, parton saturation effects in the incident nucleus, and parton energy loss through interactions with the nuclear medium of the initial state. These effects can be accessed in p+Pb collisions where a large region of hot QGP is a priori not expected to occur. At the same time, modification of quarkonia production is also expected even in p+Pb collisions to originate from final state effects, namely from the absorption of the heavy quark-antiquark pair through interactions with the co-moving hadrons. For a review of these physics mechanisms see e.g. [10].

To help disentangle the role of the above mentioned phenomena, ATLAS measured the production cross sections of five quarkonium states, J/ψ, $\psi(2S)$, and $\Upsilon(nS)$ $(n = 1, 2, 3)$ in p+Pb collisions at 5.02 TeV with an integrated luminosity of 28 nb^{-1} [11]. The resulting nuclear modification factor, R_{pPb}, is found to be con-

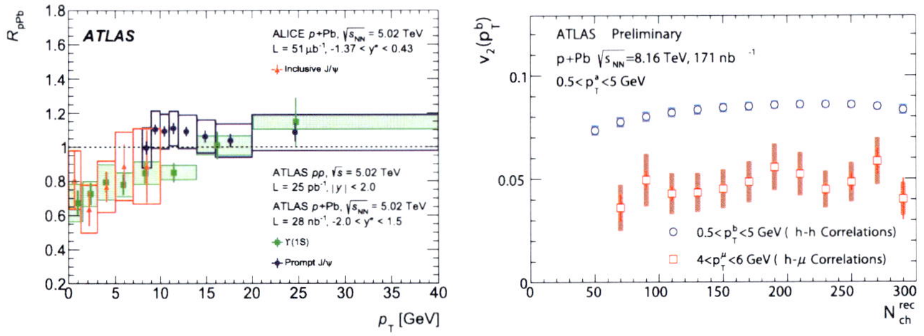

Fig. 7.2 Left: Nuclear modification factor, R_{pPb}, as a function of p_{T} for $\Upsilon(1S)$ (in green) and prompt J/ψ (in blue). The vertical error bars cover the statistical uncertainties and horizontal error bars represent the bin size. The horizontal position of data point indicates the mean of the weighted p_{T} distribution. The vertical size of colored boxes underneath the data points represent the systematic uncertainties. The R_{pPb} of inclusive J/ψ measured by ALICE from [12] is also shown in red. Figure taken from [11]. Right: The v_2 as a function of number of reconstructed charged particles ($N_{\mathrm{ch}}^{\mathrm{rec}}$) obtained from the template fits to hadron-hadron (circles) and to hadron-μ correlations (squares). The error bars and shaded bands indicate statistical and systematic uncertainties, respectively. For the hadron-hadron correlations, the statistical errors are too small to be seen. Figure taken from [13]

sistent with unity for prompt and non-prompt J/ψ for the kinematic range of the measurement, $8 < p_{\mathrm{T}} < 40$ GeV. The prompt $\psi(2S)$ production is suppressed with respect to prompt J/ψ production in p+Pb collisions with a significance of one standard deviation.

The R_{pPb} for $\Upsilon(1S)$ is measured for $p_{\mathrm{T}} < 40$ GeV and is found to be smaller than unity at $p_{\mathrm{T}} < 15$ GeV, increasing with p_{T} and becoming compatible with unity at high p_{T}. The production of $\Upsilon(2S)$ and $\Upsilon(3S)$ is found to be suppressed with respect to $\Upsilon(1S)$ in the integrated kinematic ranges of $p_{\mathrm{T}} < 40$ GeV and $-2 < y^* < 1.5$ in p+Pb collisions with significance at the level of two standard deviations. Both the prompt $\psi(2S)$ to J/ψ and $\Upsilon(2S)$ to $\Upsilon(1S)$ double ratios show decreasing trend with increasing centrality of p+Pb collisions. The left panel of Fig. 7.2 shows a compilation of results on R_{pPb} measured as a function of p_{T} for $\Upsilon(1S)$ and prompt J/ψ.

The initial stage effects can also be studied using measurements of heavy quarks. ATLAS measured prompt D^0 mesons (D^0 and $\bar{D}^0$) and D^* mesons (which are not coming from the sequential decays of b-hadrons) using p+Pb collisions at 8.16 TeV with an integrated luminosity of 76.3 $\mu\mathrm{b}^{-1}$ [14]. The D^0 and D^* mesons are reconstructed in the decay channels with kaons in the range of transverse momentum $3 < p_{\mathrm{T}} < 30$ GeV and $5 < p_{\mathrm{T}} < 30$ GeV, respectively. The relative D meson productions at forward and backward center of mass rapidities is studied via the forward ($0 < y^< 0.5$) to backward ($-0.5 < y < 0$) production ratio, R_{FB}, as a function of transverse momentum. No significant forward-backward asymmetry for prompt D^0 and D^* mesons is observed. The azimuthal correlations between inclusive D^* mesons and charged particles is also measured. The second-order harmonic coefficients for D^*-hadron correlations are extracted using template fits to remove the "non-flow" contributions. The D^*-hadron correlations are broadly consistent with the measurement of muon-hadron correlations.

A detailed measurement of long-range pseudorapidity correlations between muons and charged particles was done using p+Pb data at 8.16 TeV [13]. The template fitting method is used to obtain the second-order Fourier coefficient, $v_{2,2}$, corresponding to the genuine long-range correlations. The $v_{2,2}$ is factorized to obtain the single-particle anisotropy coefficient, v_2, of the muons. The right panel of Fig. 7.2 shows a comparison of hadron-μ and hadron-hadron v_2. Significant v_2 values are observed for muons over the transverse momentum interval $4 < p_{\mathrm{T}} < 8$ GeV and over the range of the charged particle multiplicity of $100 - 300$. The muon-v_2 values are largest at 4 GeV and decrease with increasing p_{T}. Given that nearly all the prompt muons over this p_{T} range are produced from decays of heavy-flavor particles, these results confirm a significant azimuthal anisotropy in the distribution of heavy-flavor particles produced in p+Pb collisions.

7.3 Conclusions

The ATLAS Experiment at the LHC provides a wealth of new results on quarkonia and heavy flavor measurements in Pb+Pb and p+Pb collisions. The new results from Pb+Pb collisions provide a precise quantification of strong suppression effects and flow phenomena. The new results from p+Pb collisions should allow constraints to be put on the origin of flow phenomena in small collision systems and quantify the role of initial state effects in the studied data.

Acknowledgments This work was supported by Grant Agency of the Czech Republic under grant 18-12859Y, by the Ministry of Education, Youth and Sports of the Czech Republic under grant LTT 17018, and by Charles University grant UNCE/SCI/013.

References

1. ATLAS Collaboration. The ATLAS Experiment at the CERN Large Hadron Collider. *JINST*, 3:S08003, 2008
2. T. Matsui, H. Satz, J/ψ Suppression by Quark-Gluon Plasma Formation. Phys. Lett. B **178**, 416–422 (1986)
3. Franois Arleo, Quenching of Hadron Spectra in Heavy Ion Collisions at the LHC. Phys. Rev. Lett. **119**(6), 062302 (2017)
4. Martin Spousta, On similarity of jet quenching and charmonia suppression. Phys. Lett. B **767**, 10–15 (2017)
5. ATLAS Collaboration. Prompt and non-prompt J/ψ and $\psi(2S)$ suppression at high transverse momentum in 5.02 TeV Pb+Pb collisions with the ATLAS experiment. *Eur. Phys. J.*, C78(9):762, 2018
6. Min He, Rainer J. Fries, Ralf Rapp, Heavy Flavor at the Large Hadron Collider in a Strong Coupling Approach. Phys. Lett. B **735**, 445–450 (2014)

7. Caio A. G. Prado, Jacquelyn Noronha-Hostler, Roland Katz, Alexandre A. P. Suaide, Jorge Noronha, Marcelo G. Munhoz, and Mauro R. Cosentino. Event-by-event correlations between soft hadrons and D^0 mesons in 5.02 TeV PbPb collisions at the CERN Large Hadron Collider. *Phys. Rev.*, C96(6):064903, 2017
8. ATLAS Collaboration. Measurement of the suppression and azimuthal anisotropy of muons from heavy-flavor decays in Pb+Pb collisions at $\sqrt{s_{\mathrm{NN}}} = 2.76$ TeV with the ATLAS detector. *Phys. Rev.*, C98(4):044905, 2018
9. ATLAS Collaboration. Prompt and non-prompt J/ψ elliptic flow in Pb+Pb collisions at $\sqrt{s_{NN}} = 5.02$ Tev with the ATLAS detector. *Eur. Phys. J.*, C78(9):784, 2018
10. A. Andronic et al., Heavy-flavour and quarkonium production in the LHC era: from proton proton to heavy-ion collisions. Eur. Phys. J. C **76**(3), 107 (2016)
11. ATLAS Collaboration. Measurement of quarkonium production in proton lead and protonproton collisions at 5.02 TeV with the ATLAS detector. *Eur. Phys. J.*, C78(3):171, 2018
12. ALICE Collaboration. Rapidity and transverse-momentum dependence of the inclusive J/ψ nuclear modification factor in p-Pb collisions at $\sqrt{s_{N\ N}} = 5.02$ TeV. *JHEP*, 06:055, 2015
13. ATLAS Collaboration. Measurement of the long–range pseudorapidity correlations between muons and charged-particles in $\sqrt{s_{\mathrm{NN}}}$=8.16 TeV proton-lead collisions with the ATLAS detector. ATLAS-CONF-2017-006
14. ATLAS Collaboration. *D* meson production and long-range azimuthal correlation in 8.16 TeV p+Pb collisions with ATLAS. ATLAS-CONF-2017-073

Chapter 8
Recent Results on Heavy Flavour from CMS

Ruslan Chistov

Abstract In this talk I report recent results on heavy flavor physics from pp collisions at CMS. It includes spectroscopy of open or hidden beauty hadrons from pp collision data at $\sqrt{s} = 8$ and 13 TeV. These studies are aimed at gaining new knowledge on the properties of beauty hadrons which enhances our understanding of QCD.

8.1 Introduction

The CMS [1] experiment at LHC is designed mainly for high-p_t physics including studies of the Higgs boson, precision measurements of the Standard Model processes and parameters and numerous searches for New Physics. Nevertheless, CMS is contributing intensively in heavy flavor physics. This is possible due to several features of the experiment, including:

(a) A redundant muon system with large rapidity coverage provides high-purity muon identification (muon-ID) with standalone $\delta p_t / p_t = 10\%$ up to $p_t = 200$ Gev/c;

(b) A very efficient hardware Level 1 trigger and highly flexible High Level Trigger with paths dedicated to b-hadron specific decays with muons in the final state.

(c) A good p_t resolution in tracker with high efficiency (>99%) and good secondary vertex reconstruction with impact parameter resolution down to $\approx 15\ \mu$m.

In this article I will discuss the results on b-hadrons recently obtained by CMS from the analyses of the Run 1 and Run 2 data.

R. Chistov (✉)
P. N. Lebedev Physical Institute of the Russian Academy of Sciences, Moscow, Russia
e-mail: rchistov@cern.ch

D. Elia et al. (eds.), *The XVIII International Conference on Strangeness in Quark Matter (SQM 2019)*, Springer Proceedings in Physics 250,
https://doi.org/10.1007/978-3-030-53448-6_8

8.2 Study of $B_{s1}(5830)^0$ and $B_{s2}^*(5840)$ Decaying into B^+K^- and $B^0K_s^0$ in CMS

In 2018 the CMS Collaboration published results from a study of excited B_s^0 states, the $B_{s2}^*(5840)^0$ and $B_{s1}(5830)^0$, using 19.6 fb^{-1} of pp data at $\sqrt{s} = 8$ TeV. Previously these states were observed by the CDF [2] and D0 [3] Collaborations and confirmed by the LHCb Collaboration [4] in the charged decay mode B^+K^-. CMS confirmed all previous measurements and observed the new decay mode $B_{s2}^*(5840)^0 \to B^0K_s^0$ and obtained evidence for the analogous neutral mode for the $B_{s1}(5830)^0$ state (see Fig. 8.1). Various mass and mass difference measurements along with the ratios of production cross-sections times branching fractions were measured [5]. As a by-product of this analysis, the first measurement of the mass difference between spin excited B-meson states was obtained: $M(B^{*0}) - M(B^{*+}) = (0.91 \pm 0.24 \pm 0.09 \pm 0.02(PDG))$ MeV.

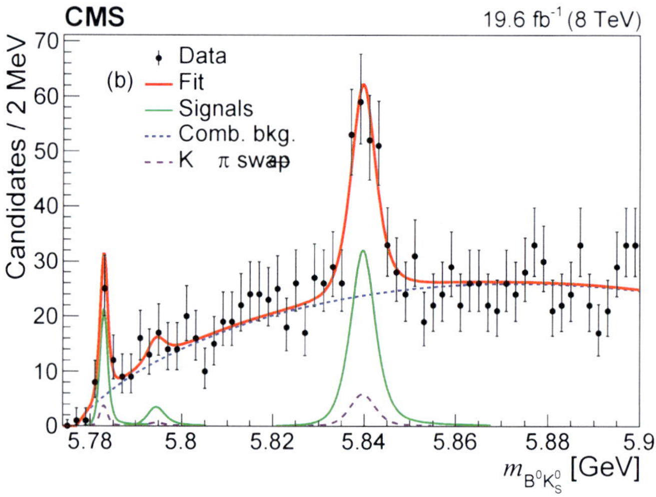

Fig. 8.1 Invariant mass distributions of $B^0K_S^0$ candidates with the results of the fit overlaid [5]. The points represent the data, the thick solid curves are the results of the overall fits, and the thin solid lines display the signal contributions. The short-dashed lines show the combinatorial background contributions. The long-dashed lines show the contributions from swapping $K^\pm \to \pi^\pm$ in the reconstruction of the B^0 mesons

8.3 Observation of Two Resolved States $\chi_{b1}(3P)$ and $\chi_{b2}(3P)$ in CMS

The bottomonium system plays very important role in our understanding of the dynamics of quarks inside the hadrons and permits thoroughly tests of various theoretical models describing the potential within the $b-\bar{b}$ system. The $\chi_{bJ}(3P)$ triplet is a special place since the masses of the triplet members are nearly at the open beauty threshold. In charmonium system such a proximity of $\chi_{c1}(2P)$ state to the $D^0\bar{D}^{*0}$ threshold results in the existence of the $X(3872)$. The question about whether this state is just the ordinary $\chi_{c1}(2P)$ with the mass modified by the interaction with the $D^0\bar{D}^{*0}$ threshold or it is a mixture of $\chi_{c1}(2P)$ and $D^0\bar{D}^{*0}$ molecular or even tetraquark, remains open. The study of the $\chi_{bJ}(3P)$ triplet states could shed new light on the existence of possible X_b state, beauty analogue of $X(3872)$. Also, this study tests theoretical models describing the spin-orbit interaction term of the inter-quark potential within the bottomonium system.

In 2018, by using the data collected at $\sqrt{s} = 13$ TeV (from 2015 to 2017 with the integrated luminosity of 80 fb^{-1}), the CMS Collaboration published the first observation of resolved $\chi_{b1}(3P)$ and $\chi_{b2}(3P)$ states along with the measurement of their masses and mass difference [6]. These measurements become possible due to the good invariant mass resolution in $\chi_{b1,2}(3P) \to \Upsilon(3S)\gamma$ decay, where $\Upsilon(3S) \to \mu^+\mu^-$ and photons are detected through the conversion into the e^+e^- pairs. Figure 8.2 shows the $\Upsilon(3S)\gamma$ invariant mass distribution after the photon energy correction as well as the overlaid fit. The $\chi_{b1,2}(3P)$ masses were measured to be $M_1 = 10513.42 \pm 0.41 \pm 0.18$ MeV, $M_2 = 10524.02 \pm 0.57 \pm 0.18$ MeV. The mass difference was

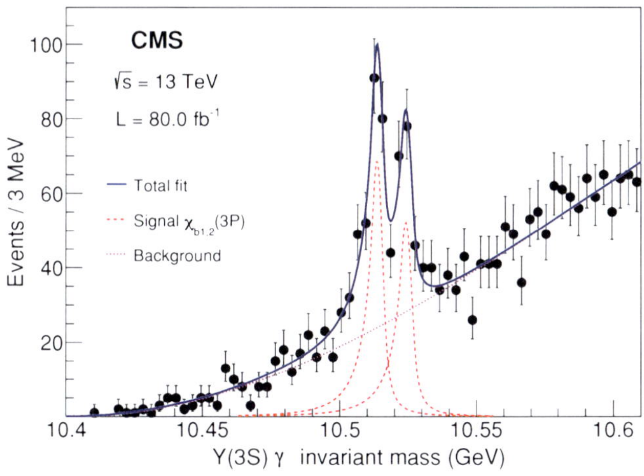

Fig. 8.2 The invariant mass distribution of the $\chi_{b1,2}(3P) \to \Upsilon(3S)\gamma$ candidates [6]. The vertical bars are the statistical uncertainties. The curves represent the fitted contributions of the two signal peaks, the background, and their sum

obtained to be $\Delta M = 10.6 \pm 0.64 \pm 0.17$ MeV. Out of twenty theoretical predictions for ΔM, eighteen of them fall into the interval [8–18] MeV and only one gives for the ΔM the value of -2 MeV. The negative sign, i.e. unnatural mass hierarchy, in this model is due to the coupling with open-beauty threshold. The obtained experimental result favors the natural mass hierarchy when the J = 2 state is heavier than the J =1 state.

8.4 Observation of Two Excited B_c^+ States

The spectrum of B_c^+ $(\bar{b}c)$ states will help to understand in a greater depth the dynamics of heavy-heavy quark systems. In 2014, using Run 1 data, the ATLAS Collaboration reported the observation (5.4 σ) of a new state decaying into $B_c^+\pi^+\pi^-$ ($B_c^+ \to J/\psi\pi^+$) with mass consistent with the predictions for $B_c(2S)^+$ [7]. Then, the LHCb Collaboration, with 8 TeV data, in the same decay chain, found no significant signal and set upper limit on the $B_c^+(2S)$ production cross-section depending on its mass [8].

In 2019, using full Run 2 statistics, the CMS Collaboration observed two well resolved peaks in $B_c^+\pi^+\pi^-$ ($B_c^+ \to J/\psi\pi^+$) invariant mass [9] (see Fig. 8.3). The lower mass peak is attributed to the reflection from the $B_c(2S)^{*+}$ meson that firstly decays into $B_c^{*+}\pi^+\pi^-$. Then the B_c^{*+} decays to B_c^+ ground state and a soft photon which is not reconstructed. The higher mass peak is just a signal from fully reconstructed $B_c(2S)^+$ in $B_c^+\pi^+\pi^-$ final state. The significance of each peak was found to be larger than 5 σ. The mass difference between two peaks was measured to be $29.1 \pm 1.5 \pm 0.7$ MeV and the mass of $B_c(2S)^+$ was measured to be $6871.0 \pm 1.2(stat.) \pm 0.8(syst.) \pm 0.8(B_c)$ MeV, in agreement with theoretical predictions.

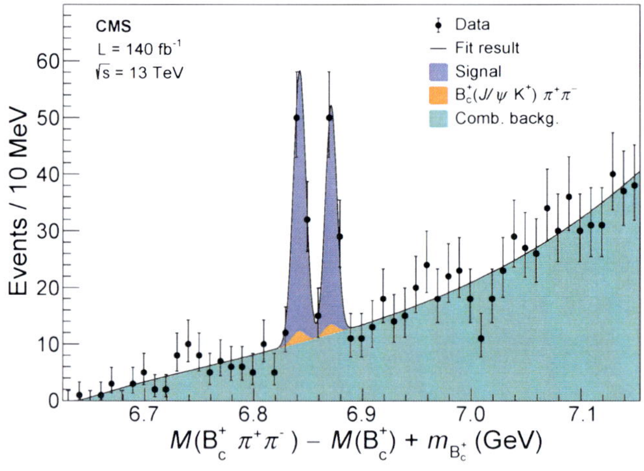

Fig. 8.3 The invariant mass distribution of $B_c^+\pi^+\pi^-$ combinations. The points are data and the curve is the result of the fit

8.5 Search for Exotic States in $B^+ \to J/\psi\bar{\Lambda}p$ Decay

Fifteen years ago the Belle Collaboration discovered the first exotic charmonium-like state, the X(3872), decaying into $J/\psi\pi^+\pi^-$. During last 15 years, many charmonium-like states were observed including charged states. There are several theoretical interpretations of these states. Among them one can mention hadrocharmonium, tetraquarks and molecular states. No one is sufficient for describing all properties of all observed states. Observation of new states will widen our understanding of dynamics of quarks in different configurations. Recently, the LHCb Collaboration by analysing the $\Lambda_b^0 \to J/\psi K^- p$ decay observed three pentaquark states in the $J/\psi p$ system. It is therefore interesting to find such b-hadron decay mode in which it is possible to search for new pentaquarks, either in the same $J/\psi p$ system or in a new system, with strangeness, e.g. $J/\psi\Lambda$. Such hidden charm strange are predicted recently close to their threshold [11].

In 2019 the CMS Collaboration released the results from the study of $B^+ \to J/\psi\bar{\Lambda}p$ decay [10]. The invariant mass distribution of the $J/\psi\bar{\Lambda}p$ combinations is shown in Fig. 8.4 demonstrated a very clean signal from this decay. The ratio $\frac{\mathcal{B}(B^+ \to J/\psi\bar{\Lambda}p)}{\mathcal{B}(B^+ \to J/\psi K^{*+})}$ was measured which corresponds to the branching fraction $\mathcal{B}(B^+ \to J/\psi\bar{\Lambda}p) = (15.1 \pm 0.8(stat.) \pm 0.5(syst.) \pm 0.9(Br))$. This is the most precise measurement of this decay to date. Also, the invariant mass distributions of the $\bar{\Lambda}p$, $J/\psi p$ and $J/\psi\bar{\Lambda}$ systems are studied and found to be inconsistent with pure phase space hypothesis. Then, the model-independent angular amplitude analysis was used to test for a presence of exotic resonances in $J/\psi p$ and $J/\psi\Lambda$ systems. It was

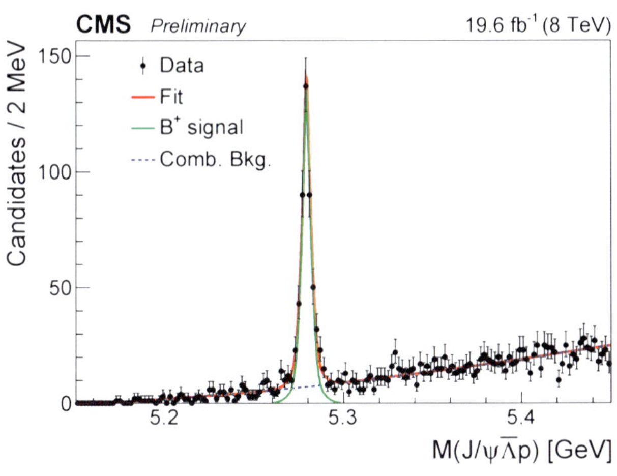

Fig. 8.4 The invariant mass distribution of $B^+ \to J/\psi\bar{\Lambda}p$ candidates. The points are data and the curve is the result of the fit

found that the inclusion of three excited kaons $K_4^*(2045)^+$ ($J^P = 4^+$), $K_2^*(2250)^+$ ($J^P = 2^-$), $K_3^*(2320)^+$ ($J^P = 3^+$) decaying into $\bar{\Lambda} p$ is enough to describe the data. The significance of incompatibility of data with this hypothesis is less than 3 σ including systematics.

8.6 Summary

CMS observed new decay modes of excited B_s^0 state, $B_{s2}^*(5840)^0 \to B^0 K_s^0$ and obtained evidence for $B_{s1}(5830)^0 \to B^0 K_s^0$. CMS observed for the first time two resolved states $\chi_{b1}(3P)$ and $\chi_{b2}(3P)$, measured their masses and mass difference. The latter supports the standard mass hierarchy (the J=2 state is heavier than the J=1 one). CMS observed for the first time two excited B_c^+ states, one of which is fully reconstructed ($B_c(2S)^+$) and the second one partially ($B_c(2S)^{*+}$) due to undetected soft photon. The measured mass of the former state is in agreement with theoretical predictions. CMS studied the $B^+ \to J/\psi \bar{\Lambda} p$ decay in which a model independent angular analysis was performed. It was shown that data, i.e. intermediate two-body masses $M(J/\psi p)$ and $M(J/\psi \bar{\Lambda})$ are well described by the presence of excited kaons decaying into $\bar{\Lambda} p$.

Although designed for high transverse momentum physics, CMS is a very suitable experiment for heavy flavor physics.

References

1. C.M.S. Collaboration, JINST **3**, S08004 (2008)
2. C.D.F. Collaboration, Phys. Rev. Lett. **100**, 082001 (2008)
3. D0 Collaboration, Phys. Rev. Lett. **100**, 082002 (2008)
4. L.H.C.B. Collaboration, Phys. Rev. Lett **110**, 151803 (2013)
5. C.M.S. Collaboration, Eur. Phys. J. C **78**, 939 (2018)
6. C.M.S. Collaboration, Phys. Rev. Lett **121**, 092002 (2018)
7. A.T.L.A.S. Collaboration, Phys. Rev. Lett. **113**, 212004 (2014)
8. LHCb Collaboration 2018 *JHEP* **01** 138
9. C.M.S. Collaboration, Phys. Rev. Lett. **122**, 132001 (2019)
10. CMS Collaboration 2019 *CERN-EP-2019-128, arXiv:1907.05461*
11. X.-Z. Weng, X.-L. Chen, W.-Z. Deng, S.-L. Zhu, Hidden-charm pentaquarks and P_c states. Phys. Rev. D **100**, 016014 (2019)

Chapter 9
Recent Results on Light Flavor from STAR

Jie Zhao

Abstract These proceedings present an overview of the recent results on light flavor by the STAR experiment at RHIC.

9.1 Introduction

Relativistic heavy-ion collisions are unique tools to study the properties of the quark-gluon plasma (QGP) in quantum chromodynamics (QCD) [1]. One important goal of the heavy-ion program in RHIC-STAR is to explore the QCD phase diagram [1, 2]. At the RHIC top energy data were collected with different species from small to large collision systems, allowing studies of the high temperature QCD medium to extract quantitative information on the QGP. The Beam Energy Scan (BES) program with the collision energies from 7.7 to 64.2 GeV extended the studies to lower temperature and higher baryon densities on the QCD phase diagram. The main goal is to search for the turn-off of QGP signatures and signals of the first-order phase transition and the critical point [2]. To further extend the coverage on the QCD phase diagram, the fixed-target program (FXT) is exploited to reach the higher baryon densities with the baryon chemical potential in the range of $\mu_B \approx 420$–720 MeV.

Starting 2010 STAR has accumulated large volume data from 200 GeV down to 7.7 GeV. A rich body of results were produced pertinent to the properties of the QCD matter. In these proceedings, we highlight selected STAR results on light flavor measurements that were presented in the "Strangeness in Quark Matter" 2019 conference. More details can be found in STAR related proceedings in [3–5].

Jie Zhao for the STAR Collaboration.

J. Zhao (✉)
Purdue University, IN 47907 West Lafayette, USA
e-mail: zhao656@purdue.edu

D. Elia et al. (eds.), *The XVIII International Conference on Strangeness in Quark Matter (SQM 2019)*, Springer Proceedings in Physics 250,
https://doi.org/10.1007/978-3-030-53448-6_9

9.2 Initial Conditions

The measurement of longitudinal decorrelation of anisotropic flow can help provide a 3D image of the QGP evolution [6]. Using the newly installed Forward Meason Spectrometer (FMS), STAR has measured longitudinal flow decorrelations in 200 GeV Au+Au collisions (Fig. 9.1). Such measurements as a function of the normalized rapidity, indicate a strong decrease with respect to LHC [7–9]. These results provide new constraints on both the initial-state geometry fluctuations and final-state dynamics of heavy-ion collisions.

The measurement of the elliptic anisotropy (v_2) in small systems can further improve our understanding of the importance of the initial geometry. Figure 9.2 shows the $V_{2,2}$ obtained by difference methods in p+Au and d+Au collisions [10]. The results for different energies show a common trend with the charged particle multiplicity, which provide important insights on the nature of collectivity in small systems.

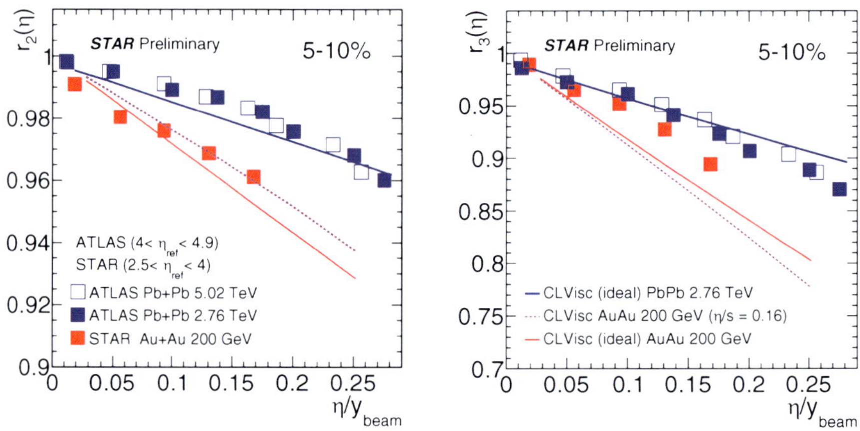

Fig. 9.1 Decorrelation parameters, r_2 (left), and, r_3 (right), as a function of the normalized rapidity in 5–10% Au+Au collisions [7] and Pb+Pb collisions [8, 9]

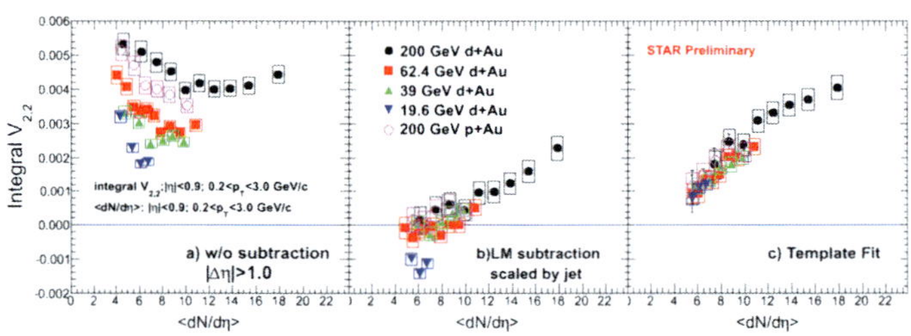

Fig. 9.2 Integral $V_{2,2}$ as function of multiplicity in p+Au and d+Au collisions [10]

9.3 Phase Transition and Critical Point

The higher-order fluctuation observables or the higher moments of conserved quantities, can be directly connected to the corresponding thermodynamic susceptibilities. It is a sensitive tool to study the criticality on the QCD phase diagram as well as to determine the freeze-out parameters [11, 12]. Figure 9.3 (left) shows the new measurements of the net-proton cumulants in Au+Au collisions at 54.4 GeV [3]. The data are compared to the results from other energies and a good agreement is found. A non-monotonic behavior as a function of the collision energy is observed. Figure 9.3 (right) shows the 6th- to 2nd-order cumulant of the net-proton multiplicity distributions [3]. The C_6/C_2 for central Au+Au collisions at 54.4 GeV is positive while that for 200 GeV is negative, although with large uncertainties. The results are in agreement with the theoretical expectation of a smooth crossover phase transition [13, 14]. STAR also measured net-Λ cumulants, which provide insights on the flavor dependence of the freeze-out parameters [4, 15].

Production of light nuclei with small binding energies, such as the triton (~8.48 MeV) and the deuteron (~2.2 MeV), formed via final-state coalescence, are sensitive to the local nucleon density [16]. The production of these nuclei can therefore be used to extract information of nucleon distributions at freeze-out, which could be associated with the QCD phase transition [17]. Figure 9.4 (left) shows that the coalescence parameter, B_2, first decreases and then increase with collision energy [18]. The extracted neutron density fluctuations [19], n, also show a non-monotonic behavior with collision energy (right panel of Fig. 9.4) [20].

One of the important QGP signatures is the nuclear modification factor, R_{CP}, being significantly smaller than unity at high energies. The strange hadron measurements from BES-I by STAR [21] show no suppression of the K_s^0 R_{cp} up to $p_{\mathrm{T}} = 3.5$ GeV/c. The particle type dependence of R_{CP} is found to be smaller at $\sqrt{s_{\mathrm{NN}}} \leq 11.5$ GeV (Fig. 9.5). These measurements point to the beam energy region below 19.6 GeV for further investigation of the deconfinement phase transition.

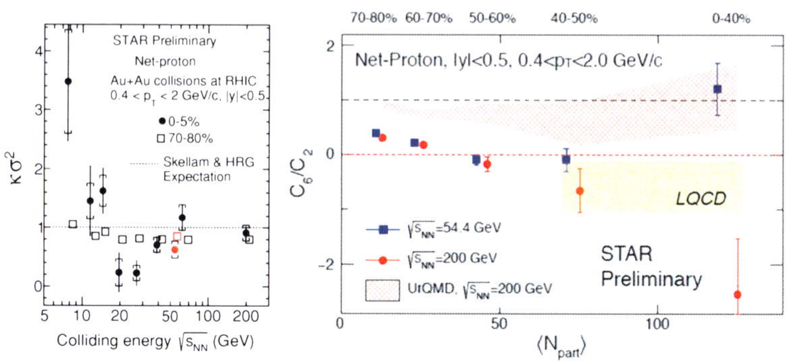

Fig. 9.3 Energy dependence of the net-proton moments products, $k\sigma^2$ (left), and the 6th- to 2nd-order cumulant ratio, C_6/C_2 (right), in 54.4 and 200 GeV Au+Au collisions [3]

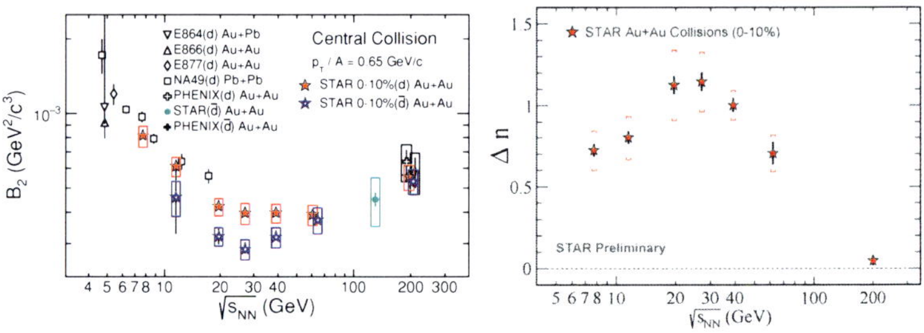

Fig. 9.4 Energy dependence of the coalescence parameter, B_2 (left), and the neutron density fluctuation, n (right), from Au+Au collisions at RHIC [18, 20]

Fig. 9.5 R_{cp} of the K_s^0, Λ, Ξ, ϕ, Ω in Au+Au collisions at $\sqrt{s_{\mathrm{NN}}}$ = 7.7–39 GeV [21]

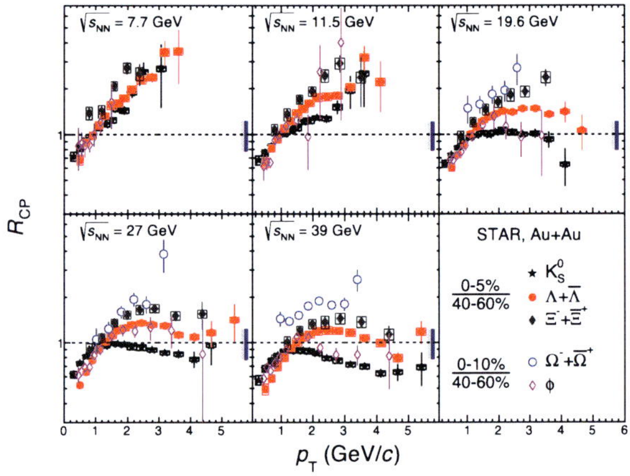

9.4 Hypertriton

The measurement of hypertriton can provide insight on hyperon-nucleon interactions [22, 23]. The heavy-flavor tracker (HFT) significantly improved the signal-to-background ratio of hypertriton, thus allowing more precise determinations of the hypertriton binding energy and mass difference between hypertriton and antihypertriton. The STAR data [24] provide the first test of the CPT symmetry in the light hypernuclei sector. No deviation from the exact symmetry is observed.

9.5 Medium Effects and Dynamics

Lifetimes of long-lived resonances are comparable to the typical lifetime of the QGP fireball created in heavy-ion collisions. Resonances can thus be used to study the properties and evolution of the hot and dense QGP medium. The K^{*0} and ϕ mesons

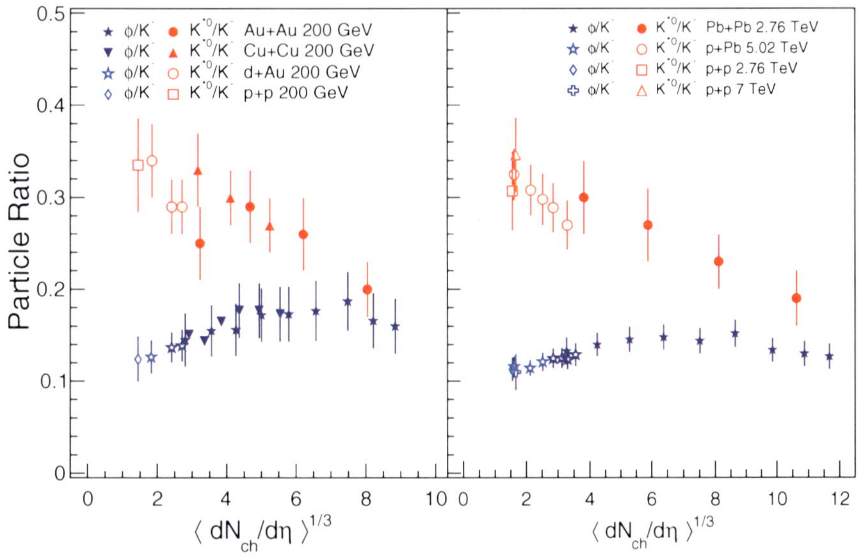

Fig. 9.6 K^{*0} and ϕ to K^- ratios as function of multiplicity from RHIC (left) [5] and LHC (right) [25]

have different hadronic cross sections and lifetimes. The comparison of ϕ/K^- and K^{*0}/K^- ratios in Fig. 9.6 indicate strong medium effects at RHIC and LHC [5, 25].

Dileptons are penetrating probe to heavy-ion collisions [26]. Recent measurements show a strong enhancement in the very low p_T region. The results point to additional physics contributions, for example contributions from photon interactions in the magnetic field trapped in the QGP [27].

9.6 Chirality, Vorticity and Polarization Effects

Due to spin-orbit coupling, particles produced in non-central heavy-ion collisions possess large orbital angular momentum and can be globally polarized along the angular momentum direction [28]. This effect was demonstrated by the global Λ polarization measurement from STAR (left panel of Fig. 9.7) [29]. The data also hint a systematic splitting between Λ and $\bar{\Lambda}$, an effect expected from the initial magnetic field. Recently, STAR reported a first observation of the Λ local polarization with a quadrupole structure (right panel of Fig. 9.7), which could be related to the elliptic flow [30].

An electric charge separation can be induced by chirality imbalance along the strong magnetic field and is predicted to occur in relativistic heavy-ion collisions because of topological charge fluctuations and the approximate chiral symmetry restoration in QCD. This effect is called the Chiral Magnetic Effect (CME) [31]. Since

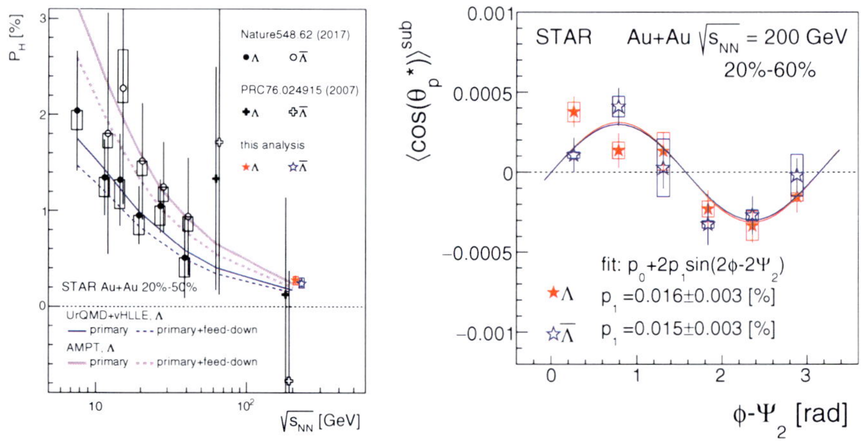

Fig. 9.7 (Left) Energy dependence of the global polarization of Λ and $\bar{\Lambda}$ in Au+Au collisions [29]. (Right) local polarization of Λ and $\bar{\Lambda}$ as a function of azimuthal angle [30]

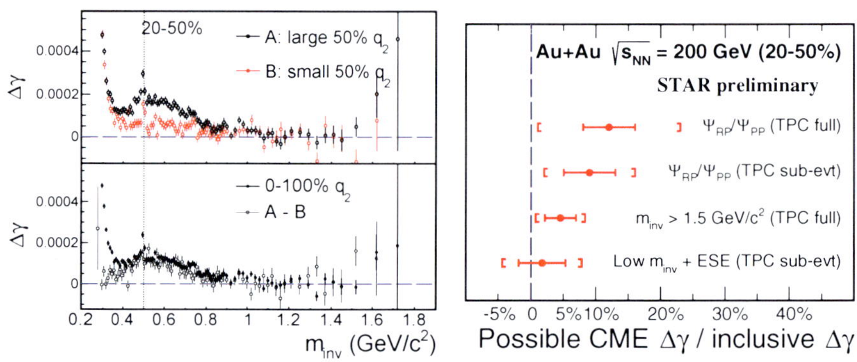

Fig. 9.8 (Left) $\Delta\gamma$ correlator as function of invariant [33]. (Right) relative contribution from possible CME signal to the measured $\Delta\gamma$ [33–35]

the first measurement of the $\Delta\gamma$ correlator in 2009 [32], there have been extensive developments to reduce or eliminate the backgrounds [31]. Figure 9.8 (left) shows the results obtained using the invariant mass method, one of the recently developed method [33]. The extracted potential CME signal relative to the inclusive $\Delta\gamma$ in 200 GeV Au+Au collisions with two novel methods [34, 35] are summarized in the right panel of Fig. 9.8. These data-driven estimates indicate that the possible CME signal is small, within 1-2 σ from zero [33].

9.7 Summary

The recent results on light flavor from the STAR experiment are overviewed. The longitudinal flow decorrelation was measured in heavy-ion data and compared to LHC data. The elliptic anisotropy is measured p+Au and d+Au collisions. These measurements will further our understanding of the importance of the initial geometry to the system evolution. The net-proton (net-Λ) cumulants, the light nuclei coalescence parameter and neutron density fluctuation are reported. All these results seem to show non-monotonic behavior with collision energy and may bear important implications to phase transitions and the possible critical point. The strange hadron production is found to be not suppressed at $\sqrt{s_{\rm NN}} \leq 11.5$ GeV, calling for further studies at low energies. The measured ratios of resonance yields to K^- indicate strong medium effects. Strong enhancement is observed in the very low $p_{\rm T}$ dielectron yield, which may be due to photon interactions. Hypertriton measurements are reported, which present the first test of the CPT symmetry in the light hypernuclei sector. The Λ local polarization with a quadrupole structure is observed for the first time, which needs further theoretical undertanding. Two novel data-driven methods are used to search for the CME signal. The present estimates indicate that the possible CME signal is small, within 1–2 σ from zero.

Acknowledgments This work was supported by the U.S. Department of Energy (Grant No. de-sc0012910).

References

1. J. Adams et al., STAR collaboration, Nucl. Phys. A 757, 102, 2005; K. Adcox et al., PHENIX Collaboration, ibid. 757, 184 2005; B. B. Back et al., PHOBOS Collaboration, ibid. 757, 28 2005; I. Arsene et al., BRAHMS Collaboration, ibid. 757, 1 2005
2. M. M. Aggarwal et al., STAR collaboration, arXiv:1007.2613
3. A. Pandav (for the STAR collaboration), contribution to these proceedings
4. R. Bellwied (for the STAR collaboration), contribution to these proceedings
5. M. Nasim (for the STAR collaboration), contribution to these proceedings
6. P. Bozek, W. Broniowski, J. Moreira, Phys. Rev. C **83**, 034911 (2011)
7. M. Nie (for the STAR collaboration), Nucl. Phys. A 982, 403, 2019
8. V. Khachatryan et al., CMS collaboration. Phys. Rev. C **92**(3), 034911 (2015)
9. M. Aaboud et al., ATLAS collaboration. Eur. Phys. J. C **78**(2), 142 (2018)
10. S. Huang (for the STAR collaboration), Nucl. Phys. A 982, 475, 2019
11. M.A. Stephanov, Phys. Rev. Lett. **107**, 052301 (2011)
12. X. Luo, N. Xu, Nucl. Sci. Tech. **28**(8), 112 (2017)
13. M. Cheng et al., Phys. Rev. D **79**, 074505 (2009)
14. B. Friman, F. Karsch, K. Redlich, V. Skokov, Eur. Phys. J. C **71**, 1694 (2011)
15. R. Bellwied, J. Hostler, P. Parotto, I. Vazquez, C. Ratti, J.M. Stafford, Phys. Rev. C **99**(3), 034912 (2019)
16. L.P. Csernai, J.I. Kapusta, Phys. Rept. **131**, 223 (1986)
17. H.H. Gutbrod, A. Sandoval, P.J. Johansen, A.M. Poskanzer, J. Gosset, W.G. Meyer, G.D. Westfall, R. Stock, Phys. Rev. Lett. **37**, 667 (1976)
18. J. Adam et al., STAR collaboration. Phys. Rev. C **99**(6), 064905 (2019)

19. K. Sun, L. Chen, C. Ko, Z. Xu, Phys. Lett. B **774**, 103 (2017)
20. D. Zhang (for the STAR collaboration), arXiv:1909.07028
21. J. Adam et al., STAR collaboration, arXiv:1906.03732
22. J. Chen, D. Keane, Y. Ma, A. Tang, Z. Xu, Phys. Rept. **760**, 1 (2018)
23. B.I. Abelev et al., STAR collaboration. Science **328**, 58 (2010)
24. J. Adam et al., STAR collaboration, arXiv:1904.10520
25. B. B. Abelev et al., ALICE collaboration, Phys. Rev. C 91, 024609, 2015; ibid. 88:044910, 2013
26. L. Adamczyk et al., STAR collaboration. Phys. Rev. Lett. **113**(2), 022301 (2014)
27. J. Adam et al., STAR collaboration. Phys. Rev. Lett. **121**(13), 132301 (2018)
28. Z. Liang, X. Wang, Phys. Rev. Lett. **94**, 102301 (2005)
29. L. Adamczyk et al., STAR collaboration. Nature **548**, 62 (2017)
30. J. Adam et al., STAR collaboration. Phys. Rev. Lett. **123**, 132301 (2019)
31. J. Zhao and Fuqiang Wang, Prog. Part. Nucl. Phys. 107, 200, 2019
32. B.I. Abelev et al., STAR collaboration. Phys. Rev. C **81**, 054908 (2010)
33. J. Zhao (for the STAR collaboration), Nucl. Phys. A 982, 535, 2019
34. H. Xu, J. Zhao, X. Wang, H. Li, Z. Lin, C. Shen, F. Wang, Chin. Phys. C 42, 084103, 2018; H. Xu, X. Wang, H. Li, J. Zhao, Z. Lin, C. Shen, F. Wang, Phys. Rev. Lett. 121, 022301, 2018
35. J. Zhao, H. Li, F. Wang, Eur. Phys. J. C **79**, 168 (2019)

Chapter 10
High Energy Hadron Production, Self-organized Criticality and Absorbing State Phase Transition

Paolo Castorina and Helmut Satz

Abstract In high energy nuclear collisions, production rates of light nuclei agree with the predictions of an ideal gas at a temperature $T = 155 \pm 10$ MeV. In an equilibrium hadronic medium of this temperature, light nuclei cannot survive. In this contribution, we suggest that the observed behavior is due to an evolution in global non-equilibrium, leading to self-organized criticality and to hadron formation as an absorbing state phase transition for color degrees of freedom. At the confinement point, the initial quark-gluon medium becomes quenched by the vacuum, breaking up into all allowed free hadronic and nuclear mass states, without (or with a very short-live) subsequent formation of thermal hadronic medium.

10.1 Introduction

The yields for deuteron, ^3He, hyper-triton, ^4He and their antiparticles have recently been measured in $Pb - Pb$ collisions by the ALICE collaboration [1–3] and are in very good agreement [4] with the statistical hadronization model (SHM) [5], with a formation temperature of $T \simeq$, 155 MeV, corresponding at the (pseudo)critical confinement temperature $T_c = 155 \pm 10$ MeV [6].

The curious feature is that the all hadron abundances are already specified once and for all at T_c and are not subsequently modified in the evolution of the hadron gas and this enigma is further enhanced by the yields for light nuclei. Indeed, these states have binding energies of a few MeV and are generally much larger than hadronic size, and therefore their survival in the assumed hot hadron gas poses an even more striking puzzle [7]. For example, the hyper-triton root-mean-square size is close to 10

P. Castorina (✉)
INFN Sezione di Catania, Catania, Italy
e-mail: paolo.castorina@ct.infn.it

Faculty of Physics and Mathematics, Charles University, Prague, Czech Republic

H. Satz
Fakultät für Physik, Universität Bielefeld, Bielefeld, Germany
e-mail: satz@physik.uni-bielefeld.de

D. Elia et al. (eds.), *The XVIII International Conference on Strangeness in Quark Matter (SQM 2019)*, Springer Proceedings in Physics 250,
https://doi.org/10.1007/978-3-030-53448-6_10

fm, about the same size of the whole fireball formed in Pb-Pb collision at $\sqrt{s} = 2.76$ TeV, and the energy needed to remove the Λ from it is 130 ± 30 KeV.

Why are the yields for the production of light nuclei determined by the rates as specified at the critical hadronization temperature, although in hot hadron gas they would immediately be destroyed?

In this contribution, following [8], we want to discuss a solution to this puzzle obtained by abandoning the idea of a thermal hadron medium existing below the confinement point. We propose that the hot quark-gluon system, when it cools down to the hadronization temperature, is effectively quenched by the cold physical vacuum.

The relevant basic mechanism for this is self-organized criticality, leading to universal scale-free behavior, based on an absorbing state phase transition for color degrees of freedom.

10.2 Self-organized Criticality and Absorbing State Phase Transition

The core hypothesis of Self-Organized Criticality (SOC) [9, 10] is that systems consisting of many interacting components will, under certain conditions, spontaneously organize into a state with properties akin to that ones observed in a equilibrium thermodynamic system, as the scale-free behavior.

The self-organized evolution indicates that the complex behavior arises spontaneously without the need for the external tuning of a control parameter (the temperature for example). In SOC the dynamics of the order parameter drives the control parameter to the critical value: natural dynamics drives the system towards and maintains it at the edge of stability.

For non-equilibrium steady states it is becoming increasingly evident that SOC is related to conventional critical behavior by the concept of absorbing-state phase transition [11, 12].

An absorbing state is a configuration that can be reached by the dynamics but cannot be left and absorbing state phase transitions are among the simplest non-equilibrium phenomena displaying critical behavior and universality [11, 12].

A clear example is given by models describing the growth of bacterial colonies or the spreading of an infectious disease among a population: once an absorbing state, e.g., a state in which all the bacteria are dead, is reached, the system cannot escape from it.

Let us now consider the hadronization dynamics where, for sake of simplicity, an initial e^+e^- annihilation produces a $\bar{q}q$ pair which evolves according to QCD dynamics. The short distances dynamics is due to local interacting color charges, with the QCD processes of parton (quarks and gluons) annihilations and creation.

The dynamics of color degrees of freedom (d.o.f.) ends up with the hadronic production, i.e. with the production of colorless clusters. The final state has no color and the evolution of the system cannot produce colored partons in the final state.

From this point of view, hadron production is a phase transition to an absorbing state for color degrees of freedom. Moreover this phase transition is a non-equilibrium one, since, by definition, the rate out of an absorbing state is zero and an absorbing state can not obey the detailed balance.

A toy model which shows how the competition between hadron (h) formation, i.e. color neutralization, and production and/or annihilation of color charges (partons P) leads to an absorbing state is easily obtained by considering a normalized quantity $\rho(t)$, proportional to color charge, as a function of time t and the processes: $P + P \to P$ with rate λ (parton annihilation); $P \to P + P$ with rate σ (parton production); $P \to h$ with rate k (color neutralization).

The mean field evolution equation is given by [11, 12]

$$\frac{d\rho}{dt} = (\sigma - k)\rho - \lambda\rho^2 = \rho(\sigma - k - \lambda\rho) \ . \tag{10.1}$$

If $\sigma < k$ the steady state is $\rho_s = 0$ and is an absorbing state. If $\sigma > k$ the steady state is $\rho_s = (\sigma - k)/\lambda$, the critical value is $\sigma_c = k$ and, as in thermal equilibrium, the critical point is governed by a power law behavior $\rho_s \simeq (\sigma - \sigma_c)^\beta$ with $\beta = 1$.

Absorbing states characterize first order phase transitions also [16] and, indeed, for pure $SU(N)$ gauge theories, where the Polyakov loop, l, is an order parameter, one can show that the dynamical evolution of the system [17] has a steady state with $l = 0$, which is an absorbing state.

According to previous discussion: (1) The Hadronization mechanism is a non equilibrium phase transition to an absorbing state for color d.o.f.; (2) The dynamical evolution is driven by color d.o.f. up to the hadronization time/temperature; (3) a natural assumption, due to the absorbing state phase transition, is that the system is essentially Afrozen at the values of the parameters at the transition.

Let us discuss the consequences of this point of view for the hadron production.

10.3 SOC in Hadron Formation

10.3.1 Self-organization and Hadronic Spectrum

The typical illustration of SOC, proposed in the pioneering work [9], is the avalanches dynamics of sandpiles, where the number $N(s)$ of avalanches of size s observed over a long period is found to vary as a power of s, $N(s) = \alpha s^{-p}$, which means that the phenomenon is scale-free.

Another useful example of self-organized criticality provided by partitioning integers [8]. Consider the *ordered* partitioning of an integer n into integers. The number $q(n)$ of such partitionings is for $n = 3$ equal to four: 3, 2+1, 1+2, 1+1+1, i.e., $q(3) = 4$. It is easily shown [14] that in general $q(n) = 2^{n-1}$, i.e. the number of partitions increases exponentially with the size of the integer.

Given an initial integer n, we would now like to know the number $N(k, n)$ specifying how often a given integer k occurs in the set of all partitionings of n. To illustrate, in the above case of $n = 3$, we have $N(3, 3) = 1$, $N(3, 2) = 2$ and $N(3, 1) = 5$. To apply the formalism of self-organized criticality, we have to attribute a strength $s(k)$ to each integer. It seems natural use the number of partitions for this, i.e., set $s(k) = q(k)$ and the desired number $N(k, n)$ in a scale-free scenario is then given by

$$N(k, n) = \alpha(n)[s(k)]^{-p}. \tag{10.2}$$

For small values of n, $N(k, n)$ is readily obtained explicitly and one finds that the critical exponent becomes $p \simeq 1.26$.

The previous example is immediately reminiscent of the statistical bootstrap model of Hagedorn [13], who had "fireballs composed of fireballs, which in turn are composed of fireballs, and so on". Indeed, its general pattern has been shown to be due to an underlying structure analogous to the partitioning of an integer into integers [14].

More precisely, Hagedorn's bootstrap approach [13] proposes that a hadronic colorless state of overall mass m can be partitioned into structurally similar colorless states, and so on. If these states were at rest, the situation would be identical to the above partioning problem. Since the constituent fireballs have an intrinsic motion, the number of states $\rho(m)$ corresponding to a given mass m is determined by the bootstrap equation which can be asympotically solved [15], giving $\rho(m) \sim m^{-a}$ $\exp(m/T_H)$ and T_H as solution of

$$\left(\frac{2}{3\pi}\right)\left(\frac{T_H}{m_0}\right) K_2(m_0/T_H) = 2\ln 2 - 1, \tag{10.3}$$

with m_0 denoting the lowest possible mass and $K_2(x)$ is a Hankel function of pure imaginary argument. For $m_0 = m_\pi \simeq 130$ Mev, this leads to the Hagedorn temperature $T_H \simeq 150$ MeV, i.e., to approximately the critical hadronization temperature found in statistical QCD. The cited solution gave $a = 3$, but other exponents could also been considered.

The previous form is an asymptotic solution of the bootstrap equation which diverges for $m \to 0$ and must be modified for small masses. Using a similar result for $\rho(m)$ obtained in the dual resonance model [18], Hagedorn proposed

$$\rho(m) = \text{const.}(1 + (m/\mu_0))^{-a} \exp(m/T_H) \tag{10.4}$$

where $\mu_0 \simeq 1 - 2$ GeV is a normalization constant.

At this point we should emphasize that the form of $\rho(m)$ is entirely due to the self-organized nature of the components and they are in no way a result of thermal behavior. We have expressed the slope coefficient of m in terms of the Hagedorn "temperature" only in reference to subsequent applications. In itself, it is totally of combinatorical origin.

10.3.2 Comparison with ALICE Data

We now apply the formalism of self-organized criticality to strong interaction physics. Our picture assumes a sudden quench of the partonic medium produced in the collision. The initial hot system of deconfined quarks and gluons rapidly expands and cools; while this system is presumably in local thermal equilibrium, the difference between transverse and longitudinal motion implies a global non-equilibrium behavior. The longitudinal expansion quickly drives the system to the hadronisation point, and it is now suddenly thrown into the cold physical vacuum. The process is not unlike that of a molten metal being dumped into cold water. In this quenching process, the system freezes out into the degrees of freedom presented by the system at the transition point and subsequently remains as such, due to the absorbing state nature of the transition, apart from possible hadron or resonance decays. There never is an evolving warm metal. In other words, in our case there is no hot (or a very short-live) interacting hadron gas. Whatever thermal features are observed, such as radial or elliptic hydrodynamic flow, must then have originated from local equilibrium in the earlier deconfined stage [19]. The mechanism driving the system rapidly to the critical point is the global non-equilibrium due to the longitudinal motion provided by the collision.

In such a scenario, high energy nuclear collisions lead to a system which at the critical point, i.e. the color absorbing state, breaks up into components of different masses m, subject to self-similar composition and hence of a strength $\rho(m)$ as given by the above Eq. (10.4). In the self-organized criticality formalism, this implies that the interaction will produce

$$N(m) = \alpha[\rho(m)]^{-p} \qquad (10.5)$$

hadrons of mass m. With $\rho(m)$ given by Eq. (10.4), the resulting powerlaw form

$$\log N(m) = -m\left(\frac{p\log e}{T_H}\right)\left[1-\left(\frac{aT_H}{m}\right)\ln(1+\frac{m}{\mu_0})\right] + \text{const.} \qquad (10.6)$$

is found to show a behavior similar to that obtained from an ideal resonance gas in equilibrium. We emphasize that it is here obtained assuming only scale-free behavior (self-organized criticality) and a mass weight determined by the number of partitions. No equilibrium thermal system of any kind is assumed.

We now consider the mentioned ALICE data [1–3]. In Fig. 10.1 the production yields for the different mass states in central $Pb - Pb$ collisions at $\sqrt{s} = 2.76$ GeV are shown; in each case, the yield is divided by the relevant spin degeneracy. We see that the yields show essentially powerlike behavior, and the light nuclei follow the same law as the elementary hadrons. The solid line in Fig. 10.1 shows the behavior obtained from eqs. (10.6), ignoring for the moment the second term in the square brackets,

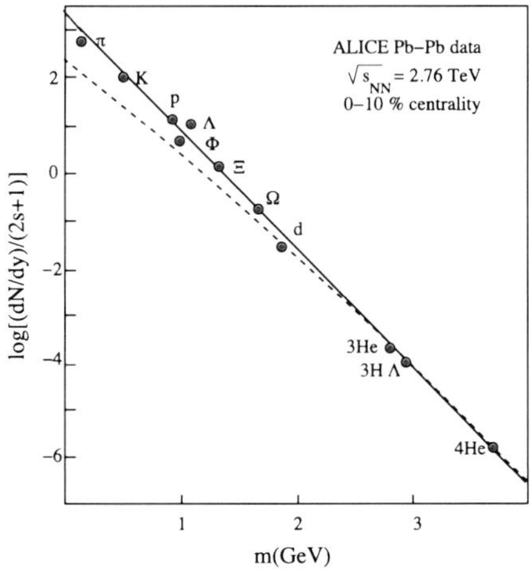

Fig. 10.1 Yield rates of species at central rapidity vs. their mass m [1–3]. The solid line corresponds to Eq. (10.7), the dashed line to Eq. (10.8)

$$\log[(dN/dy)/(2s+1)] \simeq -m\left(\frac{0.43\,p}{T_H}\right) + A, \tag{10.7}$$

with $T_H = 0.155 MeV$ and fit values $p = 0.9$, $A = 3.4$. The form is evidently in good agreement with the data.

Including the correction term to linear behavior that we had omitted above, we have

$$\log[(dN/dy)/(2s+1)] \simeq -m\left(\frac{0.43\,p}{T_H}\right) + +pa\log[1 + (m/\mu)] + A. \tag{10.8}$$

The additional term is, as indicated, rather model dependent. It will effectively turn the yield curve down for decreasing masses. This is in fact necessary, since the decay of heavier resonances will enhance the direct low mass meson yields. To illustrate the effect of the term, we choose $a = 3$, corresponding to the mentioned solution of the bootstrap equation [15], and $\mu = 2$ GeV for the normalization. The result is included in Fig. 10.1.

Acknowledgments P. C. thanks the School of Nuclear Science and Technology, Lanzhou University, for the hospitality and Marco Ruggieri and Pengming Zhang for useful comments.

References

1. J. Adam et al., ALICE. Phys. Rev. C **93**, 024917 (2016). arXiv:1506.08951 [nucl-ex]
2. J. Adam et al., ALICE. Phys. Lett. B **754**, 360 (2016). arXiv:1506.08453 [nucl-ex]
3. S. Acharya et al., ALICE. Nucl. Phys. A **971**, 1 (2018). arXiv:1710.07531 [nucl-ex]
4. A. Andronic et al., Nature **561**, 321 (2018)

5. For a comprehensive survey, see e.g. P. Braun-Munzinger, K. Redlich and J. Stachel, in Quark-Gluon Plasma 3, R. Hwa and X.-N. Wang (Eds.), World Scientific, Singapore 2003
6. A. Bazavov et al. (HotQCD), Phys. Rev. D90 (2014) 094503, and A. Bazavov et al., arXiv:1812.08235
7. For a recent review see P. Brau-Munziger and B. Donigus, Nucl. Phys. A987 (2019) 144
8. P. Castorina and H. Satz , Int.J.Mod.Phys. E28 (2019) no.04, 1950025 and references therein
9. P. Bak, C. Tang, K. Wiesenfeld, Phys. Rev. Lett. **59**, 381 (1987)
10. H.J. Jensen, *Self-Organized Criticality* (Cambridge University Press, Cambridge University Press, 1998)
11. H. Hinrichsen, Non Equilibrium Critical Phenomena and Phase Transitions into Absorbing States, arXiv:0001070
12. H. Hinrichsen, LECTURE NOTES : Non-equilibrium phase transitions
13. R. Hagedorn, Nuovo Cim. Suppl. **3**, 147 (1965)
14. Ph. Blanchard, S. Fortunato and H. Satz, Europ. Phys. J. C34 (2004) 361
15. W. Nahm, Nucl. Phys. 45 (1972) 525
16. S. si santo et al. Phys. Rev. Lett. 116, 240601 (2016)
17. A. Dumitru, R. D. Pisarski arXiv:0209001
18. K. Huang, S. Weinberg, Phys. Rev. Lett. **25**, 855 (1970)
19. For a discussion of local equilibrium in SOC, see A. Montakhab and J. M. Carlson, Phys. Rev. E58 (1998)

Chapter 11
Recent Results in Small Systems from CMS

Prabhat R. Pujahari

Abstract The observation of a wide variety of physical phenomena in the context of the formation of a strongly interacting QCD matter in heavy-ion nuclear collisions at the LHC has drawn significant attention to the high energy heavy-ion physics community. The appearance of a varieties of similar phenomena as in heavy-ion in the high multiplicity proton-proton and proton-nucleus collisions at the LHC energies has triggered further investigation to understand the dynamics of particle production mechanism in a highly dense and small QCD medium. The CMS collaboration uses many different probes in these studies ranging from the particle production cross section to multi-particle correlations. In this proceeding, I report a few selected recent CMS results from the small systems with the main focus on the measurement of collective phenomena in high multiplicity pp and pPb collisions.

11.1 Introduction

Inmulti-particle correlations the context of high energy heavy-ion physics, the collisions between protons or a proton with a nucleus is commonly referred to as small system and they can provide baseline measurements for heavy-ion collisions. Traditionally, it is thought that such small systems do not show characteristics of QGP formation a priori. However, in the recent few years, this simplistic view of a small system has been challenged at the LHC—thanks to the new frontier in energies and state-of-the-art instrumentations. The individual events in a high multiplicity pp and pPb collisions can have very high charged particle multiplicity and energy density which is comparable to that of AA collisions [1].

With the advent of the LHC, high multiplicity pp and pPb collisions show unexpected phenomena which have never been observed before in such small systems.

Prabhat R. Pujahari for the CMS Collaboration.

P. R. Pujahari (✉)
Physics Department, Indian Institute of Technology Madras, 600036 Chennai, India
e-mail: p.pujahari@cern.ch

D. Elia et al. (eds.), *The XVIII International Conference on Strangeness in Quark Matter (SQM 2019)*, Springer Proceedings in Physics 250,
https://doi.org/10.1007/978-3-030-53448-6_11

The observation of a long range rapidity ridge in the measurement of two-particle angular correlation in heavy-ion collisions is no surprise to us and this can be well explained by hydrodynamical collective flow of a strongly interacting and expanding medium [3]. However, the appearance of similar structures in a high multiplicity pp and pPb collisions has drawn a lot of attention and prompted studies to understand the cause of such behaviour in small systems. In particular, the discovery of the ridge by CMS collaboration [2] in high multiplicity pp collisions is one of such intriguing results observed in small systems [4]. The emergence of a long-range, near-side correlation in the relative azimuthal angle of the produced particles in pp and subsequently in pPb collisions motivate to investigate the existence of the collective phenomena in such systems. Further information can also be gained by focusing on multi-particle correlationsmulti-particle correlations and event-by-event fluctuations of such quantities. We observe signatures traditionally attributed to a collective behaviour not only in PbPb collisions but also in small systems. Since then, a significant amount of, unexpected phenomena has been observed in small systems with striking similarities to heavy-ion observations [5, 6].

11.2 Transverse Energy Density

The total transverse energy, E_T, is a measure of how much energy is liberated by the "stopping" of the colliding nucleons. From Fig. 11.1 it can be seen that $dE_T/d\eta\mid_{\eta=0}$ $\approx$ 22 GeV. This is nearly 1/40 of the value observed for the 2.5% most central PbPb collisions. However, since the cross sectional area of pPb collisions is much smaller than that of central PbPb collisions, this result implies that the maximum energy density in pPb collisions is comparable to that achieved in PbPb collisions [1].

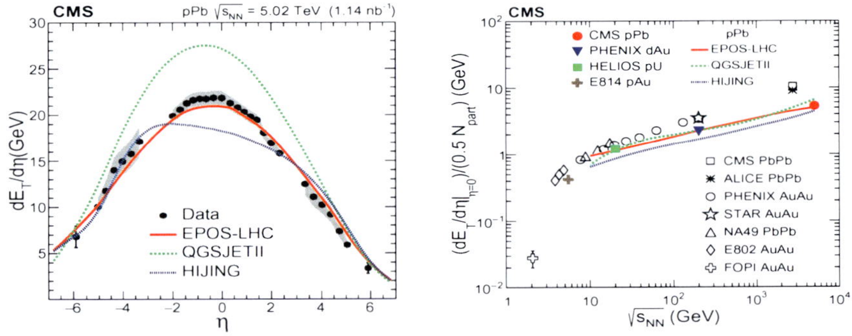

Fig. 11.1 (left) Transverse energy density versus η from minimum bias pPb collisions at $\sqrt{s_{NN}}$ = 5.02 TeV. (right) Transverse energy density per participating nucleon-nucleon pair evaluated at various $\sqrt{s_{NN}}$ for minimum bias pAu, pU, dAu, and pPb collisions [1]

11.3 Collectivity in Small Systems at the LHC

The p_T distributions of identified hadrons are one of the important tools to probe the collective behaviour of particle production. The p_T distributions in pp and pPb collisions show a clear evolution, becoming harder as the multiplicity increases [7]. As it is shown in Fig. 11.2, models including hydrodynamics describes the data better for the p_T spectra. Data-to-model agreement is good at higher charged particle multiplicity, N_{ch}. In addition, the evolution of the p_T spectra with multiplicity can be compared more directly by measuring the average transverse kinetic energy, $\langle KE_T \rangle$ [7]. Due to the presence of collective radial flow, the shape of the transverse momentum distribution depends on the mass of the particle.

The $\langle KE_T \rangle$ for $K_s{}^0$, Λ and Ξ particles as a function of multiplicity are shown in Fig. 11.3. For all particle species, $\langle KE_T \rangle$ increases with increasing multiplicity. A theoritical Blast-wave model [8] fits have also been performed to the p_T spectra of strange particles in several multiplicity bins as shown in Fig. 11.2. The interpretation of the parameters of these fits, such as kinetic freeze-out temperature, T_{kin} and transverse radial flow velocity, β_T, are model dependent. In the context of the Blast-Wave model, when comparing the parameters of different systems at similar $dN_{ch}/d\eta$, it was found that β_T is larger for small systems i.e., $\beta_T(\text{pp}) > \beta_T(\text{pPb}) > \beta_T(\text{PbPb})$. This could be an indication of a larger radial flow in more explosive small systems. However, a similar decreasing trend is observed for T_{kin} and β_T as a function of multiplicity in all three collision systems.

It needs to be further understood whether the nature of the ridge and its collectivity is due to the two-particle azimuthal correlation structures observed at large relative pseudorapidity in pp and pPb collisions, or if it is a multi-particle collective effect.

A strong hint for multi-particle correlationsmulti-particle correlations in high multiplicity pp and pPb collisions was reported by the CMS collaboration [9, 17].

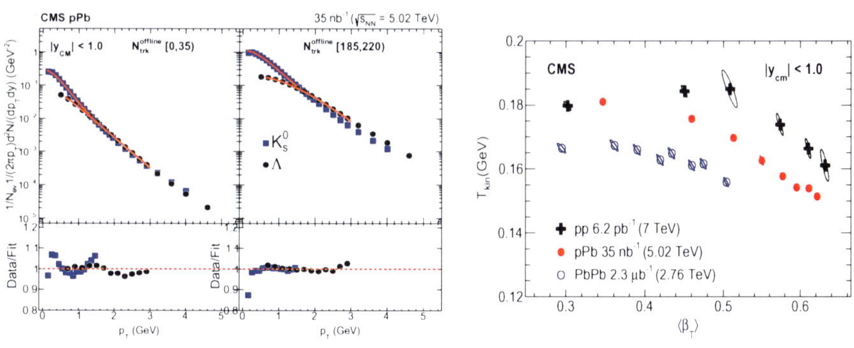

Fig. 11.2 (left) Simultaneous blast-wave fits of the p_T spectra of $K_s{}^0$ and Λ in low- and high-multiplicity pPb events. (right) The extracted kinetic freeze-out temperature, T_{kin}, versus the average radial-flow velocity, $\langle \beta_T \rangle$, from a simultaneous blast-wave fit to the $K_S{}^0$ and Λ p_T spectra at $|y_{cm}| < 1$ for different multiplicity intervals in pp , pPb , and PbPb collisions

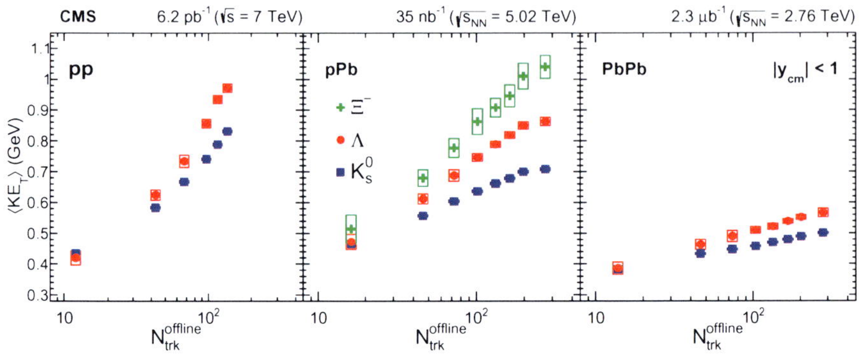

Fig. 11.3 The average transverse kinetic energy for $K_s{}^0$, Λ and Ξ particles as a function of multiplicity in pp, pPb, and PbPb collisions [7]

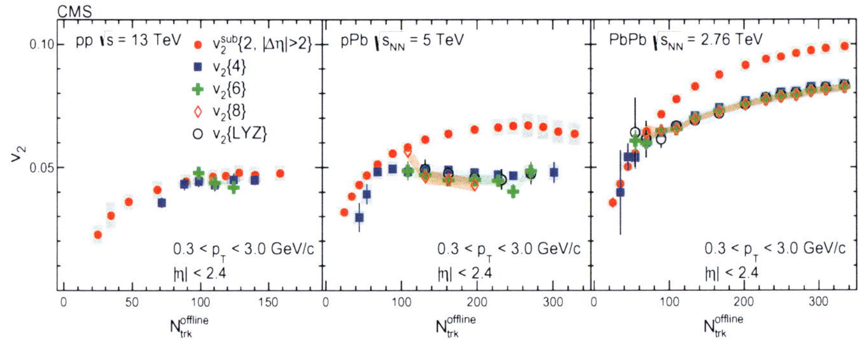

Fig. 11.4 Second-order azimuthal anisotropy Fourier harmonics, v_2 measured by CMS in pp, pPb and PbPb collisions based on multi-particlemulti-particle correlations correlations [9]

Figure 11.4 shows the second-order azimuthal anisotropy Fourier harmonics (v_2) measured in pp, pPb and PbPb collisions over a wide pseudorapidity range based on correlations calculated up to eight particles. The v_2 values stay high and show similar trends in all three systems. The v_2 computed from two-particle correlations is found to be larger than that obtained with four-, six- and eight-particle correlations, as well as the Lee-Yang zeroes method. However, the v_2 obtained from multi-particle correlationsmulti-particle correlations, all yield to similar v_2 values i.e., $v_2\{4\} \approx v_2\{6\} \approx v_2\{8\} \approx v_2\{\text{LYZ}\}$ [9]. These observations support the interpretation of a collective origin for the observed long-range correlations in high-multiplicity pp and pPb collisions.

Another useful observable in the study of collectivity is the event-by-event correlation between Fourier harmonics of different order flow coefficients. The CMS Collaboration has measured these normalized symmetric cumulants, SC(m, n), where m and n are different order flow coefficients, in pp, pPb and PbPb collisions, as a function of track multiplicity [10]. Similar observations are made in all three sys-

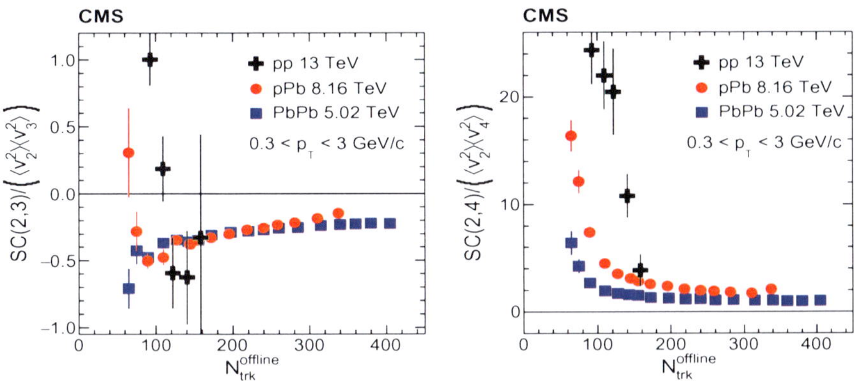

Fig. 11.5 The normalized symmetric cumulant for the second and third coefficients (left) and the second and fourth coefficients (right) are shown for pp (black cross), pPb (red circle), and PbPb (blue square). Tracks with transverse momentum between 0.3 and 3.0 GeV are used [10]

tems. In the case of SC(2, 3), which gauges the correlation between v_2 and v_3, an anti-correlation is found at high track multiplicity, as shown in Fig. 11.5. On the contrary, $SC(2, 4) > 0$: the v_2 and v_4 values are positively correlated event-by-event. Similar trends are observed in pPb and PbPb collisions, and high multiplicity pp collisions, regarding the trend of these observables as a function of track multiplicity. A long-range near-side two-particle correlation involving an identified particle is also observed [9, 11]. Results for both pPb and pp collisions are shown in Fig. 11.6. Moving to high-multiplicity events for both systems, a particle species dependence of v_2 is observed. The mass ordering of v_2 was first seen in AA collisions at RHIC and LHC energies [12, 13], which can be understood as the effect of radial flow pushing heavier particles towards higher p_T. This behavior is found to be qualita-

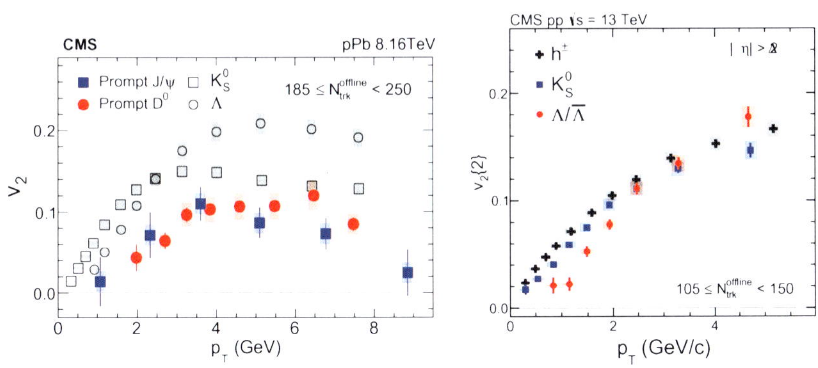

Fig. 11.6 The v_2 results for ${K_s}^0$, and Λ, prompt D^0 and prompt J/Ψ in high-multiplicity pPb (left) events. (right) The v_2 results for inclusive charged particles, ${K_s}^0$, and Λ as a function of p_T in pp collisions at $\sqrt{s} = 13$ TeV

tively consistent with both hydrodynamic models [14] and an alternative initial state interpretation [15].

A measurement of the elliptic flow of prompt J/Ψ meson in high-multiplicity pPb collisions is reported by the CMS experiment [16]. The prompt J/Ψ results are compared with the v_2 values for open charm mesons (D^0) and strange hadrons. As shown in Fig. 11.6, positive v_2 values are observed for the prompt J/Ψ meson, as extracted from long-range two-particle correlations with charged hadrons, for $2 < p_T < 8$ GeV.

The prompt J/Ψ meson results, together with results for light-flavor and open heavy-flavor hadrons, provide novel insights into the dynamics of the heavy quarks produced in small systems that lead to high final-state multiplicities.

11.4 Conclusions

Several effects, such as mass-dependent hardening of p_{T} distributions, near-side long-range correlations, multi-particle azimuthal correlations, etc, which in nuclear collisions are typically attributed to the formation of a strongly-interacting collectively-expanding quark-gluon medium, have been observed in high-multiplicity pp and pPb collisions at the LHC. The study of small collision systems at high multiplicity is undoubtedly of considerable interest. While a lot of progress has been made towards understanding the long-range correlation phenomena in small colliding systems, there are still many open questions to be addressed by the experiemental and theoritical communities.

References

1. C.M.S. Collaboration, Phys. Rev. C **100**, 024902 (2019)
2. C.M.S. Collaboration, JINST **3**, S08004 (2008)
3. C.M.S. Collaboration, Phys. Lett. B **06**, 028 (2013)
4. C.M.S. Collaboration, Phys. Lett. B **718**, 795 (2013)
5. Roberto Preghenella, arXiv:1804.03474 [hep-ex]
6. Emilien Chapon, EPJ Web Conf. **171**, 01007 (2018)
7. C.M.S. Collaboration, Phys. Lett. B **768**, 103 (2017)
8. E. Schnedermann et al., Phys. Rev. C **48**, 2462 (1993)
9. C.M.S. Collaboration, Phys. Lett. B **765**, 193 (2017)
10. C.M.S. Collaboration, Phys. Rev. Lett. **120**, 092301 (2018)
11. C.M.S. Collaboration, Phys. Rev. Lett. **121**, 082301 (2018)
12. S.T.A.R. Collaboration, Phys. Rev. C **75**, 054906 (2007)
13. PHENIX Collaboration, Phys. Rev. C **85**, 064914 (2012)
14. K. Werner et al., Phys. Rev. Lett. **112**, 232301 (2014)
15. B. Schenke et al., Phys. Rev. Lett. **117**, 162301 (2016)
16. C.M.S. Collaboration, Phys. Lett. B. **791**, 172 (2019)
17. C.M.S. Collaboration, Phys. Rev. Lett. **115**, 012301 (2015)

Part III
Heavy Flavour

Chapter 12
D Meson Sensitivity to a System Size Scan at LHC

Roland Katz, Jacquelyn Noronha-Hostler, Caio A. G. Prado, and Alexandre A. P. Suaide

Abstract Experimental measurements in pA collisions indicate no D meson suppression ($R_{\text{pPb}} \sim 1$) but a surprisingly large v_2. To better understand these results we propose a system size scan at the LHC involving ^{16}OO, ^{40}ArAr, ^{129}XeXe and ^{208}PbPb collisions. Using Trento+ v-USPhydro+DAB-MOD to make predictions, we find that the R_{AA} tends towards unity when the system size is decreased, but nonetheless, in the most central collisions $v_2\{2\}$ is almost independent of the colliding system. These results are analyzed in light of path length and initial eccentricity variations.

12.1 Introduction

Flow correlations, strangeness enhancement and suppression of hard probes are considered to be three signatures of the Quark-Gluon Plasma. The recent observation of the first two in small hadronic collisions - such as pp and pPb—raises many questions on the nature of the created "medium" in these collisions [1–3]. Jet and heavy flavor suppression is not observed in small systems [4], but the CMS collaboration has measured large D meson anisotropies in pPb collisions [5]. We still dot not understand of how such a significant v_2 in small systems could be compatible with $R_{\text{AA}} \to 1$ [6]. In order to determine the applicability of hydrodynamics in these tiny systems, it was recently proposed to run a system size scan at LHC via ArAr and OO collisions [7], on which various predictions have been made [8]. D mesons being mostly sensitive to

R. Katz (✉)
SUBATECH, Univ. de Nantes, IMT, IN2P3/CNRS, 44307 Nantes, France
e-mail: roland.katz@subatech.in2p3.fr

J. Noronha-Hostler
Department of Physics, Univ. of Illinois at Urbana-Champaign, 61801 Urbana, IL, USA

C. A. G. Prado
Institute of Particle Physics, Central China Normal University, 430079 Wuhan, Hubei, China

A. A. P. Suaide
Inst. de Física, Univ. de São Paulo, C.P. 66318, 05315-970 São Paulo, SP, Brazil

D. Elia et al. (eds.), *The XVIII International Conference on Strangeness in Quark Matter (SQM 2019)*, Springer Proceedings in Physics 250,
https://doi.org/10.1007/978-3-030-53448-6_12

equilibrium dynamics, they appear to be ideal probes of system size effects [9]. Here we investigate these effects on the R_{AA} and $v_n\{2\}$ by varying the colliding nuclei. To do so, we use Trento+v-USPhydro+DAB-MOD [10] with the same soft backgrounds as in [8] and the Langevin set up that gave us the best results in PbPb collisions [11].

12.2 Model Description

The Monte Carlo simulation DAB-MOD [9–11], developed to study open heavy favours, is coupled to 2D+1 event-by-event hydrodynamical backgrounds [8]. Heavy quarks are first sampled using FONLL distributions and then propagate with a relativistic Langevin model using the spatial diffusion coefficient from [12]. When the heavy quark / medium decoupling temperature T_d is reached, the hadronization is finally performed using a hybrid fragmentation/coalescence model. In DAB-MOD the overall magnitude of the R_{AA} is ambiguous as we usually fix the scaling parameter of the Langevin model using R_{AA} data in most central collisions. Here, we use for all colliding systems the value determined in PbPb collisions (i.e. $D/(2\pi T) = 2.2$).

12.3 Results

In Fig. 12.1, we first investigate how the system size modifies the R_{AA} as one moves towards smaller systems. First, central collisions are observed to be more sensitive to system size than mid-central collisions where there is no visible difference between OO and ArAr even though their system size is different [8]. We expect the R_{AA} to smoothly approach unity with shrinking system size, as $(1 - R_{AA})$ is approximatively proportional to the initial medium radius $\sim A^{1/3}$. However, in a future extension to

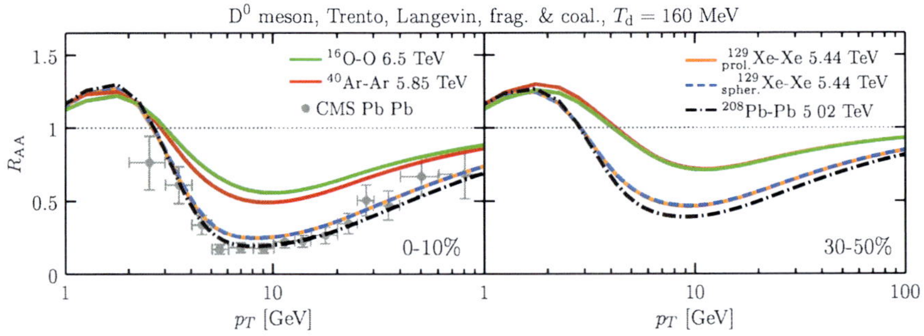

Fig. 12.1 Direct D^0 meson R_{AA} for OO, ArAr, XeXe with spherical and prolate initial nuclei, and PbPb collisions in 0–10% (left) and 30–50% (right) centrality classes. Prompt D^0 data ($|y| < 1$) from the CMS collaboration for PbPb collisions [13]

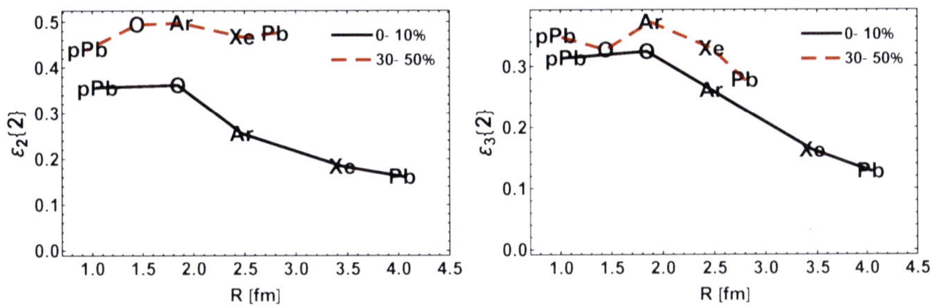

Fig. 12.2 $\varepsilon_2\{2\}$ (left) and $\varepsilon_3\{2\}$ (right) versus radius for OO, ArAr, XeXe and PbPb collisions at the LHC in 0–10% and 30–50% centrality classes

pPb collisions, we might not reach unity enough to describe the data (like in previous studies with similar frameworks [6]). Finally, the deformation of the Xe nuclei has no influence on the R_{AA}.

For the azimuthal anisotropies v_n, two factors play a significant role: the size of the system, which can be described by the typical radius of the initial conditions R, and the initial geometrical shape usually characterized by the eccentricities ε_n [8]. Their variations with the colliding system in the two centrality classes considered here are shown in Fig. 12.2. The systems resulting from OO, ArAr, XeXe and PbPb central collisions have significantly different sizes and eccentricities: the eccentricities increase while the radius decreases. In contrast, in mid-central collisions the eccentricities remain roughly constant when one varies the system size. The mid-central collisions can then be seen as probes of pure system size effects. As measured D meson data in pPb [5] correspond to central collisions, they might experience both system size and eccentricities variations compared to large AA collisions. In Fig. 12.3 (top) we show the D meson v_2 in the two different centrality classes. In the 30–50% mid-central class, the v_2 of smaller systems are significantly suppressed across all p_T. Thus, as ε_2 is $\sim$ const. in the mid-central class, the pure effect of the system size plays a dramatic role on the v_2. In the 0–10% centrality class, the v_2 is observed to be roughly independent of the colliding system across all p_T. This striking result can be understood by returning to Fig. 12.2 where we saw that for central collisions the ε_2 increases as R decreases. Thus, there are now two competing factors that influence the final v_2: a suppression effect from decreasing R, like in the mid-central class, and an enhancement effect from increasing ε_2. The similarity of the v_2 curves regardless of colliding system can therefore be explained by the two competing effects roughly compensating each other in central collisions. One can extend these ideas to pPb collisions: if they have large enough eccentricities (see Fig. 12.2) v_2 may not vanish despite the system size shrinking (other effects, e.g. the initial flow, could also contribute). Note finally that in central collisions the v_2 shows a sensitivity to the deformation of the ^{129}Xe nucleus. In Fig. 12.3 (bottom), the triangular anisotropies v_3 are observed to be more sensitive to size effects than by eccentricities, i.e. there is a consistent suppression in small systems regardless of the centrality class, even when ε_3 changes significantly. Finally, contrasting with the "universality" of $v_3(p_T)$

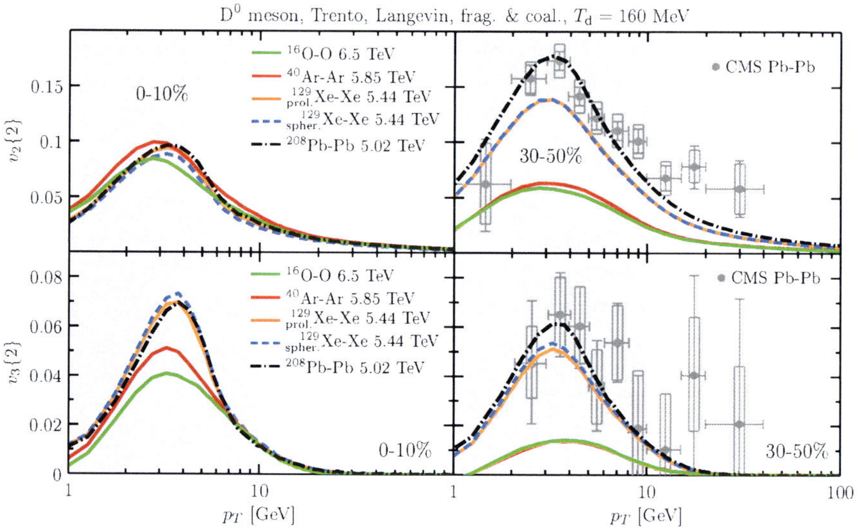

Fig. 12.3 Direct D^0 meson $v_2\{2\}$ (top) and $v_3\{2\}$ (bottom) for OO, ArAr, XeXe with spherical and prolate initial nuclei, and PbPb collisions in 0–10% (left) and 30–50% (right) centrality classes. Prompt D^0 data ($|y| < 1$) from the CMS collaboration for PbPb collisions [14]

across centralities usually observed in PbPb collisions [14], in smaller systems the v_3 vary strongly with the centrality class.

12.4 Conclusions

We made predictions for the D meson nuclear modification factors and azimuthal anisotropies for the proposed system size scan at LHC [9]. We find that the R_{AA} gradually approaches unity as the system size decreases, i.e. as the path lengths shrink. The variations of the v_2 over the colliding systems depend strongly on two competing factors: the typical system radius R and the geometry of the initial condition described by its eccentricity ε_2. In mid-central collisions, we get a clear hierarchy of the v_2 between colliding systems, showing the strong influence of the system size itself, as ε_2 is nearly constant over the different systems. In central collisions the suppression of v_2 due to the decreasing R is counterbalanced by an enhancement coming from an increasing ε_2, leading to roughly equivalent $v_2(p_T)$ across the colliding system scan. Although ε_3 increases with decreasing R in central collisions, v_3 is more sensitive to R itself and, thus, one observes a suppression following the system size hierarchy regardless of the centrality class. Finally, we find that in small systems v_3 decreases with centrality, whereas it is known to be almost constant in PbPb collisions. The latter can now be explained by a balance between a suppression effect from path length reduction and an enhancement from ε_3 increase with centrality.

Acknowledgments The authors thank Fundação de Amparo à Pesquisa do Estado de São Paulo (FAPESP) and Conselho Nacional de Desenvolvimento Científico e Tecnológico (CNPq) for support. R.K. is supported by the Region Pays de la Loire (France) under contract No. 2015-08473. C.A.G.P. is supported by the NSFC under grant No. 11890714 and 1186113100, MOST of China under Project No. 2014CB845404. J.N.H. acknowledges the support of the Alfred P. Sloan Foundation and from the US-DOE Nuclear Science Grant No. DE-SC0019175.

References

1. V. Khachatryan et al., CMS Collaboration. Phys. Lett. B **765**, 193 (2017)
2. J. Adam et al., ALICE Collaboration. Nature Phys. **13**, 535 (2017)
3. C. Aidala et al. [PHENIX Collaboration], Nature Phys. **15**, no. 3, 214 (2019)
4. S. Acharya et al. [ALICE Collaboration], arXiv:1906.03425 [nucl-ex]
5. A. M. Sirunyan et al. [CMS Collab.], Phys. Rev. Lett. **121**, no. 8, 082301 (2018)
6. Y. Xu et al., Nucl. Part. Phys. Proc. **276–278**, 225 (2016)
7. Z. Citron et al. arXiv:1812.06772 [hep-ph]
8. M.D. Sievert, J. Noronha-Hostler, Phys. Rev. C **100**(2), 024904 (2019)
9. R. Katz, C.A.G. Prado, J. Noronha-Hostler, A.A.P. Suaide, arXiv:1907.03308
10. C.A.G. Prado et al., Phys. Rev. C **96**(6), 064903 (2017)
11. R. Katz, C. A. G. Prado, J. Noronha-Hostler, J. Noronha, A. A. P. Suaide, arXiv:1906.10768 [nucl-th]
12. G.D. Moore, D. Teaney, Phys. Rev. C **71**, 064904 (2005)
13. A.M. Sirunyan et al. CMS Collaboration. Phys. Lett. B **782**, 474 (2018)
14. A. M. Sirunyan *et al.* [CMS Collab.], Phys. Rev. Lett. **120**, no. 20, 202301 (2018)

Chapter 13
Beauty Production with ALICE at the LHC

Erin F. Gauger

Abstract In this manuscript, various beauty production measurements using the ALICE detector will be presented. We will show new measurements of non-prompt D^0 mesons in pp collisions at $\sqrt{s} = 5.02$ TeV and beauty-tagged jet production in p–Pb collisions at $\sqrt{s_{NN}} = 5.02$ TeV. The R_{AA} of beauty-hadron decay electrons in central Pb–Pb collisions and the v_2 of beauty-hadron decay electrons in semi-central Pb–Pb collisions at $\sqrt{s_{NN}} = 5.02$ TeV will also be discussed.

13.1 Introduction

In hadronic collisions, beauty quarks are produced early via hard-scattering processes with large momentum transfer. Because of this early production, the beauty quark is an excellent probe of the Quark-Gluon Plasma (QGP) [1, 2] formed in heavy-ion collisions. Traveling through the QGP, beauty quarks interact with other partons via collisional and radiative processes and lose energy. The beauty quark ($m_b \simeq 4.18$ GeV/c^2 [4]) is expected to lose less energy than lighter quarks, since collisional processes depend on the mass of the particle and the dead cone effect [3] would hamper radiative energy loss. Since charm ($m_c \simeq 1.27$ GeV/c^2 [4]) and beauty quarks are both produced early in the collision but have different masses, it is useful to compare beauty and charm measurements to test our understanding of mass-dependent energy loss in heavy-ion collisions.

Measurements of beauty production in pp collisions are important to test perturbative QCD (pQCD) calculations, as well as to provide the baseline for Pb–Pb measurements. In p–Pb collisions, beauty-production measurements are crucial to isolate initial-state and cold nuclear matter effects, both of which would be present in Pb–Pb measurements.

Erin F. Gauger for the ALICE Collaboration

E. F. Gauger (✉)
The University of Texas at Austin, 78712 Austin, TX, USA
e-mail: erin.gauger@cern.ch

D. Elia et al. (eds.), *The XVIII International Conference on Strangeness in Quark Matter (SQM 2019)*, Springer Proceedings in Physics 250,
https://doi.org/10.1007/978-3-030-53448-6_13

13.2 Beauty Measurements

In ALICE, beauty production is measured via beauty-hadron decay electrons, non-prompt D^0 mesons, and beauty-tagged jets. All three measurements rely on the excellent vertex reconstruction and impact parameter resolution of the ALICE detector to isolate the decay particles and jets from beauty decay.

13.2.1 Beauty-Hadron Decay Electrons

Roughly 10% of beauty hadrons decay directly into electronic final states (e.g. $B^- \rightarrow$ e+X), and another 10% decay to charm hadrons which further decay to electrons (e.g. $B^- \rightarrow D^0 \rightarrow$ e+X) [4]. This high branching ratio coupled with the excellent electron identification of the ALICE detector makes it convenient to study beauty quarks by measuring the production of beauty-hadron decay electrons. Beauty-hadron decay electrons have been measured in pp, p–Pb, and Pb–Pb collisions [5–7].

Electrons from beauty-hadron decays must be separated from other sources of electrons, such as photon conversion, Dalitz decays, and charm-hadron decays. This separation is achieved by exploiting the relatively long decay length of beauty hadrons ($\tau_B \approx 500\ \mu\text{m}/c$) versus non-beauty hadrons (e.g. $\tau_D = 60$–$300\ \mu\text{m}/c$) [4]. The long decay length gives beauty-hadron decay electrons a longer distance of closest approach to the primary vertex (d_0), thus making the beauty-electron d_0 distribution much wider than that from other sources (Fig. 13.1, bottom left). The difference in d_0 shape allows us to fit the electron d_0 distribution with templates from Monte Carlo simulations in order to extract the beauty-hadron decay electron yield. Once the yields are obtained in both pp and Pb–Pb collisions, the nuclear modification factor $R_{\text{AA}} = \frac{\text{d}N_{\text{AA}}/\text{d}p_{\text{T}}}{T_{\text{AA}} * \text{d}\sigma_{\text{pp}}/\text{d}p_{\text{T}}}$ can be calculated. The pp reference spectrum for the beauty-hadron decay electron R_{AA} measurement in Pb-Pb at $\sqrt{s_{\text{NN}}} = 5.02$ TeV was obtained by a pQCD-driven scaling of the cross section measured at $\sqrt{s} = 7$ TeV [5]. In order to obtain a v_2 measurement, the yield is measured for electrons that lie both in and out of the event plane, and the v_2 is calculated according to $v_2 = \frac{1}{R^2}\frac{\pi}{4}\frac{N_{\text{in-plane}} - N_{\text{out-of-plane}}}{N_{\text{in-plane}} + N_{\text{out-of-plane}}}$, where N refers to the number of electrons measured, and R is the resolution correction for the event plane.

Two beauty-electron measurements in ALICE are R_{AA} in 0–10% and v_2 in 20–40% central Pb–Pb collisions at $\sqrt{s_{\text{NN}}} = 5.02$ TeV. In Fig. 13.1, top left, the beauty-electron R_{AA} is shown along with the R_{AA} of heavy-flavor decay electrons (from both charm and beauty decays). Though the systematic error bars are large, a hint of an increased R_{AA} of beauty-hadron decay electrons when compared with heavy-flavor electrons at low-p_{T} is observed. This is consistent with our expectations of the mass dependence of energy loss in the QGP medium. At high p_{T}, the two distributions overlap, in part because at higher momentum, the heavy-flavor electron sample becomes dominated by beauty-hadron decay electrons. Figure 13.1 (top right) shows a comparison of the R_{AA} of beauty-hadron decay electrons with models that include

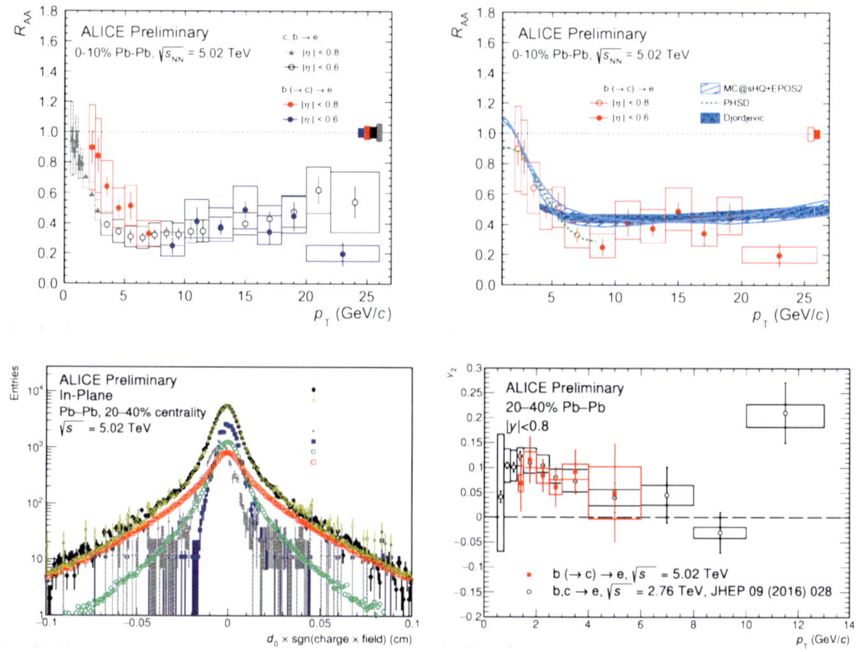

Fig. 13.1 Various results of beauty-hadron decay electrons in ALICE. The top two panels show the R_{AA} in 0–10% Pb–Pb collisions. The bottom-left panel shows an example of a MC template fit of the d_0 of electrons to extract those from beauty-decay. The bottom-right panel shows the v_2 of beauty-hadron decay electrons in 20–40% Pb–Pb collsions

both collisional and radiative energy loss. We see that the theoretical models are in good agreement with data.

Finally, the v_2 of beauty-hadron decay electrons is shown in Fig. 13.1 (bottom right). The v_2 is non-zero; in fact, between $1.3 < p_T < 4$ GeV/c, the significance of the measurement for a positive v_2 is 3.49σ. This hints that the beauty quark may participate in the collective behavior of the medium. The v_2 measurement of heavy-flavor decay electrons [8], also shown in Fig. 13.1, is similar to that of electrons from beauty decays.

13.2.2 *Non-prompt D^0 mesons*

With the ALICE detector, beauty production is also studied by measuring non-prompt D^0 mesons from beauty-hadron decays. The measurement is performed in pp collisions at $\sqrt{s} = 5.02$ TeV. The non-prompt D^0 mesons (along with their charge conjugates) are reconstructed via the decay channel to $K^-\pi^+$ (branching ratio $\sim 3.9\%$) and selected by applying various topological requirements, including a cut on the dis-

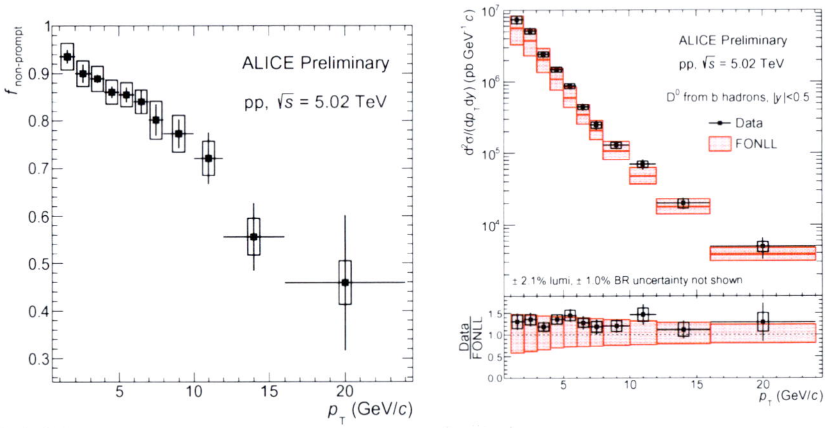

Fig. 13.2 Left: the fraction of non-prompt to inclusive D^0 mesons in the sample. Right: The cross section of non-prompt D^0 mesons in pp collisions at $\sqrt{s} = 5.02$ TeV compared to FONLL [10]

tance between the primary and secondary vertices. The topological selection criteria are optimized using boosted decision trees, allowing us to achieve a high fraction of non-prompt D^0 mesons ($f_{\text{non-prompt}}$) in our sample. At low-p_T, $f_{\text{non-prompt}}$ almost reaches 95%, an unprecedented purity for this measurement (see Fig. 13.2).

The cross-section of non-prompt D^0 mesons is compared with FONLL [10] predictions (pQCD) in Fig. 13.2. As in previous measurements [9], the two are in agreement, though the measurement lies on the upper edge of the FONLL uncertainty band.

13.2.3 Beauty-Tagged Jets

A more direct access to the initial parton kinematics is obtained by measuring beauty-tagged jets. This has been done for the first time in ALICE in p–Pb collisions at $\sqrt{s_{\text{NN}}} = 5.02$ TeV. Jets are selected using the anti-k_T algorithm [11], and a resolution parameter of $R = 0.4$. To achieve a high purity of b-jets in the sample, the long lifetime of beauty hadrons is exploited once more. Jets that contain a three-pronged secondary vertex are selected, and a number of topological requirements are applied to increase the b-jet purity. In particular, a cut is applied on the displacement significance ($SL_{xy} > 7$) of the secondary vertex. The SL_{xy} is defined as the distance between primary and secondary vertex in the xy-plane divided by the resolution of that distance.

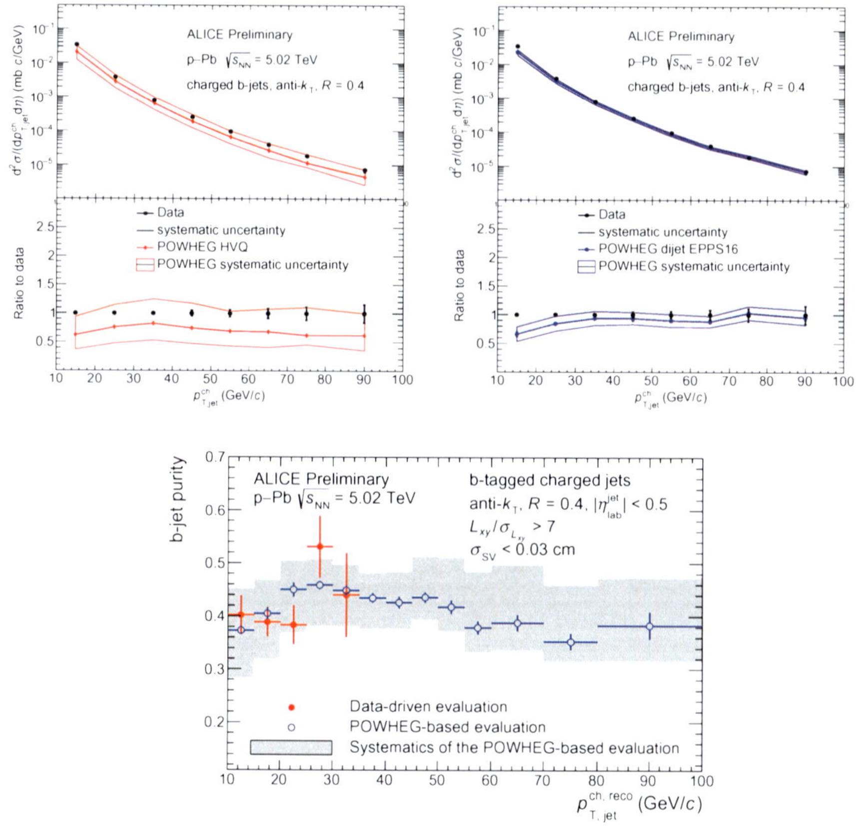

Fig. 13.3 Top left and right: the cross-section of beauty-tagged jets in p–Pb collisions compared with POWHEG. Bottom: the purity of beauty jets in the sample

The results of this measurement are shown in Fig. 13.3, where the estimated purity (∼40%) of beauty-tagged jets in the sample is shown along with the jet cross-section compared POWHEG HVQ and POWHEG dijet EPPS16. We see that both models agree with the measured cross-section.

13.3 Conclusions

In this manuscript the ALICE results on beauty production were discussed with particular focus on beauty-hadron decay electrons in Pb–Pb collisions, non-prompt D^0 mesons in pp collisions, and beauty-tagged jets in p–Pb collisions. All three analyses took advantage of the long lifetime of beauty hadrons to separate the beauty signal from background sources. The non-prompt D^0 meson measurement agrees with

FONLL predictions, and the beauty-tagged jet measurements agree with POWHEG models. In Pb–Pb collisions, we see a hint of the mass-dependent energy loss in the QGP, as well as a non-zero v_2 for beauty-hadron decay electrons.

Acknowledgments This work was supported by U.S. Department of Energy Office of Science under contract number DE–SC0013391.

References

1. S. Borsanyi et al., JHEP **09**, 073 (2010)
2. A. Bazavov et al., Phys. Rev. D **90**, 094503 (2014)
3. N. Armesto et al., Phys. Rev. D **69**, 114003 (2004)
4. M. Tanabashi et al. (Particle Data Group), Phys. Rev. D **98**, 030001 (2018) and 2019 update
5. B. Abelev et al., ALICE Collaboration. Phys. Lett. B **721**, 13–23 (2013)
6. B. Abelev et al., ALICE Collaboration. Phys. Lett. B **738**, 97 (2014)
7. J. Adam et al., ALICE Collaboration. JHEP **07**, 052 (2017)
8. J. Adam et al., ALICE Collaboration. JHEP **09**, 028 (2016)
9. A.M. Sirunyan et al., CMS Collaboration. Phys. Rev. Lett. **123**, 022001 (2019)
10. M. Cacciari, M. Greco and P. Nason, JHEP **9805** (1998) 007 [arXiv:hep-ph/9803400]; M. Cacciari, S. Frixione and P. Nason, JHEP **0103** (2001) 006 [arXiv:hep-ph/0102134]
11. M. Cacciari, G.P. Salam, G. Soyez, JHEP **0804**, 063 (2008). [arXiv:0802.1189 [hep-ph]]

Chapter 14
Measurement of Non-strange D-Meson Production and Azimuthal Anisotropy in Pb–Pb Collisions with ALICE at the LHC

Syaefudin Jaelani

Abstract Heavy quarks are effective probes of the properties of the QGP created in ultra-relativistic heavy-ion collisions. The ALICE Collaboration measured the non-strange D-meson production in Pb–Pb collisions at $\sqrt{s_{\rm NN}} = 5.02$ TeV. The in-medium energy loss can be studied via the nuclear modification factor measurement. The measurement of the D-meson elliptic-flow coefficient, v_2, allows us to investigate the participation of the heavy quarks in the collective expansion of the system at low momentum and their possible thermalization in the medium. Furthermore, sensitivity of charm quark to the light quark can be studied by the Event-Shape Engineering technique of the D-meson elliptic flow.

14.1 Introduction

Heavy quarks are effective probes of the properties of the Quark-Gluon Plasma (QGP) created in ultra-relativistic heavy-ion collisions. Charm and beauty quarks, due to their large masses, are produced in hard scattering processes on timescales shorter than the QGP formation time [1]. They experience the entire evolution of the medium, lose part of their energy interacting with its constituents via in-medium gluon radiation [2, 3] and collisional processes [4–6].

The measurement of the nuclear modification factor ($R_{\rm AA}$) of D mesons provides important information about the microscopic interactions of heavy quarks with the medium constituents, in particular on the colour-charge and parton-mass dependence of heavy-quark energy loss. Azimuthal anisotropy measurements give insight into the participation of low-momentum heavy quarks in the collective expansion of the system and their possible thermalization in the medium. At high transverse momentum, the path-length dependence of parton energy loss mechanisms can be tested [7].

Syaefudin Jaelani for the ALICE Collaboration.

S. Jaelani (✉)
Utrecht University, Princetonplein 1, 3584 CC Utrecht, Netherlands
e-mail: syaefudin.jaelani@cern.ch

D. Elia et al. (eds.), *The XVIII International Conference on Strangeness in Quark Matter (SQM 2019)*, Springer Proceedings in Physics 250,
https://doi.org/10.1007/978-3-030-53448-6_14

14.2 Non-strange D-Meson Reconstruction

The reconstruction of the non-strange D mesons (D^0, D^+ and D^{*+}) and their charge conjugates is done via haronic decay channel, given by $D^0 \rightarrow K^-\pi^+$, $D^+ \rightarrow K^-\pi^+\pi^+$ and $D^{*+} \rightarrow D^0\pi^+$, with a branching ratio of 3.93% ± 0.04% for D^0, 9.46% ± 0.24% for D^+ as well as 67.7% ± 0.5% for D^{*+} [8]. The decay topology was exploited via secondary-vertex reconstruction separated by a few hundred micrometers from the interaction point. Topological selections were applied in order to reduce the combinatorial background and enhance the signal-to-background ratio. Further background rejection was achieved by applying particle identification to the D mesons daughters using information from the Time Projection Chamber via specific energy loss $\mathrm{d}E/\mathrm{d}x$, and via the measurment of the time-of- flight in the Time Of Flight detector. Finally, the D-meson yield extraction will be done by separating the signal and background in the invariant mass spectra. The correction for acceptance and efficiency was determined using Monte Carlo simulations, including a transport code [9] which reproduces the detector response. The HIJING [10] event generator is used to simulate the underlying Pb–Pb events and D-meson signals were added using PYTHIA6 [11] event generator. The prompt yield of D mesons is obtained by subtracting from the inclusive yields the beauty-hadron decays estimated based on FONLL calculations [12, 13]. The V0 scintillators, which cover the pseudorapidity region $-3.7 < \eta < -1.7$ and $2.8 < \eta < 5.1$, provide centrality and event plane angle.

14.3 Prompt D-Meson Nuclear Modification Factor and Elliptic Flow

The R_{AA} of non-strange D mesons (D^0, D^+ and D^{*+}) is measured in two centrality classes, central 0–10% and semi-central 30–50 %, in Pb–Pb collisions at $\sqrt{s_{\mathrm{NN}}} = 5.02$ TeV. The p_{T}-differential cross section of prompt D mesons in pp collisions at $\sqrt{s} = 5.02$ TeV [17] is used as reference. Figure 14.1 shows the average non-strange D-meson R_{AA} compared to perturbative QCD model predictions in both centrality classes. The CUJET3.0 [14] and Djordjevic [15] models, which include both radiative and collisional energy loss processes, provide a fair description of the R_{AA} in both centrality classes for $p_{\mathrm{T}} > 10$ GeV/c, where radiative energy loss is expected to be the dominant interaction mechanism.

The average non-strange D-meson elliptic-flow coefficient v_2 in Pb–Pb collisions at $\sqrt{s_{\mathrm{NN}}} = 5.02$ TeV in the 30–50% centrality class is reported in Fig. 14.2 with the $\pi^{\pm}$, J/Ψ and charged particle v_2 for the same energy and centrality class. The v_2 coefficient of non-strange D-mesons is larger than zero for $p_{\mathrm{T}} > 2$ GeV/c in semi-central Pb–Pb collisions, which indicates participation of charm quark in the collective expansion dynamics.

The Event-Shape Engineering technique was used to investigate the D^0, D^+ and D^{*+} v_2. The second-harmonic reduced flow vector, $q_2 = |Q_2|/\sqrt{M}$, can be used

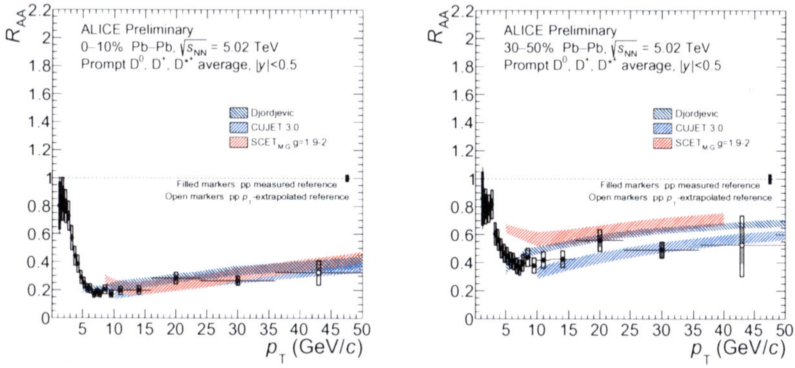

Fig. 14.1 Average non-strange D-meson R_{AA} in central 0–10% (left panel) and semi-central 30–50% (right panel) events in Pb–Pb collisions at $\sqrt{s_{\mathrm{NN}}} = 5.02$ TeV compared with pQCD model predictions [14–16]

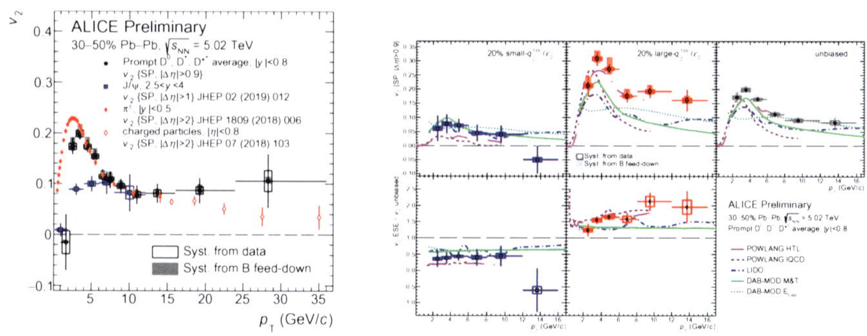

Fig. 14.2 Left: average non-strange D-meson v_2 in Pb–Pb collisions at $\sqrt{s_{\mathrm{NN}}} = 5.02$ TeV in the 30–50% centrality class compared to the $\pi^{\pm}$, J/Ψ and charged particles v_2. Right: average non-strange D-meson v_2 for the small and large q_2, and unbiased v_2 compared with some of the available models

to quantify the eccentricity of the events, where M is the multiplicity and Q_2 is the second-harmonic flow vector. The events were divided into two groups, small-q_2 class (20% of events with smallest measured $q_2^{\mathrm{TPC}}/q_2^{\mathrm{V0A}}$) and large-$q_2$ class (20% of events with largest measured $q_2^{\mathrm{TPC}}/q_2^{\mathrm{V0A}}$). The average D-meson v_2 for the small-q_2, large-q_2 and for unbiased v_2 is shown in Fig. 14.2, on the right panel, compared with model predictions. The models are based on charm-quark transport in a hydrodynamically expanding medium. The POWLANG HTL [18] model reproduces well the data for large-q_2 and unbiased D-meson v_2 below 12 GeV/c, while it underestimates the data for small-q_2. The LIDO [19] and DAB-MOD [20] models provide better description of the data for small-q_2 values, while they underestimate the data for large-q_2 and unbiased D-meson v_2.

The average R_{AA} for non-strange D mesons in the 0–10% centrality class (left) and v_2 in the 30–50% centrality class (right) are compared with Transport models

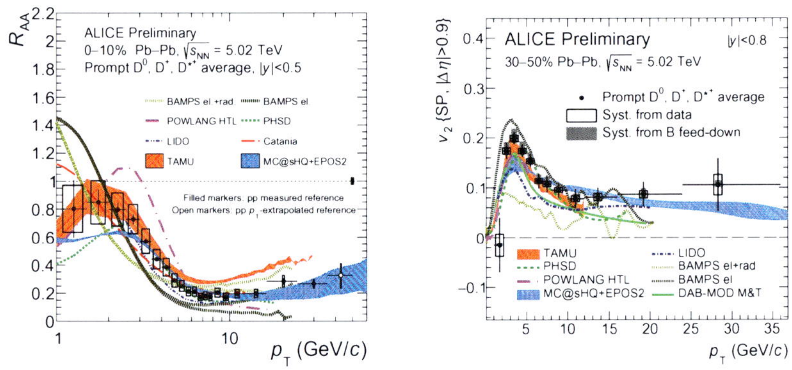

Fig. 14.3 Average non-strange D-meson R_{AA} in the 0–10% centrality class (left) and elliptic-flow coefficient v_2 in the 30–50% centrality class (right) measured in Pb–Pb collisions at $\sqrt{s_{NN}} = 5.02$ TeV, compared with the transport models [18, 19, 21–25]

in Fig. 14.3. Most of the models provide a fair description of the data in central events for $p_T < 10$ GeV/c, while POWLANG [18] and BAMPS [21], in which the interactions are only described by collisional (i.e. elastic) processes, show some tension with respect to the R_{AA} data points. The TAMU [25] model, with improved space-momentum correlations between charm quarks and underlying hydro medium, describes well the D-meson v_2 for $p_T < 12$ GeV/c. The MC@sHQ+EPOS2 [22] model provides a fair description of v_2, as PHSD [23] and TAMU [25] do for $p_T <$ 12 GeV/c, while BAMPS [21] model overestimates the maximum value of v_2. In addition, the LIDO [19] and DAB-MODE [20] models describe the shape of v_2 but underestimate its magnitude.

14.4 Conclusions

The ALICE Collaboration measured the non-strange D-meson R_{AA} and the elliptic-flow coefficient v_2 in Pb–Pb collisions at $\sqrt{s_{NN}} = 5.02$ TeV. The average R_{AA} of the non-strange D-meson shows minimum values of 0.15 in the centrality class 0–10% at p_T 6.5–8 GeV/c and 0.35 in the centrality class 30-50% at p_T 7.5–8 GeV/c. The values of the D-meson elliptic-flow coefficient v_2 are larger than zero above 2 GeV/c in mid-central Pb–Pb collisions, which provide information of the collective expansion of the system. In addition, the Event-Shape Engineering technique for the non-strange D-meson elliptic-flow coefficient v_2 to study the sensitivity of charm quark to the light quark was applied.

References

1. F.M. Liu and S.X. Liu, Phys. Rev. C **89**, no. 3, 034906 (2014) https://doi.org/10.1103/PhysRevC.89.034906 [arXiv:1212.6587 [nucl-th]]
2. M. Gyulassy, M. Plumer, Phys. Lett. B **243**, 432 (1990). https://doi.org/10.1016/0370-2693(90)91409-5
3. R. Baier, Y. L. Dokshitzer, A. H. Mueller, S. Peigne and D. Schiff, Nucl. Phys. B **484**, 265 (1997) https://doi.org/10.1016/S0550-3213(96)00581-0 [hep-ph/9608322]
4. M.H. Thoma, M. Gyulassy, Nucl. Phys. B **351**, 491 (1991). https://doi.org/10.1016/S0550-3213(05)80031-8
5. E. Braaten, M.H. Thoma, Phys. Rev. D **44**, 1298 (1991). https://doi.org/10.1103/PhysRevD.44.1298
6. E. Braaten, M.H. Thoma, Phys. Rev. D **44**(9), R2625 (1991). https://doi.org/10.1103/PhysRevD.44.R2625
7. S. Jaelani [ALICE], Int. J. Mod. Phys. Conf. Ser. **46** (2018), 1860018 https://doi.org/10.1142/S2010194518600182 [arXiv:1802.09245 [nucl-ex]]
8. C. Patrignani *et al.* [Particle Data Group], Chin. Phys. C **40**, no. 10, 100001 (2016). https://doi.org/10.1088/1674-1137/40/10/100001.
9. R. Brun, F. Bruyant, F. Carminati, S. Giani, M. Maire, A. McPherson, G. Patrick and L. Urban, https://doi.org/10.17181/CERN.MUHF.DMJ1.
10. X.N. Wang, M. Gyulassy, Phys. Rev. D **44**, 3501 (1991). https://doi.org/10.1103/PhysRevD.44.3501
11. T. Sjostrand, S. Mrenna and P. Z. Skands, JHEP **0605**, 026 (2006) https://doi.org/10.1088/1126-6708/2006/05/026 [hep-ph/0603175]
12. M. Cacciari, M. Greco and P. Nason, JHEP **9805**, 007 (1998) https://doi.org/10.1088/1126-6708/1998/05/007 [hep-ph/9803400]
13. M. Cacciari, S. Frixione and P. Nason, JHEP **0103**, 006 (2001) https://doi.org/10.1088/1126-6708/2001/03/006 [hep-ph/0102134]
14. J. Xu, J. Liao, M. Gyulassy, JHEP **1602**, 169 (2016). https://doi.org/10.1007/JHEP02(2016)169. [arXiv:1508.00552 [hep-ph]]
15. M. Djordjevic and M. Djordjevic, Phys. Rev. C **92**, no. 2, 024918 (2015) https://doi.org/10.1103/PhysRevC.92.024918 [arXiv:1505.04316 [nucl-th]]
16. Z.B. Kang, F. Ringer, I. Vitev, JHEP **1703**, 146 (2017). https://doi.org/10.1007/JHEP03(2017)146. [arXiv:1610.02043 [hep-ph]]
17. S. Acharya *et al.* [ALICE Collaboration], Eur. Phys. J. C **79**, no. 5, 388 (2019) https://doi.org/10.1140/epjc/s10052-019-6873-6 [arXiv:1901.07979 [nucl-ex]]
18. A. Beraudo, A. De Pace, M. Monteno, M. Nardi and F. Prino, Eur. Phys. J. C **75**, no. 3, 121 (2015) https://doi.org/10.1140/epjc/s10052-015-3336-6 [arXiv:1410.6082 [hep-ph]]
19. W. Ke, Y. Xu and S. A. Bass, Phys. Rev. C **98**, no. 6, 064901 (2018) https://doi.org/10.1103/PhysRevC.98.064901 [arXiv:1806.08848 [nucl-th]]
20. C. A. G. Prado, J. Noronha-Hostler, R. Katz, A. A. P. Suaide, J. Noronha, M. G. Munhoz and M. R. Cosentino, Phys. Rev. C **96**, no. 6, 064903 (2017) https://doi.org/10.1103/PhysRevC.96.064903 [arXiv:1611.02965 [nucl-th]]
21. J. Uphoff, O. Fochler, Z. Xu and C. Greiner, J. Phys. G **42**, no. 11, 115106 (2015) https://doi.org/10.1088/0954-3899/42/11/115106 [arXiv:1408.2964 [hep-ph]]
22. M. Nahrgang, J. Aichelin, P. B. Gossiaux and K. Werner, Phys. Rev. C **89**, no. 1, 014905 (2014) https://doi.org/10.1103/PhysRevC.89.014905 [arXiv:1305.6544 [hep-ph]]
23. T. Song, H. Berrehrah, D. Cabrera, W. Cassing and E. Bratkovskaya, Phys. Rev. C **93**, no. 3, 034906 (2016) https://doi.org/10.1103/PhysRevC.93.034906 [arXiv:1512.00891 [nucl-th]]
24. S. Plumari, V. Minissale, S. K. Das, G. Coci and V. Greco, Eur. Phys. J. C **78**, no. 4, 348 (2018) https://doi.org/10.1140/epjc/s10052-018-5828-7 [arXiv:1712.00730 [hep-ph]]
25. M. He, R.J. Fries, R. Rapp, Phys. Lett. B **735**, 445 (2014). https://doi.org/10.1016/j.physletb.2014.05.050. [arXiv:1401.3817 [nucl-th]]

Chapter 15
Transport Properties of Heavy Quarks and Their Correlations to the Bulk Dynamics and the Initial Electromagnetic Field

Salvatore Plumari, Gabriele Coci, Vincenzo Minissale, Santosh K. Das, Lucia Oliva, Maria Lucia Sambataro, and Vincenzo Greco

Abstract We study the propagation of heavy quarks (HQs) in the quark-gluon plasma (QGP) by means of a event-by-event relativistic Boltzmann transport (RBT) approach. We discuss the correlations between light flavor and heavy quarks (HQs) flow harmonics and the relative fluctuations of anisotropic flows $\sigma_{v_n}/\langle v_n \rangle$ Moreover, we investigate the role of QCD interaction in developing these flows correlations. As recently recognized, strong electro-magnetic (E.M.) fields are created in Ultra-relativistic Heavy-Ion Collision (HIC). We show that these fields are responsible for a splitting of directed flow v_1 of D and anti-D mesons. Moreover, we discuss the effect of the initial large bulk vorticity on the build up of rapidity odd HQs directed flow v_1.

S. Plumari · G. Coci · M. L. Sambataro · V. Greco
Department of Physics and Astronomy 'Ettore Majorana', University of Catania, Via S. Sofia 64, I-95125 Catania, Italy

S. Plumari (✉) · G. Coci · V. Minissale · M. L. Sambataro · V. Greco
Laboratori Nazionali del Sud, INFN-LNS, Via S. Sofia 62, 95123 Catania, Italy
e-mail: salvatore.plumari@ct.infn.it

S. K. Das
School of Physical Science, Indian Institute of Technology Goa, Ponda-403401, Goa, India

L. Oliva
GSI Helmholtzzentrum für Schwerionenforschung GmbH, Planckstr. 1, 64291 Darmstadt, Germany

Institut für Theoretische Physik, Johann Wolfgang Goethe-Universität, Max-von-Laue-Str. 1, 60438 Frankfurt am Main, Germany

D. Elia et al. (eds.), *The XVIII International Conference on Strangeness in Quark Matter (SQM 2019)*, Springer Proceedings in Physics 250,
https://doi.org/10.1007/978-3-030-53448-6_15

15.1 Introduction

Heavy quarks (HQs), charm and bottom quarks, are excellent probes of the system created in a ultra-Relativistic Heavy Ion Collision (uRHIC). They can probe both for the initial stages of uRHIC and the thermalized QGP evolution because their formation time is very small and the their large masses. In their final state the charm quarks appear as constituent of charmed hadrons mainly D mesons and Λ_c baryons. The main observables studied in HQ sector were the heavy mesons nuclear suppression factor R_{AA} and the elliptic flow $v_2(p_T)$. Several theoretical efforts have been made in order to describe both these observables [1–5]. A more realistic modelling requires to take into account initial state fluctuations. Theoretical studies including event-by-event fluctuations in the initial geometry, have shown that the triangular flow $v_3(p_T)$ of D mesons is finite and it could provide constraint on heavy quark transport coefficient [6, 7]. In uRHIC very strong Electromagnetic (EM) fields are created and because HQ are produced in the very early stages of uRHIC they will be directly affected by such a strong EM field. This results in a rapidity-odd directed flow v_1 for D^0 and $\bar{D}^0$ [8]. Moreover, recently, it was shown that the D meson directed flow measurement can probes to the initial conditions of the initial system after the collision and providing information about its tilted profile in the reaction plane and the initial e.m. field [9]. In this contribution we will discuss these aspects in the framework of a transport calculation.

15.2 Transport Equation for Charm Quarks

The time evolution of the charm quark distribution function in QGP is obtained by solving the RBT equations (for more details see [5]).

$$
\begin{aligned}
p^{\mu}\partial_{\mu} f_Q(x, p) &= \mathcal{C}[f_q, f_g, f_Q](x, p) \\
p_j^{\mu}\partial_{\mu} f_j(x, p) &= \mathcal{C}[f_q, f_g](x_j, p_j) \;\; for \;\; j = q, g
\end{aligned}
\tag{15.1}
$$

where $f_i(x, p)$ is the on-shell phase space distribution function for the i parton and $\mathcal{C}[f_q, f_g, f_Q](x, p)$ is the relativistic collision integral. The the phase-space distribution function of the bulk medium (quarks and gluons) enters in the equation for charm quarks as an external quantities in $\mathcal{C}[f_q, f_g, f_Q]$ therefore f_q and f_g are independent of $f_Q(x, p)$ and we discard collisions between charm quarks which is by far a solid approximation. The evolution of the bulk of quarks and gluons is instead given by the solution of the other two transport equations where the $\mathcal{C}[f_q, f_g]$ is tuned to a fixed $\eta/s(T)$, for details see [10]. For the heavy quark bulk interaction we consider a quasi-particle model (QPM) [11]. Finally, for charm quark hadronization in to D mesons, we consider an hybrid approach of hadronization by coalescence plus fragmentation [12]. In order to extend the analysis to high order harmonic flows $v_n(p_T)$ we need to include the initial state fluctuations. Recently, we have developed

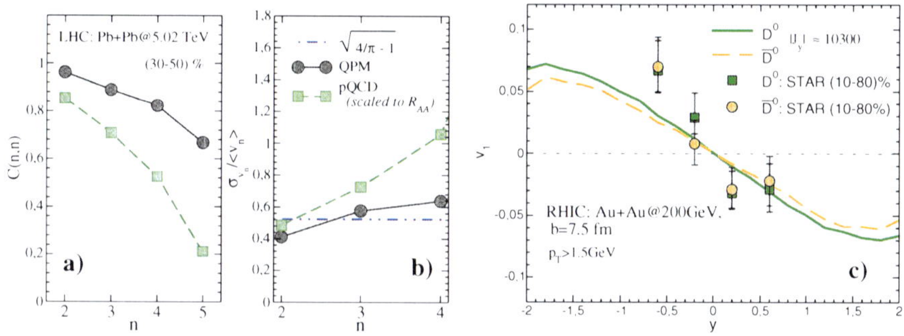

Fig. 15.1 Left panel: **a** correlation coefficient $C(n, n)$ and **b** $\sigma_{v_n}/\langle v_n \rangle$ as a function of the order of harmonics n. The circles refer to the case of QPM model while the squares to the case corresponding to a pQCD interaction. The dashed line indicate the value $\sqrt{4/\pi - 1}$ expected for a 2D Gaussian distribution. These results are for $Pb + Pb$ at $\sqrt{s_{NN}} = 5.02\,TeV$. Right panel: Rapidity dependence of v_1 for D^0 for $Au + Au$ collisions at $\sqrt{s_{NN}} = 200\,GeV$ for $(10 - 80\%)$ centrality cut. Data taken from [15]

an event-by-event RBT approach for the bulk in order to study the role of finite η/s on the anisotropic flows $v_n(p_T)$ [13, 14]. Therefore, within this approach it is possible to study the correlation between different anisotropic flow harmonics. The novelty of this contribution is the extension of these studies to the heavy quark sector where we study the correlations between charm quarks and light quarks. For this purpose a measure of the strength of the correlation is given by the correlation coefficient $C(v_n^{(light)}, v_n^{(heavy)})$. We have studied the impact of the temperature dependence of the drag coefficient on both correlation and v_n distribution. For this scope we have considered two different models having different T dependent drag coefficients: one with a pQCD like interaction with constant $\alpha_s = 0.4$ that leads a T^2 dependence similar to AdS/CFT approach and another one with an interaction coming from the QPM model with a weakly T dependent drag coefficient. Notice that in the pQCD interaction we rescale the interaction in order to reproduce the same $R_{AA}(p_T)$ obtained in the QPM. As shown in Fig. 15.1a, the correlation coefficient decrease with respect to the order of harmonics for both pQCD and QPM. The specific T dependence of the drag strongly modify the soft-hard anisotropic flows correlations, even if the models are tuned to reproduced the same experimental D meson $R_{AA}(p_T)$. As shown, the correlation is stronger for QPM than pQCD for all the harmonics considered.

Interesting properties of HQs can be inferred by studying the relative fluctuations $\sigma_{v_n}/\langle v_n \rangle$ where σ_{v_n} are the standard deviation for v_n. As shown in Fig. 15.1b the $\sigma_{v_n}/\langle v_n \rangle$ are increasing functions with the order of the harmonics and are sensitive to the temperature dependence of the transport coefficients for $n \geq 3$. The upcoming experimental results will help to constrain heavy quark transport coefficients.

Another new observable that has raised a growing attention is the directed flow $v_1 =< p_x/p_T >$ of D meson. There are two sources for a finite v_1 of HQs: (i) the initial large vorticity coming from a tilted initial distribution in the reaction plane that produce a v_1 of D meson several times larger than the one of charged particles [9] (ii) the initial strong E.M. field produced in a HIC that give a sizeable v_1 for charm (anti-charm) quarks which is odd respect to the charge. We have studied both these effects solving the RBT equation coupled to the e.m. field. The time evolution of E.M. field produce in a HIC is calculated by solving Maxwell equations for a single charge and then they are folded with the nuclear transverse density and summed over forward (η) and backward ($-\eta$) rapidity. For details about the e.m. field configuration used see [8]. Due to the collision geometry the generated magnetic field **B** is dominated by the y-component, this results in a Lorentz force that acts on the expanding medium in the xz plane (orthogonal to **B**). On the other hand, time variation of **B** induces an electric field **E**, whose dominant component is E_x, and results in a Faraday current in the xz plane. The net combination of the two effects leads to the formation of a finite direct flow. In realistic simulations for RHIC energies we distribute charm quarks in momentum space according to Fixed Order + Next-to-Leading-Order (FONLL) [16]. For bulk partons we employ thermal distribution plus minijet tail at high p_T while in coordinate space we use the standard Glauber model with a slight modification. The initial conditions are a modification of longitudinal boost invariant to take into account that the initial longitudinal energy density profile is no longer symmetric respect to $\eta \rightarrow -\eta$. These initial condition have been implemented by using the same parametrization in [17] where the parameter have been fixed in order to reproduce the experimental data for the v_1 of charged particles. The initial tilted fireball develops a rapidity-odd directed flow v_1 of charged particles. In Fig. 15.1c we show results for the v_1 under the combined effect of the initial tilt and e.m. field. Since the Lorentz force acts in opposite directions for oppositely charged particles, the v_1 from the e.m. field is expected to generate a non-zero split in the v_1 of D and $\bar{D}$ mesons. On the other hand, the tilt mechanism can generate only finite v_1 without splitting. As shown we predict a large v_1 consistent with the recent experimental data and splitting between particle and anti-particle of about 1%.

References

1. P.B. Gossiaux, J. Aichelin, Phys. Rev. C **78**, 014904 (2008). https://doi.org/10.1103/PhysRevC.78.014904
2. W.M. Alberico et al., Eur. Phys. J. C **71**, 1666 (2011). https://doi.org/10.1140/epjc/s10052-011-1666-6
3. T. Song et al., Phys. Rev. C **92**(1), 014910 (2015). https://doi.org/10.1103/PhysRevC.92.014910
4. S. Cao, G.Y. Qin, S.A. Bass, Phys. Rev. C **92**(2), 024907 (2015). https://doi.org/10.1103/PhysRevC.92.024907
5. F. Scardina, S.K. Das, V. Minissale, S. Plumari, V. Greco, Phys. Rev. C **96**(4), 044905 (2017). https://doi.org/10.1103/PhysRevC.96.044905
6. M. Nahrgang, et al., **C91**(1), 014904 (2015). https://doi.org/10.1103/PhysRevC.91.014904
7. A. Beraudo, et al., **C79**(6), 494 (2019). https://doi.org/10.1140/epjc/s10052-019-6996-9
8. S.K. Das, S. Plumari, S. Chatterjee, J. Alam, F. Scardina, V. Greco, Phys. Lett. B **768**, 260 (2017). https://doi.org/10.1016/j.physletb.2017.02.046

9. S. Chatterjee, P. Bożek, Phys. Rev. Lett. **120**(19), 192301 (2018). https://doi.org/10.1103/PhysRevLett.120.192301
10. M. Ruggieri, F. Scardina, S. Plumari, V. Greco, Phys. Rev. C **89**(5), 054914 (2014). https://doi.org/10.1103/PhysRevC.89.054914
11. S. Plumari, W.M. Alberico, V. Greco, C. Ratti, Phys. Rev. D **84**, 094004 (2011). https://doi.org/10.1103/PhysRevD.84.094004
12. S. Plumari, V. Minissale, S.K. Das, G. Coci, V. Greco, Eur. Phys. J. C **78**(4), 348 (2018). https://doi.org/10.1140/epjc/s10052-018-5828-7
13. S. Plumari, G.L. Guardo, F. Scardina, V. Greco, Phys. Rev. C **92**(5), 054902 (2015). https://doi.org/10.1103/PhysRevC.92.054902
14. S. Plumari, Eur. Phys. J. C **79**(1), 2 (2019). https://doi.org/10.1140/epjc/s10052-018-6510-9
15. J. Adam, et al., (STAR) (2019). arXiv:1905.02052 [nucl-ex]
16. M. Cacciari, P. Nason, R. Vogt, Phys. Rev. Lett. **95**, 122001 (2005). https://doi.org/10.1103/PhysRevLett.95.122001
17. P. Bozek, I. Wyskiel, Phys. Rev. C **81**, 054902 (2010). https://doi.org/10.1103/PhysRevC.81.054902

Chapter 16
Measurements of Open-Charm Hadrons in Au+Au Collisions at $\sqrt{s_{NN}} = 200$ GeV by the STAR Experiment

Jan Vanek

Abstract At RHIC energies, charm quarks are primarily produced at early stages of ultra-relativistic heavy-ion collisions. This makes them an excellent probe of the Quark-Gluon Plasma (QGP) produced in these collisions since they experience the whole evolution of the medium. STAR is able to study the production of charm quarks through direct reconstruction of hadronic decays of open-charm hadrons. This is possible thanks to an excellent vertex resolution provided by the Heavy Flavor Tracker. In these proceedings, we present a selection of the most recent results on open-charm hadron production, in particular the nuclear modification factors of $D^{\pm}$ and D^0, elliptic and triangular flow of D^0, the $\Lambda_c^{\pm}/D^0$ yield ratio, and the directed flow of D^0 mesons.

16.1 Introduction

One of the main goals of the heavy-ion program at the STAR experiment is to study properties of the Quark-Gluon Plasma (QGP). Charm quarks are an excellent probe of the QGP as they are produced at very early stages of ultra-relativistic heavy-ion collisions and so experience the whole evolution of the hot and dense medium. STAR is able to study production of charm quarks through a precise topological reconstruction of open-charm hadron decays utilizing the Heavy Flavor Tracker (HFT) [1].

Various measurements are used to study interactions of charm quarks with the QGP. In these proceedings, we present a selection of the most recent results on open-charm hadron production from the STAR experiment. In particular, we discuss the nuclear modification factors of $D^{\pm}$ and D^0 mesons which give access to the charm quark energy loss in the QGP, and also D^0 elliptic (v_2) and triangular flow

Jan Vanek for the STAR Collaboration.

J. Vanek (✉)
Nuclear Physics Institute, Czech Academy of Sciences, Řež, Czech Republic
e-mail: vanek@ujf.cas.cz

D. Elia et al. (eds.), *The XVIII International Conference on Strangeness in Quark Matter (SQM 2019)*, Springer Proceedings in Physics 250,
https://doi.org/10.1007/978-3-030-53448-6_16

(v_3) coefficients which can probe the charm quark transport in the QGP. We show the $\Lambda_c^{\pm}/D^0$ yield ratio as a function of transverse momentum (p_T) and collision centrality that helps us better understand the charm quark hadronization process in heavy-ion collisions. In addition, we present the rapidity-odd directed flow of D^0 mesons, which can be used to probe the initial tilt of the QGP bulk and the effects of the early-time magnetic field.

16.2 Results

Figure 16.1 shows the nuclear modification factors (R_{AA}) of D^0 and $D^{\pm}$ mesons as a function of transverse momentum in 0–10% central Au+Au collisions. Both open-charm mesons show a significant suppression at high p_T which suggests strong interactions of the charm quarks with the QGP. The R_{AA} evolution in low to intermediate p_T region suggests a large collective flow of charm quarks [2] which can also be seen in Fig. 16.2.

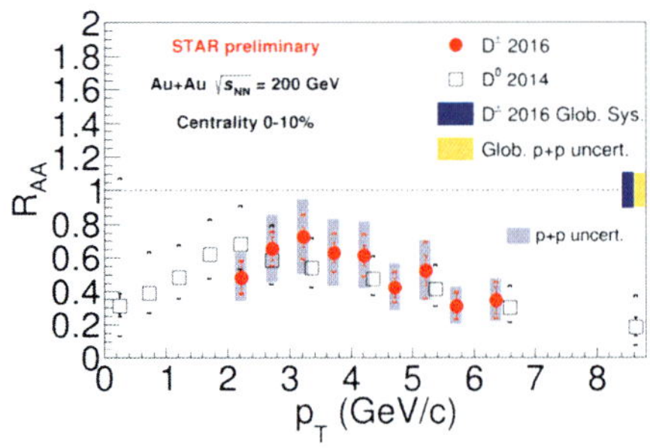

Fig. 16.1 Nuclear modification factor of D^0 [2] and $D^{\pm}$ mesons as a function of p_T in 0–10% central Au+Au collisions at $\sqrt{s_{NN}} = 200$ GeV

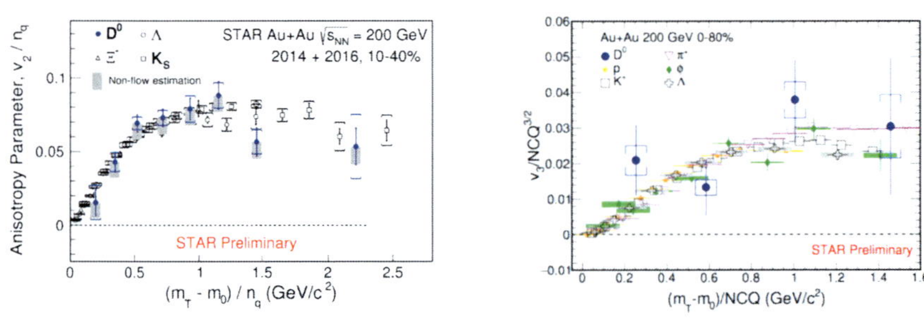

Fig. 16.2 The NCQ-scaled elliptic **(left)** and triangular **(right)** flow of D^0 mesons and light-flavor hadrons [4] in Au+Au collisions at $\sqrt{s_{NN}} = 200$ GeV

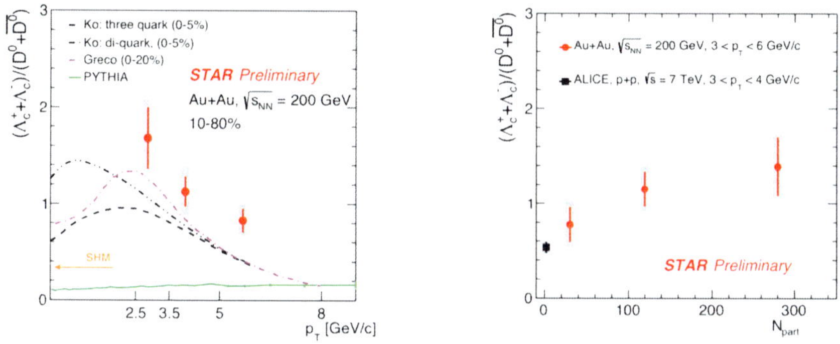

Fig. 16.3 **(left)** $\Lambda_c^{\pm}/D^0$ yield ratio as a function of transverse momentum p_T in Au+Au collisions at $\sqrt{s_{NN}} = 200$ GeV. The data are compared to PYTHIA, Statistical Hadronization Model [5] and coalescence model calculations [6, 7]. **(right)** $\Lambda_c^{\pm}/D^0$ yield ratio as a function of number of participants N_{part}. The ALICE experiment measurement of the ratio in p+p collisions at $\sqrt{s} = 7$ TeV [8] is shown for comparison

Figure 16.2 demonstrates a test of the Number of Constituent Quarks (NCQ, or n_q) scaling [3] for elliptic flow (left panel) and triangular flow (right panel) for both D^0 mesons and light-flavor hadrons. The STAR data show that charm quarks acquire similar level of collectivity as the light quarks in the QGP medium.

The presence of the QGP may also influence the charm quark hadronization. In order to study that, STAR has measured the $\Lambda_c^{\pm}/D^0$ yield ratio as a function of p_T (Fig. 16.3, left panel) and number of participants N_{part} (Fig. 16.3, right panel). The ratio shows an enhancement with respect to p+p collisions and PYTHIA calculation, and is reasonably reproduced by models incorporating coalescence hadronization of the charm quarks [6, 7].

Theoretical calculations predict that the charm quarks might also be sensitive to the initial tilt of the QGP bulk and the electromagnetic (EM) field induced by the passing spectators [9]. The former leads to a large negative slope of the directed flow versus rapidity (dv_1/dy) of open-charm mesons, and the latter to a negative slope for D^0 and a positive slope for $\overline{D^0}$. When combined, the slope is predicted to be negative for both D^0 and $\overline{D^0}$ but larger for D^0 than for $\overline{D^0}$ in Au+Au collisions at $\sqrt{s_{NN}} = 200$ GeV. The STAR result on D^0 and $\overline{D^0}$ v_1 are shown in Figure 16.4. The current precision of the measurement is not sufficient to conclude on the EM induced splitting, but the dv_1/dy slopes are indeed negative and significantly larger that of light-flavor mesons, as discussed in [10].

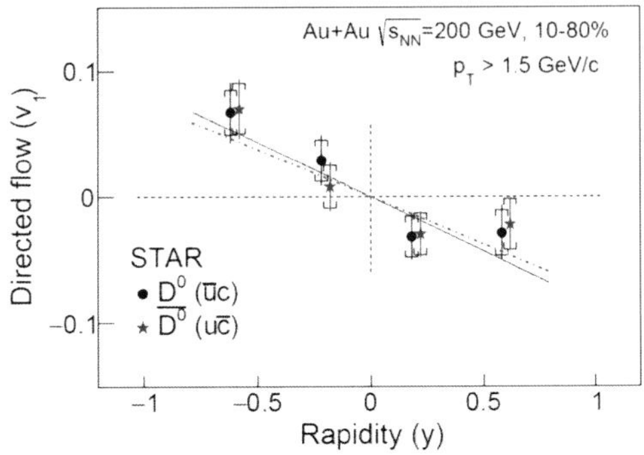

Fig. 16.4 Directed flow D^0 and $\overline{D^0}$ mesons as a function of rapidity in 10-80% central Au+Au collisions at $\sqrt{s_{\rm NN}} = 200$ GeV. The solid black and blue dashed lines are fits to the data. Taken from [10]

16.3 Summary

The STAR experiment has extensively studied the production of open-charm hadrons in Au+Au collisions at $\sqrt{s_{NN}} = 200$ GeV through a precise topological reconstruction of their hadronic decays, utilizing the HFT. The latest results show that the D^0 and $D^{\pm}$ mesons are suppressed in central Au+Au collisions, suggesting a substantial energy loss of the charm quarks in the QGP. The charm quarks also exhibit a significant collective motion as suggested by the observed large elliptic and triangular flow of D^0 mesons. The QGP seems to influence the charm quark hadronization. The STAR results on the $\Lambda_c^{\pm}/D^0$ yield ratio are in qualitative agreement with theoretical models incorporating coalescence hadronization of charm quarks. The measured D^0 dv_1/dy slope is qualitatively consistent with hydrodynamical model calculations with tilted QGP bulk [9].

Acknowledgments These proceedings and presentation are supported by OPVVV grant CZ.02.1.01/0.0/0.0/16_013/0001569 and the project LTT18002 of the Ministry of Education, Youth and Sports of the Czech Republic.

References

1. D. Beavis, et al., The STAR Heavy-flavor tracker, technical design report (2011). https://drupal.star.bnl.gov/STAR/starnotes/public/sn0600. Accessed 6 Sep 2019
2. J. Adam et al., [STAR Collaboration]: centrality and transverse momentum dependence of D^0-meson production at mid-rapidity in Au+Au collisions at $\sqrt{s_{\rm NN}}$= 200 GeV. Phys. Rev. C **99**, 034908 (2019). https://doi.org/10.1103/PhysRevC.99.034908
3. S. Afanasiev, et al., [PHENIX Collaboration]: elliptic flow for phi mesons and (anti)deuterons in Au+Au collisions at $\sqrt{s_{\rm NN}} = 200$ GeV. Phys. Rev. Lett. **99**, 052301 (2007). https://doi.org/10.1103/PhysRevLett.99.052301
4. B.I. Abelev et al., [STAR Collaboration]: centrality dependence of charged hadron and strange hadron elliptic flow from $\sqrt{s_{\rm NN}}$= 200 GeV Au+Au collisions. Phys. Rev. C **77**, 054901 (2008). https://doi.org/10.1103/PhysRevC.77.054901

5. A. Andronic et al., Statistical hadronization of charm in heavy-ion collisions at SPS. RHIC and LHC. Phys. Lett. B **571**, 36–44 (2003). https://doi.org/10.1016/j.physletb.2003.07.066
6. Y. Oh et al., Ratios of heavy baryons to heavy mesons in relativistic nucleus-nucleus collisions. Phys. Rev. C **79**, 044905 (2009). https://doi.org/10.1103/PhysRevC.79.044905
7. S. Plumari, V. Minissale, S.K. Das, G. Coci, V. Greco, Charmed hadrons from coalescence plus fragmentation in relativistic nucleus-nucleus collisions at RHIC and LHC. https://doi.org/10.1140/epjc/s10052-018-5828-7
8. S. Acharya, et al., [The ALICE collaboration]: c^+ production in pp collisions at $\sqrt{s} = 7\,\mathrm{TeV}$ and in p-Pb collisions at $\sqrt{s_{\mathrm{NN}}}= 5.02\,\mathrm{TeV}$. J. High Energ. Phys. **2018**, 108 (2018). https://doi.org/10.1007/JHEP04(2018)108
9. S. Chatterjee, P. Bozek, Interplay of drag by hot matter and electromagnetic force on the directed flow of heavy quarks. **2018**. arXiv:1804.04893
10. J. Adam, et al., [STAR Collaboration]: first observation of the directed flow of D^0 and $\overline{D^0 \mathrm{inAu} + \mathrm{Aucollisionsat}\sqrt{s_{\mathrm{NN}}}} = 200\,\mathrm{GeV}$. Phys. Rev. Lett. **123**, 162301 (2019). https://doi.org/10.1103/PhysRevLett.123.162301

Chapter 17
Overview on Hidden Heavy Avour Results

Roberta Arnaldi

Abstract A long road has been made since when quarkonium was proposed for the first time as a signature for the formation of a plasma of quarks and gluons (QGP). Now, thirty years later after the first SPS measurements, the high precision data obtained from both RHIC and LHC experiments confirm its central role in the investigation of the properties of the QGP created in heavy-ion collisions. In this paper an overview of the most recent quarkonium results will be presented, focussing on the production of the J/ψ , $\psi(2S)$ and Υ states in pp, p-A and A-A collisions.

17.1 Introduction

The existence of a phase transition between the hadronic matter and the Quark Gluon Plasma (QGP), a state where quarks and gluons are no longer confined into hadrons, is predicted by QCD to occour when high temperatures and/or baryonic densities are reached. The formation and the properties of such a state can be investigated experimentally in ultrarelativistic heavy-ion collisions. These studies were performed for the first time thirty years ago at the CERN SPS, at a centre-of-mass energy ($\sqrt{s_{\rm NN}}$) of 17 GeV. From year 2000, the higher energies reached at RHIC ($\sqrt{s_{\rm NN}}$ = 200 GeV) and, ten years later, at LHC (up to $\sqrt{s_{\rm NN}}$ = 5.02 TeV) allowed to get further insight on the QGP properties, investigating a system where higher temperatures and energy densities were reached.

Quarkonia are well-known probes of the formation of the QGP since the high colour-charge density reached in such a medium can screen the binding force between the Q and $\overline{\rm Q}$, leading to a temperature-dependent melting of the quarkonium states according to their binding energies [1]. Quarkonium production is usually investigated in three colliding systems, i.e. in proton-proton, proton-nucleus and nucleus-nucleus collisions. Proton-proton interactions have a twofold interest: on one side, they allow to address the production mechanisms, on the other side they represent

R. Arnaldi (✉)
INFN Sez. Torino, Via P. Giuria 1, I-10125 Torino, Italy
e-mail: arnaldi@to.infn.it

D. Elia et al. (eds.), *The XVIII International Conference on Strangeness in Quark Matter (SQM 2019)*, Springer Proceedings in Physics 250,
https://doi.org/10.1007/978-3-030-53448-6_17

a reference for the production in p-A and A-A collisions. In fact, the modification induced by medium effects on the quarkonium yield in heavy-ion collisions is usually quantified through the nuclear modification factor $R_{AA} = Y_{AA}/\langle N_{coll}\rangle \times Y_{pp}$, defined as the ratio of yields in A-A (or p-A) and pp collisions, scaled by the number of interactions $\langle N_{coll}\rangle$ computed within the Glauber model. Proton-nucleus interactions allow the investigation of quarkonia in a cold nuclear medium, where the effects which may modify the production are not related to the formation of a hot system [2]. Among the cold nuclear matter (CNM) effects, the dominant ones are due to the modification of the quarks and gluons structure functions for nucleons inside nuclei, modeled either via the nuclear parton distribution functions (nPDF) or through a Glass Condensate effective theory, or the initial/final energy loss of the $Q\overline{Q}$ pair during its path through the cold nuclear medium. A-A collisions allow the study of the modifications induced by a hot medium on the probes. In a hot and deconfined medium the production of quarkonium is expected to be significantly suppressed with respect to the pp yield scaled by the number of collisions. However, at very high energies as those reached at LHC, quarkonium can also be produced by recombination of charm quarks, due to their large abundance in the medium. This mechanism, particularly important for charmonium, i.e. states formed by c and $\overline{c}$, may counterbalance partly or totally, the suppression process. Finally, it should be noted that, even if a priori the three colliding systems are well distinguished, the separation between their properties is in reality not so sharp. Recent results at LHC energies indicate, in fact, that some of the behaviours observed in the smaller system, as pp and p-A, are reminiscent of observations done in the larger A-A systems (see e.g. [3]) and hence the role of small systems in the understanding of heavy-ion collisions is now deeply investigated.

17.2 Quarkonium in Proton–Proton Collisions

The study of quarkonium production in pp collisions has now reached a high level of precision. The most recent RHIC [4] and LHC [5] measurements have significantly extended, at various energies and in several rapidity (y) regions, the transverse momentum (p_T) coverage allowing not only a more stringent comparison to theory models, but also an improvement of the reference for p-A and A-A studies. Calculations based on the color glass condensate approach, on an improved color evaporation model and on NRQCD show a reasonable agreement with the experimental results over the full kinematic range [6, 7]. Quarkonium production is also investigated as a function of the charged particle multiplicity. The self-normalised yields show an increase with event activity which is weakly dependent on energy and on the quarkonium state, while it is stronger for high-p_T (Fig. 17.1 (left)). It can be observed that the increase of the self-normalised yields is more than linear when no rapidity gap is present between the y regions where quarkonium and multiplicity are measured. Most of the theory predictions, even if based on different underlying processes, fairly describe the experimental observations.

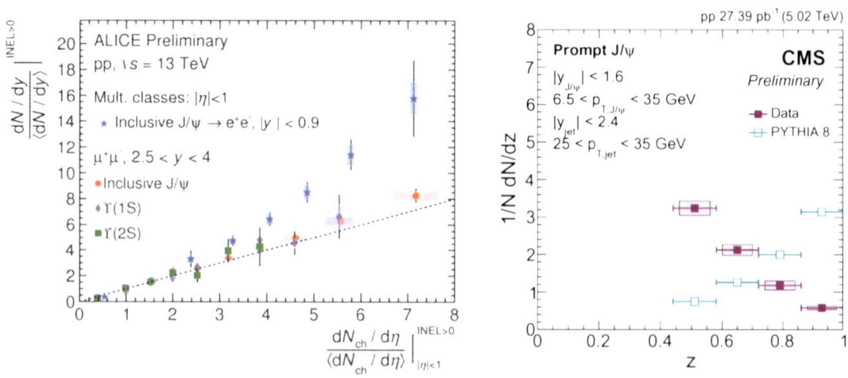

Fig. 17.1 Left: Self-normalised quarkonium yields as a function of the event activity. Right: Self-normalised z distribution for prompt J/ψ, both in data and in PYTHIA8 [8]

Finally, the production of J/ψ is also studied in jets [8, 9]. The measurements in Fig. 17.1 (right) show that prompt J/ψ carry a small fraction of the jet momentum (quantified through the momentum fraction $z = p_{\mathrm{T}}^{J/\psi}/p_{\mathrm{T}}^{jet}$), indicating that their production is accompanied by a large jet activity, much larger than the one predicted by PYTHIA8 [10]. These observations may help to put further constraints to the quarkonium production models.

17.3 Quarkonium in Proton-Nucleus Collisions

The role of CNM effects on quarkonium production has been studied in several systems at RHIC energies. Results from the lightest systems, as p-Al, show a very little modification of the yields as a function of p_{T}, both in the forward and in the backward y regions accessible by the PHENIX experiment. When heavier nuclei are involved, a reduction of the quarkonium production with respect to the one measured in pp interactions is observed. Such a reduction is more important at low p_{T} and can be reproduced by theoretical models based on shadowing, with a possible inclusion of a further contribution related to the break-up of the resonance due to inelastic interactions with the surrounding nucleons. CNM effects have been studied in details also at LHC energies, over a broad kinematic range, exploiting the complementarity of the four experiments. As shown in Fig. 17.2 (left) the J/ψ R_{pA} shows a suppression going from the backward-y region to the most forward one [11]. Models based on shadowing, color glass condensate implementations, energy loss or transport calculations (see [11] and references therein) nicely describe the observed trend, even if the uncertainties, mainly due to the shadowing contribution, do not allow to discriminate between the various approaches.

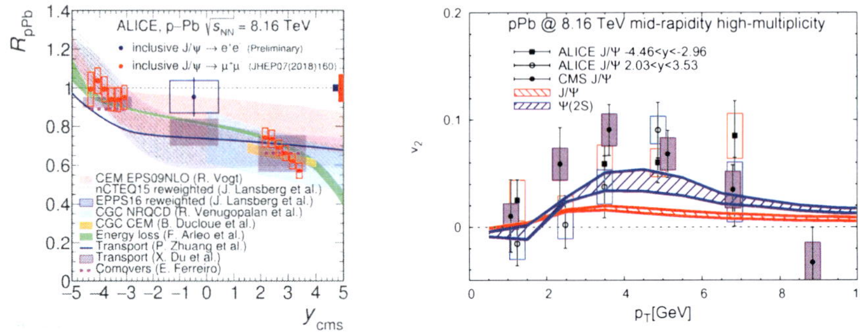

Fig. 17.2 Left: J/ψ R_{pA} as a function of rapidity [11]. Right: J/ψ v_2 as a function of p_{T} [19, 20]. Both results are compared to theoretical calculations

Similar conclusions can be drawn also for the $\Upsilon(1\mathrm{S})$ production in p-A collisions [12, 13], since the R_{pA} shows a suppression, dominant at low p_{T}, which increases from backward to forward y and which can be described by theoretical models based on shadowing or energy loss. On the contrary, a different behaviour is observed for the quarkonium excited states. The $\psi(2\mathrm{S})$ resonance show a stronger suppression than the J/ψ, in particular in the backward-y region [14]. Such a suppression cannot be described by models based on the aforementioned CNM effects, but final state mechanisms, as the interactions with comoving particles [15, 16], affecting mainly the more loosely bound quarkonium states, have to be added. Same conclusions can be drawn for the $\Upsilon(2\mathrm{S})$ and $\Upsilon(3\mathrm{S})$ resonances [12, 13, 17, 18]. The R_{pA} studies are complemented by the investigation of the quarkonium elliptic flow v_2, which measures the participation of the resonance to the collective flow of the medium. As shown in Fig. 17.2 (right) a non-zero v_2 is observed for the J/ψ as a function of p_{T}, by both ALICE [19] and CMS [20]. The size of the v_2 is similar to the one observed in Pb-Pb collisions, which will be discussed later on, even if the mechanisms involved should be different. Furthermore, it can be observed that models including final state mechanisms still underestimate the observed effect [16].

17.4 Quarkonium in Nucleus–Nucleus Collisions

It is now well assessed that, for low p_{T} J/ψ (Fig. 17.3 left), the smaller R_{AA} suppression observed at LHC, with respect to RHIC measurements, can be interpreted as due to a different balance of the suppression and recombination mechanisms. The latter is more important at higher energies because of the larger amount of $c\bar{c}$ pairs in the medium [21].

On the contrary, at high p_{T}, where the recombination contribution is expected to play a minor role, the suppression observed by RHIC and LHC experiments is of the same order [22, 23]. A similar suppression at the two energies is also observed for

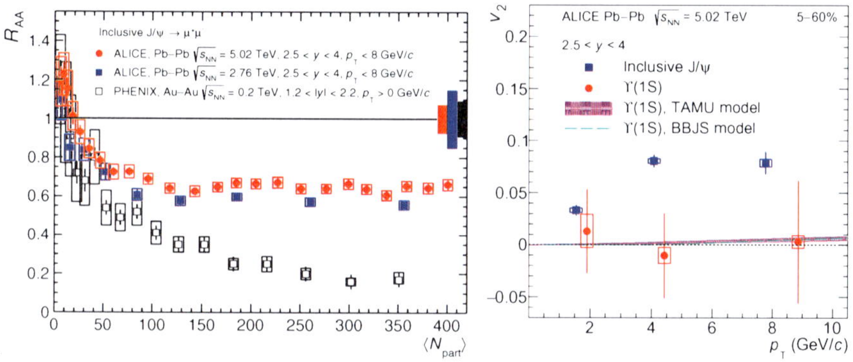

Fig. 17.3 Left: centrality-dependence of J/ψ R_{AA}. Right: p_{T}-dependence of the Υ(1S) and J/ψ v_2

the Υ(1S) [24]. Even if uncertainties are still rather large, this result seems to imply a little modification induced by the hot medium on the strongly bound Υ(1S). On the contrary, the Υ(2S) state shows a stronger suppression at LHC energies, as expected in a sequential suppression scenario. The measurement of the R_{AA} is complemented by the observation of the v_2, whose measurement is now extended over a broad kinematic range. For the J/ψ, a significant v_2 is observed, reaching a 7σ effect at intermediate p_{T} and remaining different from zero up to p_{T} ~20 GeV/c [23, 25, 26]. In the intermediate p_{T} region, the results can be described by theoretical models which assume that the J/ψ is inheriting the flow of the charm quarks, as a further confirmation of the role played by recombination. At high p_{T}, where recombination is not expected to be present, the models so far underestimate the experimental data. For the first time the v_2 of the Υ(1S) has also been measured [23] with values close to zero over the full p_{T} interval, contrarily to the J/ψ case, suggesting a different production mechanisms (see Fig. 17.3 (right)). Finally, for the first time ALICE has evaluated the J/ψ polarization in Pb-Pb collisions, in various reference frames. All polarization parameters turn out to be zero, when studied as a function of p_{T}, similarly to what was observed in pp collisions.

17.5 Conclusions

The new results presented at this Conference confirm the crucial role of quarkonia in the understanding of QGP properties. LHC and RHIC results have now reached a remarkable precision, allowing a detailed study of several observables as the R_{AA} and the v_2 for charmonium and bottomonium states. A simultaneous description of all the available measurements is now challenging for theoretical models, but mandatory to get further insight into the QGP characterization.

References

1. T. Matsui, H. Satz, J/ψ suppression by quark-gluon plasma formation. Phys. Lett. B **178**, 416–422 (1986)
2. J.L. Albacete et al., Predictions for cold nuclear matter effects in p+Pb collisions at $\sqrt{s_{NN}}$ = 8.16 TeV. Nucl. Phys. A **972**, 18–85 (2018)
3. CMS Coll., S. Chatrchyan, et al., Observation of long-range near-side angular correlations in pPb collisions at the LHC. Phys. Lett. B **718**, 795–814 (2013)
4. STAR Coll., J. Adam, et al., Measurements of the p_T-dependent cross sections of J/ψ production at mid-rapidity in proton+proton collisions at $\sqrt{s}$ = 510 and 500 GeV with the STAR detector. Phys. Rev. D **100**(5), 052009 (2019)
5. ALICE Coll., A. Acharya, et al., Inclusive J/ψ production at mid-rapidity in pp collisions at $\sqrt{s}$ = 5.02 TeV. arXiv:1905.07211 [nucl-ex]
6. Y.Q. Ma, R. Venugopalan, Comprehensive description of J/ψ production in pp collisions at collider energies. Phys. Rev. Lett. **113**(19), 192301 (2014)
7. Y.-Q. Ma, R. Vogt, Quarkonium production in an improved color evaporation model. Phys. Rev. **D94**(11), 114029 (2016). arXiv:1609.06042
8. CMS Coll., Production of prompt and nonprompt J/ψ mesons in jets in pp collisions at $\sqrt{s}$ =5.02 TeV. https://cds.cern.ch/record/2318344
9. LHCb Coll., R. Aaij, et al., Study of J/ψ production in Jets. Phys. Rev. Lett. **118**(19), 192001 (2017). arXiv:1701.05116 [hep-ex]
10. T. Sjostrand et al., An introduction to PYTHIA 8.2. Comput. Phys. Commun. **191**, 159–177 (2015). arXiv:1410.3012 [hep-ph]
11. ALICE Coll., S. Acharya, et al., Inclusive J/ψ production at forward and backward rapidity in p-Pb collisions at $\sqrt{s_{\rm NN}}$ = 8.16 TeV, JHEP 07 160 (2018)
12. ATLAS Coll., M. Aaboud, et al., Measurement of quarkonium production in p-Pb and pp collisions at 5.02 TeV with the ATLAS detector Eur. Phys. J. **C78**(3), 171 (2018)
13. ALICE Coll., Inclusive Υ production in p-Pb collisions at $\sqrt{s_{NN}}$= 8.16 TeV. https://cds.cern.ch/record/2317189
14. ALICE Coll., Abelev B *et al.* (2014), Suppression of ψ(2S) production in p-Pb collisions at $\sqrt{s_{\rm NN}}$ = 5.02 TeV, JHEP 12 073, arXiv:1405.3796
15. E. Ferreiro, Excited charmonium suppression in proton-nucleus collisions as a consequence of comovers. Phys. Lett. B **749**, 98–103 (2015). arXiv:1411.0549
16. X. Du, R. Rapp, In-medium charmonium production in proton-nucleus collisions, JHEP03 015 (2019). arXiv:1808.10014 [nucl-th]
17. CMS Coll., S. Chatrchyan, et al., Event activity dependence of Y(nS) production in $\sqrt{s_{NN}}$=5.02 TeV pPb and $\sqrt{s}$=2.76 TeV pp collisions, JHEP 04 103 (2014)
18. LHCb Coll., R. Aaij, et al., Study of Υ production in pPb collisions at $\sqrt{s_{NN}}$ = 8.16 TeV, JHEP 11 194 (2018), arXiv:1810.07655 [hep-ex]
19. ALICE Coll., S. Acharya, et al., Search for collectivity with azimuthal J/ψ-hadron correlations in high multiplicity p-Pb collisions at $\sqrt{s_{\rm NN}}$ = 5.02 and 8.16 TeV. Phys. Lett. B **780**, 7–20 (2018)
20. CMS Coll., Observation of prompt J/ψ meson elliptic flow in high-multiplicity pPb collisions at $\sqrt{s_{NN}}$=8.16 TeV. https://cds.cern.ch/record/2318326
21. ALICE Coll., J. Adam, et al., J/ψ suppression at forward rapidity in Pb-Pb collisions at $\sqrt{s_{\rm NN}}$ = **5.02** TeV. Phys. Lett. B **766**, 212–224 (2017)

22. STAR Coll., J. Adam, et al., Measurement of inclusive J/ψ suppression in Au+Au collisions at $\sqrt{s_{NN}} = 200$ GeV through the dimuon channel at STAR. Phys. Lett. B **797**, 134917 (2019). arXiv:1905.13669 [nucl-ex]
23. CMS Coll., V. Khachatryan, et al., Suppression and azimuthal anisotropy of prompt and non-prompt J/ψ production in PbPb collisions at $\sqrt{s_{NN}}$ =2.76 TeV, Eur. Phys. J. **C77**(4), 252 (2017). arXiv:1610.00613 [nucl-ex]
24. CMS Coll., A.M. Sirunyan, et al., Measurement of nuclear modification factors of $\Upsilon(1S)$, $\Upsilon(2S)$, and $\Upsilon(3S)$ mesons in PbPb collisions at $\sqrt{s_{\rm NN}} = 5.02$ TeV. Phys. Lett. B **790**, 270–293 (2019). arXiv:1805.09215 [hep-ex]
25. ALICE Coll., S. Acharya, et al., Study of J/ψ azimuthal anisotropy at forward rapidity in Pb-Pb collisions at $\sqrt{s_{\rm NN}} = 5.02$ TeV, JHEP 02 012 (2019)
26. ATLAS Coll., M. Aaboud, et al., Prompt and non-prompt J/ψ elliptic flow in Pb+Pb collisions at $\sqrt{s_{NN}} = 5.02$ TeV with the ATLAS detector. Eur. Phys. J. **C78**(9), 784 (2018). arXiv:1807.05198 [nucl-ex]

Chapter 18
J/ψ Production Measurements in pp, p–Pb and Pb–Pb Collisions at Midrapidity at the LHC

Minjung Kim

Abstract Measurements of J/ψ production are a valuable probe to study the properties of the hot and dense medium created in heavy-ion collisions. We report on the latest measurements of J/ψ production in pp, p–Pb and Pb–Pb collisions at midrapidity by ALICE at the LHC.

Suppression of J/ψ production in ultra-relativistic heavy-ion collisions induced by color screening of the $c\bar{c}$ pair was proposed as a signature of the Quark–Gluon Plasma (QGP) created in the collisions [1]. The in-medium dissociation probability of different charmonium states was further expected to provide an estimate of the QGP temperature [2, 3]. At LHC energies, as the $c\bar{c}$ cross section increases, the recombination of abundantly produced $c\bar{c}$ pairs into J/ψ, (re)generation, is expected to be the dominant contribution to the J/ψ production and is therefore a direct probe of the charm quark deconfinement and thermalization in the QGP [4, 5].

The ALICE detector is dedicated to study the QGP properties [6]. J/ψ production is measured via the dielectron decay channel at midrapidity down to zero transverse momentum (p_T) with a separation of prompt and non-prompt J/ψ down to p_T = 1 GeV/c. The Inner Tracking System (ITS) and the Time Projection Chamber (TPC) are used for track and collision vertex reconstruction. Electrons are identified with the specific energy loss ($\mathrm{d}E/\mathrm{d}x$) in the TPC. The Electromagnetic Calorimeter (EMCal) and the Transition Radiation Detector (TRD) are used to trigger for the events with high p_T electrons and for further electron identification. In these proceedings, a review of recent results is presented on J/ψ production at midrapidity in pp collisions at $\sqrt{s}$ = 5.02 TeV and $\sqrt{s}$ = 13 TeV, in p–Pb collisions at $\sqrt{s_{NN}}$ = 5.02 TeV and $\sqrt{s_{NN}}$ = 8.16 TeV, and in Pb–Pb collisions at $\sqrt{s_{NN}}$ = 5.02 TeV.

Minjung Kim for the ALICE Collaboration.

M. Kim (✉)
Physikalisches Institut, Ruprecht-Karls-Universität Heidelberg, Heidelberg, Germany
e-mail: minjung.kim@cern.ch

D. Elia et al. (eds.), *The XVIII International Conference on Strangeness in Quark Matter (SQM 2019)*, Springer Proceedings in Physics 250,
https://doi.org/10.1007/978-3-030-53448-6_18

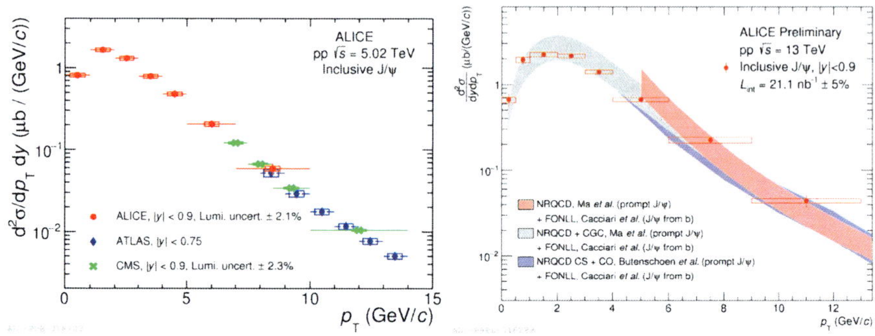

Fig. 18.1 p_T—differential inclusive J/ψ cross section at midrapidity in pp collisions at $\sqrt{s}$ = 5.02 TeV [7] compared with ATLAS [8] and CMS [9] measurements (left) and at $\sqrt{s}$ = 13 TeV compared with theoretical model calculations [10–12] (right)

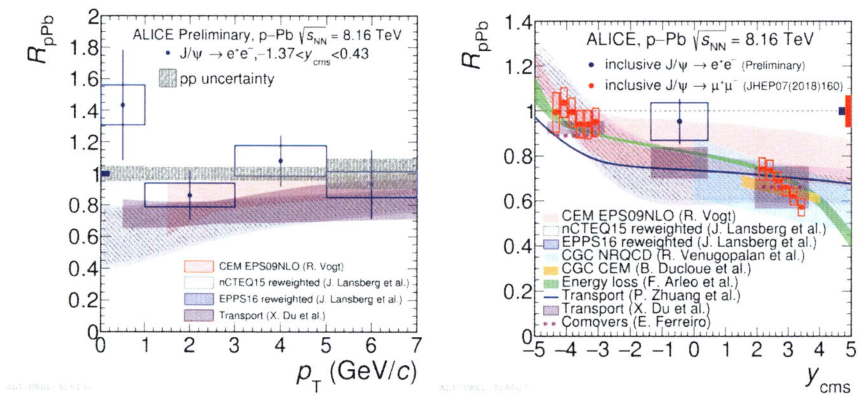

Fig. 18.2 R_{pPb} of inclusive J/ψ as a function of p_T at midrapidity (left) and of rapidity (right) in p–Pb collisions at $\sqrt{s_{\text{NN}}}$ = 8.16 TeV compared with model calculations [14–21]

In pp collisions, the p_T–differential invariant cross section of inclusive J/ψ production for rapidity $|y| < 0.9$ at $\sqrt{s}$ = 5.02 TeV [7] and 13 TeV is measured as shown in Fig. 18.1. The measurement at $\sqrt{s}$ = 5.02 TeV (Fig. 18.1, left) agrees well with ATLAS [8] and CMS [9] measurements in the p_T region of overlap. The Non-Relativistic QCD (NRQCD) [10] calculation coupled to a Color Glass Condensate (CGC) description of the low-x gluons in the proton for prompt J/ψ production [12] together with Fixed—Order Next-To-Leading-Logarithm (FONLL) calculations for the non-prompt contribution J/ψ [11] describe the measurements well [7, 13] as illustrated in Fig. 18.1, right.

In order to quantify possible cold-nuclear-matter (CNM) effects present in Pb–Pb collisions, we measure the nuclear modification factor R_{pPb}, the ratio of the cross section in p–Pb collisions to that measured in pp collisions scaled by the number of nucleons in the nucleus, in p–Pb collisions where we do not expect significant

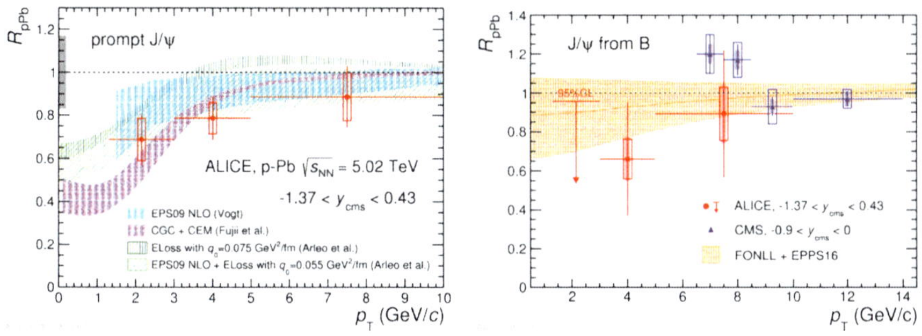

Fig. 18.3 R_{pPb} of prompt (left) and non-prompt (right) J/ψ at midrapidity in p–Pb collisions at $\sqrt{s_{\text{NN}}}$ = 5.02 TeV [24] compared with model calculations [9, 25–28]

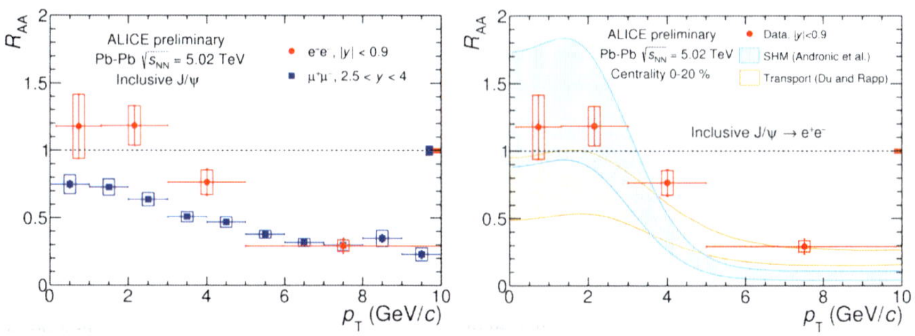

Fig. 18.4 Inclusive J/ψ R_{AA} in 0–20% centrality as a function of p_{T} in Pb–Pb collisions at $\sqrt{s_{\text{NN}}}$ = 5.02 TeV at midrapidity compared with measurements at forward–rapidity [30, 31] (left) and model calculations [20, 32] (right)

hot–medium effects. Figure 18.2 shows the R_{pPb} of inclusive J/ψ in p–Pb collisions at $\sqrt{s_{\text{NN}}}$ = 8.16 TeV as a function of p_{T} at midrapidity and as a function of centre-of-mass rapidity, y_{CMS}. Various theoretical calculations considering different initial state effects, nuclear shadowing [14, 15], CGC effects [16, 17] and parton coherent energy loss [18], and assuming additional final state interactions [19–21] describe the p_{T} and y_{CMS} dependence of the R_{pPb}.

Based on the long lifetime of beauty hadrons, we can separate the prompt and non-prompt J/ψ production at midrapidity using the pseudo-proper decay length [22–24]. Figure 18.3 shows the R_{pPb} of prompt J/ψ (left) and non-prompt J/ψ (right) as a function of p_{T}. Within experimental and theoretical uncertainties, models including CNM effects [25–28] describe the measured R_{pPb} of prompt J/ψ which is smaller than unity at low p_{T}. The R_{pPb} of non-prompt J/ψ is consistent with FONLL calculations [11] including EPPS16 nPDFs [29] and matches CMS result [24] at high p_{T} within uncertainties.

The modification of J/ψ production in heavy–ion collisions can be studied via the nuclear modification factor R_{AA} which is defined as the ratio between the yield in

Pb–Pb collisions and the J/ψ cross section in pp collisions scaled by the nuclear overlap function. The R_{AA} of inclusive J/ψ in Pb–Pb collisions at midrapidity in 0–20% centrality is shown as a function of p_T in Fig. 18.4 together with the measurements at forward rapidity [30] and with model calculations [20, 32]. At low p_T ($p_T < 5$ GeV/c), the R_{AA} of inclusive J/ψ is larger at midrapidity compared to forward rapidity while both results show comparable suppression at high p_T. This behavior can be understood within (re)generation scenario of J/ψ since the $c\bar{c}$ density is larger at midrapidity. The p_T dependence of the J/ψ R_{AA} is described by the statistical hadronization model (SHM) [32] and a transport model [20] within experimental and theoretical uncertainties.

Selected measurements of J/ψ production in pp, p–Pb and Pb–Pb collisions at midrapidity were presented. In p–Pb collisions, suppression of prompt J/ψ production is observed due to nuclear effects. The measurements in Pb–Pb collisions indicate a sizable contribution of J/ψ production from (re)generation. The additional data samples collected in Run 2 in 2018 and the much larger data samples with even better detector performance expected in Run 3 and Run 4 [33] will further improve precision of the measurements.

References

1. T. Matsui, H. Satz, Phys. Lett. B **178**, 416 (1986)
2. F. Karsch, M.T. Mehr, H. Satz, Z. Phys. C **37**, 617 (1988)
3. S. Digal, P. Petreczky, H. Satz, Phys. Rev. D **64**, 094015 (2001)
4. P. Braun-Munzinger, J. Stachel, Phys. Lett. B **490**, 196 (2000)
5. R.L. Thews, M. Schroedter, J. Rafelski, Phys. Rev. C **63**, 054905 (2001)
6. K. Aamodt et al., ALICE collaboration. JINST **3**, S08002 (2008)
7. S. Acharya et al., [ALICE Collaboration]. arXiv:1905.07211 [nucl-ex]
8. M. Aaboud et al., [ATLAS Collaboration]. Eur. Phys. J. C **78**(3), 171 (2018)
9. A.M. Sirunyan et al., [CMS Collaboration]. Eur. Phys. J. C **77**(4), 269 (2017)
10. Y.Q. Ma, K. Wang, K.T. Chao, Phys. Rev. Lett. **106**, 042002 (2011)
11. M. Cacciari et al., JHEP **1210**, 137 (2012)
12. Y.Q. Ma, R. Venugopalan, Phys. Rev. Lett. **113**(19), 192301 (2014)
13. S. Acharya et al., [ALICE Collaboration]. Eur. Phys. J. C **77**(6), 392 (2017)
14. J.L. Albacete et al., Nucl. Phys. A **972**, 18 (2018)
15. A. Kusina et al., Phys. Rev. Lett. **121**(5), 052004 (2018)
16. Y.Q. Ma et al., Phys. Rev. C **97**(1), 014909 (2018)
17. B. Ducloué, T. Lappi, H. Mäntysaari, Phys. Rev. D **94**(7), 074031 (2016)
18. F. Arleo, S. Peigné, JHEP **1410**, 073 (2014)
19. B. Chen et al., Phys. Lett. B **765**, 323 (2017)
20. X. Du, R. Rapp, Nucl. Phys. A **943**, 147 (2015)
21. E.G. Ferreiro, Phys. Lett. B **749**, 98 (2015)
22. B. Abelev et al., ALICE Collaboration. JHEP **1211**, 065 (2012)
23. J. Adam et al., ALICE Collaboration. JHEP **1507**, 051 (2015)
24. S. Acharya et al., [ALICE Collaboration]. Eur. Phys. J. C **78**(6), 466 (2018)
25. R. Vogt, Phys. Rev. C **81**, 044903 (2010)
26. J.L. Albacete et al., Int. J. Mod. Phys. E **22**, 1330007 (2013)
27. H. Fujii, K. Watanabe, Nucl. Phys. A **915**, 1 (2013)
28. F. Arleo et al., JHEP **1305**, 155 (2013)

29. K.J. Eskola et al., Eur. Phys. J. C **77**(3), 163 (2017)
30. J. Adam et al., ALICE Collaboration. Phys. Lett. B **766**, 212 (2017)
31. S. Acharya et al., [ALICE Collaboration]. arXiv:1909.03158 [nucl-ex]
32. A. Andronic et al., Phys. Lett. B **797**, 134836 (2019)
33. Z. Citron et al., arXiv:1812.06772 [hep-ph]

Chapter 19
Quarkonia and Its Fate in the Anisotropic Hot QGP Medium

Mohammad Yousuf Jamal

Abstract The dissociation of the heavy quarkonia states (s-wave bottomonia specifically, for radial quantum numbers n = 1 and 2) have been determined considering the (momentum) anisotropic hot QCD medium. To do that, the in-medium (corrected) complex potential has been obtained. The non- ideal hot QCD medium effects have been included employing a quasi-particle description and found to suppress the dissociation temperature as compared to the ideal case.

19.1 Introduction

After the discovery of J/ψ (a bound state of $c\bar{c}$), [1], in 1974, both the experimental as well as the theoretical studies of heavy quarkonia has become an interesting topic for the researchers to investigate. Later, the quarkonia ($Q\bar{Q}$) suppression has become one among the most reliable signatures of the Quark-Gluon-Plasma (QGP) production in the relativistic heavy-ion collision experiments.

As the bottom quark mass, $m_b \gg \Lambda_{QCD}$ (QCD scale), the velocity of the bound state ($b\bar{b}$) remain small and hence, the Non-Relativistic QCD (NRQCD) approach [2, 3], using non-relativistic potential models can be exploited in this context. Here, we have considered the medium as a hot thermal bath and employed the quasi-particle distribution functions to incorporate the hot medium effects, using the effective fugacity quasi-particle model (EQPM) [4, 5]. The anisotropy has been introduced at the level of distribution function by stretching/squeezing it in one of the direction. Thereafter, the complex dielectric permittivity in the presence of anisotropy has been obtained. The in-medium complex Cornell potential has been obtained by modifying it using complex dielectric permittivity. The thermal width/binding energy of quarkonia bound states is then determined by the imaginary/real part of the modified potential [6]. The dissociation temperatures have been calculated by exploiting

M. Y. Jamal (✉)
School of Physical Sciences, National Institute of Science Education and Research, HBNI, Jatni 752050, India
e-mail: mohammad.yousuf@niser.ac.in

D. Elia et al. (eds.), *The XVIII International Conference on Strangeness in Quark Matter (SQM 2019)*, Springer Proceedings in Physics 250,
https://doi.org/10.1007/978-3-030-53448-6_19

the criterion [7], that says, at the dissociation temperature, the thermal width equals twice the binding energy. To examine the hot QCD medium effects using EQPM [4, 5], the hot QCD equation of state (EoS) have been updated with the recent lattice [8, 9].

19.2 Heavy-Quark Potential, Thermal Width and the Quarkonia Binding Energy in the Anisotropic Hot QCD Medium

In the present analysis, we shall employ the Cornell potential [10, 11], that contains the Coulombic as well as the string part, given as

$$V(r) = -\frac{\alpha}{r} + \sigma r, \tag{19.1}$$

modifying it in the presence of dissipative medium using the dielectric permittivity, $\epsilon(k)$, in the Fourier space. Here, r is the effective radius of the corresponding quarkonia state, α is the strong coupling constant and σ is the string tension. Let us now briefly discuss the model [4, 5] that has been employed in the analysis.

19.2.1 Effective Fugacity Quasi-Particle Model (EQPM) and Debye Screening

EQPM, maps the hot QCD medium effects with the effective equilibrium distribution function, $f_{g,q}(p)$, of quasi-partons [4, 5], that describes the strong interaction effects in terms of effective fugacities, $z_{g,q}$. Where the quasi-parton equilibrium distribution functions for gluon and quark/anti-quark, respectively read as,

$$f_g(p) = \frac{1}{z_g^{-1} e^{\beta E_p} - 1}, \quad f_{q/\bar{q}}(p) = \frac{1}{z_{q/\bar{q}}^{-1} e^{\beta E_p} + 1}. \tag{19.2}$$

The Debye mass, m_D can be obtained using the distribution functions given in 19.2 as,

$$m_D^2 = -4\pi\alpha\left(2N_c \int \frac{d^3p}{(2\pi)^3} \partial_p f_g(\mathbf{p}) + 2N_f \int \frac{d^3p}{(2\pi)^3} \partial_p f_q(\mathbf{p})\right), \tag{19.3}$$

Let us now discuss the modification of the potential, considering the presence of anisotropy in the hot QCD medium.

19.2.2 Medium Modified Heavy-Quark Potential in the Presence of Anisotropy

The anisotropic distribution function has been obtained from isotropic one by rescaling (stretching/squeezing) it in one of the direction in the momentum space as [12],

$$f(\mathbf{p}) \rightarrow f_{\xi}(\mathbf{p}) = C_{\xi}\ f(\sqrt{\mathbf{p}^2 + \xi(\mathbf{p}\cdot\hat{\boldsymbol{n}})^2}), \tag{19.4}$$

where, $\hat{\boldsymbol{n}}$ is a unit vector ($\hat{\boldsymbol{n}}^2 = 1$), showing the direction of momentum anisotropy. The parameter ξ, gives the anisotropic strength in the medium, and describes the amount of squeezing ($\xi > 0$, or oblate form) in the $\hat{\boldsymbol{n}}$, direction. The normalization constant, C_{ξ} by normalizing the Debyee mass comes out as,

$$C_{\xi} = \begin{cases} \frac{\sqrt{\xi}}{\tan^{-1}\sqrt{\xi}} & \text{if} \quad \xi \geq 0. \end{cases} \tag{19.5}$$

Now, the medium modified potential can be obtained in the Fourier space by dividing the heavy-quark potential using the medium dielectric permittivity, $\epsilon(\mathbf{k})$ as,

$$\grave{V}(k) = \frac{\bar{\mathrm{V}}(k)}{\epsilon(k)}. \tag{19.6}$$

19.2.3 Binding Energy and Thermal Width

In the small anisotropy limit, the binding energy (BE) can be obtained by using the real part of the modified potential and solving the Schrödinger equation, considering the isotropic part and the first order perturbation in anisotropy parameter, ξ, [6] as,

$$\mathrm{Re}[BE(T)] = \left(\frac{m_Q\,\sigma^2}{m_D^4\,n^2} + \alpha\,m_D + \frac{\xi}{3}\Big(\frac{m_Q\,\sigma^2}{m_D^4\,n^2} + \alpha\,m_D + \frac{2\,m_Q\,\sigma^2}{m_D^4\,n^2}\Big)\right). \tag{19.7}$$

Using the imaginary part of the potential, the thermal width for a particular resonance state can be found as,

$$\Gamma(T) = -\int d^3\mathbf{r}\ |\Psi(r)|^2\ \mathrm{Im}\ V(\mathbf{r}), \tag{19.8}$$

where, $\Psi(r)$ is the Coulombic wave function for ground state (1 s, corresponding to $n = 1\,(\Upsilon)$) and the first excited state (2 s, corresponding to $n = 2\,(\Upsilon')$), respectively.

Table 19.1 Dissociation temperature for isotropic and anisotropic cases

Temperatures are in the unit of $T_c (= 0.16$ GeV)		
Leading Order		
States ↓	$\xi = 0.0$	$\xi = 0.4$
Υ	2.96	3.09
Υ'	1.51	1.56
(2 + 1)-Lattice EoS		
Υ	2.56	2.69
Υ'	1.14	1.21

19.3 Results and Discussion

As discussed earlier, the dissociation temperature has been obtained by employing the criterion that says, the temperature at which twice the binding energy (real part) equals the thermal width, causes dissociation of quarkonia, is the dissociation temperature. The results for leading order and the lattice EoS are shown in Table 19.1. One can observe that the dissociation temperature is higher for Υ (1s-state), as compared to the excited states, $\Upsilon^{'}$ (2s-state). The numbers for dissociation temperatures are found to be consistent with those given in [7]. Specifically, while implementing the non-ideal (lattice) EoS, the numbers are observed to be closer.

19.4 Summary and Conclusion

The dissociation temperatures for the bottonium states have been obtained using the medium modified inter-quark potential in the anisotropic hot QCD medium. Using the real/imaginary part of the medium modified potential the binding energy/thermal dissociation width and hence, the dissociation temperatures have been obtained. The momentum-space anisotropy, $\xi > 0$ increases the dissociation temperature whereas non-ideal medium interaction effects are found to suppress the results.

Acknowledgments We would like to acknowledge people of INDIA for their generous support for the research in fundamental sciences in the country. Specifically would like to acknowledge DST-SERB (Young Scientist) travel grant for attending the international conference.

References

1. J.J. Aubert et al., E598 Collaboration. Phys. Rev. Lett. **33**, 1404 (1974)
2. A. Pineda, Prog. Part. Nucl. Phys. **67**, 735 (2012)
3. H.S. Shao et al., JHEP **1505**, 103 (2015)

4. V. Chandra, A. Ranjan, V. Ravishankar, Eur. Phys. J. A **40**, 109–117 (2009)
5. V. Chandra, V. Ravishankar, Phys. Rev. D **84**, 074013 (2011)
6. M. Margotta et al., Phys. Rev. D **83**, 105019 (2011)
7. A. Mocsy, P. Petreczky, Phys. Rev. Lett. **99**, 211602 (2007)
8. A. Bazabov et al., Phys. Rev D **90**, 094503 (2014)
9. S. Borsanyi et al., Phys. Lett. B **370**, 99–104 (2014)
10. E. Eichten et al., Phys. Rev. D **17**, 3090 (1978)
11. E. Eichten et al., Phys. Rev. D **21**, 203 (1980)
12. M. Yousuf, S. Mitra, V. Chandra, Phys. Rev. D **95**, 094022 (2017)

Chapter 20
Quarkonium Measurements at Forward Rapidity with ALICE at the LHC

Wadut Shaikh

Abstract Heavy quarks are produced at the first instant of ultra-relativistic nucleus–nucleus collisions and therefore are an important tool to study the subsequent high energy-density medium formed in such collisions. A series of experimental efforts for understanding the properties of the Quark–Gluon Plasma (QGP), a medium consisting of a deconfined state of quarks and gluons, are based on measuring the bound states of heavy quark-antiquark pairs known as quarkonia. However, the medium modification of heavy-flavour hadron production includes also the contribution of Cold Nuclear Matter (CNM) effects such as shadowing or nuclear breakup in addition to the QGP effects. Proton–nucleus collisions, where no QGP is expected, are used to measure CNM effects on quarkonium production. Finally, quarkonium measurements in proton–proton collisions are used as reference for both heavy-ion and proton–ion collisions. ALICE measurements of quarkonia at forward rapidity for various energies and colliding systems (pp, p–Pb, Pb–Pb and Xe–Xe) during the LHC Run 1 and Run 2 periods will be discussed.

20.1 Introduction

Quarkonia are a fruitful probe to investigate the properties of the deconfined medium, called Quark–Gluon Plasma (QGP), created in ultra-relativistic heavy-ion collisions. At LHC energies the modification of the J/ψ ($c\bar{c}$) yield in heavy-ion collisions with respect to the binary-scaled yield in pp collisions is explained as an interplay of suppression [1] and (re)generation [2, 3]. For Υ ($b\bar{b}$), on the other hand, (re)generation effects are expected to be negligible due to the small number of produced b quarks. In addition, the regenerated quarkonia inherit the flow of their constituting c quarks and thus participate to the collective motion in the QGP. The CNM effects (shadowing,

Wadut Shaikh for the ALICE Collaboration.

W. Shaikh (✉)
Saha Institute of Nuclear Physics, HBNI, Kolkata 700064, India
e-mail: wadut.shaikh@cern.ch

D. Elia et al. (eds.), *The XVIII International Conference on Strangeness in Quark Matter (SQM 2019)*, Springer Proceedings in Physics 250,
https://doi.org/10.1007/978-3-030-53448-6_20

parton energy loss, interaction with hadronic medium) which are not related to the deconfined medium may also lead to a modification of quarkonium production. In order to disentangle the CNM effects from the hot nuclear matter effects, quarkonium production is studied in p–Pb collisions in which the QGP is not expected to be formed. In pp collisions, the quarkonium production can be understood as the creation of a heavy-quark pair ($q\bar{q}$) (perturbative) followed by its hadronization into a bound state (non-perturbative). None of the existing models fully describe the quarkonium production in pp collisions and more differential measurements will further constrain the quarkonium production models in elementary hadronic collisions.

20.2 Analysis and Results

The ALICE Collaboration has studied quarkonium production in various collision systems (pp, p–Pb, Pb–Pb, and Xe–Xe) at different center-of-mass energies per nucleon pair $\sqrt{s_{\mathrm{NN}}}$ down to zero transverse momentum (p_{T}) and at forward rapidity ($2.5 < y < 4$) with the Muon Spectrometer [6] through the dimuon decay channel.

pp collisions: In Fig. 20.1 (left), the inclusive J/ψ production cross section at $\sqrt{s}$ = 13 TeV is compared to different sets of Non-Relativistic QCD (NRQCD) predictions [5]. Recent theoretical developments, e.g. combining NRQCD with Color Glass Condensate (GCG+NRQCD) [7] reproduce the J/ψ p_{T} shape. In Fig. 20.1 (right), the quarkonium production shows a linear increase with charged-particle multiplicity.

The ALICE Collaboration has also measured the inclusive J/ψ polarization in pp collisions at $\sqrt{s}$ = 7 and 8 TeV [9]. The measured J/ψ polarization is compatible with zero within uncertainties in the different studied p_{T} intervals. The Color-Singlet Model (CSM) and the Next-to-Leading Order (NLO) NRQCD [10] predictions do not describe the polarization parameters.

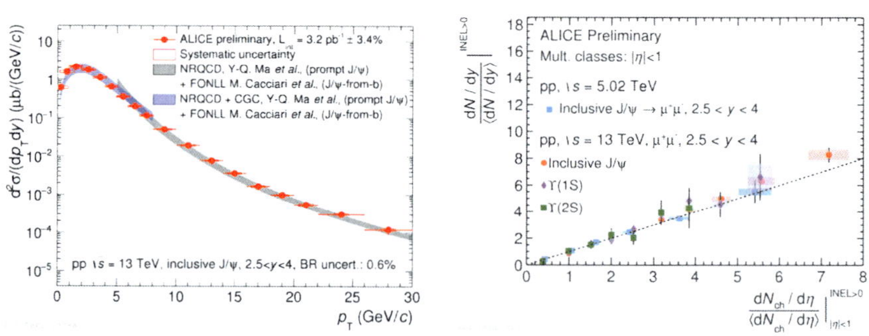

Fig. 20.1 p_{T} differential inclusive J/ψ cross section measured at forward rapidity in pp collisions at $\sqrt{s}$ = 13 TeV (left). Relative quarkonium yield as a function of the relative charged-particle density in pp collisions at $\sqrt{s}$ = 13 and 5.02 TeV (right)

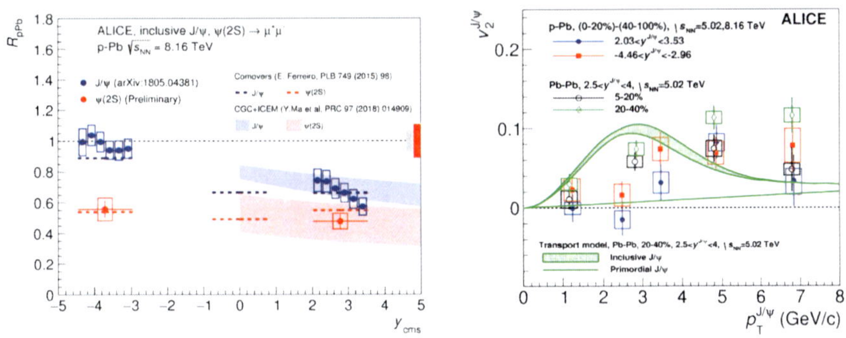

Fig. 20.2 The $\psi(2S)$ and J/ψ R_{pA} as a function of y_{cms} with different model predictions in p–Pb collisions at $\sqrt{s_{NN}}$ = 8.16 TeV (left). J/ψ v_2 coefficients in p–Pb and Pb–p collisions compared to the results in Pb–Pb at $\sqrt{s_{NN}}$ = 5.02 TeV and the transport model calculations for semi-central Pb–Pb collisions at $\sqrt{s_{NN}}$ = 5.02 TeV (right)

p–Pb collisions: The CNM effects can be studied in p–Pb collisions via the nuclear modification factor (R_{pA}), defined as:

$$R_{pPb} = \frac{\sigma_{pPb}}{A_{Pb} \cdot \sigma_{pp}},$$

where σ_{pPb} and σ_{pp} are the production cross sections in p–Pb and pp collisions respectively. A_{Pb} is the atomic mass number (208) of the Pb nucleus. Inclusive J/ψ production in p–Pb collisions at $\sqrt{s_{NN}}$ = 8.16 TeV [11] was measured by ALICE and a suppression was observed at positive y_{cms} corresponding to low Bjorken-x gluons in the Pb nucleus. The models which include various combinations of cold nuclear matter effects describe the J/ψ p–Pb data.

The R_{pA} of $\psi(2S)$ together with J/ψ is shown in Fig. 20.2 (left). Contrary to the J/ψ case [11], models must include final-state interactions with the surrounding medium in order to describe the $\psi(2S)$ results.

The ALICE Collaboration measured azimuthal correlations between J/ψ emitted at forward and backward rapidities with mid-rapidity charged particles at $\sqrt{s_{NN}}$ = 5.02 and 8.16 TeV [12]. The data indicate persisting long-range correlation structures at $\Delta\varphi \approx 0$ and $\Delta\varphi \approx \pi$. The elliptic-flow is defined as the 2nd order coefficient v_2 of the Fourier expansion of the azimuthal distribution. Figure 20.2 (right) shows the elliptic-flow coefficient v_2 of J/ψ for p–Pb and Pb–p collisions [12] together with measurements [8] and model calculations for Pb–Pb collisions at $\sqrt{s_{NN}}$ = 5.02. The J/ψ v_2 in $3 < p_T < 6$ GeV/c is found to be positive in both rapidity intervals and of the same order as that measured in Pb–Pb collisions.

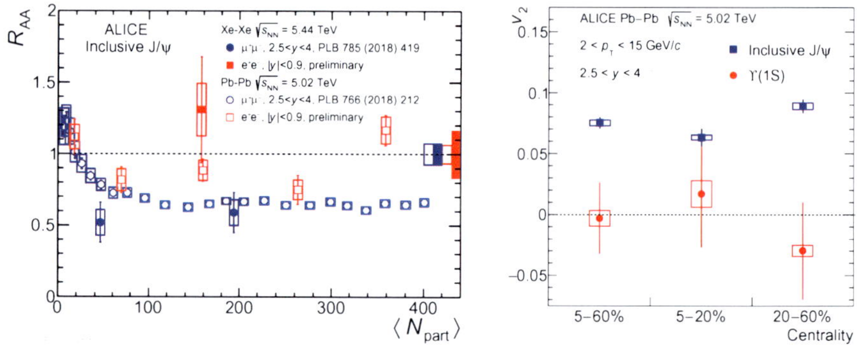

Fig. 20.3 J/ψ R_{AA} in Pb–Pb (Xe–Xe) at $\sqrt{s_{NN}} = 5.02$ (5.44) TeV (left). The $\Upsilon(1S)$ v_2 coefficient integrated over the transverse momentum range $2 < p_T < 15$ GeV/c in three centrality intervals compared to that of inclusive J/ψ at $\sqrt{s_{NN}} = 5.02$ TeV (right)

Pb–Pb (Xe–Xe) collisions: The nuclear modification factor for a given centrality class i in A–A collisions can be defined as:

$$R^i_{AA} = \frac{d^2 N_i^{AA}/dy dp_T}{< T^i_{AA} > \, . \, d^2\sigma^{pp}/dy dp_T},$$

where $d^2 N_i^{AA}/dy dp_T$ is the yield in nucleus-nucleus collisions, $< T^i_{AA} >$ is the nuclear overlap function and $d^2\sigma^{pp}/dy dp_T$ is the production cross section in pp collisions. In Fig. 20.3 (left), the J/ψ R_{AA} measured in Xe–Xe collisions [13] as a function of centrality is compared to the one measured in Pb–Pb collisions [14] both at forward rapidity and mid-rapidity ($-0.5 < y < 0.5$). The ALICE Collaboration also measured the R_{PbPb} of Υ states for Pb–Pb collisions at $\sqrt{s_{NN}} = 5.02$ TeV [15]. A strong suppression in central collisions is observed for $\Upsilon(1S)$. A larger suppression of $\Upsilon(2S)$ compared to $\Upsilon(1S)$ is also observed.

The data samples recorded by ALICE during the 2015 and 2018 LHC Pb–Pb runs at $\sqrt{s_{NN}} = 5.02$ TeV were used for $\Upsilon(1S)$ v_2 measurement [16]. In Fig. 20.3 (right), the v_2 coefficient of $\Upsilon(1S)$ in three centrality intervals is shown, together with that of the J/ψ [8]. The measured $\Upsilon(1S)$ v_2 coefficient is compatible with zero within current uncertainties and consistent with no recombination expectations in the bottomonium sector. The R_{AA} and v_2 results suggest that (re)generation is the dominant contribution to the production of J/ψ, but it only gives a marginal contribution, if any, to the production of $\Upsilon(1S)$ in heavy-ion collisions at LHC collision energies.

References

1. T. Matsui, H. Satz, Phys. Lett. B **178**, 416 (1986)
2. P. Braun-Munzinger, J. Stachel, Phys. Lett. B **490**, 196 (2000)
3. R.L. Thews, M. Schroedter, J. Rafelski, Phys. Rev. C **63**, 054905 (2001)
4. A. Andronic et al., Eur. Phys. J. C **76**(3), 107 (2016)
5. S. Acharya et al., [ALICE Collaboration]. Eur. Phys. J. C **77**(6), 392 (2017)
6. ALICE Collaboration, ALICE-TDR-5-add-1, CERN-LHCC-2000-046
7. Y.Q. Ma, R. Venugopalan, Phys. Rev. Lett. **113**(19), 192301 (2014)
8. S. Acharya et al., [ALICE Collaboration]. Phys. Rev. Lett. **119**(24), 242301 (2017)
9. S. Acharya et al., [ALICE Collaboration]. Eur. Phys. J. C **78**(7), 562 (2018)
10. M. Butenschoen, B.A. Kniehl, Phys. Rev. Lett. **108**, 172002 (2012)
11. S. Acharya et al., ALICE Collaboration. JHEP **07**, 160 (2018)
12. S. Acharya et al., ALICE Collaboration. Phys. Lett. B **780**, 7 (2018)
13. S. Acharya et al., ALICE Collaboration. Phys. Lett. B **785**, 419 (2018)
14. J. Adam et al., ALICE Collaboration. Phys. Lett. B **766**, 212 (2017)
15. S. Acharya et al., ALICE Collaboration. Phys. Lett. B **790**, 89 (2019)
16. S. Acharya et al., [ALICE Collaboration]. arXiv:1907.03169 [nucl-ex]

Chapter 21
AdS/CFT Calculations for Rapidity Dependence of Non-prompt J/Ψ Suppression at LHC

R. Hambrock and William A. Horowitz

Abstract We compare the rapidity dependence of non-prompt J/Ψ suppression of data from CMS at $\sqrt{s} = 2.76\text{TeV}$ with the calculations from two AdS/CFT based energy loss models. When computed in the aMC@NLO framework matched to Herwig++, we fail to reproduce the increase in suppression with increasing suppression that is observed in data. Using FONLL for production and fragmentation, we however reproduce this relationship.

21.1 Introduction

Heavy flavour observables are a crucial probe of the quark gluon plasma. By virtue of their high mass, heavy quarks effectively only enter the picture via hard production, and are thus participants for the entire lifespan of a QGP.

To perform heavy flavour phenomenology, we require observables that allow us to distinguish between different theoretical approaches. One line that can be drawn between theoretical models is the coupling strength of the plasma they assume. Our past work has shown that both weak- and strong-coupling exhibit agreement with suppression of **B**- and **D**-mesons [1], but that the more differential observable of azimuthal heavy flavour correlations reveals stark differences between the results at low p_T [2]. While their mutual agreement with suppression data suggests maturity of both frameworks, the azimuthal correlations constitute a promising observable for distinguishing weakly- and strongly coupled plasmas.

R. Hambrock (✉) · W. A. Horowitz
Department of Physics, University of Cape Town, Private Bag X3, Rondebosch, Cape Town 7701, South Africa
e-mail: roberthambrock@gmail.com

W. A. Horowitz
e-mail: wa.horowitz@uct.ac.za

D. Elia et al. (eds.), *The XVIII International Conference on Strangeness in Quark Matter (SQM 2019)*, Springer Proceedings in Physics 250,
https://doi.org/10.1007/978-3-030-53448-6_21

In this paper, we show calculations for another more differential observable, namely non-prompt J/Ψ suppression with respect to rapidity. We compare these calculations with data from CMS [3]. Our energy loss model interfaces with both aMC@NLO [4] and FONLL [5], and we'll show that the choice of production and hadronization/fragmentation frameworks is critical for this observable.

21.2 Energy Loss Model

21.2.1 Overview

The following will outline our computational procedure and its background. Subsequent to initializing the momenta of heavy quark pairs either to leading order with FONLL [5] or to next-to-leading order with aMC@NLO [4] using Herwig++ [6] for the showering, the production points of the heavy quarks are weighted by the Glauber binary distribution [7]. The particles are propagated through the plasma via the energy loss mechanism described in Sect. 21.2.2, either until the temperature in their local fluid cell drops below a critical threshold where hadronization is presumed to occur, or until the maximum time the VISHNU background [8] is calculated for has passed. Depending on the initialization mechanism used, the bottom quarks are now hadronized to **B** mesons and then fragmented to J/Ψs using either Herwig++ or FONLL.

21.2.2 Langevin Energy Loss

The stochastic equation of motion for a heavy quark in the fluid's rest frame is [9]

$$\frac{dp_i}{dt} = -\mu p_i + F_i^L + F_i^T \tag{21.1}$$

where F_i^L and F_i^T are longitudinal and transverse momentum kicks with respect to the quark's direction of propagation and μ is the drag loss coefficient, given by $\mu = \pi\sqrt{\lambda}T^2/2M_Q$ [10] where M_Q is the mass of a heavy quark in a plasma of temperature T with 't Hooft coupling constant λ. The correlations of momentum kicks at time t_1 and t_2 are given by

$$\langle F_i^T(t_1)F_j^T(t_2)\rangle = \kappa_T\left(\delta ij - \frac{\mathbf{p}_i\mathbf{p}_j}{|p|^2}\right)g(t_2 - t_1)$$

$$\langle F_i^L(t_1)F_j^L(t_2)\rangle = \kappa_L\frac{p_i p_j}{|p|^2}g(t_2 - t_1)$$

where the function g is only known numerically [7] and with

$$\kappa_T = \pi\sqrt{\lambda}T^3\gamma^{1/2} \tag{21.2}$$

$$\kappa_L = \gamma^2\kappa_T = \pi\sqrt{\lambda}T^3\gamma^{5/2} \tag{21.3}$$

$$\hat{q} = \langle p_\perp(t)^2\rangle\lambda \approx \kappa_T t/\lambda = \gamma(2\pi T^3\sqrt{\lambda})/v \tag{21.4}$$

where γ is the speed of the quark. The coupling of the longitudinal fluctuations to velocity grows as $\gamma^{5/2}$, thus growing significant extremely quickly [7]. The fluctuations are thus important to include for finite $\lambda \sim \mathcal{O}(10)$, where $\gamma_{crit}^{fluc} = \frac{M_Q^2}{4T^2}$ is lower than the speed limit on a quark, $\gamma_{crit}^{sl} = (1 + \frac{2M_Q}{\sqrt{\lambda}T})^2$ where $g(0) = 1$, since any kick will be fully correlated to itself. If the time scale of momentum kick correlations is small compared the time scale determined by the drag coefficient, we can model the colouring as white noise, hence treat g as a Dirac delta [7]. Since we require $\gamma < \gamma_{crit}^{sl}$, it follows that $t_{corr}\mu \sim \frac{1}{2}\sqrt{\lambda}\sqrt{\gamma}\frac{T}{M_Q} < \frac{1}{2}\sqrt{\lambda}\sqrt{\frac{4M_Q^2}{\lambda T^2}}\frac{T}{M_Q} = 1$, and we may thus safely approximate the colouring as white noise. It should be noted that this construction does not obey the fluctuation-dissipation theorem [7]. The computations based on this model will be labeled *D(p)*.

21.2.3 Development on Energy Loss Model

The problem with the energy loss mechanism described in 21.2.2 is that, since the longitudinal momentum fluctuations grow as $\gamma^{\frac{5}{2}}$, our setup breaks down for high momenta, where in a perturbative QCD setting, Brehmstrahlung would restrict the momentum growth of the quark. Via a novel calculation presented in [11], we instead consider a stationary string in AdS_d hanging into a black hole horizon and calculate $s^2(t, a, d)$, the average transverse distance squared travelled by the string's free endpoint, where t is the time, d the dimension of the setup, and a parametrizes between a heavy quark for $a = 0$ and a light quark for $a = 1$. Crucially, $s^2(t, a, d = 3)$ can be determined analytically for small string lengths, which is identical to the asymptotically late time behavior of a string with arbitrary initial length. We thus find the asymptotically late time behavior of a string in d dimensions by

$$s^2(t \gg \beta, a, d) = s^2_{\text{small}}(t \gg \beta, a, d)$$
$$= (\frac{d-1}{2})^2 s^2_{\text{small}}\ (t \gg \beta, a, d = 3) = \frac{(d-1)^2}{8\pi\sqrt{\lambda}}\beta(1 - \frac{a}{2}) \tag{21.5}$$

where $\beta = T^{-1}$. At late times, the motion is diffusive, thus we can extract the diffusion coefficient

$$D(a, d) \sim \frac{1}{2} s^2 (t \gg \beta, a, d) \tag{21.6}$$

which in AdS_5 for a heavy quark reads $2\beta/\pi\sqrt{\lambda}$. From this, we obtain

$$\kappa_T = 2T^2/D = \pi\sqrt{\lambda}T^2/\beta = \pi\sqrt{\lambda}T^3 \tag{21.7}$$

$$\hat{q} = \langle p_\perp(t)^2 \rangle \lambda \approx \kappa_T t/\lambda = (2\pi T^3 \sqrt{\lambda})/v \tag{21.8}$$

Requiring these fluctuations to obey the fluctuation-dissipation theorem (which the construction in Sect. 21.2.2 could not), we attain $\mu = \pi\sqrt{\lambda}T^2/2E$. The computations based on this model will be labeled *D=const.*

21.3 R_{AA}(y)

We compare non-prompt J/Ψ suppression calculations with data from CMS [3] (Fig. 21.1). All of our calculations overpredict the suppression of non-prompt J/Ψ. However, the FONLL based calculations qualitatively capture the increase in suppression with increasing rapidity that is seen in the data. On the other hand, the aMC@NLO based calculations show no rapidity dependence of the suppression at all.

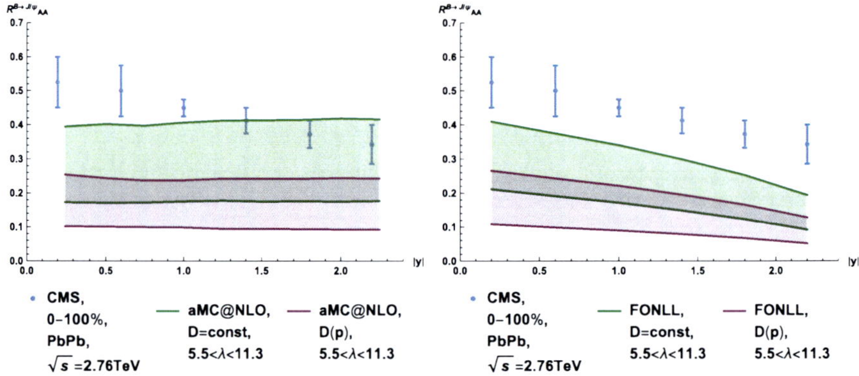

Fig. 21.1 Comparison of $R_{AA}^{J/\Psi}$ (y) CMS data [3] [$6.5 < p_T < 30$GeV] with our energy loss models using MC@NLO (Left) and FONLL (Right) respectively

21.4 Conclusion and Outlook

We have compared AdS/CFT based non-prompt J/Ψ suppression with CMS data (Fig. 21.1). We find that the choice of production and hadronization/fragmentation frameworks leads to starkly different results. In particular, aMC@NLO does not appear to have rapidity dependence in the production spectrum of bottom quarks. This lead us to henceforth switch to FONLL for production and hadronization/fragmentation for observables that are rapidity sensitive. Unfortunately, this comes at the loss of next-to leading order production, which is critical for realistic azimuthal correlation predictions. For these, using Herwig++ appears to be the better choice [12], if restricted to mid-rapidity.

Acknowledgments The authors wish to thank the NRF and SA-CERN for their generous financial support.

References

1. R. Hambrock, W.A. Horowitz, *EPJ Web of Conferences*, vol. 171, p. 18002 (2018). arXiv:1802.02442
2. R. Hambrock, W.A. Horowitz, Nucl. Part. Phys. Proc. **289–290** 233–236 (2017). arXiv:1703.05845
3. V. Khachatryan et al., (CMS). Eur. Phys. J. **C77** 252 (2017). arXiv:1610.00613
4. J. Alwall, R. Frederix, S. Frixione, V. Hirschi, F. Maltoni, O. Mattelaer, H.S. Shao, T. Stelzer, P. Torrielli, M. Zaro, JHEP **07** 079 (2014). arXiv:1405.0301
5. M. Cacciari, S. Frixione, N. Houdeau, M.L. Mangano, P. Nason, G. Ridolfi, JHEP **10** 137 (2012). arXiv:1205.6344
6. M. Bahr et al., Eur. Phys. J. **C58** 639–707 (2008). arXiv:0803.0883
7. W.A. Horowitz, Phys. Rev. **D91** 085019 (2015). arXiv:1501.04693
8. C. Shen, U. Heinz, P. Huovinen, H. Song, Phys. Rev. **C84** 044903 (2011). arXiv:1105.3226
9. G.D. Moore, D. Teaney, Phys. Rev. **C71** 064904 (2005). arXiv:hep-ph/0412346
10. C.P. Herzog, A. Karch, P. Kovtun, C. Kozcaz, L.G. Yaffe, JHEP **07**, 013 (2006). arXiv:hep-th/0605158
11. R.W. Moerman, W.A. Horowitz, (2016). arXiv:1605.09285
12. R. Hambrock, W. Horowitz, J. Phys. Conf. Ser. **889**, 012015 (2017)

Chapter 22
Heavy-Flavour Jet Production and Charm Fragmentation with ALICE at LHC

Auro Mohanty

Abstract Heavy quarks, produced in hard parton scatterings in the early stage of ultra-relativistic heavy-ion collisions, are ideal probes to investigate the properties of the Quark–Gluon Plasma (QGP) produced in such collisions. Measurements of heavy-flavour jets can provide constraints on energy-loss models. In particular, they add information on how the radiated energy is dissipated in the medium. Studies of angular correlations between heavy-flavour and charged particles allow us to characterize the heavy-quark fragmentation process and its possible modification in a hot nuclear matter environment. This manuscript will focus on the latest results on heavy-flavour jets and D-meson correlations with charged particles studied with the ALICE detector in pp, p–Pb and Pb–Pb collisions.

22.1 Physics Motivation

Fragmentation function of photon-triggered mesons was studied by Kang and Vitev [1] and a flavour dependence of energy loss in QGP medium was predicted. Measurements in pp collisions provide essential reference to interpret those in proton–nucleus (p–A) and nucleus–nucleus (A–A) collisions. They also provide an excellent test of the perturbative quantum chromodynamics (pQCD) because heavy-flavour observables are calculable in pQCD down to $p_{\mathrm{T}} \approx 0$. Anderle et al. [2] presented a global QCD analysis of $\mathrm{D}^{*\pm}$-meson fragmentation functions in pp scatterings.

ALICE is uniquely placed to play a significant role in the low and intermediate p_{T} (p_{T}: transverse momentum) sector.

Auro Mohanty on behalf of the ALICE Collaboration.

A. Mohanty (✉)
Institute for Subatomic Physics, Utrecht University/Nikhef, Utrecht, Netherlands
e-mail: auro.mohanty@cern.ch

D. Elia et al. (eds.), *The XVIII International Conference on Strangeness in Quark Matter (SQM 2019)*, Springer Proceedings in Physics 250,
https://doi.org/10.1007/978-3-030-53448-6_22

22.2 Procedure and Physics Results

Heavy-flavour jets are studied by means of two different methods: by reconstructing jets with a heavy-flavour tag ('heavy-flavour jets') and by studying correlations between heavy-flavour hadrons with other hadrons ('D-meson-hadron correlations').

22.2.1 *Heavy-Flavour Jets*

Jets are reconstructed using the anti-k_T algorithm [3]. They are tagged as heavy-flavour jets if they have within their constituents: heavy-flavour electrons, D mesons, or beauty mesons (by an indirect measurement).

22.2.1.1 Heavy-Flavour Electron Jets

Electrons resulting from the semi-leptonic decay of heavy-flavour hadrons are used to tag the jets, called heavy-flavour electron (HFe) jets. First, jets are reconstructed using charged tracks. Then a constituent track in each jet is searched for, having the same momentum as the heavy-flavour electrons identified separately (see [4] for a detailed description of the ALICE apparatus).

Figure 22.1 shows the nuclear modification factor R_{pPb} of HFe jets with jet radii $R = 0.3$, 0.4, and 0.6 in p–Pb collisions at a center of mass energy per nucleon pair, $\sqrt{s_{\mathrm{NN}}} = 5.02$ TeV. No cold nuclear matter (CNM) effects are observed.

D-Meson Tagged Jets. Jets are tagged if they contain a D^0 meson within the jet cone. D^0 mesons are reconstructed in the $\mathrm{D}^0 \rightarrow \mathrm{K}^- \pi^+$ [5] decay channel. The daughter kaon and pion tracks are replaced by an equivalent D^0 constituent, which is then used together with the other charged tracks to reconstruct the jets.

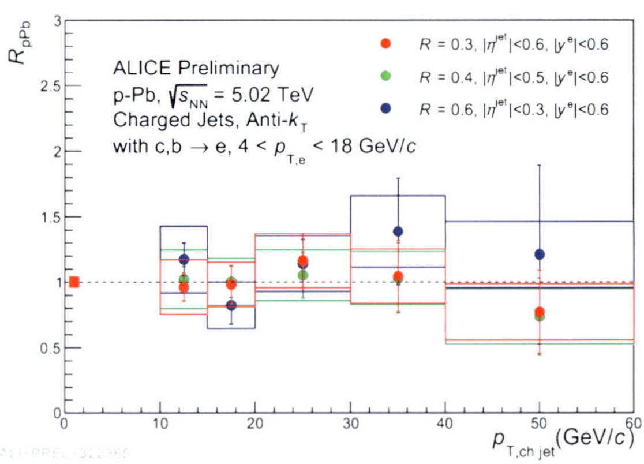

Fig. 22.1 R_{pPb} of HFe jets with $R = 0.3$, 0.4, and 0.6

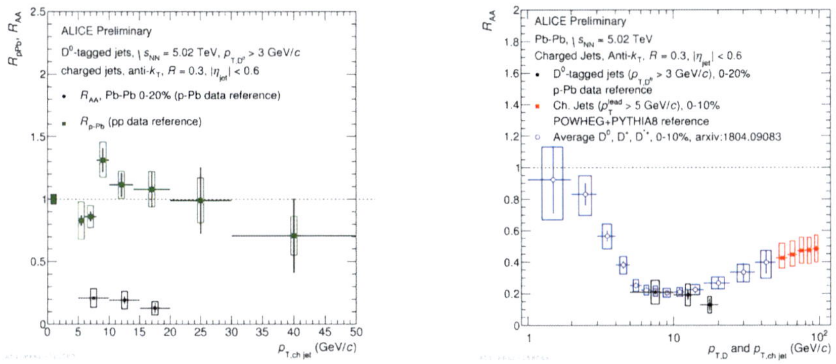

Fig. 22.2 R_{pPb} and R_{AA} of D-jets (left) and R_{AA} of D-jets, D mesons, and charged jets (right) at $\sqrt{s_{\mathrm{NN}}} = 5.02$ TeV

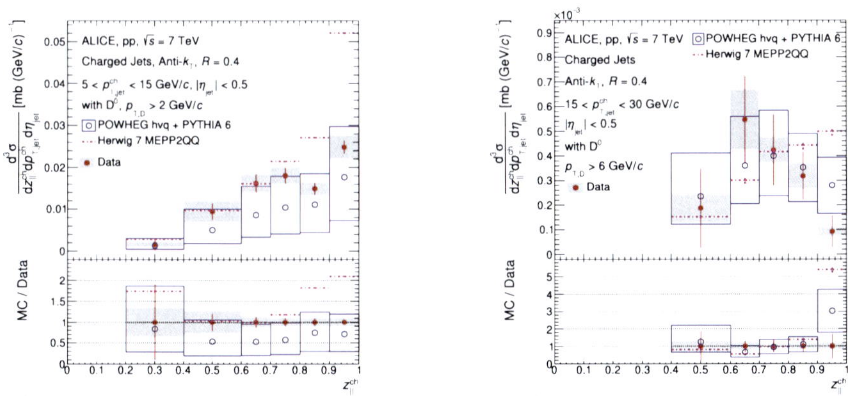

Fig. 22.3 $z_{||}^{\mathrm{ch}}$-differential cross section of D^0-jets in pp collisions at $\sqrt{s} = 7$ TeV

Jet-p_{T} differential cross section of D^0-jets was measured in pp collisions at $\sqrt{s} = 13$ TeV with $R = 0.4$. The D^0-jet production cross section was also measured at $\sqrt{s_{\mathrm{NN}}} = 5.02$ TeV in pp, p–Pb, and Pb–Pb collisions in 0–20% centrality with $R = 0.3$. The nuclear modification factor is shown for p–Pb and Pb–Pb in Fig. 22.2 (left panel). R_{pPb} is consistent with unity within uncertainties, while R_{AA} is 0.2 at $p_{\mathrm{T}} \sim 10$ GeV/c . On the right panel, it can be seen that the R_{AA} of D^0-jets is compatible with that of D mesons.

Fractional momentum ($z_{||}^{\mathrm{ch}}$) carried by the constituent D^0 meson along the jet axis was measured for D^0-jets [6] in pp collisions at $\sqrt{s} = 7$ TeV, with $R = 0.4$. Hard fragmentation is observed in $5 < p_{\mathrm{T}}^{\mathrm{ch.\,jet}} < 15$ GeV/c, compatible with leading order (LO) and next-to-leading order (NLO) pQCD predictions, as seen on the left in Fig. 22.3. However, in $15 < p_{\mathrm{T}}^{\mathrm{ch.\,jet}} < 30$ GeV/c, there is a hint of softer fragmentation observed in data, seen on the right panel.

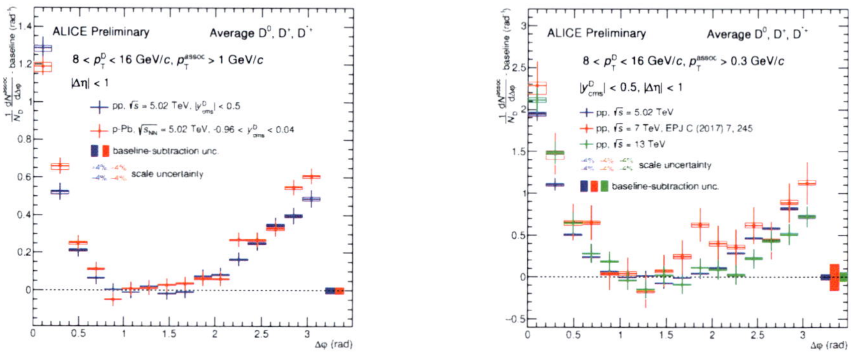

Fig. 22.4 D-meson-hadron azimuthal correlations measured in pp and p–Pb collisions at $\sqrt{s_{\rm NN}}$ = 5.02 TeV (left), and in pp collisions at $\sqrt{s}$ = 5.02, 7, and 13 TeV (right)

b-Tagged Jets. The property of B mesons having a longer lifetime is exploited here to identify jets originating from b quarks without explicitly reconstructing the B mesons. Jets containing a 3-pronged secondary vertex within the cone are selected since B mesons tend to decay into at least three daughters. Our measurements of jet-$p_{\rm T}$ differential production cross section of b-jets in p–Pb collisions at $\sqrt{s_{\rm NN}}$=5.02 TeV are in agreement with POWHEG+PYTHIA predictions.

22.2.2 *D-Meson-Hadron Correlations*

Charm jet fragmentation is also studied using azimuthal ($\Delta\phi$) correlations between D mesons and associated charged hadrons. Two peaks are observed in the $\Delta\phi$ distribution, one at $\Delta\phi \approx 0$, called the near side, and a broader peak at the away side $\Delta\phi \approx \pi$, signifying two leading jets emitted in opposite directions in a collision. No evidence of CNM effects were found (see Fig. 22.4, left panel) when production of associated tracks was compared across pp and p–Pb collisions, at $\sqrt{s_{\rm NN}}$ = 5.02 TeV. No energy dependence could also be observed in pp collisions as seen on the right panel.

22.3 Summary

ALICE obtained new measurements of HFe jets in pp and p–Pb collisions at $\sqrt{s_{\rm NN}}$ = 5.02 TeV. Also shown is a first measurement of D^0-jets in Pb–Pb collisions at $\sqrt{s_{\rm NN}}$ = 5.02 TeV in 0–20% centrality with a suppression by a factor of five at $p_{\rm T} \sim 10$ GeV/c. At $\sqrt{s}$ = 7 TeV, hard fragmentation for D^0-jets was seen in pp collisions in $5 < p_{\rm T} < 15$ GeV/c along with a hint of softer fragmentation in

$15 < p_{\mathrm{T}} < 30$ GeV/c in data when compared to theoretical predictions. Our new measurements of b-tagged jets in p–Pb collisions at $\sqrt{s_{\mathrm{NN}}} = 5.02$ TeV were also reported. No energy dependence was observed for D-meson-hadron correlations in pp collisions. And there was no evidence of CNM effects in p–Pb collisions for any study reported in this manuscript.

References

1. Z.B. Kang, I. Vitev, Photon-tagged heavy meson production in high energy nuclear collisions. Phys. Rev. **D84**, 014,034 (2011). https://doi.org/10.1103/PhysRevD.84.014034
2. D.P. Anderle, T. Kaufmann, M. Stratmann, F. Ringer, I. Vitev, Using hadron-in-jet data in a global analysis of D^* fragmentation functions. Phys. Rev. **D96**(3), 034,028 (2017). https://doi.org/10.1103/PhysRevD.96.034028
3. M. Cacciari, G.P. Salam, G. Soyez, The anti-k_t jet clustering algorithm. JHEP **04**, 063 (2008). https://doi.org/10.1088/1126-6708/2008/04/063
4. K. Aamodt, et al., The ALICE experiment at the CERN LHC. J. Instrum. **3**(08), S08,002–S08,002 (2008). https://doi.org/10.1088/1748-0221/3/08/s08002
5. M. Tanabashi, et al., Review of particle physics. Phys. Rev. D **98**, 030,001 (2018). https://doi.org/10.1103/PhysRevD.98.030001
6. S. Acharya et al., Measurement of the production of charm jets tagged with D^0 mesons in pp collisions at $\sqrt{s} = 7$ TeV. JHEP **08**, 133 (2019). https://doi.org/10.1007/JHEP08(2019)133

Part IV
Strangeness and Light Flavour

Chapter 23
Strangeness Enhancement from Dynamical Core–Corona Initialisation Model

Yuuka Kanakubo, Michito Okai, Yasuki Tachibana, and Tetsufumi Hirano

Abstract We estimate the fraction of energy of the quark gluon plasma (QGP) fluids in high-energy proton–proton and nuclear collisions with the dynamical core–corona initialisation (DCCI) model. The fraction of the QGP fluids turn out to be the key to interpret the strangeness enhancement in high-multiplicity small colliding systems.

23.1 Introduction

Strangeness enhancement reported from ALICE Collaboration [1] implies existence of the quark gluon plasma (QGP) fluids in small colliding systems such as p + p and p + Pb collisions. Strange hadron yield ratios increase with multiplicity and saturates at high multiplicity in which local chemical equilibrium is archived. One of the possible ideas to give insight into these data is the core–corona picture [2, 3]: The system formed in relativistic heavy-ion collisions can be separated into two components, "core" and "corona". The core is the matter under thermal equilibrium and the corona is like a system in p + p collisions in which thermalisation is not supposed to be archived.

In this study, we introduce the concept of the core–corona picture into a dynamical initialisation model [4], analyse strange hadron yield ratios as functions of multiplicity using this model and estimate the fraction of energy of the QGP fluids from small to large colliding systems [5].

Y. Kanakubo (✉) · M. Okai · Y. Tachibana · T. Hirano
Department of Physics, Sophia University, Tokyo 102-8554, Japan
e-mail: y-kanakubo-75t@eagle.sophia.ac.jp

Y. Tachibana
Department of Physics and Astronomy, Wayne State University, Detroit, MI 48201, USA

D. Elia et al. (eds.), *The XVIII International Conference on Strangeness in Quark Matter (SQM 2019)*, Springer Proceedings in Physics 250,
https://doi.org/10.1007/978-3-030-53448-6_23

23.2 Dynamical Core–Corona Initialisation (DCCI) Model

In this study initial conditions of the QGP fluids are dynamically generated as a consequence of four-momentum deposition from partons which are formed just after the collision of two nuclei. We use relativistic hydrodynamic equations with source terms to describe the dynamical generation of the QGP fluids,

$$\partial_\mu T^{\mu\nu}_{\mathrm{fluid}}(x) = J^\nu(x)\,. \tag{23.1}$$

Assuming the dynamics of partons in phase space, we derive the source terms in (23.1) as

$$J^\mu(x) = -\sum_i \frac{dp_i^\mu(t)}{dt} G\left(\boldsymbol{x} - \boldsymbol{x}_i(t)\right), \tag{23.2}$$

where p_i^μ is four-momentum of the i th parton and G is the Gaussian function. Here we call dp_i^μ/dt the "fluidisation rate". We assume that the dynamical initialisation takes place from formation time of partons, $\tau = \tau_{00}(= 0.10\ \mathrm{fm})$, to initial time of fluids, $\tau = \tau_0(= 0.60\ \mathrm{fm})$. Here we note that energy-momentum conservation is satisfied within the combined system of the QGP fluids and the traversing partons during the dynamical initialisation.

Under the basic concept of the core–corona picture, we assume the fluidisation rate is proportional to density of initially produced partons,

$$\frac{dp_i^\mu}{dt}(t) = -a_0 \frac{\rho(\boldsymbol{x}_i(t))}{p_{\mathrm{T,i}}{}^2} p_i^\mu(t)\,, \tag{23.3}$$

where $p_{\mathrm{T,i}}$ and p_i^μ are transverse momentum and four-momentum of the i th parton. The dimensionless parameter a_0 is a free parameter to control the intensity of energy-momentum deposition. ρ is the density of partons surrounding the i th parton defined as

$$\rho(\boldsymbol{x}_i(t)) = \sum_{j\neq i} G\left(\boldsymbol{x}_i(t) - \boldsymbol{x}_j(t)\right). \tag{23.4}$$

Here we assume that each parton has a Gaussian profile.

We briefly summarize the entire model flow in the following. First we generate the initial partons with the Monte Carlo event generator PYTHIA ver 8.230 [6] in which hadronisation is switched off. Next we perform the dynamical core–corona initialisation to obtain initial conditions of the QGP fluids. We describe space-time evolution of the QGP fluids with ideal (3 + 1)-dimensional hydrodynamics. The core and corona undergo the different hadronisation processes. Hadron yields from the QGP fluids (core) are obtained by integrating Cooper–Frye formula at hypersurface of decoupling temperature. On the other hand, hadron production from the surviving partons (corona) is described by string fragmentation with PYTHIA.

23.3 Results

Figure 23.1 shows the strange hadron yield ratios in p + p, p + Pb and Pb + Pb collisions as functions of multiplicity and are compared with the ALICE experimental data [1]. The yield ratios monotonically increase with multiplicity and saturate in high-multiplicity events. The yield ratio of cascades to pions $(\Xi^- + \bar{\Xi}^+)/(\pi^+ + \pi^-)$ purely from the corona is ~ 0.002 at low multiplicity, while it saturates at ~ 0.005 at high multiplicity, which is almost the same as the value purely obtained from the core. The behaviour in these results is similar to that in the experimental data. Thus, as a result of competition between the core and the corona in final hadron productions, we successfully describe the continuous change of strange hadron yield ratios as seen in the experimental data.

We study how much energy and momentum produced at the initial stage of the collisions are turned into the fluids. To see this, we show in Fig. 23.2 the fraction of the fluidised energy at midrapidity,

$$R = \left. \frac{dE_{\text{fluids}}/d\eta_s}{dE_{\text{tot}}/d\eta_s} \right|_{\eta_s=0}. \tag{23.5}$$

The fluidized energy can be obtained by integrating the time component of source terms from $\tau = \tau_{00}$ to τ_0 in the transverse plane as

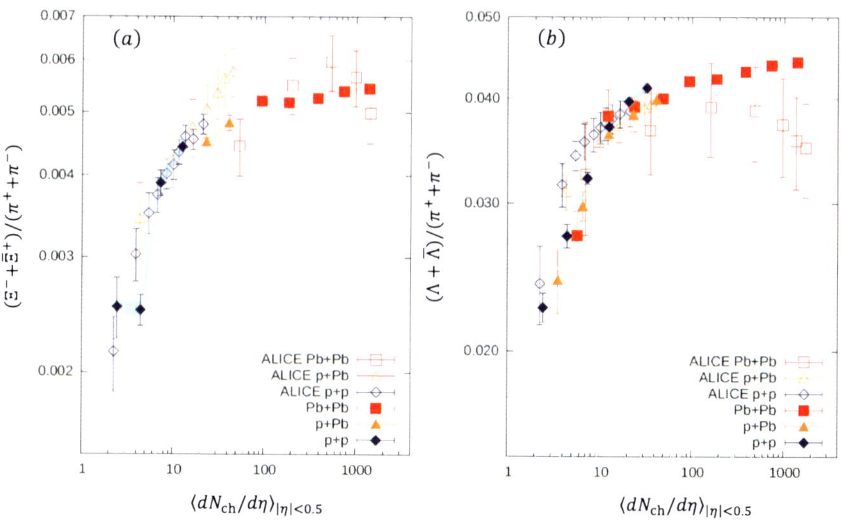

Fig. 23.1 (Color online) Yield ratio of **a** cascades (Ξ^- and $\bar{\Xi}^+$), **b** lambdas (Λ and $\bar{\Lambda}$) as a function of multiplicity at mid-rapidity, $\langle dN_{\text{ch}}/d\eta\rangle$, in p + p (diamonds) at $\sqrt{s_{NN}} = 7$ TeV, p + Pb (triangles) at 5.02 TeV and Pb + Pb (squares) at 2.76 TeV collisions at the LHC energies. Results from the DCCI model (closed symbols) are compared with the ALICE data (open symbols) [1]

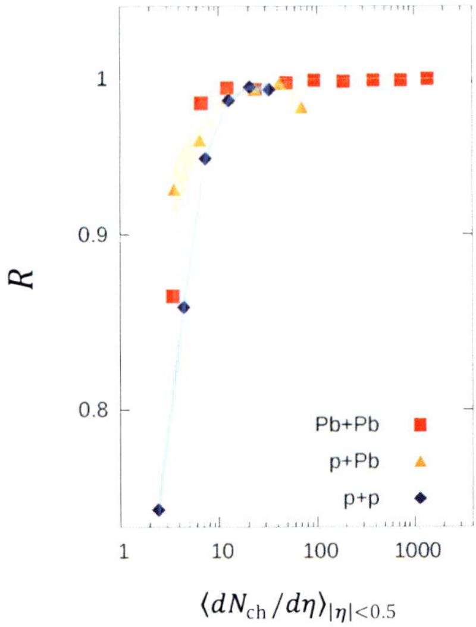

Fig. 23.2 (Color online) Fraction of the fluidized energy to the total energy at $\eta_s = 0$ as a function of multiplicity in p + p (diamonds) at $\sqrt{s_{NN}} = 7$, p + Pb (triangles) at 5.02 TeV, and Pb + Pb (squares) at 2.76 TeV.

$$\left.\frac{dE_{\text{fluids}}}{d\eta_s}\right|_{\eta_s=0} = \int_{\tau_{00}}^{\tau_0} d\tau \int d^2\mathbf{x}_\perp \tau J^\tau(\tau, \mathbf{x}_\perp, \eta_s = 0). \tag{23.6}$$

On the other hand, the total energy density $dE_{\text{tot}}/d\eta_s$ is calculated by taking the sum of energy of all the initial partons at $\tau = \tau_{00}$.

Fraction of the fluidized energy, R, increases monotonically and saturates around $\langle dN_{\text{ch}}/d\eta\rangle_{|\eta|<0.5} \sim 10$. The most notable thing is that the fraction of fluidized energy, R, is finite in $\langle dN_{\text{ch}}/d\eta\rangle \sim$ 3–10. This means that, within the DCCI model calculations, the contribution of hadron production from the QGP fluids is needed to reproduce the yield ratio of cascades reported in the experimental data. Thus, this is the strong indication of the partial QGP generation even in averaged events in p + p collisions at $\sqrt{s_{NN}} = 7$ TeV.

23.4 Summary

We investigated multiplicity dependence of hadron yield ratios from p + p to A + A collision systems with the DCCI model. Modeling the dynamical generation of initial conditions of the QGP fluids from initial partons, we describe the strangeness enhancement in high-multiplicity small colliding systems reported by ALICE Collaboration. We found that there exists contribution from the QGP fluids to final hadron yields even in averaged p + p events within the DCCI model.

References

1. J. Adam et al., (ALICE), Enhanced production of multi-strange hadrons in high-multiplicity proton-proton collisions. Nat. Phys. **13**, 535 (2017)
2. P. Bozek, Size of the thermal source in relativistic heavy-ion collisions. Acta Phys. Pol. ***B36***, 3071 (2005)
3. K. Werner, Core–corona separation in ultra-relativistic heavy ion collisions. Phys. Rev. Lett. ***98***, 152301 (2007)
4. M. Okai, K. Kawaguchi, Y. Tachibana, T. Hirano, New approach to initializing hydrodynamic fields and mini-jet propagation in quark-gluon fluids. Phys. Rev. C **95**(5), 054914 (2017)
5. Y. Kanakubo, M. Okai, Y. Tachibana, T. Hirano, Enhancement of strange baryons in high-multiplicity proton–proton and proton–nucleus collisions. PTEP ***2018***, 121D01 (2018)
6. T. Sjöstrand, S. Mrenna, P.Z. Skands, A brief introduction to PYTHIA 8.1, Comput. Phys. Commun. ***178***, 852 (2008)

Chapter 24
Production of Light Flavor Hadrons Measured by PHENIX at RHIC

Iurii Mitrankov

Abstract Light flavor hadrons are copiously produced in hadronic and heavy-ion interactions and bring a wealth of information about properties of the produced medium and reaction dynamics. Having different masses, quark content and lifetimes, light flavour hadrons do not only serve as general observables in the soft sector, but also play an important role as high transverse momentum probes and signatures of the onset of collectivity in collisions of small systems. We present review of the most recent PHENIX results on the production of π^0, η, K_S, φ and ω mesons in p (d, ^{3}He) + Au, Cu + Cu, Cu + Au, Au + Au and U + U collisions at top RHIC energies with emphasis on study of the parton energy loss in heavy ion collisions, cold nuclear matter effects in small systems.

24.1 Introduction

The ultra-relativistic heavy-ion collisions provide a capability to investigate the properties of quark-gluon plasma (QGP)-state of matter at very high temperature and/or density where quarks and gluons are in the deconfined state [1]. The QGP study is one of the main goals of PHENIX experiment [2].

One of the ways to investigate QGP properties in experiments is to measure the yields of final state particles. The properties of hot nuclear matter can be studied by measuring level of hadron suppression or so-called jet quenching effect in high transverse momentum p_T range in heavy-ion collisions. Particles, containing strange quarks, are considered to be a convenient tool to investigate QGP. Strangeness enhancement in QGP may cause an increase in the production of hadrons containing $^s/_{\bar{s}}$ quarks with increasing number of participants from proton-proton to nucleus-nucleus collisions [1].

Iurii Mitrankov for the PHENIX Collaboration.

I. Mitrankov (✉)
Peter the Great St.Petersburg Polytechnic University, St.Petersburg 195251, Russia
e-mail: mitrankovy@gmail.com

D. Elia et al. (eds.), *The XVIII International Conference on Strangeness in Quark Matter (SQM 2019)*, Springer Proceedings in Physics 250,
https://doi.org/10.1007/978-3-030-53448-6_24

The light hadron production was well studied in symmetric Au + Au and Cu + Cu collisions at $\sqrt{s_{NN}} = 200$ GeV at RHIC [3, 4]. Systematical study of light hadron production is continued in asymmetric Cu + Au collisions at $\sqrt{s_{NN}} = 200$ GeV and in collisions of deformed Uranium nuclei at $\sqrt{s_{NN}} = 192$ GeV, which can provide more information about jet quenching and therefore the characteristics of the QGP.

The study of light hadron production with different masses, quark content, lifetimes and flavor in small collision systems like p + Au, d + Au and He + Au can help to explore onset of collectivity (mini QGP) in such collisions as well as cold nuclear matter effects (CNM), which are important for interpreting of heavy-ion results.

24.2 Data Analysis

The studying of hadrons production was done using the two PHENIX central arms each covering 90° in azimuth at the mid rapidity ($|y| < 0.35$) [1]. The measurement of particles momentum is done using drift chambers and the first layer of the pad chambers. Kaons identification is performed by the time-of-flight detector (ToF). Electromagnetic calorimeter provides measurements of photon kinematic properties.

Light mesons $\pi^0 \rightarrow \gamma\gamma$, $\eta \rightarrow \gamma\gamma$, $K_S \rightarrow \pi^0\pi^0$, $\varphi \rightarrow K^+K^-$, $\omega \rightarrow \pi^0$ were reconstructed by handling respective invariant mass distributions.

Nuclear modification factors R_{AB} are often used to study collective effects that affect hadrons invariant transverse momentum spectra and are calculated according to the relation

$$R_{AB} = \frac{\sigma_{pp}^{inel}}{N_{coll}} \frac{dN_{AB}}{d\sigma_{pp}},$$

where $dN_{AB} = 1/(2\pi p_T) \cdot \left(d^2 N_{AB}\right)/(dp_T dy)$—inclusive hadron's production spectrum in A + B collisions, $d\sigma_{pp} = 1/(2\pi p_T) \cdot \left(d^2\sigma_{pp}\right)/(dp_T dy)$—inclusive differential cross section for hadron production in p + p collisions at the same center-of-mass energy, N_{coll}—number of nucleon-nucleon collisions in the A + B system for selected centrality intervals are estimated with the Glauber Monte-Carlo simulation and $\sigma_{pp}^{inel} = 42.2mb$ is the total inelastic proton-proton cross sections.

24.2.1 *Large Systems*

The π^0, η, K_S, φ and ω mesons nuclear modification factors in Cu + Au collisions at $\sqrt{s_{NN}} = 200$ GeV as a function of transverse momentum is shown in Fig. 24.1. In all figures vertical bars correspond to statistical uncertainties and rectangles—to systematic ones.

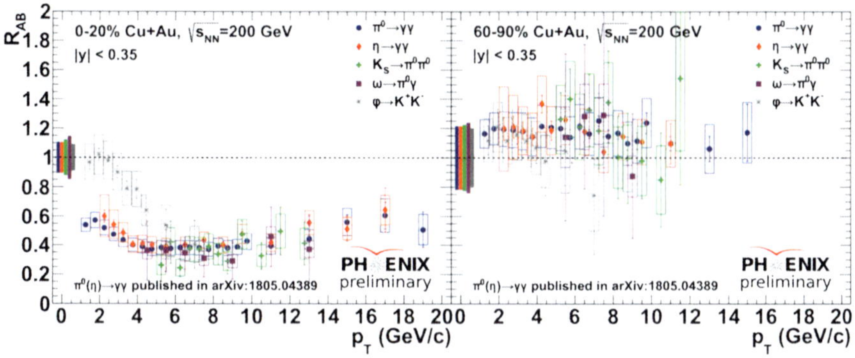

Fig. 24.1 Nuclear modification factors R_{AB} as a function of p_T measured for φ, π^0, η, ω and K_S in Cu + Au collisions

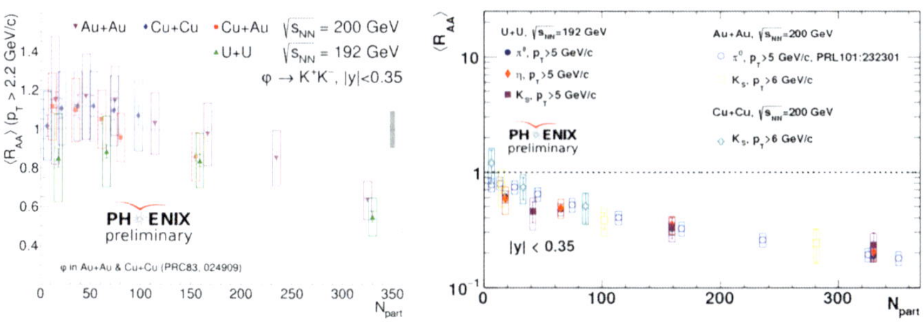

Fig. 24.2 **Left**: R_{AB} for φ-meson integrated at $p_T > 2.2$ GeV/c in Cu + Au, U + U, Au + Au and Cu + Cu collisions as a function of N_{part}. **Right**: R_{AB} for π^0, η, K_S -mesons integrated at $p_T >$ 5 GeV/c in U + U, Au + Au and Cu + Cu collisions as a function of N_{part}

In contrast to π^0, η, K_S, and ω, φ shows less suppression at $p_T < 5\text{GeV}/c$, while at $p_T > 5\text{GeV}/c$ it exhibits the same suppression pattern as π^0, η, K_S, and ω.

In peripheral Cu + Au and U + U collisions all considered mesons R_{AB} exhibit similar shape within systematic uncertainties. The same light mesons behaviour was observed in Cu + Cu and Au + Au collisions, as presented at Fig. 24.2.

There is no difference in the integrated R_{AB} of φ-meson between Au + Au, Cu + Cu, Cu + Au and U + U collisions and of π^0, η, K_S-mesons between U + U, Au + Au and Cu + Cu collisions. This might indicate that the suppression level scales with the average size of the nuclear overlap region.

24.2.2 Small Systems

The Fig. 24.3 shows φ meson R_{AB} measured in p + Au, d + Au and ^{3}He + Au collisions at the $\sqrt{s_{NN}} = 200$ GeV. From comparison of p + Au, d + Au and ^{3}He + Au results, an ordering for φ R_{AB} can be seen in the intermediate p_T in most

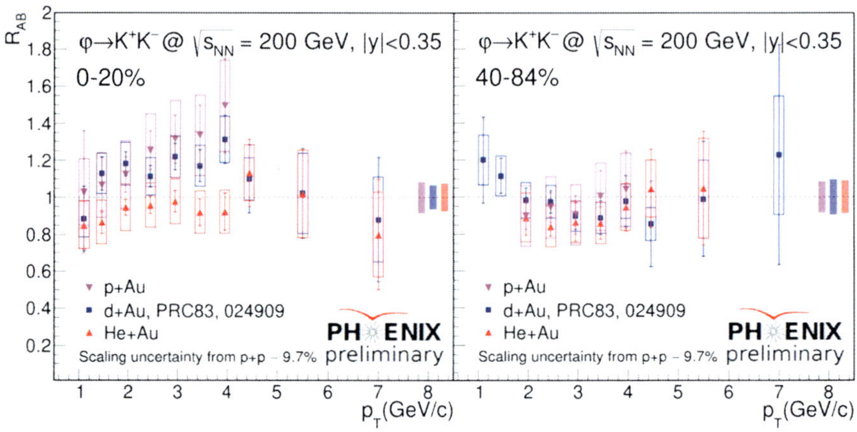

Fig. 24.3 The φ-meson R_{AB} in p + Au, d + Au and ^{3}He + Au collisions at 200 GeV

central collisions: $R_{He+Au} < R_{d+Au} < R_{p+Au}$. In other centralities φ-meson nuclear modification factors for all three systems exhibit similar shape.

The R_{AB} of φ and π^0 mesons in central p + Au and ^{3}He + Au collisions at 200 GeV are shown in Fig. 24.4. It seems that φ and π^0 R_{AB} exhibit similar shape within systematic uncertainties. Therefore, the production of light mesons shows similar behavior, it can be assumed that CNM effects are not responsible for the differences between φ and π^0 seen in Au + Au, Cu + Cu, Cu + Au, and U + U central collisions.

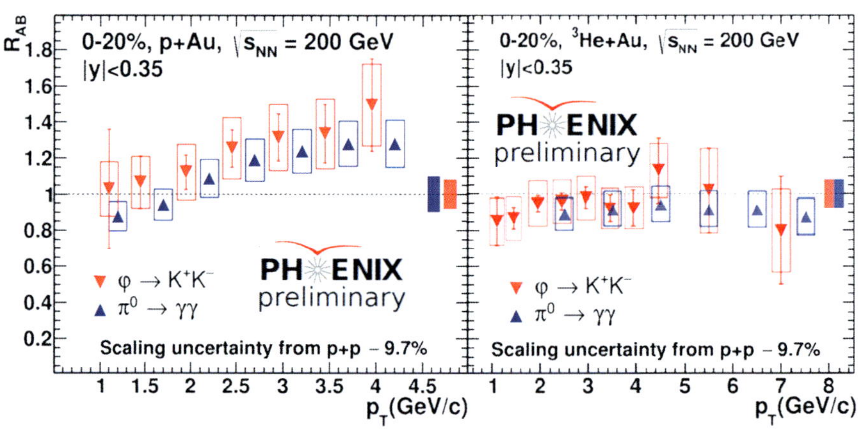

Fig. 24.4 The φ and π^0 meson R_{AB} measured in central p(^{3}He) + Au collisions at 200 GeV

24.3 Conclusions

PHENIX experiment has measured invariant yields and R_{AB} of π^0, η, K_S, φ and ω mesons in p(d,^{3}He) + Au, Cu + Cu, Cu + Au, Au + Au, U + U collisions at top RHIC energies.

The φ-meson exhibits a different suppression pattern, being an outlier, compared to π^0, η, K_S, ω mesons in large systems collisions. These results provide a contribution to the understanding of the strangeness enhancement competing with energy loss as well as additional examination of fragmentation and recombination models.

There is no difference in the integrated R_{AB} of light mesons between heavy-ion collisions systems. This might indicate that the suppression level scales with the average size of the nuclear overlap region.

The φ and π^0 R_{AB} shows similar values in central p(^{3}He) + Au collisions within uncertainties, which might indicate that CNM effects are not responsible for the differences between π^0 and φ observed in large systems. In the intermediate p_T range in most central collisions φ R_{AB} seem to depend on the collision system size.

Acknowledgments We acknowledge support from Russian Ministry of Education and Science, state assignment 3.1498.2017/4.6.

References

1. K. Adcox et al., Formation of dense partonic matter in relativistic nucleus-nucleus collisions at RHIC: Experimental evaluation by the PHENIX collaboration. Nucl. Phys. A **757** (2005)
2. K. Adcox et al, PHENIX detector overview. Nucl. Instrum. Meth. A **499** (2003)
3. A. Adare et al., Nuclear modification factors of phi mesons in d + Au, Cu + Cu and Au + Au collisions at sqrt(sNN) = 200 GeV. Phys. Rev. C **83**(2), 024909 (2011)
4. K. Adcox et al., Suppression of Hadrons with Large Transverse Momentum in Central Au + Au Collisions at sqrt(s) = 130 GeV. Phys. Rev. Lett. **88** (2002)

Chapter 25
Strange and Non-strange Light-Flavour Hadron Production in Pb–Pb and p–Pb Collisions at LHC Energies with ALICE

Michal Šefčík

Abstract The ALICE experiment is dedicated to the study of strongly interacting matter at the extremely high temperatures and energy densities reached at the LHC. We report on the production of π, K and p measured with ALICE in p–Pb and Pb–Pb collisions at the top LHC energies of $\sqrt{s_{\mathrm{NN}}} = 8.16$ and 5.02 TeV, respectively, as well as on the production of Λ, Ξ and Ω in Pb–Pb collisions at $\sqrt{s_{\mathrm{NN}}} = 5.02$ TeV. Excellent tracking and particle-identification capabilities of the ALICE experiment allow characterising the hot nuclear matter via detailed measurements of particle production in nucleus-nucleus collisions. In addition, the study of proton-nucleus collisions provides a fundamental benchmark for initial state and cold nuclear matter effects.

25.1 Introduction

Measurement of light flavour hadron observables provides important information about the colliding system. In these proceedings we briefly discuss methods for particle reconstruction and identification with ALICE. Then we describe the study of the production of π, K and p in p–Pb collisions at $\sqrt{s_{\mathrm{NN}}} = 8.16$ TeV and in Pb–Pb collisions at $\sqrt{s_{\mathrm{NN}}} = 5.02$ TeV, as well as the study of the production of Λ, Ξ and Ω in Pb–Pb collisions at $\sqrt{s_{\mathrm{NN}}} = 5.02$ TeV. The results include transverse momentum spectra ratios, Blast-Wave model fits, and yields integrated in transverse momentum as a function of multiplicity.

Michal Šefčík for the ALICE Collaboration.

M. Šefčík (✉)
Pavol Jozef Šafárik University, Košice, Slovakia
e-mail: michal.sefcik@cern.ch

D. Elia et al. (eds.), *The XVIII International Conference on Strangeness in Quark Matter (SQM 2019)*, Springer Proceedings in Physics 250,
https://doi.org/10.1007/978-3-030-53448-6_25

25.2 Particle Detection in ALICE

The ALICE detector was designed for the study of heavy-ion collisions at the LHC. The main sub-detectors used to measure identified hadrons are the Inner Tracking System (ITS), the Time Projection Chamber (TPC), the Time-Of-Flight detector (TOF) and the High Momentum Particle Identification Detector (HMPID). The V0 detector is used for triggering and for the estimation of multiplicity and centrality. A full description of the ALICE sub-detectors can be found in [1, 2].

Charged light-flavour hadrons (π, K, p) are identified using combined information from ITS, TPC, TOF and HMPID detectors. Weakly decaying strange particles (K^0_S, Λ, Ξ, Ω) are reconstructed via their characteristic decay topologies. For a detailed description, see similar analyses [3, 4].

25.3 Results

The transverse momentum spectra of pions, kaons and protons have been fitted simultaneously with a Blast-Wave function [5]. The resulting fitted parameters are presented in Fig. 25.1. The evolution with multiplicity in Pb–Pb collisions at $\sqrt{s_{\rm NN}}$ = 5.02 TeV follows a trend similar to that observed at the lower energy $\sqrt{s_{\rm NN}}$ = 2.76 TeV, with further increase in radial expansion velocity. In p–Pb collisions at $\sqrt{s_{\rm NN}}$ = 8.16 TeV, we observe higher kinetic freeze-out temperature than at the lower energy $\sqrt{s_{\rm NN}}$ = 5.02 TeV.

The K/π and p/π ratios are shown in Fig. 25.2 for p–Pb collisions and in Fig. 25.3 for Pb–Pb collisions. The p/π ratio exhibits a pronounced enhancement at intermediate $p_{\rm T}$ in central Pb–Pb collisions w.r.t. peripheral collisions. The K/π ratio enhancement is comparatively negligible.

Figure 25.3 also shows a comparison of spectra ratios to hydrodynamic models. The models used are: iEBE-VISHNU hybrid model [6, 7], McGill [8] and

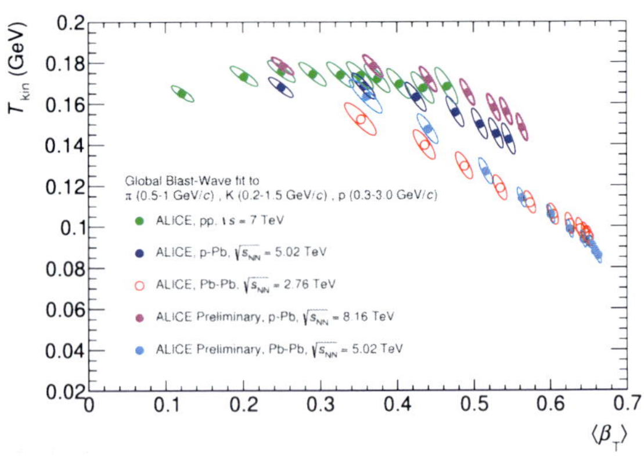

Fig. 25.1 Blast-wave parameters resulting from a simultaneous fit of π, K and p spectra. $T_{\rm kin}$ represents the kinetic freeze-out temperature and $\beta_{\rm T}$ the radial expansion velocity. The fit ranges are specified in the plot legend. Various collision systems and energies are shown

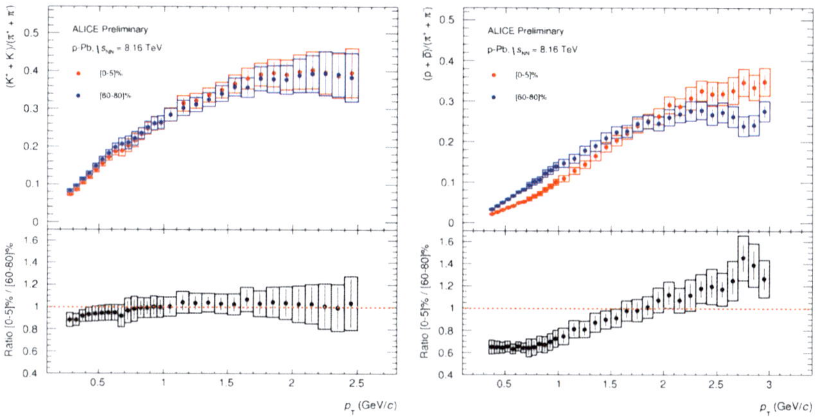

Fig. 25.2 $(K^+ + K^-)/(\pi^+ + \pi^-)$ (left) and $(p + \bar{p})/(\pi^+ + \pi^-)$ yield ratios as a function of transverse momentum in 0–5% and 80–90% centrality classes in p–Pb collisions at $\sqrt{s_{NN}} = 8.16$ TeV. In the lower part, the ratio of central to peripheral collisions is shown

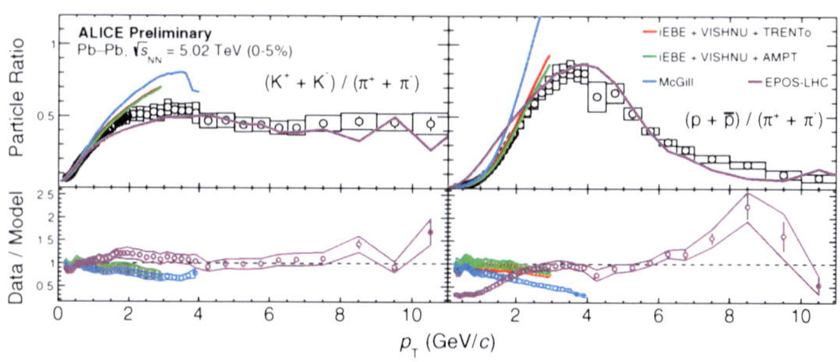

Fig. 25.3 $(K^+ + K^-)/(\pi^+ + \pi^-)$ (left) and $(p + \bar{p})/(\pi^+ + \pi^-)$ yield ratios as a function of transverse momentum in 0–5% centrality class in Pb–Pb collisions at $\sqrt{s_{NN}} = 5.02$ TeV. Comparison with hydrodynamic models is shown. In the lower part, the ratio of data to models is shown

EPOS [9]. All the hydrodynamic models describe the behaviour at low p_T. EPOS, which includes radial flow and jet-medium interactions, describes the dependence over the entire p_T range.

The p/π and Ω/π ratios of p_T-integrated yields are shown in Fig. 25.4 as a function of multiplicity. All measured collision systems and energies follow a common trend. The increase in small systems observed for strange particles is more pronounced for particles with greater strangeness content. For strange particles, there is a flat trend at high multiplicities. The p/π ratio exhibits a hint of decrease at high multiplicities.

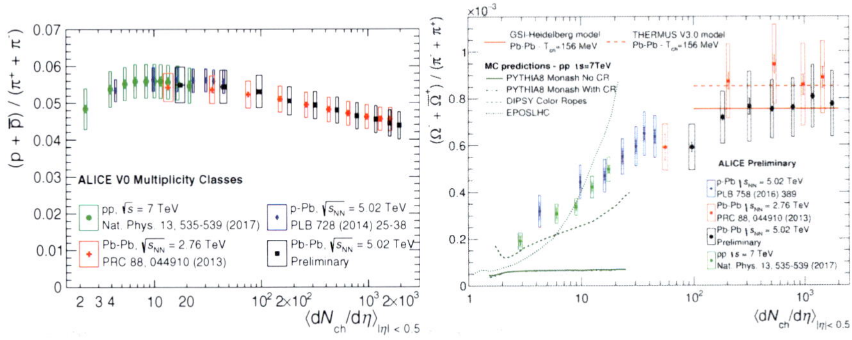

Fig. 25.4 Ratio of proton (left) and Ω (right) yields to pion yield as a function of charged particle multiplicity at medium pseudo-rapidity. Data from pp collisions at $\sqrt{s} = 7$ TeV, p–Pb collisions at $\sqrt{s_{NN}} = 5.02$ TeV and Pb–Pb collisions at $\sqrt{s_{NN}} = 2.76$ TeV and $\sqrt{s_{NN}} = 5.02$ TeV are shown

25.4 Conclusions

The production of light-flavour hadrons in p–Pb and Pb–Pb collisions was measured by ALICE. Blast-wave fits of the p_T spectra indicate that with the increase in energy, the radial expansion velocity increases in central Pb–Pb collisions and the kinetic freeze-out temperature increases in p–Pb collisions. The yield ratio to pions versus multiplicity has a continuous behaviour across colliding systems and energies.

References

1. K. Aamodt et al., (ALICE Colaboration): the ALICE experiment at the CERN LHC. JINST **3**, S08002 (2008)
2. B. Abelev et al., (ALICE Colaboration): performance of the ALICE experiment at the CERN LHC. Int. J. Mod. Phys. A **29**, 1430044 (2014)
3. B. Abelev et al., (ALICE Collaboration): multiplicity dependence of pion, kaon, proton and lambda production in p–Pb collisions at $\sqrt{s_{NN}} = 5.02$ TeV. Phys. Lett. B **728**, 25–38 (2014)
4. B. Abelev et al., (ALICE Collaboration): multi-strange baryon production at mid-rapidity in Pb–Pb collisions at $\sqrt{s_{NN}} = 2.76$ TeV. Phys. Lett. B **728**, 216–227 (2014)
5. E. Schnedermann, J. Sollfrank, U.W. Heinz, Thermal phenomenology of hadrons from 200 A/GeV S+S collisions. Phys. Rev. C **48**, 2462–2475 (1993)
6. C. Shen, Z. Qiu, H. Song, J. Bernhard, S. Bass, U. Heinz, The iEBE-VISHNU code package for relativistic heavy-ion collisions. Comput. Phys. Commun. **199**, 61–85 (2016)
7. W. Zhao, H.-J. Xu, H. Song, Collective flow in 2.76 A TeV and 5.02 A TeV Pb + Pb collisions. Eur. Phys. J. C **77**(9), 645 (2017)
8. S. McDonald, C. Shen, F. Fillion-Gourdeau, S. Jeon, C. Gale, Hydrodynamic predictions for Pb + Pb collisions at 5.02 TeV. Phys. Rev. C **95**(6), 064913 (2017)
9. K. Werner, B. Guiot, I. Karpenko, T. Pierog, Analysing radial flow features in p–Pb and p–p collisions at several TeV by studying identified particle production in EPOS3. Phys. Rev. C **89**(6), 064903 (2014)

Chapter 26
Chemical Equilibration of QGP in Hadronic Collisions

Aleksi Kurkela and Aleksas Mazeliauskas

Abstract We performed state-of-the-art QCD effective kinetic theory simulations of chemically equilibrating QGP in longitudinally expanding systems. We find that chemical equilibration takes place after hydrodynamization, but well before local thermalization. By relating the transport properties of QGP and the system size we estimate that hadronic collisions with final state multiplicities $dN_{\text{ch}}/d\eta > 10^2$ live long enough to reach approximate chemical equilibrium for all collision systems. Therefore we expect the saturation of strangeness enhancement to occur at the same multiplicity in proton-proton, proton-nucleus and nucleus-nucleus collisions.

The experimental measurements of the final-state particles in the ultra-relativistic proton–proton, proton-nucleus and nucleus-nucleus collisions at hadron colliders show a qualitative change of hadron chemistry with the increasing particle multiplicity [1–3]. At central nucleus-nucleus collisions the observed ratios of long-lived hadrons is consistent with chemical equilibrium hadron resonance gas models at temperature $T_{\text{ch}} \approx 155\,\text{MeV}$ [4], while the lowest multiplicity proton-proton collisions are reproduced by perturbative event generators. One interpretation behind this change is the formation of deconfined state of Quark-Gluon Plasma (QGP), which reaches approximate thermal and chemical equilibrium in the collision fireball, from which the thermal hadron productions ensues. Indeed, the hydrodynamic models of the QGP expansion have been very successful in describing small transverse momentum particle spectra and multi-particle correlations. However the initial state created in high-energy nuclear collisions even locally is far from equilibrium and the study of thermalization has been a very active topic [5]. In this work [6, 7] we describe

A. Kurkela · A. Mazeliauskas (✉)
Theoretical Physics Department, CERN, 1211 Genève 23, Meyrin, Switzerland
e-mail: aleksas.mazeliauskas@cern.ch

A. Kurkela
Faculty of Science and Technology, University of Stavanger, 4036 Stavanger, Norway

A. Mazeliauskas
Institut für Theoretische Physik, Universität Heidelberg, 69120 Heidelberg, Germany

D. Elia et al. (eds.), *The XVIII International Conference on Strangeness in Quark Matter (SQM 2019)*, Springer Proceedings in Physics 250,
https://doi.org/10.1007/978-3-030-53448-6_26

the QGP approach to chemical equilibrium within the framework of QCD effective kinetic theory [8].

At high energy, weak coupling limit of QCD, the mid-rapidity interaction region is dominated by strong non-equilibrium gluonic fields [5]. The microscopic description to local thermal equilibrium from such initial configuration is described by the "bottom-up" thermalization scenario [9], which was explicitly realized by classical-statistical Yang-Mills [10] and gluonic kinetic theory [11] simulations. Here for the first time we use the complete leading order QCD kinetic theory of quarks and gluons [8] to simulate the hydrodynamic, chemical and kinetic equilibration of QGP. We solve the coupled Boltzmann equations for quark and gluon distribution functions undergoing homogeneous boost-invariant expansion

$$\partial_\tau f_{g,q} - \frac{p_z}{\tau}\partial_{p_z} f_{g,q} = -\mathcal{C}_{2\leftrightarrow 2}[f] - \mathcal{C}_{1\leftrightarrow 2}[f]. \tag{26.1}$$

We include elastic $2 \leftrightarrow 2$ and colinear $1 \leftrightarrow 2$ collision processes at leading order in the coupling constant $\lambda = 4\pi\alpha_s N_c$. We consider $N_c = 3$ colors and $N_f = 3$ flavours of massless fermions with equal quark-antiquark content, i.e., at vanishing baryon chemical potential $\mu_B = 0$. For the explicit expressions of collision kernels and the regularization procedure of the infrared divergences see the published work [7].

Following the previous work [11], initially at $\tau = 1\,Q_s^{-1}$ the fermion distribution function $f_q = 0$ is set to zero, while the initial gluon density is

$$f_0 = \frac{2A}{\lambda}\frac{Q_0}{\sqrt{p_\perp^2 + p_z^2\xi^2}}e^{-\frac{2}{3}\frac{p_\perp^2+\xi^2 p_z^2}{Q_0^2}}, \tag{26.2}$$

where the values of $A = 5.24$ and $Q_0 = 1.8\,Q_s$ are adjusted to match transverse momentum and energy density extracted from the lattice. The parameter ξ determines the initial anisotropy and is set to $\xi = 10$ to reflect highly anisotropic initial conditions in heavy ion collisions. We then follow the evolution of quark and gluon distribution functions to equilibrium at different values of the coupling constant λ and find the corresponding values of shear viscosity over entropy ratio η/s. In the following we discuss the time evolution of energy density obtained by momentum integration of the non-equilibrium quark and gluon distribution functions.

It was observed in previous works [11–14], that the parametrically large differences in the equilibration rates for different coupling constants λ can be largely scaled out by measuring time in units of *relaxation time*

$$\tau_R(\tau) = \frac{4\pi\eta/s}{T_{\text{id.}}(\tau)}, \tag{26.3}$$

where η/s is the specific shear viscosity. Note that for expanding system the local thermalization temperature and the relaxation time itself is time dependent. Here

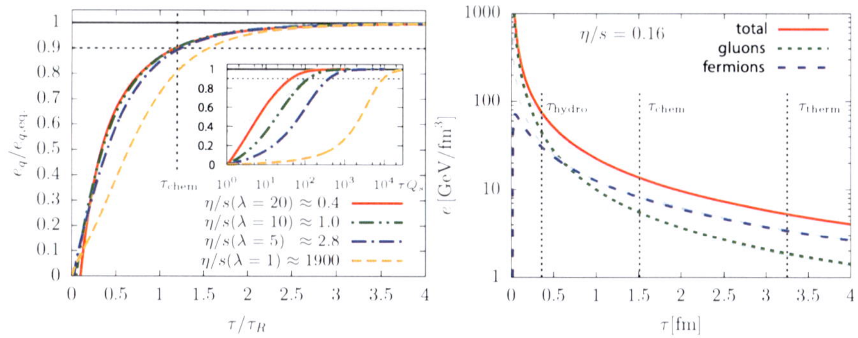

Fig. 26.1 **a** Fermion energy density fraction $e_q(\tau)/e_{q,\text{eq}}(\tau)$ as a function of rescaled time for different coupling constants. The inset shows un-rescaled time dependence. **b** Evolution of total energy density and its gluonic and fermion components in kinetic theory converted to physical units using universality of τ/τ_R scaling and phenomenological values of $\eta/s = 0.16$, $\langle s\tau\rangle = 4.1\,\text{GeV}^2$ and $\nu_{\text{eff}} = 40$. The grey lines correspond to ideal, viscous and chemically equilibrated energies. Figures taken from [6]

we define the effective temperature $T_{\text{id.}}(\tau) = \frac{(T(\tau)\tau^{1/3})|_{\tau\to\infty}}{\tau^{1/3}}$, which asymptotically coincides with the temperature of near equilibrium QGP.

In chemical equilibrium, $N_f = 3$ fermions constitute $e_{q,\text{eq}}/e_{\text{total}} \approx 0.66$ of the total equilibrium energy density. In Fig. 26.1a we present the time evolution of quark energy density relative to equilibrium expectation as a function of rescaled time time τ/τ_R. We consider different values of the coupling constant $\lambda = 1, 5, 10, 20$, which corresponds to $\eta/s \approx 1900, 2.8, 1.0, 0.4$ and vastly different relaxation times τ_R. Remarkably, the large separation of equilibration time-scales shown in the inset of Fig. 26.1a collapses to near λ-independent equilibration curve shown in the main panel of Fig. 26.1a. We find that at the time $\tau/\tau_R \approx 1.2$ and for larger values of the coupling, the fermions reach 90% of their chemical equilibrium energy density. We checked that this is not affected my small non-zero fermion density at early times indicated by grey dashed line in Fig. 26.1a.

We can convert the scaled chemical equilibration time $\tau_{\text{chem}}/\tau_R$ to physical time by solving (26.3) and noting that the asymptotic constant $T^3_{\text{id.}}\tau = (T^3\tau)_\infty$ is proportional to the local entropy density per rapidity $(s\tau)_\infty$:

$$\tau_{\text{chem}} = \underbrace{(\tau_{\text{chem}}/\tau_R)^{3/2}}_{\text{scaled time variable}} \times \underbrace{(4\pi\eta/s)^{3/2} \times (s\tau)_\infty^{-1/2} \times (4\pi^2\nu_{\text{eff}}/90)^{1/2}}_{\text{phenomenological input}}. \qquad (26.4)$$

Substituting the empirical estimate $\tau_{\text{chem}}/\tau_R = 1.2$ we obtain a pocket formula for chemical equilibration time as a function of entropy density, specific shear viscosity and number of degrees of freedom. It is important to emphasize that although the parametric dependence of equilibration time on, say, entropy density, can be inferred from general physics arguments, (26.4) has all relevant numerical factors and is therefore a quantitative formula. The approximate independence of η/s seen in Fig. 26.1a, gives us motivation of plugging in phenomenological relevant parameter values of $\eta/s = 2/4\pi \approx 0.16$, $\langle s\tau\rangle = 4.1\,\mathrm{GeV}^2$ and $\nu_{\text{eff}} = 40$. In Fig. 26.1b we show the converted time evolution of total, gluon and fermion energy density. We observe that at $\tau \approx 0.6\,\text{fm}$ fermion and gluon energy contributions become equal and at $\tau_{\text{chem}} \approx 1.5\,\text{fm}$ the chemical equilibration defined above is achieved. In the paper [6, 7] we also studied hydrodynamic and kinetic equilibrations and found the following ordering of equilibration scales (also shown in Fig. 26.1b)

$$\underbrace{\tau_{\text{hydro}}}_{\pm 10\%\ \text{viscous}\ e(\tau)} < \underbrace{\tau_{\text{chem}}}_{\pm 10\%\ \text{fermion eq.}\ e(\tau)} < \underbrace{\tau_{\text{therm}}}_{\pm 10\%\ \text{ideal}\ e(\tau)} . \tag{26.5}$$

Strikingly, the empirical weak coupling equilibration formula (26.4) produces realistic equilibration timescales compatible with heavy-ion phenomenology if extrapolated to realistic values of η/s. To what extent such extrapolation captures the dynamics of relatively strongly coupled system is debatable, but it is clear that the weak coupling equilibration baseline is not in contradiction with rapid thermalization of QGP.

Finally, (26.4) can be reformulated as a bound on charged particle multiplicities necessary to achieve chemical equilibrium by freeze-out. We relate the multiplicity $dN_{\text{ch}}/d\eta$ in terms of scaled variables using $\langle \tau s\rangle \approx (S/N_{\text{ch}})\ 1/(\pi R^2)\ dN_{\text{ch}}/d\eta$ (where $S/N_{\text{ch}} \approx 7$ [15]) and (26.4), so that

$$\frac{dN_{\text{ch}}}{d\eta} = \tfrac{4\pi^3}{90}\nu_{\text{eff}}\,(S/N_{\text{ch}})^{-1}\,(4\pi\eta/s)^3\,(\tau/\tau_R)^3\,(\tau/R)^{-2}\,. \tag{26.6}$$

Assuming that the system disintegrates once its lifetime exceeds the system size $\tau \sim R$ [14], the minimum multiplicity needed for that time to be at or above chemical equilibration time $\tau_{\text{chem}}/\tau_R = 1.2$ is

$$\frac{dN_{\text{ch}}}{d\eta} \gtrsim 110\left(\frac{\tau_{\text{chem}}}{1.2\tau_R}\right)^3\left(\frac{\eta/s}{0.16}\right)^3\left(\frac{\tau_{\text{chem}}}{R}\right)^{-2}. \tag{26.7}$$

This bound is roughly compatible with experimentally observed trends of strangeness enhancement [1–3], which gives theoretical ground for the assumed formation of chemically equilibrated QGP at high multiplicity collisions.

In summary, we presented the detailed picture of QGP equilibration within QCD kinetic theory framework. Strikingly, extrapolated to phenomenological values of η/s, this picture gives realistic equilibration timescales and connects in a novel way the transport properties of QGP to experimental observations of hadron chemistry in proton–proton, proton-nucleus and nucleus-nucleus collisions.

Acknowledgments This work was supported in part by the German Research Foundation (DFG) Collaborative Research Centre (SFB) 1225 (ISOQUANT) (A.M.).

References

1. J. Adam et al., Enhanced production of multi-strange hadrons in high-multiplicity proton–proton collisions. Nature Phys. **13**, 535–539 (2017). https://doi.org/10.1038/nphys4111
2. B.B. Abelev, et al., Multiplicity dependence of pion, kaon, proton and lambda production in p–Pb collisions at $\sqrt{s_{NN}}$ = 5.02 TeV. Phys. Lett. **B728**, 25–38 (2014). https://doi.org/10.1016/j.physletb.2013.11.020
3. B.B. Abelev et al., Multi-strange baryon production at mid-rapidity in Pb–Pb collisions at $\sqrt{s_{NN}}$ = 2.76 TeV. Phys. Lett. B **728**, 216–227 (2014). https://doi.org/10.1016/j.physletb.2014.05.052, https://doi.org/10.1016/j.physletb.2013.11.048.; Erratum: Phys. Lett. B **734**, 409 (2014)
4. A. Andronic, P. Braun-Munzinger, K. Redlich, J. Stachel, Decoding the phase structure of QCD via particle production at high energy. Nature **561**(7723), 321–330 (2018). https://doi.org/10.1038/s41586-018-0491-6
5. S. Schlichting, D. Teaney, The first fm/c of heavy-ion collisions (2019)
6. A. Kurkela, A. Mazeliauskas, Chemical equilibration in hadronic collisions. Phys. Rev. Lett. **122**, 142,301 (2019). https://doi.org/10.1103/PhysRevLett.122.142301
7. A. Kurkela, A. Mazeliauskas, Chemical equilibration in weakly coupled QCD. Phys. Rev. D **99**(5), 054,018 (2019). https://doi.org/10.1103/PhysRevD.99.054018
8. P.B. Arnold, G.D. Moore, L.G. Yaffe, Effective kinetic theory for high temperature gauge theories. JHEP **01**, 030 (2003). https://doi.org/10.1088/1126-6708/2003/01/030
9. R. Baier, A.H. Mueller, D. Schiff, D.T. Son, 'Bottom up' thermalization in heavy ion collisions. Phys. Lett. B **502**, 51–58 (2001). https://doi.org/10.1016/S0370-2693(01)00191-5
10. J. Berges, K. Boguslavski, S. Schlichting, R. Venugopalan, Turbulent thermalization process in heavy-ion collisions at ultrarelativistic energies. Phys. Rev. D **89**(7), 074,011 (2014). https://doi.org/10.1103/PhysRevD.89.074011
11. A. Kurkela, Y. Zhu, Isotropization and hydrodynamization in weakly coupled heavy-ion collisions. Phys. Rev. Lett. **115**(18), 182,301 (2015). https://doi.org/10.1103/PhysRevLett.115.182301
12. L. Keegan, A. Kurkela, P. Romatschke, W. van der Schee, Y. Zhu, Weak and strong coupling equilibration in nonabelian gauge theories. JHEP **04**, 031 (2016). https://doi.org/10.1007/JHEP04(2016)031
13. A. Kurkela, A. Mazeliauskas, J.F. Paquet, S. Schlichting, D. Teaney, Effective kinetic description of event-by-event pre-equilibrium dynamics in high-energy heavy-ion collisions. Phys. Rev. C **99**(3), 034,910 (2019). 10.1103/PhysRevC.99.034910
14. A. Kurkela, A. Mazeliauskas, J.F. Paquet, S. Schlichting, D. Teaney, Matching the nonequilibrium initial stage of heavy ion collisions to hydrodynamics with QCD kinetic theory. Phys. Rev. Lett. **122**(12), 122,302 (2019). https://doi.org/10.1103/PhysRevLett.122.122302
15. P. Hanus, A. Mazeliauskas, K. Reygers, Entropy production in pp and Pb–Pb collisions at the LHC (2019)

Chapter 27
Strangeness and Light Fragment Production at High Baryon Density

David Blaschke, Gerd Röpke, Yuri Ivanov, Marina Kozhevnikova, and Simon Liebing

Abstract We discuss medium effects on light cluster production in the QCD phase diagram within a generalized Beth-Uhlenbeck (GBU) approach by relating Mott transition lines to those for chemical freeze-out. We find that in heavy-ion collisions at highest energies provided by the LHC light cluster abundances should follow the statistical model because of low baryon densities. At low energies in the nuclear fragmentation region, where the freeze-out interferes with the liquid-gas phase transition, selfenergy and Pauli blocking effects are important. At intermediate energies the HADES, FAIR and NICA experiments can give new information. The GBU approach provides new insights to strange hadron production in this energy domain for explaining the "horn" effects.

D. Blaschke
University of Wroclaw, 50-204 Wroclaw, Poland

D. Blaschke · Y. Ivanov · M. Kozhevnikova
JINR Dubna, BLTP, Joliot-Curie Str. 6, 141980 Dubna, Russia

D. Blaschke (✉) · G. Röpke
NRNU (MEPhI), Kashirskoe Shosse 31, 115409 Moscow, Russia
e-mail: david.blaschke@uwr.edu.pl
URL: http://www.ift.uni.wroc.pl/~blaschke

G. Röpke
University of Rostock, Albert-Einstein-Str. 23-24, 18051 Rostock, Germany

S. Liebing
TU Bergakademie Freiberg, ITP, Leipziger Str. 23, 09599 Freiberg, Germany

D. Elia et al. (eds.), *The XVIII International Conference on Strangeness in Quark Matter (SQM 2019)*, Springer Proceedings in Physics 250,
https://doi.org/10.1007/978-3-030-53448-6_27

27.1 Chemical Freeze-Out in the QCD Phase Diagram

The beam energy scan (BES) programs of heavy-ion collisions (HIC) provide insights into the systematics of particle production under varying conditions of temperature and density of the evolving hadronic fireball created in these experiments. A remarkable fact is that the thermal statistical model of hadron production makes excellent predictions for particle yields with just two free parameters: the temperature T_{fo} and the baryon chemical potential $\mu_{B,fo}$ at the chemical freeze-out [1]. The chemical freeze-out concept assumes that the system is approximately described by the equilibrium as long as collisions are frequent to establish the corresponding distributions. For an expanding fireball, this is no longer the case at a critical density so that the chemical equilibrium freezes out at the corresponding parameter values for temperature T and baryon number density n_B. An empirical relation for T_{fo} as function of μ_B has been given in [2]

$$T_{fo}[\text{GeV}] = 0.166 - 0.139(\mu_B/\text{GeV})^2 - 0.053(\mu_B/\text{GeV})^4, \tag{27.1}$$

with $\mu_B[\text{GeV}] = 1.308/[1 + 0.273\sqrt{s_{NN}}/\text{GeV}]$ and $\sqrt{s_{NN}} = \sqrt{2m_N E_{\text{lab}} + 2m_N^2}$, $m_N = 0.939$ GeV. It has been discussed in [3] that (27.1) apparently works well down to $T_{fo} \sim 10$ MeV corresponding to HIC at moderate laboratory energies of $E_{\text{lab}} = 35$ AMeV [4] which have been analyzed in the quantum statistical freeze-out scheme in [5], but also up to highest energies provided by heavy-ion collisions at the LHC [1].

In this contribution, we would like to focus on the region between these two extremes, where the transition from baryon stopping to nuclear transparency takes place and the highest baryon densities at freeze-out are reached. It is the region of c.m.s. energies of the future NICA facility [6], $\sqrt{s_{NN}} = 2 \ldots 11$ GeV, which was partly addressed already by AGS and CERN-SPS experiments as well as the RHIC BES I program and the recent HADES experiment at GSI. Besides the MPD and BM@N experiments at NICA it will be covered in future by the low-energy RHIC and the RHIC fixed target programs as well as the FAIR CBM experiment. We want to elucidate that the yields of light fragments (clusters) like neutrons (n), protons (p), deuterons (^{2}H, d), tritons (^{3}H, t), helions (^{3}He, h), and α-particles (^{4}He) undergo a change from being strongly affected by medium effects at lowest energies to free-streaming quasiparticle behaviour at highest energies and that a similar transition governs the appearance of the "horn" effects for the ratios of strange to nonstrange hadron production in this energy range.

In Table 27.1 we show freeze-out parameter values according to (27.1) for different collision energies from the available and planned beam energy scan programs together with the baryon densities for these parameters, calculated within the statistical model with and without light clusters. The results are shown by the plus symbols in the QCD phase diagram of Fig. 27.1, where also the coexistence regions for the nuclear gas-liquid transition and for two examples of the hadron-quark matter transi-

Table 27.1 Freeze-out temperatures T_{fo} and chemical potentials μ_B according to (27.1) by Cleymans et al. [2] for heavy-ion collisions with beam energies $E_{\rm lab}$ in the fixed target mode and the corresponding center of mass energy $\sqrt{s}$ in the nucleon-nucleon system relevant for the collider mode. The last two columns display the total baryon density according to the statistical model without and with light clusters

$E_{\rm lab}$ [A GeV]	$\sqrt{s_{NN}}$ [GeV]	T_{fo} [MeV]	μ_B [MeV]	$n^{\rm nocluster}$[fm^{-3}]	$n^{\rm cluster}$[fm^{-3}]
1.23	2.0	39.34	846.0	0.0231	0.0529
2.0	2.35	56.38	796.8	0.0371	0.0699
3.85	3.0	79.956	719.1	0.0636	0.0938
8.0	4.3	108.7	601.7	0.1045	0.1238
10	4.7	114.7	572.9	0.1122	0.1288
15	5.6	125.0	517.2	0.1225	0.1347
20	6.4	131.8	476.1	0.1260	0.1354
30	7.7	139.6	421.6	0.1251	0.1316
43	9.2	145.7	372.5	0.1196	0.1242
70	11.6	151.8	313.9	0.1086	0.1158
158	17.2	158.6	229.0	0.0875	0.0890
204	19.6	160.0	205.9	0.0813	0.0825
387	27	162.6	156.2	0.0680	0.0687
809	39	164.2	112.3	0.0566	0.0571
2194	64.2	165.3	72.5	0.0470	0.0474
21298	200	165.9	23.5	0.0364	0.0366

tion from [7] and [8] are displayed as grey shaded regions together with their critical endpoints.

Accounting for an excluded volume of the hadrons according to [9] (version I) results in a reduction of the maximal densities that can be reached at chemical freeze-out (cross symbols), see also [10]. Note that the account of light clusters increases the freeze-out densities at low temperatures by about a factor two when compared with a result where no light clusters have been taken into account (triangle symbols) [10]. The freeze-out at high temperatures proceeds from the quark-gluon plasma (QGP) and basically coincides with the behaviour of the pseudocritical temperatures of the deconfinement transition T_c obtained in lattice QCD simulations [11, 12]. For $T_{fo} <$ 100 MeV the freeze-out proceeds from a subphase within hadronic matter, such as the quarkyonic phase characterized by confinement and partial chiral symmetry restoration, resulting in a triple point in the QCD phase diagram [13, 14].

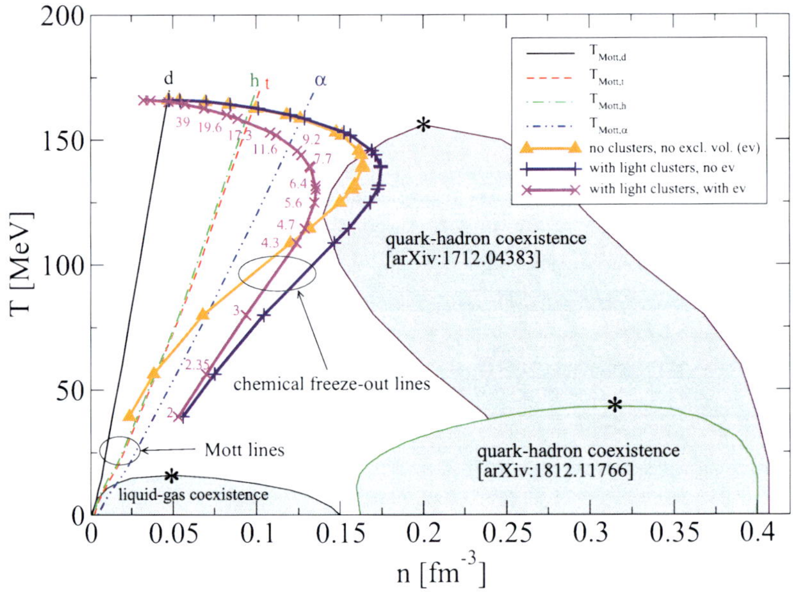

Fig. 27.1 Chemical freezeout lines in the temperature density plane (phase diagram) together with Mott lines for light clusters. The coexistence regions for the nuclear gas-liquid transition and for two examples of the hadron-quark matter transition are shown as grey shaded regions together with their critical endpoints. For details, see text

27.2 Freeze-Out and Mott Transition for Light Clusters

The particle production measured in heavy ion collisions (HIC) is of interest to infer the properties of dense matter. The time evolution of the fireball produced in HIC demands a nonequilibrium approach to describe the time dependence of the distribution function of the observed products, which are mainly neutrons, protons, and clusters at low energies, but also mesons, hyperons and antiparticles at high energies. Different transport codes have been developed to describe the time evolution of the fireball, but the formation of bound states (clusters) remains an open problem where some semi-empirical assumptions such as the coalescence model are applied. The freeze-out concept can only be considered as an approximation to describe disassembling matter. It has the advantage that correlations and bound state formation are correctly described within a quantum statistical approach. For a nonequilibrium theory, the equilibrium is a limiting case, and even more, the quasi equilibrium (generalized Gibbs ensemble) serves to define the boundary conditions for the nonequilibrium evolution, see [15].

A topical question is the debate about the origin of light (anti-)hypernuclear clusters at LHC in the ALICE experiment which follow thermal statistical model predictions with $T_{fo} \sim 156.5$ MeV [1] despite the fact that their binding energy is about two orders of magnitude smaller. Are they formed at the hadronization of the QGP

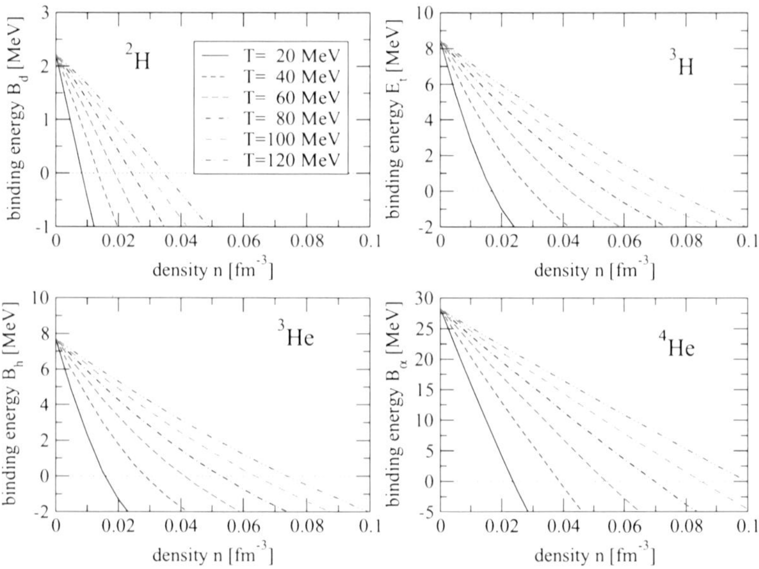

Fig. 27.2 Binding energies of light clusters in nuclear matter as a function of density for different temperatures. Vanishing binding energy defines the Mott transition density. For details of the calculation, see [17, 18]

and survive because of the sudden character of the freeze-out or are they formed by coalescence of baryons at a much later stage? Here we add the notion of the Mott transition [16] to this discussion. Due to an interplay of selfenergy and Pauli-blocking effects in the Bethe-Salpeter equation for light nuclear clusters their binding energy becomes medium dependent, see Fig. 27.2 and vanishes (Mott effect) along the Mott lines in the phase diagram, see Fig. 27.1. For details of the calculation, see [18]. To the left of the Mott lines the corresponding cluster is a bound state (at rest in the medium) while to the right of it the cluster is partially dissociated. For highest energies (and temperatures) the medium does not lead to a Mott effect and clusters are formed like in free space. When lowering the energy there is a crossing between freeze-out and Mott lines which strongly depends on details of the calculation as, e.g., the account for a finite momentum of the clusters relative to the medium which would stabilize them, see Fig. 27.3. The account for light clusters in calculating the freeze-out line removes the second crossing with the Mott lines at low energies. It is very important to study light (hyper)nuclear cluster production in an experiment like HADES, BM@N or MPD to analyse the transition from high to low T_{fo} systematically.

Here we want to report first results of applying the sudden freeze-out scheme to light cluster formation in the three-fluid hydrodynamics simulation program THESEUS [21], by including scalar and vector selfenergies for nucleons as medium effects in the particle distribution functions on the freeze-out surface. Calculations for the deuteron rapidity distribution in Pb+Pb collisions at $E_{\rm lab} = 20$ A GeV and 30 A GeV are shown in Fig. 27.4 and compared to experimental data from the NA49

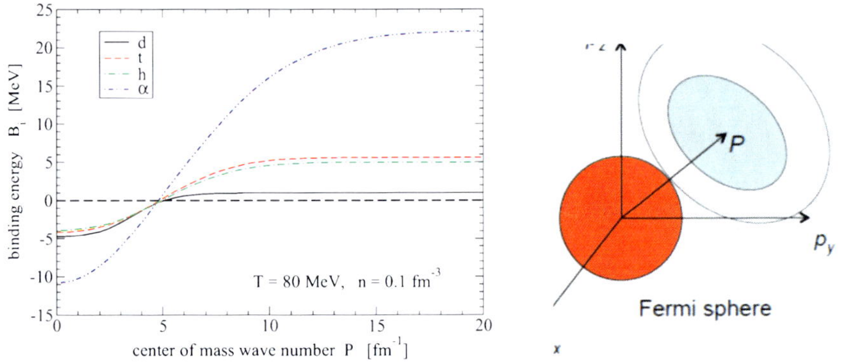

Fig. 27.3 Left: Binding energies of light clusters in nuclear matter as a function of the center of mass momentum P relative to the nuclear medium (Fermi sphere), see right panel. Vanishing binding energy defines the Mott momentum above which the clusters are bound states, see [18]

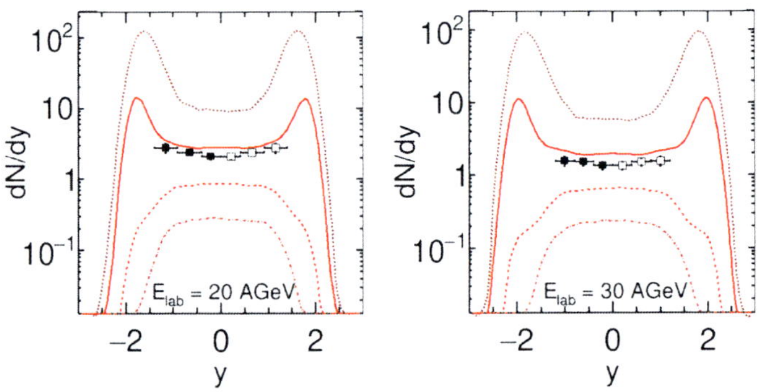

Fig. 27.4 Results of the NA49 collaboration for the deuteron rapidity distribution in Pb+Pb collisions at $E_{\text{lab}} = 20$ A GeV (left panel) and $E_{\text{lab}} = 30$ A GeV (right panel) compared with results from the three-fluid hydrodynamics simulation (THESEUS) for impact parameter $b = 3$ fm using a crossover equation of state model with a thermal statistical model with (broken lines) and without (solid lines) selfenergy effects

collaboration. We observe that the best description is obtained when the selfenergy effects are discarded. It is necessary to include also the momentum-dependent Pauli-blocking effects in a further development of this study and compare results with recent data from the HADES collaboration at lower energies $E_{\text{lab}} = 1.23$ A GeV [22] to find out at which energy the importance of nuclear medium effects for cluster production sets in. These new results can be compared with previous results for light clusters [23] obtained within THESEUS by applying the coalescence scheme [24] for cluster formation in the final state of the collision.

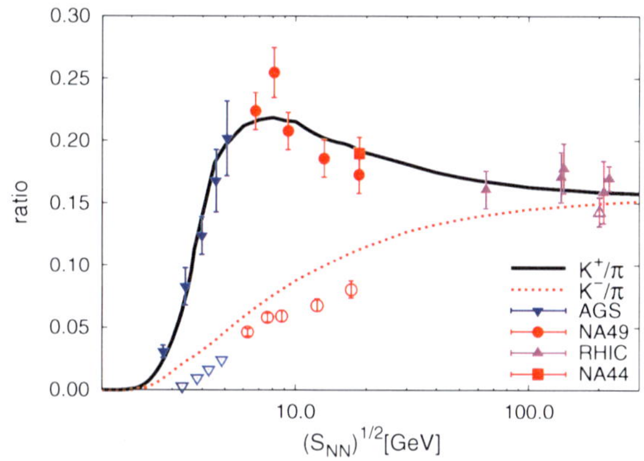

Fig. 27.5 The "horn" in the collision energy dependence of the ratio K^+/π^+ from different experiments (full symbols) compared to the results from a GBU approach to pions and kaons in dense quark matter (solid line) [26]. The results for K^-/π^- are shown as dotted line, to be compared with the experimental data (open symbols)

27.3 "Horn" Effects in Strangeness Production

The effect of a "horn" structure for the ratio of strange to nonstrange particle production as a function of the collision energy has been suggested by Gazdzicki and Gorenstein [19] and was established experimentally by the NA49 collaboration [20] for the particle ratio K^+/π^+ to be located between the AGS and the SPS domains of collision energy at $\sqrt{s_{NN}} \sim 8$ GeV. A similar structure has also been found in the energy scan of the ratio Λ/π^- [1] and they have been attributed to a tricritical point in the QCD phase diagram [13, 14]. Within standard HIC simulations this horn effect has not been reproduced (see, e.g., [21]), but when accounting for effects of partial chiral symmetry restoration on strangeness production already in the hadronic phase the horn effect is interpreted as an enhancement of strangeness production on the rising branch of the ratio [25]. Another interesting aspect may be the occurrence of a plasmon-like mode in the K^+ channel at the hadronisation transition that can be described within a PNJL model for the quark-gluon plasma [26], see Fig. 27.5. It is also possible that a threshold-like enhancement of the pion production for collisions at $\sqrt{s_{NN}} > 8$ GeV due to the onset of Bose condensation of pions [27] contributes to the pronouncedness of the effect [28]. A systematic approach should also explain the strong system-size dependence observed by the NA61/SHINE experiment [29].

27.4 Conclusions

We have argued for the importance to combine quantum statistical analyses of low-energy heavy-ion collisions with simulations the NICA parameter region. Light clusters are of interest to determine the parameter values at freeze-out, in particular the density that is not well known and the role of the in-medium effects, for instance for the deuteron yield. A second topical goal is the study of strange hadron production in this context, with inputs from chiral quark models of hadrons within the generalized

Beth-Uhlenbeck approach in order to pin down the nature of the "horn" effects for strange-to-nonstrange particle ratios.

Acknowledgments Discussions with Manuel Lorenz and M. Szala at the SQM-2019 conference and at the ECT* Trento workshop on "Light clusters in nuclei and nuclear matter" on the preliminary results of the HADES collaboration are gratefully acknowledged. I. Karpenko has provided an update of the particlization routine in the THESEUS programme to include selfenergy effects on light clusters. We thank N.-U. Bastian and S. Typel for providing the data for the coexistence regions in Fig. 27.3 and A. Radzhabov for the preparation of Fig. 27.5. S.L. acknowledges the Heisenberg-Landau program for supporting his visit to JINR Dubna. Y.B.I. was supported by the Russian Foundation for Basic Research, Grants No. 18-02-40084 and No. 18-02-40085 for work related to the upgrade of the three-fluid hydrodynamics code. The other work was supported by the Russian Science Foundation under grant No. 17-12-01427.

References

1. A. Andronic, P. Braun-Munzinger, K. Redlich, J. Stachel, Nature **561**(7723), 321 (2018)
2. J. Cleymans, H. Oeschler, K. Redlich, S. Wheaton, Phys. Rev. C **73**, 034905 (2006)
3. G. Röpke, D. Blaschke, Y.B. Ivanov, I. Karpenko, O.V. Rogachevsky, H.H. Wolter, Phys. Part. Nucl. Lett. **15**(3), 225 (2018)
4. S. Kowalski et al., Phys. Rev. C **75**, 014601 (2007)
5. J. Natowitz et al., Phys. Rev. Lett. **108**, 172701 (2010); L. Qin et al., Phys. Rev. Lett. **108**, 172701 (2012); L. Qin et al., Phys. Rev. Lett. **108**, 172701 (2012)
6. V.D. Kekelidze, R. Lednicky, V.A. Matveev, I.N. Meshkov, A.S. Sorin, G.V. Trubnikov, Eur. Phys. J. A **52**(8), 211 (2016); K. Hagel et al., Eur. Phys. J. A **50**, 39 (2014)
7. S. Typel, D. Blaschke, Universe **4**(2), 32 (2018)
8. N.U.F. Bastian, D.B. Blaschke. arXiv:1812.11766 [nucl-th]
9. M. Albright, J. Kapusta, C. Young, Phys. Rev. C **90**(2), 024915 (2014)
10. J. Randrup, J. Cleymans, Eur. Phys. J. **52**, 218 (2016)
11. S. Borsanyi, Z. Fodor, C. Hoelbling, S.D. Katz, S. Krieg, K.K. Szabo, Phys. Lett. B **730**, 99 (2014)
12. A. Bazavov et al., HotQCD Collaboration. Phys. Rev. D **90**, 094503 (2014)
13. A. Andronic et al., Nucl. Phys. A **837**, 65 (2010)
14. K.A. Bugaev et al., Phys. Part. Nucl. Lett. **15**(3), 210 (2018)
15. G. Röpke, Il Nuovo Cimento C **39**, 392 (2016)
16. G. Röpke, M. Schmidt, L. Münchow, H. Schulz, Nucl. Phys. A **399**, 587 (1983)
17. S. Typel, G. Röpke, T. Klähn, D. Blaschke, H.H. Wolter, Phys. Rev. C **81**, 015803 (2010)
18. G. Röpke, Correlations and clustering in dilute matter, in W.U. Schröder (ed.), *Nuclear Particle Correlations and Cluster Physics* (World Scientific, 2017); [[nucl-th]]
19. M. Gazdzicki, M.I. Gorenstein, Acta Phys. Polon. B **30**, 2705 (1999)
20. S.V. Afanasiev et al., NA49 Collaboration. Phys. Rev. C **66**, 054902 (2002)
21. P. Batyuk et al., Phys. Rev. C **94**, 044917 (2016)
22. M. Szala (HADES Collaboration), contribution at this conference (2019)
23. N.-U. Bastian et al., Eur. Phys. J. A **52**(8), 244 (2016)
24. V.N. Russkikh, Y.B. Ivanov, Y.E. Pokrovsky, P.A. Henning, Nucl. Phys. A **572**, 749 (1994)
25. A. Palmese, W. Cassing, E. Seifert, T. Steinert, P. Moreau, E.L. Bratkovskaya, Phys. Rev. C **94**(4), 044912 (2016)
26. D. Blaschke, A. Dubinin, A. Radzhabov, A. Wergieluk, Phys. Rev. D **96**(9), 094008 (2017)
27. V. Begun, W. Florkowski, Phys. Rev. C **91**, 054909 (2015)
28. D. Blaschke, A.V. Friesen, A. YuL Kalinovsky, Radzhabov. arXiv:1912.13162 [hep-ph]
29. P. Podlaski (NA61/SHINE Collaboration), contribution at this conference (2019)

Chapter 28
Cross-Correlators of Conserved Charges in QCD

Rene Bellwied, Szabolcs Borsanyi, Zoltan Fodor, Jona N. Guenther, Jacquelyn Noronha-Hostler, Paolo Parotto, Attila Pásztor, Claudia Ratti, and Jamie M. Stafford

Abstract We present cross-correlators of QCD conserved charges at $\mu_B = 0$ from lattice simulations and perform a Hadron Resonance Gas (HRG) model analysis to break down the hadronic contributions to these correlators. We construct a suitable hadronic proxy for the ratio $-\chi_{11}^{BS}/\chi_2^S$ and discuss the dependence on the chemical potential and experimental cuts. We then perform a comparison to preliminary STAR results and comment on a possible direct comparison of lattice and experiment.

28.1 Introduction and Setup

The transition between hadronic matter and deconfined Quark Gluon Plasma (QGP) is a smooth crossover at vanishing baryon chemical potential [1–4], and is believed to turn into a first order transition for larger values of the chemical potential. Among

R. Bellwied · P. Parotto · C. Ratti · J. M. Stafford
Department of Physics, University of Houston, Houston, TX 77204, USA

S. Borsanyi · Z. Fodor · J. N. Guenther · P. Parotto (✉)
Department of Physics, University of Wuppertal, Wuppertal D-42119, Germany
e-mail: parotto@uni-wuppertal.de

Z. Fodor · A. Pásztor
Eötvös University, Budapest 1117, Hungary

Z. Fodor
Jülich Supercomputing Centre, Jülich D-52425, Germany

Physics Department, UCSD, San Diego, CA 92093, USA

J. N. Guenther
Department of Physics, University of Regensburg, Regensburg D-93053, Germany

J. Noronha-Hostler
Department of Physics, University of Illinois at Urbana-Champaign, Urbana, IL 61801, USA

D. Elia et al. (eds.), *The XVIII International Conference on Strangeness in Quark Matter (SQM 2019)*, Springer Proceedings in Physics 250,
https://doi.org/10.1007/978-3-030-53448-6_28

the best suited observables for the study of the QCD phase diagram in the transition region are fluctuations of conserved charges. They can be studied in theory through first principles lattice QCD calculations (see e.g. [5–9]), as well as being closely connected to experimentally available measurements of net-particle fluctuations and correlations [10–14]. Due to the fact that some hadrons cannot be detected in experiment, a sizable share of B, Q, S is lost. Historically, the hadronic proxies used for B, Q and S are protons, the sum $p + \pi + K$ and the kaons themselves, respectively. More recently, the attention has moved towards non-diagonal correlators between conserved charges [13, 14].

In this contribution we build a bridge between lattice-QCD-calculated correlators of conserved charges and experimentally accessible fluctuations and correlations of hadronic species, focusing on the correlator between baryon number and strangeness χ_{11}^{BS}. We employ the HRG model in order to include the effect of resonance decays and cuts on the kinematics, which are present in the experiment. Most importantly, the HRG model allows us to isolate single particle-particle correlations and connect them to the correlators of conserved charges.

The ideal HRG model partition function is a sum over the single-state partition functions. Fluctuations of conserved charges are expressed as derivatives of the grand partition function with respect to the different chemical potentials:

$$\chi_{ijk}^{BQS}(T, \hat{\mu}_B, \hat{\mu}_Q, \hat{\mu}_S) = \frac{\partial^{i+j+k}\left(p/T^4\right)}{\partial \hat{\mu}_B^i \partial \hat{\mu}_Q^j \partial \hat{\mu}_S^k} = \sum_R B_R^i\, Q_R^j\, S_R^k\, I_{i+j+k}^R\left(T, \hat{\mu}_B, \hat{\mu}_Q, \hat{\mu}_S\right), \quad (28.1)$$

where $\hat{\mu}_i = \mu_i/T$, and the phase space integral at order $i + j + k$ reads (note that it is completely symmetric in all indices, hence $i + j + k = l$):

$$I_l^R\left(T, \hat{\mu}_B, \hat{\mu}_Q, \hat{\mu}_S\right) = \frac{\partial^l p_R/T^4}{\partial \hat{\mu}_R^l}. \quad (28.2)$$

It is possible to recast the sum in (28.1) as a sum over the fewer states which are stable under strong interactions:

$$\sum_R B_R^l Q_R^m S_R^n I_p^R \rightarrow \sum_{i \in \text{stable}} \sum_R \left(P_{R\rightarrow i}\right)^p B_i^l Q_i^m S_i^n I_p^R, \quad (28.3)$$

where $(P_{R\rightarrow i})^p$ is the average number of particles i produced by the decay of particle R.

The advantage of expressing the fluctuations in (28.1) in term of stable particles, is that we can further distinguish by particles which can be—or usually are—detected in experiment, and those which are not. In this work we employ the hadronic list labeled as PDG2016+ in [17], with the list of decays described and first utilized in [18]. We will hereafter consider the following species as the commonly measured ones:

$$\pi^{\pm},\ K^{\pm},\ p\,(\overline{p}),\ \Lambda(\overline{\Lambda}),\ \Xi^-(\overline{\Xi}^+),\ \Omega^-(\overline{\Omega}^+).$$

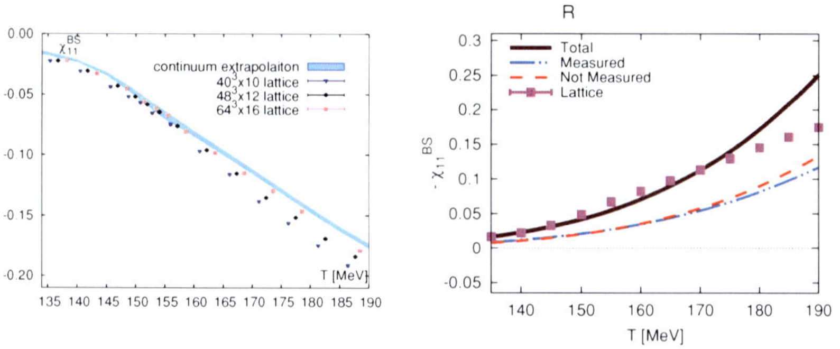

Fig. 28.1 The correlator χ_{11}^{BS} as a function of the temperature at $\mu_B = 0$. (Left panel) results for the ω_0-based scale setting [15] at finite lattice spacing and continuum extrapolation. (Right panel) the continuum extrapolated results compared to HRG model calculations (solid black line), with the contribution from measured (dotted-dashed blue line) and non-measured (dashed reline) hadronic species. Figure from [16]

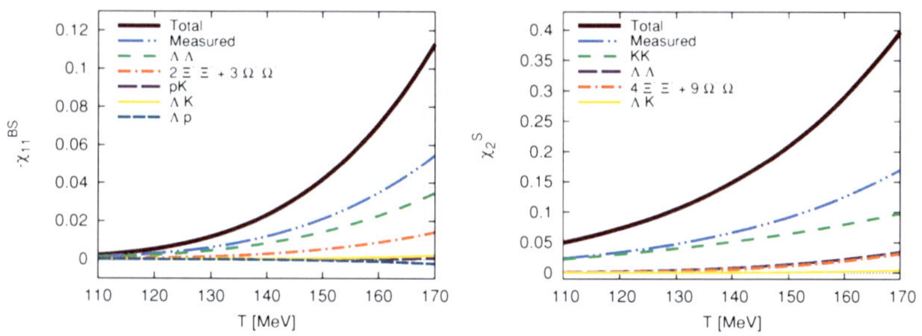

Fig. 28.2 Breakdown of the contributions from *measured* particle-particle correlators to $-\chi_{11}^{BS}$ (left panel) and χ_2^S (right panel), at $\mu_B = 0$, as a function of the temperature. Figure from [16]

In Fig. 28.1 we show the χ_{11}^{BS} correlator as a function of the temperature for vanishing chemical potential, calculated from the lattice (left panel) at different finite spacings, as well as its continuum extrapolation. In the right panel, along with the continuum extrapolation, we show the results from our HRG model analysis, where we separate the contribution to this correlator from measured and non-measured hadronic species. We see that the contributions roughly correspond to the same amount. Moreover, we can see in Fig. 28.2 the breakdown of the main contributions from *measured* particle-particle correlations to $-\chi_{11}^{BS}$ and χ_2^S at vanishing chemical potential.

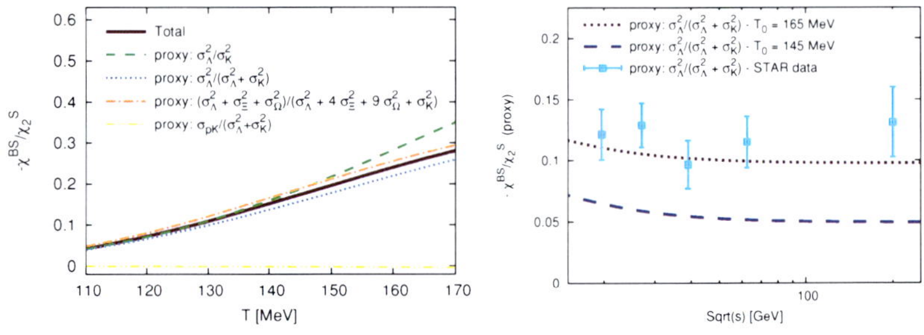

Fig. 28.3 (Left panel): comparison of different proxies (the proxy $\widetilde{C}^{\Lambda,\Lambda K}_{BS,SS}$ is shown as a blue dotted line) and the total contribution (black solid line) for the ratio $-\chi^{BS}_{11}/\chi^S_2$ at $\mu_B = 0$. (Right panel): comparison of our proxy with the kinematic cuts from [12, 13], along parametrized freeze-out lines with $T(\mu_B = 0) = 145, 165$ MeV (black dotted and blue dashed). The STAR preliminary data are shown in light blue. Figure from [16]

Proxy for $-\chi^{BS}_{11}/\chi^S_2$

In order to perform a comparison to experiment and potentially to lattice QCD results, we consider the ratio $-\chi^{BS}_{11}/\chi^S_2$. Exploiting the information in Fig. 28.2, we construct the following proxy for this ratio:

$$\widetilde{C}^{\Lambda,\Lambda K}_{BS,SS} = \sigma^2_\Lambda/(\sigma^2_K + \sigma^2_\Lambda), \tag{28.4}$$

which is shown in the left panel of Fig. 28.3 as a blue dotted line, alongside the ratio $-\chi^{BS}_{11}/\chi^S_2$ (black solid line). This quantity well reproduces the full contribution for all temperatures around the QCD transition.

In Fig. 28.4 we show the same comparison for finite chemical potential, along parametrized chemical freeze-out lines with $T(\mu_B = 0) = 145, 165$ MeV. In the left panel we show the comparison in the absence of cuts on the kinematics, while in the central panel we introduce "exemplary" cuts, which are the same for all the hadronic species; in both cases we see that the proxy works well for a broad range of collision energies. In the right panel we compare our proxy in the case with and without the cuts, and notice that the effect of the cuts is quite modest. This hints at the possibility of directly comparing lattice QCD calculations and experimental measurements for this particular ratio.

In the right panel of Fig. 28.3 we compare preliminary STAR results to our proxy, where we have utilized the same cuts as present in the experimental analysis [12, 13]. We see that a higher chemical freeze-out temperature is preferred, which is in line with previous findings [19, 20]. A direct comparison to lattice QCD results is however premature, since it is essential that the same cuts are applied to all hadronic species for the proxy we constructed to reproduce the ratio $-\chi^{BS}_{11}/\chi^S_2$.

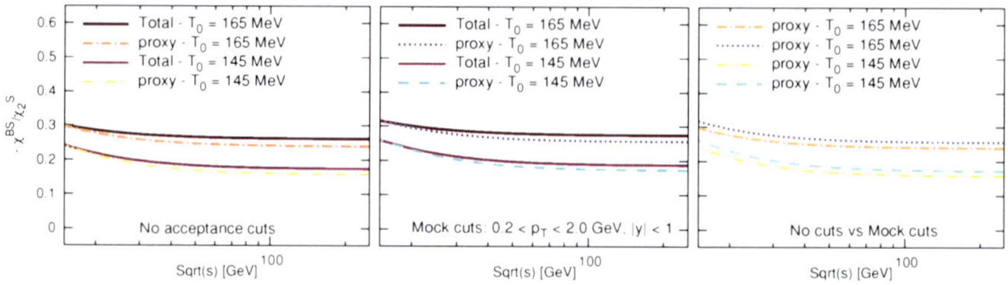

Fig. 28.4 Comparison of the proxy $\widetilde{C}^{\Lambda,\Lambda K}_{BS,SS}$ to the ratio $-\chi^{BS}_{11}/\chi^{S}_{2}$ for finite chemical potential, along the same freeze-out lines as in Fig. 28.3. The left panel shows the comparison without cuts, while the central panel with "exemplary" cuts. In the right panel the proxy is shown both with and without cuts. Figure from [16]

Acknowledgments This project was partly funded by the DFG grant SFB/TR55 and also supported by the Hungarian National Research, Development and Innovation Office, NKFIH grants KKP126769 and K113034. The project also received support from the BMBF grant 05P18PXFCA. Parts of this work were supported by the National Science Foundation under grant no. PHY-1654219 and by the U.S. Department of Energy, Office of Science, Office of Nuclear Physics, within the framework of the Beam Energy Scan Theory (BEST) Topical Collaboration. A.P. is supported by the János Bolyai Research Scholarship of the Hungarian Academy of Sciences and by the ÚNKP-19-4 New National Excellence Program of the Ministry of Innovation and Technology. The authors gratefully acknowledge the Gauss Centre for Supercomputing e.V. (www.gauss-centre.eu) for funding this project by providing computing time on the GCS Supercomputer JUWELS and JURECA/Booster at Jülich Supercomputing Centre (JSC), and on SUPERMUC-NG at LRZ, Munich as well as on HAZELHEN at HLRS Stuttgart, Germany. C.R. also acknowledges the support from the Center of Advanced Computing and Data Systems at the University of Houston. J.N.H. acknowledges the support of the Alfred P. Sloan Foundation, support from the US-DOE Nuclear Science Grant No. de-sc0019175. R.B. acknowledges support from the US DOE Nuclear Physics Grant No. DE-FG02-07ER41521.

References

1. Y. Aoki, G. Endrodi, Z. Fodor, S.D. Katz, K.K. Szabo, Nature **443**, 675 (2006). https://doi.org/10.1038/nature05120
2. S. Borsanyi et al., Wuppertal-budapest collaboration. JHEP **1009**, 073 (2010). https://doi.org/10.1007/JHEP09(2010)073
3. A. Bazavov et al., Phys. Rev. D **85**, 054503 (2012). https://doi.org/10.1103/PhysRevD.85.054503
4. A. Bazavov et al., HotQCD collaboration. Phys. Lett. B **795**, 15 (2019). https://doi.org/10.1016/j.physletb.2019.05.013
5. S. Borsanyi, Z. Fodor, S.D. Katz, S. Krieg, C. Ratti, K. Szabo, JHEP **1201**, 138 (2012). https://doi.org/10.1007/JHEP01(2012)138
6. A. Bazavov et al., HotQCD cllaboration. Phys. Rev. D **86**, 034509 (2012). https://doi.org/10.1103/PhysRevD.86.034509
7. A. Bazavov et al., Phys. Rev. Lett. **109**, 192302 (2012). https://doi.org/10.1103/PhysRevLett.109.192302
8. S. Borsanyi, Z. Fodor, S.D. Katz, S. Krieg, C. Ratti, K.K. Szabo, Phys. Rev. Lett. **111**, 062005 (2013). https://doi.org/10.1103/PhysRevLett.111.062005

9. M. D'Elia, G. Gagliardi, F. Sanfilippo, Phys. Rev. D **95**(9), 094503 (2017). https://doi.org/10.1103/PhysRevD.95.094503
10. L. Adamczyk et al., STAR collaboration. Phys. Rev. Lett. **112**, 032302 (2014). https://doi.org/10.1103/PhysRevLett.112.032302
11. L. Adamczyk et al., STAR collaboration. Phys. Rev. Lett. **113**, 092301 (2014). https://doi.org/10.1103/PhysRevLett.113.092301
12. L. Adamczyk et al., STAR collaboration. Phys. Lett. B **785**, 551 (2018). https://doi.org/10.1016/j.physletb.2018.07.066
13. T. Nonaka [STAR Collaboration], Nucl. Phys. A **982**, 863 (2019). https://doi.org/10.1016/j.nuclphysa.2018.10.092
14. J. Adam et al., [STAR Collaboration], Phys. Rev. C **100**(1), 014902 (2019). https://doi.org/10.1103/PhysRevC.100.014902
15. S. Borsanyi et al., JHEP **1209**, 010 (2012). https://doi.org/10.1007/JHEP09(2012)010
16. R. Bellwied et al.,. arXiv:1910.14592 [hep-lat]
17. P. Alba et al., Phys. Rev. D **96**(3), 034517 (2017). https://doi.org/10.1103/PhysRevD.96.034517
18. P. Alba, V. Mantovani Sarti, J. Noronha, J. Noronha-Hostler, P. Parotto, I. Portillo Vazquez, C. Ratti, Phys. Rev. C **98**(3), 034909 (2018). https://doi.org/10.1103/PhysRevC.98.034909
19. R. Bellwied, J. Noronha-Hostler, P. Parotto, I. Portillo Vazquez, C. Ratti, J.M. Stafford, Phys. Rev. C **99**(3), 034912 (2019). https://doi.org/10.1103/PhysRevC.99.034912
20. M. Bluhm, M. Nahrgang, Eur. Phys. J. C **79**(2), 155 (2019). https://doi.org/10.1140/epjc/s10052-019-6661-3

Chapter 29
Parton Hadron Quantum Molecular Dynamics (PHQMD)—A Novel Microscopic N-Body Transport Approach For Heavy-Ion Dynamics and Hypernuclei Production

Elena Bratkovskaya, Jörg Aichelin, Arnaud Le Fèvre, Viktar Kireyeu, Vadim Kolesnikov, Yvonne Leifels, and Vadim Voronyuk

Abstract We present the novel microscopic n-body dynamical transport approach PHQMD (Parton-Hadron-Quantum-Molecular-Dynamics) for the description of particle production and cluster formation in heavy-ion reactions at relativistic energies. The PHQMD extends the established PHSD (Parton-Hadron-String-Dynamics) transport approach by replacing the mean field by density dependent two body interactions in a similar way as in the Quantum Molecular Dynamics (QMD) models. This allows for the calculation of the time evolution of the n-body Wigner density and therefore for a dynamical description of clusters and hypernuclei formation. The clusters are identified with the MST (Minimum Spanning Tree) or the SACA ('Simulated Annealing Cluster Algorithm') algorithm which—by regrouping the nucleons in single nucleons and noninteracting clusters—generates the most bound configuration of nucleons and clusters. Collisions among particles in PHQMD are treated in the same way as in PHSD. In [1] we presented the first results from the PHQMD for general 'bulk' observables such as rapidity distributions and transverse mass spectra for hadrons as well as for clusters production, including hypernuclei, at SIS and

---

E. Bratkovskaya (✉) · A. Le Fèvre · Y. Leifels
GSI Helmholtzzentrum für Schwerionenforschung GmbH, Planckstr. 1, 64291 Darmstadt, Germany
e-mail: E.Bratkovskaya@gsi.de

E. Bratkovskaya
Institut für Theoretische Physik, Johann Wolfgang Goethe-Universität, Max-von-Laue-Str. 1, 60438 Frankfurt am Main, Germany

J. Aichelin
SUBATECH, Université de Nantes, IMT Atlantique, IN2P3/CNRS 4 rue Alfred Kastler cedex 3, 44307 Nantes, France

Frankfurt Institute for Advanced Studies, Ruth Moufang Str. 1, 60438 Frankfurt, Germany

V. Kireyeu · V. Kolesnikov · V. Voronyuk
Joint Institute for Nuclear Research, Joliot-Curie 6, 141980 Dubna, Moscow region, Russia

D. Elia et al. (eds.), *The XVIII International Conference on Strangeness in Quark Matter (SQM 2019)*, Springer Proceedings in Physics 250,
https://doi.org/10.1007/978-3-030-53448-6_29

FAIR/NICA/BES RHIC energies. The selected results on clusters and hypernuclei production from [1] are discussed in this contribution.

29.1 Introduction

The study of cluster and hypernucleus production, which reflects the phase space density during the expansion phase, is of particular interest from experimental as well as from theoretical side. The production mechanisms of hypernuclei may shed light on the theoretical understanding of the dynamical evolution of heavy-ion reactions which cannot be addressed by other probes. In particular, the formation of heavy projectile/target like hypernuclei elucidates the physics at the transition region between spectator and participant matter. Since hyperons are produced in the overlap region, multiplicity as well as rapidity distributions of hypernuclei formed in the target/projectile region depend crucially on the interactions of the hyperons with the hadronic matter, e.g. cross sections and potentials. On the other hand, midrapidity hypernuclei test the phase space distribution of baryons in the expanding participant matter, especially whether the phase space distributions of strange and non-strange baryons are similar and whether they are in thermal equilibrium.

The description of cluster and hypernuclei formation is a challenging theoretical task which requires (I) the microscopic dynamical description of the time evolution of heavy-ion collisions; (II) the modelling of the mechanisms for the clusters formation.

Cluster formation has often been described either by a coalescence model [2, 3] or statistical methods [4] assuming that during the heavy-ion reaction at least a subsystem achieves thermal equilibration. Both of these models have serious drawbacks. The most essential is that they are not able to address the question of how the clusters are formed and what we can learn from the cluster formation about the reaction dynamics.

29.2 PHQMD: Basic Ideas

In order to overcome the limitations of existing models for the clusters formation, we advance the novel Parton-Hadron-Quantum-Molecular Dynamics (PHQMD) [1] approach which is based on the collision integrals of the Parton-Hadron-String Dynamics approach [5, 6] and density dependent 2-body potential interactions of QMD type models [7, 8]. These clusters can be identified by two methods: either by the minimum spanning tree (MST) procedure [7] or by a cluster finding algorithm based on the simulated annealing technique, the Simulated Annealing Clusterization Algorithm (SACA) [9, 10]. Presently an extended version—the 'Fragment Recognition In General Application' (FRIGA) [11] is under development which includes symmetry and pairing energy as well as hyperon-nucleon interactions. The MST algorithm is based on spatial correlations and it is effective in finding the clusters at the end of the reaction. In order to identify the cluster formation already at early times of the reaction, when the collisions between the nucleons are still on-going and the nuclear density is high, the SACA approach is used. It is based on the idea of Dorso and Randrup [12] that the most bound configuration of nuclei and nucleons evolves in time towards the final cluster distribution.

29.3 Results

First results from combined PHSD/SACA approach have been reported in [13]. Recently we presented the first results from the PHQMD approach on 'bulk' dynamics, covering the energy range from SIS to RHIC, as well as the results on dynamical cluster formation, including hypernuclei, based on the MST and SACA models [1]. In this contribution we show the selected results from [1].

In Fig. 29.1 we display our results for Au+Au at 600 *A*MeV calculated with a hard EoS in comparison with minimum bias ALADIN data [14]. The clusters identified by SACA are stable for times larger than 50 fm/c as shown in Fig. 29.1. One can see clearly that PHQMD with a hard EoS reproduces quite nicely the experimentally observed 'rise and fall'. The rise and fall of the intermediate mass cluster multiplicity depends strongly on the nuclear equation-of-state. As shown in [1], the rise and fall for a soft EoS overpredicts the data at large $Z_{\text{bound }2}$. There in semi-peripheral and peripheral collisions, where $Z_{\text{bound }2}$ is large, the spectator matter is much less stable and fragments into a much larger number of intermediate mass clusters as compared to a hard EoS (Fig. 29.1). Thus, the fragment pattern in semi-peripheral reactions can serve as an additional observable to determine the hadronic EoS experimentally.

A special interest is related to the production of hypernuclei in heavy-ion collisions. In Fig. 29.2 we show the multiplicity of light and heavy hypercluster as a

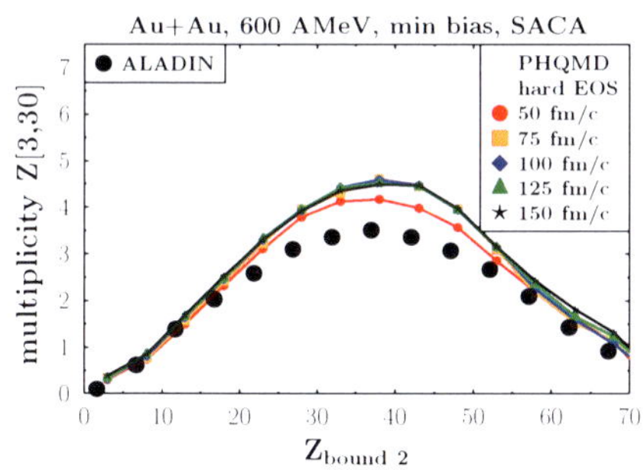

Fig. 29.1 (left) 'Rise and fall' of the multiplicity of clusters with $Z \in [3]$ as a function of the total bound charge Z_{bound}. 2 Both quantities are measured for forward emitted clusters. The experimental data of the ALADIN Collaborations are from [11, 14]. The plot shows the PHQMD results with hard EoS using cluster identification by SACA for 600 *A*GeV at different times—50 (red line), 75(orange line), 100 (blue line), 125 (green line), 150 (black line) fm/c

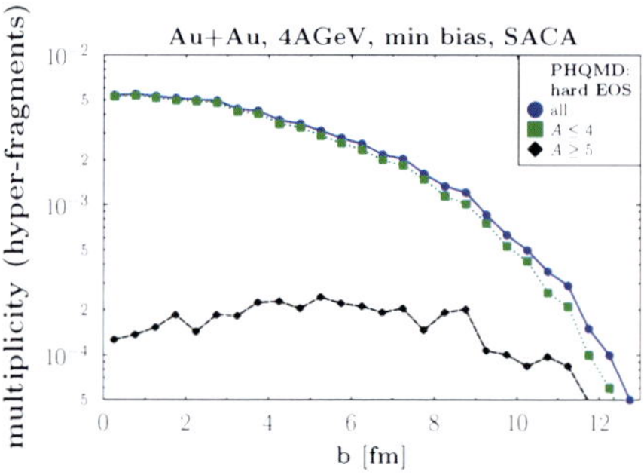

Fig. 29.2 (right) The multiplicity of hyperclusters as a function of the impact parameter for Au+Au collisions at 4 *A*GeV calculated with the PHQMD using the SACA cluster recognition algorithm. The blue dots show the multiplicity of all hyper-nuclei, while the green squares and black rhombus stand for $A \leq 4$ and $A \geq 5$, respectively

function of the impact parameter for Au+Au collisions at 4 *A*GeV. As seen from this figure, the yield of light hyper-clusters decreases with the impact parameter, mainly because the overlap region between projectile and target gets smaller and hence less hyperons are produced. In central collisions, mainly small hypernuclei ($A \leq 4$) are formed while mid-central collisions are better suited for a study of heavier hypernuclei ($A \geq 5$). Hypernuclei with $A \geq 5$ are dominantly produced by hyperons which enter the spectator matter and get caught there. Therefore, for heavy hyper-nuclei production there is a competition between the hyperon production which decreases with impact parameter and the spectator matter whose size increases with impact parameter.

Summarizing, we present the PHQMD transport approach which can be used for the dynamical cluster identification including the hypernuclei production from low to ultrarelativistic energies.

Acknowledgments The authors acknowledge inspiring discussions with C. Blume, Wolfgang Cassing, C. Hartnack, Pierre Moreau, Taesoo Song, Io. Vassiliev and M. Winn. Furthermore, we acknowledge support by the Deutsche Forschungsgemeinschaft (DFG, German Research Foundation), by the Russian Science Foundation grant 19-42-04101 and by the GSI-IN2P3 agreement under contract number 13–70. Also we thank the COST Action THOR, CA15213. The computational resources have been provided by the LOEWE-Center for Scientific Computing and the "Green Cube" at GSI.

References

1. J. Aichelin, E. Bratkovskaya, A. Le Fèvre, V. Kireyeu, V. Kolesnikov, Y. Leifels, V. Voronyuk, arXiv:1907.03860 [nucl-th]
2. L. Zhu, C.M. Ko, X. Yin, Phys. Rev. C **92**, 064911 (2015)
3. Z. Fecková, J. Steinheimer, B. Tomášik, M. Bleicher, Phys. Rev. C **93**, 054906 (2016)

4. A.S. Botvina, K.K. Gudima, J. Steinheimer, M. Bleicher, I.N. Mishustin, Phys. Rev. C **84**, 064904 (2011)
5. W. Cassing, E.L. Bratkovskaya, Nucl. Phys. A **831**, 215 (2009)
6. O. Linnyk, E.L. Bratkovskaya, W. Cassing, Prog. Part. Nuclear Phys. **87**, 50 (2016)
7. J. Aichelin, Phys. Rep. **202**, 233 (1991)
8. R. Marty, J. Aichelin, Phys. Rev. C **87**, 034912 (2013)
9. R.K. Puri, C. Hartnack, J. Aichelin, Phys. Rev. C **54**, R28 (1996)
10. R.K. Puri, J. Aichelin, J. Comput. Phys. **162**, 245 (2000)
11. A. Le Fèvre, J. Aichelin, C. Hartnack, Y. Leifels, Phys. Rev. C **100**, 034904 (2019)
12. C. Dorso, J. Randrup, Phys. Lett. B **301**, 328 (1993)
13. A. Le Fèvre, Y. Leifels, J. Aichelin, C. Hartnack, V. Kireyev, E. Bratkovskaya, J. Phys. Conf. Ser. **668**, 012021 (2016)
14. A. Le Fèvre, Y. Leifels, J. Aichelin, C. Hartnack, V. Kireyev, E. Bratkovskaya, Nuovo Cim. C **39**, 399 (2017)

Chapter 30
ΛK Femtoscopy in Pb–Pb Collisions at $\sqrt{s_{NN}} = 2.76$ TeV Measured with ALICE

Jesse T. Buxton

Abstract The first measurements of the scattering parameters of ΛK pairs in all three charge combinations (ΛK^+, ΛK^-, and ΛK^0_S) are presented. The results are achieved through a femtoscopic analysis of ΛK correlations in Pb–Pb collisions at $\sqrt{s_{NN}} = 2.76$ TeV recorded by ALICE at the LHC. The femtoscopic correlations result from strong final-state interactions, and are fit with a parametrization allowing for both the characterization of the pair emission source and the measurement of the scattering parameters for the particle pairs.

30.1 Introduction

Femtoscopy is an experimental method used to study the space–time characteristics of the particle emitting sources in relativistic particle collisions. With this method, two- (or many-) particle relative-momentum correlation functions are used to connect the final-state momentum distributions to the space–time distributions of particle emission at freeze-out [1]. The correlation functions are sensitive to quantum statistics, as well as strong and Coulomb final-state interactions (FSI). Thus, femtoscopy can offer a unique environment in which to measure nuclear scattering parameters, many of which are difficult, if not impossible, to measure otherwise.

30.2 Data Analysis

This work reports on the analysis of Pb–Pb collisions at $\sqrt{s_{NN}} = 2.76$ TeV produced by the LHC and measured by the ALICE experiment in 2011. Charged particle tracking was performed using the Time Projection Chamber (TPC) and the Inner

On behalf of the ALICE Collaboration.

J. T. Buxton (✉)
The Ohio Statue University, Columbus, OH 43210, USA
e-mail: jesse.thomas.buxton@cern.ch

D. Elia et al. (eds.), *The XVIII International Conference on Strangeness in Quark Matter (SQM 2019)*, Springer Proceedings in Physics 250,
https://doi.org/10.1007/978-3-030-53448-6_30

Tracking System (ITS). Particle identification for reconstructed tracks was carried out using both the TPC and Time-Of-Flight (TOF) detectors in the pseudorapidity range $|\eta| < 0.8$. The purities of the $K^{\pm}$ collections are estimated to be $P_{K^{\pm}} \approx 97\%$ in the range $0.14 < p_T < 1.5$ GeV/c. Electrically neutral Λ ($\overline{\Lambda}$) and K^0_S particles were reconstructed through their weak decays: $\Lambda \rightarrow p\pi^-$ and $K^0_S \rightarrow \pi^+\pi^-$. The Λ and $\overline{\Lambda}$ purities are estimated to be $P_{\Lambda(\overline{\Lambda})} \approx 95\%$ ($p_T > 0.4$ GeV/c), and that of the K^0_S is $P_{K^0_S} \approx 98\%$ ($p_T > 0.2$ GeV/c). When forming particle pairs, a shared daughter restriction is applied to ensure the first particle in the pair is unique from the second. Furthermore, an average separation constraint is imposed to remove splitting and merging effects.

30.3 Analysis Methods

The two-particle correlation function is defined as the ratio of the covariant two-particle and single-particle spectra. In practice, the correlation function is formed experimentally as $C(k^*) = \mathcal{N}\frac{A(k^*)}{B(k^*)}$, where $A(k^*)$ is the signal distribution, $B(k^*)$ is the reference distribution, and $\mathcal{N}$ is a normalization parameter. The signal distribution is the same-event distribution of particle pairs, and the reference distribution is obtained using mixed-event pairs [2], i.e., particles from a given event are paired with those from another event.

Theoretically, the ΛKcorrelation function can be described analytically with a model derived by Lednický and Lyuboshitz [3],

$$
\begin{aligned}
C(k^*)_{\text{Lednický}} = 1 + \sum_S \rho_S \Bigg[& \frac{1}{2}\left|\frac{f^S(k^*)}{R_{\text{inv}}}\right|^2 \left(1 - \frac{d_0^S}{2\sqrt{\pi}R_{\text{inv}}}\right) \\
& + \frac{2\Re f^S(k^*)}{\sqrt{\pi}R_{\text{inv}}} F_1(2k^*R_{\text{inv}}) - \frac{\Im f^S(k^*)}{R_{\text{inv}}} F_2(2k^*R_{\text{inv}}) \Bigg],
\end{aligned}
\tag{30.1}
$$

where $f(k^*)$ is the complex scattering amplitude, F_1 and F_2 are analytic functions, and R_{inv} is the radius of the spherically symmetric Gaussian distribution assumed for the pair emission source in the pair rest frame. The complex scattering amplitude is evaluated via the effective range approximation, $f(k^*) = \left(\frac{1}{f_0} + \frac{1}{2}d_0 k^{*2} - ik^*\right)^{-1}$, where f_0 is the complex s-wave scattering length and d_0 is the effective range of the interaction.

Residual correlations The finally measured correlation function is a combination of the genuine ΛKcorrelation with contributions from both impurities and residual correlations resulting from resonance feed-down [4],

$$
C_{\text{measured}}(k^*_{\Lambda K}) = 1 + \lambda'_{\Lambda K}[C_{\Lambda K}(k^*_{\Lambda K}) - 1] + \sum_{ij} \lambda'_{ij}[C_{ij}(k^*_{\Lambda K}) - 1],
\tag{30.2}
$$

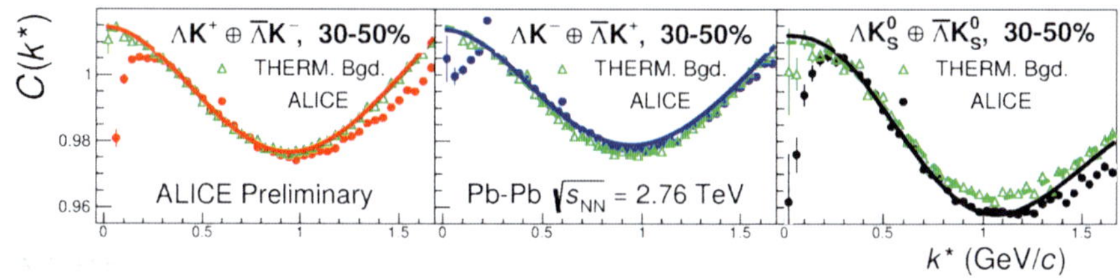

Fig. 30.1 THERMINATOR 2 simulation (open triangles) with experimental data (closed circles) for the 30–50% centrality range. Results are shown for ΛK^+ (left), ΛK^- (middle), and ΛK^0_S (right). A 6th-order polynomial fit to the simulation is shown as a dashed curve. This polynomial scaled to match the experimental data is drawn as a solid curve

where $\lambda'_{ij} = \lambda_{Fit}\lambda_{ij}$, the ΛK term represents the genuine ΛK correlation, and the ij terms denote the contributions from impurities and residual correlations. The λ_{ij} parameters serve as weights dictating the relative strength of each component's contribution to the observed signal, and are estimated using the THERMINATOR 2 and HIJING simulations [4, 5]. The net correlation signal from the small subset of pairs containing an impurity (i.e., pairs with at least one misidentified member) is assumed to average to unity. The main sources of residual correlations in the ΛK systems result from Λ hyperons which have decayed from Σ^0, Ξ^0, and Ξ^- parents. When modeling the $\Sigma^0 K$ and $\Xi^0 K$ systems, for which the interactions are not known, the source radii and scattering parameters are assumed equal to those of the daughter ΛK system. For modeling the $\Xi^- K^\pm$ contribution, the available experimental $\Xi^- K^\pm$ data are used. Each residual component, $C_{ij}(k^*_{\Lambda K})$ in (30.2), is the parent correlation function expressed in terms of the relative momentum of the daughter ΛKpair, and is obtained using a transform matrix generated with the THERMINATOR 2 [6] simulation (see [4] for more details).

Non-femtoscopic background A significant non-femtoscopic background is observed in all studied ΛK correlations, which is primarily due to particle collimation associated with elliptic flow, and results from mixing events with unlike event planes [7]. Each background is modeled using a 6th-order polynomial fit to the THERMINATOR 2 simulation, as shown in Fig. 30.1, which is then applied as a scale factor in the final fit function.

30.4 Results

Figure 30.2 shows experimental ΛK correlation functions with fits for the 0–10% centrality percentile interval. All six ΛK systems are fit simultaneously across all centralities, with a single radius and λ_{Fit} parameter for each centrality interval. Scattering parameters ($\Re f_0$, $\Im f_0$, d_0) are shared between pair-conjugate systems, but they are assumed unique among the different ΛK charge combinations. Figure 30.3 sum-

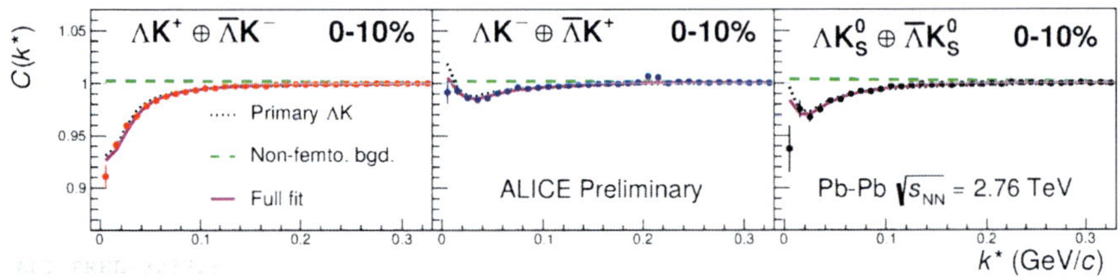

Fig. 30.2 Fit results for the ΛK data in the 0–10% centrality range; $\Lambda K^+ \oplus \bar{\Lambda} K^-$ are shown in the left column, $\Lambda K^- \oplus \bar{\Lambda} K^+$ in the middle, and $\Lambda K^0_S \oplus \bar{\Lambda} K^0_S$ in the right. The curves show the primary ΛK contribution to the fit, i.e., $1 + \lambda'_{\Lambda K} C_{\Lambda K}$ in (30.2) (dotted), the fit to the non-femtoscopic background (dashed), and final fit (solid)

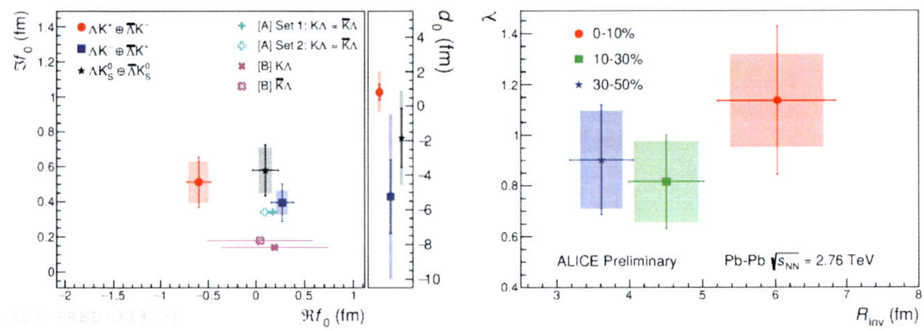

Fig. 30.3 Extracted fit parameters for all of the ΛK systems. The cross ([A] = [8]) and X (B = [9]) points show theoretical predictions made using chiral perturbation theory

marizes the extracted ΛK fit parameters. For all ΛK systems, positive imaginary parts of the scattering lengths, $\Im(f_0)$, describing the inelastic scattering channels, are extracted. More interestingly, the results show that the ΛK^+ and ΛK^- systems differ in the sign of the real part, $\Re(f_0)$, of their scattering lengths, with a negative value for ΛK^+ and positive value for ΛK^- (and ΛK^0_S). The real part of the scattering length describes the effect of the strong interaction, with a positive value signifying an attraction and a negative value signifying a repulsion. Therefore, the results from this analysis indicate that the strong force is repulsive in the ΛK^+ interaction and attractive in the ΛK^- and ΛK^0_S interactions.

30.5 Summary

Results from a femtoscopic analysis of ΛK correlations in Pb–Pb collisions at $\sqrt{s_{NN}}$ = 2.76 TeV measured by the ALICE experiment at the LHC have been presented. The femtoscopic radii, λ parameters, and scattering parameters were extracted from one-dimensional correlation functions in terms of the invariant momentum difference. Striking differences are observed in the ΛK^+, ΛK^-, and ΛK^0_S correlation functions, and the extracted scattering parameters indicate that the strong force is repulsive in the ΛK^+ interaction and attractive in the ΛK^- and ΛK^0_S interactions. This effect could be due to different quark–antiquark interactions between the pairs, or from different net strangeness for each system.

References

1. M.A. Lisa, S. Pratt, R. Soltz, U. Wiedemann, Ann. Rev. Nucl. Part. Sci. **55**, (2005)
2. G.I. Kopylov, Phys. Lett. B **50**, 472 (1974)
3. R. Lednický, V.L. Lyuboshitz, Sov. J. Nuclear Phys. **35**, 770 (1982)
4. A. Kisiel, H. Zbroszczyk, M. Szymański, Phys. Rev. C **89**(5), 054916 (2014)
5. S. Acharya et al., Phys. Rev. C **99**(2), 024001 (2019)
6. M. Chojnacki, A. Kisiel, W. Florkowski, W. Broniowski, Comput. Phys. Commun. **183**, 746 (2012)
7. A. Kisiel, Acta Physica Polonica B **48**, 717 (2017)
8. Y.R. Liu, S.L. Zhu, Phys. Rev. D **75**, 034003 (2007)
9. M. Mai, P.C. Bruns, B. Kubis, U.G. Meissner, Phys. Rev. D **80**, 094006 (2009)

Chapter 31
Geometry and Dynamics of Particle Production Seen by Femtoscopic Probes in the STAR Experiment

Paweł Szymański

Abstract The main goal of studying heavy-ion collisions is to understand the properties of the matter under extreme conditions. The spatial and temporal characteristics of particle emission can be extracted using the femtoscopy technique. From non-identical particle correlations one can obtain information about asymmetry in emission process between two kind of particles. Such asymmetry can provide information about which type of particles are emitted earlier and/or closer to the center of the source. In these proceedings, results on femtoscopic observables of pion, kaon and proton non-identical particle combinations and transverse mass dependence of three-dimensional femtoscopic observables for charged kaons in Au+Au collisions at energy $\sqrt{s_{NN}} = 39$ GeV will be reported.

31.1 Introduction

The Solenoidal Tracker at RHIC (STAR) is a large detector system at Relativistic Heavy Ion Collider (RHIC) [1]. A comprehensive program called Beam Energy Scan (BES) designed to study the phase diagram of nuclear matter, has been launched at RHIC. It uses gold ion collisions at collision energy ($\sqrt{s_{NN}}$) in the range from 7.7 up to 200 GeV. The program studies the phase transition signatures and it is aimed at finding a localization of the critical point between cross over and the first-order phase transitions [2].

Pawel Szymanski for the STAR Collaboration.

For the STAR Collaboration.

P. Szymański (✉)
Faculty of Physics, Warsaw University of Technology,
Koszykowa 75, 00-662 Warsaw, Poland
e-mail: pawel.szymanski.dokt@pw.edu.pl

D. Elia et al. (eds.), *The XVIII International Conference on Strangeness in Quark Matter (SQM 2019)*, Springer Proceedings in Physics 250,
https://doi.org/10.1007/978-3-030-53448-6_31

31.2 Femtoscopy

Femtoscopy is the only tool to perform studies of the particle-emitting source which has a size and life time of the order 10^{-15} m and 10^{-23} s, respectively. Through two particle correlations at low relative momentum, which measure their statistical effects and interactions between them in the final state, one can study space-time characteristics of the source.

The correlation function ($C(\mathbf{q})$), which is used in femtoscopy, is defined as a ratio of probability of observing two particles with momenta $\mathbf{p_1}$ and $\mathbf{p_2}$ to a product of probabilities of observing such particles independently. This function is determined by a pair wave function ($\psi(\mathbf{q}, \mathbf{r})$), which includes information about quantum-statistical effects and interactions, and emission function ($S(\mathbf{q}, \mathbf{r})$), which contains space-time information about the source:

$$C(\mathbf{q}) = \int d^3 S(\mathbf{q}, \mathbf{r}) |\psi(\mathbf{q}, \mathbf{r})|^2 \qquad (31.1)$$

where $\mathbf{q}$ is a difference between momenta $\mathbf{p_1}$ and $\mathbf{p_2}$, and $\mathbf{r}$ is a difference between the position of the first and second particle in the pair, respectively.

Kaon femtoscopy provides complementary information to pions. Strange particles provide cleaner signal as compared to pions, because they are less affected by resonance decays. Such correlation depends on quantum-statistical effects and final state interactions, which are Coulomb and strong forces. In case of identical kaon pairs strong interaction is assumed to be negligible [3].

Non-identical particle femtoscopy is a useful tool to study geometry and dynamics of particle production, like measurements of asymmetries in emission process [4].

31.3 Identical Kaon Femtoscopy

Geometrical source characteristics are determined through fitting procedure using Bowler-Sinyukov approach [5, 6]:

$$C(q_o, q_s, q_l) = 1 - \lambda + \lambda K(q_{inv})(1 + exp[-R_o^2 q_o^2 - R_s^2 q_s^2 - R_l^2 q_l^2]) \qquad (31.2)$$

where λ is the correlation strength, $R_{o,s,l}$ are the radii of the particle-emitting source in out, side and long directions, respectively (using Bertsch-Pratt parametrization [7, 8]), and $K(q_{inv})$ is the Coulomb factor.

Figure 31.1 shows the transverse mass ($m_T = \sqrt{k_T^2 + m^2}$, $k_T = |\mathbf{p_{T,1}} + \mathbf{p_{T,2}}|/2$) dependence of the extracted femtoscopic radii for pions and charged kaons measured in Au+Au collision at $\sqrt{s_{NN}} = 39$ GeV. Obtained sizes are smaller for peripheral collisions. The radii decrease with increasing m_T. The radii in *long* and *out* directions

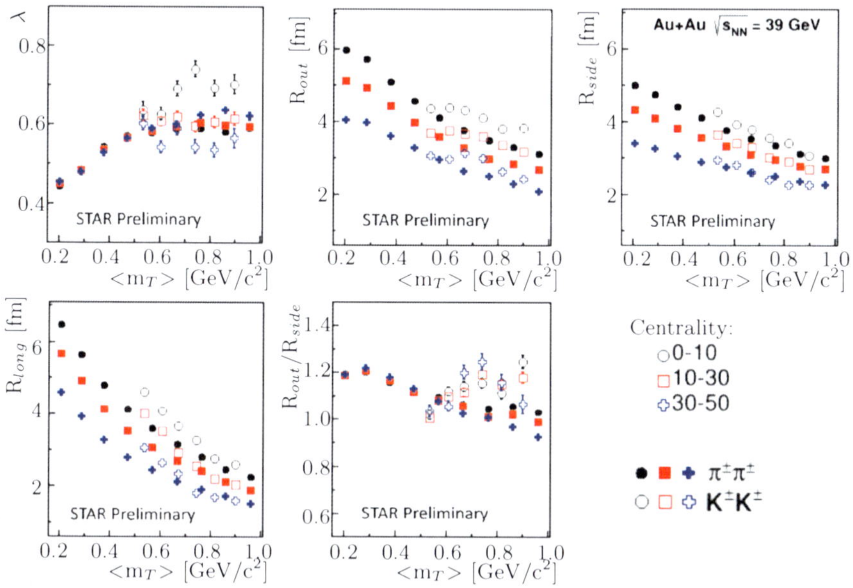

Fig. 31.1 Transverse mass dependence of sizes of kaon and pion source at $\sqrt{s_{NN}} = 39$ GeV. Only stat. uncertainties

are larger for kaons than for pions at the same transverse mass, that indicate breaking of m_T-scaling.

31.4 Non-identical Particle Femtoscopy

The correlation function, $C(k^*)$, can be represented as the decomposition into spherical harmonic components [9]:

$$C(k^*) = \sum_{l,m} C_l^m(q)Y_l^m(\theta, \phi), \qquad C_l^m(q) = \int_{\Omega} C(k^*, \theta, \phi)Y_l^m(\theta, \phi)d\Omega \quad (31.3)$$

where k^* is the particle momentum in Pair Rest Frame, θ and ϕ are polar and azimuthal angles, respectively. The C_0^0 component is sensitive to the system size and C_1^1 is sensitive to the emission asymmetry.

Figure 31.2a presents C_0^0 components for different particle system with like-sign particle combinations for 0–10% of the most central Au+Au collisions at $\sqrt{s_{NN}}$ = 39 GeV. These correlations are dominated by the Coulomb interaction. Interactions between unlike-sign particle combinations (Fig. 31.2b) are more complicated. For pion-proton unlike-sign pairs there is a visible peak around $k^* = 0.1$ GeV/c that

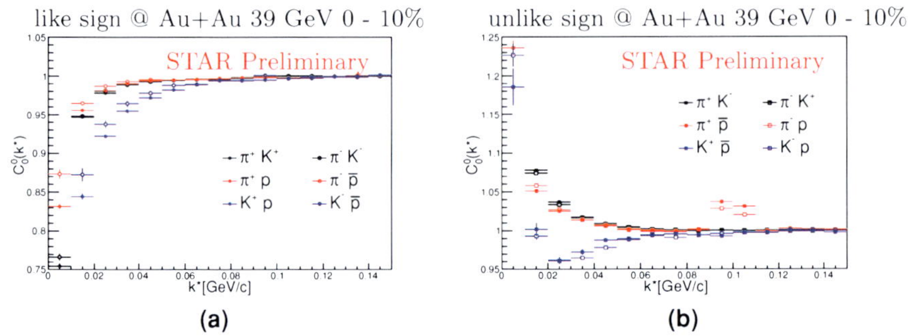

Fig. 31.2 Spherical harmonics C_0^0 components for like-sign pairs **a** and unlike-sign pairs **b** for different particle combinations. Only stat. uncertainties

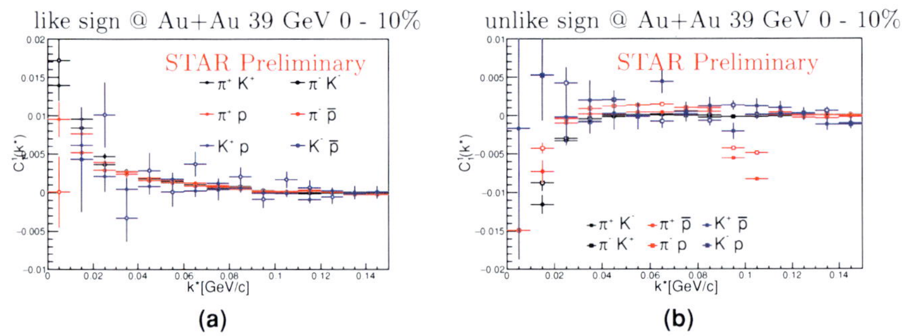

Fig. 31.3 Spherical harmonics C_1^1 components for like-sign pairs **a** and unlike-sign pairs **b** for different particle combinations. Only stat. uncertainties

corresponds to the Λ hyperon decay. The shape of the correlation functions of kaon-proton pairs is determined by the non-negligible contribution from strong interaction. Figure 31.3a, b show that asymmetries in the emission process existed for each analysed pair combination.

31.5 Summary

For Au+Au collisions at $\sqrt{s_{NN}} = 39$ GeV pion and kaon source sizes seem to follow different transverse mass dependence, which indicate breaking of the transverse mass scaling. There is a different shape of correlation functions for various non-identical particle combinations. In case of Kp the strong interaction is not negligible and requires further investigation.

An emission asymmetry is observed for particle combinations with different masses produced in Au+Au collisions at $\sqrt{s_{NN}} = 39$ GeV. The clear asymmetry signal implies collective (and dynamical) effects. Shape of both components of spherical harmonics (C_0^0 and C_1^1) suggest that lighter particles are emitted closer to the center of the source and/or later than heavier particles.

Acknowledgments This work was supported by the grant: UMO-2017/27/B/ST2/01947 supported by the National Science Centre, Poland.

References

1. K. Ackermann et al., STAR detector overview. Nucl. Instr. Meth. A **499**, 624–632 (2003). https://doi.org/10.1016/S0168-9002(02)01960-5
2. G. Odyniec, The RHIC Beam Energy Scan program in STAR and what's next. J. Phys.: Conf. Ser. **455**, 012037 (2013). https://doi.org/10.1088/1742-6596/455/1/012037
3. R. Lednicky, Correlation femtoscopy of multiparticle processes. Phys. Atom. Nuclear **67**, 72 (2004). https://doi.org/10.1134/1.1644010
4. R. Lednicky et al., How to measure which sort of particles was emitted earlier and which later. Phys. Lett. B **373**, 30–34 (1996). https://doi.org/10.1016/0370-2693(96)00124-4
5. M. Bowler, Coulomb corrections to Bose-Einstein correlations have been greatly exaggerated. Phys. Lett. B **270**, 69–74 (1991). https://doi.org/10.1016/0370-2693(91)91541-3
6. Y. Sinyukov et al., Coulomb corrections for interferometry analysis of expanding hadron systems. Phys. Lett. B **432**, 248–257 (1998). https://doi.org/10.1016/S0370-2693(98)00653-4
7. S. Pratt, Pion interferometry of quark-gluon plasma. Phys. Rev. D **33**, 1314 (1986). https://doi.org/10.1103/PhysRevD.33.1314
8. G. Bertsch et al., Pion interferometry in ultrarelativistic heavy-ion collisions. Phys. Rev. C **37**, 1896 (1988). https://doi.org/10.1103/PhysRevC.37.1896
9. P. Danielewicz, S. Pratt, Analysis of low-momentum correlations with Cartesian harmonics. Phys. Lett. B **618**, 60 (2005). https://doi.org/10.1016/j.physletb.2005.05.019

Chapter 32
The Spatial Sub-separation of Strangeness from Antistrangeness in Heavy-Ion Collisions at Energies of FAIR and NICA

Larisa Bravina, Oleksandra Panova, Oleksandr Vitiuk, Evgeny Zabrodin, and Horst Stöcker

Abstract The heavy-ion collisions at energies of FAIR and NICA are studied within the microscopic transport model. The whole interaction area is subdivided into the small cells. We perform the analysis of the space-time evolution of all particles in all cells, in the $T - \mu_B$ and $T - \mu_S$ planes, and the analysis of the finally emitted strange and non-strange particles in the $x - t$ plane. Following the time evolution of all distributions, we clearly see the spacial separation of strangeness from anti-strangeness, as well as earlier freeze-out times of kaons and pions compared to those of protons and Lambdas. The latter appear to be frozen out at lower temperature and larger strangeness chemical potential.

32.1 Introduction

Experiments with heavy-ion collisions at center-of-mass energies $3.5 \le \sqrt{s} \le 11$ GeV accessible for Beam Energy Scan at RHIC and for coming facilities FAIR (GSI) and NICA (JINR), aim to study properties of hot and dense nuclear matter at temperatures above 100 MeV and baryon chemical potentials about 400–500 MeV.

L. Bravina (✉) · O. Panova · O. Vitiuk · E. Zabrodin
Department of Physics, University of Oslo, Oslo, Norway
e-mail: larissa.bravina@fys.uio.no

O. Panova · O. Vitiuk
Taras Shevchenko National University of Kyiv, Kyiv, Ukraine

E. Zabrodin
Skobeltzyn Institute for Nuclear Physics, Moscow State University, Moscow, Russia

H. Stöcker
Frankfurt Institute for Advanced Studies, Frankfurt a.M., Germany

Institute for Theoretical Physics, J.W. Goethe University, Frankfurt a.M., Germany

D. Elia et al. (eds.), *The XVIII International Conference on Strangeness in Quark Matter (SQM 2019)*, Springer Proceedings in Physics 250,
https://doi.org/10.1007/978-3-030-53448-6_32

The measured final state distributions are compared to calculations of various models in order to recognize new phenomena, most likely associated with the transition to new state of matter, quark-gluon plasma (QGP). These models can be roughly subdivided into microscopic transport models and macroscopic models, which include hydrodynamic and thermal statistical models. Among the differences between the microscopic and macroscopic approaches is the treatment of particle freeze-out. Whereas the macroscopic models employ the ad hoc assumption of sudden freeze-out of hadrons, microscopic models allow the particles to leave the interaction area from the very beginning of the collision. Instead of sharp simultaneous freeze-out of all hadrons microscopic models favor continuous sequential particle freeze-out [1, 2]. But the hot fireball rapidly expands, therefore, temperatures of hadrons emitted at different times should be different. To check the consequences of such scenario for heavy-ion collisions at energies of FAIR and NICA we employ the UrQMD model [3, 4].

32.2 Space-Time Freeze-Out of (Non) Strange Hadrons

Our strategy is as follows. We run about $5 \cdot 10^4$ central Au + Au collisions at each of four energies, $E_{lab} = 10, 20, 30$, and 40 AGeV. The whole volume of the fireball is subdivided into small cubic cells with volume 3 fm^3. The snapshots of the system evolution are made with the time step $\Delta t = 1$ fm/c. After determination of energy density ε, net baryon density ρ_B, and net strangeness density ρ_S, we fit the data to statistical model (SM) of ideal hadron gas to find temperature T, baryon chemical potential μ_B and strangeness chemical potential μ_S of the cell (for details see, e.g. [5, 6]). Then, every particle from the final spectrum of hadrons is traced back to its last interaction point, elastic or inelastic. Knowing the space-time coordinates of the area from which the hadron was emitted, we get both the space-time freeze-out picture of hadron species and their freeze-out distributions in the $T - \mu_B$ plane. Figure 32.1 shows that main part of kaons is emitted within $2 \leq t \leq 10$ fm/c, whereas for Λ this period lasts from $t \approx 8$ fm/c to $t \approx 20$ fm/c. Temperature and baryochemical

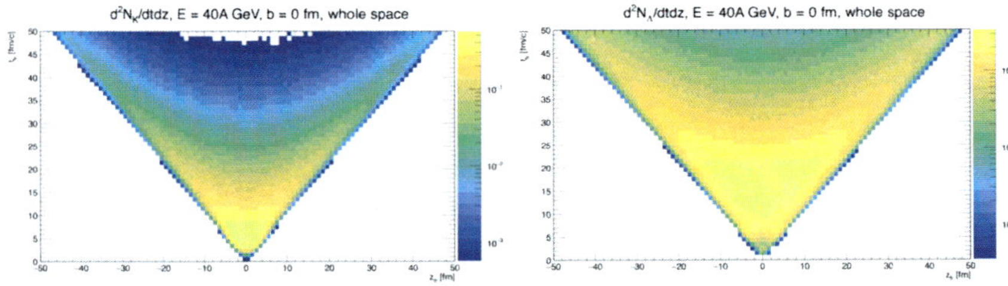

Fig. 32.1 $d^2N/dtdz$ distribution of kaons (left) and Lambdas (right) produced in UrQMD calculations of central Au + Au collisions at $E_{lab} = 40$ AGeV

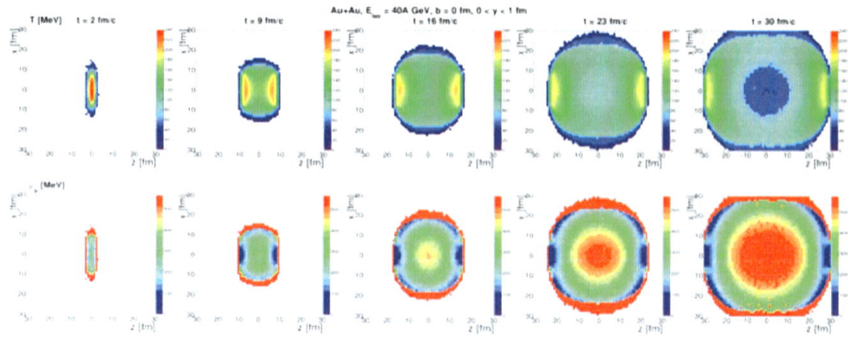

Fig. 32.2 Time evolution of temperature (upper row) and baryon chemical potential (bottom row) in the central layer $0 \le y \le 1$ fm of central Au + Au collisions at $E_{lab} = 40$ AGeV

potential are also distributed non-uniformly even within thin layer $0 \le y \le 1$ fm, as displayed in Fig. 32.2. The hottest areas are the remnants of colliding nuclei, while the central area has the largest μ_B. Therefore, we restrict our consideration to central area with volume $V = 125$ fm^3. It looks homogeneous w.r.t. T and μ_B after $t \ge 8$ fm/c. Time evolution of T and μ_B in the central cell in Au + Au collisions at all 4 energies is depicted in Fig. 32.3. Here average freeze-out times of π, N, $\bar{N}$, K, $\bar{K}$, Λ, $\bar{\Lambda}$, Σ, and $\bar{\Sigma}$ are plotted also. We see that mesons are frozen earlier than baryons. Antikaons are frozen earlier than kaons, and $\bar{\Lambda}$ are frozen earlier than Λ. The sequential freeze-out of hadrons (i) influences directed and elliptic flow of these particles [7] and (ii) helps to explain the difference in Λ and $\bar{\Lambda}$ polarization at intermediate energies [8, 9].

32.3 Conclusions

The following conclusions can be drawn from our study. (1) Microscopic models favor continuous freeze-out of hadrons. Each hadron specie i has its own average T^i_{FO} and μ^i_{FO}, which complicates the comparison with the standard statistical model of ideal hadron gas. (2) The sequential chemical freeze-out of hadrons influences development of their anisotropic flow, particularly, directed flow. (3) Different freeze-out conditions both in time and in space for Λ and $\bar{\Lambda}$ can explain the experimentally observed difference in polarization of both hyperons.

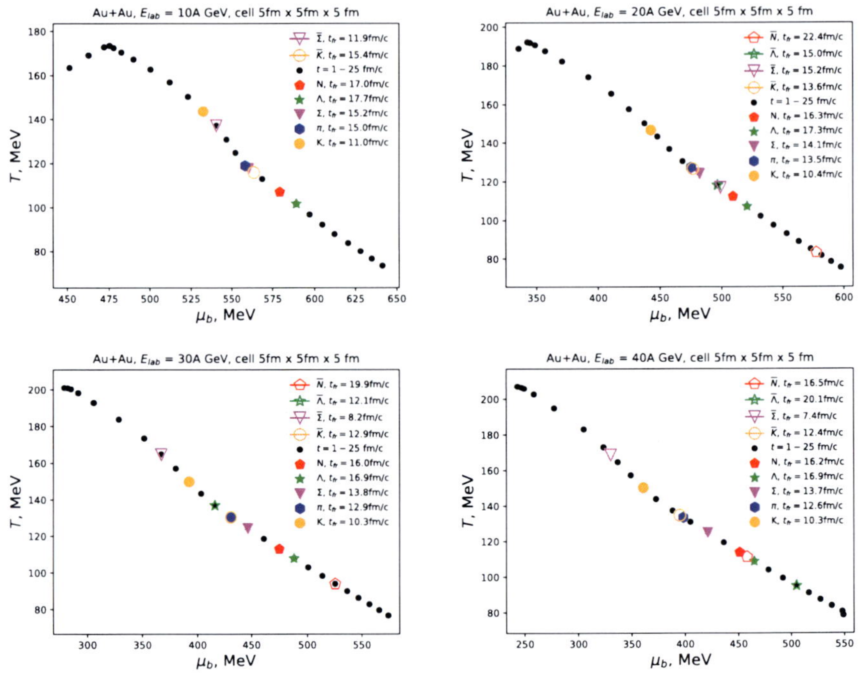

Fig. 32.3 Evolution of T versus μ_B and average freeze-out temperatures and baryon chemical potentials of different hadrons in central cell with $V = 125$ fm^3 in central Au + Au collisions at $E_{lab} = 10, 20, 30$, and 40 AGeV

Acknowledgments The work of L. B. and E. Z. was supported by Russian Foundation for Basic Research (RFBR) under Grants No. 18-02-40084 and No. 18-02-40085, and by the Norwegian Research Council (NFR) under Grant No. 255253/F50. O. P. and O. V. acknowledge the financial support of the Norwegian Centre for International Cooperation in Education (SIU) under Grant CPEA-LT-2016/10094. Computer calculations were made at Abel (UiO, Oslo) and Govorun (JINR, Dubna) computer cluster facilities.

References

1. S.A. Bass et al., Hadronic freezeout following a first order hadronization phase transition in ultrarelativistic heavy ion collisions. Phys. Rev. **C60**, 021902 (1999). https://doi.org/10.1103/PhysRevC.60.021902
2. L.V. Bravina et al., Microscopic study of freezeout in relativistic heavy ion collisions at SPS energies. Phys. Rev. **C60**, 044905 (1999). https://doi.org/10.1103/PhysRevC.60.044905
3. S.A. Bass et al., Microscopic models for ultrarelativistic heavy ion collisions. Prog. Part. Nuclear Phys. **41**, 255 (1998). https://doi.org/10.1016/S0146-6410(98)00058-1
4. M. Bleicher et al., Relativistic hadron-hadron collisions in the ultra-relativistic quantum molecular dynamics model. J. Phys. **G25**, 1859 (1999). https://doi.org/10.1088/0954-3899/25/9/308

5. L.V. Bravina et al., Local equilibrium in heavy ion collisions: microscopic model versus statistical model analysis. Phys. Rev. **C60**, 024904 (1999). https://doi.org/10.1103/PhysRevC.60.024904
6. L.V. Bravina et al., Microscopic models and effective equation of state in nuclear collisions in the vicinity of $E_lab = 30A$ GeV at the GSI Facility for Antiproton and Ion Research (FAIR) and beyond. Phys. Rev. **C78**, 014907 (2008). https://doi.org/10.1103/PhysRevC.78.014907
7. L. Bravina et al., Directed flow in microscopic models in relativistic A+A collisions. Universe **5**, 69 (2019). https://doi.org/10.3390/universe5030069
8. O. Vitiuk, L. Bravina, E. Zabrodin, Different space-time freeze-out picture—an explanation of different Λ and $\bar{\Lambda}$ polarization? eprint: arXiv:1910.06292 [hep-ph]
9. O. Vitiuk et al., Vorticity and $\Lambda/\bar{\Lambda}$ polarization in heavy-ion collisions at FAIR and NICA energies, in *These Proceedings*

Chapter 33
Strangeness Flow in Au + Au Collisions at 1.23 AGeV Measured with HADES

Lukáš Chlad

Abstract We present preliminary results on anisotropic flow of particles with strangeness content (K_s^0 and K^+) in Au + Au collisions at $\sqrt{s_{NN}} = 2.42$ GeV measured with HADES. The strange particle flow in heavy-ion collisions is a good probe for the nuclear equation-of-state. Kaon flow was rarely measured at such low centre-of-mass energies, due to the sub-threshold production of strangeness. Thanks to the large statistics of 2.6 billion 40% most central collisions, this study is now possible. The obtained flow parameters (differential measurement of directed and elliptic flow w.r.t. centre-of-mass rapidity and transverse momentum) are compared with previously published world data as well as with flow of non-strange particles.

33.1 Introduction

In order to understand the extreme state of nuclear matter that exists for example in neutron stars, kaon properties in dense nuclear matter are of particular interest [1]. In simulations [2] it was shown that kaon flow can differentiate between various applied potentials thanks to the large mean free path of kaons in matter. Using the SIS18 accelerator at GSI, several measurements have been done by the KaoS and FOPI collaborations [3, 4] resulting in an almost negligible directed kaon flow. At around the same time it was shown in [5] that zero transverse flow of nucleons can be caused by the integration of nucleons over a wide range of transverse momentum, due to the fact that nucleons with high transverse momentum have positive flow and those with low p_T have negative flow. A similar effect can be observed for kaons.

Lukáš Chlad for the HADES collaboration.

L. Chlad (✉)
Nuclear Physics Institute CAS, Řež 130, 25068 Řež Husinec, Czech Republic
e-mail: chlad@ujf.cas.cz

Faculty of Mathematics and Physics, Charles University,
Ke Karlovu 2027/3, 12116 Prague 2, Prague, Czech Republic

D. Elia et al. (eds.), *The XVIII International Conference on Strangeness in Quark Matter (SQM 2019)*, Springer Proceedings in Physics 250,
https://doi.org/10.1007/978-3-030-53448-6_33

33.2 HADES Au + Au Experiment

The analysis presented in this contribution is based on a data set collected by the High-Acceptance Dielectron Spectrometer (HADES) in 2012, measuring Au + Au collisions with $\sqrt{s_{NN}} = 2.42$ GeV. The HADES detector is located at the GSI Helmholtzzentrum für Schwerionenforschung near Darmstadt, Germany. It is a multi-purpose, fixed target, charged particle detector providing the possibility to measure both elementary and heavy-ion collisions thanks to the broad range of projectiles accelerated by the SIS18. The spectrometer consists of six superconducting coils which divide the detector into six identical sectors covering almost the full azimuthal angle and a polar angle interval of 18–85°. Each sector contains a Ring Imaging Cherenkov detector (RICH) for electron identification, four planes of Mini-Drift Chambers (MDC) for momentum determination, scintillator based Time-Of-Flight (TOF) and Resistive Plate Chamber (RPC) detectors for time measurement. In addition, there is a diamond START detector in front of the target for trigger purposes. Behind the spectrometer the Forward Wall scintillator hodoscope is located, which is used to reconstruct the reaction plane, for more details about the detector see [6]. During the 2012 campaign we collected billions of events thanks to very fast data acquisition system (DAQ), from which 2.6×10^9 events with 0–40% centrality were selected for analysis. One can find an overview of existing results in [7].

33.3 Particle Identification and Flow Analysis

The analysis of kaon flow presented here consists of several steps: identification of K^0_s and K^+, reaction plane reconstruction and event centrality determination. The identification of charged kaons is based on a combined momentum and velocity measurement with additional criteria on the specific energy loss measurement in the MDCs, see [8]. Neutral kaons are detected based on the reconstruction of the weak decay $K^0_s \rightarrow \pi^+ + \pi^-$ with a branching ratio of 69.2%. To extract the yield of neutral kaons topological cuts as well as a mixed-event technique to remove the combinatorial background are needed, for more details see [9]. Using the comparison of simulations with the Glauber model and recorded data it is possible to estimate the centrality of each event as described in [10]. Finally, the reaction plane reconstruction and the determination of its resolution follow the procedures given in [11]. To obtain a differential measurement of the flow of kaons using the event plane method we need to know the number of kaons for each centrality, rapidity, transverse momentum interval and for several bins in $\Delta\phi = \psi_{RP} - \phi_K$, where ψ_{RP} denotes the azimuthal angle of reaction plane and ϕ_K is azimuthal angle of a given kaon. Angular distributions of kaons are then fitted using the first two terms of Fourier expansion $\frac{dN}{d\Delta\phi} \propto \frac{1}{2\pi}\left(1 + 2v_1\cos(\Delta\phi) + 2v_2\cos(2\Delta\phi)\right)$, where the coefficients v_1 and v_2 have to be corrected for the reaction plane resolution in order to obtain the correct flow parameters.

33.4 Results

Our preliminary results of the first differential kaon flow analysis at sub-threshold energy show that there is no significant difference between neutral and positively charged kaons which might be expected since both contain anti-strange quark, see Fig. 33.1b. From Fig. 33.1a one can see that there is no strong dependence of the directed flow coefficients on the centrality of collisions, but there is a visible dependence of v_1 on the transverse momentum, see Fig. 33.1c. On the other hand, the elliptic flow coefficients are increasing towards more central events (Fig. 33.1b). Finally, v_2 is approximately constant as a function of rapidity and with increasing transverse momentum it is decreasing towards more negative values as displayed in Fig. 33.1d. For the average value of $p_T \approx 400\,\text{MeV}/c$ determined by our acceptance $v_2 \approx -0.1$, i.e. kaons tend to be squeezed out of the dense nuclear matter.

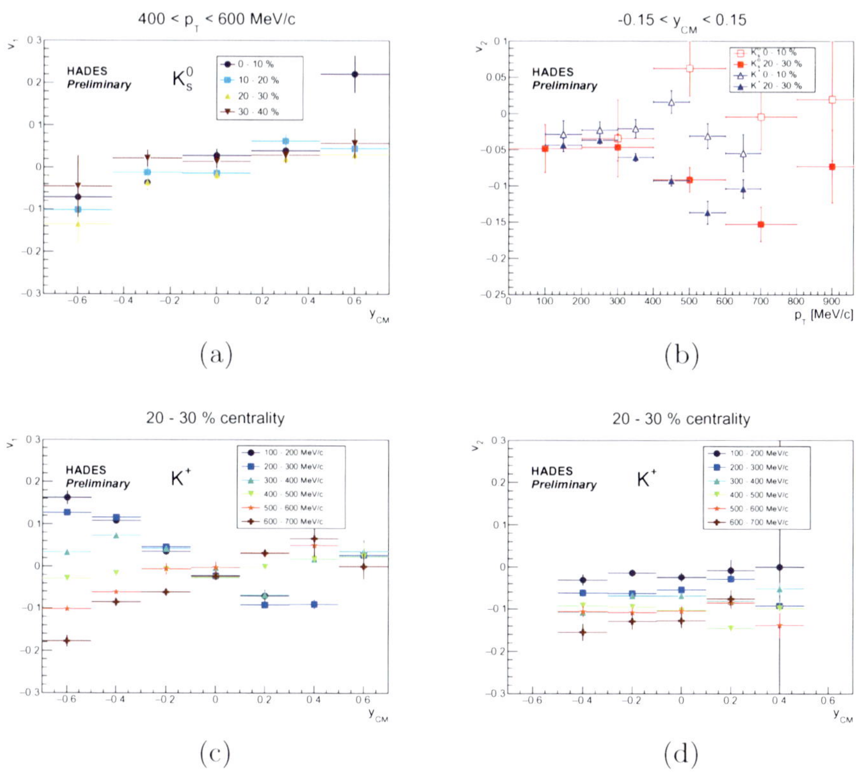

Fig. 33.1 **a** Directed flow for K^0_S in Au + Au collisions at different centralities at 1.23 AGeV as a function of the centre-of-mass rapidity y_{CM}. **b** Elliptic flow as a function of transverse momentum p_{T} comparing K^0_S and K^+. **c** Directed flow for K^+ in semi-central collisions as a function of the centre-of-mass rapidity y_{CM} for 100 MeV/c intervals of transverse momentum. **d** The same for elliptic flow. On all pictures only statistical errors are shown

33.5 Summary and Outlook

In the previous section we presented our preliminary results on the differential flow of kaons. Comparisons with published data are only possible if we do an integration over bigger transverse momentum intervals, like it was done in [3, 4], resulting in a good agreement within error bars. It is also interesting to mention that kaons and pions show many similarities, as the reader can convince himself by looking to [12].

For the future, we would like to extend this analysis to the newly collected data on Ag + Ag collisions at 1.58 AGeV measured by HADES during March 2019. Last but not least, a comparison with relativistic transport models will be of interest.

Acknowledgments The author gratefully acknowledges the support by AS CR M100481202, GACR 13-06759S, MSMT LM2015049, PO VVV CZ.02.1.01/0.0/0.0/16_013/0001677 and MSMT LTT17003 NPI CAS Rez.

References

1. V. Thorsson, M. Prakash, J.M. Lattimer, Composition, structure and evolution of neutron stars with kaon condensates. Nucl. Phys. A **572**, 693 (1994). https://doi.org/10.1016/0375-9474(94)90407-3
2. G.Q. Li, C.M. Ko, B.-A. Li, Kaon flow as a probe of the kaon potential in nuclear medium. Phys. Rev. Lett. **74**, 235 (1995). https://doi.org/10.1103/PhysRevLett.74.235
3. Y. Shin et al., (KaoS Collaboration): Enhanced out-of-plane emission of K^+ mesons observed in Au + Au collisions at 1 AGeV. Phys. Rev. Lett. **81**, 1576 (1998). https://doi.org/10.1103/PhysRevLett.81.1576
4. V. Zinyuk et al., (FOPI Collaboration): Azimuthal emission patterns of K^+ and of K^- mesons in Ni + Ni collisions near the strangeness production threshold. Phys. Rev. C **90**, 025210 (2014). https://doi.org/10.1103/PhysRevC.90.025210
5. B.A. Li, A.T. Sustich, Differential flow in heavy-ion collisions at balance energies. Phys. Rev. Lett. **82**, 5004 (1999). https://doi.org/10.1103/PhysRevLett.82.5004
6. G. Agakishiev et al., (HADES Collaboration): The high-acceptance dielectron spectrometer HADES. Eur. Phys. J. A **41**, 243 (2009). https://doi.org/10.1140/epja/i2009-10807-5
7. M. Lorenz, Overview of recent results from HADES. Nucl. Phys. A **967**, 27 (2017). https://doi.org/10.1016/j.nuclphysa.2017.06.010
8. J. Adamczewski-Musch et al., (HADES Collaboration): Deep sub-threshold φ production in Au + Au collisions. Phys. Lett. B **778**, 403–407 (2018). https://doi.org/10.1016/j.physletb.2018.01.048
9. J. Adamczewski-Musch et al., (HADES Collaboration): Sub-threshold production of K^0_s mesons and Λ hyperons in Au + Au collisions at $\sqrt{s_{NN}} = 2.4$ GeV. Phys. Lett. B **793**, 457-463 (2019). https://doi.org/10.1016/j.physletb.2019.03.065
10. J. Adamczewski-Musch et al., (HADES Collaboration): Centrality determination of Au + Au collisions at 1.23 AGeV with HADES. Eur. Phys. J. A **54**, 85 (2018). https://doi.org/10.1140/epja/i2018-12513-7
11. J.-Y. Ollitrault, Reconstructing azimuthal distributions in nucleus-nucleus collisions. arXiv:nucl-ex/9711003v2
12. Y. Liu, Y, Wang, Q. Li, L. Liu, Collective flows of pions in Au + Au collisions at energies 1.0 and 1.5 GeV/nucleon. Phys. Rev. C **97**, 034602 (2018). https://doi.org/10.1103/PhysRevC.97.034602

Chapter 34
Strangeness Production with Respect to High Momentum Hadrons in pp and p-Pb Collisions with ALICE at the LHC

Justin T. Blair

Abstract In order to understand strangeness production mechanisms, one can study the correlations of hadrons with hidden strangeness (e.g. the ϕ meson) and open strangeness (K^0_S, Λ and $\bar{\Lambda}$) in hard (jet) processes and in soft (bulk) processes. Two-particle angular correlations with K^0_S triggers allow for the study of fragmentation in pp collisions as a function of multiplicity. Similarly, correlations with associated ϕ mesons in p-Pb collisions allow both the jet and the underlying event components of strange particle production to be probed. Presented are K^0_S-h correlations measured by ALICE [1] in pp collisions at $\sqrt{s} = 13$ TeV, and h-ϕ correlations measured in p-Pb collisions at $\sqrt{s_{\rm NN}} = 5.02$ TeV.

34.1 Introduction

Historically, an enhanced production of multi-strange hadrons in heavy-ion collisions was seen as a clear signature of the presence of a QGP [2]. Recent results have shown that this increase in multi-strange hadrons is also present in high multiplicity pp and p-A collisions with a smooth increase in strange hadron production as a function of charged particle multiplicity [3–5]. This increase is seen across all collision systems, and is more prominent for multi-strange hadrons.

In order to probe the origin of this strangeness enhancement in small systems, it is necessary to study the different production mechanisms of strange quarks. Jet production can be separated from production in the underlying event by measuring angular correlations [6], allowing for the study of strangeness production in both hard and soft scatterings. Using a high momentum strange hadron (K^0_S) as a jet proxy, the fragmentation of jets containing strange quarks can be studied as a function

Justin T. Blair for the ALICE Collaboration.

J. T. Blair (✉)
University of Texas at Austin, Austin, TX 78712, USA
e-mail: jtblair@utexas.edu

D. Elia et al. (eds.), *The XVIII International Conference on Strangeness in Quark Matter (SQM 2019)*, Springer Proceedings in Physics 250,
https://doi.org/10.1007/978-3-030-53448-6_34

of multiplicity. Similarly, measuring associated strange hadrons (ϕ) correlated to an high momentum trigger hadron can separate strange quarks produced in hard processes (jet-like) from those produced in soft processes (underlying event).

34.2 K^0_S-h Angular Correlations

To study jets containing strangeness, high momentum K^0_S triggers are selected as a jet proxy in pp collisions at $\sqrt{s} = 13$ TeV. The trigger K^0_S are reconstructed from the $K^0_S \rightarrow \pi^+\pi^-$ decay channel (B. R $\approx 69\%$). These decay pions can be identified in the ALICE detector using particle identification (PID) cuts on signals from both the Time Projection Chamber (TPC) and Time of Flight (TOF) detectors, as well as topological cuts to account for the displaced decay vertex of the K^0_S. Event multiplicity is determined using the V0 detectors at large forward and backward rapidity (V0M estimator). Yields of associated hadrons are measured in both the near-side and away-side jet peaks for different event multiplicities: 0–10% (highest multiplicity), 10–50, 50–100, and 0–100% inclusive. Results are compared to predictions from simulated PYTHIA [7] events.

These yields are then compared to measurements from h-h correlations to look for differences in jet fragmentation. In all multiplicity classes, yields in both the near and away-side for K^0_S triggers were lower than for unidentified hadron triggers across all associate p_T (Fig. 34.1). However, simulated PYTHIA events have a ratio consistent with the measured data for both near and away-side yields.

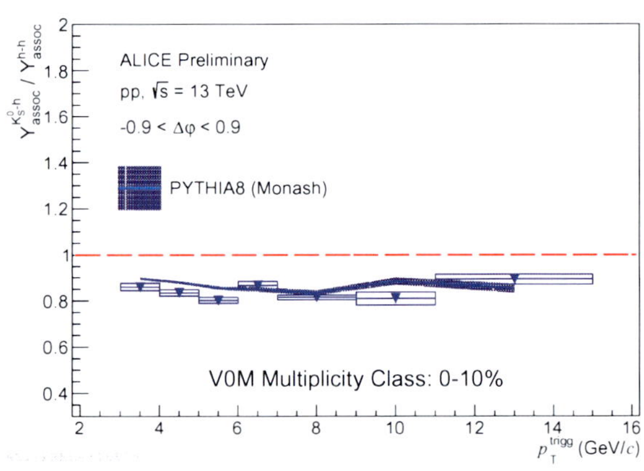

Fig. 34.1 Ratio of yields of (K^0_S-h/h-h) correlated pairs in the near-side jet peak compared to PYTHIA predictions (*shaded band*) for 0–10% V0M multiplicity class pp events

34.3 h-ϕ Angular Correlations

Hadron-ϕ angular correlations are measured in p-Pb collisions at $\sqrt{s_{NN}} = 5.02$ TeV to separate the ϕ produced within jets from ϕ produced in the underlying event. Candidate ϕ's are reconstructed from the $\phi \rightarrow K^+K^-$ decay channel (B. R. $\approx 49\%$). Trigger hadrons are selected with a transverse momentum of $4 < p_T^h < 8$ GeV/c and correlated with (K^+K^-) pairs with momentum $2 < p_T^{KK} < 4$ GeV/c. Decay kaons are identified using PID cuts on signals from both the TPC and TOF detectors [5]. The h-ϕ correlation is then obtained by subtracting combinatoric h-(K^+K^-) correlated pairs (measured in the invariant mass sideband region) from the h-(K^+K^-) correlation measured in the ϕ mass signal region. These correlations are measured for three different event multiplicity classes: 0–20% (highest mult.), 20–50%, and 50–80% (Fig. 34.2). Event multiplicity is determined using the A-side V0 detector (V0A) at large forward rapidity. Both the trigger hadron and associated ϕ are corrected for detector efficiency and acceptance effects.

After the h-ϕ correlation is measured, the yields of associated ϕ can be separated into a jet-like component and an underlying event component. To separate these, an assumed flat non-jet contribution is calculated as the average of the correlation in the regions away from the jet peaks. Once this underlying event is defined, the jet-like yields can be further separated into a near-side yield ($|\Delta\varphi| < \frac{\pi}{2}$) and an away-side yield ($\frac{\pi}{2} < \Delta\varphi < \frac{3\pi}{2}$).

Taking the ratio of h-ϕ pair yields to measured h-h pair yields gives a proxy measurement for ϕ/h in jets and ϕ/h in the underlying event (Fig. 34.3). The ratio of ϕ/h in jets is seen to be systematically lower than the inclusive ratio, while the ratio in the underlying event is systematically higher. Further, as multiplicity increases so does the fraction of total particles coming from the underlying event. Together, these observations show that the rise of the inclusive ratio is in part due to higher multiplicity events being dominated by production in the underlying event regime.

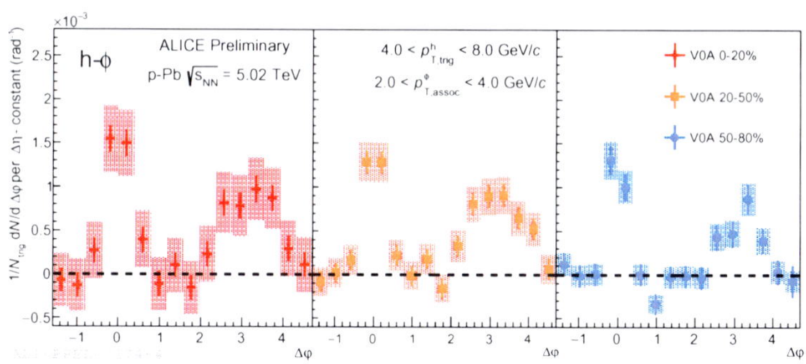

Fig. 34.2 Angular correlations between a high momentum hadron and intermediate momentum associated ϕ meson for different multiplicity classes

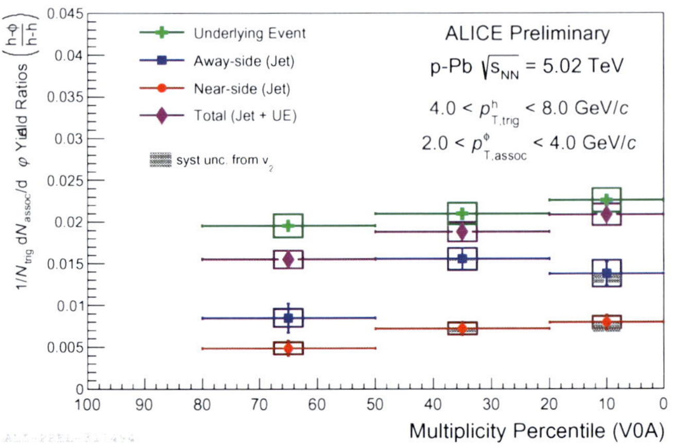

Fig. 34.3 Ratio of (h-ϕ)/(h-h) pair yields in p-Pb events as a function of event multiplicity. Yield ratios are calculated both for inclusive total pairs, as well as for just the near and away-side jet regions and the underlying event region separately. Additional systematic uncertainties (*gray shaded boxes*) are included for the presence of a non-zero v_2 [8]

34.4 Conclusion

The ALICE collaboration has measured K^0_S-h angular correlations as a function of multiplicity in pp events at $\sqrt{s} = 13$ TeV. This measurement compares the fragmentation of jets containing strangeness with inclusive charged jets. For all multiplicities, yields of associated hadrons with K^0_S triggers were measured lower than predicted by simulated PYTHIA events. Taking the ratio to correlated pairs with an inclusive hadron trigger shows that K^0_S triggers gave lower associated yields accross all multiplicities, but the ratio of (K^0_S-h)/(h-h) pairs is consistent with PYTHIA.

Hadron-ϕ angular correlations are measured as a function of multiplicity in p-Pb events at $\sqrt{s_{NN}} = 5.02$ TeV. By measuring the yield of ϕ mesons associated with a high momentum trigger hadron, it is possible to separately measure the ratio of ϕ/h within jets and ϕ/h in the underlying event. The ϕ/h ratio in jets (hard production) is systematically lower than the inclusive ratio, while the ϕ/h ratio in the underlying event (softer production) is systematically higher for all multiplicities.

Acknowledgments This work was supported by U. S. Department of Energy Office of Science under contract number DE–SC0013391.

References

1. A.L.I.C.E. Collaboration, The ALICE experiment at the CERN LHC. JINST **3**, S08002 (2008). https://doi.org/10.1088/1748-0221/3/08/S08002
2. J. Rafelski, B. Muller, Strangeness Production in the Quark-Gluon Plasma. Phys. Rev. Lett. **48**, 1066–1069 (1982). Erratum: [Phys. Rev. Lett. **56**, 2334]. https://doi.org/10.1103/PhysRevLett.48.1066, https://doi.org/10.1103/PhysRevLett.56.2334
3. A.L.I.C.E. Collaboration, Enhanced production of multi-strange hadrons in high-multiplicity proton-proton collisions. Nat. Phys. **13**, 535–539 (2017). https://doi.org/10.1038/nphys4111
4. A.L.I.C.E. Collaboration, Multi-strange baryon production in p-Pb collisions at $\sqrt{s_{NN}} = 5.02$ TeV. Phys. Lett. B **758**, 389–401 (2016). https://doi.org/10.1016/j.physletb.2016.05.027
5. ALICE Collaboration, Production of K^* $(892)^0$ and ϕ (1020) in p-Pb collisions at $\sqrt{s_{NN}} = 5.02$ TeV. Eur. Phys. J. C **76**, 245 (2016). https://doi.org/10.1140/epjc/s10052-016-4088-7
6. A.L.I.C.E. Collaboration, Multiplicity dependence of jet-like two-particle correlation structures in p-Pb collisions at $\sqrt{s_{NN}} = 5.02$ TeV. Phys. Lett. B **741**, 38–50 (2015). https://doi.org/10.1016/j.physletb.2014.11.028
7. T. Sjöstrand et al., An Introduction to PYTHIA 8.2. Comput. Phys. Commun .**191**, 159–177 (2015). https://doi.org/10.1016/j.cpc.2015.01.024
8. A.L.I.C.E. Collaboration, Long-range angular correlations of π, K and p in p-Pb collisions at $\sqrt{s_{NN}} = 5.02$ TeV. Phys. Lett. B **726**, 164–177 (2013). https://doi.org/10.1016/j.physletb.2013.08.024

Part V
Collectivity in Small Systems

Chapter 35
Hadrochemistry of Particle Production in Small Systems with ALICE at the LHC

Emily Willsher

Abstract The results of multi-strange particle production as a function of transverse momentum and event multiplicity in p–Pb collisions at $\sqrt{s_{NN}} = 8.16$ TeV measured with ALICE at the LHC are reported. A comparison to other collision energies and systems such as pp, Pb–Pb and Xe–Xe is also reported and discussed. These measurements help further the understanding of the differences and similarities between small and large collision systems.

35.1 Introduction

Historically, small collision systems, such as proton-proton and proton-lead, have been used as a reference for measurements in heavy-ion collisions in order to separate Quark-Gluon Plasma (QGP) effects from hadronic or cold nuclear matter effects. However, recent measurements of high-multiplicity pp and p–Pb collisions at the LHC have led to the observation of effects previously only seen in heavy-ion collisions [1, 2]. This includes results of strange particle production as a function of charged-particle multiplicity [3]. In this contribution, new results on the production of $\Xi^{\pm}$ and $\Omega^{\pm}$ in p–Pb collisions at the energy of $\sqrt{s_{NN}} = 8.16$ TeV are presented.

35.1.1 Strangeness Production

The enhanced production of strange hadrons was one of the earliest proposed signatures of QGP formation [4]. The enhancement was predicted to follow a hierarchy dependent on the strangeness content of the particle, meaning the level of enhance-

---

Emily Willsher for the ALICE Collaboration.

E. Willsher (✉)
University of Birmingham, Birmingham, UK
e-mail: emily.willsher@cern.ch

D. Elia et al. (eds.), *The XVIII International Conference on Strangeness in Quark Matter (SQM 2019)*, Springer Proceedings in Physics 250,
https://doi.org/10.1007/978-3-030-53448-6_35

ment is stronger for multi-strange particles. This has been observed in heavy-ion collisions at SPS, RHIC, and the LHC [5–7]. ALICE has also measured this effect in small systems by considering the ratio of strange particle to pion yields as a function of the number of charged particles produced in the collisions. The measured enhancement is strongly dependent on charged-particle multiplicity and independent of the collision system [3, 8]. This smooth strange particle enhancement across collision systems can be explained with statistical hadronisation models [9], core-corona models [10] or by using energy density as a universal scaling constant [11]. However, particle production mechanisms in these small collision systems are still unclear.

35.2 Multi-strange Hadron Detection with ALICE

The ALICE apparatus and its performance are described in detail in [12]. The main detectors used in this analysis were: the inner tracking system (ITS), which provides detailed tracking and vertex information that allows the primary and secondary vertices to be distinguished; the time projection chamber (TPC), which provides further tracking information and particle identification (PID) through the measurement of the specific energy loss (dE/dx) in the gas; and the V0 detectors which are used for triggering, as well as estimating the multiplicity of charged particles in the events as described in [13]. Multi-strange hadrons ($\Xi^{\pm}$ and $\Omega^{\pm}$) are reconstructed by utilising the characteristic topology of their weak decays into three identifiable charged particles. The reconstructed momenta of the charged decay tracks are combined in order to compute the invariant mass of the cascade candidates. The significance of the signal peak in the invariant mass distribution is optimised by applying selections on the decay topology. The data are divided into different transverse momentum (p_T) and multiplicity bins and the yields are extracted by using bin counting in the signal region and a fit of the background. The results are corrected with acceptance and efficiency values which are calculated using Monte Carlo simulations of the ALICE detector using DPMJET [14] and EPOS-LHC [15] generators.

35.3 Results

This analysis was carried out on a 16M Minimum Bias (MB) triggered event sample of p–Pb collisions at $\sqrt{s_{NN}} = 8.16$ TeV, where the MB trigger is given by a hit in both V0 detectors. The transverse momentum distributions for $\Xi^{\pm}$ and $\Omega^{\pm}$ in different multiplicity classes are shown in Fig. 35.1. The yields can be extracted from the spectra by integrating the data points and using a Lévy-Tsallis [16] fit to extrapolate to the unmeasured p_T regions. The corrected yields plotted as a function

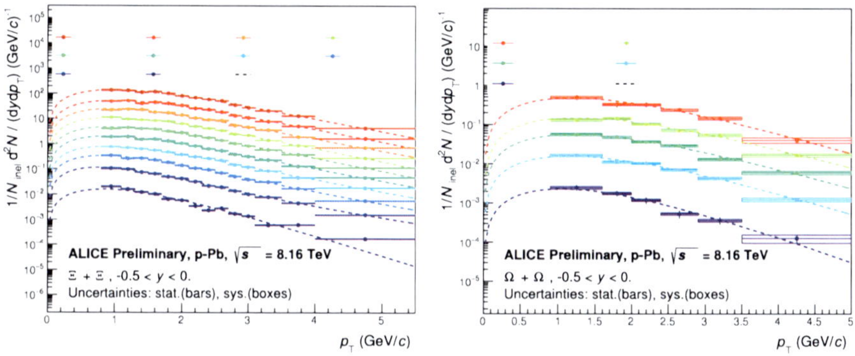

Fig. 35.1 Tranverse momentum spectra for $\Xi^{\pm}$ (left) and $\Omega^{\pm}$ (right) for different multiplicity intervals in p–Pb collisions at $\sqrt{s_{NN}} = 8.16$ TeV. The dashed lines correspond to a Lévy-Tsallis fit

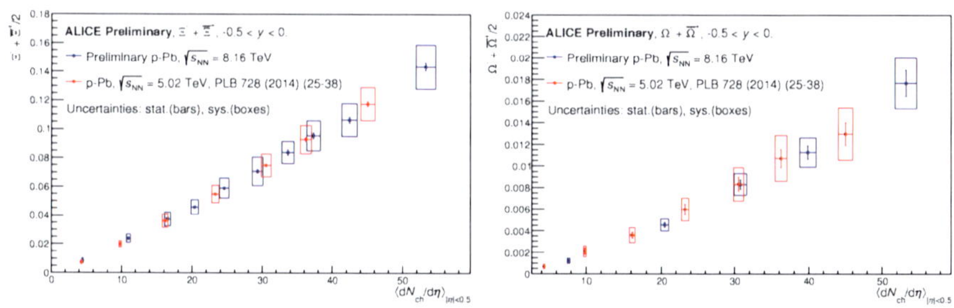

Fig. 35.2 Yields of $\Xi^{\pm}$ (left) and $\Omega^{\pm}$ (right) as a function of $\mathrm{d}N_{\mathrm{ch}}/\mathrm{d}\eta$ in p–Pb collisions at $\sqrt{s_{NN}} = 8.16$ TeV (blue) and $\sqrt{s_{NN}} = 5.02$ TeV (red)

of $\mathrm{d}N_{\mathrm{ch}}/\mathrm{d}\eta$ are shown in Fig. 35.2, where they are compared to the previous results obtained at $\sqrt{s_{NN}} = 5.02$ TeV. It can be observed that the multiplicity dependence of these yields follows a common linear trend at both energies, as was seen in pp collisions [17]. The multiplicity reach of the present measurement has also been extended to higher values of $\mathrm{d}N_{\mathrm{ch}}/\mathrm{d}\eta$. The strangeness enhancement was studied by calculating the ratio of strange particle to pion yields, where pions measured at $\sqrt{s_{NN}} = 8.16$ TeV have been used. The results as a function of $\mathrm{d}N_{\mathrm{ch}}/\mathrm{d}\eta$ are shown in Fig. 35.3 for $\Xi^{\pm}$ and $\Omega^{\pm}$ in p–Pb collisions at $\sqrt{s_{NN}} = 8.16$ TeV by the filled circle points, alongside the ratios in pp, Pb–Pb and Xe–Xe. The measurements show no significant dependence on the collision energy or system and scale with charged-particle multiplicity at LHC energies.

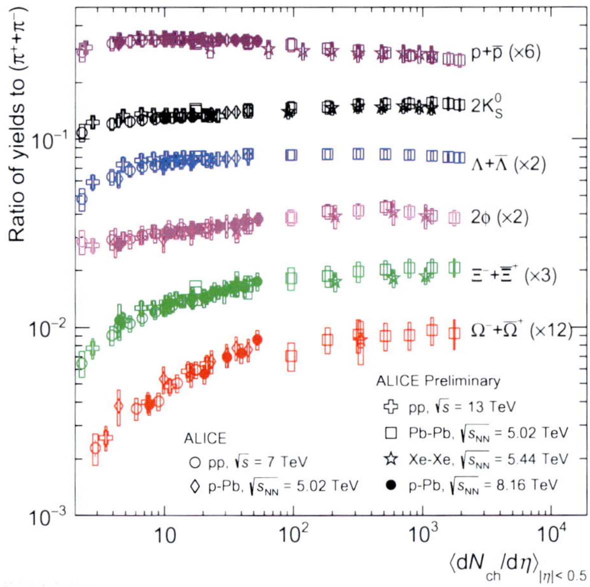

Fig. 35.3 Ratio of particle yields to pion yields as a function of $dN_{ch}/d\eta$. Results in p–Pb collisions at $\sqrt{s_{NN}}$ = 8.16 TeV are shown by filled in circle points. Results are compared to those from other centre-of-mass energies and collision systems

35.4 Conclusion

In summary, measurements of multi-strange hadrons as a function of charged-particle multiplicity in p–Pb collisions at ALICE have been reported. The results add to the picture of a smooth evolution of strangeness enhancement across collision systems. The latest data reported here also show that the particle yields as a function of charged-particle multiplicity in p–Pb collisions at $\sqrt{s_{NN}}$ = 5.02 TeV and $\sqrt{s_{NN}}$ = 8.16 TeV have a common linear trend. These results are important to improve the current understanding of the microscopic mechanisms behind particle production in small systems. Future measurements will extend pp and p–Pb results to $dN_{ch}/d\eta$ values reached in central Pb–Pb collisions so we can understand these small systems better.

References

1. CMS Collaboration, Observation of long-range, near-side angular correlations in proton-proton collisions at the LHC. JHEP **1009**, 091 (2010). https://doi.org/10.1007/JHEP09(2010)091
2. CMS Collaboration, Evidence for collectivity in pp collisions at the LHC. Phys. Lett. B **765**, 193 (2017). https://doi.org/10.1016/j.physletb.2016.12.009

3. ALICE Collaboration, Enhanced production of multi-strange hadrons in high-multiplicity proton-proton collisions. Nat. Phys. **13**, 535 (2017). https://doi.org/10.1038/nphys4111
4. J. Rafelski, B. Müller, Strangeness Production in the Quark-Gluon Plasma. Phys. Rev. Lett. **48**, 1066 (1982). https://doi.org/10.1103/PhysRevLett.48.1066
5. WA97 Collaboration, Strangeness enhancement at mid-rapidity in Pb-Pb collisions at 158 A GeV/c. Phys. Lett. B **449**, 401 (1999). https://doi.org/10.1016/S0370-2693(99)00140-9
6. STAR Collaboration, Strangeness enhancement in Cu-Cu and Au-Au collisions at $\sqrt{s_{NN}} = 200$ GeV. Phys. Rev. Lett. **108**, 072301 (2012). https://doi.org/10.1103/PhysRevLett.108.072301
7. ALICE Collaboration, Multi-strange baryon production at mid-rapidity in PbPb collisions at $\sqrt{s_{NN}} = 2.76$ TeV. Phys. Lett. B **728**, 216 (2014). https://doi.org/10.1016/j.physletb.2013.11.048
8. ALICE Collaboration, Multi-strange baryon production in pPb collisions at $\sqrt{s_{NN}} = 5.02$ TeV. Phys. Lett. B **758**, 389 (2016). https://doi.org/10.1016/j.physletb.2016.05.027
9. ALICE Collaboration, Multiplicity dependence of light-flavor hadron production in $\sqrt{s} = 7$ TeV. Phys. Rev. C **99**, 024906 (2019). https://doi.org/10.1103/PhysRevC.99.024906
10. Y. Kanakubo, M. Okai, Y. Tachibana, T. Hirano, Strangeness enhancement in p+p, p+Pb and Pb+Pb collisions at LHC energies (2019). arXiv:1901.08189v2 [nucl-th]
11. E. Cuautle, G. Paić, The energy density representation of the strangeness enhancement from p+p to Pb+Pb (2016). arXiv:1608.02101 [hep-ph]
12. ALICE Collaboration, The ALICE experiment at the CERN LHC. JINST **3**, S08002 (2008). https://doi.org/10.1088/1748-0221/3/08/S08002
13. ALICE Collaboration, Charged-particle pseudorapidity density at mid-rapidity in pPb collisions at $\sqrt{s_{NN}} = 8.16$ TeV. Eur. Phys. J. C **79**, 307 (2019). https://doi.org/10.1140/epjc/s10052-019-6801-9
14. S. Roesler, R. Engel, J. Ranft, The Monte Carlo Event Generator DPMJET-III (2000). arXiv:0012252 [hep-ph]
15. T. Pierog, I. Karpenko, J. Katzy, E. Yatsenko, K. Werner, EPOS LHC: test of collective hadronization with data measured at the CERN Large Hadron Collider. Phys. Rev. C **92**, 034906 (2015). https://doi.org/10.1103/PhysRevC.92.034906
16. C. Tsallis, Possible generalization of Boltzmann-Gibbs statistics. J. Stat. Phys. **52**, 479 (1988). https://doi.org/10.1007/BF01016429
17. ALICE Collaboration, Multiplicity dependence of (multi-)strange hadron production in proton-proton collisions at $\sqrt{s} = 13$ TeV (2019). arXiv:1908.01861 [nucl-ex]

Chapter 36
Collectivity and Electromagnetic Fields in Proton-Induced Collisions

Lucia Oliva, Pierre Moreau, Vadim Voronyuk, and Elena Bratkovskaya

Abstract We investigate the influence of the intense electromagnetic fields generated in p+Au collisions at top RHIC energy on the collective anisotropies of charged mesons, performing simulations with the Parton-Hadron-String Dynamics (PHSD) transport approach. We show that the electric field E_x along the impact parameter axis and the magnetic field B_y orthogonal to the reaction plane are comparable in magnitude, with a maximum strength of $\simeq 2.5\ \mathrm{m}_\pi^2$. We found that those fields affect the directed flow v_1 of pions and kaons: a splitting in the rapidity dependence of v_1 of particles with opposite charge is generated in the Au-going side and is mainly driven by the huge E_x component. The effect is stronger for the strange mesons and for larger impact parameters. Moreover, we demonstrate that the main contribution to the electromagnetically-induced splitting comes from the deconfined phase.

36.1 Introduction

In recent years the small systems, such as proton-nucleus collisions, has stimulated a lot of interest due to the new experimental and theoretical findings indicating the formation of Quark-Gluon Plasma (QGP) droplets in high-multiplicity events [1], leading to final momentum anisotropies comparable to those found in large colliding systems. While many works has been devoted to investigate the huge electromagnetic fields (EMF) produced in non-central heavy-ion collisions [2, 3] and their influence on final observables, such extended analysis is missing in the case of relativistic

L. Oliva (✉) · E. Bratkovskaya
GSI Helmholtzzentrum für Schwerionenforschung GmbH, Darmstadt, Germany
e-mail: oliva@fias.uni-frankfurt.de

L. Oliva · P. Moreau · E. Bratkovskaya
Institut für Theoretische Physik, Goethe-Universität, Frankfurt am Main, Germany

V. Voronyuk
Joint Institute for Nuclear Research, Dubna, Moscow region, Russia

Bogolyubov Institute for Theoretical Physics, Kiev, Ukraine

D. Elia et al. (eds.), *The XVIII International Conference on Strangeness in Quark Matter (SQM 2019)*, Springer Proceedings in Physics 250,
https://doi.org/10.1007/978-3-030-53448-6_36

proton-induced collisions. The generation of collective flow patterns in relativistic nuclear collisions is experimentally described by the Fourier expansion in momentum space of the azimuthal particle distribution. Among the flow anisotropies the first Fourier coefficient, namely the directed flow v_1, is one of the most compelling probes of the presence and the impact of the EMF in such collisions [4–6]. We aim at studying the impact of the EMF on the collective behaviour of the matter created in small colliding systems on the basis of a microscopic transport approach: by means of the Parton-Hadron-String Dynamics (PHSD) model [7, 8] we quantify the electromagnetically-induced directed flow of the most abundant charged mesons produced in p+Au collisions at $\sqrt{s_{NN}} = 200$ GeV [9].

36.2 Reminder of PHSD and The Electromagnetic Fields

The PHSD approach is a microscopic description of the dynamical evolution of heavy-ion collisions and small systems at relativistic energy [7, 8]. In the PHSD model the initial scatterings between nucleons in the two colliding nuclei lead to formation of strings, which fragment into hadrons or partons depending on whether the local energy density is respectively below or above the critical value for the QCD transition; the transition from partonic to hadronic degrees of freedom is described by dynamical hadronization. The evolution of the strongly interacting particles both in the QGP and in the hadronic phase is governed by off-shell transport equations derived from the Kadanoff-Baym equations for Green functions in phase-space representation. In the deconfined stage the QGP properties, namely the parton spectral functions and the self-generated mean-field potentials, are described by the Dynamical Quasi-Particle Model (DQPM), whose parameters are determined by a fit of lattice QCD thermodynamics.

PHSD includes the dynamical formation and evolution of the retarded EMF, its influence on quasi-particles and the back-reaction of particle dynamics on the fields [3, 4]: the transport equations are coupled to the Maxwell equations for the electric field $\mathbf{E}$ and the magnetic field $\mathbf{B}$ and the quasiparticle propagation in the EMF is determined by the Lorentz force $\mathbf{F} = q\,(\mathbf{E} + \mathbf{v} \times \mathbf{B})$, being q and $\mathbf{v}$ respectively charge and velocity of the particle.

36.3 Results and Discussion

We have performed simulations with PHSD of p + Au collisions at $\sqrt{s_{NN}} = 200$ GeV with fixed impact parameter b.

In Fig. 36.1 we show the distribution in the transverse plane of E_x (left panel) and B_y (right panel) produced in collisions at $b = 4$. The field strengths are computed in the plane $z = 0$ at the maximum overlap time of the collision. While in symmetric nucleus-nucleus collisions (e.g. Au + Au) the EMF produced in the early stage is

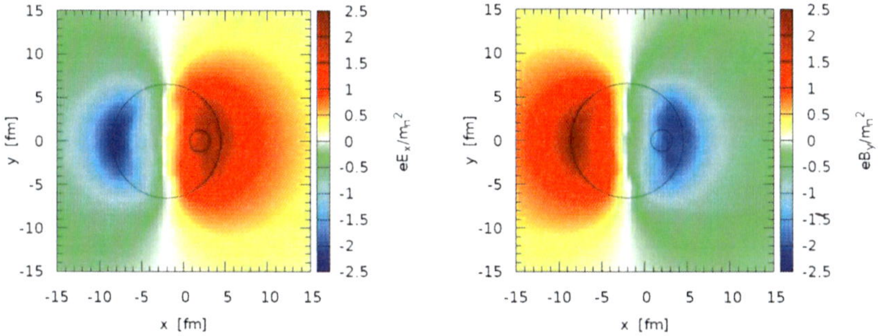

Fig. 36.1 Distribution of the EMF components E_x (*left panel*) and B_y (*right panel*) in the transverse plane at $z = 0$ at the maximum overlap time of p + Au collisions with $b = 4$ fm at $\sqrt{s_{NN}} = 200$ GeV. The circles are drawn to guide the eye

dominated by the magnetic field perpendicular to the reaction plane B_y and the other components are very small [2, 3], in asymmetric systems (e.g. Cu + Au) a significant electric field along the impact parameter axis E_x directed from the heavier nucleus towards the lighter nucleus is generated in the overlap area due to the different number of protons in the two colliding nuclei [4]. In the extremely asymmetric case of proton-induced reactions the field distributions are essentially that generated by the heavy nucleus and E_x reaches values comparable to B_y; we see from Fig. 36.1 that the maximum magnitude of both components is about $|eB_y| \simeq |eE_x| \simeq 2.5\,\mathrm{m}_\pi^2$ and for collisions at $b = 4$ the interaction area is affected by field strengths close to the maximum value. Regarding the temporal evolution of the two EMF components, both $|B_y|$ and $|E_x|$ decrease very fast, becoming close to zero after ~ 0.25 fm/c from the first nucleon-nucleon collisions; see [9] for more details.

In order to disentangle the influence of the EMF on particle dynamics, we computed the directed flow $v_1 = \langle \cos\phi \rangle$, being ϕ the azimuthal angle of the particle momentum. In Fig. 36.2 we plot the rapidity dependence of v_1 (in percentage) of pions (left panel) and kaons (right panel) obtained with PHSD simulations at $b = 2$ fm with and without the inclusion of EMF, labelled by solid and dashed curves

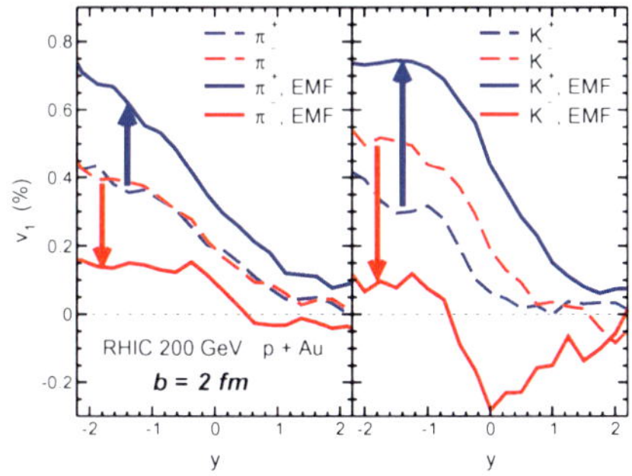

Fig. 36.2 Directed flow of pions (*left panel*) and kaons (*right panel*) versus rapidity for $b = 2$ fm p + Au collisions at $\sqrt{s_{NN}} = 200$ GeV obtained with PHSD simulations with (solid curves) and without (dashed curves) EMF, whose effect is highlighted by arrows

respectively. For what concern pions, we see that in simulations that do not include the EMF π^+ (blue lines) and π^- (red lines) show basically the same v_1. Switching on the EMF in PHSD, π^+ and π^- are pushed respectively in the positive and negative x−direction, this resulting in a splitting of the v_1 of the two particles. The effect on kaons is very interesting since K^+ (blue lines) and K^- (red lines) have a different v_1 even in the case without EMF; indeed, K^+ receive more contributions from u and d quarks from the initial colliding nuclei with respect to K^-. The EMF turn over the splitting in the kaon v_1 with respect to that generated by the vortical flow mentioned above, leading to a maximal flipping at backward rapidities. We highlight with arrows how the EMF modify the v_1 of pions and kaons: indeed, the direction of the splitting in v_1 results from the contrast between sideways kicks on charged particles by electric and magnetic forces and from the arrows and Fig. 36.1 it becomes clear that in proton-nucleus reactions the EMF effect on v_1 is mainly driven by E_x.

Therefore, the EMF lead to a separation of hadrons of the same species but with opposite charge along the impact parameter axis. The magnitude of the splitting of such partners can be identified by means of the quantity $\Delta v_1 \equiv v_1^+ - v_1^-$, where v_1^+ and v_1^- are the directed flow of the positively and negatively charged particles respectively. Then we compute the quantity

$$\Delta v_1^{emf} \equiv \Delta v_1^{(PHSD+EMF)} - \Delta v_1^{(PHSD)}. \tag{36.1}$$

i.e., the difference of Δv_1 in PHSD simulations with and without the inclusion of EMF; this way we pinpoint the magnitude of the directed flow splitting induced by the EMF as the one coming from other sources such as vortical motions and fluctuations is removed. In Fig. 36.3 we plot in percentage the quantity (36.1) for pions (left panel) and kaons (right panel) in collisions at $b = 2$ fm and $b = 6$ fm. We notice that the effect increases for collisions with larger impact parameter following the increasing trend of E_x. Moreover, the electromagnetically-induced splitting is bigger in magnitude for kaons respect to pions, at least in the rapidity region $|y| < 2$.

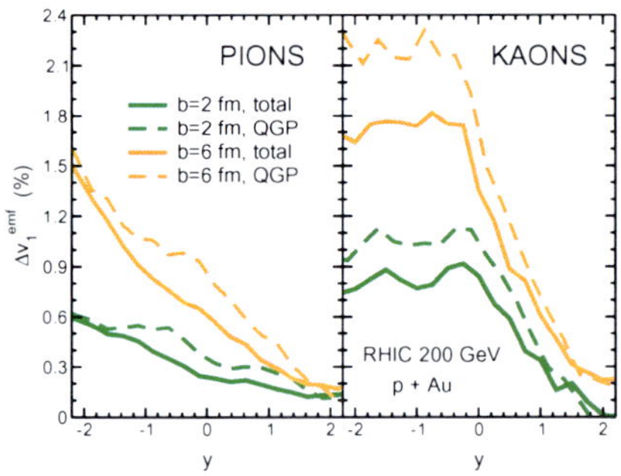

Fig. 36.3 Electromagnetically-induced splitting in $v_1(y)$ of pion (*left panel*) and kaons (*right panel*) in p + Au collisions at $\sqrt{s_{NN}} = 200$ with $b = 2$ fm and $b = 6$ fm. The dashed lines are the contribution of mesons coming directly from QGP hadronization

The reason is the different mass and hence velocity of the two meson species: for the slower moving kaons the influence of the EMF lasts for a longer time and the magnetic force that tends to compensate the push of the electric force is smaller. In Fig. 36.3, along with the total result (solid lines), we plot the v_1 of mesons coming directly from hadronization of QGP without suffering further rescattering in the hadronic phase (dashed lines); we see that for $|y| < 2$ the v_1 splitting induced by the EMF in the partonic phase is higher than that generated in the confined stage and this difference is remarkable for kaons.

Acknowledgments The authors acknowledge support by the DFG grant CRC-TR 211, the COST Action THOR CA15213 and the Humboldt Foundation. The computational resources were provided by LOEWE-CSC and GSI Green Cube.

References

1. C. Aidala et al., [PHENIX Collaboration]. Nat. Phys. **15**(3), 214 (2019)
2. V. Skokov, A.Y. Illarionov, V. Toneev, Int. J. Mod. Phys. A **24**, 5925 (2009)
3. V. Voronyuk et al., Phys. Rev. C **83**, 054911 (2011)
4. V. Voronyuk, V.D. Toneev, S.A. Voloshin, W. Cassing, Phys. Rev. C **90**, 064903 (2014)
5. U. Gursoy, D. Kharzeev, K. Rajagopal, Phys. Rev. C **89**, 054905 (2014)
6. S.K. Das et al., Phys. Lett. B **768**, 260 (2017)
7. W. Cassing, E.L. Bratkovskaya, Nucl. Phys. A **831**, 215 (2009)
8. E.L. Bratkovskaya, W. Cassing, V.P. Konchakovski, O. Linnyk, Nucl. Phys. A **856**, 162 (2011)
9. L. Oliva, P. Moreau, V. Voronyuk, E. Bratkovskaya, arXiv:1909.06770 [nucl-th]

Chapter 37
Particle Production as a Function of Underlying-Event Activity Measured with ALICE at the LHC

Valentina Zaccolo

Abstract ALICE has performed several measurements aimed at understanding the collective-like effects observed in small collision systems. New approaches may be needed to clarify particle-production mechanisms in high-multiplicity pp collisions. Transverse momentum (p_T) spectra as a function of charged-particle multiplicity show intriguing features. For example, data exhibit a stronger-than-linear increase of the self-normalised high-p_T particle yields versus multiplicity. In order to understand the role of auto-correlations on these effects, it has been proposed to use the underlying event as a multiplicity estimator to factorise the hardest and the softest components of the events. This approach can also be used to study collective effects in events with exceptionally large activity in the underlying-event region with respect to the event-averaged mean. In these proceedings, p_T spectra as a function of underlying-event activity in pp collisions measured with the ALICE detector are presented. Results are compared with PYTHIA 8.2 event generator.

37.1 Introduction

In hadronic interactions at high energies, as the ones achieved at the Large Hadron Collider (LHC), there are significant contributions from hard processes, which can be described by perturbative QCD precise calculations. Nevertheless, particle production at LHC is still dominated by soft-QCD processes. Soft QCD is characterised by non-perturbative phenomenology and requires accurate modelling. The Underlying Event is constituted by multiple semi-hard parton interactions, initial and final state radiation and beam remnants, therefore, all the event components but the hardest scattering.

Valentina Zaccolo for the ALICE Collaboration.

V. Zaccolo (✉)
University of Trieste and INFN, Via Alfonso Valerio 2, 34127 Trieste, Italy
e-mail: valentina.zaccolo@ts.infn.it

D. Elia et al. (eds.), *The XVIII International Conference on Strangeness in Quark Matter (SQM 2019)*, Springer Proceedings in Physics 250,
https://doi.org/10.1007/978-3-030-53448-6_37

The results presented in the following sections are motivated by two observations. Firstly, the p_{T} spectra versus multiplicity results measured recently by ALICE [1] highlighted a stronger than linear increase with multiplicity, which grows with p_{T}. Using a forward multiplicity estimator, that is well separated in rapidity from the region of the measurement, the increase is reduced. One could argue that the reason of the increase with multiplicity is due to the presence of jets which bias the estimator. This last point can be understood better using a jet-free multiplicity estimator, which could then help characterising the correlation effects between low- and high-p_{T} particle production. The second observation relates to multiplicity-dependent studies for heavy-flavours and high-p_{T} particle production [1–3], where ALICE observes, again, a non-linear particle production increase with multiplicity. This effect is often attributed to multiplicity saturation due to coherent hadronisation effects. Anyway, recently, the non-linear particle production increase with multiplicity was explained with auto-correlations effects [4]. The authors argue that removing them would lead to a weaker-than-linear increase.

The measurements are performed with ALICE, one of the four main experiments at the LHC. It is constituted by 18 different detector systems and its peculiarities are a very high momentum resolution and excellent particle identification capabilities [5]. The central barrel detectors are embedded in a solenoidal magnet of $B = 0.5$ T nominal field. The tracking and vertexing detectors used for results shown in the following are the Inner Tracking System (ITS) and the Time Projection Chamber (TPC). The events used in the analysis were collected in 2016 proton–proton (pp) collisions at $\sqrt{s} = 13$ TeV selected using a charged-particle signal coincidence in V0A and V0C arrays of scintillator counters.

37.2 Analysis and Results

The analysis was performed considering two regions defined by the relative azimuthal angle with respect to the leading charged particle: $|\Delta\phi| = \phi - \phi_{\mathrm{leading}}$. The toward and transverse regions are defined by $|\Delta\phi| < \pi/3$ and $\pi/3 < |\Delta\phi| < 2\pi/3$, respectively. The relative transverse activity classifier, R_{T}, is the self-normalised particle density in the transverse region [6]. An additional selection is done to characterise only the UE plateau region (the region of the jet pedestal): $5 < p_{\mathrm{T}}^{\mathrm{leading}} < 40$ GeV/c. Several intervals of R_{T} were selected in order to distinguish between low and high UE activity. From PYTHIA 8 simulations, one expects that the R_{T} distributions exhibit a Koba-Nielsen-Olesen scaling [8].

Results for the transverse region are shown in Fig. 37.1 (left). The p_{T} spectra of charged particles reconstructed in the transverse region above 0.6 GeV/c exhibit a strong R_{T} dependence. The evolution of the p_{T}-dependent spectra normalised to the inclusive p_{T} spectrum ($R_{\mathrm{T}} \geq 0$) are reminiscent of the observed behaviour when the inclusive multiplicity estimator was used [1]. A different behaviour is observed for analogous measurements considering the toward region (Fig. 37.1, right), where the

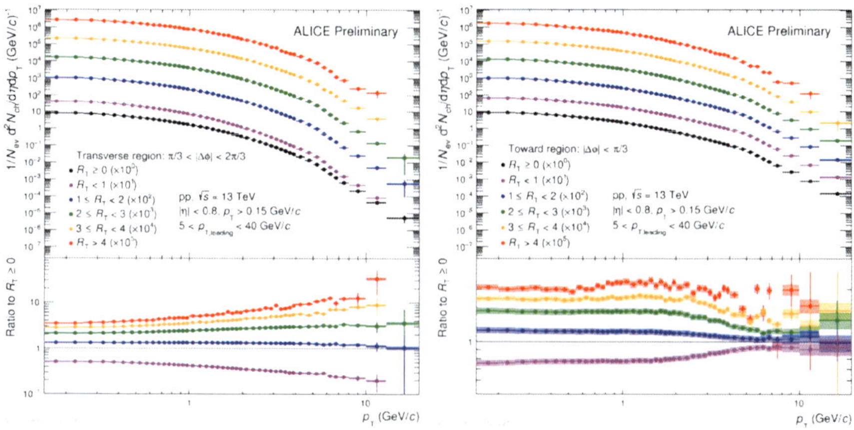

Fig. 37.1 Charged-particle p_T spectra as a function of R_T in pp collisions at $\sqrt{s} = 13$ TeV. Results for the transverse (left) and toward (right) regions are compared with their corresponding p_T spectrum for the R_T-integrated event class. Vertical bars are statistical errors uncertainties, while boxes are systematic uncertainties

Fig. 37.2 Self-normalised charged-particle yields in the transverse (green) and toward (red) regions as a function of R_T for pp collisions at $\sqrt{s} = 13$ TeV in the p_T range from 2 to 4 GeV/c. The data are compared with calculations of PYTHIA 8.2 with the Monash tune

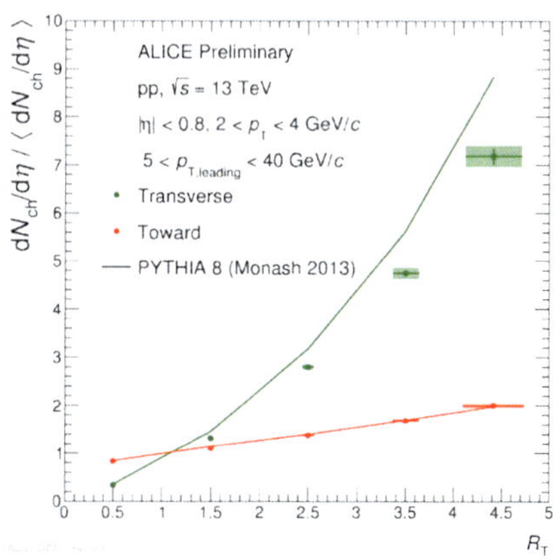

ratios to the inclusive p_T spectrum in the toward region converge to unity at high p_T. One can conclude that we have achieved an almost complete separation between the soft (UE) and hard (jet) parts of the event at high p_T. Moreover, the auto-correlation effects are significantly reduced.

Figure 37.2 shows the self-normalised particle yields, considering particles within $2 < p_T < 4$ GeV/c, as a function of R_T. On one hand, the self-normalised yield in the transverse region, in green, shows the same behaviour of the one observed with the full midrapidity-based multiplicity estimator [1]. On the other hand, the self-normalised yields in the toward region, in red, as a function of R_T do not converge to zero showing that at $R_T \sim 0$ one can still have the presence of a jet. This opens the possibility of studying a hard object with almost no UE activity. Moreover, for the toward region the self-normalised yields exhibit a weaker dependence on R_T than that observed for the transverse region. The results are compared to PYTHIA 8 (Monash tune) calculations [7] which reproduce well the observed trends.

37.3 Summary and Outlook

The charged-particle p_T-spectra in the transverse and toward regions as a function of the relative transverse activity classifier R_T have been measured using the ALICE detector. Results for pp collisions at $\sqrt{s} = 13$ TeV have been presented. The charged-particle p_T-spectra in the transverse region show a hardening at high R_T, which confirms the trend observed using the mid-pseudorapidity multiplicity estimator in the full azimuthal region. The charged-particle p_T-spectra in the toward region, instead, exhibit a weaker dependence with R_T, suggesting that the auto-correlation effects are at play when p_T spectra and multiplicity are both determined within the same pseudorapidity and $|\Delta\phi|$ regions. Finally, it was observed that it is still possible to identify a jet with zero activity in the UE, giving the opportunity to relate pp collisions to elementary colliding systems like e^+e^-. Therefore, R_T is an effective instrument to disentangle jet and UE components of the particle spectra and is a promising instrument for light and heavy-flavour particle-production studies.

References

1. ALICE Collaboration, Charged-particle production as a function of multiplicity and transverse spherocity in pp collisions at $\sqrt{s} = 5.02$ and 13 TeV. Eur. Phys. J. C **79**(10), 857 (2019). https://doi.org/10.1140/epjc/s10052-019-7350-y
2. ALICE Collaboration, J/ψ production as a function of charged particle multiplicity in *pp* collisions at $\sqrt{s} = 7$ TeV. Phys. Lett. B **712**, 165–175 (2012). https://doi.org/10.1016/j.physletb.2012.04.052
3. ALICE Collaboration, Measurement of charm and beauty production at central rapidity versus charged-particle multiplicity in proton-proton collisions at $\sqrt{s} = 7$ TeV. JHEP **1509**, 148 (2015). https://doi.org/10.1007/JHEP09(2015)148
4. S.G. Weber, A. Dubla, A. Andronic, A. Morsch, Elucidating the multiplicity dependence of J/ψ production in proton–proton collisions with PYTHIA8. Eur. Phys. J. C **79**(1), 36 (2019). https://doi.org/10.1140/epjc/s10052-018-6531-4
5. ALICE Collaboration, The ALICE experiment at the CERN LHC. JINST **3**, S08002 (2008). https://doi.org/10.1088/1748-0221/3/08/S08002
6. T. Martin, P. Skands, S. Farrington, Probing collective effects in hadronisation with the extremes of the underlying event. Eur. Phys. J. C **76**(5), 299 (2016). https://doi.org/10.1140/epjc/s10052-016-4135-4
7. T. Sjöstrand, S. Ask, J.R. Christiansen, R. Corke, N. Desai, P. Ilten, S. Mrenna, S. Prestel, C.O. Rasmussen, P.Z. Skands, An introduction to PYTHIA 8.2. Comput. Phys. Commun. **191**, 159–177 (2015). https://doi.org/10.1016/j.cpc.2015.01.024
8. A. Ortiz, L. Valencia Palomo, Universality of the underlying event in pp collisions. Phys. Rev. D **96**(11), 114019 (2017). https://doi.org/10.1103/PhysRevD.96.114019

Chapter 38
Bottomonium Production in pp, pPb and PbPb Collisions at 5.02 TeV with the CMS Detector

Daniele Fasanella

Abstract Bottomonia are important probes for the study of properties of the Quark-Gluon Plasma (QGP): they are produced at early times and then propagate in the strongly interacting medium of deconfined quarks and gluons. The CMS experiment is particularly well suited for the study of bottomonia in their dimuon final state. Measurements of bottomonium production are reported for the Υ(1S), Υ(2S), and Υ(3S) mesons in pp, pPb and PbPb collisions at 5.02 TeV. The results are compared with theory models, which can help to improve and constrain the theoretical calculations.

38.1 Introduction

It is expected that matter at high energy density and temperature forms the Quark-Gluon Plasma (QGP), a strongly interacting medium of deconfined quarks and gluons [1, 2]. One of the most remarkable signatures of quarkonium state interactions with the medium is the sequential suppression in heavy ion collisions compared to the production in proton-proton (pp) collisions. The dissociation temperatures for the states, above which suppression occurs, are expected to be correlated with their binding energies, and are predicted to be $T_{dissoc} \approx 2\ T_c$, $1.2\ T_c$ and $1\ T_c$ for the Υ(1S), Υ(2S), and Υ(3S) states, respectively, where T_c is the critical temperature for deconfinement [3]. The suppression of the Υ(1S), Υ(2S), and Υ(3S) yields produced in heavy-ion collisions relative to proton-proton (pp) collisions was first measured by the Compact Muon Solenoid (CMS) experiment, at the Large Hadron Collider (LHC), in PbPb collisions at a centre-of-mass energy per nucleon pair of $\sqrt{s_{NN}} =$ 2.76 TeV [4]. The tightest bound state, Υ(1S), was observed to be less suppressed than the more loosely bound excited states, Υ(2S) and Υ(3S).

Daniele Fasanella for the CMS Collaboration.

D. Fasanella (✉)
CERN, Espl. des Particules 1, 1211 Meyrin, Switzerland
e-mail: daniele.fasanella@cern.ch

D. Elia et al. (eds.), *The XVIII International Conference on Strangeness in Quark Matter (SQM 2019)*, Springer Proceedings in Physics 250,
https://doi.org/10.1007/978-3-030-53448-6_38

38.2 Bottomonia Production Results

In this report the more recent CMS results using data samples at a center-of-mass energy of 5.02 TeV in pp, pPb and PbPb collisions [5–7] are presented . The availability of data coming from different kinds of collision and different energies opens interesting opportunities in the understanding of bottomonia production. The events are selected online with a hardware-based trigger requiring two muon candidates in the muon detectors. The yields of Upsilon mesons are extracted using unbinned maximum-likelihood fits to the invariant mass spectra, where signals of each state are modelled by two Crystal-Ball (CV) functions, and the chosen background PDF is usually an error function multiplied by an exponential.

Different observables are used to study bottomonium production characteristics between the collision types. Double ratios, defined as

$$\frac{(Y(2S)/Y(1S))_{PbPb}}{Y(2S)/Y(1S))_{pp}}, \tag{38.1}$$

are very useful to quantify the relative modification of yields of the excited states, where reconstruction, trigger, and muon identification efficiencies largely cancel. In Fig. 38.1 (left) double ratios for Y(2S) are reported, as a function of the average number of nucleons (N_{part}) that participate in the interaction. Predictions of suppression from Krouppa and Strickland [8], incorporating color-screening effects on the bottomonium family and reflecting feed-down contributions from decays of heavy quarkonia, are in overall agreement with the Y(2S) double ratio results. Another theoretical curve from Du et al. [9], based on a kinetic-rate equation and containing a small component of regenerated bottomonia, shows a similar level of agreement with the data. Within this model, our data rule out a scenario where the regeneration component is absent.

The excited-to-ground-states cross section ratio, Y(2S)/Y(1S) are also studied in Pb, pp and PbPb collisions, as a function of event activity variables, like the number of tracks ($N_{tracks}^{|\eta|<2.4}$). This is evaluated as the number of tracks reconstructed in the tracker, not including the two muons, originating from the same vertex as the Y, with $p_T > 400$ MeV/c at $|\eta| < 2.4$. The pp and pPb activity-integrated ratio results far above the PbPb ones. In Fig. 38.1 (right) the Y(2S)/Y(1S) is shown as a function of $N_{tracks}^{|\eta|<2.4}$. This ratio is found to decrease with increasing $N_{tracks}^{|\eta|<2.4}$ for pp and Pb collisions, and is not visible a pronounced PbPb centrality dependence (within large uncertainties). The unexpected dependence in pp and pPb suggests novel phenomena in quarkonium production that could arise from a larger number of charged particles being systematically produced with the ground state, or from a stronger impact of the growing number of nearby particles on the more weakly bound states.

Normalized cross sections are also obtained, correcting for acceptance and efficiency the yields extracted from the fits to the dimuon invariant mass spectra. These cross sections are used to evaluate the modifications of particle production in nucleus-nucleus (AA) interactions, thought the nuclear modification factor, R_{AA}:

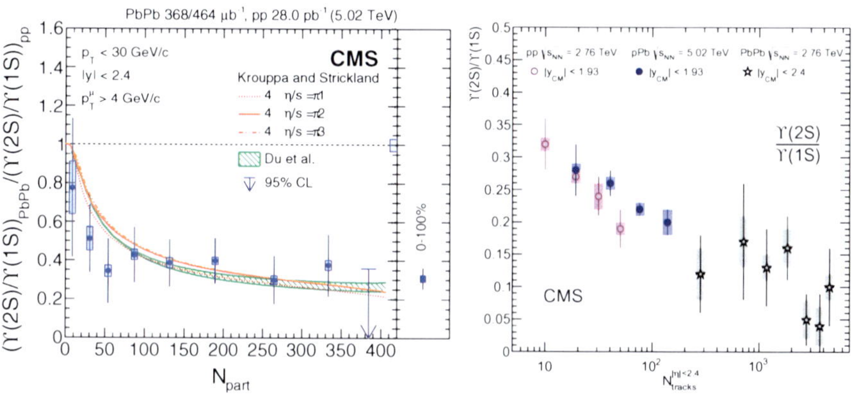

Fig. 38.1 On the left, the double ratio of the Υ(2S) as a function of centrality is shown [7]. The centrality-integrated value is shown in the right panel. The error bars represent the statistical uncertainty in the PbPb data while the boxes represent the systematic uncertainty. The global systematic uncertainty is represented as a grey box drawn around the line at unity. Calculations by Krouppa and Strickland (orange curves [8]) and by Du et al. (green hatchedband [9]) are also shown. On the right, single cross section ratios Υ(2S)/Υ(1S) versus $N_{tracks}^{|\eta|<2.4}$ is shown, measured, for pp and PbPb collisions at $\sqrt{s_{NN}} = 2.76$ TeV (open circles and open stars respectively) and for pPb collisions at $\sqrt{s_{NN}} = 5.02$ TeV(closed circles) [5]. The error bars in the figures indicate the statistical uncertainties, and the boxes represent the point-to-point systematic uncertainties

$$R_{AA}(p_{\mathrm{T}}, y) = \frac{N^{AA}(p_{\mathrm{T}}, y)}{\langle T_{AA}\rangle \sigma^{pp}(p_{\mathrm{T}}, y)}. \tag{38.2}$$

The quantity $\langle T_{AA}\rangle$ is the average nuclear overlap function, N^{AA} is the normalized corrected yield of Upsilon mesons in PbPb collisions, and σ_{pp} refers to the pp cross section. The dependence of R_{AA} on PbPb collision centrality, as quantified using the average number of participating nucleons $\langle N_{part}\rangle$, is depicted in Fig. 38.2, where a gradual decrease in R_{AA} for the Υ(1S) and Υ(2S) states is observed. Within the current theoretical and experimental uncertainties, both the considered theoretical models in [8, 9] are in agreement with the results. The R_{AA} of the Υ(3S) state is measured to be below 0.096 at 95% confidence level, making this the strongest suppression observed for a quarkonium state in heavy ion collisions to date. The obtained centrality-integrated R_{AA} value has been compared to those measured by CMS at $\sqrt{s_{NN}} = 2.76$ TeV, resulting in an increase by a factor of $\sim 1.20 \pm 0.15$, although the two R_{AA} values are compatible within the uncertainties. This is in agreement with the expectation of larger suppression at higher medium temperatures. The models predict increases in the medium temperature for PbPb collisions of ~16% [8] and ~7% [9] between $\sqrt{s_{NN}} = 2.76$ TeV and $\sqrt{s_{NN}} = 5.02$ TeV.

In conclusion, CMS has produced a large amount of results on bottomonium production in pp, pPb and PbPb collision. All the three states are suppressed in PbPb collisions with respect to pp, with a gradual decrese of R_{AA} with $\langle N_{part}\rangle$.

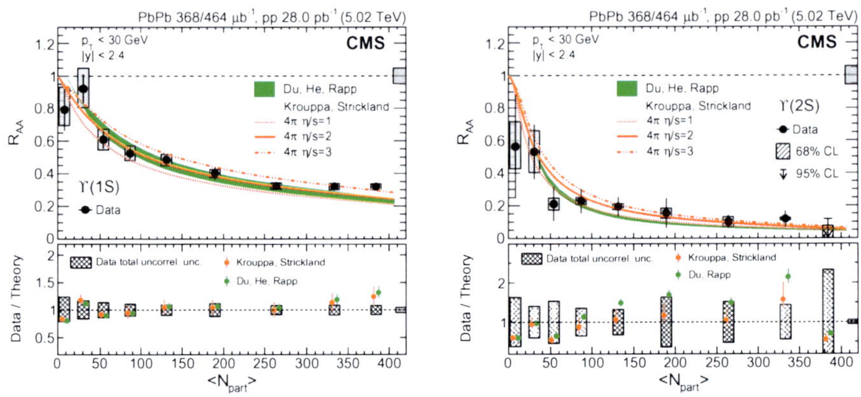

Fig. 38.2 Nuclear modification factors for the Y(1S) (left) and Y(2S) (right) mesons as a function of $\langle N_{part} \rangle$ [6] compared to calculations from Krouppa and Strickland [8], and Du, He, and Rapp [9]. The box at the dashed line at unity represents the global uncertainty. The data to theory ratios are shown in the bottom panels. For [8], the points correspond the $4\pi \eta/s = 2$ curve (where η is the viscosity and s the entropy), while the error bars show the difference between this one and the other two η/s curves. For [9], the points and error bars correspond to the center and width of the published theory band, respectively

Good agreement has been found with theoretical models. The excited-to-ground-states cross section ratios, Y(2S)/Y(1S), are found to unexpectedly decrease with increasing charged-particle multiplicity.

References

1. F. Karsch and E. Laermann, Thermodynamics and in-medium hadron properties from lattice QCD, in Quark-Gluon Plasma III, arXiv:hep-lat/0305025
2. E.V. Shuryak, Theory of hadronic plasma. Sov. Phys. JETP **47**, 212 (1978)
3. A.Mocsyand, P.Petreczky, "Color screening melts quarkonium", Phys. Rev. Lett. 99, https://doi.org/10.1103/PhysrevLett.99.211602
4. CMS Collaboration, "Observation of sequential Upsilon suppression in PbPb collisions", https://doi.org/10.1103/PhysRevLett.109.222301
5. CMS Collaboration, "Event activity dependence of Y(nS) production in $\sqrt{s_{NN}} = 5.02$ TeV pPb and $\sqrt{s} = 2.76$ TeV pp collisions", https://doi.org/10.1007/JHEP04(2014)103
6. CMS Collaboration, "Measurement of nuclear modification factors of (1S), (2S), and (3S) mesons in PbPb collisions at $\sqrt{s_{NN}} = 5.02$ TeV", https://doi.org/10.1016/j.physletb.2019.01.006
7. CMS Collaboration, "Suppression of Excited ϒ States Relative to the Ground State in Pb-Pb Collisions at $\sqrt{s_{\rm NN}} = 5.02$ TeV", https://doi.org/10.1103/PhysRevLett.120.142301
8. B. Krouppa and M. Strickland, "Predictions for bottomonia suppression in 5.023 TeV Pb–Pb collisions", https://doi.org/10.3390/universe2030016
9. X. Du, R. Rapp and M. He, "Color Screening and Regeneration of Bottomonia in High-Energy Heavy-Ion Collisions", https://doi.org/10.1103/PhysRevC.96.054901

Chapter 39
Study of Open Heavy-Flavour Hadron Production in pp and p–Pb Collisions with ALICE

Preeti Dhankher

Abstract Measurement of heavy-flavour production in small systems can be used to test the Quantum ChromoDynamic (QCD) models. In this manuscript, the production cross section of D mesons at midrapidity and open heavy-flavour decay muons (HFM) measured at forward rapidity in pp collisions at $\sqrt{s} = 5.02$ TeV, and open heavy-flavour decay electrons (HFE) measured at midrapidity in pp collisions at $\sqrt{s} = 13$ TeV with ALICE detector will be presented. The self-normalized yield of open heavy-flavour decay electrons and muons as a function of multiplicity in pp and p–Pb collisions will be presented. Finally, the latest result on nuclear modification factor ($Q_{\rm pPb}$) of D mesons and the v_2 of open heavy-flavour decay electrons and muons in p–Pb collisions will be discussed.

39.1 Introduction

Heavy quarks (charm and beauty) are effective probes to test perturbative QCD (pQCD) based calculations in pp collisions and to study cold nuclear matter (CNM) effects such as gluon saturation, shadowing, $k_{\rm T}$ broadening and energy loss in p–Pb collisions. Recent observations in pp and p–Pb collisions shown remarkable similarities with Pb–Pb collisions, which might suggest the presence of collectivity. To further explore the origin of the collective-like effects observed in pp and p–Pb collisions, the study of open heavy-flavour production as a function of the charged-particle multiplicity naturally links soft and hard processes that occur in the collision and allows one to study their interplay.

In ALICE [1], open heavy-flavour hadrons are either detected directly via the reconstruction of hadronic decays of D mesons (D^0, D^+, D^{*+}, and D_s^+) and Λ_c^+ baryon at midrapidity, or indirectly by reconstructing a single electron at midrapidity,

Preeti Dhankher for the ALICE Collaboration.

P. Dhankher (✉)
Indian Institute of Technology Bombay, Mumbai 400076, India
e-mail: preeti.dhankher@cern.ch

D. Elia et al. (eds.), *The XVIII International Conference on Strangeness in Quark Matter (SQM 2019)*, Springer Proceedings in Physics 250,
https://doi.org/10.1007/978-3-030-53448-6_39

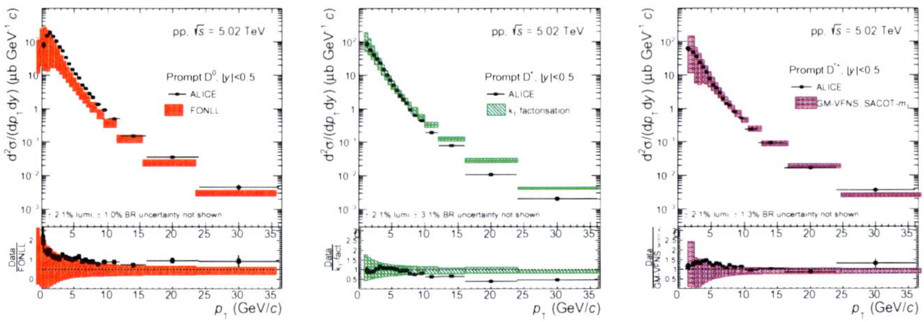

Fig. 39.1 D-meson production cross sections measured in pp collisions at $\sqrt{s} = 5.02$ TeV compared with pQCD models

or a muon produced at forward rapidity via semi-leptonic decay channels. In addition, Λ_c^+ and Ξ_c^0 are also reconstructed via semi-leptonic decays at midrapidity.

39.1.1 Heavy-Flavour Production in pp *Collisions*

The non-strange D-meson (D^0, D^+ and D^{*+}) production cross sections measured down to low p_T (0 GeV/c for D^0) in pp collision at $\sqrt{s} = 5.02$ TeV [2] are shown in Fig. 39.1. The measurement is compared with different pQCD models using the Fixed Order plus Next-to-Leading Logarithms approach (FONLL) [3], k_T-factorization [4] and General-Mass Variable-Flavour-Number Scheme (GM-VFNS) [5]. All the models describe well the experimental data within their large uncertainties. The data are more precise than the model predictions, therefore providing strong constraints on their parameters. The production cross section of HFE and HFM (c, b $\rightarrow$ e/μ) in pp collisions at $\sqrt{s} = 13$ TeV and 5.02 TeV is shown in Fig. 39.2 along with the FONLL calculations. The data lie at the upper edge of the theoretical prediction for both measurements. The semi-leptonic decay measurement for muons (right) also shows separately c $\rightarrow$ μ and b $\rightarrow$ μ cross section predictions from FONLL, which indicate that beauty is the main contributor for $p_T \gtrsim 5$ GeV/c, whereas at low p_T the charm contribution dominates.

39.1.2 Multiplicity Dependence in Heavy-Flavour Production

The study of heavy-flavour production as a function of multiplicity correlates the hard and soft processes of particle production. The self-normalized yields of HFE in pp collisions at $\sqrt{s} = 13$ TeV (left) and HFM measured in pp collisions at $\sqrt{s} = 8.16$ TeV (right) are shown in Fig. 39.3, in which they both exhibit a faster-than-linearly increasing trend, where the increase is steeper for high-p_T intervals.

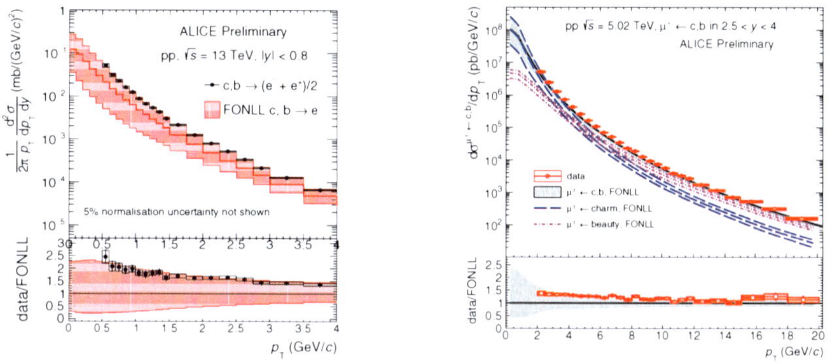

Fig. 39.2 Production cross section of HFE in pp collisions at $\sqrt{s} = 13$ TeV (left) and HFM in pp collisions at $\sqrt{s} = 5.02$ TeV [6] (right)

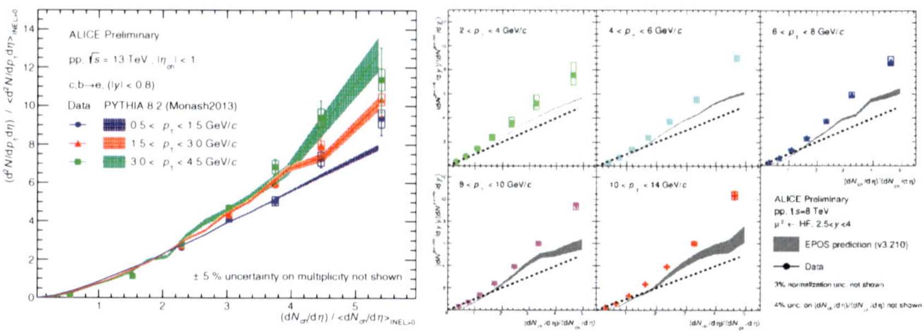

Fig. 39.3 Multiplicity-dependent self-normalized yields of HFE in pp collisions at $\sqrt{s} = 13$ TeV compared with PYTHIA8.2 predictions (left) and of HFM in pp collisions at $\sqrt{s} = 5.02$ TeV compared with EPOS3.2 predictions without hydrodynamics

The left panel shows that the HFE measurement is consistent with the PYTHIA8.2 [7] predictions, which incorporates Multiple Parton Interactions (MPI) and indicates that the increasing trend may be linked to MPIs. The right panel shows the HFM measurement compared with the EPOS3.2 [8] predictions without hydrodynamic treatment of the collision, which also include MPIs. The data are underpredicted by the model at high multiplicity which suggests that there can be some collective-like effects along with the MPIs. Figure 39.4, left panel, demonstrates that the similar steeper-than-linear increasing trend is observed also for the average D-meson relative yields measured in p–Pb collisions and in pp collisions and the right panel shows that the D-meson and HFE measurements are consistent within the uncertainties.

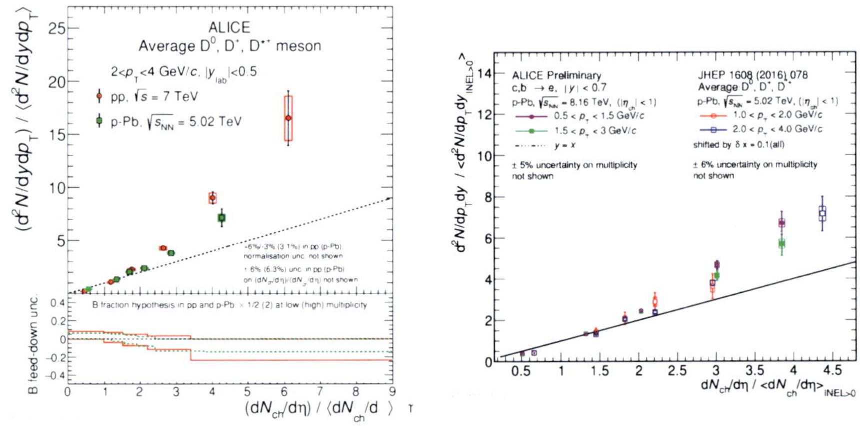

Fig. 39.4 Multiplicity-dependent self-normalized yield of average D-mesons in pp collisions at $\sqrt{s} = 7$ TeV (left) and of HFE in p–Pb collisions at $\sqrt{s_{\rm NN}} = 8.16$ TeV (right). Both results are compared with average D-meson measurement in p–Pb collisions at $\sqrt{s_{\rm NN}} = 5.02$ TeV

39.1.3 Nuclear Matter Effects and v_2 in p–Pb Collisions

Nuclear modification factor measurements in p–Pb collision provide access to CNM effects and possible collective-like effects. Figure 39.5, left and middle panels show the nuclear modification factor ($R_{\rm pPb}$) of prompt D-mesons given by (39.1), comparing it to model predictions as well.

$$R_{\rm pPb} = \frac{1}{A}\frac{{\rm d}^2\sigma_{\rm pPb}^{\rm prompt\ D}/{\rm d}p_{\rm T}dy}{{\rm d}^2\sigma_{\rm pp}^{\rm prompt\ D}/{\rm d}p_{\rm T}{\rm d}y},\quad Q_{\rm cp} = \frac{({\rm d}^2\sigma_{\rm pPb}^{\rm prompt\ D}/{\rm d}p_{\rm T}dy)^{i}_{\rm pPb}}{({\rm d}^2\sigma_{\rm pPb}^{\rm prompt\ D}/{\rm d}p_{\rm T}dy)^{60-100\%}_{\rm pPb}} \tag{39.1}$$

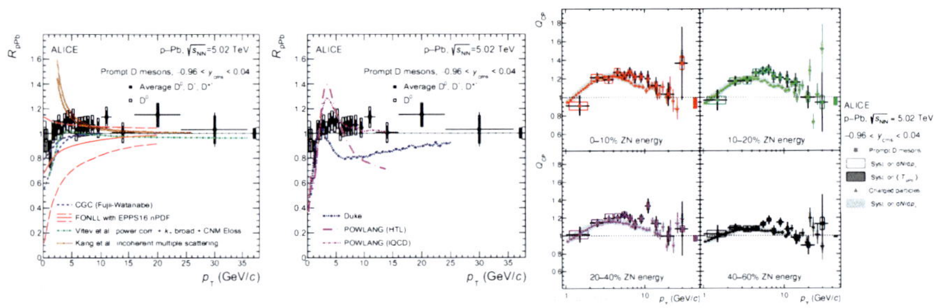

Fig. 39.5 Nuclear modification factor $R_{\rm pPb}$ (left and middle) and $Q_{\rm cp}$ (right) of prompt D-meson measured in p–Pb collisions at $\sqrt{s_{\rm NN}} = 5.02$ TeV [14] $R_{\rm pPb}$ measurement compared with models including CNM effects (left) and assuming formation of QGP (right)

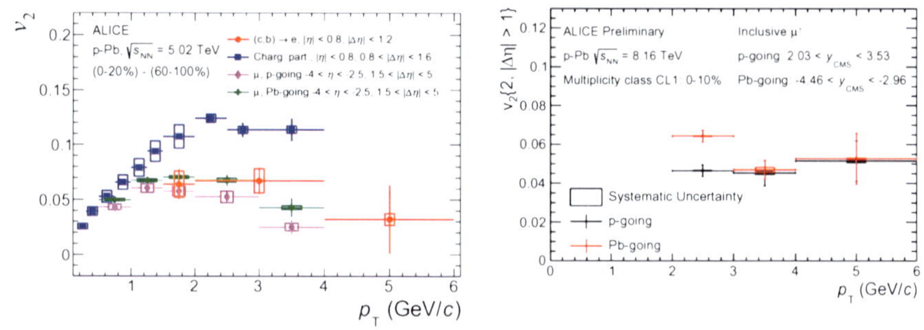

Fig. 39.6 Elliptic-flow coefficient (v_2) of HFE in p–Pb collisions at $\sqrt{s_{NN}} = 5.02$ TeV compared with the one of charged particles and inclusive muons [15] (left) and v_2 of inclusive muons measured in p–Pb collisions at $\sqrt{s_{NN}} = 8.16$ TeV (right)

The data is described well by models including CNM effects [9–12] (left), whereas the middle panel shows the comparison of data with models which predict a small QGP droplet formation [13] and a suppression at high p_T, which is disfavoured by the data. The right panel shows the measurement of Q_{cp} given by (39.1) which is comparable to the measurement of inclusive charged particles. The enhancement ($\sim 3\sigma$) observed at intermediate p_T could be interpreted as larger radial flow in most central p–Pb collisions.

As shown in Fig. 39.6, a positive elliptic-flow coefficient (v_2) is observed in the heavy-flavour sector, which potentially can be interpreted as the existence of collective-like behaviour in small systems. The left panel of Fig. 39.6 shows the v_2 measurement obtained from a Fourier decomposition of the correlation distribution between HFE and charged particles in p–Pb collisions at $\sqrt{s_{NN}} = 5.02$ TeV [15]. The measured v_2 is positive with a significance of more than 5σ in the $1.5 < p_T < 4$ GeV/c range. The right panel shows the positive v_2 of inclusive muons obtained with Q-cumulant method in p–Pb collisions at $\sqrt{s_{NN}} = 8.16$ TeV.

39.2 Summary and Conclusions

In summary, the results of D-meson, HFE, and HFM production cross section measured in pp collisions at 5.02 TeV and 13 TeV are discussed. Various pQCD theoretical models describe the pp data within their uncertainties. The self-normalized yields of D mesons, HFE, and HFM demonstrate stronger-than-linear dependence on charged-particle multiplicity and this trend can be qualitatively explained by models including MPIs. The R_{pPb} measurement of prompt D mesons is consistent with no modification of the spectra over the whole momentum range within the current uncertainties. The Q_{cp} measurement of prompt D mesons and the positive v_2 of HFE and HFM in p–Pb collisions suggest potential collectivity in small systems.

References

1. K. Aamodt et al., ALICE Collaboration. JINST **3**, S08002 (2008)
2. Acharya, S. et al., (ALICE Collaboration). Eur. Phys. J. C **79**(5), 388 (2019)
3. M. Cacciari, S. Frixione, N. Houdeau et al., JHEP **10**, 137 (2012)
4. R. Maciuła, A. Szczurek, Phys. Rev. D **87**, 094022 (2013)
5. I. Helenius, H. Paukkunen, JHEP **05**, 196 (2018)
6. S. Acharya et al., ALICE Collaboration. JHEP **1909**, 008 (2019). arXiv:1905.07207 [nucl-ex]
7. Sjostrand, T. et al., Comput. Phys. Commun. **191**, 159 (2015). arXiv:1410.3012 [hep-ph]
8. K. Werner, B. Guiot, Iu. Karpenko, T. Pierog, Phys. Rev. C **89**, 064903 (2014)
9. H. Fujii, K. Watanabe, Nucl. Phys. A **920**, 78 (2013)
10. M.L. Mangano, P. Nason, G. Ridolfi, Nucl. Phys. B **373**, 295 (1992)
11. R. Sharma, I. Vitev, B.W. Zhang, Phys. Rev. C **80**, 054902 (2009)
12. Z.B. Kang, I. Vitev, E. Wang, H. Xing, C. Zhang, Phys. Lett. B **740**, 23 (2015)
13. M. Nardi, A. Beraudo, A. De Pace, M. Monteno, F. Prino, arXiv:1509.04083 [hep-ph]
14. Acharya, S. et al., (ALICE Collaboration), arXiv:1906.03425 [nucl-ex]
15. S. Acharya et al., ALICE Collaboration. PRL **122**, 072301 (2019). arXiv:1805.04367 [nucl-ex]

Chapter 40
Experimental Overview on Strangeness and Collectivity in Small Systems

Roberto Preghenella

Abstract Since the discovery of the "ridge" in high-multiplicity proton-proton collisions at the LHC, the investigation of collectivity in small systems has become an increasingly large subject. In these proceedings, I report a brief overview of the past and recent experimental results addressing the physics of strange particle production and the search for collective phenomena in small collision systems.

40.1 Introduction

Soon after the startup of the LHC, the analysis of high-multiplicity events in proton-proton collisions has revealed the existence of unexpected phenomena in small collision systems. The CMS Collaboration discovered the so-called "ridge" in pp collisions [1], which was the first of a list of unexpected results that followed suit. Studies of particle production have uncovered the presence of long-range, near-side angular correlations in pp and in p–Pb collisions [2], paving the way for a systematic investigation of the existence of collective phenomena, known since long in heavy-ion collisions [3], in the much smaller pp and p–Pb collision systems. A wealth of new, unexpected phenomena have been observed so far in small collision systems at the LHC and at RHIC, with striking similarities to heavy-ion phenomenology [4].

40.2 Collective Phenomena

The initial discovery of the ridge in pp [1] and its subsequent observation in p–Pb collisions [2] was the trigger for more detailed searches for collective phenomena in small collision systems. Two-particle correlations results in p–Pb collisions have been reported by the ALICE Collaboration, showing the presence of the so-called

R. Preghenella (✉)
Istituto Nazionale di Fisica Nucleare, Sezione di Bologna, Italy
e-mail: preghenella@bo.infn.it

D. Elia et al. (eds.), *The XVIII International Conference on Strangeness in Quark Matter (SQM 2019)*, Springer Proceedings in Physics 250,
https://doi.org/10.1007/978-3-030-53448-6_40

"double ridge" [5, 6] and a large number of results on two-particle correlations followed, especially on correlation of identified hadrons [7–13] and of Bose-Einstein (HBT) correlations [14, 15]. The similarities between pp, p–Pb and Pb–Pb collisions, consistent with the hydrodynamic picture of a particle-production source expanding collectively and more explosively along the collision event plane, were striking.

The LHC discoveries were complemented by results from RHIC experiments that followed closely the investigation of small systems with a set of reanalyses of previously recorded d–Au data [16, 17]. A significant addition to this exploration comes from a recent work reported by the PHENIX Collaboration at RHIC that has reported on the observation of elliptic and triangular flow patterns of charged particles produced in p–Au, d–Au and ^{3}He–Au collisions [18]. With the unique combination of three distinct initial geometries and two flow patterns PHENIX provided a powerful way to discriminate between different models, showing that hydrodynamical ones, which include the formation of a short-lived Quark-Gluon Plasma (QGP) droplet, provide the best simultaneous description of the measurements.

Measurements of two-particle angular correlations of charged particles emitted in hadronic Z^0 decays have been recently performed using the archived e^+e^- annihilation data collected with the ALEPH detector at LEP [19]. No significant long-range correlation is observed in either the beam-axis analysis or the thrust-axis one, where the latter is sensitive to a medium expanding transverse to the color string between the outgoing $q\bar{q}$ pair. Similar negative results from the analysis of archived ZEUS data in electron-proton DIS collisions indicate that flow-like correlations are a feature unique to hadronic collision systems.

One question one needs to ask for the case of small collision systems is whether the observed correlation structures actually result from genuine multi-particle collective effects. Strong hints for collective multi-particle correlations in pp and p–Pb collisions were reported already [7, 20], supporting the interpretation of a collective origin for the observed long-range correlations in high-multiplicity pp and p–Pb collisions. But one has to be careful and find methods to reliably suppress the correlations that result from non-flow effects present in pp collision in order to be able to unveil the actual underlying collectivity. As a matter of fact, a significant fraction of non-flow correlations that are present in small systems in a broad range of N_{ch} have been revealed [21–27].

The measurement of the production of identified hadrons and their p_{T} distributions provides another tool to look from a difference perspective at the collective behaviour. The shape of the transverse momentum distributions encode in their characteristics dependence on the particle mass the signature of the development on collective radial flow. A clear evolution in the p_{T} distributions in pp and p–Pb collisions can be observed, with spectra becoming harder with increasing charged-particle multiplicity and with a more pronounced modification for heavier particles [28–32]. The p_{T}-dependent ratios, such as the Λ/K^0_S ratio measured by the CMS Collaboration [29] can better reveal the stronger multiplicity dependence of the spectral shapes of heavier particles. The ratios in pp and p–Pb collisions show a significant enhancement at intermediate p_{T} ($\sim$3 GeV/c), quantitatively similar to that measured in Pb–Pb collisions at similar multiplicity. Results from the ALICE Collaboration show that the modification of the p_T-dependent ratios of p/π and of Λ/K^0_S follows a rather smooth evolution at low and mid p_T across different systems, pointing towards a universal soft mechanism driven by final state $dN_{ch}/d\eta$ [33]. The high-p_T part of the ratio, which is dominated by hard fragmentation, is unchanged. The spectral shapes of identified hadrons can be characterised with the Blast-Wave model [34], that has proven to be a useful tool to test the data against the radial flow picture. In this context, when comparing the Bast-Wave parameters of different systems at similar $dN_{ch}/d\eta$, T_{kin} values are similar, whereas β_T are larger for small systems. Stronger pressure gradients in smaller systems could be at play and result in a larger observed radial flow, as predicted in [35]. Both T_{kin} and β_T parameters obtained from pp and p–Pb fits are consistent for similar $dN_{ch}/d\eta$ when looking at ALICE results. On the other hand, a different conclusion emerges from CMS results [29], where pp collisions exhibit a larger β_T for similar $dN_{ch}/d\eta$.

40.3 Strangeness Production

The study of the production of strange hadrons in high-energy hadronic interactions provides an important means to investigate the properties of QCD. String fragmentation, as an example of non-perturbative QCD process, dominates the production of strange hadrons at low p_T. As a matter of fact, given that the mass of the strange quark is larger than the one of up and down quarks, production of strange hadrons in fragmentation is suppressed relative to hadrons containing only light quarks. An enhanced production of strange hadrons was one of the earliest proposed indicators for the formation of a Quark-Gluon Plasma (QGP) state [36–38]. Strangeness enhancement was expected to be more pronounced for multi-strange baryons and was indeed observed in collisions of heavy nuclei at the SPS, RHIC and LHC [39–48].

The first observation of strangeness enhancement in high-multiplicity pp collisions has been reported by the ALICE Collaboration with measurements of the production of primary strange and multi-strange hadrons [30]. Similar measurements have been previously performed in p–Pb collisions [33, 49] and in Pb–Pb collisions [48, 50]. When increasing the charged-particle multiplicity in pp collisions, a significant enhancement of strange to non-strange hadron production is observed. The values of the ratios and their evolution with $dN_{ch}/d\eta$ resembles those observed in p–Pb collisions at a slightly lower centre-of-mass energy and suggests that if there exists a common driver for the strangeness production that is not given by the initial-state collision system or energy but rather by the final-state event activity [51]. The observation of the constant production of protons relative to pions along with the multiplicity-dependent enhancement of the production of strange hadrons could not be simultaneously reproduced by the Monte Carlo event generator models commonly used at the moment of the discovery [52–56]. This has raised a challenge for the general purpose pp event generators and their underlying models that has significantly stimulated and attracted the corresponding theoretical community towards a better understanding of the physics of heavy-ion collisions from a microscopical point of view [57–60].

Hadron abundances in heavy-ion collisions, including those of strange particles, are compatible with the expectations from chemical equilibrium and can be described using statistical hadronisation models in the grand-canonical ensamble [61, 62]. To effectively produce a suppression of strangeness production in small systems one needs to extend the grand-canonical statistical description and introduce new concepts, like the ones employed in the strangeness canonical suppression [63] and the core-corona superposition [64, 65] models. The study reported in [66] has analysed the prodution of light-flavour hadrons as a function of the multiplicity at LHC energies in the strangeness canonical suppression picture using THERMUS [67]. The model provides a good description of the data, but with a notable exception for the case of the ϕ meson. Given the fact that a strangeness canonical suppression model is only sensitive to the hadronic quantum numbers, the strangeness in this case, the observation of an enhancement (suppression) of ϕ meson production with increasing (decreasing) multiplicity is inconsistent with the canonical suppression interpretation. The indication is that the ϕ meson behaviour is closer to the behaviour expected from a hadron formed by two strange quarks. The increase of the relative ϕ meson production with $\mathrm{d}N_{\mathrm{ch}}/\mathrm{d}\eta$ does not allow one to proceed along with the canonical suppression interpretation of the observed strangeness enhancement in high-multiplicity pp and p–Pb collisions in a fully consistent way. Further analyses of hadron production in small collision systems employing statistical models in several configurations have been performed in [68]. The authors conclude that high-multiplicity proton-proton collisions are close to the thermodynamic limit and estimate that the final state in pp collisions could reach the thermodynamic limit when $\mathrm{d}N_{\mathrm{ch}}/\mathrm{d}\eta$ is larger than twenty per unit of rapidity. Interestingly, results from the study of the PHENIX experiment at RHIC on the measurement of direct photon emission from heavy ion collisions seem to imply that there must be a transition at a similar $\mathrm{d}N_{\mathrm{ch}}/\mathrm{d}\eta \approx 2$ to 20 from the small yield of direct photons in pp collisions to an enhanced regime like the one observed in Au–Au collisions [69].

40.4 Conclusions

A multitude of features which in nuclear collisions are typically attributed to the formation of a strongly-interacting collectively-expanding quark-gluon medium, like near-side long-range correlations and mass-dependent hardening of p_{T} distributions, have been observed in high-multiplicity events with small collision systems at the LHC and at RICH. An enhanced production of strange particles as a function of the charged-particle multiplicity density ($\mathrm{d}N_{\mathrm{ch}}/\mathrm{d}\eta$), originally considered to be another signature of QGP formation in nuclear collisions, has also been observed in small collision systems.

It is unequivocal that the study of the features emerging from small collision systems at high multiplicity is of significant interest. Further studies are essential to assess whether the combination of these observations can be interpreted as signal of the progressive onset of a QGP medium, which seems to start developing already in small systems.

References

1. V. Khachatryan et al., JHEP **09**, 091 (2010)
2. S. Chatrchyan et al., Phys. Lett. B **718**, 795 (2013)
3. B. Abelev et al., Phys. Rev. C **80**, 064912 (2009)
4. R. Preghenella, EPJ Web Conf. **171**, 11003 (2018)
5. B. Abelev et al., Phys. Lett. B **719**, 29 (2013)
6. G. Aad et al., Phys. Rev. Lett. **110**(18), 182302 (2013)
7. V. Khachatryan et al., Phys. Lett. B **765**, 193 (2017)
8. B. Abelev et al., Phys. Lett. B **726**, 164 (2013)
9. V. Khachatryan et al., Phys. Lett. B **742**, 200 (2015)
10. B. Abelev et al., JHEP **06**, 190 (2015)
11. V. Pacík, Nucl. Phys. A **982**, 451 (2019)
12. A.M. Sirunyan et al., Phys. Rev. Lett. **121**(8), 082301 (2018)
13. A.M. Sirunyan et al., Phys. Lett. B **791**, 172 (2019)
14. ATLAS Collaboration, ATLAS-CONF-2017-008 (2017)
15. D. Adamova et al., Phys. Rev. Lett. **118**(22), 222301 (2017)
16. A. Adare et al., Phys. Rev. Lett. **111**(21), 212301 (2013)
17. A. Adare et al., Phys. Rev. Lett. **114**(19), 192301 (2015)
18. C. Aidala et al., Nat. Phys. **15**(3), 214 (2019)
19. A. Badea, A. Baty, P. Chang, G.M. Innocenti, M. Maggi, C. Mcginn, M. Peters, T.A. Sheng, J. Thaler, Y.J. Lee (2019). arXiv:hep-ex/1906.00489
20. V. Khachatryan et al., Phys. Rev. Lett. **115**(1), 012301 (2015)
21. J. Adam et al., Phys. Rev. Lett. **117**, 182301 (2016)
22. M. Aaboud et al., Phys. Rev. C **97**(2), 024904 (2018)
23. A.M. Sirunyan, et al. (2019). arXiv:hep-ex/1905.09935
24. M. Aaboud et al., Phys. Lett. B **789**, 444 (2019)
25. A.M. Sirunyan, et al. (2019). arXiv:1904.11519
26. A.M. Sirunyan et al., Phys. Rev. Lett. **120**(9), 092301 (2018)
27. S. Acharya et al., Phys. Rev. Lett. **123**(14), 142301 (2019)
28. S. Acharya et al., Phys. Rev. C **99**(2), 024906 (2019)
29. V. Khachatryan et al., Phys. Lett. B **768**, 103 (2017)
30. J. Adam et al., Nat. Phys. **13**, 535 (2017)
31. A.M. Sirunyan et al., Phys. Rev. D **96**(11), 112003 (2017)
32. CMS Collaboration, CMS-PAS-HIN-16-013 (2018)
33. B. Abelev et al., Phys. Lett. B **728**, 25 (2014)
34. E. Schnedermann et al., Phys. Rev. C **48**, 2462 (1993)
35. E. Shuryak, I. Zahed, Phys. Rev. C **88**(4), 044915 (2013)
36. J. Rafelski et al., Phys. Rev. Lett. **48**, 1066 (1982)
37. P. Koch, J. Rafelski, W. Greiner, Phys. Lett. B **123**, 151 (1983)
38. P. Koch, B. Muller, J. Rafelski, Phys. Rep. **142**, 167 (1986)
39. E. Andersen et al., Phys. Lett. B **433**, 209 (1998)
40. E. Andersen et al., Phys. Lett. B **449**, 401 (1999)
41. F. Antinori et al., Phys. Lett. B **595**, 68 (2004)
42. S.V. Afanasiev et al., Phys. Lett. B **538**, 275 (2002)
43. T. Anticic et al., Phys. Rev. Lett. **93**, 022302 (2004)
44. J. Adams et al., Phys. Rev. Lett. **92**, 182301 (2004)

45. J. Adams et al., Phys. Rev. Lett. **98**, 062301 (2007)
46. B. Abelev et al., Phys. Rev. C **77**, 044908 (2008)
47. F. Antinori et al., J. Phys. **G37**, 045105 (2010)
48. B. Abelev et al., Phys. Lett. B **728**, 216 (2014)
49. J. Adam et al., Phys. Lett. B **758**, 389 (2016)
50. B. Abelev et al., Phys. Rev. Lett. **111**, 222301 (2013)
51. S. Acharya et al. (2019). arXiv:nucl-ex/1908.01861
52. T. Sjöstrand et al., Comput. Phys. Commun. **178**, 852 (2008)
53. T. Pierog, I. Karpenko, J. Katzy, E. Yatsenko, K. Werner (2013). arXiv:hep-ph/1306.0121
54. C. Flensburg, G. Gustafson, L. Lonnblad, JHEP **08**, 103 (2011)
55. C. Bierlich, G. Gustafson, L. Lönnblad, A. Tarasov, JHEP **03**, 148 (2015)
56. C. Bierlich, J.R. Christiansen, Phys. Rev. D **92**, 094010 (2015)
57. C. Bierlich, G. Gustafson, L. Lönnblad, H. Shah, JHEP **10**, 134 (2018)
58. C. Bierlich, G. Gustafson, L. Lönnblad (2016). arXiv:hep-ph/1612.05132
59. S. Ferreres-Solé, T. Sjöstrand, Eur. Phys. J. C **78**(11), 983 (2018)
60. T. Sjöstrand, Nucl. Phys. A **982**, 43 (2019)
61. J. Cleymans et al., Phys. Rev. C **74**, 034903 (2006)
62. A. Andronic et al., Phys. Lett. B **673**, 142 (2009)
63. K. Redlich, A. Tounsi, Eur. Phys. J. C **24**, 589 (2002)
64. F. Becattini, J. Manninen, J. Phys. **G35**, 104013 (2008)
65. J. Aichelin et al., Phys. Rev. C **79**, 064907 (2009)
66. V. Vislavicius, A. Kalweit (2016). arXiv:nucl-ex/1610.03001
67. S. Wheaton, J. Cleymans, J. Phys. **G31**, S1069 (2005)
68. N. Sharma, J. Cleymans, B. Hippolyte, Adv. High Energy Phys. **2019**, 5367349 (2019)
69. A. Adare et al., Phys. Rev. Lett. **123**(2), 022301 (2019)

Chapter 41
Experimental Overview on Heavy Flavour in Small Systems

Émilien Chapon

Abstract Heavy flavour production in small systems is sensitive to a variety of effects, from initial to final state. It can be studied using leptons from heavy flavour hadron decays, or meson or baryon exclusive hadronic decays. The yields are compared to models including initial state effects, such as nuclear modifications of the parton distribution functions or in the colour glass condensate framework. These data can also be used to look for signals of a quark gluon plasma. Additional information is gained from studies of the heavy flavour elliptic flow and correlations.

41.1 Introduction

The large mass of the heavy quarks make them excellent probes of the physics of small collision systems (proton-proton and proton-nucleus). Their production can be computed in a perturbative way, which helps study the initial state: either the nuclear modifications of the parton distribution functions (nPDF), or gluon saturation in the colour glass condensate (CGC) framework. Measurements in small systems also provide a comparison point for the interpretation of results in nucleus-nucleus collisions: in these intermediate size systems, a different interplay of initial state and other cold nuclear matter (CNM) effects is expected, as well as a smaller amount of quark gluon plasma (QGP) effects, if any. They provide insights on collectivity, where again comparing light and heavy flavours sheds light on the importance of the quark mass. We will see whether heavy flavour (HF) hadrons exhibit elliptic flow in proton-nucleus collisions, and how it compares with that of light flavours. Finally, since qualitative similarities are found with nucleus-nucleus collisions, all these measurements help look for a possible QGP created in small systems, in proton-nucleus or even proton-proton collisions.

Experimentally, several methods can be used to measure heavy flavour. The first one is to measure charged leptons, produced in the weak decays of HF hadrons (the

É. Chapon (✉)
CERN, Geneva, Switzerland
e-mail: emilien.chapon@cern.ch

D. Elia et al. (eds.), *The XVIII International Conference on Strangeness in Quark Matter (SQM 2019)*, Springer Proceedings in Physics 250,
https://doi.org/10.1007/978-3-030-53448-6_41

decay chain of beauty hadrons can involve a charm hadron). Heavy flavour electrons and muons can be separated from other sources (such as decays from light flavour hadrons or resonances, and photon conversions for electrons) by the fact that they are displaced with respect to the primary vertex: the lifetime of HF hadrons translates into measurable flight distances, from $c\tau \approx 150\,\mu$m for charm to $c\tau \approx 500\,\mu$m for beauty. This difference is also used for extracting separately the charm and beauty contributions. A second method is to focus on the exclusive decay of specific hadrons, such as $D^0 \rightarrow K^-$ß$^+$ or $B^+ \rightarrow J/\ K^+$. Also for these hadronic final states, the presence of a displaced vertex is used experimentally for suppressing the combinatorial background. Finally, further information on heavy quark fragmentation can be gained from the study of HF-tagged jets, defined experimentally either from the presence of a HF hadron (such as D^0) inside the jet, or from other algorithms based on jet properties (presence of a secondary vertex or a decay lepton). In a similar fashion, observables can also be defined from the correlation between the HF hadron and light hadrons.

41.2 Production

Measurements of HF electrons and muons in pPb collisions give access to inclusive HF hadrons in semi-leptonic decays. Recent results at $\sqrt{s_{NN}} = 5.02\,$TeV by the ALICE Collaboration of beauty electrons [1], at midrapidity, and HF muons [2], at forward rapidity, are consistent with expectations from EPS09 [3] shadowing, though with large uncertainties in the former case. The PHENIX Collaboration has also reported the nuclear modification factor of inclusive $c\bar{c}$ and $b\bar{b}$ production in deuteron-gold collisions using e^+e^- production [4], found to be consistent with one, with large uncertainties and some dependence on the model assumed in the analysis procedure. Very precise cross sections are also reported in pp [5], with much smaller uncertainties than models (such as FONLL [6]).

The measurement of identified hadrons, such as $D^0 \rightarrow K^-$ß$^+$, leads to several advantages, despite the reduction in number of events because of the fragmentation and branching ratio. It gives access to the original hadron kinematics (including p_T), and to the study of possible differences between mesons and baryons, for instance. In pp collisions, the recent ALICE measurement of 4 different D meson species at $\sqrt{s} = 5.02\,$TeV [7] is very precise, lying at the upper edge of the large uncertainty band from FONLL [6]: these data may be used to further constrain this HF production model and others. In addition, such measurements can also be used to extract the total charm cross section [8], which is an important input to several calculations such as quarkonium production in heavy ion collisions.

In pPb collisions, the measurement of D meson production cross sections allows the direct comparison of the data [9–11] to models, though the corresponding theoretical uncertainties are usually large, as discussed above, from scale uncertainties in FONLL and gluon shadowing at low Bjorken x. This motivates the direct use of pp measurements as a reference, building nuclear modification factors R_{pPb} (Fig. 41.1).

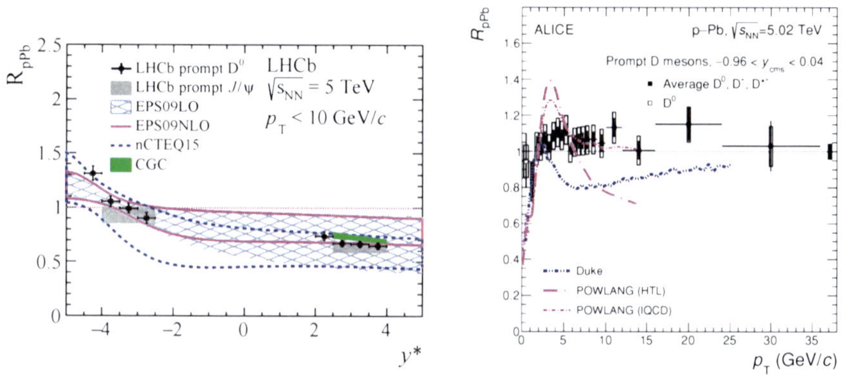

Fig. 41.1 Nuclear modifications factor of D^0 mesons from the LHCb [10] (left) and ALICE [9] (right) experiments, compared to initial state models (left) and transport models including a QGP (right)

Several observations can be made from these ratios. First, results for different D meson species are found to be consistent, between D^+, D^0, D^{*+}, and even D_s^+ [9, 11]. Further, modifications are found to be consistent with expectations from nPDF models [10], but also with CGC expectations in the corresponding domain of applicability (at small x_{Pb}, i.e. at forward rapidity). The data, especially from LHCb, are found to be very precise, much more than nPDFs, especially at small x_{Pb}, in the shadowing region, but also at larger x_{Pb} in the antishadowing region. It is then natural to try to constrain nPDFs using these data. A couple of theory groups have discussed recently the use of HF data for constraining low x gluon shadowing and the potentially large reductions of uncertainties that could be obtained. The authors of a first paper [12] have used a data-driven parametrisation of matrix elements, and a Bayesian reweighting of MC replicas. On the other hand, the second group [13] has obtained the partonic cross sections in a SACOT-m_T scheme in the GM-VFNS approach, with a Hessian reweighting for constraining nPDFs. Very recent LHCb D^0 meson results in pPb [14] suggest CNM effects on top of nPDF, however, showing that some care is needed if planning to use these data as inputs to an nPDF analysis.

By injecting noble gases in the interaction region (using SMOG), the LHCb experiment has the capability to measure charm production in a fixed-target mode, which is probing the large x region. In order to remove "normal" beam-beam interactions, events are selected when only beam 1 is present. The D^0 results in proton-helium collisions [15], reported as a function of rapidity in the centre-of-mass frame (spanning backward to midrapidity), show no evidence for substantial valence-like intrinsic charm contribution, since the data are below the expectation in the most backward (largest x) bin, where intrinsic charm would lead to an enhancement [16].

The ALICE Collaboration has compared their D meson data in pPb (averaging D^+, D^0, and D^{*+}) [9] to several models of initial state (CGC, nPDF) and CNM effects (k_T broadening, energy loss, incoherent multiple scattering), or even transport models assuming the creation of a QGP droplet in pPb collisions. The latter predict a bump in the R_{pPb} around 2–4 GeV and a suppression at higher p_{T}, but neither of these features is found in the data, disfavouring these models, even in the most "central" (higher activity) pPb collisions.

Further studies of D meson production as a function of event activity in pPb collisions have been reported by the ALICE Collaboration [17]. Different trends are found depending on the event activity estimator: when charged particle tracks are used, measured at midrapidity, a faster than linear increase is observed, while an almost linear correlation between D meson yields and event activity is found when the V0A detector is used. In all cases, EPOS 3 [18] provides a better description when including hydrodynamics than without. In addition, the nuclear modification factor as a function of p_{T} in different centrality classes, Q_{pPb}, has also been studied by the ALICE Collaboration [9]. In all centrality classes, a similar trend is found as for light hadrons (charged particles), including a hint of suppression at low p_{T} and a hint of enhancement at high p_{T} in central events.

Heavy flavour hadronisation in a heavy ion environment can be studied by comparing meson to baryon production. The $\Lambda_{\mathrm{c}}^{+}/\mathrm{D}^{0}$ ratio has been measured by the ALICE and LHCb [19] Collaborations, and found to be consistent with the one in pp collisions, decreasing with p_{T}. Nuclear PDF uncertainties almost full cancel in this ratio, and a better description is found by a model using a parametrisation of pp cross section than from event generators. Similar observations have been made in pp by the ALICE Collaboration for the $\Xi_{\mathrm{c}}^{0}/\mathrm{D}^{0}$ ratio [20].

Heavy flavour tagged jets have been compared in pp and pPb collisions by the CMS [21] and ALICE [22] Collaborations, respectively for charm jets, and bottom and heavy flavour electron jets. In all cases, results are found to be consistent in pp and pPb. The ALICE Collaboration has also compared D^{0} mesons in jets between pp and pPb collisions, with two measurements: D^{0}-tagged jets (jets including a D^{0} meson), and D^{0}-hadron correlations (near and away side production of charged hadrons) [23]. In both cases, no evidence has been found for CNM effects, within uncertainties.

Inclusive beauty production can be accessed in nonprompt J/ψ production, measured by all four LHC experiments [24–27]. Results are consistent with nPDF expectations, with an additional hint of higher suppression in high event activity events. Similarly to charm production, identified B mesons, such as $\mathrm{B}^{+} \rightarrow \mathrm{J}/\psi\mathrm{K}^{+}$, allow the reconstruction of the original hadron kinematics and particle identification, at the cost of smaller yields compared to nonprompt J/ψ. Results from the CMS [28] and LHCb [29] Collaborations are found to be consistent with nonprompt J/ψ, as well as nPDF expectations. Recently, the LHCb Collaboration has reported a measurement of Λ_{b}^{0} in pPb collisions [29], the first b baryon measurement in heavy ion collisions, constraining the b quark fragmentation in a nuclear environment. The baryon-to-meson ratio is found to be compatible with the pp value.

Going to even higher masses, the CMS Collaboration has reported the first measurement of top quarks in pPb collisions [30]. Top quarks are too short-lived to hadronise, and measured by CMS in the lepton+jets final state. This process is sensitive to high x gluons, providing a new handle for nPDFs in this kinematic region.

41.3 Flow and Correlations

Measurements of heavy flavour hadron anisotropic flow in small systems are confronted with some experimental challenges. The so-called nonflow contribution, due to jet-like back-to-back topologies, has to be accounted for, usually using an estimation based on low multiplicity events, where it is expected to be dominating. In addition, the heavy flavour signal usually sits on top of a background, which can be large, and the signal v_2 needs to be extracted from the signal + background v_2. Different strategies are chosen by the experiments to solve these challenges. The ALICE and CMS Collaborations use two-particle correlations (Fourier fits), with nonflow subtraction. The ATLAS Collaboration relies on template fits of $\Delta\phi$ between the particle of interest and charged hadrons, where a template accounting for nonflow is built using low multiplicity events. Finally, the event plane method is used by the PHENIX Collaboration.

The elliptic flow of heavy flavour leptons has been measured by the ALICE [31], ATLAS [32], and PHENIX Collaborations. A non-zero v_2 is consistently found, also smaller than that of charged hadrons, consistently for all three experiments. The CMS experiment has also reported the elliptic flow of D^0 mesons in pPb collisions [33], scaling results by the number of constituent quarks n_q as motivated by the quark coalescence model: the hadron v_2 should reflect that initially carried by its constituent quarks. In pPb collisions, the v_2/n_q of D^0 mesons is smaller than strange hadrons for small kinematic energy KE_T/n_q, while in PbPb, both particles follow the same trend.

The ALICE [34] and CMS [35] Collaborations have also measured the J/ψ meson v_2 in pPb collisions. Large values are found, comparable to those in PbPb collisions, and they are similar in the p-going and Pb-going directions. The v_2 is lower for J/ψ than for strange hadrons, though both are consistent at low KE_T/n_q. Finally, it is similar to the D^0 meson v_2. These indications of the collective behaviour of J/ψ mesons in pPb are a challenge to transport models, which usually predict a smaller v_2 (Fig. 41.2).

41.4 Future

Finally, projections for the expected performance and precision of the heavy flavour measurements in the upcoming Runs 3 and 4 at the LHC are presented in the High-Luminosity LHC Yellow Report [36]. A large gain with respect to current LHC results will be brought both from the larger data sets (especially for the ALICE experiment) and detector upgrades. New ways to understand the heavy quark dynamics in small system and look for the QGP will include precise nuclear modification factor measurements (including B^+, currently suffering from large uncertainties), elliptic flow for heavy flavour leptons, D^0 and J/ψ, as well as $D^0\bar{D}^0$ correlations.

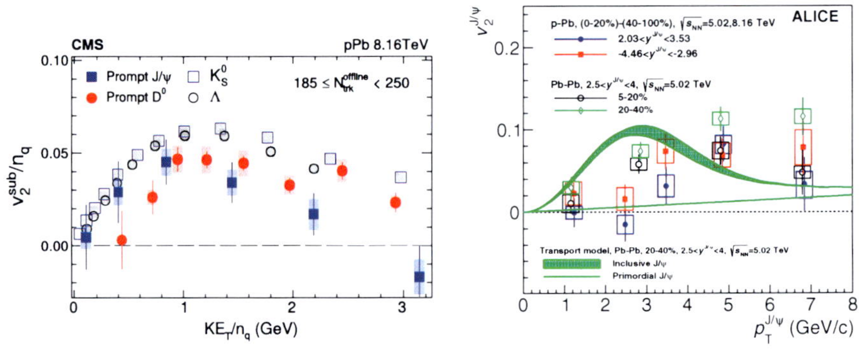

Fig. 41.2 Left: elliptic flow of J/ψ mesons in pPb collisions, compared to D^0 mesons and strange hadrons, from the CMS experiment [35]. Right: elliptic flow of J/ψ mesons in pPb collisions, separately for the p-going and Pb-going direction, compared to PbPb collisions, from the ALICE experiment [34]

41.5 Summary

Heavy flavour hadron production in pPb collisions shows modest modifications with respect to pp collisions, overall consistent with expectations from initial state modifications (nPDF or CGC). These modifications could then, in turn, be potentially used to constrain nPDFs. The data show no sign of the expectations from transport models assuming the creation of a QGP in pPb collisions, namely a flow peak at low pT and suppression at high pT. On the other hand, a large heavy flavour flow is found in pPb collisions, larger than predicted by transport models. Initial state fluctuations, expected to play a significant role in small systems, are however missing from models. Heavy flavour production in small systems sheds light on the initial state modifications and final state interactions, and shows no clear sign of QGP up to now.

References

1. ALICE Collaboration, Measurement of electrons from beauty-hadron decays in p-Pb collisions at $\sqrt{s_{NN}} = 5.02$ TeV and Pb-Pb collisions at $\sqrt{s_{NN}} = 2.76$ TeV. JHEP **1707**, 052 (2017). https://doi.org/10.1007/JHEP07(2017)052
2. ALICE Collaboration, Production of muons from heavy-flavour hadron decays in p-Pb collisions at $\sqrt{s_{NN}} = 5.02$ TeV. Phys. Lett. B **770**, 459 (2017). https://doi.org/10.1016/j.physletb.2017.03.049
3. K.J. Eskola, H. Paukkunen, C.A. Salgado, EPS09: A new generation of NLO and LO nuclear parton distribution functions. JHEP **04**, 065 (2009). https://doi.org/10.1088/1126-6708/2009/04/065
4. PHENIX Collaboration, Measurements of e^+e^- pairs from open heavy flavor in $p+p$ and $d+A$ collisions at $\sqrt{s_{NN}} = 200$ GeV. Phys. Rev. C **96**(2), 024907 (2017). https://doi.org/10.1103/PhysRevC.96.024907

5. ALICE Collaboration, Production of muons from heavy-flavour hadron decays in pp collisions at $\sqrt{s}$ = 5.02 TeV. JHEP **09**, 008 (2019). https://doi.org/10.1007/JHEP09(2019)008
6. M. Cacciari, M. Greco, P. Nason, The P(T) spectrum in heavy flavor hadroproduction. JHEP **05**, 007 (1998). https://doi.org/10.1088/1126-6708/1998/05/007
7. ALICE Collaboration, Measurement of D^0, D^+, D^{*+} and D_s^+ production in pp collisions at $\sqrt{s}$ = 5.02 TeV with ALICE. Eur. Phys. J. C **79**(5), 388 (2019). https://doi.org/10.1140/epjc/s10052-019-6873-6
8. ALICE Collaboration, D-meson production in p-Pb collisions at $\sqrt{s_{NN}}$ =5.02 TeV and in pp collisions at $\sqrt{s}$ =7 TeV. Phys. Rev. C **94**(5), 054908 (2016). https://doi.org/10.1103/PhysRevC.94.054908
9. ALICE Collaboration, Measurement of prompt D^0, D^+, D^{*+}, and D_s^+ production in p−Pb collisions at $\sqrt{s_{NN}}$ = 5.02 TeV. arXiv:1906.03425 [nucl-ex]
10. LHCb Collaboration, Study of prompt D^0 meson production in pPb collisions at $\sqrt{s_{NN}}$ = 5 TeV. JHEP **10**, 090 (2017). https://doi.org/10.1007/JHEP10(2017)090
11. ATLAS Collaboration, D meson production and long-range azimuthal correlation in 8.16 TeV p+Pb collisions with ATLAS, ATLAS-CONF-2017-073 (2017). http://cds.cern.ch/record/2285811
12. A. Kusina, J.P. Lansberg, I. Schienbein, H.S. Shao, Gluon shadowing in heavy-flavor production at the LHC. Phys. Rev. Lett. **121**(5), 052004 (2018). https://doi.org/10.1103/PhysRevLett.121.052004
13. K.J. Eskola, I. Helenius, P. Paakkinen, H. Paukkunen, A QCD analysis of LHCb D-meson data in p+Pb collisions. arXiv:1906.02512 [hep-ph]
14. LHCb Collaboration, Study of prompt D^0 meson production in pPb at $\sqrt{s_{NN}}$ = 8.16 TeV at LHCb, LHCb-CONF-2019-004 (2019). https://cds.cern.ch/record/2700244
15. LHCb Collaboration, First measurement of charm production in its fixed-target configuration at the LHC. Phys. Rev. Lett. **122**(13), 132002 (2019). https://doi.org/10.1103/PhysRevLett.122.132002
16. J. Pumplin, H.L. Lai, W.K. Tung, The charm parton content of the nucleon. Phys. Rev. D **75**, 054029 (2007). https://doi.org/10.1103/PhysRevD.75.054029
17. ALICE Collaboration, Measurement of D-meson production versus multiplicity in p-Pb collisions at $\sqrt{s_{NN}}$ = 5.02 TeV. JHEP **08**, 078 (2016). https://doi.org/10.1007/JHEP08(2016)078
18. H.J. Drescher, M. Hladik, S. Ostapchenko, T. Pierog, K. Werner, Parton based Gribov-Regge theory. Phys. Rep. **350**, 93 (2001). https://doi.org/10.1016/S0370-1573(00)00122-8
19. LHCb Collaboration: Prompt Λ_c^+ production in pPb collisions at $\sqrt{s_{NN}}$ = 5.02 TeV. JHEP **02**, 102 (2019). https://doi.org/10.1007/JHEP02(2019)102
20. ALICE Collaboration, First measurement of Ξ_c^0 production in pp collisions at $\sqrt{s}$ = 7 TeV. Phys. Lett. B **781**, 8 (2018). https://doi.org/10.1016/j.physletb.2018.03.061
21. CMS Collaboration, Measurements of the charm jet cross section and nuclear modification factor in pPb collisions at $\sqrt{s_{NN}}$ = 5.02 TeV. Phys. Lett. B **772**, 306 (2017). https://doi.org/10.1016/j.physletb.2017.06.053
22. ALICE Collaboration, These proceedings
23. ALICE Collaboration, Azimuthal correlations of prompt D mesons with charged particles in pp and p-Pb collisions at $\sqrt{s_{NN}}$ = 5.02 TeV. arXiv:1910.14403 [nucl-ex]
24. ATLAS Collaboration, Measurement of quarkonium production in protonlead and protonproton collisions at 5.02 TeV with the ATLAS detector. Eur. Phys. J. C **78**(3), 171 (2018). https://doi.org/10.1140/epjc/s10052-018-5624-4
25. ALICE Collaboration, Prompt and non-prompt J/ψ production and nuclear modification at mid-rapidity in pPb collisions at $\sqrt{s_{NN}}$ = **5.02** TeV. Eur. Phys. J. C **78**(6), 466 (2018). https://doi.org/10.1140/epjc/s10052-018-5881-2
26. LHCb Collaboration, Prompt and nonprompt J/ψ production and nuclear modification in pPb collisions at $\sqrt{s_{NN}}$ = 8.16 TeV. Phys. Lett. B **774**, 159 (2017). https://doi.org/10.1016/j.physletb.2017.09.058
27. CMS Collaboration, Measurement of prompt and nonprompt J/ψ production in pp and pPb collisions at $\sqrt{s_{NN}}$ = 5.02 TeV. Eur. Phys. J. C **77**(4), 269 (2017). https://doi.org/10.1140/epjc/s10052-017-4828-3

28. CMS Collaboration, Study of B meson production in p+Pb Collisions at $\sqrt{s_{NN}} = 5.02$ TeV using exclusive hadronic decays. Phys. Rev. Lett. **116**(3), 032301 (2016). https://doi.org/10.1103/PhysRevLett.116.032301
29. LHCb Collaboration, Measurement of B^+, B^0 and Λ_b^0 production in pPb collisions at $\sqrt{s_{\mathrm{NN}}} =$ 8.16 TeV. Phys. Rev. D **99**(5), 052011 (2019). https://doi.org/10.1103/PhysRevD.99.052011
30. CMS Collaboration, Observation of top quark production in proton-nucleus collisions. Phys. Rev. Lett. **119**(24), 242001 (2017). https://doi.org/10.1103/PhysRevLett.119.242001
31. ALICE Collaboration, Azimuthal anisotropy of heavy-flavor decay electrons in p-Pb collisions at $\sqrt{s_{\mathrm{NN}}} = 5.02$ TeV. Phys. Rev. Lett. **122**(7), 072301 (2019). https://doi.org/10.1103/PhysRevLett.122.072301
32. ATLAS Collaboration, Measurement of the long–range pseudorapidity correlations between muons and charged-particles in $\sqrt{s_{\mathrm{NN}}}$=8.16 TeV proton-lead collisions with the ATLAS detector. ATLAS-CONF-2017-006 (2017). https://cds.cern.ch/record/2244808
33. CMS Collaboration, Elliptic flow of charm and strange hadrons in high-multiplicity pPb collisions at $\sqrt{s_{_{\mathrm{NN}}}} = 8.16$ TeV. Phys. Rev. Lett. **121**(8), 082301 (2018). https://doi.org/10.1103/PhysRevLett.121.082301
34. ALICE Collaboration, Search for collectivity with azimuthal J/ψ-hadron correlations in high multiplicity p-Pb collisions at $\sqrt{s_{\mathrm{NN}}} = 5.02$ and 8.16 TeV. Phys. Lett. B **780**, 7 (2018). https://doi.org/10.1016/j.physletb.2018.02.039
35. CMS Collaboration, Observation of prompt J/ψ meson elliptic flow in high-multiplicity pPb collisions at $\sqrt{s_{\mathrm{NN}}} = 8.16$ TeV. Phys. Lett. B **791**, 172 (2019). https://doi.org/10.1016/j.physletb.2019.02.018
36. Z. Citron et al., Future physics opportunities for high-density QCD at the LHC with heavy-ion and proton beams. arXiv:1812.06772 [hep-ph]

Part VI
Hadron Resonances, Hadronization and Coalescence

Chapter 42
Sequential Coalescence with Charm Conservation

Jiaxing Zhao, Shuzhe Shi, Nu Xu, and Pengfei Zhuang

Abstract In the frame of sequential coalescence, we discuss heavy flavor production in high energy nuclear collisions. The charm conservation during the evolution of the hot medium enhances the earlier hadron production and suppresses the later production. This relative enhancement (suppression) changes significantly the ratios between charmed hadrons in heavy ion collisions at RHIC and LHC energies.

The hadronization mechanism is a fundamental problem in Quantum Chromodynamics (QCD). Different from the parton hadronization in vacuum, the statistics plays an important role in the hadronization of the Quark-Gluon Plasma (QGP) formed in high energy nuclear collisions. If the produced hadrons are in thermal and chemical equilibrium state at the hadonization, their yield can be described by a statistical model with temperature and chemical potentials as parameters [1, 2]. More dynamical approaches are the coalescence models which describe well the quark number scaling of the elliptic flow and the enhancement of the baryon to meson ratios [3–5] for light hadrons in heavy ion collisions. However, there are two assumptions in these coalescence models. One is the coalescence probability or Wigner function which is usually taken as a Gaussian distribution with the width as a free parameter, the other is the simultaneous hadronization for all the hadrons.

The hadronization of heavy quarks provides a way to understand the hadronization mechanism. Due to the large mass of heavy quarks, their bound states, mesons and baryons, can be described by two- or three-body Schrödinger or Dirac equations in both vacuum and hot medium [6–9]. In this case, the eigenvalue of the equation, namely the binding energy, determines the dissociation or production temperature of the bound state, which leads to a sequential coalescence process for all the heavy

J. Zhao · P. Zhuang (✉)
Physics Department, Tsinghua University, Beijing 100084, China
e-mail: zhuangpf@mail.tsinghua.edu.cn

S. Shi
Department of Physics, McGill University, Montréal H3A 2T8, Canada

N. Xu
Institute of Modern Physics, Chinese Academy of Sciences, Lanzhou, Gansu 730000, China

D. Elia et al. (eds.), *The XVIII International Conference on Strangeness in Quark Matter (SQM 2019)*, Springer Proceedings in Physics 250,
https://doi.org/10.1007/978-3-030-53448-6_42

flavor hadrons. On the other hand, the eigenstate of the equation, namely the wave function of the bound state, controls the coalescence probability of the corresponding quarks. In addition, the heavy quark mass is much larger than the typical temperature of the fireball produced in heavy ion collisions at the Relativistic Heavy Ion Collider (RHIC) and the Large Hadron Collider (LHC), the thermal production of heavy quarks can be safely neglected [10], and the initially produced heavy quarks are conserved during the evolution of the colliding system. This charm conservation effect changes remarkably the heavy flavor yields [11].

By solving the two-body Dirac equation [8] with lattice simulated heavy quark potential at finite temperature [12], we obtain the charmed meson binding energy $\epsilon(T)$ and the radial wave function $\psi(r, T)$. From the condition for meson melting $\epsilon(T_m) = 0$ or $\langle r\rangle(T_m) \to \infty$, we extract the meson dissociation (production) temperatures $T_m = 1.2T_c$ for $m = D_s^+$ and $1.15T_c$ for $m = D_s^{*+}$, D^0, D^{*0} and D^{*+}, where T_c is the deconfinement temperature for light quarks. Considering the relation $V_{qq} \simeq V_{q\bar{q}}/2$ between the quark-quark and quark-anti-quark potentials, the dissociation temperatures for charmed baryons like Λ_c, Σ_c, Ξ_c and Ω_c should be lower than that for charmed mesons [7, 9]. To simplify the calculation, we take the dissociation or production temperature as $T_b = T_c$ for all the charmed baryons. The evolution of the hot medium created in heavy ion collisions at RHIC and LHC can be described by the energy-momentum conservation law $\partial_\mu T^{\mu\nu} = 0$, which determines the space-time dependence of the medium temperature T and baryon chemical potential μ_B. During the evolution, the medium cools down continuously due to the expansion of the system. When the local temperature $T(x)$ reaches the production temperature T_h for charmed hadron h, the hadron will be produced via coalescence mechanism at time $t_h(\mathbf{x})$. Since different hadrons have different production temperatures, the hadronziation of charmed quarks happens sequentially. D_s^+s are produced first, then D_s^{*+}, D^0, D^{*0} and D^{*+}, and finally the charmed baryons Λ_c, Σ_c, Ξ_c and Ω_c.

The spectra of charmed hadrons produced through sequential coalescence are controlled by both the dynamics (Wigner function W) and statistics (quark distribution f_i),

$$\frac{dN}{d^2p_T d\eta} = C \int p^\mu d\sigma_\mu \prod_{i=1}^{n} \frac{d^4x_i d^4p_i}{(2\pi)^3} f_i(x_i, p_i) W(x_1, ..., x_n, p_1, ..., p_n) \quad (42.1)$$

where p_T and η are the charmed hadron transverse momentum and longitudinal rapidity, the summation is over the constituent quarks with $n = 2$ for mesons and 3 for baryons, the integration is on the coalescence hypersurface $\sigma_\mu(t_h, \mathbf{x})$ at production time t_h, and the constant C is a statistical factor to take into account the internal quantum numbers in forming a colorless hadron. The Wigner function or coalescence probability $W(x_1, ..., x_n, p_1, ..., p_n)$ for n quarks to combine into a hadron is the Fourier transformation of the probability density $\psi^\dagger(x_1, ..., x_n)\psi(x_1, ..., x_n)$ with the wave function $\psi(x_1, ..., x_n)$ determined by the two- or three-body Dirac equation. Considering thermal and chemical equilibrium for light quarks, non-chemical equilibrium for strange quarks, and non-thermal equilibrium for charm quarks, the quark distributions can be taken as

$$\begin{aligned} f_q(x, p) &= \frac{N_q}{e^{p^\mu u_\mu(x)/T(x)} + 1}, \quad q = u, d \\ f_s(x, p) &= \frac{N_s \gamma_s}{e^{p^\mu u_\mu(x)/T(x)} + 1}, \\ f_c(x, p) &= r_h \rho_c(x) \left[\alpha f_{th}(p) + \beta f_{pp}(p)\right], \end{aligned} \tag{42.2}$$

where the constants N_q and N_s are determined by the inner quantum numbers of the quarks. Since strange quarks may not reach fully chemical equilibrium at RHIC energy, a fugacity factor is included in the distribution which is $\gamma_s = 0.85$ at RHIC and 1 at LHC from the comparison with the experimental data [13]. For charm quarks, they lose energy during the transportation in the quark-gluon plasma and become partially thermalized at the hadronization. We take, as a first approximation, a linear combination of the pQCD distribution f_{pp} determined by experimental p+p collisions and a thermal distribution f_{th} as the charm quark distribution. The coefficients α and β control the degree of the thermalization. The space density $\rho_c(x)$ is determined by the initial p+p collisions and nuclear geometry. The charm quark number fraction r_h describes the charm conservation during the hadronization. More charm quarks are involved in the earlier coalescence, and less charm quarks join the later hadronization. Suppose the produced D_s^+ number and all the charmed meson number are $N_{D_s^+}$ and N_m, the charm quark number fraction is then $r_h = 1$ for D_s^+, $1 - N_{D_s^+}/N_c \sim 0.9$ for the other charmed mesons, and $1 - N_m/N_c \sim 0.6$ for charmed baryons. Note that, if all the charmed hadrons are simultaneously produced at the phase transition temperature T_c or coalescence time t_c in usual coalescence models, the charm fraction is the same for all the singly charmed hadrons, and therefore the charm conservation will not affect the yield ratios in this case.

Recently, the yield ratio of D_s^+/D^0 is measured in heavy ion collisions at RHIC [15] and LHC [16] energies. In comparison with the corresponding p+p collisions simulated via PYTHIA[17], the ratio in A+A collisions is strongly enhanced. One of the reasons for this enhancement is the strangeness enhancement [18] in heavy ion collisions which leads to D_s^+ enhancement. There is another reason in our calculation: The enhanced charm quark number used in the coalescence for D_s^+ results in a D^0 suppression due to the charm conservation. These two reasons make a very strong D_s^+/D^0 enhancement which explains well the experimental data [11]. Figure 42.1 shows the centrality dependence of the ratio at RHIC energy. Since the fugacity is not sensitive to the centrality, the ratio is almost a constant in semi-central and central collisions.

In summary, we have established a sequential coalescence model with charm conservation and applied it to the charmed hadron production in heavy ion collisions. The charm conservation leads to an enhancement for earlier produced hadrons and

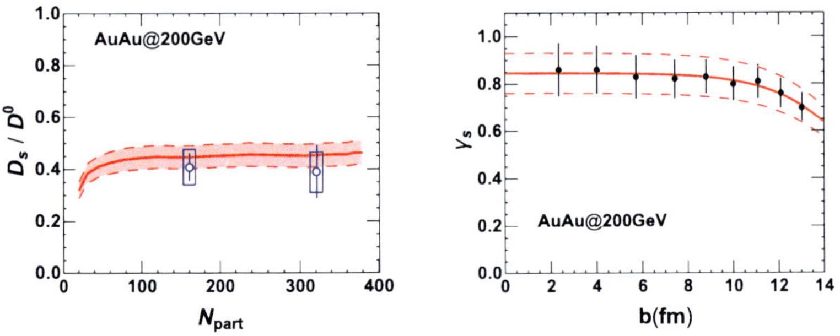

Fig. 42.1 The centrality dependence of the yield ratio D_s/D^0 (left panel) and strangeness fugacity γ_s (right panel) for Au+Au collisions at RHIC energy. b is the impact parameter, N_p is the number of the participant nucleons, and the data are from [14] for γ_s and [15] for D_s/D^0

a suppression for later produced hadrons. The strangeness enhancement and charm conservation in the sequential hadronization model explain well the strong D_s/D^0 enhancement observed in heavy ion collisions at RHIC and LHC energies.

References

1. P. Braun-Munzinger, K. Redlich, J. Stachel, in *Quark Gluon Plasma*, ed. by R.C. Hwa, et al., pp. 491–599
2. P. Braun-Munzinger, J. Stachel, Phys. Lett. B **490**, 196 (2000)
3. D. Molnar, S. Voloshin, Phys. Rev. Lett. **91**, 092301 (2003)
4. V. Greco, C. Ko, P. Levai, Phys. Rev. C **68**, 034904 (2003)
5. R. Fries, B. Muller, C. Nonaka, S. Bass, Phys. Rev. Lett. **90**, 202303 (2003)
6. H. Satz, J. Phys. G **32**, R25 (2006)
7. J. Zhao, P. Zhuang, Few Body Syst. **58**, 100 (2017); J. Zhao, H. He, P. Zhuang, Phys. Lett. B **771**, 349 (2017)
8. H.W. Crater, J.H. Yoon, C.Y. Wong, Phys. Rev. D **79**, 034011 (2009); S. Shi, X. Guo, P. Zhuang, Phys. Rev. D **88**(1), 014021 (2013)
9. S. Shi, J. Zhao, P. Zhuang, arXiv:1905.10627 [nucl-th]
10. K. Zhou, Z. Chen, C. Greiner, P. Zhuang, Phys. Lett. B **758**, 434 (2016)
11. J. Zhao, S. Shi, N. Xu, P. Zhuang, arXiv:1805.10858 [hep-ph]d
12. O. Kaczmarek, Eur. Phys. J. C **61**, 811 (2009)
13. P. Castorina, S. Plumari, H. Satz, Int. J. Mod. Phys. E **26**(12), 1750081 (2017)
14. B.I. Abelev et al., STAR Collaboration. Phys. Rev. C **79**, 034909 (2009)
15. L. Zhou, [STAR Collaboration]. Nucl. Phys. A **967**, 620 (2017)
16. J. Adam et al., ALICE Collaboration. JHEP **1603**, 082 (2016)
17. T. Sjostrand, S. Mrenna, P.Z. Skands, JHEP **0605**, 026 (2006)
18. J. Rafelski, B. Muller, Phys. Rev. Lett. **48**, 1066 (1982). Erratum: [Phys. Rev. Lett. **56**, 2334 (1986)]

Chapter 43
Λ_c^+ Production in pp and PbPb Collisions at 5.02 TeV with the CMS Detector

Rui Xiao

Abstract Due to the large masses of heavy quarks, their interactions with the quark-gluon plasma (QGP) may be different from those of light quarks. The lightest charm baryon is Λ_c^+, composed of a charm quark and two light quarks. Measurements of Λ_c^+ production in both pp and PbPb collisions can provide important inputs to the understanding of heavy quark transport in the QGP and the creation of heavy quark mesons and baryons via coalescence. Models involving quark coalescence predict a large enhancement of Λ_c^+ production in PbPb collisions compared to pp collisions. The high luminosity datasets collected at a nucleon-nucleon center-of-mass energy of 5.02 TeV using the CMS detector have been used to measure Λ_c^+ production in both pp and PbPb collisions via the $\Lambda_c^+ \rightarrow pK^-\pi^+$ decay channel. Results for differential cross sections for Λ_c^+ and ratios of Λ_c^+ over D^0 yields in pp and PbPb collisions, as well as the nuclear modification factor for Λ_c^+, are presented.

43.1 Introduction

Measurements of heavy-quark production provide unique inputs in understanding the properties of quark-gluon plasma (QGP) [1]. In relativistic heavy ion collisions, besides the fragmentation process present in proton-proton (pp) collisions, hadron production can also occur via quark coalescence. Due to the larger number of constituent quarks, the relative coalescence contribution to baryon production is expected to be more significant than for mesons. In particular, models involving coalescence of charm and light-flavor quarks predict a large enhancement in the Λ_c^+/D^0 production ratio in heavy ion collisions relative to pp collisions [2–4]. All discussions of Λ_c^+ and D^0 also include the corresponding charge conjugate states. In this contribution, we report a study of inclusive Λ_c^+ baryon production in pp and lead-lead (PbPb)

Rui Xiao for the CMS Collaboration.

R. Xiao (✉)
Purdue University, West Lafayette, IN 47906, USA
e-mail: xiao147@purdue.edu

D. Elia et al. (eds.), *The XVIII International Conference on Strangeness in Quark Matter (SQM 2019)*, Springer Proceedings in Physics 250,
https://doi.org/10.1007/978-3-030-53448-6_43

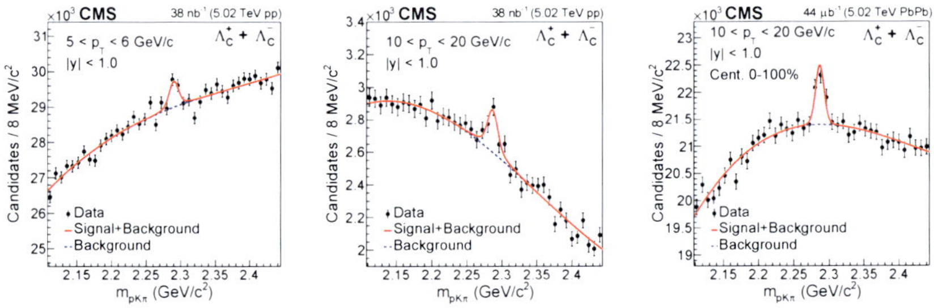

Fig. 43.1 Invariant mass distribution of Λ_c^+ candidates with p_T = 5–6 GeV/c (left), 10–20 GeV/c (middle) in pp collisions, and p_T = 10–20 GeV/c in PbPb collisions within the centrality range 0–100% (right). The solid line represents the full fit and the dashed line represents the background component [7]

collisions. The data were collected at $\sqrt{s_{NN}} = 5.02$ TeV in 2015 using the CMS detector [5]. The Λ_c^+/D^0 production ratios use the corresponding CMS measurements of prompt D^0 production [6].

43.2 Analysis Procedure

The $\Lambda_c^+ \to pK^-\pi^+$ candidates are reconstructed by selecting three charged tracks with $|\eta| < 1.2$ and a net charge of $+1$. During the invariant mass reconstruction, both possibilities for the mass assignments of the same-sign tracks are considered, while the kaon mass is assigned to the opposite-signed track. The incorrect assignment results in a broad distribution in the invariant mass and is indistinguishable from the combinatorial background. The Λ_c^+ baryon yields in each transverse momentum (p_T) interval are obtained from unbinned maximum likelihood fits to the invariant mass distribution. The signal shape is modeled by the sum of two Gaussian functions. Representative invariant mass distributions in pp and PbPb collisions are shown in Fig. 43.1.

43.3 Results

Figure 43.2 shows the p_T-differential cross section of Λ_c^+ baryon in pp and PbPb collisions. The shape of the p_T distribution in pp collisions is consistent with the PYTHIA8 calculation [8], while the data are systematically higher than PYTHIA8.

The nuclear modification factor R_{AA} for Λ_c^+ baryons is shown in Fig. 43.3 as a function of the number of participating nucleons $\langle N_{part}\rangle$ for PbPb collisions. There is a hint that the production of Λ_c^+ is suppressed in PbPb collisions for $p_T > 10$ GeV/c, but no conclusion can be drawn due to the large uncertainty in the pp differential cross

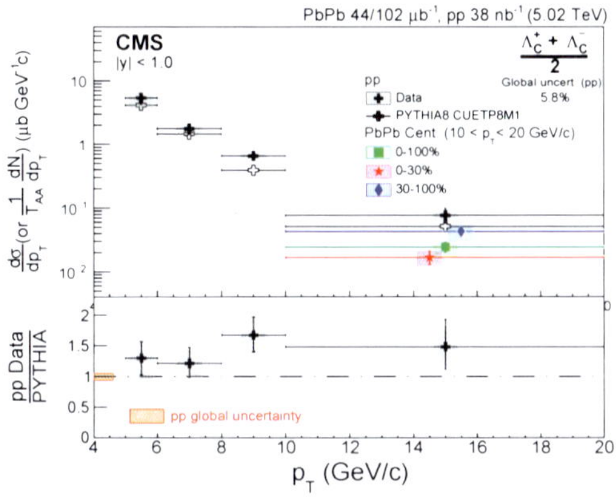

Fig. 43.2 The p_T-differential cross sections for Λ_c^+ baryon in pp collisions and in three centrality regions of PbPb collisions, along with the PYTHIA8 calculation for pp collisions [7]. The boxes and error bars represent the systematic and statistical uncertainties, respectively. The PbPb data points are shifted in the horizontal axis for clarity. The bottom panel shows the ratio of the measured p_T-differential cross section in pp data to the PYTHIA8 calculation. The box at unity indicates the 5.8% normalization uncertainty for pp collision arising from the integrated luminosity and branching fraction

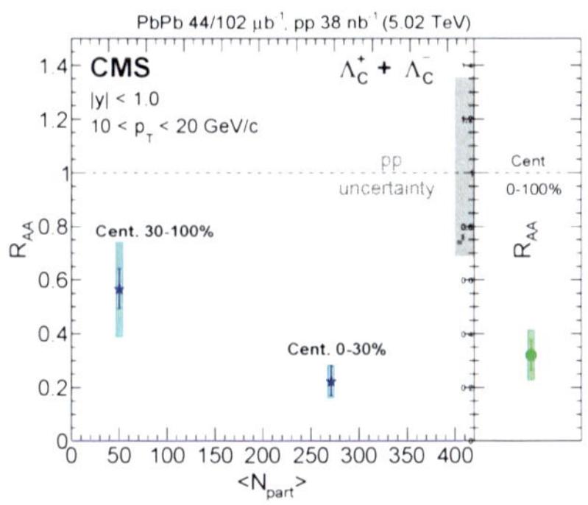

Fig. 43.3 The nuclear modification factor R_{AA} versus $\langle N_{part} \rangle$ [7]. The boxes and error bars on the data points represent the systematic uncertainties, which also contain the track reconstruction efficiency and T_{AA} uncertainties, and statistical uncertainties in the PbPb invariant yield, respectively. The box at unity indicates the total uncertainty in the pp differential cross section, which is common to all three R_{AA} values

section. However, the ratio of the R_{AA} values for the 0–30% and 30–100% centrality ranges shows evidence for more suppression in the more central PbPb collisions.

Figure 43.4 shows the Λ_c^+/D^0 production ratio as a function of p_T for pp and 0–100% centrality PbPb collisions. The production ratio in pp collisions, is similar in shape versus p_T but systematically higher than the calculation from PYTHIA8.212 tune CUETP8M1. The calculations of PYTHIA8 adding a color reconnection (CR) mechanism [9], shown in Fig. 43.4, are consistent with the Λ_c^+/D^0 production ratio

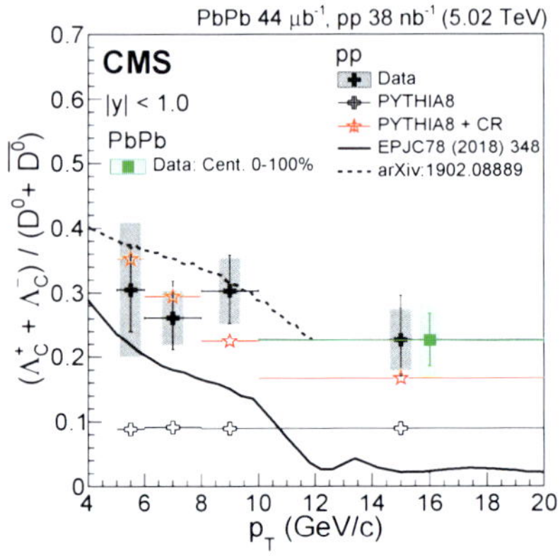

Fig. 43.4 The Λ_c^+/D^0 production cross section ratio versus p_T in pp collisions and for minimum bias PbPb collisions (0–100% centrality) [7]. The PbPb data point is shifted in the horizontal axis for clarity. The open crosses and open stars represent the predictions of PYTHIA8 with the CUETP8M1 tune and with color reconnection [9], respectively. The solid and dashed lines are the calculations from [10] and [11], respectively

results in pp collisions. The Λ_c^+/D^0 production ratio in pp collisions is also compared to two models. The black solid line shows calculations using a model that includes both coalescence and fragmentation in pp collisions [10]. Compared to the data, this model predicts a stronger dependence on p_T. Another model includes enhanced contributions from the decay of excited charm baryon states not included in PYTHIA [11] also shown in Fig. 43.4 by the dashed line. This model provides a reasonable description of the data in the range where it is available. The PbPb measurement in the p_T range 10–20 GeV/c is consistent with the pp result. This lack of an enhancement may suggest that there is no significant contribution from the coalescence process for $p_T > 10$ GeV/c in PbPb collisions.

43.4 Summary

In summary, this report presents the Λ_c^+ spectra in pp and PbPb collisions at 5.02 TeV with the CMS detector. A hint of suppression of Λ_c^+ production for $10 < p_T < 20$ GeV/c is observed in PbPb when compared to pp data, with central PbPb events showing stronger suppression. The Λ_c^+/D^0 production ratios in pp collisions are consistent with the predictions of PYTHIA8 adding color reconnection in hadronization, and also with a model that includes enhanced contributions from the decay of excited charm baryons. The Λ_c^+/D^0 production ratios in pp and PbPb collisions for $p_T = 10$–20 GeV/c are found to be consistent with each other. This may suggest that the coalescence process does not play a significant role in Λ_c^+ baryon production in this p_T range.

References

1. E.V. Shuryak, Theory of hadronic plasma. Sov. Phys. JETP **47**, 212 (1978)
2. Y. Oh, C.M. Ko, S.H. Lee, S. Yasui, Heavy baryon/meson ratios in relativistic heavy ion collisions. Phys. Rev. C **79**, 044905 (2009). https://doi.org/10.1103/PhysRevC.79.044905
3. S.H. Lee et al., Λ_c enhancement from strongly coupled quark-gluon plasma. Phys. Rev. Lett. **100**, 222301 (2008). https://doi.org/10.1103/PhysRevLett.100.222301
4. S. Ghosh et al., Diffusion of Λ_c in hot hadronic medium and its impact on Λ_c/D ratio. Phys. Rev. D **90**, 054018 (2014). https://doi.org/10.1103/PhysRevD.90.054018
5. CMS Collaboration, The CMS Experiment at the CERN LHC. JINST **3**, S08004 (2008)
6. CMS Collaboration, Nuclear modification factor of D^0 mesons in PbPb collisions at $\sqrt{s_{NN}} =$ 5.02 TeV. Phys. Lett. B **782**, 474 (2018). https://doi.org/10.1016/j.physletb.2018.05.074
7. CMS Collaboration, Production of Λ_c^+ baryons in proton-proton and lead-lead collisions at $\sqrt{s_{NN}} = 5.02$ TeV. Phys. Lett. B. (submitted to), arXiv:1906.03322
8. T. Sjöstrand, S. Ask, J. R. Christiansen, R. Corke, N. Desai, P. Ilten, S. Mrenna, S. Prestel, C. O. Rasmussen, P.Z. Skands, An introduction to PYTHIA 8.2. Comput. Phys. Commun. **191**, 159 (2015). https://doi.org/10.1016/j.cpc.2015.01.024
9. J.R. Christiansen, P.Z. Skands, String formation beyond leading colour. JHEP **08**, 003 (2015). https://doi.org/10.1007/JHEP08(2015)003
10. S. Plumari et al., Charmed hadrons from coalescence plus fragmentaion in relativistic nucleus-nucleus collisions at RHIC and LHC. Eur. Phys. J. C **78**, 348 (2018). https://doi.org/10.1140/epjc/s10052-018-5828-7
11. M. He, R. Rapp, Charm-baryon production in proton-proton collisions. Phys. Lett. B **795**, 117 (2019). https://doi.org/10.1016/j.physletb.2019.06.004

Chapter 44
Charmed Hadron Production by Recombination in Heavy Ion Collisions

Sungtae Cho

Abstract Starting from the investigation on recent experiments about charmed hadrons, e.g., nuclear modification factor ratios between charmoinum states and measurements of doubly charmed hadrons, we discuss the production of those charmed hadrons by recombination in heavy ion collisions. We adopt the coalescence model, and evaluate transverse momentum distributions of not only charmonium states but also charmed hadrons such as Ξ_{cc} baryons and $X(3872)$ mesons produced from quark-gluon plasma. We discuss the important characteristics of charmed hadron production in heavy ion collisions by showing the transverse momentum distribution ratio between various charmed hadrons.

44.1 Introduction

Heavy quark hadrons have been considered to be useful probes in understanding the properties of the quark-gluon plasma. Since heavy quarks are expected to be more produced than before as the energies available in heavy ion collisions are increased, the possibility of producing heavy quark hadrons from the quark-gluon plasma is expected to increase as well [1, 2]. Therefore it is necessary to study the production of hadrons with heavy quarks in high energy heavy ion collisions from a point of view of the coalescence model in which the hadron production is described as the process of coalescing constituents into hadrons [3, 4].

Recently, the production of various kinds of exotic hadrons with or without heavy quarks in heavy ion collision have been investigated based on the coalescence model [5–7]. Nevertheless, the production of many normal hadrons with multi-heavy quarks, e.g., Ξ^*_{cc}, Ω_{scc}, Ω^*_{scc}, and Ω_{ccc} baryons have not been studied yet in heavy ion collision experiments. We discuss here the production of those multi-charmed hadrons in heavy ion collisions, mostly by focusing on the production yield and the yield distribution as a function of transverse momenta.

S. Cho (✉)
Division of Science Education, Kangwon National University, Chuncheon 24341, Korea
e-mail: sungtae.cho@kangwon.ac.kr

D. Elia et al. (eds.), *The XVIII International Conference on Strangeness in Quark Matter (SQM 2019)*, Springer Proceedings in Physics 250,
https://doi.org/10.1007/978-3-030-53448-6_44

We investigate also the transverse momentum distribution of the $X(3872)$ meson. As attempts to understand the $X(3872)$ meson, e.g., the structure and spin, the $X(3872)$ meson has been studied in heavy ion collisions [8]. Here we try to understand the internal structure of the $X(3872)$ meson by evaluating the transverse momentum distribution of the $X(3872)$ in both a four-quark and a two-quark state. Since it has been found that the yield depends on the structure of hadrons [5–7], transverse momentum distributions are also expected to be dependent on the structure of hadrons as well as the wave function through the Wigner function [9]. Therefore by evaluating two different transverse momentum distribution of the $X(3872)$ meson we expect to obtain information about the dependence of the transverse momentum distribution on their constituents.

We also discuss transverse momentum distribution ratios between various multi-charmed hadrons. Studying the transverse momentum distribution ratio, especially between the $X(3872)$ and multi-charmed hadrons will shed some lights in investigating roles of heavy quarks in the heavy quark hadron production.

44.2 Production of Multi-charmed Hadrons from the Quark-Gluon Plasma

We first focus on the production yields of Ξ_{cc}, Ξ_{cc}^*, Ω_{scc}, Ω_{scc}^*, and Ω_{ccc} baryons in heavy ion collisions. Using the same method introduced in [7] we evaluate the production yields of the above multi-charmed hadrons in both the coalescence and statistical hadronization model [10]. When evaluating yields and transverse momentum distributions of Ξ_{cc} and Ω_{scc} baryons, we have not included decay contributions from Ξ_{cc}^* and Ω_{scc}^* baryons. We show results in Table 44.1.

Secondly, we calculate transverse momentum distributions of Ξ_{cc}, Ω_{scc}, Ω_{ccc} baryons, and $X(3872)$ mesons in the coalescence model. Especially, we consider two transverse momentum distributions for the $X(3872)$ meson, one in a four-quark state, and the other in a two-quark state [10]. In evaluating transverse momentum

Table 44.1 The Ξ_{cc}, Ξ_{cc}^*, Ω_{scc}, Ω_{scc}^*, and Ω_{ccc} yields at mid-rapidity in both the statistical and coalescence model expected at RHIC in $\sqrt{s_{NN}} = 200$ GeV Au+Au collisions and at LHC in $\sqrt{s_{NN}} = 2.76$ TeV Pb+Pb collisions

	RHIC		LHC	
	Stat.	Coal.	Stat.	Coal.
Ξ_{cc}	3.7×10^{-3}	4.5×10^{-4}	1.0×10^{-2}	1.6×10^{-3}
Ξ_{cc}^*	6.4×10^{-3}	9.0×10^{-4}	1.8×10^{-2}	3.3×10^{-3}
Ω_{scc}	1.3×10^{-3}	8.2×10^{-5}	3.7×10^{-3}	3.0×10^{-4}
Ω_{scc}^*	1.5×10^{-3}	1.6×10^{-4}	4.3×10^{-3}	6.0×10^{-4}
Ω_{ccc}	1.1×10^{-4}	1.1×10^{-6}	4.0×10^{-4}	5.3×10^{-6}

distributions of multi-charmed hadrons we adopt the charm quark transverse momentum distribution developed in [11], and use the oscillator frequency, ω_c, leading to consumption of all charm quarks at zero transverse momentum by quark coalescence [11, 12]. We show results in Fig. 44.1. A superscript $R(L)$ represents the transverse momentum distribution at RHIC(LHC) $\sqrt{s_{NN}} = 200\,\text{GeV}$ ($\sqrt{s_{NN}} = 2.76\,\text{TeV}$).

We also show in Fig. 44.1 some transverse momentum distribution ratios between multi-charmed hadrons. We see different shape of peaks at different transverse momentum regions in Fig. 44.1b–d. We argue that the peak appearing here in each transverse momentum distribution ratio is related to the type and number of quark constituents participating in hadron production. The peak is located at higher transverse momentum when more heavier quarks are involved; the peak at the higher transverse momentum for hadrons with heavier quarks supports the argument that the momentum of heavy quark hadrons is mostly carried by heavy quarks due to their heavier mass [11, 12].

We show in Fig. 44.2 transverse momentum distribution ratios between the $X(3872)$ in a four-quark state and the Ξ_{cc}, $c\bar{c}q\bar{q}/ccq$ and those between the $X(3872)$ in a two-quark state and the Ξ_{cc}, $c\bar{c}/ccq$ at both RHIC $\sqrt{s_{NN}} = 200\,\text{GeV}$ and LHC $\sqrt{s_{NN}} = 2.76\,\text{TeV}$. We see clear difference between two ratios, thereby we can obtain

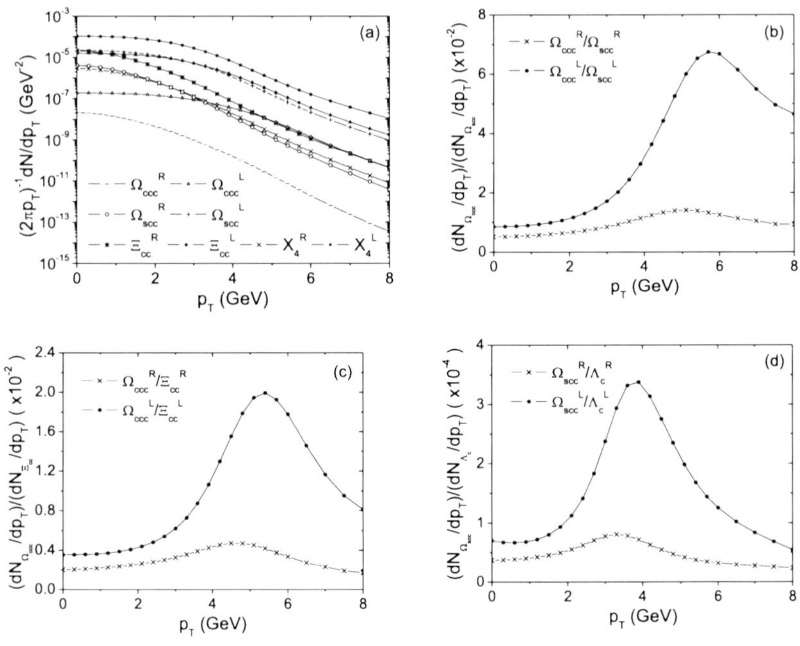

Fig. 44.1 **a** Transverse momentum distributions of Ξ_{cc}, Ω_{scc}, Ω_{ccc} baryons, and a $X(3872)$ meson in a four-quark state, X_4 and transverse momentum distribution ratios between the Ω_{ccc} and the Ω_{scc} (**b**), between the Ω_{ccc} and the Ξ_{cc} (**c**), and between the Ω_{scc} and the Λ_c (**d**) at both RHIC $\sqrt{s_{NN}} = 200\,\text{GeV}$ and LHC $\sqrt{s_{NN}} = 2.76\,\text{TeV}$

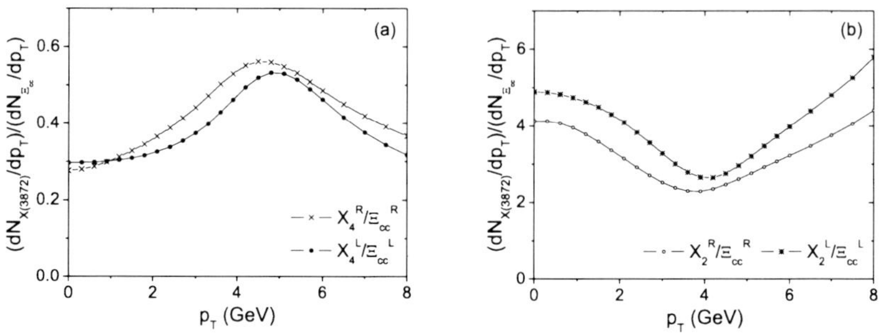

Fig. 44.2 **a** Transverse momentum distribution ratios between the $X(3872)$ in a four-quark state and the Ξ_{cc}, and **b** those between the $X(3872)$ in a two-quark state and the Ξ_{cc} at both RHIC $\sqrt{s_{NN}} = 200\,\text{GeV}$ and LHC $\sqrt{s_{NN}} = 2.76\,\text{TeV}$

the information on the quark structure of $X(3872)$ mesons by evaluating the transverse momentum distribution ratio between the $X(3872)$ meson and various multi-charmed hadrons.

44.3 Conclusion

We have investigated the yield and transverse momentum distribution of $X(3872)$ mesons as well as those of multi-charmed baryons. We find that estimated yields of multi-charmed hadrons produced in heavy ion collisions are large compared to those of the $X(3872)$, and thereby we expect to observe sufficient amount of multi-charmed hadrons in heavy ion collision. We also see that transverse momentum distributions of multi-charmed hadrons keep the information on their constituent quarks at the moment of hadron production very well, especially charm quarks carrying most of the momentum of multi-charmed hadrons due to their heavier masses. Moreover we note that the transverse momentum distribution ratio between various multi-charmed hadrons reflects the interplay between quark contents of corresponding hadrons, enabling us to infer the internal structure of the $X(3872)$ meson.

References

1. P. Braun-Munzinger, J. Stachel, Phys. Lett. B **490**, 196 (2000)
2. R.L. Thews, M. Schroedter, J. Rafelski, Phys. Rev. C **63**, 054905 (2001)
3. V. Greco, C.M. Ko, P. Levai, Phys. Rev. C **68**, 034904 (2003)
4. R.J. Fries, B. Muller, C. Nonaka, S.A. Bass, Phys. Rev. C **68**, 044902 (2003)
5. S. Cho et al. (ExHIC Collaboration), Phys. Rev. Lett. **106**, 212001 (2011)
6. S. Cho et al. (ExHIC Collaboration), Phys. Rev. C **84**, 064910 (2011)
7. S. Cho et al. (ExHIC Collaboration), Prog. Part. Nucl. Phys. **95**, 279 (2017)
8. S. Cho, S.H. Lee, Phys. Rev. C **88**, 054901 (2013)

9. S. Cho, Phys. Rev. C **91**(5), 054914 (2015)
10. S. Cho, S.H. Lee, arXiv:1907.12786 [nucl-th]
11. S. Plumari, V. Minissale, S.K. Das, G. Coci, V. Greco, Eur. Phys. J. C **78**(4), 348 (2018)
12. Y. Oh, C.M. Ko, S.H. Lee, S. Yasui, Phys. Rev. C **79**, 044905 (2009)

Chapter 45
Heavy Quark Baryon and Meson Production in *pp* and *AA* at RHIC and LHC Within a Coalescence Plus Fragmentation Model

Vincenzo Minissale, Salvatore Plumari, Gabriele Coci, Giuseppe Galesi, and Vincenzo Greco

Abstract The hadronization process of heavy hadrons with bottom and charm quarks, especially for baryons Λ_c, in a dense QGP medium is largely not understood. We present predictions obtained with a coalescence plus fragmentation model, for D^0, D_s, Λ_c spectra, the related baryon to meson ratios and the D_s/D^0 ratio, both at RHIC and LHC energies in a wide range of transverse momentum. We discuss how our model can naturally predict values of the order of O(1) for Λ_c/D^0 as recently measured at both RHIC and LHC. Moreover assuming that at the LHC top energies there can be the formation of QGP, we show that in the same scheme due to considerable volume effect a still large $\Lambda_c/D^0 \approx 0.5$ is predicted as seen by ALICE in *pp* collisions.

45.1 Introduction

A new state of matter composed of a strongly interacting plasma of deconfined quark and gluons called Quark-Gluon Plasma (QGP) is created and studied in Ultra-relativistic heavy ion collision at LHC and at RHIC. In recent years a lot of efforts have been made to study the heavy quark dynamics in the QGP [1–8]. The final observable states that contain charm quarks are the charmed hadrons, mainly D mesons and Λ_c, Σ_c baryons. Recent experimental results from STAR and ALICE collaborations have shown a baryon/meson ratio in the heavy flavor sector like the one observed for light and strange hadrons [9]. The explanation of the hadronization process trough the recombination mechanism starts from the idea that the final state

V. Minissale (✉) · S. Plumari · G. Coci · G. Galesi · V. Greco
Laboratori Nazionali del Sud, INFN-LNS, Via S. Sofia 62, 95123 Catania, Italy
e-mail: vincenzo.minissale@lns.infn.it

S. Plumari · G. Coci · G. Galesi · V. Greco
Department of Physics and Astronomy 'Ettore Majorana',
University of Catania, Via S. Sofia 64, 95125 Catania, Italy

D. Elia et al. (eds.), *The XVIII International Conference on Strangeness in Quark Matter (SQM 2019)*, Springer Proceedings in Physics 250,
https://doi.org/10.1007/978-3-030-53448-6_45

particles are formed by comoving partons in the QGP that combine their transverse momentum to produce a meson or a baryon with an higher transverse momentum [10–13].

45.2 Coalescence Plus Fragmentation Model

Our coalescence model is based on the Wigner formalism, the resolution of the coalescence integral gives the momentum spectrum of hadrons that can be written as:

$$\frac{d^2 N_H}{dP_T^2} = g_H \int \prod_{i=1}^{n} \frac{d^3 p_i}{(2\pi)^3 E_i} p_i \cdot d\sigma_i \; f_{q_i}(x_i, p_i) f_H(x_1 \ldots x_n, p_1 \ldots p_n)\, \delta^{(2)}\left(P_T - \sum_{i=1}^{n} p_{T,i}\right) \tag{45.1}$$

where $d\sigma_i$ denotes an element of a space-like hypersurface, g_H is the statistical factor to form a colorless hadron while f_{q_i} are the quark phase-space distribution functions. f_H is the Wigner function and describes the spatial and momentum distribution of quarks within an hadron. The Wigner distribution function used has a Gaussian shape in space and momentum, with a normalization fixed to guarantee that in the limit $p \to 0$ we have all the charm hadronizing. While the covariant width parameter an be related to the oscillator frequency ω by $\sigma = 1/\sqrt{\mu\omega}$ where μ is the reduced mass. The width of f_M is related to the root mean square charge radius of the hadron. For D^+ meson $\langle r^2 \rangle_{ch} = 0.184\, fm^2$ corresponding to a $\sigma_p = \sigma_r^{-1} = 0.283\,\mathrm{GeV}$; for Λ_c^+ the mean square charge radius is $\langle r^2 \rangle_{ch} = 0.15\,\mathrm{fm}^2$. We compute the coalescence probability P_{coal} for each charm quark then we assign a fragmentation probability $P_{frag}(p_T) = 1 - P_{coal}(p_T)$. Therefore the hadron spectra comes from the charm spectrum that do not undergo to coalescence, and evaluates by the convolution with the fragmentation function, that in our case is the Peterson fragmentation function $D_{had}(z, Q^2) \propto [z[1 - \frac{1}{z} - \frac{\epsilon_c}{1-z}]^2]^{-1}$ (see [3]). The relative ratios between different hadron channels are properly calculated and normalized according to the ratio of fragmentation fraction in [14]. Our approach is based on a fireball where the bulk of particles is a thermalized system of gluons and u, d, s quarks and anti-quarks at a temperature of $T_C = 165\,\mathrm{MeV}$. The fireball is considered at $\tau = 7.8\,\mathrm{fm/c}$, for LHC Pb+Pb at 2.76 TeV, and $\tau = 4.5\,\mathrm{fm/c}$, for RHIC Au+Au at 200 GeV. To take into account for the collective flow, we assume a radial flow profile as $\beta_T(r_T) = \beta_{\max}\frac{r_T}{R}$, where R is the transverse radius of the fireball. For partons at low transverse momentum we consider a thermal distribution, instead for $p_T > 2.5\,\mathrm{GeV}$, we consider the minijets that have undergone the jet quenching mechanism. For heavy quarks we use the transverse momentum distribution obtained by solving the relativistic Boltzmann equation [3]. The heavy quark numbers are estimated to be $dN_c/dy \simeq 2$ at RHIC and $dN_c/dy \simeq 15$ at LHC.

45.3 Results

In Fig. 45.1a (left) are shown the transverse momentum spectra at midrapidity for *Au+Au* collisions at 200 GeV and for 0–10% centrality for D^0 meson (left) and for Λ_c^+ baryon (right). We have also considered the feed-down for the Λ_c^+ and D^0 that comes from the decay of the main hadronic channels, i.e. D^{*0}, D^{*+}, $\Sigma_c^*(2520)$ and $\Sigma_c(2455)$. The contribution from coalescence (black solid) and fragmentation (dashed) for D^0 production is similar at momenta for $p_T < 3$ GeV. The two production differs at higher momenta where fragmentation becomes dominant for the yield. The coalescence is the dominant mechanism for the Λ_c^+ production for $p_T < 7$ GeV and it is mainly related to the fragmentation fraction of charmed hadrons formed via fragmentation from charm quarks, as showed in the analysis in [14], where this fraction is about the 6% of the total produced heavy hadrons. In Fig. 45.1b (right) are shown the transverse momentum spectra at midrapidity for $Pb + Pb$ collisions at 2.76 TeV and for 0–10% centrality for D^0 meson (left) and for Λ_c^+ baryon (right). For D^0 the contribution from coalescence is smaller than the one from fragmentation in all the momentum region considered, thus fragmentation is dominant for the overall yield. For the Λ_c^+ production the coalescence is the dominant mechanism up to $p_T = 7$ GeV. In Fig. 45.2a we show the Λ_c^+/D^0 ratio in comparison with the STAR experimental data shown by squares [17] (left). Coalescence by itself predicts a rise and fall of the baryon/meson ratio, the inclusion of fragmentation reduces the ratio, and we can see a quite good agreement with the experimental data in the peak region (blue line). The two panels in Fig. 45.2 show the comparison between RHIC and LHC for this ratio. Even if the only coalescence and the only fragmentation ratio remain similar, the combined ratio is different because, for each species, the production ratio between coalescence and fragmentation is smaller at LHC than at RHIC as can be seen from the spectra in Fig. 45.1 and so at LHC the final ratio is smaller than at RHIC. Recently have been released new experimental data of Λ_c^+/D^0 in pp collisions at LHC [18–20], and show an unexpected excess of production of Λ_c. We have applied our model assuming the formation of a QGP system in the case of *pp*

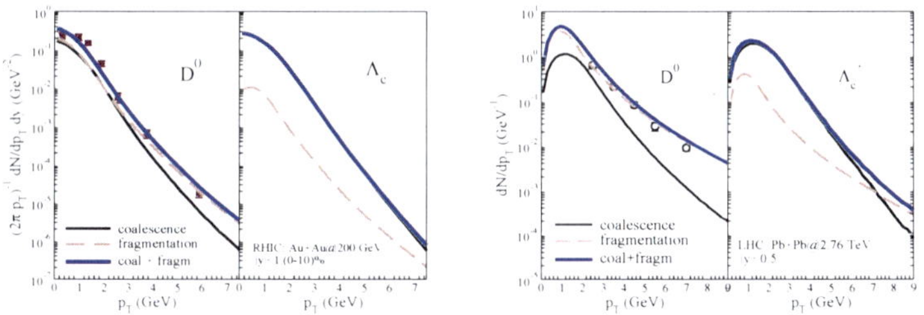

Fig. 45.1 (Color online) **a** (left) Transverse momentum spectra for $Au + Au$ collisions at $\sqrt{s} =$ 200 GeV and for (0−10%) centrality. Experimental data from [15]. **b** (right) Transverse momentum spectra for $Pb + Pb$ collisions at $\sqrt{s} = 2.76$ TeV and for (0−10%) centrality. Experimental data from [16]

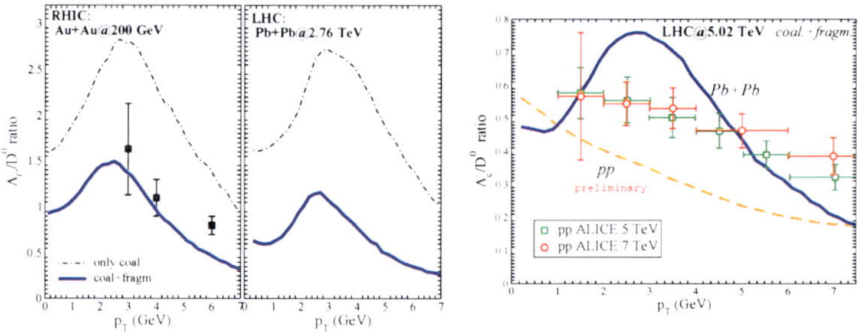

Fig. 45.2 (Color online) **a** Λ_c^+ to D^0 ratio as a function of p_T and at mid-rapidity for AA collisions. [17] (**b**) and pp collisions with a coalescence plus fragmentation model. Experimental data from [18, 19]

collisions. In Fig. 45.2b is shown with the red dashed line the Λ_c^+/D^0 ratio obtained for this kind of system. Our calculations predict a significantly different ratio respect to the only fragmentation. Moreover, the presence of a coalescence mechanism can have a deep impact on the pp baseline used to evaluate theoretically the R_{AA} for the different species, in particular for Λ_c the presence of coalescence implies a different behavior especially at low momenta. Recently released data from ALICE for the Λ_c nuclear modification factor [19] can lead to a better understanding of the heavy hadrons production in both pp and heavy ion collision.

References

1. M. He, R.J. Fries, R. Rapp, Phys. Rev. Lett. **110**(11), 112301 (2013). https://doi.org/10.1103/PhysRevLett.110.112301
2. S. Cao et al., Phys. Rev. C **92**(2), 024907 (2015). https://doi.org/10.1103/PhysRevC.92.024907
3. F. Scardina, S.K. Das, V. Minissale, S. Plumari, V. Greco, Phys. Rev. C **96**(4), 044905 (2017). https://doi.org/10.1103/PhysRevC.96.044905
4. S.K. Das et al., Phys. Lett. B **768**, 260 (2017). https://doi.org/10.1016/j.physletb.2017.02.046
5. S.K. Das, F. Scardina, S. Plumari, V. Greco, Phys. Lett. B **747**, 260 (2015). https://doi.org/10.1016/j.physletb.2015.06.003
6. S.K. Das, F. Scardina, S. Plumari, V. Greco, Phys. Rev. C **90**, 044901 (2014). https://doi.org/10.1103/PhysRevC.90.044901
7. S.K. Das et al., Phys. Rev. D **94**(11), 114039 (2016). https://doi.org/10.1103/PhysRevD.94.114039
8. S. Cao et al., Phys. Rev. C **94**(1), 014909 (2016). https://doi.org/10.1103/PhysRevC.94.014909
9. X. Dong, Nucl. Phys. A **967**, 192 (2017). https://doi.org/10.1016/j.nuclphysa.2017.04.023
10. V. Greco, C. Ko, P. Levai, Phys. Rev. C **68**, 034904 (2003). https://doi.org/10.1103/PhysRevC.68.034904
11. V. Minissale, F. Scardina, V. Greco, Phys. Rev. C **92**(5), 054904 (2015). https://doi.org/10.1103/PhysRevC.92.054904
12. Y. Oh, C.M. Ko, S.H. Lee, S. Yasui, Phys. Rev. C **79**, 044905 (2009). https://doi.org/10.1103/PhysRevC.79.044905
13. S. Plumari, V. Minissale, S.K. Das, G. Coci, V. Greco, Eur. Phys. J. C **78**(4), 348 (2018). https://doi.org/10.1140/epjc/s10052-018-5828-7

14. M. Lisovyi et al., Eur. Phys. J. C **76**(7), 397 (2016). https://doi.org/10.1140/epjc/s10052-016-4246-y
15. L. Adamczyk et al., Phys. Rev. Lett. **113**(14), 142301 (2014). https://doi.org/10.1103/PhysRevLett.121.229901, https://doi.org/10.1103/PhysRevLett.113.142301. [Erratum: Phys. Rev. Lett. **121**(22), 229901 (2018)]
16. B. Abelev et al., JHEP **09**, 112 (2012). https://doi.org/10.1007/JHEP09(2012)112
17. L. Zhou, Nucl. Phys. A **967**, 620 (2017). https://doi.org/10.1016/j.nuclphysa.2017.05.114
18. S. Acharya et al., JHEP **04**, 108 (2018). https://doi.org/10.1007/JHEP04(2018)108
19. C. Zampolli, SQM2019 talk (2019)
20. A.M. Sirunyan et al. (2019), arXiv:1906.03322

Chapter 46
Proton and Light Nuclei from Au+Au Collisions at $\sqrt{s_{NN}} = 2.4$ GeV Measured with HADES

Melanie Szala

Abstract The HADES (**H**igh **A**cceptance **D**i**E**lectron **S**pectrometer) experiment at the SIS18 accelerator of the GSI Helmholtzzentrum für Schwerionenforschung investigates heavy-ion collisions at kinetic beam energies of a few GeV per nucleon. In this energy regime a large fraction of protons are bound in light nuclei. In order to unterstand the production mechanism of this nuclei, we present details of the analysis for Au+Au collisions with 1.23A GeV ($\sqrt{s_{NN}} = 2.4$ GeV).

46.1 Motivation

The further one moves in the QCD phase diagram to the high μ_B region, the more abundant light nuclei become. In heavy ion collisions at $\sqrt{s_{NN}} = 2.4$ GeV almost 50% of all protons are bound in light nuclei [1]. This is in strong contrast to ultra-relativistic heavy ion collisions at LHC, where light nuclei are rare [2]. The HADES experiment is investigating rare and penetrating probes such as strangeness production [3] and dilepton radiation [4] in order to bridge towards lower and higher energy regions [1, 5]. Now, we extend our measurements also to light nuclei. Due to the very high statistics and clear separation of charged hadrons by the time-of-flight and energy-less measurements, we can study p, d, t and ^{3}He.

Melanie Szala for the HADES Collaboration.

M. Szala (✉)
Institut für Kernphysik, Goethe Universität, 60438 Frankfurt, Germany
e-mail: M.Szala@gsi.de

D. Elia et al. (eds.), *The XVIII International Conference on Strangeness in Quark Matter (SQM 2019)*, Springer Proceedings in Physics 250,
https://doi.org/10.1007/978-3-030-53448-6_46

46.2 Experimental Details

46.2.1 HADES Experiment

The setup of the HADES experiment is described in detail in [6]. The HADES spectrometer is a fixed target experiment located at the heavy-ion synchrotron SIS18 of the Helmholtzzentrum für Schwerionenforschung GSI in Darmstadt, Germany. Since 2002, the HADES experiment operates with the aim of investigating dense matter in the energy regime of a few GeV per nucleon. The spectrometer was constructed to have a large geometrical acceptance covering polar angles from 18 to 85° and almost the full azimuthal angle (see Fig. 46.1). Further characteristics of the spectrometer are its good momentum resolution and high read-out rate. In 2012 HADES has recorded about seven billion Au+Au collisions at a collision energy of 1.23A GeV ($\sqrt{s_{NN}} = 2.4$ GeV).

46.2.2 Particle Identification

The particle identification is based on the velocity vs. momentum/Z correlation (Fig. 46.2) where the velocity is determined from the time-of-flight measurement in the TOF and RPC scintillators with respect to the time-zero information delivered by the START detector and the tracked flight path. If needed, additional particle discrimination power is gained from the energy loss (*dE/dx*) information in the MDC and TOF. P, d, t and ^{3}He as charged particles can be selected by the HADES detector directly. Protons and the light nuclei are identified by applying pre-cuts on the correlation between β-momentum and the energy loss. As shown in Fig. 46.2 protons and light nuclei are well described by the expected correlation in these observables (black lines).

By cutting around the expected time-of-flight or energy loss identified hadron candidates can be selected. To select the pure signal out of this candidates, the mass

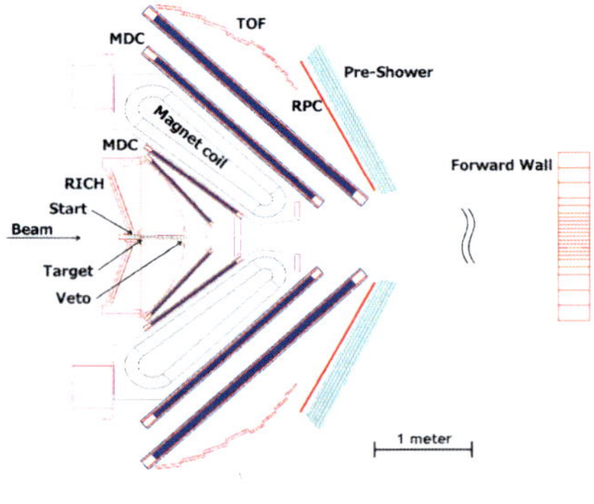

Fig. 46.1 Cross section of the HADES spectrometer. The spectrometer is divided into six identical sectors that are placed symmetrically around the beam axis

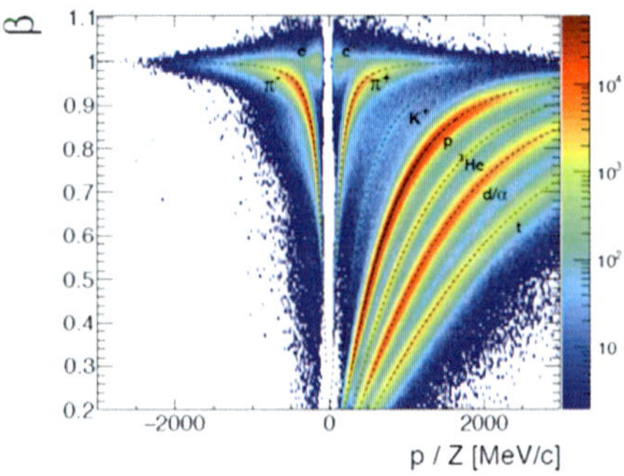

Fig. 46.2 Correlation between the momentum/Z and β of all selected tracks. Black lines correspond to the expected values for the different particle species

projection is fitted in the expected mass region and the background is subtracted using the ROOT class Interpolator. Afterwards, the phase space is divided into bins of reduced transverse mass $m_t - m_0$ and rapidity y and the reconstructed protons and light nuclei are counted in each cell of the phase space, in order to obtain the reconstructed raw yields of protons and light nuclei inside the HADES acceptance as a function of $(m_t - m_0) - y$.

46.2.3 *Efficiency Correction*

The measured particle spectra are corrected for geometric acceptance of the detector as well as the detector and track reconstruction efficiency. The acceptance and efficiency matrices are generated based on simulations. The events are generated with PLUTO [7] or the UrQMD [8] and then passed through a HGeant simulation that mimics a realistic detector response. The simulated data can thus be analyzed exactly in the same way as the real data.

46.3 Results

46.3.1 $m_t - m_0$ *Spectra*

We provide multi-differential spectra based on our large data sample of 2.2×10^9 events used in the analysis. We investigate the 0–10% most central collisions [9] in order to study only the mid-rapidity source. The relative fraction of protons and light nuclei originating from the break-up spectator regions inside the HADES acceptance gets larger going to more peripheral collisions. The number of events for the 0–10% most central collisions is about 5×10^8. The phase space is divided in reduced transverse mass bins of 25 MeV/c and in rapidity bins of 0.1, giving a total number of

52 bins for protons. The acceptance and efficiency corrected transverse mass spectra of p, d, t and ^{3}He for bins of rapidity are presented in Fig. 46.3. The coverage for protons is in total 1.5 units in rapidity. For comparison, a thermal source of protons with a temperature of 120 MeV has a total coverage of 1.8 units in rapidity. Hence, we cover the majority of protons and hence have small uncertainties due to extrapolation. The same applies to d, however the coverage in $m_t - m_0$ is slightly lower than for protons. For t and ^{3}He the coverage is further reduced to mostly the backward region as well as midrapidity and only small coverage in forward regions.

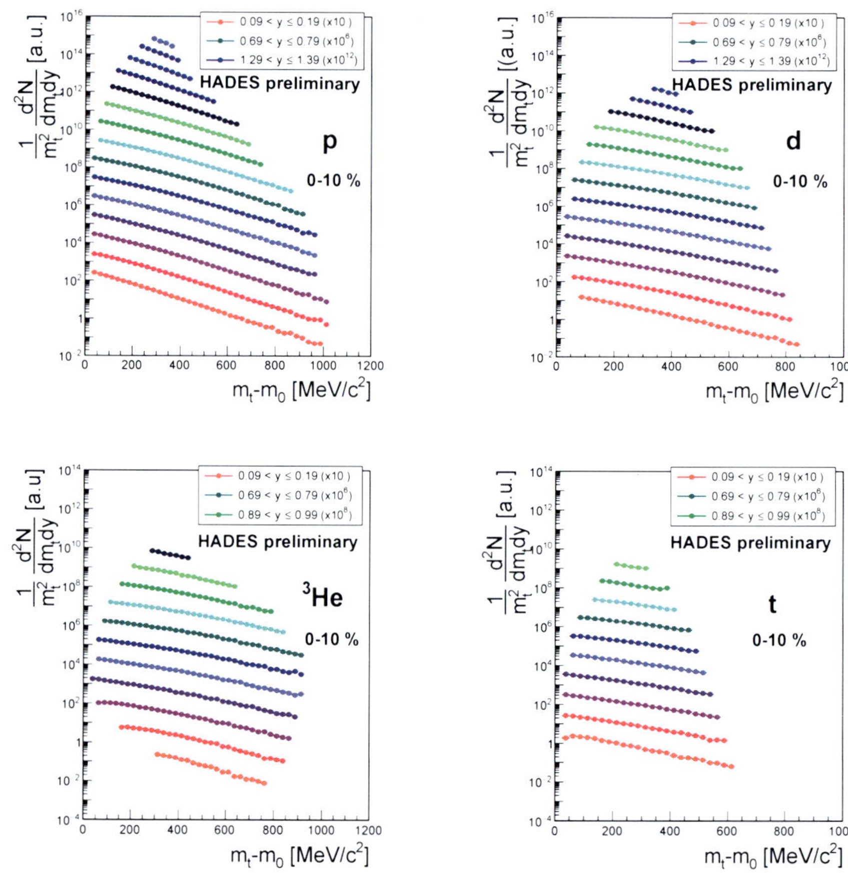

Fig. 46.3 Efficiency and acceptance corrected reduced transverse mass spectra of protons, deuterons, triton and ^{3}He for 0–10% most central collisons. For visibilty, the spectra are scaled by transverse mass squared $\frac{1}{m_t^2}$ and by different powers of 10 for the different measured rapidity regions

46.4 Summary and Outlook

We present high statistic, multi-differential data of p, d, t and ^{3}He. The results shown are only preliminary results and the analysis of light nuclei is still on-going. The future analysis will involve comparisons of $m_t - m_0$ spectra with a blast wave model for extraction of kinetic freeze-out parameters and the extraction of the yields. Furthermore, we can test possible production scenarios, e.g. thermal model versus coalescence, and will provide multi-differential B_A parameter. This measurements are supplemented by data of Ag+Ag collisions at $\sqrt{s_{NN}} = 2.55$ GeV taken as part of the FAIR Phase-0 in spring 2019. The first analysis of this data shows very promising prospects for new even more precise measurements.

References

1. W. Reisdorf et al. (FOPI Collaboration), Nucl. Phys. A **848**, 366 (2010)
2. P. Braun-Munzinger, B. Doenigus, Nucl. Phys. A **987**, 144 (2019)
3. J. Adamczewski-Musch et al. (HADES Collaboration), Phys. Lett. B **793**, 457 (2019)
4. J. Adamczewski-Musch et al., Nat. Phys. **15**(10), 1040 (2019)
5. A. Andronic, P. Braun-Munzinger, K. Redlich, J. Stachel, Nature **561**(7723), 321 (2018)
6. G. Agakichiev et al. (HADES Collaboration), Eur. Phys. J. A **41**, 243 (2009)
7. I. Froehlich et al., PoS. ACAT 076 (2007)
8. S.A. Bass et al., Prog. Part. Nucl. Phys. **41**, 255 (1998)
9. J. Adamczewski-Musch et al. (HADES Collaboration), Eur. Phys. J. A **54**, 85 (2018)

Chapter 47
Testing Coalescence and Thermal Models with the Production Measurement of Light (Anti-)Nuclei as Function of the Collision System Size with ALICE at the LHC

Luca Barioglio

Abstract High energy pp, p–Pb, and Pb–Pb collisions at the LHC offer a unique tool to study the production of light (anti-)nuclei. The study of the production yield of (anti-)nuclei in heavy-ion collisions at the LHC energy probes the late stages in the evolution of the hot, dense nuclear matter created in the collision. The same measurements performed in smaller collision systems are crucial to understand how the particle production mechanism evolves going from small to large systems. The latest ALICE results on the production yields of light (anti-)nuclei in pp, p–Pb, and Pb–Pb collisions at energies going from 5.02 to 13 TeV centre-of-mass energies are presented. A critical comparison of the experimental results with the predictions of the statistical (thermal) model and baryon coalescence approach is given to provide insight into the production mechanisms of light anti-nuclei in ultra-relativistic collisions.

47.1 Nuclear Matter Production

At the LHC, light (anti-)nuclei are abundantly produced in different collision systems and at different energies. Light nuclei are characterized by a low binding energy ($E_B \sim 1$ MeV) compared to the temperature of the chemical freeze-out, which is the time when the abundances of particle species are fixed ($T_{ch} \sim 160$ MeV). Therefore, in principle one would not expect to observe any nucleus. The study of nuclear (anti-)matter production mechanisms is crucial to understand how these loosely bound objects can form and survive in such extreme conditions.

Luca Barioglio for the ALICE Collaboration.

L. Barioglio (✉)
University and INFN of Turin, via P. Giuria 1, 10125 Turin, Italy
e-mail: luca.barioglio@cern.ch

D. Elia et al. (eds.), *The XVIII International Conference on Strangeness in Quark Matter (SQM 2019)*, Springer Proceedings in Physics 250,
https://doi.org/10.1007/978-3-030-53448-6_47

The experimental results on light-nuclei production are compared with the predictions of the available phenomenological models, namely the statistical hadronization [1] and the coalescence [2] models.

According to the statistical hadronisation model (SHM), (anti-)nuclei are produced at the chemical freeze-out in statistical equilibrium, along with all the other hadrons. The SHM can describe the production yields $\mathrm{d}N/\mathrm{d}y$ in central Pb–Pb collisions [3], including nuclei. In Pb–Pb collisions a grand canonical approach is used, since the condition $VT^3 > 1$ is satisfied, where V and T are the system temperature and volume respectively. In pp and p–Pb collisions, characterized by a smaller volume, this condition is not met and hence a canonical approach is used. In the canonical statistical model (CSM) the local conservation of quantum numbers is constrained. In these proceedings data are compared with the prediction of THERMAL-FIST package [4], in which baryon number, strangeness content and electric charge are exactly conserved. For increasing system size the canonical approach tends to the grand canonical one.

In the coalescence picture, nucleons that are close to each other in phase space after chemical freeze-out can merge and form a nucleus via coalescence [2]. The key concept of this class of models is the overlap between the nuclear wave function and the phase space of the constituent nucleons and the main observable is the coalescence parameter, defined as:

$$B_A\left(p_{\mathrm{T}}^{p}\right) = \frac{1}{2\pi p_{\mathrm{T}}^{A}} \frac{\mathrm{d}^2 N_A}{\mathrm{d}y \mathrm{d}p_{\mathrm{T}}^{A}} \Big/ \left(\frac{1}{2\pi p_{\mathrm{T}}^{p}} \frac{\mathrm{d}^2 N_p}{\mathrm{d}y \mathrm{d}p_{\mathrm{T}}^{p}}\right)^A \tag{47.1}$$

where the invariant spectra of the (anti-)protons are evaluated at the transverse momentum of the nucleus divided by its mass number A, so that $p_{\mathrm{T}}^{p} = p_{\mathrm{T}}^{A}/A$. The coalescence parameter is related to the probability to form a nucleus via coalescence.

47.2 The Coalescence Parameter

The B_2 has been measured in different collision systems at different energies. The dependence of the B_2 on the system size can be studied by measuring for each collision system and energy the value of the B_2 at a fixed p_{T}/A value. The B_2 measurements for $p_{\mathrm{T}}/A = 0.75$ GeV/c are shown in Fig. 47.1. The results for other values of p_{T}/A are similar. The data are compared with the calculations [5] for the coalescence model for two different parameterizations of the system size concerning the radius found in HBT studies. The B_2 does not show any discontinuity between different collision systems and different energies. Moreover, it evolves smoothly with the multiplicity of charged particles created at mid rapidity in the collision. The B_2 slightly decreases with increasing multiplicity in pp and p–Pb collisions, where the

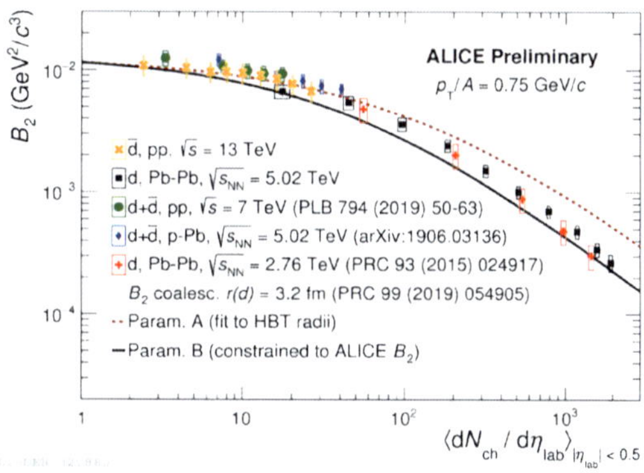

Fig. 47.1 B_2 as a function of the average event multiplicity $\langle dN_{ch}/d\eta\rangle$ at $p_T/A = 0.75$ GeV/c. Data are compared with the predictions of the coalescence model [5]. The red dashed line and the black continuous one correspond to different parameterizations of $\langle dN_{ch}/d\eta\rangle$ with respect to the HBT radius

system size is smaller than the deuteron size. At high multiplicities, where the system size is larger than the deuteron size, the decrease is more noticeable. The global trend is described by the coalescence model.

47.3 The Ratio Between Nucleus and Proton Yields

In Fig. 47.2 the ratio between the p_T-integrated yields of deuterons and protons (d/p) and the ratio between the yields of ^{3}He nuclei and protons (^{3}He/p) are shown. Data are compared with the prediction of Thermal-FIST package and coalescence calculations [6]. For ^{3}He, results from both the two-body and three-body coalescence are reported. For both deuterons and ^{3}He nuclei the ratio evolves smoothly as a function of the event multiplicity, hence of the system size. The drop at low multiplicity can be interpreted in the CSM as the result of canonical suppression and for coalescence as a consequence of the small phase-space of the nucleons in small systems. The flatness of the ratio at high multiplicity follows the prediction of both the CSM and the coalescence model.

47.4 Conclusions

The measurements of the coalescence parameter B_2 and of the yield ratios d/p and ^{3}He/p as a function of the event multiplicity suggest a common production mechanism that depends only on the system size. Coalescence can describe both the B_2 and the yield ratios as a function of multiplicity. CSM can describe the evolution of d/p and ^{3}He/p with the system size. However, in the current state more data and more precise model calculations are needed to discern the production mechanisms of light nuclei.

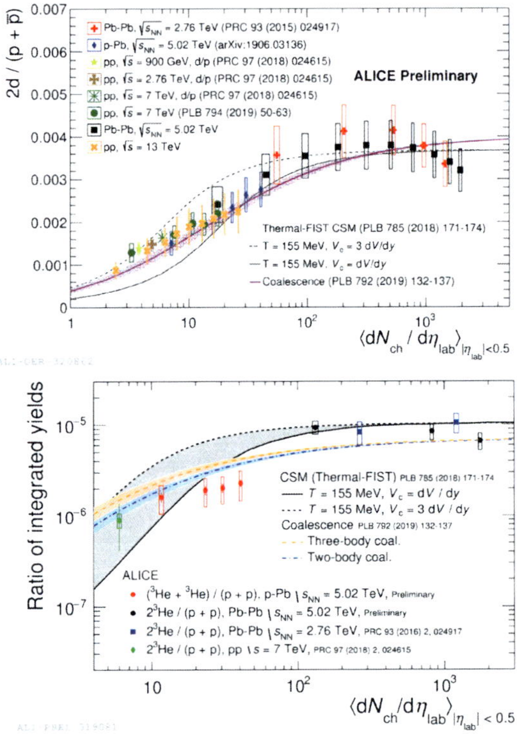

Fig. 47.2 Ratio between the p_T-integrated yields of deuterons and protons (top) and of ^{3}He nuclei and protons (bottom). Data are compared with the predictions of the THERMAL-FIST CSM [4] and the coalescence model [6]

References

1. A. Andronic, P. Braun-Munzinger, J. Stachel, H. Stöcker, Production of light nuclei, hypernuclei and their antiparticles in relativistic nuclear collisions. Phys. Lett. B **697**(3), 203–207 (2011)
2. J.I. Kapusta, Mechanisms for deuteron production in relativistic nuclear collisions. Phys. Rev. C **21**(4), 1301 (1980)
3. ALICE Collaboration et al., Production of ^{4}He and (anti-)^{4}He in pb–pb collisions at $\sqrt{s_{\rm NN}}$ = 2.76 tevat the LHC. Nucl. Phys. A **971**, 1–20 (2018)
4. V. Vovchenko, B. Dönigus, H. Stoecker, Multiplicity dependence of light nuclei production at LHC energies in the canonical statistical model. Phys. Lett. B **785**, 171–174 (2018)
5. F. Bellini, A.P. Kalweit, Testing production scenarios for (anti-)(hyper-)nuclei and exotica at energies available at the cern large hadron collider. Phys. Rev. C **99**, 054905 (2019). May
6. K.-J. Sun, C.M. Ko, B. Dönigus, Suppression of light nuclei production in collisions of small systems at the large hadron collider. Phys. Lett. B **792**, 132–137 (2019)

Chapter 48
Light Nuclei Production in Ultra-Relativistic Heavy Ion Collisions

Dmytro Oliinychenko, Long-Gang Pang, Hannah Elfner, and Volker Koch

Abstract We briefly overview the motivations of recent studies of light nuclei production in ultra-relativistic heavy ion collisions: anti-nuclei in space, and search for the critical point of the strongly-interacting matter. Then we focus on a particular recent development—the possible solution of the "snowballs in hell" puzzle—why the nuclei with binding energies of few MeV apparently survive at temperatures of around 155 MeV. Recent simulations within a hydrodynamics + hadronic transport approach, where deuterons are produced and destroyed mainly in $\pi pn \leftrightarrow \pi d$ reactions show that deuterons do not survive. They are rather created and disintegrated with approximately equal rates during certain period of time. Complementing previous publications, we show that these simulations reproduce not only deuteron spectra, but also the deuteron flow v_2 in PbPb collisions at 2.76 TeV.

48.1 Overview: Anti-nuclei in Space and Relation Between Light Nuclei Production and Critical Fluctuations

In the few past years light nuclei production (d, t, ^{3}He, $^3_\Lambda$H, ^{4}He, and their antiparticles) in ultra-relativistic heavy ion collisions has been actively explored, both theoretically and experimentally. There are at least two major motivations. First, AMS-02

D. Oliinychenko (✉) · L.-G. Pang · V. Koch
Lawrence Berkeley National Laboratory, 1 Cyclotron Rd, Berkeley, CA 94720, USA
e-mail: oliiny@fias.unifrankfurt.de

L.-G. Pang
Physics Department, University of California, Berkeley, CA 94720, USA

H. Elfner
GSI Helmholtzzentrum für Schwerionenforschung, Planckstr. 1, 64291 Darmstadt, Germany

Institute for Theoretical Physics, Goethe University, Max-von-Laue-Strasse 1, 60438 Frankfurt am Main, Germany

Frankfurt Institute for Advanced Studies, Ruth-Moufang-Strasse 1, 60438 Frankfurt am Main, Germany

D. Elia et al. (eds.), *The XVIII International Conference on Strangeness in Quark Matter (SQM 2019)*, Springer Proceedings in Physics 250,
https://doi.org/10.1007/978-3-030-53448-6_48

experiment at the International Space Station [1] has measured anti-nuclei in space [2], possibly including few unpublished events of $^3\overline{\text{He}}$ and $^4\overline{\text{He}}$. These events may originate from exotic sources such as antimatter clouds in space or dark matter annihilations, and from ordinary sources, such as pp, pA, and AA collisions. Presently the estimates of the ordinary pp background vary by an order of magnitude [3, 4], even though models are calibrated using the same data [5]. This stimulated interest to building more precise models of light nuclei production and obtaining new data to constrain them.

Another reason to study light nuclei is related to the search of the critical point of strongly-interacting matter [6]. A generic feature of this critical point is that in its vicinity spatial fluctuations of the baryon density are expected to increase. Experimentally only momentum space is accessible, therefore spatial fluctuations cannot be measured directly. However, based on a simple coalescence model it was recently suggested, that light nuclei production is sensitive to these fluctuations [7, 8]. In this model the density is separated into average and fluctuating contributions

$$\rho_n(x) = \langle \rho_n \rangle + \delta\rho_n(x) \tag{48.1}$$
$$\rho_p(x) = \langle \rho_p \rangle + \delta\rho_p(x) \tag{48.2}$$

and the triton and deuteron yields are expressed as

$$N_d \approx \frac{3}{2^{1/2}} \left(\frac{2\pi}{mT}\right)^{3/2} \int d^3x\, \rho_p(x)\rho_n(x) \sim \langle \rho_n \rangle N_p(1 + C_{np}) \tag{48.3}$$
$$N_t \approx \frac{3^{1/2}}{4} \left(\frac{2\pi}{mT}\right)^{3} \int d^3x\, \rho_p(x)\rho_n^2(x) \sim \langle \rho_n \rangle^2 N_p(1 + 2C_{np} + \Delta\rho_n)\,, \tag{48.4}$$

where

$$C_{np} \equiv \langle \delta\rho_n(x)\delta\rho_p(x) \rangle / (\langle \rho_n \rangle \langle \rho_p \rangle) \tag{48.5}$$
$$\Delta\rho_n \equiv \langle \delta\rho_n(x)^2 \rangle / \langle \rho_n^2 \rangle\,. \tag{48.6}$$

Here C_{np} represents the spatial correlations between neutrons and protons, and $\Delta\rho_n$ corresponds to the spatial fluctuations of neutron density. Therefore the ratio $N_t N_p / N_d^2$ becomes

$$\frac{N_t N_p}{N_d^2} = \frac{1}{2\sqrt{3}} \frac{1 + 2C_{np} + \Delta\rho_n}{(1 + C_{np})^2} \tag{48.7}$$

Here the factor $\frac{1}{2\sqrt{3}}$ originates from spin degeneracies and masses, $\frac{g_t g_p}{g_d^2}\left(\frac{3m \cdot m}{(2m)^2}\right)^{3/2} = \frac{1}{2\sqrt{3}} \approx 0.29$, and the second factor characterizes spatial correlations and fluctuations. In absence of fluctuations and correlations the ratio $\frac{N_t N_p}{N_d^2}$ is expected to be independent of collision energy, centrality, baryon or isospin content of the system. Therefore a non-flat behaviour of the ratio $N_t N_p / N_d^2$ against collision energy characterizes spatial fluctuations and may help to locate the critical point. The experimentally

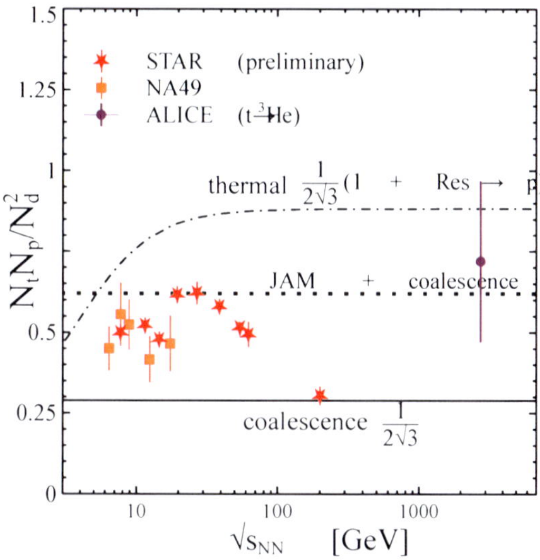

Fig. 48.1 Comparison of the $\frac{N_t N_p}{N_d^2}$ ratio in central PbPb or AuAu collisions at midrapidity between data from NA49 [10–12] (squares), STAR [13–15] (stars), and ALICE [16] collaborations (circle) and models (lines): uniform coalescence (solid) [7], thermal (dash-dotted) [17], and JAM hadronic transport with coalescence afterburner [9]

measured $N_t N_p/N_d^2$ ratio indeed exhibits a non-flat behaviour against energy, as shown in Fig. 48.1. However, a theoretical interpretation of this behaviour is currently not possible, because different models of light nuclei production do not agree even without critical point effects. This disagreement can be traced back to the model assumptions. The coalescence model assumes that light nuclei are produced at the late stage of the collision, when all resonances, such as Δ^* and N^* have decayed into nucleons. Then nucleons form light nuclei as described above, resulting in $\frac{N_t N_p}{N_d^2} = \frac{1}{2\sqrt{3}}$. Correlations between protons and neutrons and $\Delta\rho_n$ from thermal fluctuations are neglected. In the thermal model nuclei are formed from nucleons early, at the hadronic chemical freeze-out. After this resonances decay into protons, increasing their amount by a feed-down factor of $(1 + Res \to p)$, which varies from roughly 1.5 to 3 depending on collision energy. The intermediate case is represented by the JAM + coalescence model [9], where the JAM transport code is run until time 40 fm/c, allowing some resonances to decay into protons and neutrons and then coalescing nucleons that lie close in the phase space. Interestingly, the ratio $N_t N_p/N_d^2$ in this model is almost independent on collision energy, even though it takes into account correlations between protons and neutrons, as well as $\Delta\rho_n$ from thermal fluctuations.

The disagreement between the assumptions of these models can be resolved within a dynamical approach, where light nuclei are produced and destroyed continually as a result of reactions. Such an approach was recently suggested, but presently has been explored only for deuterons [18, 19]. It allows to determine, where and when deuterons are produced and resolve the tension between the extreme assumptions of the thermal and coalescence models—early production at hadronic chemical freeze-out versus late production at the stage, when nucleons received the full feed-down from resonances.

Of course, in addition to a realistic space-time picture any reasonable model of light nuclei production should properly take the known nuclear effects into account. In particular, recently it has been pointed out that the feed-down to tritons and deuterons from the numerous excited states of ^{4}He is important, especially at lower energies [20, 21]. Even if instead of the $\frac{N_t N_p}{N_d^2}$ ratio one considers a similar proton-free ratio

$$\frac{N_{^3\mathrm{He}} N_t}{N_d N_\alpha}, \tag{48.8}$$

which is insensitive to resonances feed-down to protons (also insensitive to weak decays, for which the published STAR proton spectra are not corrected!), the decays of ^{4}He* should still be taken into account.

48.2 Snowballs from Hell

The dynamical approach mentioned above suggests a solution to the following "snowballs in hell" puzzle [22]: midrapidity yields of light nuclei at LHC are described well by a thermal model with the temperature of hadronic chemical freeze-out $T_{chem} \approx 155$ MeV, while their spectra follow a blast-wave model with kinetic freeze-out temperature around $T_{kin} \approx 115$ MeV [16]. This poses the question, how light nuclei with binding energies of only few MeV ("the snowballs") survive between chemical and kinetic freeze-outs ("the hell"). The puzzle is relevant for deuterons at a wide range of energies of $\sqrt{s_{NN}} = 5 - 5000$ GeV, and for other (anti-)nuclei (t, ^{3}He, $^3_\Lambda$H, ^{4}He, and their antiparticles) at 2.76 GeV. In contrast, as one can see in Fig. 48.1, tritons at lower energies are not described by the thermal model, therefore the puzzle is not persistent there. A possible solution of this puzzle was suggested for deuterons in a dynamical model [18, 19] and for other light nuclei in analytical models [23, 24]: the nuclei are destroyed and created, while their yield remains approximately constant over time.

The approach used in [18] is a hybrid approach combining relativistic hydrodynamics for the earlier stage of the reaction and hadronic transport for the later stage. The transformation between those, called particlization, is performed at $T = 155$ MeV, the temperature of the hadronic chemical freeze-out according to the thermal model. Thermal deuterons are sampled from hydrodynamics and allowed to rescatter in the hadronic afterburner, mainly via $\pi pn \leftrightarrow \pi d$ reactions. The amounts of these reactions in forward and reverse directions turns out to differ by less than 5% after $t > 15$ fm/c, and deuteron yield remains approximately unchanged. Nevertheless, the majority of the final deuterons are originating from the afterburner stage, not from hydrodynamics. Another option is tested in [18], where no deuterons are produced from the hydrodynamics at all, and therefore all deuterons originate from the afterburner. The resulting deuteron yield is then less by ≈20%, but is still consistent with experimental data.

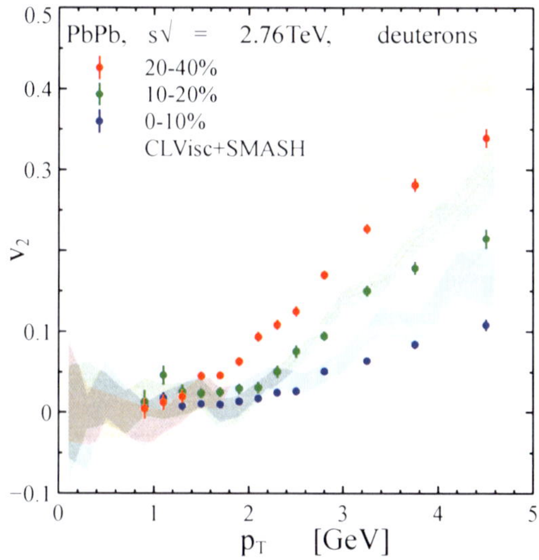

Fig. 48.2 Elliptic flow of deuterons from the dynamical approach [18, 19] is compared to the data from ALICE [25]

As demonstrated in [18, 19], this approach is able to describe deuteron spectra in PbPb collisions at $\sqrt{s_{NN}} = 2.76\,\text{TeV}$ at different centralities. Here, in Fig. 48.2 we show that the deuteron elliptic flow $v_2(p_T)$ is also reproduced. This calculation was performed with deuterons being sampled at the particlization. It would be interesting to see, if these results change considerably in case of no deuterons sampled at particlization. An ongoing work is devoted to this question, as well as to testing the same mechanism at lower energies and for different nuclei.

48.3 Summary and Perspectives

In the last two years the interest to better understand light nuclei production in heavy ion collisions has increased considerably. There are at least two motivating challenges: precisely estimate the anti-nuclei flux in space from pp, pA, and AA collisions; and the search for the critical point of the strongly-interacting matter by measuring spatial density fluctuations using light nuclei production. The first challenge can in principle be resolved experimentally, but the second one needs precise models of nuclei production, where it would be possible to switch critical point effects on and off. No such model is currently available. Moreover, even models without critical point disagree on where and how the light nuclei are produced, see Fig. 48.1. This disagreement can be alleviated by dynamical simulations. The latter have already proven successful, resolving the "snowballs in hell" puzzle. To improve the dynamical models one needs more data on inclusive and exclusive inelastic cross sections of hadrons with light nuclei, such as $\pi + d$, $\pi + t$, $\pi + {}^3\text{He}$, $p + d$, $p + t$,

$p + {}^3$He, etc. The rates of reverse reactions that produce light nuclei can be computed from these cross sections using the detailed balance principle.

Simulations with some (but not all) critical point effects can be performed by supplying transport or hybrid simulation with an adjustable equation of state. In case of transport this can be a mean field potential tuned to provide an equation of state with a critical point. In hydrodynamics one can directly choose the equation of state with a phase transition, such as [26]. Certain care should be taken at the particlization to preserve the spatial fluctuations. A recently developed local microcanonical sampling approach [27] allows to address this difficulty. It is possible, that spatial fluctuations generated by the critical point and preserved by the proper particlization are still vanishing during the afterburner stage and are not influencing the light nuclei production. In this case the experimentally measured bumps in Fig. 48.1 may be related to the collision dynamics or to nuclear liquid-gas phase transition. These possibilities should not be dismissed and can be addressed with dynamical models in future.

Acknowledgments We would like to thank X. Luo for sharing the STAR data shown in Fig. 48.1, and to the hepdata project (hepdata.net), that allowed to seamlessly obtain NA49 and ALICE data shown in the same Figure. D.O. thanks B. Dönigus, V. Vovchenko, and E. Shuryak for fruitful discussions. D. O. and V. K. were supported by the U.S. Department of Energy, Office of Science, Office of Nuclear Physics, under contract number DE-AC02-05CH11231 and received support within the framework of the Beam Energy Scan Theory (BEST) Topical Collaboration. L.-G.P. was supported by the National Science Foundation (NSF) within the framework of the JETSCAPE collaboration, under grant number ACI-1550228. Computational resources were provided by the Center for Scientific Computing in Frankfurt.

References

1. A. Kounine, Int. J. Mod. Phys. E **21**(08), 1230005 (2012). https://doi.org/10.1142/S0218301312300056
2. S. Ting, The first five years of the alpha magnetic spectrometer on the ISS (2016), https://indico.cern.ch/event/592392/
3. V. Poulin, P. Salati, I. Cholis, M. Kamionkowski, J. Silk, Phys. Rev. D **99**(2), 023016 (2019). https://doi.org/10.1103/PhysRevD.99.023016, arXiv:1808.08961 [astro-ph.HE]
4. K. Blum, K.C.Y. Ng, R. Sato, M. Takimoto, Phys. Rev. D **96**(10), 103021 (2017). https://doi.org/10.1103/PhysRevD.96.103021, arXiv:1704.05431 [astro-ph.HE]
5. S. Acharya et al. [ALICE Collaboration], Phys. Rev. C **97**(2), 024615 (2018). https://doi.org/10.1103/PhysRevC.97.024615, arXiv:1709.08522 [nucl-ex]
6. A. Bzdak, S. Esumi, V. Koch, J. Liao, M. Stephanov, N. Xu, arXiv:1906.00936 [nucl-th]
7. K.J. Sun, L.W. Chen, C.M. Ko, Z. Xu, Phys. Lett. B **774**, 103 (2017). https://doi.org/10.1016/j.physletb.2017.09.056, arXiv:1702.07620 [nucl-th]
8. K.J. Sun, L.W. Chen, C.M. Ko, J. Pu, Z. Xu, Phys. Lett. B **781**, 499 (2018). https://doi.org/10.1016/j.physletb.2018.04.035, arXiv:1801.09382 [nucl-th]
9. H. Liu, D. Zhang, S. He, N. Yu, X. Luo, arXiv:1909.09304 [nucl-th]
10. T. Anticic et al. [NA49 Collaboration], Phys. Rev. C **83**, 014901 (2011). https://doi.org/10.1103/PhysRevC.83.014901, arXiv:1009.1747 [nucl-ex]
11. T. Anticic et al. [NA49 Collaboration], Phys. Rev. C **94**(4), 044906 (2016). https://doi.org/10.1103/PhysRevC.94.044906, arXiv:1606.04234 [nucl-ex]

12. C. Blume [Na49 Collaboration], J. Phys. G **34**, S951 (2007). https://doi.org/10.1088/0954-3899/34/8/S133 [nucl-ex/0701042]
13. J. Adam et al. [STAR Collaboration], Phys. Rev. C **99**(6), 064905 (2019). https://doi.org/10.1103/PhysRevC.99.064905, arXiv:1903.11778 [nucl-ex]
14. D. Zhang [STAR Collaboration], arXiv:1909.07028 [nucl-ex]
15. X. Luo, private communication
16. J. Adam et al. [ALICE Collaboration], Phys. Rev. C **93**(2), 024917 (2016). https://doi.org/10.1103/PhysRevC.93.024917, arXiv:1506.08951 [nucl-ex]
17. V. Vovchenko, H. Stoecker, Comput. Phys. Commun. **244**, 295 (2019). https://doi.org/10.1016/j.cpc.2019.06.024, arXiv:1901.05249 [nucl-th]
18. D. Oliinychenko, L.G. Pang, H. Elfner, V. Koch, Phys. Rev. C **99**(4), 044907 (2019). https://doi.org/10.1103/PhysRevC.99.044907, arXiv:1809.03071 [hep-ph]
19. D. Oliinychenko, L.G. Pang, H. Elfner, V. Koch, MDPI Proc. **10**(1), 6 (2019). https://doi.org/10.3390/proceedings2019010006, arXiv:1812.06225 [hep-ph]
20. E. Shuryak, J.M. Torres-Rincon, arXiv:1910.08119 [nucl-th]
21. V. Vovchenko, in preparation (2019)
22. P. Braun-Münzinger, B. Dönigus, N. Löher, CERN Courier, August 2015
23. X. Xu, R. Rapp, Eur. Phys. J. A **55**(5), 68 (2019). https://doi.org/10.1140/epja/i2019-12757-7, arXiv:1809.04024 [nucl-th]
24. V. Vovchenko, K. Gallmeister, J. Schaffner-Bielich, C. Greiner, arXiv:1903.10024 [hep-ph]
25. S. Acharya et al. [ALICE Collaboration], Eur. Phys. J. C **77**(10), 658 (2017). https://doi.org/10.1140/epjc/s10052-017-5222-x, arXiv:1707.07304 [nucl-ex]
26. P. Parotto et al., arXiv:1805.05249 [hep-ph]
27. D. Oliinychenko, V. Koch, Phys. Rev. Lett. **123**(18), 182302 (2019). https://doi.org/10.1103/PhysRevLett.123.182302, arXiv:1902.09775 [hep-ph]

Chapter 49
$K^*(892)^0$ Production in p+p Interactions from NA61/SHINE

Angelika Tefelska

Abstract The measurement of $K^*(892)^0$ resonance production via its $K^+\pi^-$ decay mode in inelastic p+p collisions at beam momenta 40–158 GeV/c ($\sqrt{s_{NN}} = 8.8 - 17.3\,\text{GeV}$) is presented. The data were recorded by the NA61/SHINE hadron spectrometer at the CERN Super Proton Synchrotron. The analysis of $K^*(892)^0$ was done with the template method. The results include the double differential spectra $d^2n/(dydp_T)$, $d^2n/(m_T dm_T dy)$ as well as dn/dy spectra.

49.1 Introduction

The study of short-lifetime resonances are unique tools to understand the less known aspects of high energy collisions, especially its time evolution. The measurement of $K^*(892)^0$ meson production may help to distinguish between two possible scenarios for the fireball freeze-out: the sudden and the gradual one [1]. The ratio of $K^*(892)^0$ to charged kaon production may allow to determine the time between chemical and kinetic freeze-outs [1, 2].

The transverse mass spectra and yields of $K^*(892)^0$ mesons are also important inputs for Blast-Wave models and Hadron Resonance Gas models. Moreover, resonance spectra and yields provide an important reference for tuning Monte Carlo string-hadronic models.

In this paper we report measurements of $K^*(892)^0$ resonance production via its $K^+\pi^-$ decay mode in inelastic p+p collisions at beam momenta 40–158 GeV/c ($\sqrt{s_{NN}} = 8.8 - 17.3\,\text{GeV}$). The data were recorded by the NA61/SHINE hadron spectrometer [3] at the CERN SPS. The template fitting method was used to extract the $K^*(892)^0$ signal. This analysis method is also known as the cocktail fit method and was used by many other experiments such as ALICE, ATLAS, CDF, and CMS.

Angelika Tefelska for the NA61/SHINE Collaboration.

A. Tefelska (✉)
Faculty of Physics, Warsaw University of Technology, Warsaw, Poland
e-mail: angelika.tefelska@cern.ch

D. Elia et al. (eds.), *The XVIII International Conference on Strangeness in Quark Matter (SQM 2019)*, Springer Proceedings in Physics 250,
https://doi.org/10.1007/978-3-030-53448-6_49

49.2 Methodology

The $K^*(892)^0$ analysis was done for p+p interactions based on data sets recorded in years 2009, 2010 and 2011 which contained about 5.26×10^6 (beam momentum 40 GeV/c), 4.78×10^6 (beam momentum 80 GeV/c) and 56.65×10^6 (beam momentum 158 GeV/c) good-quality collisions of the proton beam with a 20 cm long liquid hydrogen target. The NA61/SHINE calibration, track and vertex reconstruction procedures and simulations are discussed in [4–6].

The template method was used to extract raw $K^*(892)^0$ signals. In this method the background is described as a sum of two components: mixed events and Monte Carlo generated templates which describe the contribution of $K^+\pi^-$ pairs coming from sources other than the $K^*(892)^0$. For the studied resonance in the small p+p system, the template method was found to be much more effective in estimating the background than the standard procedure relying on mixed events only. More information about the methodology is available in [7].

49.3 Results

Figure 49.1 presents the mid-rapidity transverse mass spectra of $K^*(892)^0$ mesons produced in inelastic p+p collisions at 40, 80, and 158 GeV/c. The rapidity spectra obtained by integrating and extrapolating the transverse momentum distributions are shown in Fig. 49.2.

The mean multiplicities of $K^*(892)^0$ mesons in full phase-space (158 GeV/c) or in $0 < p_T < 1.5$ GeV/c (40 and 80 GeV/c) were obtained from summing the measured points and adding the contribution from a Gaussian fit in the unmeasured region (lines in Fig. 49.2). For 158 GeV/c the point with $y < 0$ was calculated only to check the symmetry of the rapidity distribution and was not included in the procedure of mean multiplicity determination. The numerical values of the mean multiplicities of $K^*(892)^0$ mesons are presented in Table 49.1.

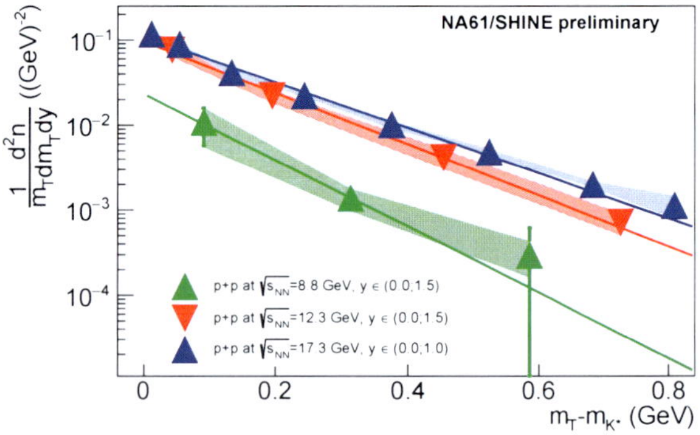

Fig. 49.1 Preliminary results on mid-rapidity transverse mass spectra of $K^*(892)^0$ mesons produced in inelastic p+p collisions at 40, 80, and 158 GeV/c

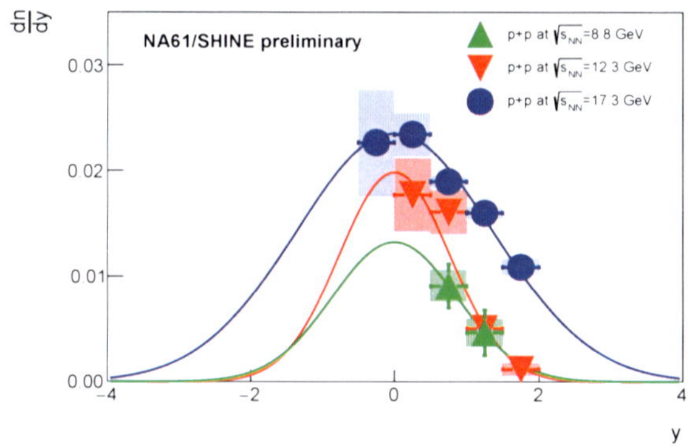

Fig. 49.2 Preliminary results on rapidity spectra of $K^*(892)^0$ mesons produced in inelastic p+p collisions at 40, 80, and 158 GeV/c. For 40 and 80 GeV/c the results were obtained in a wide transverse momentum range ($0 < p_T < 1.5$ GeV/c), whereas for 158 GeV/c p_T-extrapolated and integrated ($0 < p_T < \infty$) results are shown

Table 49.1 Mean multiplicities of $K^*(892)^0$ mesons produced in inelastic p+p collisions at 40, 80, and 158 GeV/c from NA61/SHINE, as well as from NA49 [8] at 158 GeV/c. The first uncertainty is statistical and the second is systematic

$\sqrt{s_{NN}}$	NA61/SHINE preliminary	NA49 [8]
8.8	$0.0285 \pm 0.0031 \pm 0.0046$	–
12.3	$0.0381 \pm 0.0054 \pm 0.0037$	–
17.3	$0.0806 \pm 0.0006 \pm 0.0026$	$0.0741 \pm 0.0015 \pm 0.0067$

49.4 Comparison with Hadron Resonance Gas Model

The new $K^*(892)^0$ results of NA61/SHINE were compared with predictions of the statistical Hadron Resonance Gas model [9, 10] in Canonical (CE) and Grand Canon-

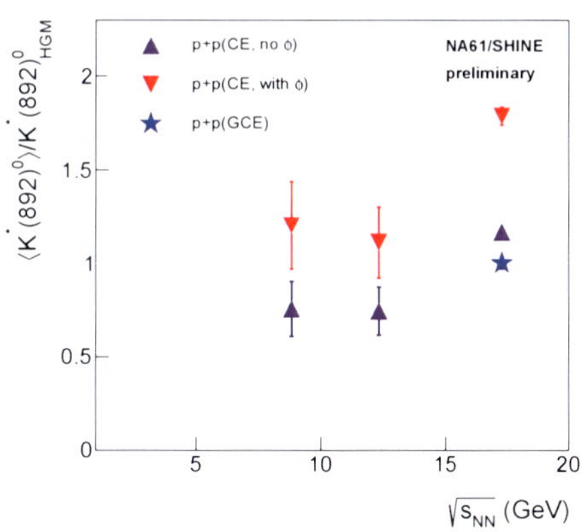

Fig. 49.3 Preliminary results on mean multiplicities of $K^*(892)^0$ mesons produced in p+p collisions compared to Hadron Gas Model predictions [9, 10]. Statistical and systematic uncertainties of $\langle K^*(892)^0\rangle$ were added in quadrature

ical (GCE) formulations (Fig. 49.3). At 158 GeV/c the GCE model provides a very good description of $K^*(892)^0$ production in the small p+p system. The CE model also agrees provided that the ϕ meson is excluded from the fits. More comparisons of the 158 GeV/c measurements with Hadron Resonance Gas model predictions of other authors can be found in [7, 11].

Acknowledgments This work was supported by the National Science Centre, Poland (grant 2017/25/N/ST2/02575) and partially supported by the Ministry of Science and Higher Education, Poland (DIR/WK/2016/2017/10-1).

References

1. C. Markert, G. Torrieri, J. Rafelski, AIP Conf. Proc. **631**, 533 (2002)
2. C. Blume, Acta Phys. Polon. B **43**, 577 (2012)
3. N. Abgrall et al. (NA61/SHINE Collaboration), J. Inst. **9**, P06005 (2014)
4. N. Abgrall et al. (NA61/SHINE Collaboration), Eur. Phys. J. C **74**(3), 2794 (2014)
5. A. Aduszkiewicz et al. (NA61/SHINE Collaboration), Eur. Phys. J. C **76**(11), 635 (2016)
6. A. Aduszkiewicz et al. (NA61/SHINE Collaboration), Eur. Phys. J. C **77**(2), 59 (2017)
7. A. Tefelska (for the NA61/SHINE Collaboration), PoS CORFU2018, 203 (2019)
8. T. Anticic et al. (NA49 Collaboration), Phys. Rev. C **84**, 064909 (2011)
9. V.V. Begun, V. Vovchenko, M.I. Gorenstein, H. Stoecker, Phys. Rev. C **98**(5), 054909 (2018)
10. V.V. Begun, private communication.
11. K. Grebieszkow (for the NA61/SHINE Collaboration), PoS CORFU2018, 152 (2019)

Chapter 50
Studying the Effect of the Hadronic Phase in Nuclear Collisions with PYTHIA and UrQMD

A. S. Vieira, C. Bierlich, D. D. Chinellato, and J. Takahashi

Abstract In this work, we couple Pb-Pb events simulated with the PYTHIA Angantyr event generator at $\sqrt{s_{\rm NN}} = 2.76$ and 5.02 TeV with the hadronic cascade simulator UrQMD to study the effect of the hadronic phase on observables such as charged-particle multiplicity densities, transverse momentum spectra and identified particle ratios, giving special emphasis to short-lived resonances.

The extreme conditions reached in ultra-relativistic heavy-ion collisions at the LHC are expected to produce a state of matter in which quarks and gluons are deconfined, the quark-gluon plasma (QGP). As a consequence, several features, such as elliptic flow and chemically equilibrated particle production, are expected and observed in these collision systems. However, it has to be noted that, once hadronization takes place, inelastic and elastic interactions may still take place. A proper distentanglement of the effects of this final hadronic phase and any features emerging from previous stages of the system evolution is fundamental to the understanding of heavy-ion collisions.

50.1 The Used Models

To characterize the hadronic phase, we couple the PYTHIA Angantyr model [1, 2] to the hadronic cascade simulator UrQMD [3]. The Angantyr model is based on the wounded nucleon model [4], and as such on a Glauber model. In Angantyr, several

CB was supported by Swedish Research Council, contract number 2017-003. AV, DDC and JT were supported by FAPESP grant 2017/05685-2.

A. S. Vieira · D. D. Chinellato · J. Takahashi
Universidade Estadual de Campinas, São Paulo, Brazil

C. Bierlich (✉)
Niels Bohr Institute, University of Copenhagen, Copenhagen, Denmark
e-mail: bierlich@nbi.dk

D. Elia et al. (eds.), *The XVIII International Conference on Strangeness in Quark Matter (SQM 2019)*, Springer Proceedings in Physics 250,
https://doi.org/10.1007/978-3-030-53448-6_50

additions to the original model has been made. The full nucleon–nucleon scattering amplitude is parametrized, allowing for a distinction between nucleons which are elastically scattered, diffractively excited or participating with colour exchange from both nucleons, denoted absorptively scatterred. All parameters in the parametrization of the scattering amplitude are fitted to the total and semi-inclusive cross sections in pp collisions. The parameters of the PYTHIA models for multiparton interactions (MPIs) [5], hadronization, parton showers as well as parton density functions, are all fixed in e^+e^-, ep and pp collisions. In a pp collision, MPIs are selected by the $2 \rightarrow 2$ perturbative parton–parton scattering cross section. Since this cross section has a $1/p_\perp^4$ divergence, it is regulated by one of the aforementioned parameters of the MPI model. In a heavy ion collision, a single projectile nucleon can interact absorptively with several target nucleons. In such cases, shadowing effects must be taken into account. In Angantyr this is done by ordering absorptively wounded nucleons into two categories, using the following strategy. All *possible* nucleon–nucleon interactions are ordered in increasing nucleon–nucleon impact parameter. Going from smallest to largest impact parameter, the nucleon pairs are labelled *primary* if neither of the nucleons have participated before, and *secondary* if one of the nucleons have previously participated in an interaction. The primary interactions are modelled as a normal inelastic non-diffractive collision. A secondary interaction is treated as only a single wounded nucleon, with inspiration from the Fritiof model [6], where a wounded nucleon contributes to the final state as a string with a mass distribution $\propto dM^2/M^2$, similar as the modelling of diffractive excitation. Instead of using a single string, as the original Fritiof program did, Angantyr allows for MPIs in secondary collisions, by treating the system as a collision between a nucleon and an Ingelman–Schlein Pomeron [7], followed by a (perturbative) parton shower. The nucleon–Pomeron scatterings are adjusted to look as normal inelastic non–diffractive nucleon–nucleon scatterings in the direction of the wounded nucleon. Furthermore, secondary collisions are rejected in cases where they would break overall energy–momentum conservation. This has the effect that not all possible nucleon–nucleon interactions from the Glauber calculation, are in fact realized in the generation of a final state.

As such, the Angantyr framework produces hadronic final states of heavy ion collisions (pA and AA), without any free parameters, as all are fixed by data from smaller collision systems. It should be noted that this treatment generates a heavy ion final state with no effects from a potential QGP, thus providing a baseline for tests of various models for collectivity. The recent models for "Rope hadronization" [9] and "String shoving" [10] implemented in Pythia, are not included in any results of these proceedings. The final state hadrons are assigned spatial vertices calculated in the string model [11]. These spatial vertices are then passed on to UrQMD, leading to a complete model of a heavy-ion collision in which no equilibrium is assumed.

50.2 Results

In order to test if this simulation chain describes heavy-ion collision data adequately, we compare predictions of the basic observables $\mathrm{d}N_{\mathrm{ch}}/\mathrm{d}\eta$ and $\langle p_{\mathrm{T}}\rangle$ to ALICE measurements [12]. While the $\mathrm{d}N_{\mathrm{ch}}/\mathrm{d}\eta$ is within 10% of the measured values, the $\langle p_{\mathrm{T}}\rangle$ exhibits stronger deviations due to the absence of radial flow. The corresponding plots have been omitted for brevity.

To further characterize the effect of the hadronic phase, we study how the p_{T}-differential spectra of charged particles are altered by hadronic scattering in UrQMD by calculating the ratio of the p_{T} spectra obtained when scattering is allowed to the spectra obtained when only decays occur in the hadronic phase. As can be seen in Fig. 50.1, hadronic interactions are seen to reduce high-p_{T} yields rather significantly, with the effect being most pronounced around 5 GeV/c. In addition, a minor radial-flow-like effect can be seen at low-p_{T}, where very low momentum particles are pushed to a momentum of about 1.5 GeV/c.

The observation that high-p_{T} is suppressed in this way prompts the calculation of the nuclear modification factor R_{AA}, shown in Fig. 50.1 for two selected centralities and for PYTHIA+UrQMD calculations with and without hadronic interactions. For this calculation, the same framework was used to determine particle spectra in pp collisions at $\sqrt{s}$ = 2.76 TeV and the number of binary collisions is taken to be the one calculated by the ALICE experiment [14]. While for low p_{T} the R_{AA} from simulations fails at describing the measurement, the model matches data above a p_{T} of 6 GeV/c. Quantitatively, two contributions lead to this correct description of the R_{AA}:

(a) Angantyr does not follow normal N_{coll} scaling. As explained in Sect. 50.1, the number of binary collisions does not enter as a scaling parameter, as shadowing effects are included by treatment of secondary wounded nucleons.
(b) UrQMD further modifies the R_{AA}, through rescattering effects, to precisely reproduce the minimum at around 5–6 GeV/c.

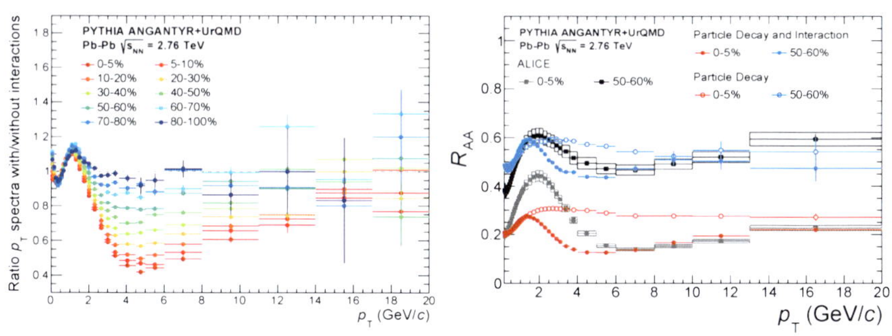

Fig. 50.1 (left) Ratio of p_{T} spectra with hadronic interactions and without interactions in various centrality classes and (right) R_{AA} in two selected centrality classes in Pb-Pb at 2.76 TeV. Data on the right plot are from [8]

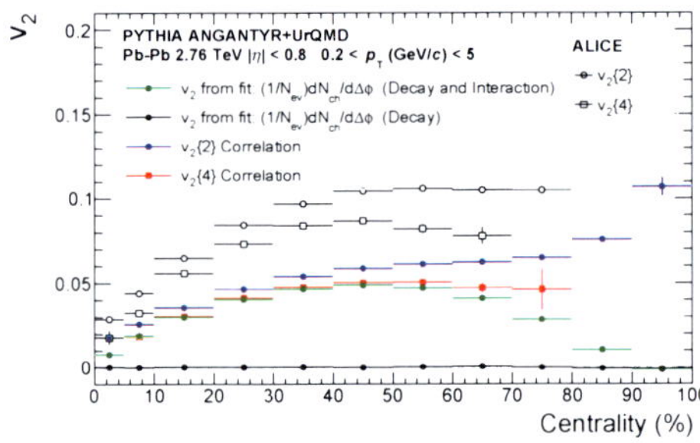

Fig. 50.2 The flow coefficients $v_2\{2\}$ and $v_2\{4\}$ from Pythia8/Angantyr + UrQMD (blue and red), as well as v_2 extracted from fit of the $dN_{ch}/d\Delta\phi$ distribution, all as function of centrality. Compared to data from ALICE [13]

Furthermore, the model is also seen to exhibit a progressively smaller high-$p_{\rm T}$ suppression for increasing $p_{\rm T}$. In the framework of the hadron vertex model, this is understood to be a consequence of high-momentum particles originating from boosted jet-like structures that hadronize progressively further away from the bulk of particles created in a heavy-ion collision. Another observable commonly used to characterize heavy-ion collisions is the anisotropic flow, quantified via the flow coefficients v_n. The leading elliptic flow term, v_2, is shown in Fig. 50.2 as a function of centrality in Pb-Pb collisions at 2.76 TeV simulated with the PYTHIA+UrQMD chain. The v_2 has been calculated in two distinct ways: first, with a fit to the azimuthal particle distribution with respect to the simulated event plane $dN_{ch}/d\Delta\phi$, and then also using 2- and 4-particle cumulants following [15]. If no hadronic interactions take place, particle production is seen to exhibit no correlation with the event plane as expected. However, if hadronic scattering is enabled, a substantial amount of elliptic flow builds up in the hadronic phase, reaching around 50–60% of the values measured by ALICE [13], which is consistent with observations from earlier studies using UrQMD [16]. The elliptic flow estimated via 4-particle cumulants $v_2\{4\}$ matches the one calculated via a fit to the $dN_{ch}/d\Delta\phi$ distribution in 0–50% collisions, indicating that it is well correlated with initial hadronic geometry, while for more peripheral events this correlation with the simulated event plane becomes weaker, indicating that the non-zero $v_2\{2\}$ and $v_2\{4\}$ in peripheral events is due to non-flow contributions such as dijet structures. Therefore, even in the absence of initial hadronic flow, hadronic interactions may build up a substantial amount of elliptic flow. Further studies are needed to determine the exact level of flow that would have to be present at hadronization time so as to be able to reproduce measurements from ALICE.

References

1. C. Bierlich et al., JHEP **10**, 139 (2016), arXiv:1607.04434
2. C. Bierlich et al. (2018), arXiv:1806.10820
3. M. Bleicher et al., J. Phys. **G25**, 1859–1896 (1999), arXiv:hep-ph/9909407
4. A. Bialas et al., Nucl. Phys. B **111**, 461–476 (1976)
5. T. Sjöstrand, M. van Zijl, Phys. Rev. D **36**, 2019–2041 (1987)
6. B. Andersson et al., Nucl. Phys. B **281**(1), 289–309 (1987)

7. G. Ingelman, P.E. Schlein, Phys. Lett. **152B**, 256–260 (1985)
8. S. Acharya et al. (ALICE), JHEP **11**, 013 (2018), arXiv:1802.09145
9. C. Bierlich et al., JHEP **03**, 148 (2015), arXiv:1412.6259
10. C. Bierlich et al., Phys. Lett., **B779**, 58–63 (2018), arXiv:1710.09725
11. S. Ferreres-Solé, T. Sjöstrand, Eur. Phys. J. **C78**(11), 983(2018), arXiv:1808.04619
12. K. Aamodt (ALICE) et al., Phys. Rev. Lett. **106**, 032301 (2011), arXiv:1012.1657
13. K. Aamodt, (ALICE) et al., Phys. Rev. Lett. **105**, 252302 (2010), arXiv:1011.3914
14. ALICE (2018), https://cds.cern.ch/record/2636623
15. A. Bilandzic et al., Phys. Rev. C **83**, 044913 (2011)
16. Y. Lu et al., J. Phys. G: Nucl. Part. Phys. **32**(8), 1121–1129 (2006)

Chapter 51
Probing QCD Matter via K^{*0}(892) and ϕ(1020) Resonance Production at RHIC

Md Nasim

Abstract We present some measurements of K^{*0} and ϕ resonances at midrapidity in Au+Au collisions at $\sqrt{s_{NN}} = 7.7 - 200$ GeV using the STAR detector. The transverse momentum (p_T) spectra and p_T-integrated yields of K^{*0} and ϕ have been studied. The ratios between resonance (K^{*0} and ϕ) to non-resonance particles (K) are presented as a function of centrality. It is found that K^{*0}/K^- ratios are suppressed in the most central collisions as compared to peripheral ones for all studied collision energies. On the other hand, ϕ/K^- ratios are weakly dependent on centrality. These results can be understood by considering the effect of more hadronic rescattering for K^{*0} (lifetime $\sim$ 4 fm/c) as compared to ϕ (lifetime $\sim$ 42 fm/c). We have also presented the measurement of the first-order azimuthal anisotropy (known as directed flow, v_1) of ϕ meson as a function of rapidity (y) at midrapidity in Au+Au collisions at $\sqrt{s_{NN}} = 7.7 - 200$ GeV. The slope of ϕ-meson v_1 (dv_1/dy) has been compared to the dv_1/dy of other identified particles. We have found that all particles that consist of produced quarks show similar behaviour for $\sqrt{s_{NN}} > 14.5$ GeV.

51.1 Introduction

The aim of the STAR experiment at Relativistic Heavy Ion Collider (RHIC) is to study the QCD matter by colliding nuclei at ultra-relativistic speeds [1]. The study of K^{*0} (lifetime $\sim$ 4 fm/c) and ϕ (lifetime $\sim$ 42 fm/c) production in heavy-ion collisions can be used to probe the medium created after the collision [2]. The K^{*0} resonance has a short lifetime, therefore it may decay during the hadronic phase and its decay products may undergo elastic or pseudoelastic scatterings [3]. Due to elastic or pseudoelastic scatterings, the final yield of K^{*0} resonance may get changed. The K^{*0} resonance yields may get reduced due to rescattering of its daughters through elastic scattering

Md Nasim for the STAR Collaboration.

M. Nasim (✉)
Indian Institute of Science Education and Research Berhampur, Berhampur 760010, Odisha, India
e-mail: nasim@iiserbpr.ac.in

D. Elia et al. (eds.), *The XVIII International Conference on Strangeness in Quark Matter (SQM 2019)*, Springer Proceedings in Physics 250,
https://doi.org/10.1007/978-3-030-53448-6_51

or the yields may be regenerated through pseudoelastic scattering between chemical and kinetic freeze-out [2]. However, due to longer lifetime as compared to the fireball, ϕ mesons mostly decay outside of the fireball and its daughters are not affected by late-stage hadronic scatterings. Hence, the study of K^{*0} and ϕ resonances provides an important information about the late-stage hadronic scatterings.

In high-energy heavy-ion collisions, particles are produced with an azimuthally anisotropic momentum distribution. Directed flow (v_1) is a measure of azimuthal angular anisotropy of the produced particles with respect to the first-order event plane [4]. The v_1 is an initial state effect and expected to be sensitive to the equation of state of the system formed in the collision. The ϕ meson freezes out early and is expected to have small hadronic interaction cross section. The measured v_1 of ϕ meson can be used as a clean probe to study the QCD matter [5].

51.2 Data Sets and Methods

The results presented here are based on data collected at $\sqrt{s_{NN}} = 7.7, 11.5, 14.5, 19.6, 27, 39, 62.4$ and 200 GeV in Au+Au collisions by the STAR detector using a minimum-bias trigger. The Time Projection Chamber (TPC) [6] and Time of Flight (TOF) [7] detectors with full 2π coverage are used for particle identification in the central pseudorapidity (η) region ($|\eta| < 1.0$). The K^{*0} and ϕ resonances are reconstructed from the following hadronic decay channels: $K^{*0} \longrightarrow K^{\pm} + \pi^{\mp}$ and $\phi \longrightarrow K^{+} + K^{-}$. Event mixing technique has been used for combinatorial background estimation [8]. Figure 51.1 shows invariant mass distribution of $K^{\pm}\pi^{\mp}$ and $K^{+}K^{-}$ pairs after mixed event background subtraction.

The first harmonic coefficient of the Fourier decomposition of azimuthal distribution with respect to the first-order event plane angle (ψ_1) can be expressed as $v_1 = \langle \cos(\varphi - \psi_1) \rangle$, where φ is the azimuthal angle of the produced particle. The

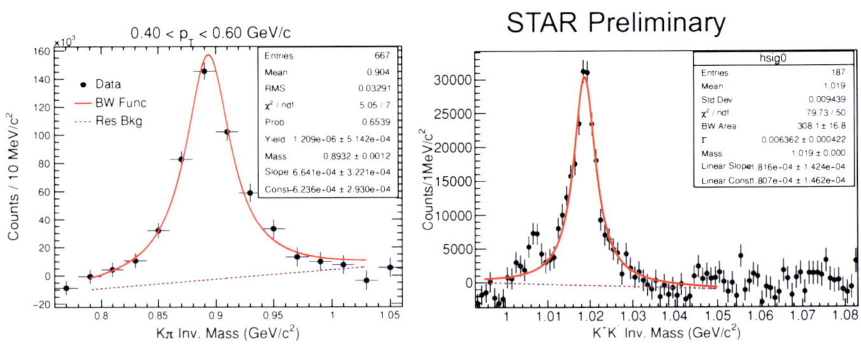

Fig. 51.1 Invariant mass distribution of $K^{\pm}\pi^{\mp}$ and $K^{+}K^{-}$ pairs after mixed event background subtraction for miminum bias (0–80%) Au+Au collisions at $\sqrt{s_{NN}} = 11.5$ GeV for $0.4 < p_T < 0.6$ GeV/c

first-order event plane angles are calculated using a forward rapidity detectors (Beam-Beam Counters or Zero-Degree Calorimeters) [9]. The measured v_1 with respect to the first-order event plane has been corrected for the finite event plane resolution. More details on the v_1 measurement in STAR can be found in [10].

51.3 Results

Figure 51.2 shows K^{*0}/K^- and ϕ/K^- ratios as a function of $(dN_{ch}/d\eta)^{1/3}$ measured at midrapidity (the cubic root of the charged-particle multiplicity density is proportional to the system radius) in Au+Au collisions at various center-of-mass energies measured by STAR experiment [8, 11–14]. The results are compared to the measurements performed at $\sqrt{s_{NN}} = 2.76$ TeV in Pb+Pb collision by the ALICE collaboration. The K^{*0}/K^- ratios decrease with increasing $(dN_{ch}/d\eta)^{1/3}$ for all studied collisions energies. On the other-hand, ϕ/K^- ratios are almost independent of $(dN_{ch}/d\eta)^{1/3}$. The observed suppression of the K^{*0}/K^- ratios could be due to rescattering effect as discussed earlier.

Directed flow slope, dv_1/dy, versus beam energy for ϕ, $\bar{\Lambda}$, p and $\bar{p}$ is presented in Fig. 51.3 for 10–40% central Au+Au collisions [10]. The slope for $\bar{\Lambda}$ is negative for all energies and is consistent within errors with that for $\bar{p}$. The $\bar{p}$, ϕ and $\bar{\Lambda}$ are seen to have similar $v_1(y)$ for $\sqrt{s_{NN}} > 14.5$ GeV. All these species consist from quarks that are produced in the collision. For $\sqrt{s_{NN}} < 14.5$ GeV, the current statistical uncertainty is too large to make any conclusion.

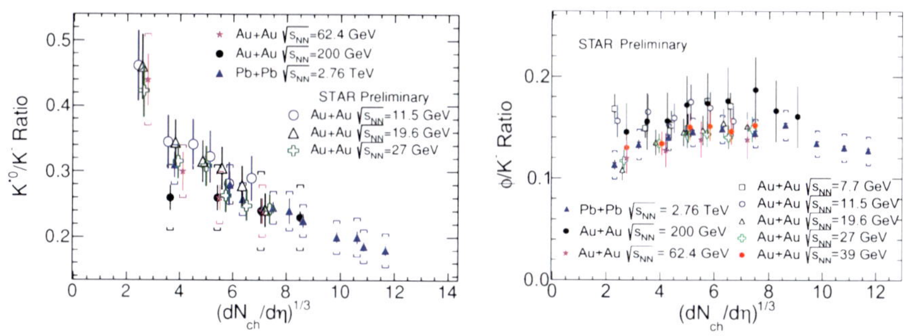

Fig. 51.2 Ratios K^{*0}/K^- and ϕ/K^- as a function of $(dN_{ch}/d\eta)^{1/3}$ in Au+Au and Pb+Pb collisions at various $\sqrt{s_{NN}}$ [8, 11–14]

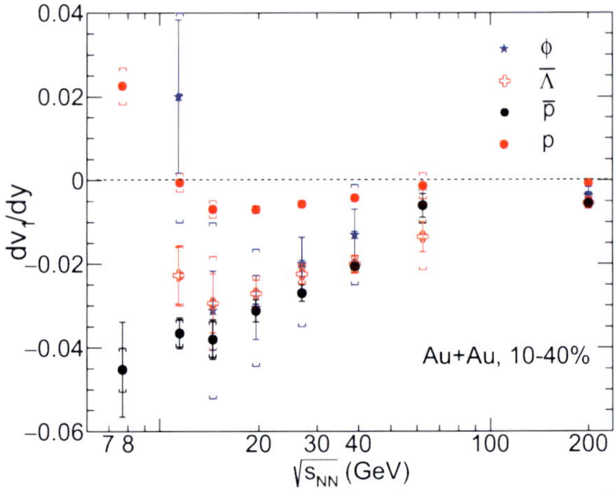

Fig. 51.3 Directed flow slope (dv_1/dy) versus beam energy for 10–40% centrality in Au+Au collisions for ϕ, $\bar{\Lambda}$, p and $\bar{p}$ [10]

51.4 Summary

We report the measurement of K^{*0} and ϕ resonance production at midrapidity in Au+Au collisions at $\sqrt{s_{NN}} = 7.7 - 200\,\text{GeV}$ recorded by the STAR detector. The K^{*0}/K^- and ϕ/K^- ratios as a function of $(dN_{ch}/d\eta)^{1/3}$ is presented for different center-of-mass energies. The K^{*0}/K^- ratios are found to decrease with increasing $(dN_{ch}/d\eta)^{1/3}$. Whereas, ϕ/K^- ratios are nearly independent of $(dN_{ch}/d\eta)^{1/3}$. The observed suppression of K^{*0}/K^- ratios in central collisions could be due to the effect of hadronic rescattering which reduce the measured yield of short-lived resonances. The directed flow of ϕ meson is presented for different collision energies. Directed flow slope (dv_1/dy) of ϕ meson is found to be consistent within errors with that of $\bar{\Lambda}$ and $\bar{p}$ for $\sqrt{s_{NN}} > 14.5\,\text{GeV}$.

References

1. J. Adams et al. (STAR Collaboration), Nucl. Phys. A **757**, 102 (2005)
2. J. Adams et al. (STAR Collaboration), Phys. Rev. C **71**, 64902 (2005)
3. M. Bleicher et al., Phys. Lett. B **530**, 81 (2002)
4. A.M. Poskanzer, S.A. Voloshin, Phys. Rev. C **58**, 1671 (1998)
5. A. Shor, Phys. Rev. Lett. **54**, 1122 (1985)
6. M. Anderson et al., Nucl. Instrum. Methods A **499**, 659 (2003)
7. W.J. Llope (STAR TOF Group), Nucl. Instrum. Methods B 241, 306 (2005)
8. L. Adamczyk et al. (STAR Collaboration), Phys. Rev. C **93**, 21903 (2016)
9. C. Adler et al., Nucl. Instrum. Methods A **470**, 488 (2001)
10. L. Adamczyk et al. (STAR Collaboration), Phys. Rev. Lett. **120**, 62301 (2018)
11. L. Adamczyk et al. (STAR Collaboration), Phys. Rev. C **96**, 44904 (2017)
12. M.M. Aggarwal et al. (STAR Collaboration), Phys. Rev. C **84**, 034909 (2011)
13. B.I. Abelev et al. (STAR Collaboration), Phys. Rev. C **79**, 64903 (2009)
14. J. Adam et al. (ALICE Collaboration), Phys. Rev. C **91**, 024609 (2015)

Chapter 52
Hadronic Resonances Production with ALICE at the LHC

Sushanta Tripathy

Abstract Measurements of the production of short-lived hadronic resonances are used to probe the properties of the late hadronic phase in ultra-relativistic heavy-ion collisions. Since these resonances have lifetimes comparable to that of the fireball, they are sensitive to the competing effects of particle re-scattering and regeneration in the hadronic gas, which modify the observed particle momentum distributions and yields after hadronisation. Having different masses, quantum numbers and quark content, hadronic resonances carry a wealth of information on different aspects of ion-ion collisions, including the processes that determine the shapes of particle momentum spectra, insight into strangeness production and collective effects in small collision systems. We present the most recent ALICE results on $\rho(770)^0$, $K^*(892)^0$, $\phi(1020)$, $\Sigma(1385)^{\pm}$, $\Lambda(1520)$ and $\Xi(1530)^0$ production at the LHC. They include measurements performed in pp, p–Pb and Pb–Pb collisions at different energies, as well as the latest results from the LHC Run 2 with Xe–Xe collisions at $\sqrt{s_{NN}} = 5.44\,\mathrm{TeV}$ and with Pb–Pb collisions at $\sqrt{s_{NN}} = 5.02\,\mathrm{TeV}$. Collision energy, centrality and multiplicity differential measurements integrated yields and particle ratios are discussed in detail. A critical overview of these results are given through comparisons to measurements from other experiments and theoretical models.

52.1 Introduction

Hadronic resonances are very interesting probes to study the properties of the hadronic medium formed in ultra-relativistic heavy-ion collisions, as the yield ratios of resonances to stable hadrons provide information about the re-scattering and regeneration effects in the hadronic medium [1, 2]. Resonances with shorter lifetime are

Sushanta Tripathy for the ALICE Collaboration.

S. Tripathy (✉)
Discipline of Physics, School of Basic Sciences, Indian Institute of Technology Indore, Simrol, Indore 453552, India
e-mail: Sushanta.Tripathy@cern.ch

D. Elia et al. (eds.), *The XVIII International Conference on Strangeness in Quark Matter (SQM 2019)*, Springer Proceedings in Physics 250,
https://doi.org/10.1007/978-3-030-53448-6_52

expected to decay inside the hadronic phase formed in ultra-relativistic heavy-ion collisions and if the elastic or pseudo-elastic scattering of their decay products (re-scattering) is dominant over regeneration, the resonance yield after kinetic freeze-out would be smaller than the one originally produced at the chemical freeze-out. The re-scattering and regeneration processes may also cancel each other. If the lifetimes of resonances are more than the lifetime of the hadronic phase, their yields will not be affected by any such processes. Along with the insight into the hadronic phase of the system, resonances with open and hidden strangeness shed light on strangeness production in different collision systems. It is expected that the particles with open strangeness may be subject to canonical suppression in small collision systems with respect to large systems. However the ϕ meson, a hidden-strangeness particle, is not expected to be canonically suppressed [3].

We present the most recent ALICE results on $\rho(770)^0$, $K^*(892)^0$, $\phi(1020)$, $\Sigma(1385)^{\pm}$, $\Lambda(1520)$ and $\Xi(1530)^0$ production at the LHC. They include measurements performed in pp, p–Pb and Pb–Pb collisions at different energies, as well as the latest results from the LHC Run 2 with Xe–Xe collisions at $\sqrt{s_{NN}} = 5.44$ TeV and with Pb–Pb collisions at $\sqrt{s_{NN}} = 5.02$ TeV.

52.2 Results and Discussions

Figure 52.1 shows the particle yield ratios of different resonances to their ground states (ordered by increasing resonance lifetime from top to bottom) as a function of multiplicity for pp, p–Pb, Xe–Xe and Pb–Pb collisions. Comparisons with EPOS3 [4] and STAR data are also shown in Fig. 52.1. A significant suppression of the ρ/π, K^{*0}/K and Λ^*/Λ ratios has been observed with increasing charged particle multiplicity. This indicates the dominance of re-scattering over regeneration for ρ, K^{*0} and Λ. However, the long-lived resonance ratios like Ξ^*/Ξ and ϕ/K are nearly constant as a function of multiplicity. This suggests the long-lived resonances are not significantly affected by the re-scattering or regeneration processes. It indicates that the long lived resonances decay predominantly outside the hadronic medium. The EPOS3 calculations without UrQMD seem to describe only the ϕ/K while they fail to explain ρ/π and K^{*0}/K ratios. However, we observe that EPOS3 with UrQMD, which includes a modelling of re-scattering and regeneration in the hadronic phase, seems to explain all the particle ratios qualitatively.

Figures 52.2 and 52.3 show the p_T-integrated $\phi/(\pi^+ + \pi^-)$ and Ξ/ϕ ratios, respectively in pp, p–Pb and Pb–Pb collisions as a function of charged-particle multiplicity. These ratios give insight into strangeness production using the ϕ meson, a hidden-strangeness particle. The ratio ϕ/π increases as a function of multiplicity in small collision systems and approaches the thermally predicted value [7] at high multiplicity

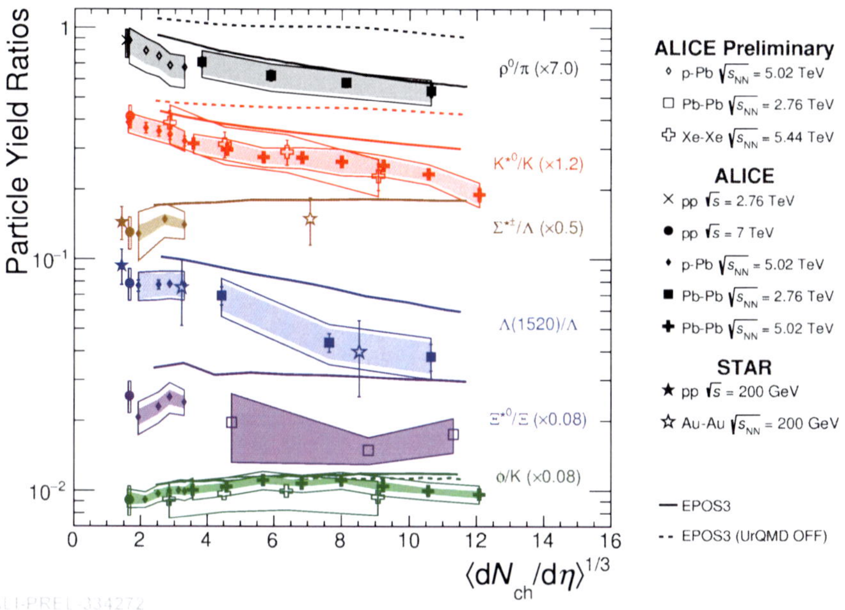

Fig. 52.1 Summary of particle yield ratios of different resonances to their respective ground-state particles as a function of multiplicity for pp, p–Pb, Xe–Xe and Pb–Pb collisions along with comparisons to EPOS3 predictions and STAR data

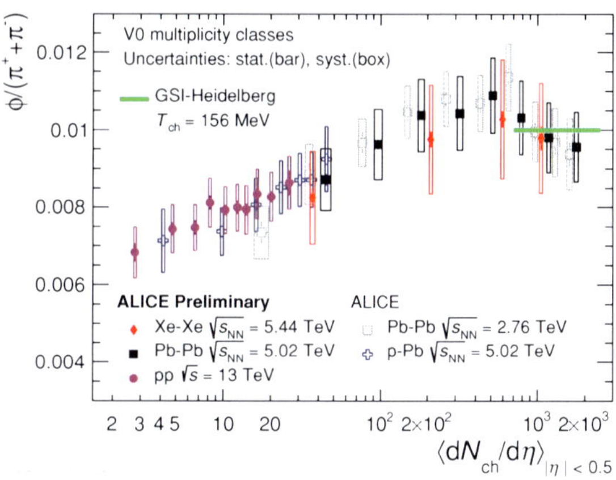

Fig. 52.2 p_{T}-integrated $\phi/(\pi^+ + \pi^-)$ ratio in pp, p–Pb and Pb–Pb collisions as a function of charged-particle multiplicity [5, 6]. Statistical uncertainties are represented by bars and total systematic uncertainties by open boxes. The Grand Canonical thermal model prediction [7] is shown as solid green line

in large collision systems. Even though the ϕ meson has hidden strangeness, this behavior is consistent, against the expectations with its canonical suppression in small systems [3] and it favors the non-equilibrium production of ϕ and/or strange particles. The ϕ/K (Fig. 52.1) ratio remains fairly flat and the Ξ/ϕ remains flat or slightly increases across a wide multiplicity range. These ratios suggest that the ϕ behaves similarly to particles with open strangeness.

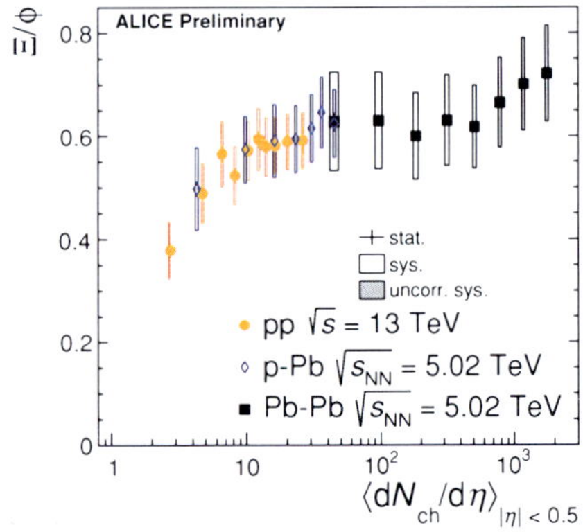

Fig. 52.3 p_T-integrated Ξ/ϕ ratio in pp, p–Pb and Pb–Pb collisions as a function of charged-particle multiplicity. Statistical uncertainties are represented by bars and total systematic uncertainties by open boxes

52.3 Summary

In summary, the ratios of ρ/π, K^{*0}/K and Λ^*/Λ show a decreasing trend from pp and peripheral Pb–Pb to central Pb–Pb collisions. This suggests the dominance of re-scattering over regeneration processes for short-lived resonances in the hadronic phase of the system formed in heavy-ion collisions. However, yield ratios for long-lived resonances remain flat as a function of multiplicity. This suggests that long-lived resonances are not significantly affected by re-scattering or regeneration processes, or that those processes may cancel each other out. It indicates that the long lived resonances decay predominantly outside the hadronic medium. EPOS3 with UrQMD seems to reproduce these trends qualitatively. The ϕ meson, a hidden-strangeness particle, shows similar behavior to particles with open strangeness and seems to have an effective strangeness between 1 and 2 [8, 9].

References

1. M. Bleicher, J. Aichelin, Phys. Lett. B **530**, 81 (2002)
2. M. Bleicher, H. Stoecker, J. Phys. G **30**, S111 (2004)
3. V. Vislavicius, A. Kalweit, arXiv:1610.03001 [nucl-ex]
4. A.G. Knospe et al., Phys. Rev. C **93**(1), 014911 (2016)
5. B.B. Abelev et al. (ALICE Collaboration), Phys. Rev. C **91**, 024609 (2015)
6. J. Adam et al. (ALICE Collaboration), Eur. Phys. J. C **76**(5), 245 (2016)
7. A. Andronic et al., Nature **561**(7723), 321 (2018)
8. S. Tripathy (ALICE Collaboration), Nucl. Phys. A **982**, 180 (2019)
9. J. Song (ALICE Collaboration), PoS ICHEP **2018**, 096 (2019)

Part VII
QCD Phase Diagram and Critical Point

Chapter 53
Influence of Centrality Definition and Detector Efficiency on the Net-Proton Kurtosis

Sukanya Sombun, Jan Steinheimer, Christoph Herold, Ayut Limphirat, Yupeng Yan, and Marcus Bleicher

Abstract We study the influence of the centrality definition and detector efficiency on the net-proton kurtosis for minimum bias Au+Au collisions at a beam energy of $\sqrt{s_{\mathrm{NN}}} = 7.7\,\mathrm{GeV}$ by using the UrQMD model. We find that different ways of defining the centrality lead to different cumulant ratios. Moreover, we demonstrate that the kurtosis is suppressed for central collisions when a wider transverse momentum acceptance is used. Finally, the influence of a detector efficiency on the measured cumulant ratios is estimated.

53.1 Introduction

One aim of heavy ion collisions is to study the phase structure of the strong interaction or quantum chromodynamics (QCD) which has been widely investigated both in experiment and theory. QCD based models predict a first-order phase transition at large μ_B which ends in a QCD critical point (CP) [1]. The QCD phase structure can be disclosed from the study of event-by-event fluctuations of conserved quantities which can be defined in the form of cumulants. Especially cumulants of the

S. Sombun (✉) · C. Herold · A. Limphirat · Y. Yan
School of Physics and Center of Excellence in High Energy Physics & Astrophysics, Suranaree University of Technology, Nakhon Ratchasima 30000, Thailand
e-mail: sukanya_joy33@hotmail.com

J. Steinheimer · M. Bleicher
Frankfurt Institute for Advanced Studies, Ruth-Moufang-Str. 1, 60438 Frankfurt am Main, Germany

M. Bleicher
Institut für Theoretische Physik, Goethe Universität Frankfurt, Max-von-Laue-Strasse 1, 60438 Frankfurt am Main, Germany

GSI Helmholtzzentrum für Schwerionenforschung GmbH, Planckstr. 1, 64291 Darmstadt, Germany

John von Neumann-Institut für Computing, Forschungzentrum Jülich, 52425 Jülich, Germany

D. Elia et al. (eds.), *The XVIII International Conference on Strangeness in Quark Matter (SQM 2019)*, Springer Proceedings in Physics 250,
https://doi.org/10.1007/978-3-030-53448-6_53

net-proton number are predicted to be sensitive to the presence of a QCD phase transition [2, 3] and CP [4–7]. Higher order cumulants are particularly sensitive to a divergent correlation length close to the CP. The higher order susceptibilities of baryon number, electric charge and strangeness have been calculated theoretically in [8–16]. Experimentally, event-by-event fluctuations of higher order cumulants of net-proton, net-pion and net-kaon number were measured by RHIC [17–22] and LHC [23–25]. Hereby, a non-monotonic behaviour of higher order cumulant ratios as a function of beam energy might disclose critical behaviour. Although critical fluctuations have been widely studied, there are still difficulties in understanding the interplay of different effects and their impact on the measured observables. Some uncertainties that we are going to address in this paper using the UrQMD model are the centrality determination, the importance of volume fluctuations, the transverse momentum (p_T) acceptance, and efficiency corrections.

53.1.1 The UrQMD Model

For the present study, we use the Ultrarelativistic Quantum Molecular Dynamics (UrQMD) model to study cumulant ratios of the net-proton number distribution. The UrQMD model is a microscopic transport model which is able to simulate p+p, p+A and A+A collisions. It is based on the covariant propagation of constituent quarks and anti-quarks accompanied by mesonic and baryonic degrees of freedom, binary elastic and inelastic scattering of hadrons, resonance excitations as well as string dynamics and strangeness exchange reactions [26–28]. The elementary cross-sections are interpreted geometrically and are taken from experimental data [29].

53.2 Method

53.2.1 Calculation of Cumulants

The fluctuations of the net-proton number distribution can be characterized by the corresponding cumulants. These are calculated as

$$C_1 = M = \langle \mathrm{N} \rangle \tag{53.1}$$

$$C_2 = \sigma^2 = \left\langle (\delta \mathrm{N})^2 \right\rangle \tag{53.2}$$

$$C_3 = S\sigma^3 = \left\langle (\delta \mathrm{N})^3 \right\rangle \tag{53.3}$$

$$C_4 = \kappa\sigma^4 = \left\langle (\delta \mathrm{N})^4 \right\rangle - 3\left\langle (\delta \mathrm{N})^2 \right\rangle^2 \tag{53.4}$$

where $(\delta \mathrm{N}) = \mathrm{N} - \langle \mathrm{N} \rangle$, and $\langle \mathrm{N} \rangle$ is the event-averaged value of the net-proton number N. We consider the following ratios of these cumulants,

$$\frac{C_2}{C_1} = \frac{\langle \delta N^2 \rangle}{\langle N \rangle} = \frac{\sigma^2}{M} \tag{53.5}$$

$$\frac{C_3}{C_2} = \frac{\langle \delta N^3 \rangle}{\langle \delta N^2 \rangle} = S\sigma \tag{53.6}$$

$$\frac{C_4}{C_2} = \frac{\langle \delta N^4 \rangle}{\langle \delta N^2 \rangle} - 3\langle \delta N^2 \rangle = \kappa\sigma^2 \tag{53.7}$$

with combinations of mean (M), variance (σ^2), skewness (S) and kurtosis (κ) of measured event-by-event fluctuations. For the statistical error calculations, we use the delta theorem [30] which states that the error of cumulant ratios is proportional to a certain power of the standard deviation as:

$$error\left(\frac{C_r}{C_2}\right) \propto \frac{\sigma^{r-2}}{\sqrt{n}} . \tag{53.8}$$

Here, r denotes the order of the cumulant, and n the number of events.

53.2.2 Centrality Definition

In heavy-ion experiments, there is no unique definition of centrality. As the impact parameter (b) is experimentally not accessible, observables like the number of participants ($N_{\rm part}$) or the number of charged particles ($N_{\rm charge}$) are used. These can be related to the impact parameter e.g. by a Glauber model. Therefore, the centrality is practically determined by particle multiplicities. In our work, the centrality will be defined by $N_{\rm part}$, $N_{\rm charge}$ or the number of participants in the projectile, $N_{\rm part-projectile}$. Calculation of cumulants as a function of $N_{\rm part}$ and $N_{\rm charge}$ is believed to minimize volume fluctuations which can have an influence on the extracted ratio of cumulants. We study the dependence of the net-proton number on the different centrality definitions. For this, we simulate Au+Au collision at a beam energy of $\sqrt{s_{\rm NN}} = 7.7\,\text{GeV}$ with impact parameters in the range $0 \leq b \leq 15\,\text{fm}$. We define the following quantities,

- $N_{\rm charge}$: The number of all charged particles with $|\eta| \leq 1$ and $0.15 < p_T < 2.0\,\text{GeV}$ minus the number of protons and anti-protons in this specific acceptance range.
- $N_{\rm part}$: 394 minus all spectator protons and neutrons defined by $|y| > 1.5$ and $p_T < 0.3\,\text{GeV}$.
- $N_{\rm part-projectile}$: 197 minus all projectile spectator protons and neutrons defined by $y > 1.5$ and $p_T < 0.3\,\text{GeV}$.

We determine the distribution of these three different quantities as shown in Fig. 53.1. The three different methods give different distributions. We find that the participant distribution of $N_{\rm part}$ shows a sharp cutoff at the maximum number of participants,

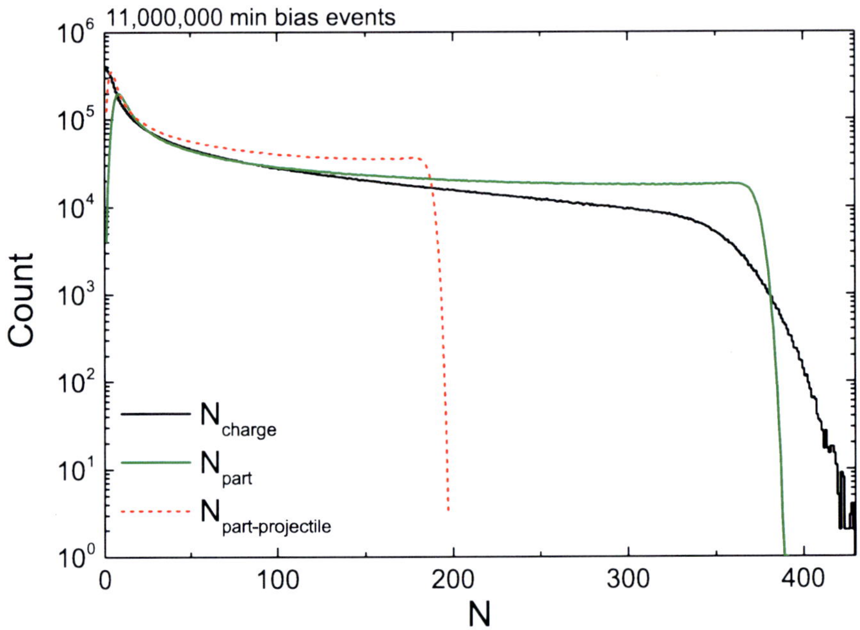

Fig. 53.1 [Color online] Distributions of N_{charge}, N_{part} and $N_{\text{part}-\text{projectile}}$ in Au+Au collision at a beam energy of $\sqrt{s_{\text{NN}}} = 7.7\,\text{GeV}$ with impact parameter $0 \leq b \leq 15$ fm. Figure from [31]

whereas the distribution of N_{charge} shows a much smoother drop. The centrality is classified into 10 centrality bins of 0–10, 10–20, 20–30, 30–40, 40–50, 50–60, 60–70, 70–80, 80–90 and 90–100%. We calculate the cumulants C_n for a fixed N_{part} (N_{charge}, $N_{\text{part}-\text{projectile}}$) and then average those cumulants over all N_{part} in a given centrality bin. The ratios of cumulants are then determined as ratios of averages.

53.3 Results

53.3.1 Dependence on Centrality Definition

We first assume that the efficiency of the detector and its acceptance are perfect (100%) for all particles. Figure 53.2 shows the result of the net-proton number kurtosis as a function of centrality for the three centrality definitions, at mid-rapidity $|y| < 0.5$ and within transverse momentum $0.4 < p_T < 0.8\,\text{GeV}$. For the most central collisions, we find that the value of the kurtosis does not depend on the centrality definition. On the other hand, for mid-central collision, the difference between the values of $\kappa\sigma^2$ becomes larger. Moreover, the centrality defined by N_{charge} yields only

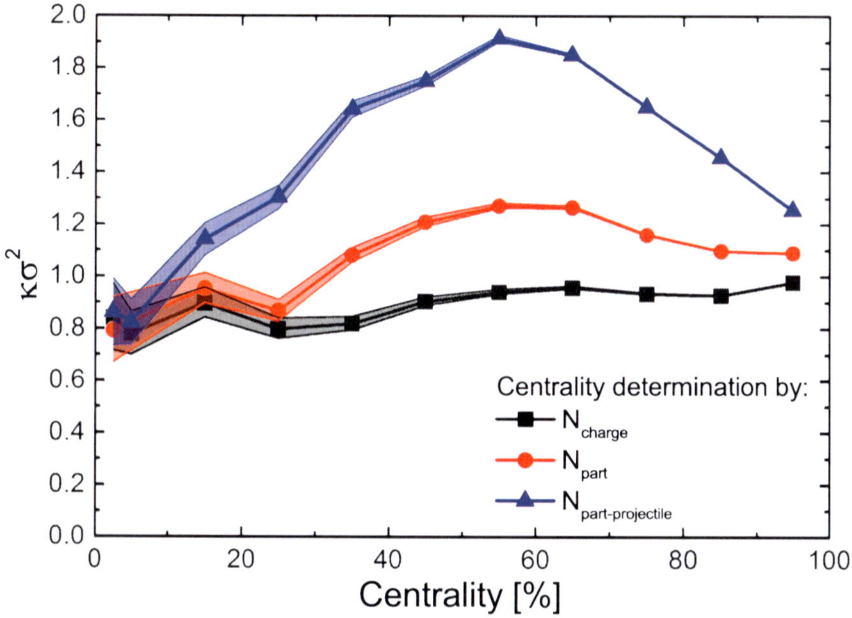

Fig. 53.2 [Color online] The kurtosis of the net-proton number as a function of centrality which is defined by three different quantities, the number of charged particles, the number of participants and the number of participants in the projectile. Figure from [31]

a mild dependence of the kurtosis on the centrality because it not subject to large volume fluctuations. Thus, we use N_{charge} in what follows to define the centrality to investigate other effects.

53.3.2 Effect of Transverse Momentum Range

We now study the kurtosis of the net-proton number as a function of centrality which is defined by N_{charge} for two different transverse momentum ranges, namely $(0.4 < p_T < 0.8\,\text{GeV})$ and $(0.4 < p_T < 2.0\,\text{GeV})$. The result is shown in Fig. 53.3. It can be seen that for the most central collisions, the value of the kurtosis is strongly suppressed for the larger acceptance range due to baryon number conservation. At mid-central collisions, the value of the kurtosis is larger for the wider transverse momentum range which indicates that volume fluctuations affect the extracted kurtosis.

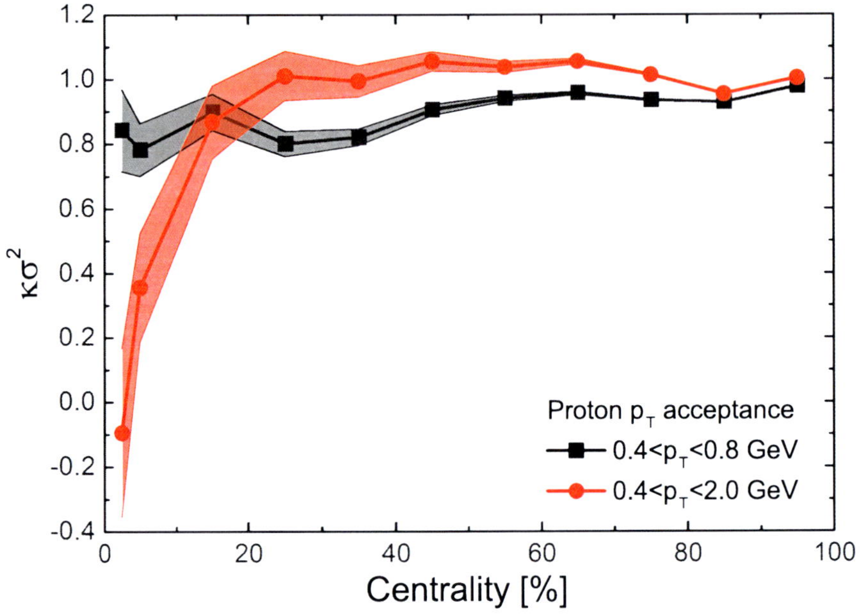

Fig. 53.3 [Color online] The kurtosis of the net-proton number as a function of centrality (defined by N_{charge}) for two different transverse momentum p_T ranges. Figure from [31]

53.3.3 Effect of Efficiency

In practice, particle detectors are not ideal systems but suffer from a finite particle detection efficiency. The efficiency is the number of produced particles that are recorded in the detector divided by the overall yield, see [32]. Therefore, detectors measuring particle number fluctuations will never perform perfectly. We now show how an efficiency below 100% influences the measurement of cumulants. We begin by studying the effect of efficiency on the fluctuations. In Fig. 53.4, we show the result of the net-proton number kurtosis in two different p_T ranges as a function of centrality at 100 and 70% N_{charge} efficiency. It is found that the reduced (realistic) efficiency leads to an overall increase of the kurtosis.

Second, we study the effect of a proton efficiency. The result of the variance, skewness and kurtosis of the net-proton number as a function of centrality defined by 70% N_{charge} efficiency is shown in Fig. 53.5. We compare the results for 100 and 75% proton efficiency. The circles represent calculations with 100% proton efficiency, the squares are 75% proton efficiency. We find that the cumulant ratios of the net-proton number for the 75% proton efficiency is smaller than for the 100% proton efficiency.

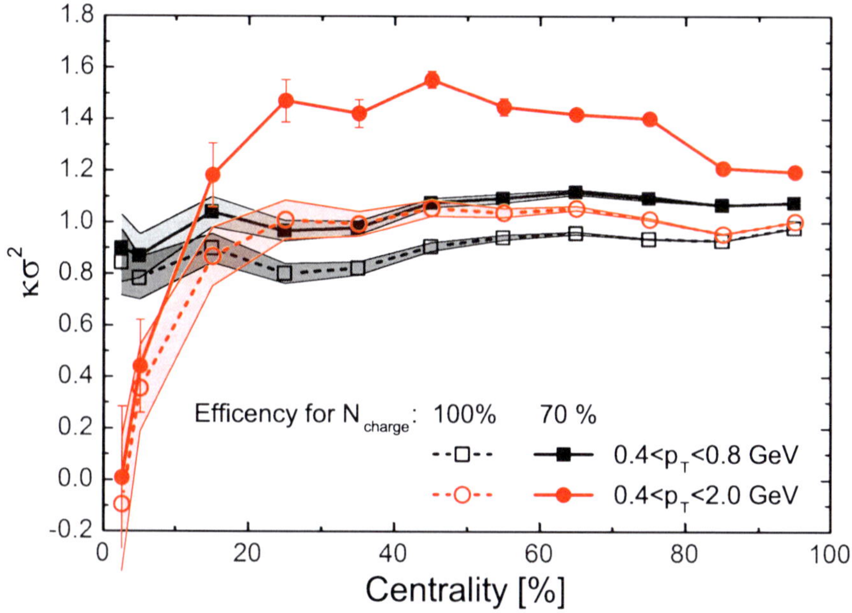

Fig. 53.4 [Color online] The kurtosis of the net-proton number in two different p_T acceptance bins (black squares and red circles) as a function of centrality with 100% (open symbols) and 70% (full symbols) N_{charge} efficiency. Figure from [31]

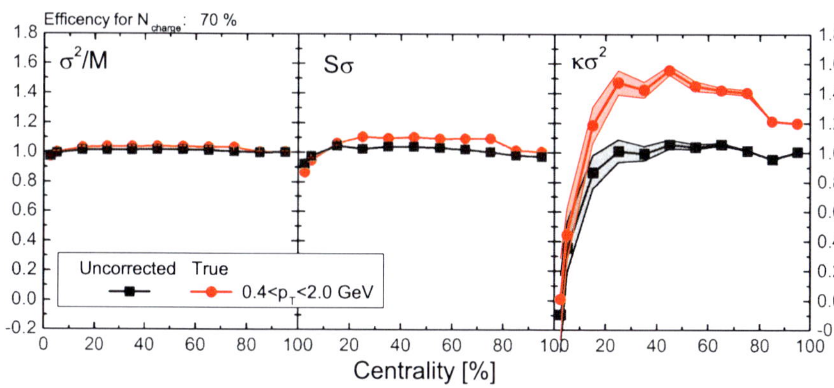

Fig. 53.5 [Color online] The result of variance, skewness and kurtosis of the net-proton number in two different p_T acceptance bins for a 70% N_{charge} efficiency. We compare the result between 100% (red circle symbol) and 75% (black squares symbol) proton efficiency. Figure from [31]

53.4 Conclusions

We have studied the effect of different methods for centrality determination on the measured net-proton kurtosis. We have found that different centrality definitions give different results for the kurtosis. Using a centrality defined by N_{charge} reduces the effect of volume fluctuations. Moreover, we have studied the effect of two different transverse momentum ranges accepting net-protons in the measurement. We have seen that the wider transverse momentum range leads to a strongly suppressed kurtosis at central collisions. Finally, we have observed the effect of centrality determination, finding a clear impact on the value of the kurtosis.

Acknowledgments The computational resources have been provided by the LOEWE Frankfurt Center for Scientific Computing (LOEWE-CSC) and the Center for Computer Services at SUT. This work is supported by the German Academic Exchange Service (DAAD), HIC for FAIR and the Thailand Research Fund (TRF). SS and AL acknowledge support from TRF-RGJ (PHD/0185/2558). CH, AL and YY acknowledge support from Suranaree University of Technology and the Office of the Higher Education Commission under NRU project of Thailand.

References

1. M.A. Stephanov, QCD phase diagram and the critical point. Prog. Theor. Phys. Suppl. **153**, 139 (2004). https://doi.org/10.1142/S0217751X05027965
2. J. Randrup, Spinodal decomposition during the hadronization stage at RHIC? Phys. Rev. Lett. **92**, 122301 (2004). https://doi.org/10.1103/PhysRevLett.92.122301
3. C. Sasaki, B. Friman, K. Redlich, Density fluctuations in the presence of spinodal instabilities. Phys. Rev. Lett. **99**, 232301 (2007). https://doi.org/10.1103/PhysRevLett.99.232301
4. M.A. Stephanov, Non-Gaussian fluctuations near the QCD critical point. Phys. Rev. Lett. **102**, 032301 (2009). https://doi.org/10.1103/PhysRevLett.102.032301
5. C. Herold, M. Bleicher, M. Nahrgang, J. Steinheimer, A. Limphirat, C. Kobdaj, Y. Yan, Broadening of the chiral critical region in a hydrodynamically expanding medium. Eur. Phys. J. A **54**(2), 19 (2018). https://doi.org/10.1140/epja/i2018-12438-1
6. M. Nahrgang, C. Herold, M. Bleicher, Influence of an inhomogeneous and expanding medium on signals of the QCD phase transition. Nucl. Phys. A **904–905**, 899c (2013). https://doi.org/10.1016/j.nuclphysa.2013.02.160
7. M. Szymański, M. Bluhm, K. Redlich, C. Sasaki, Net-proton number fluctuations in the presence of the QCD critical point
8. M. Cheng et al., Baryon number, strangeness and electric charge fluctuations in QCD at high temperature. Phys. Rev. D **79**, 074505 (2009)
9. W.J. Fu, Y.X. Liu, Y.L. Wu, Fluctuations and correlations of conserved charges in QCD at finite temperature with effective models. Phys. Rev. D **81**, 014028 (2010)
10. L. Chen, X. Pan, F.B. Xiong, L. Li, N. Li, Z. Li, G. Wang, Y. Wu, Statistical and dynamical fluctuations in the ratios of higher net-proton cumulants in relativistic heavy-ion collisions. J. Phys. G **38**, 115004 (2011)
11. F. Karsch, K. Redlich, Has Tc been measured by heavy ion experiments? Phys. Rev. D **84**, 051504 (2011)
12. B.J. Schaefer, M. Wagner, QCD critical region and higher moments for three flavor models. Phys. Rev. D **85**, 034027 (2012)
13. X. Wang, C.B. Yang, On the energy and centrality dependence of higher order moments of net-proton distributions in relativistic heavy ion collisions. Phys. Rev. C **85**, 044905 (2012)

14. D.M. Zhou, A. Limphirat, Y.I., Yan, C. Yun, Y.P., Yan, X. Cai, L.P. Csernai, B.H. Sa, Higher moment singularities explored by the net proton non-statistical fluctuations. Phys. Rev. C **85**, 064916 (2012)
15. P. Rau, J. Steinheimer, S. Schramm, H. Stöcker, Conserved charge fluctuations in a chiral hadronic model including hadrons and quarks. Phys. Lett. B **733**, 176 (2014). https://doi.org/10.1016/j.physletb.2014.04.035
16. W. Fan, X. Luo, H. Zong, Probing the QCD phase structure with higher order baryon number susceptibilities within the NJL model. Chin. Phys. C **43**(3), 033103 (2019). https://doi.org/10.1088/1674-1137/43/3/033103
17. M.M. Aggarwal et al. (STAR Collaboration), Higher moments of net-proton multiplicity distributions at RHIC. Phys. Rev. Lett. **105**, 022302 (2010)
18. X.F. Luo (STAR Collaboration), Probing the QCD critical point with higher moments of net-proton multiplicity distributions. J. Phys. Conf. Ser. **316**, 012003 (2011)
19. L. Adamczyk et al. (STAR Collaboration), Energy dependence of moments of net-proton multiplicity distributions at RHIC. Phys. Rev. Lett. **112**, 032302 (2014)
20. L. Adamczyk et al. (STAR Collaboration), Beam energy dependence of moments of the net-charge multiplicity distributions in Au+Au collisions at RHIC. Phys. Rev. Lett. **113**, 092301 (2014)
21. A. Adare et al. (PHENIX Collaboration), Measurement of higher cumulants of net-charge multiplicity distributions in Au+Au collisions at $\sqrt{s_{NN}} = 7.7 - 200$ GeV. Phys. Rev. C **93**(1), 011901 (2016)
22. J. Xu, Energy dependence of moments of net-proton, net-kaon, and net-charge multiplicity distributions at STAR. J. Phys. Conf. Ser. **736**(1), 012002 (2016). https://doi.org/10.1088/1742-6596/736/1/012002
23. B. Abelev et al. (ALICE Collaboration), Net-charge fluctuations in Pb-Pb collisions at $\sqrt{s}_{NN} =$ 2.76 TeV. Phys. Rev. Lett. **110**(15), 152301 (2013)
24. A. Rustamov, (ALICE Collaboration), Net-baryon fluctuations measured with ALICE at the CERN LHC, arXiv:1704.05329 [nucl-ex]
25. A. Ohlson (ALICE Collaboration), Measurements of the fluctuations of identified particles in ALICE at the LHC. PoS CPOD **2017**, 031 (2018). https://doi.org/10.22323/1.311.0031
26. S.A. Bass et al., Microscopic models for ultrarelativistic heavy ion collisions. Prog. Part. Nucl. Phys. **41**, 255 (1998)
27. M. Bleicher et al., Relativistic hadron hadron collisions in the ultrarelativistic quantum molecular dynamics model. J. Phys. G **25**, 1859 (1999)
28. G. Graef, J. Steinheimer, F. Li, M. Bleicher, Deep sub-threshold Ξ and Λ production in nuclear collisions with the UrQMD transport model. Phys. Rev. C **90**, 064909 (2014)
29. C. Patrignani et al. (Particle Data Group), Rev. Part. Phys. Chin. Phys. C **40**(10), 100001 (2016)
30. X. Luo, Error estimation for moments analysis in heavy ion collision experiment. J. Phys. G **39**, 025008 (2012)
31. S. Sombun, J. Steinheimer, C. Herold, A. Limphirat, Y. Yan, M. Bleicher, Higher order net-proton number cumulants dependence on the centrality definition and other spurious effects. J. Phys. G **45**(2), 025101 (2018). https://doi.org/10.1088/1361-6471/aa9c6c
32. B.I. Abelev et al. (STAR Collaboration), Systematic measurements of identified particle spectra in pp, d^+ Au and Au+Au collisions from STAR. Phys. Rev. C **79**, 034909 (2009)

Chapter 54
Direct Photon and Light Neutral Meson Production in the Era of Precision Physics at the LHC

Meike Charlotte Danisch

Abstract We present the latest results from ALICE on direct photon and light neutral meson production in p–Pb and Pb–Pb collisions. The direct photon excess ratio R_γ in different charged particle multiplicity classes of p–Pb collisions at $\sqrt{s_{\rm NN}} = 5.02$ TeV is shown. In addition, we present the direct photon elliptic flow coefficient v_2 in central and semicentral events of Pb–Pb collisions at $\sqrt{s_{\rm NN}} = 2.76$ TeV. An outlook on ongoing and future measurements is given.

54.1 Introduction

Direct photons, being defined as photons not originating from hadron decays, are a valuable tool to investigate the space-time evolution of the medium created in heavy-ion collisions. The measured elliptic flow coefficient of photons reflects the momentum anisotropy of the source, convoluted with the photon emission rate, integrated over time. Direct photons can also contribute to the effort of testing bulk effects in high multiplicity collisions of small systems.

There are different sources of photons in heavy-ion collisions. Prompt photons are created in initial hard scatterings. In addition, we expect thermal photons from the QGP and the hadron gas phases, which are sensitive to the medium temperature [1]. There will also be photons from hadron decays, which are the vast majority of all photons. The latter originate mostly from decays of the neutral mesons π^0 and η into two photons. Therefore we need to measure their spectra precisely in order to be able to obtain the excess of direct photons over this decay photon background. Due to the different shapes of transverse momentum ($p_{\rm T}$) distributions of prompt and thermal photons (power-law and close-to-exponential respectively) they can be distinguisted on a statistical basis. Prompt photons dominate direct photons at high

Meike Charlotte Danisch for the ALICE Collaboration.

M. Charlotte Danisch (✉)
Physikalisches Institut, Universität Heidelberg, Heidelberg, Germany
e-mail: meike.charlotte.danisch@cern.ch

D. Elia et al. (eds.), *The XVIII International Conference on Strangeness in Quark Matter (SQM 2019)*, Springer Proceedings in Physics 250,
https://doi.org/10.1007/978-3-030-53448-6_54

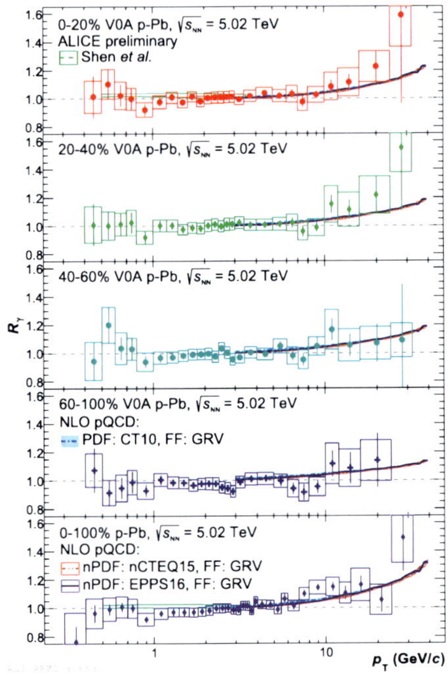

Fig. 54.1 Direct photon R_γ, labeled by the percentile of the multiplicity distribution

$p_T \gtrsim 4$ GeV/c and thermal photons at low $p_T \lesssim 3$ GeV/c. Therefore, a direct photon excess at low p_T can be interpreted as thermal photon signal. So far it was assumed that thermal photons are relevant only in AA collisions but after collective effects have been observed also in high-multiplicity pp and p–A collisions (see for example [2]) we can question this assumption.

54.2 Direct Photon R_γ in p–Pb Collisions at $\sqrt{s_{NN}}$ = 5.02 TeV

Observation of a diret photon signal implies that the ratio of inclusive (γ_{inc}) over decay photons (γ_{dec}) is larger than one. In order to eliminate parts of the uncertainty, we define the double ratio as $R_\gamma = \frac{\gamma_{\text{inc}}}{\gamma_{\text{dec}}} \equiv \frac{\gamma_{\text{inc}}}{\pi^{0,\text{p}}} / \frac{\gamma_{\text{dec}}}{\pi^{0,\text{p}}}$, where $\pi^{0,\text{p}}$ stands for the parametrisation of the measured π^0 spectrum. In the analysis presented here, inclusive photons are measured with three different techniques: PCM, EMCal and PHOS [3]. The EMCal is a sampling calorimeter composed of alternating layers of lead and plastic scintillators. It is placed at a radius of $R = 4.3$ m from the beam and covers a range of $\Delta\varphi = 100°$ in azimuthal angle and $|\eta| < 0.7$ in pseudorapidity. The cell size is about $6 \times 6\,\text{cm}^2$.

The Photon Spectrometer PHOS is a homogeneous calorimeter made from $PbWO_4$ crystals. The acceptance of $\Delta\varphi = 60°$ and $|\eta| < 0.125$ is smaller than the one of EMCal but PHOS has a finer granularity with a cell size of about $2.2 \times 2.2\,\text{cm}^2$ at $R = 4.6\,\text{m}$. With the photon conversion method (PCM) we exploit the fact that photons can convert to an electron positron pair in the detector material with a probability of about 8% (for $R < 1.8\,\text{m}$). We search for secondary vertices characterized by two oppositely charged emerging electron tracks which we detect in the Inner Tracking System (ITS, $|\eta| < 1.2$) and Time Projection Chamber (TPC, $|\eta| < 0.9$). This methods benefits from the large acceptance and the good momentum resolution especially at low p_{T}. In addition to the inclusive photons, we measure π^0 and η mesons via their decay to two photons, performing an invariant mass analysis of photon pairs. For this purpose, the photon samples are taken from the same methods as mentioned above, as well as from one additional hybrid method (PCM-EMC), where one photon is detected with PCM and one with EMCal. The decay photon spectra are then obtained using a Monte Carlo simulation including all relevant hadron decays, based on the measured neutral meson spectra and m_{T} scaling. For each of the four methods, the R_γ is calculated. In case of the PCM-EMC method, inclusive photons are taken from PCM. After checking that the results from all methods are consistent, they were combined. The analysis was performed in event multiplicity classes of data from p–Pb collisions recorded in 2013. The results are shown in Fig. 54.1. The dotted blue, red and purple lines, starting at $p_{\mathrm{T}} = 3\,\text{GeV}/c$ show results of different pQCD calculations. They are all well compatible with the measured points at high p_{T}. For 0–20 and 0–100% samples a green band is drawn in addition, which shows a prediction from a hydrodynamic model [4] including thermal photon emission at low p_{T}. The current data are not sensitive to the predicted very small thermal photon signal.

54.3 Direct Photon v_2 in $\sqrt{s_{\mathrm{NN}}}$ = 2.76 TeV Pb–Pb Collisions

The inclusive photon v_2 is obtained using the scalar product method [5]. Reference particles are measured in the V0 scintillation detectors placed in a different pseudorapidity region ($2.8 < \eta < 5.1$ and $-3.7 < \eta < -1.7$) [3]. Results from the independent methods PCM and PHOS are combined after they were found to be consistent. In the analysis presented here, Pb–Pb collisions were analyzed in two centrality classes, 0–20 and 20–40%. For PCM, 13.6×10^6 events were available and 18.8×10^6 for PHOS. The direct photon $v_2^{\gamma,\mathrm{dir}}$ can be calculated by subtracting the $v_2^{\gamma,\mathrm{dec}}$ of decay photons from the $v_2^{\gamma,\mathrm{inc}}$ of inclusive photons using the following formula: $v_2^{\gamma,\mathrm{dir}} = \frac{v_2^{\gamma,\mathrm{inc}} R_\gamma - v_2^{\gamma,\mathrm{dec}}}{R_\gamma - 1}$, where the R_γ measured with PCM and PHOS [7] was used. $v_2^{\gamma,\mathrm{dec}}$ is obtained from a MC simulation including all relevant hadron decays. The simulation is based on hadron spectra and v_2 measurements and uses KE_{T} scaling [8] when necessary. The results for decay photons and direct photons are shown in Fig. 54.2. At low p_{T}, where thermal photons should dominate, we measure a pos-

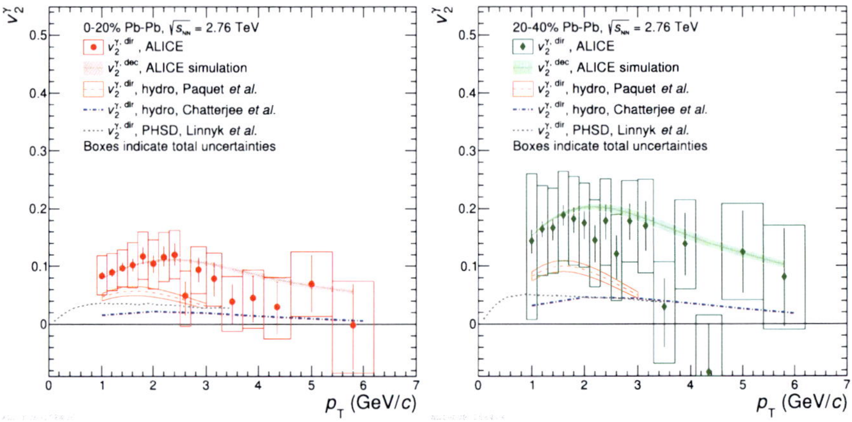

Fig. 54.2 Direct and decay photon v_2 in central (left) and semicentral (right) collisions compared with hydrodynamic and cascade model predictions [6]

itive $v_2^{\gamma,\mathrm{dir}}$ which is close to $v_2^{\gamma,\mathrm{dec}}$. This indicates an already developed momentum anisotropy of the medium at direct photon production times. At higher p_T, where the prompt photon contribution increases, the $v_2^{\gamma,\mathrm{dir}}$ decreases. In more peripheral events, the thermal photon v_2, and therefore also the direct photon v_2 at low p_T, are expected to be larger than in central events because of the more pronounced initial spatial anisotropy of the medium. Because the direct photon signal is smaller in more peripheral events [7] the uncertainties are larger in this case and therefore we cannot yet make a conclusive statement on how the direct photon v_2 changes with centrality. Calculations from different theoretical models are illustrated by the dashed lines. They tend to underestimate the v_2 with respect to the measured values.

54.4 Summary and Outlook

In summary, ALICE has measured the direct photon elliptic flow coefficient v_2 in Pb–Pb collisions at $\sqrt{s_\mathrm{NN}}$ = 2.76 TeV. It was found to be consistent with the current knowledge of the space-time-evolution and photon emission rates but smaller uncertainties will be needed to confirm or exclude given model predictions. The direct photon R_γ in high multiplicity p–Pb collisions at $\sqrt{s_\mathrm{NN}}$ = 5.02 TeV is not yet sensitive to the predicted very small thermal photon signal. Analyses of $\sqrt{s_\mathrm{NN}}$ = 5.02 TeV Pb–Pb collisions are ongoing. Neutral pion spectra are a crucial input for the decay photon simulation which is required for direct photon measurements. They are presented in Fig. 54.3 in different centrality classes. The increased statistics of the $\sqrt{s_\mathrm{NN}}$ = 5.02 TeV dataset will enable us to reduce statistical uncertainties of R_γ and v_2 by one order of magnitude compared to the measurements at $\sqrt{s_\mathrm{NN}}$ = 2.76 TeV. It was investigated that also the systematic uncertainties, which dominate over parts of the p_T range,

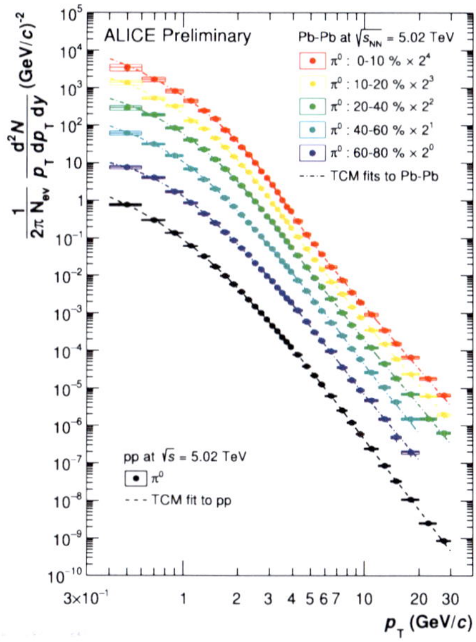

Fig. 54.3 Neutral pion spectra in pp and Pb–Pb collisions at $\sqrt{s_{NN}} = 5.02$ TeV

can be reduced by employing new analysis techniques. This allows to exploit the advantages of the ALICE upgrade in future runs.

References

1. J. Kapusta, P. Lichard, D. Seibert, Phys. Rev. D **47**, 4171 (1993)
2. C. Loizides, Nucl. Phys. A **956**, 200 (2016)
3. ALICE Collaboration, Int. J. Mod. Phys. A **29**(24), 1430044 (2014)
4. C. Shen, J.F. Paquet, G.S. Denicol, S. Jeon, C. Gale, Phys. Rev. C **95**, 014906 (2017)
5. STAR Collaboration, Phys. Rev. C **66**, 034904 (2002)
6. ALICE Collaboration, Phys. Lett. B **789**, 308 (2019)
7. ALICE Collaboration, Phys. Lett. B **754**, 235 (2016)
8. STAR Collaboration, Phys. Rev. Lett. **98**, 162301 (2007)

Chapter 55
Time-Evolution of Net-Baryon Density Fluctuations Across the QCD Critical Region

Marcus Bluhm and Marlene Nahrgang

Abstract We investigate the role of a finite surface tension during the time-evolution of fluctuations in the net-baryon density. The systems in this study undergo a temperature evolution across the phase transition in the critical region of the QCD phase diagram. The occuring non-equilibrium effects are discussed.

55.1 Introduction

The search for the conjectured critical point associated with the deconfinement and chiral phase transition in the QCD phase diagram has received tremendously increasing attention in recent years, both theoretically and experimentally. In order to identify potential signatures of the critical point in experimental data a firm understanding of the dynamics of critical fluctuations during the evolution of matter created in heavy-ion collisions is crucial. For the net-baryon density n_B as the slow critical mode [1] this dynamics is governed by diffusion processes. The time-evolution of fluctuations in n_B, which are expected to be enhanced in the critical region [2, 3], can therefore be modeled by a stochastic diffusion equation. Numerical simulations of the diffusion of critical fluctuations both on the crossover [4–6] and first-order phase transition sides of the critical point have been performed recently for one spatial dimension. In these studies, non-equilibrium effects [7–10] such as retardation or critical slowing down have been observed quantitatively.

In this talk, we study the time evolution of net-baryon density fluctuations across the phase transition in the crossover domain near the critical point. We look at the local variance, i.e. the variance of event-by-event fluctuations over the size of a fluid cell. In continuum it is related to the zero-distance value of the two-point correlation function. We numerically solve the stochastic diffusion equation in the form

M. Bluhm (✉) · M. Nahrgang
SUBATECH UMR 6457 (IMT Atlantique, Université de Nantes, IN2P3/CNRS), 4 rue Alfred Kastler, 44307 Nantes, France
e-mail: bluhm@subatech.in2p3.fr

ExtreMe Matter Institute EMMI, GSI, Planckstr. 1, 64291 Darmstadt, Germany

D. Elia et al. (eds.), *The XVIII International Conference on Strangeness in Quark Matter (SQM 2019)*, Springer Proceedings in Physics 250,
https://doi.org/10.1007/978-3-030-53448-6_55

$$\partial_t n_B(x,t) = \frac{D}{n_c}\left(m^2\nabla_x^2 n_B - K\nabla_x^4 n_B\right) + \sqrt{2Dn_c/A}\,\nabla_x\zeta_x(x,t)$$
$$+ D\nabla_x^2\left(\frac{\lambda_3}{n_c^2}(\Delta n_B)^2 + \frac{\lambda_4}{n_c^3}(\Delta n_B)^3 + \frac{\lambda_6}{n_c^5}(\Delta n_B)^5\right) . \quad (55.1)$$

Here, $\Delta n_B = n_B - n_c$ with a critical density of $n_c = 1/(3\ \text{fm}^3)$. Equation (55.1) describes the dynamics of the critical fluctuations in one spatial, the longitudinal, dimension which is not coupled to the physics in the transverse area A. Accordingly, the covariance of the white noise reads $\langle\zeta_x(x,t), \zeta_x(x',t')\rangle = \delta(x-x')\delta(t-t')$. This ensures that the long-time equilibrium distribution of n_B is governed by the free energy of the system in agreement with the fluctuation-dissipation theorem.

The particular form of (55.1) is based on a free energy functional which contains apart from a Gaussian mass term, where m^2 is inversely connected with the correlation length ξ, also a kinetic energy term with surface tension K and non-linear coupling terms which are proportional to λ_i. This is valid in the vicinity of the critical point and may still be improved by including further regular contributions. The dependence of the coupling coefficients on ξ is defined by universality [1] through the mapping of the 3-dimensional Ising spin model onto an effective potential [11] and the dependence of ξ on temperature T and baryo-chemical potential μ_B in QCD can be determined by a matching to the susceptibility of the spin model scaling equation of state [12] as explained in [13, 14]. The numerical results presented in this work depend on the values of the dimensionless couplings $\tilde{\lambda}_i$ and the diffusion coefficient D.

The stochastic diffusion equation (55.1) allows us to contrast three distinct physical cases: the Gauss model with $K = \lambda_i = 0$ and $m^2 \neq 0$ which was studied in detail in [5], the Gauss+surface model for which also $K \neq 0$, and the Ginzburg–Landau model as considered in [6] for which we exemplarily choose $\tilde{\lambda}_3 = 1$, $\tilde{\lambda}_4 = 10$ and $\tilde{\lambda}_6 = 3$ for the dimensionless couplings. In the numerical implementation we apply a semi-implicit scheme for the Ginzburg–Landau model and an implicit scheme for the other two. Exact net-baryon number conservation is ensured by imposing periodic boundary conditions.

55.2 Time Evolution of the Variance

To study the time evolution of fluctuation observables such as the local variance σ^2, we consider a system of temporally evolving but spatially constant T which is otherwise static, i.e. of a fixed finite size $L = 20$ fm. The temperature cools down following $T(\tau) = T_0\,(\tau_0/\tau)$ where we start from $T_0 = 0.5$ GeV with equilibrated initial conditions at $\tau_0 = 1$ fm/c. Correspondingly, $D(\tau) = D(\tau_0)T(\tau)/T_0$. At $\tau - \tau_0 = 2.33$ fm/c the pseudo-critical temperature is reached.

Let us first focus on the Gauss+surface model. For this model analytic expressions for the long-time equilibrium value of the local variance at fixed T can be derived both in the continuum and in discretized space-time. In the latter case the result depends

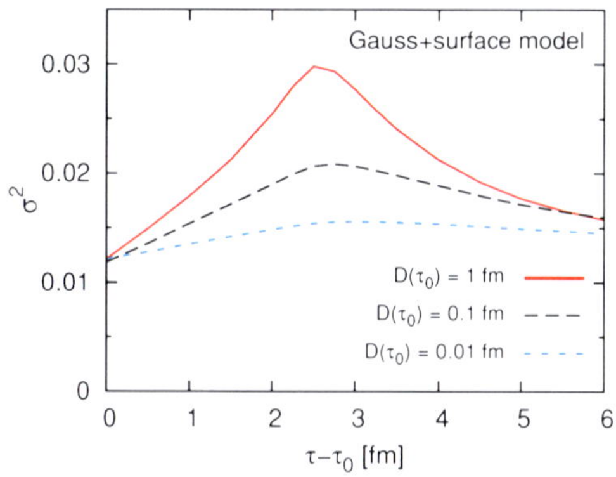

Fig. 55.1 Comparison of the time-evolution of the local variance σ^2 for the Gauss+surface model for different diffusion coefficients $D/\text{fm} = 1, 0.1, 0.01$ at $\tau_0 = 1$ fm/c. The critical temperature T_c is reached at $\tau - \tau_0 = 2.33$ fm/c

on the employed resolution $\Delta x = L/N_x$ with N_x considered lattice sites and reads $\sigma^2 = n_c^2 e^{-\Delta x\sqrt{m^2/K}}/(2A\sqrt{m^2 K})$. In equilibrium, physical observables such as σ^2 are independent of the diffusion coefficient D. As the temperature cools down, the system falls out of equilibrium as a consequence of this dynamics and the time-evolution of σ^2 becomes sensitive to the value of D. This is illustrated in Fig. 55.1. One observes that with decreasing diffusion coefficient finite-range fluctuations are not formed fast enough and as a consequence their magnitude is reduced during the time evolution. Similar observations have been reported for the Gauss model in [5]. Here, even for $D(\tau_0) = 1$ fm the maximum of σ^2 is reduced by a factor of 2 compared to the equilibrium value at T_c due to non-equilibrium effects. The temporal evolution of T is much larger than the diffusion rate. Moreover, this maximum is shifted towards later times, i.e. smaller T than T_c, where this retardation effect becomes stronger with decreasing D.

In Fig. 55.2 we contrast the impact of the time-evolution of T on the behavior of the local variance for the three different physical models. One observes that the non-equilibrium effects are larger once a finite surface tension is included because it takes more time to build up the finite correlation length associated with K. The local variance in the Gauss model reaches almost its equilibrium value at $T = T_c$, see [5], while it is significantly reduced once $K \neq 0$. Moreover, it is interesting to point out another effect of finite K: the local variance at a temperature away from T_c is not suppressed as much as it is the case in the Gauss model. The retardation effect is larger for the models with $K \neq 0$, as the maximal value of the local variance is shifted to $T < T_c$.

55.3 Conclusions and Outlook

In this work, we have outlined the importance of finite-range correlations associated with the presence of a surface tension K in models describing the diffusion of net-baryon density fluctuations. The time needed to build up these correlations can be

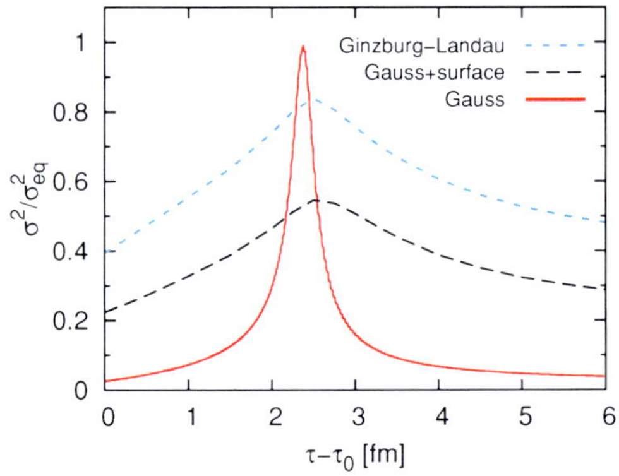

Fig. 55.2 Comparison of the time-evolution of the local variance σ^2 scaled by its equilibrium value σ^2_{eq} at $T = T_c$ for different forms of the stochastic diffusion equation (Gauss, Gauss+surface and Ginzburg–Landau models) and $D(\tau_0) = 1$ fm. The equilibrium value has been calculated for a static box of length $L = 20$ fm at constant temperature in the long-time limit

important and of the same order as the cooling time, which leads to important non-equilibrium effects.

In future work, the expansion of the system, e.g. in a Bjorken geometry, will be added and second-order relaxation equations will be applied in order to guarantee causality and prepare a future embedding in full fluid dynamical simulations of heavy-ion collisions.

Acknowledgments The authors acknowledge the support of the program "Etoiles montantes en Pays de la Loire 2017". The work was in parts supported by the ExtreMe Matter Institute (EMMI) at the GSI Helmholtzzentrum für Schwerionenforschung, Darmstadt, Germany. The authors thank T. Schäfer, S. Bass, M. Kitazawa and the members of the BEST Topical Collaboration for many stimulating discussions.

References

1. P.C. Hohenberg, B.I. Halperin, Rev. Mod. Phys. **49**, 435 (1977)
2. M.A. Stephanov, Phys. Rev. Lett. **102**, 032301 (2009)
3. M. Asakawa, S. Ejiri, M. Kitazawa, Phys. Rev. Lett. **103**, 262301 (2009)
4. M. Sakaida, M. Asakawa, H. Fujii, M. Kitazawa, Phys. Rev. C **95**(6), 064905 (2017)
5. M. Nahrgang, M. Bluhm, T. Schäfer, S.A. Bass, Nucl. Phys. A **967**, 824 (2017)
6. M. Nahrgang, M. Bluhm, T. Schaefer, S.A. Bass, Phys. Rev. D **99**(11), 116015 (2019)
7. B. Berdnikov, K. Rajagopal, Phys. Rev. D **61**, 105017 (2000)
8. M. Nahrgang, S. Leupold, C. Herold, M. Bleicher, Phys. Rev. C **84**, 024912 (2011)
9. M. Kitazawa, M. Asakawa, H. Ono, Phys. Lett. B **728**, 386 (2014)
10. S. Mukherjee, R. Venugopalan, Y. Yin, Phys. Rev. C **92**(3), 034912 (2015)
11. M.M. Tsypin, Phys. Rev. B **55**, 8911 (1997)
12. R. Guida, J. Zinn-Justin, Nucl. Phys. B **489**, 626 (1997)
13. M. Bluhm, M. Nahrgang, S.A. Bass, T. Schaefer, Eur. Phys. J. C **77**(4), 210 (2017)
14. M. Bluhm, M. Nahrgang, S.A. Bass, T. Schäfer, J. Phys. Conf. Ser. **779**(1), 012074 (2017)

Chapter 56
Measurement of Higher Moments of Net-Proton Distributions in Au+Au Collisions at $\sqrt{s_{NN}}$ = 54.4 GeV at RHIC

Ashish Pandav

Abstract We report the measurement of cumulants (C_n) of event-by-event net-proton distributions up to the sixth order in Au+Au collisions for center-of-mass energy $\sqrt{s_{NN}}$ = 54.4 GeV using the STAR detector at RHIC. Measurements are performed as a function of collision centrality. Protons and antiprotons are selected within the rapidity range $|y| < 0.5$ and within transverse momentum range $0.4 < p_T < 2.0$ GeV/c. We observe a strong dependence of cumulants ($C_n, n \leq 4$) on collision centrality whereas the cumulant ratios C_2/C_1, C_3/C_2 and C_4/C_2 exhibit a weak collision-centrality dependence. The cumulant ratio (C_6/C_2) of net-proton distributions for most central gold nuclei collisions at $\sqrt{s_{NN}}$ = 54.4 GeV shows positive value while at $\sqrt{s_{NN}}$ = 200 GeV it is negative for the same collision system. The expectation from the UrQMD and HIJING models calculated in the STAR acceptance are compared to the measured cumulant ratios and found not to describe the measurements.

56.1 Introduction

Cumulants of conserved charges in high-energy heavy-ion collisions are excellent probes for the study of phase structure in the QCD phase diagram, nature of quark-hadron phase transition and freeze-out dynamics. The cumulants and their ratios are related to the correlation length of the system and the thermodynamic number susceptibilities that are calculable in the lattice QCD and various QCD-based models [1–3]. Cumulants up to the 4th order ($C_n, n \leq 4$) of event-by-event distributions of net charge, net proton (proxy for net baryon) and net kaon (proxy for net strangeness) was measured by the STAR experiment in the phase I of Beam

Ashish Pandav for the STAR Collaboration.

A. Pandav (✉)
School of Physical Sciences, National Institute of Science Education and Research, HBNI, Jatni 752050, India
e-mail: apandav10@gmail.com

D. Elia et al. (eds.), *The XVIII International Conference on Strangeness in Quark Matter (SQM 2019)*, Springer Proceedings in Physics 250,
https://doi.org/10.1007/978-3-030-53448-6_56

Energy Scan (BES) program at RHIC [4]. The cumulant ratios C_3/C_2 and C_4/C_2 of net-proton distribution in the most central (0–5%) gold nuclei collisions show non-monotonic dependence as a function of beam energy [5] which has important implication vis-a-vis the critical point search. Furthermore, the ratio of sixth- to second-order cumulants (C_6/C_2) can also provide insights into the nature of phase transition. The QCD-based model calculations predict a negative value of C_6/C_2 of net-baryon distributions for crossover phase transitions if the chemical freeze-out is close to the chiral phase transition [6].

56.2 Analysis Details

In order to obtain the cumulants of net-proton distributions, protons and antiprotons are selected within the rapidity coverage $|y| < 0.5$ and within p_T range 0.4–2.0 GeV/c. The collision centrality is determined from the charged particle multiplicity within pseudorapidity range $|\eta| < 1$, excluding the protons and antiprotons to avoid autocorrelation effect. The centrality bin width correction is applied to the measurement of the cumulants and their ratios in order to suppress the volume fluctuation effects [7]. Cumulants are corrected for the efficiency and acceptance effects of the detector assuming the distribution of the detector response to be binomial. For estimation of statistical uncertainties of cumulants and their ratios, a resampling method called the bootstrap was used [8]. Systematic uncertainties of the C_n's are estimated varying track selection and particle identification criteria.

56.3 Results and Discussions

In this section, new results on the measurement of cumulants up to the 6th order of the event-by-event net-proton distributions for Au+Au collisions at $\sqrt{s_{NN}} = 54.4$ GeV are presented as a function of collision centrality. Figure 56.1 shows the event-by-event net-proton multiplicity distributions from Au+Au collisions at $\sqrt{s_{NN}} = 54.4$ GeV for 0–5, 5–10, 30–40, 60–70 and 70–80% most central collisions. The distributions are uncorrected for the efficiency and acceptance effects. The shape of the net-proton distribution broadens and its mean increases as one approaches central collsions from peripheral collisions. Cumulants are obtained for these distributions. They are corrected for finite centrality bin width, the detector efficiency and acceptance effects [9, 10]. Figure 56.2 shows the cumulants (up to the 4th order) of net-proton distribution as a function of collision centrality (given by the average number of participant nucleons, <Npart>) for Au+Au collisions at $\sqrt{s_{NN}} = 54.4$ GeV. The cumulants increase with number of participant nucleons.

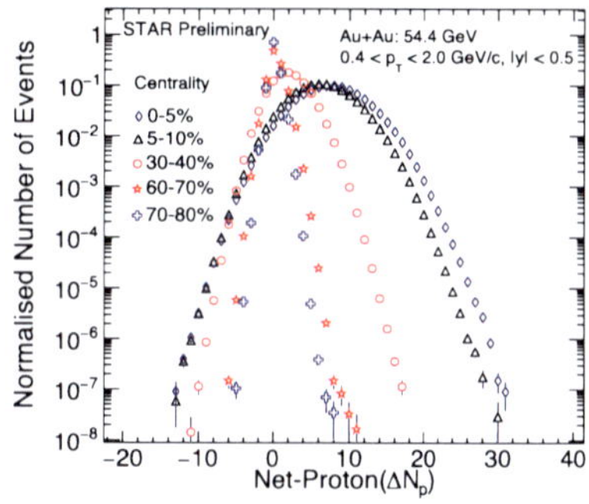

Fig. 56.1 Event-by-event net-proton multiplicity distributions in Au+Au collisions at $\sqrt{s_{NN}} = 54.4\,\text{GeV}$ for 0–5, 5–10, 30–40, 60–70 and 70–80% collision centralities at mid-rapidity. The distributions are uncorrected for the efficiency and acceptance effects

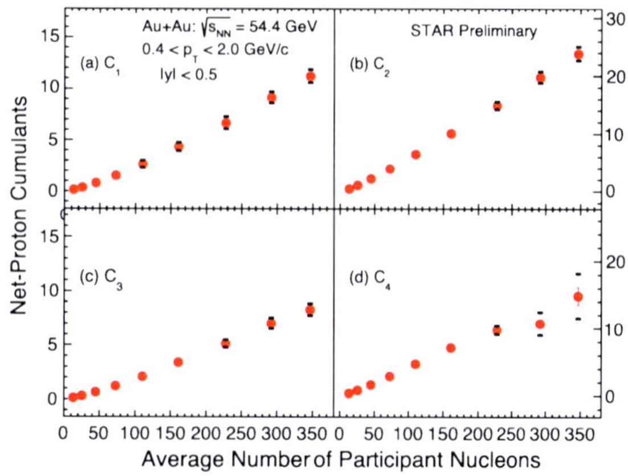

Fig. 56.2 Cumulants of net-proton distribution up to the 4th order as a function of average number of participant nucleons for Au+Au collisions at $\sqrt{s_{NN}} = 54.4\,\text{GeV}$

Collision-centrality dependence of cumulant ratios C_2/C_1, C_3/C_2 and C_4/C_2 is shown in Fig. 56.3. The C_2/C_1 decreases with collision centrality. The ratios C_3/C_2 and C_4/C_2 exhibit a weak dependence on collision centrality. The cumulant ratios obtained from the UrQMD [11] and HIJING [12] models are also compared and found to qualitatively reproduce the measured centrality dependence, whereas quantitative differences exist. The Skellam baseline for C_4/C_2, which is the expected value if protons and antiprotons follow Poisson distribution, fails to describe the measured values. The non-monotonic $\sqrt{s_{NN}}$ dependence of C_4/C_2 of net-proton distribution is shown in Fig. 56.4 for peripheral (70–80%) and most central (0–5%) collisions with inclusion of the results from the current measurement (open and solid red markers respectively). The new measurement with high statistics (~550 million minimum-bias events) gives a precise baseline for the possible critical fluctuations studies which is expected to be observed at lower center-of-mass energies.

Figure 56.5 shows the ratio of the sixth- to second-order cumulants (C_6/C_2) of net-proton distribution as a function of collision centrality for Au+Au collisions at $\sqrt{s_{NN}} = 54.4\,\text{GeV}$ and $\sqrt{s_{NN}} = 200\,\text{GeV}$ [13]. The value of C_6/C_2 for central collisions (0–40%) at $\sqrt{s_{NN}} = 54.4\,\text{GeV}$ is positive. This is in contrast to the value of C_6/C_2 measured in Au+Au collisions at $\sqrt{s_{NN}} = 200\,\text{GeV}$, which is negative

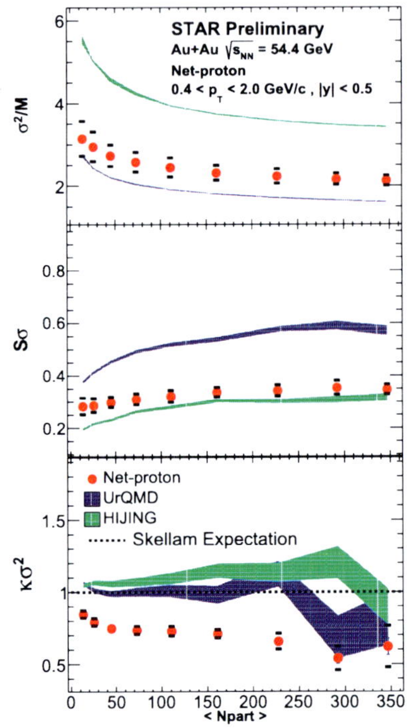

Fig. 56.3 Ratio of cumulants C_2/C_1, C_3/C_2 and C_4/C_2 of net-proton distribution as a function of average number of participant nucleons (<Npart>) for Au+Au collisions at $\sqrt{s_{NN}} = 54.4\,\text{GeV}$. Blue and green band are the UrQMD and HIJING expectations, respectively

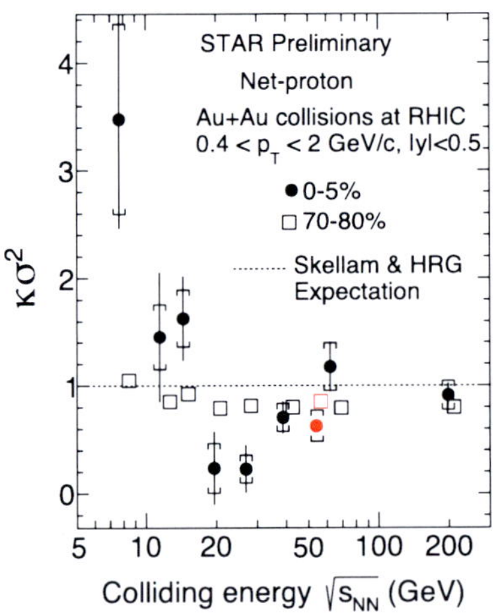

Fig. 56.4 Collision-energy dependence of C_4/C_2 for net-proton distributions for 0–5 and 70–80% most central collisions for Au+Au collisions with inclusion of the results from the C_4/C_2 measurement at $\sqrt{s_{NN}} = 54.4\,\text{GeV}$

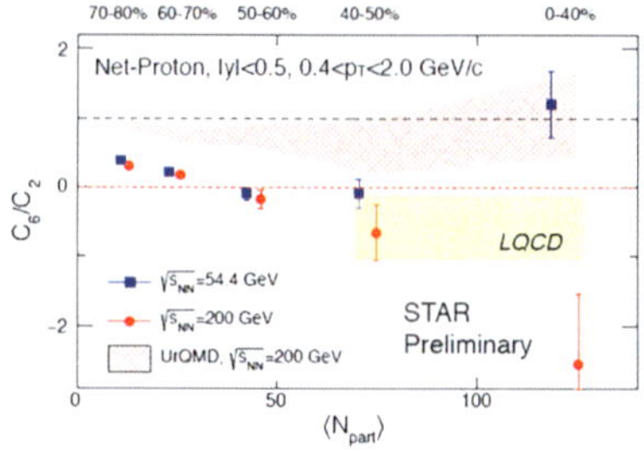

Fig. 56.5 Cumulant ratio C_6/C_2 of net-proton distribution for Au+Au collisions at $\sqrt{s_{NN}} = 54.4\,\text{GeV}$ (blue) and $200\,\text{GeV}$ (red) as a function of collision centrality. Red band is the UrQMD expectation for Au+Au: $200\,\text{GeV}$ and yellow band is the Lattice QCD prediction [14]

for central (0–40%) collisions. A negative value of C_6/C_2 is predicted to occur for a cross over transition between hadronic matter and quark-gluon plasma in QCD-based calculations [6]. The UrQMD model expectation for Au+Au collisions at $\sqrt{s_{NN}} = 200\,\text{GeV}$ are found to be positive and consistent with the Skellam baseline for (0–40%) collision centrality.

Acknowledgments We acknowledge the support by Department of Atomic Energy, Govt. of India.

References

1. M.A. Stephanov, Phys. Rev. Lett. **102**, 032301 (2009)
2. R.V. Gavai, S. Gupta, Phys. Lett. B **696**, 459 (2011)
3. S. Gupta, X. Luo, B. Mohanty, H.G. Ritter, N. Xu, Science **332**, 1525 (2011)
4. X. Luo, N. Xu, Nucl. Sci. Tech. **28**(8), 112 (2017)
5. X. Luo (STAR Collaboration), PoS CPOD 2014, 019 (2015)
6. B. Friman et al., Eur. 346 Phys. J. C **71**, 1694 (2011)
7. X. Luo, J. Xu, B. Mohanty, N. Xu, J. Phys. G **40**, 105104 (2013)
8. A. Pandav, D. Mallick, B. Mohanty, Nucl. Phys. A **991**, 121608 (2019)
9. X. Luo, Phys. Rev. C **91**, 034907 (2015)
10. T. Nonaka, M. Kitazawa, S. Esumi, Phys. Rev. C **95**(6), 064912 (2017)
11. S.A. Bass et al., Prog. Part. Nucl. Phys. **41**, 255 (1998)
12. M. Gyulassy et al., Comput. Phys. Commun. **83**, 307 (1994)
13. T. Nonaka (STAR Collaboration), ATHIC (2018)
14. M. Cheng et al., Phys. Rev. D **79**, 074505 (2009)

Chapter 57
Equation of State of QCD Matter Within the Hagedorn Bag-Like Model

Volodymyr Vovchenko, Mark I. Gorenstein, Carsten Greiner, and Horst Stoecker

Abstract The QCD equation of state at finite temperature and densities of conserved charges is considered in the framework of a Hagedorn bag-like model, incorporating both the finite sizes of hadrons as well as their exponential mass spectrum. Augmented with non-zero masses of quarks and gluons in the bag spectrum, the model yields a fair quantitative description of lattice data on thermodynamic functions, the conserved charges susceptibilities, and Fourier coefficients of net-baryon density. Both at zero and finite baryon densities a broad crossover transition between hadronic and quark-gluon matter is observed. The model thus provides a thermodynamically consistent construction of a crossover equation of state for finite baryon number, electric charge and strangeness chemical potentials, which can be used in fluid dynamical simulations of heavy-ion collisions.

57.1 Introduction

The known hadron mass spectrum [1] is consistent with an exponential increase at large masses, as first suggested by Hagedorn long time ago [2]. Other evidence for Hagedorn states comes from analyses of QCD thermodynamics [3, 4] and transport properties [5] in the vicinity of the chiral pseudocritical temperature T_{pc}, as well as from their role as a possible explanation of fast equilibration of hadrons in heavy-ion

V. Vovchenko · C. Greiner · H. Stoecker
Institut für Theoretische Physik, Goethe Universität Frankfurt, 60438 Frankfurt am Main, Germany

V. Vovchenko (✉) · Mark I. Gorenstein · H. Stoecker
Frankfurt Institute for Advanced Studies, Giersch Science Center, 60438 Frankfurt am Main, Germany
e-mail: vovchenko@fias.uni-frankfurt.de

Mark I. Gorenstein
Bogolyubov Institute for Theoretical Physics, Kyiv 03680, Ukraine

H. Stoecker
GSI Helmholtzzentrum für Schwerionenforschung GmbH, 64291 Darmstadt, Germany

D. Elia et al. (eds.), *The XVIII International Conference on Strangeness in Quark Matter (SQM 2019)*, Springer Proceedings in Physics 250,
https://doi.org/10.1007/978-3-030-53448-6_57

collisions [6, 7]. The Hagedorn states have recently been implemented into transport codes, as an alternative to the string mechanism of hadron production [8, 9].

The class of Hagedorn bag-like models for the QCD equation of state (EoS) has been constructed starting from the early 80s [10], offering perhaps the first example of a framework which realizes a transition between hadronic matter and QGP. One crucial ingredient there is an exponential spectrum of Hagedorn states, realized in the framework of the MIT bag model [11, 12]. Coupled with an introduction of finite eigenvolumes [13, 14], this allows to obtain a first-order, second-order, or a crossover transition in the gas of spatially extended quark-gluon bags, with thermodynamic properties at high temperatures being similar to the MIT bag model equation of state [15–18]. In contrast to many common constructions of the parton-hadron transition (see e.g. [19, 20]), the transition in the Hagedorn model is described within a single partition function.

A comparison of the Hagedorn model crossover EoS with lattice QCD thermodynamics at $\mu_B = 0$ has first been considered in [21]. General qualitative features were found to be compatible with lattice QCD, although a quantitative description is lacking. A more recent study [22] has achieved a semi-quantitative description of the lattice data by introducing quasiparticle-type quarks and gluons with non-zero masses into the bag spectrum. This work reports on the recent advances in that direction, with a focus on finite chemical potentials of conserved charges.

57.2 Model

The Hagedorn bag-like model comprises of a statistical mechanics treatment of a multi-component system of non-overlapping particles which have finite sizes. The particles carry integer values of the three QCD conserved charges: baryon number, electric charge, and strangeness. The number of particle species with a particular mass m, eigenvolume v, and conserved charges b, q, and s is characterized by a mass-volume density $\rho(m, v; b, q, s)$. A generalized mass-volume density of states which depends on the chemical potentials is given by a Fourier transform of $\rho(m, v; b, q, s)$:

$$\rho(m, v; \mu_B, \mu_Q, \mu_S) = \sum_{b,q,s=-\infty}^{\infty} e^{\frac{b\mu_B}{T}} e^{\frac{q\mu_Q}{T}} e^{\frac{s\mu_S}{T}} \rho(m, v; b, q, s). \tag{57.1}$$

The density of states consists of two contributions, $\rho = \rho_H + \rho_Q$. Here

$$\rho_H(m, v; \mu_B, \mu_Q, \mu_S) = \sum_{i \in \text{pdg}} e^{\frac{b_i\mu_B}{T}} e^{\frac{q_i\mu_Q}{T}} e^{\frac{s_i\mu_S}{T}} d_i\, \delta(m - m_i)\, \delta(v - v_i). \tag{57.2}$$

is the discrete part of the particle spectrum, consisting of the low-mass, established hadrons and resonances listed in Particle Data Tables [1]. m_i and d_i are the ith species' mass and degeneracy, respectively, and $v_i = m_i/(4B)$ is its bag-model motivated eigenvolume. B is the bag constant, which will be specified later.

The continuum part of the particle spectrum is ρ_Q. It corresponds to the exponential spectrum of quark-gluon bags, evaluated assuming bags filled with non-interacting quarks and gluons with constant masses [12, 22, 23]:

$$\rho_Q(m, v; \mu_B, \mu_Q, \mu_S) = C\, v^{\gamma}\, (m - Bv)^{\delta} \exp\left\{\tfrac{4}{3}[\sigma_Q(\mu_B, \mu_Q, \mu_S)\, v]^{\frac{1}{4}}\, (m - Bv)^{\frac{3}{4}}\right\} \times \theta(v - V_0)\, \theta(m - Bv - M_0). \tag{57.3}$$

Here $\sigma_Q(\mu_B, \mu_Q, \mu_S)$ is three times the pressure of an ideal gas of massive quarks and gluons. C, γ, δ, V_0 and M_0 are model parameters. The values of the exponents γ and δ determine the nature of the transition between hadrons and QGP. The different possibilities were categorized in [24]. In the following we only consider the crossover case.

To evaluate the system pressure one needs to incorporate the eigenvolume effects. This is achieved through the isobaric ensemble [10]. For the case of a crossover transition the resulting pressure is given by a transcendental equation [22]:

$$p(T, \mu_B, \mu_Q, \mu_S) = T \sum_{i \in \mathrm{pdg}} d_i\, \phi(T, m) \exp\left(\frac{b_i \mu_B + q_i \mu_Q + s_i \mu_S}{T}\right) \exp\left(-\frac{m_i p}{4BT}\right) + \frac{C}{\pi} T^{5+4\delta} [\sigma_Q]^{\delta+\frac{1}{2}} [B + \sigma_Q T^4]^{\frac{3}{2}} \left(\frac{T}{p - p_B}\right)^{\gamma+\delta+3} \Gamma\left[\gamma + \delta + 3, \frac{(p - p_B)V_0}{T}\right]. \tag{57.4}$$

Here Γ is the incomplete Gamma function. Equation (57.4) is solved numerically.

57.3 Results

For calculations we use the following parameter set:

$$\gamma = 0, \quad \delta = -2, \quad B^{1/4} = 250\ \mathrm{MeV}, \quad C = 0.03\ \mathrm{GeV}^4, \quad V_0 = 4\ \mathrm{fm}^3, \tag{57.5}$$

$$m_u = m_d = 300\ \mathrm{MeV}, \qquad m_s = 350\ \mathrm{MeV}, \qquad m_g = 800\ \mathrm{MeV}. \tag{57.6}$$

These parameter values were obtained in [22] by constraining the model to lattice QCD thermodynamics at $\mu_B = 0$, yielding a reasonable description of the lattice data on thermodynamic functions, conserved charges susceptibilities, and Fourier coefficients of net-baryon density (see [22] for the agreement level).

We emphasize now, that the present Hagedorn model is defined not only for zero chemical potentials, but also for finite values of μ_B, μ_Q, and μ_S. Here we explore the structure and behavior of the model at finite baryon density, something which was not done before in this framework quantitatively. In the present work we let $\mu_Q = \mu_S = 0$ and perform calculations for finite μ_B values only. The left panel of Fig. 57.1 depicts the μ_B/T dependence of the pressure along the $T = 155$ MeV isotherm. The pressure is scaled by the corresponding value in the Stefan-Boltzmann (SB) limit of 3-flavor

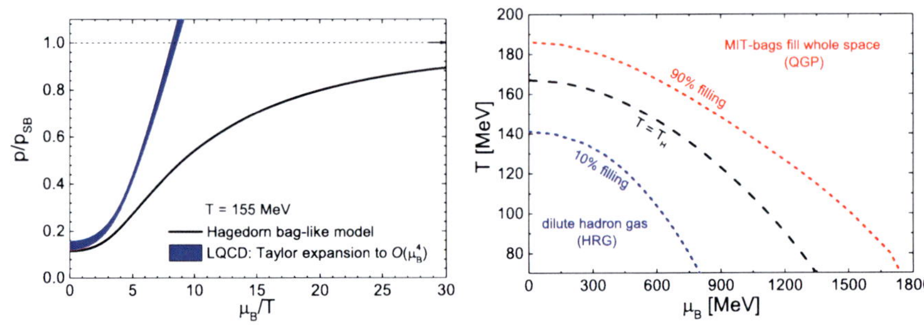

Fig. 57.1 *Left panel*: Dependence of the pressure on the baryochemical potential along the $T =$ 155 MeV isotherm. The pressure is scaled by the corresponding value in the Stefan-Boltzmann limit of 3-flavor QCD. *Right panel*: The phase structure of the Hagedorn bag-like model in the μ_B-T plane. The dashed black line corresponds to the μ_B-dependent Hagedorn temperature whereas the blue and red dashed lines depict the contours corresponding to 10% and 90% eigenvolume filling fractions, respectively

QCD – the expected $\mu_B \to \infty$ limit. The pressure of the Hagedorn model exhibits a consistent approach towards the SB limit at large μ_B/T. The behavior is smooth in the entire range of μ_B/T values considered. For comparison we also depict the pressure by using a Taylor expansion truncated at $\mathcal{O}(\mu_B^4)$, with expansion coefficients taken from the lattice data [25]. The truncated Taylor expansion behaves similarly to the Hagedorn model at moderate $\mu_B/T < 3$, but becomes inaccurate at larger chemical potentials, overshooting the SB limit at $\mu_B/T \approx 8$.

To further clarify the phase structure of the Hagedorn model we consider the $\mu_B - T$ plane (right panel of Fig. 57.1). The dashed black line depicts the μ_B-dependent of the Hagedorn temperature of the quark-gluon bag spectrum in (57.3). This line, defined through a transcendental equation $T_H = [3B/\sigma_Q(T_H)]^{1/4}$, corresponds to the region of the phase diagram where heavy Hagedorn states start significantly contributing to the thermodynamics. The line can roughly be viewed as a crossover line for the transition between normal hadronic matter and the QGP. Another interesting quantity is the eigenvolume filling fraction which corresponds to the fraction of the whole space occupied by finite-sized particles. The blue and red dashed lines in Fig. 57.1 depict the contours corresponding to 10% and 90% eigenvolume filling fractions, respectively. The matter with less than 10% filling is dilute and very similar to a hadron resonance gas. In the opposite side of the phase diagram one has large filling fractions, meaning that most of the space is occupied by bags filled with QGP. The results shown in Fig. 57.1 suggest that the transition between these two regimes is rather broad, not only at $\mu_B = 0$, but also in the baryon-rich region of the phase diagram.

57.4 Summary

The Hagedorn bag-like model with quasiparticle-type parton masses provides a reasonable description of parton-hadron crossover at zero and finite densities of conserved charges within a single partition function. It is thus suitable as an input for fluid dynamical simulations of heavy-ion collisions. One possible future extension is incorporating a hypothetical first-order phase transition in the baryon-rich region, e.g. as outlined in [16]. Such a generalization can then be used to predict testable signatures of a first-order transition at finite baryon densities.

References

1. C. Patrignani et al., Particle data group. Chin. Phys. C **40**, 100001 (2016)
2. R. Hagedorn, Nuovo Cim. Suppl. **3**, 147 (1965)
3. A. Majumder, B. Muller, Phys. Rev. Lett. **105**, 252002 (2010)
4. P.M. Lo, M. Marczenko, K. Redlich, C. Sasaki, Phys. Rev. C **92**, 055206 (2015)
5. J. Noronha-Hostler, J. Noronha, C. Greiner, Phys. Rev. Lett. **103**, 172302 (2009)
6. J. Noronha-Hostler, C. Greiner, I.A. Shovkovy, Phys. Rev. Lett. **100**, 252301 (2008)
7. J. Noronha-Hostler, M. Beitel, C. Greiner, I. Shovkovy, Phys. Rev. C **81**, 054909 (2010)
8. M. Beitel, C. Greiner, H. Stoecker, Phys. Rev. C **94**, 021902 (2016)
9. K. Gallmeister, M. Beitel, C. Greiner, Phys. Rev. C **98**, 024915 (2018)
10. M.I. Gorenstein, V.K. Petrov, G.M. Zinovev, Phys. Lett. **106B**, 327 (1981)
11. A. Chodos, R.L. Jaffe, K. Johnson, C.B. Thorn, V.F. Weisskopf, Phys. Rev. D **9**, 3471 (1974)
12. J.I. Kapusta, Phys. Rev. D **23**, 2444 (1981)
13. J. Baacke, Acta Phys. Polon. B **8**, 625 (1977)
14. R. Hagedorn, J. Rafelski, Phys. Lett. **97B**, 136 (1980)
15. M.I. Gorenstein, W. Greiner, S.N. Yang, J. Phys. G **24**, 725 (1998)
16. M.I. Gorenstein, M. Gazdzicki, W. Greiner, Phys. Rev. C **72**, 024909 (2005)
17. I. Zakout, C. Greiner, J. Schaffner-Bielich, Nucl. Phys. A **781**, 150 (2007)
18. I. Zakout, C. Greiner, Phys. Rev. C **78**, 034916 (2008)
19. P.F. Kolb, J. Sollfrank, U.W. Heinz, Phys. Rev. C **62**, 054909 (2000)
20. M. Albright, J. Kapusta, C. Young, Phys. Rev. C **90**, 024915 (2014)
21. L. Ferroni, V. Koch, Phys. Rev. C **79**, 034905 (2009)
22. V. Vovchenko, M.I. Gorenstein, C. Greiner, H. Stoecker, Phys. Rev. C **99**, 045204 (2019)
23. M.I. Gorenstein, S.I. Lipskikh, V.K. Petrov, G.M. Zinovev, Phys. Lett. **123B**, 437 (1983)
24. V.V. Begun, M.I. Gorenstein, W. Greiner, J. Phys. G **36**, 095005 (2009)
25. A. Bazavov et al., Phys. Rev. D **95**, 054504 (2017)

Chapter 58
Determination of Chemical Freeze-Out Parameters from Net-Kaon Fluctuations at RHIC

Jamie M. Stafford, Paolo Alba, Rene Bellwied, Valentina Mantovani-Sarti, Jacquelyn Noronha-Hostler, Paolo Parotto, Israel Portillo-Vazquez, and Claudia Ratti

Abstract We calculate the mean-over-variance ratio of the net-kaon fluctuations in the Hadron Resonance Gas (HRG) Model for the five highest energies of the RHIC Beam Energy Scan (BES) for different particle data lists. We compare these results with the latest experimental data from the STAR collaboration in order to extract sets of chemical freeze-out parameters for each list. We focused on the PDG2012 and PDG2016+ particle lists, which differ largely in the number of resonant states. Our analysis determines the effect of the amount of resonances included in the HRG on the freeze-out conditions.

58.1 Introduction

Characterizing the transition region of the QCD phase diagram is a focal point for current theoretical and experimental investigations in nuclear physics. Lattice QCD calculations have shown that there is a cross-over transition at T $\simeq$ 155 MeV in the low μ_B region of the phase diagram [1–3]. In addition, models have indicated that the

J. M. Stafford (✉) · R. Bellwied · P. Parotto · I. Portillo-Vazquez · C. Ratti
University of Houston, Houston, TX 77204, USA
e-mail: jmstafford@uh.edu

P. Alba
Lucht Probst Associates GmbH, Grosse Gallusstrasse 9, 60311 Frankfurt am Main, Germany

V. Mantovani-Sarti
Technische Universität München, James Franck Strasse 1, 85748 Garching, Germany

J. Noronha-Hostler
University of Illinois at Urbana-Champaign, Urbana, IL 61801, USA

P. Parotto
University of Wuppertal, 42097 Wuppertal, Germany

D. Elia et al. (eds.), *The XVIII International Conference on Strangeness in Quark Matter (SQM 2019)*, Springer Proceedings in Physics 250,
https://doi.org/10.1007/978-3-030-53448-6_58

transition becomes first order at high baryonic chemical potential, thus implying the presence of a critical point [4]. The search for this proposed critical point represents the main goal of the Beam Energy Scan II (BES-II) program at RHIC.

The evolution of the system in a heavy-ion collision (HIC) can be broken into several stages; two important stages of HICs, the chemical and kinetic freeze-outs, can be related to the experimental results. The chemical freeze-out corresponds to the point in the evolution of the system at which inelastic collisions cease, and at this time the chemical composition is fixed, corresponding to the measured particle yields and fluctuations. The next freeze-out stage, the kinetic freeze-out occurs when the system is so sparse that elastic collisions can no longer occur, and this corresponds to information on the particle spectra and correlations. Within this framework it is clear that the study of the chemical freeze-out is an important aspect of phenomenological explorations in heavy-ion collisions.

Chemical freeze-out parameters are typically obtained by treatment of the particle yields or fluctuations in a thermal model, such as the Hadron Resonance Gas (HRG) model [5–7]. Thermal fits of particle yields can be used to determine the temperature, baryonic chemical potential, and volume at freeze-out (T_f, $\mu_{B,f}$, V_f). In addition, the fluctuations of conserved charges can be used to identify freeze-out parameters by comparing experimental results for the particle fluctuations to a thermal model. The HRG Model has been used to determine the chemical freeze-out conditions in this way in works such as [6]. This study produced sets of freeze-out parameters for a range of collision energies of the BES by performing a combined fit of net-p and net-Q. Thus, these correspond to the freeze-out conditions of the light particles. Combining this information with the isentropic trajectories from Lattice QCD, we study the net-kaon fluctuations in the HRG model in order to determine if the kaons freeze-out at the same temperature as the light hadrons. This addresses an important effect that has been seen in data at both RHIC and the LHC, which is the tension between the light and strange particles in thermal fits [5, 8]. One explanation that attempts to resolve the difference in temperatures is that there could be missing resonances in the thermal model [9, 10]. While recent lattice QCD calculations [10, 11] indicate that several resonances are indeed missing, their full decay channels need to be implemented in order to determine their influence on the freeze-out temperature. Other explanations include a higher freeze-out temperature for strange hadrons [12, 13], pion-nucleon interactions in the S-matrix formalism [14], and large annihilation cross-sections that lead to a lower (anti)proton freeze-out [15–17] In this study, we seek to characterize the chemical freeze-out during heavy-ion collisions by utilizing the HRG Model to calculate fluctuations of conserved charges.

58.2 Methodology

The Hadron Resonance Gas (HRG) Model describes an interacting gas of ground state hadrons as a system of non-interacting particles and their resonant states. The pressure of such a system of hadrons can be treated in the grand canonical formalism:

$$\frac{P}{T^4} = \frac{1}{VT^3} \sum_i lnZ_i(T, V, ^-),$$

$$\ln Z_i^{M/B} = \mp \frac{Vd_i}{(2\pi)^3} \int d^3k \, \ln\left(1 \mp exp\left[-\left(\epsilon_i - \mu_a X_a^i\right)/T\right]\right)$$

where, the index i runs over all the particles included in the HRG model from the Particle Data Group listing, the energy $\epsilon_i = \sqrt{k^2 + m_i^2}$, conserved charges $\mathbf{X_i} = (B_i, S_i, Q_i)$, degeneracy d_i, mass m_i, and volume V. The fluctuations of conserved charges are defined as the derivative of the pressure with respect to the chemical potentials of the conserved charges of interest.

$$\chi_{ijk}^{BSQ} = \frac{\partial^{i+j+k}\left(P/T^4\right)}{\partial(\mu_B/T)^i \, \partial(\mu_S/T)^j \, \partial\left(\mu_Q/T\right)^k}$$

In order to compare the HRG model results to the experimental data, it is important to also consider the experimental cuts in rapidity and transverse momentum. Taking this into account, the fluctuations of net-kaons, the difference of kaons minus anti-kaons, can be written as:

$$\chi_n^{net-K} = \sum_i^{N_{HRG}} \frac{(Pr_{i\to net-K})^n}{T^{3-(n-1)}} \frac{S_i^{1-n} d_i}{4\pi^2} \frac{\partial^{n-1}}{\partial \mu_S^{n-1}} \left\{ \int_{-0.5}^{0.5} dy \int_{0.2}^{1.6} dp_T \right.$$

$$\left. \times \frac{p_T \sqrt{p_T^2 + m_i^2} Cosh[y]}{(-1)^{B_i+1} + exp((Cosh[y]\sqrt{p_T^2 + m_i^2} - (B_i\mu_B + S_i\mu_S + Q_i\mu_Q))/T)} \right\}$$

An important goal of this study is to determine the effect of the amount of resonances included in the HRG model on the net-kaon fluctuations. We do so by utilizing different particle lists. In a previous analysis [11], it was determined that the most experimentally well-known states are not enough to complete the hadronic spectrum in HRG model calculations as compared to results from Lattice QCD. Therefore, we utilize the measurements of all observed hadronic states from the Particle Data Group over all confidence levels [18]. This compilation of resonances was found to be the one which best reproduces the Lattice QCD results. We will compare the net-kaon fluctuations calculated using this optimal PDG list to the results from [19], which utilized a different particle list. These lists differ largely in the number of resonant states, particularly in the strange sector.

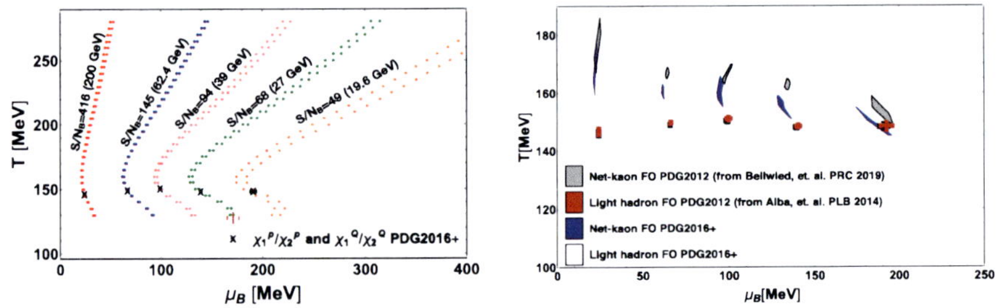

Fig. 58.1 Left: Isentropic trajectories for the five highest energies of the BES. Entropy per baryon number is conserved along these trajectories, which start from the freeze-out parameters determined by the combined fit of χ_1^p/χ_2^p and χ_1^Q/χ_2^Q. Right: Comparison of net-kaon and light hadron freeze-out parameters for two different HRG particle lists, PDG2012 and PDG2016+, over a range of collision energies at RHIC

58.3 Results

In order to compare the results from these two different particle lists, we followed the same treatment from [6, 19, 20]. First the freeze-out parameters for the light particles were calculated via a combined fit of χ_1^p/χ_2^p and χ_1^Q/χ_2^Q in the HRG model with the PDG2016+ particle list. Isospin randomization was taken into account for the determination of the freeze-out parameters from net-proton and net-electric charge, as a similar analysis has shown previously [6]. Then, isentropic trajectories were developed by starting from these new freeze-out points obtained with the PDG2016+ list for the various BES energies and fixing the entropy per baryon number along the path in the phase diagram, consistent with [20]. They are shown in Fig. 58.1 (left).

By calculating the net-kaon fluctuations along these isentropes and comparing to the experimental results for the mean-over-variance, we have identified sets of freeze-out parameters, $(T_{FO}, \mu_{B,FO})$, for the five highest energies of the Beam Energy Scan. The freeze-out parameters for the kaons with the two different PDG lists are compared to the freeze-out parameters for the light particles determined by the combined fit of net-proton and net-electric charge as shown in Fig. 58.1 (right). Even with the inclusion of more strange resonances, the separation between freeze-out conditions for net-kaons and light particles remains.

58.4 Conclusions

By utilizing different compilations of resonances from the Particle Data Group, we study the effect of the number of particles included in the HRG model on the results for net-kaon fluctuations. We find a slightly lower freeze-out temperature for the kaons when more strange resonances are included in the particle list. However, at

$\sqrt{s_{NN}}$=200 GeV there is a separation of about 10 MeV between these two sets of freeze-out conditions for net-kaons and light hadrons.

Acknowledgments This material is based upon work supported by the National Science Foundation under Grant No. PHY-1654219 and by the U.S. Department of Energy, Office of Science, Office of Nuclear Physics, within the framework of the Beam Energy Scan Theory (BEST) Topical Collaboration. We also acknowledge the support from the Center of Advanced Computing and Data Systems at the University of Houston. P.P. also acknowledges support from the DFG grant SFB/TR55. J.N.H. acknowledges the support of the Alfred P. Sloan Foundation, support from the US-DOE Nuclear Science Grant No. de-sc0019175. R.B. acknowledges support from the US DOE Nuclear Physics Grant No. DE-FG02-07ER41521.

References

1. Y. Aoki, G. Endrődi, Z. Fodor, S.D. Katz, K.K. Szabó, Nature **443**, 675–678 (2006)
2. S. Borsányi et al., J. High Energy Phys. **09**, 73 (2010)
3. A. Bazavov et al., Phys. Rev. D **85**, (2012)
4. Bzdak, A. et al., arXiv:1906.00936 (2019)
5. S. Acharya et al., Nucl. Phys. A **971**, 1–20 (2018)
6. P. Alba et al., Phys. Lett. B **738**, 305–310 (2014)
7. A. Andronic, P. Braun-Munzinger, J. Stachel, Phys. Lett. B **673**, 142–145 (2009)
8. L. Adamczyk et al., Phys. Rev. C **96**, 1–35 (2017)
9. J. Noronha-Hostler, C. Greiner, Nucl. Phys. A **931**, 1108–1113 (2014)
10. A. Bazavov et al., Phys. Rev. Lett. **113**, 1–5 (2014)
11. P. Alba et al., Phys. Rev. D **96**, (2017)
12. R. Bellwied, S. Borsányi, Z. Fodor, S.D. Katz, C. Ratti, Phys. Rev. Lett. **111**, (2013)
13. M. Bluhm, M. Nahrgang, Eur. Phys. J. C **79**, 155 (2019)
14. A. Andronic et al., Phys. Lett. B **792**, 304–309 (2019)
15. J. Steinheimer, J. Aichelin, M. Bleicher, Phys. Rev. Lett. **110**, (2013)
16. F. Becattini et al., Phys. Rev. Lett. **111**, (2013)
17. F. Becattini, J. Steinheimer, R. Stock, M. Bleicher, Phys. Lett. B **764**, 241–246 (2017)
18. C. Patrignani et al., Chin. Phys. C **40**, 1048–1061 (2016)
19. R. Bellwied et al., Phys. Rev. C **99**, (2019)
20. J.N. Guenther et al., Nucl. Phys. A **967**, 720–723 (2017)

Chapter 59
Update on BEST Collaboration and Status of Lattice QCD

Claudia Ratti

Abstract The Beam Energy Scan Topical (BEST) Collaboration was formed to support the Relativistic Heavy Ion Collider (RHIC) experimental program in search for the QCD critical point. I will report on the status of the BEST collaboration, mainly focusing on the lattice QCD effort but also touching on the other main topics.

59.1 Introduction and Structure of the Collaboration

There are many open questions on the QCD phase diagram that can be answered by studying finite density QCD. The most pressing one is whether the deconfinement phase transition, an analytical crossover at chemical potential $\mu_B = 0$ [1], can turn into first order as the chemical potential is increased. Other relevant questions are the location of the transition line and the nature of the QCD phases at large densities. The only way to produce the high density phase of matter in the laboratory is to bring more net-baryon number in the mid-rapidity region, by systematically decreasing the collision energy so that some of the primordial baryons are left in the collision area. The RHIC facility at BNL is devoted to this purpose: the second Beam Energy Scan (BESII) is scheduled for 2019–2021. The foreseen runs will take place both in the collider and fixed target modes, so that higher values of the baryonic chemical potential can be reached. After RHIC, other facilities will study dense QCD matter: NICA, CBM and JPARC will pursue the study of critical point, onset of deconfinement and dense hadronic matter at least till 2025. For reviews on these topics see [2–4].

Fundamental theory and phenomenology need to provide adequate support to such a rich experimental program. The main purpose and goal of the project pursued by the Beam Energy Scan Topical (BEST) collaboration is to develop a dynamical framework which allows for a quantitative description of the heavy ion reaction for collision energies relevant to the RHIC BES. This framework will then enable a

C. Ratti (✉)
Department of Physics, University of Houston, Houston, TX 77204, USA
e-mail: cratti@uh.edu

D. Elia et al. (eds.), *The XVIII International Conference on Strangeness in Quark Matter (SQM 2019)*, Springer Proceedings in Physics 250,
https://doi.org/10.1007/978-3-030-53448-6_59

quantitative assessment of the data from the BES in order to either claim discovery or rule out the presence of a critical point and the onset of chiral restoration in the region of the QCD phase diagram accessible by the RHIC beam energy scan. In this contribution I will review the current status of the BEST collaboration, mainly focusing on the lattice QCD effort but also touching on other relevant topics.

To achieve the proposed goal, several elements need to be developed and eventually merged into the final framework. Following essentially the time evolution of the heavy ion collision, from initial conditions to hydrodynamic and transport evolution to the final analysis of the data, these are:

- **Initial conditions**: Calculate and model the initial distribution of all conserved charges as well as of axial charges.
- **Hydrodynamic evolution**: Development of (3+1)-D viscous relativistic (anomalous-magneto) hydrodynamics with transport coefficients for all charges which includes hydro fluctuations and coupling to a fluctuating critical field.
- **Equation of state and transport coefficients**: Carry out lattice QCD calculations of the EoS at large densities, needed for the hydrodynamic evolution, and determine some of the needed transport coefficients. Extend the EoS to regions beyond the reach of lattice calculations by careful modeling and matching to lattice results.
- **Transport evolution**: Development of a transport model which matches onto the hydrodynamic phase and is able to propagate fluctuations and anomalous currents.
- **Data analysis**: Develop and apply a data analysis framework which will enable a comparison of all observables, to constrain the model parameters.

The left panel of Fig. 59.1 shows a schematics of the above steps and how they are linked together. The right panel shows the twelve Universities and two national laboratories that belong to BEST.

Fig. 59.1 Left: The various components of the BEST program and how they are linked together. Right: BEST collaboration institutes

59.2 Lattice QCD and Equation of State with 3D-Ising Critical Point

The lattice effort in the BEST collaboration is conducted completely independently by the group at Brookhaven National Laboratory and the University of Houston group, using two different lattice discretizations and two different methodologies. The goals until now were to obtain the QCD transition temperature at finite μ_B, extract the equation of state up to power six in μ_B/T, to use as an input in hydro simulations and as a baseline in the EoS with critical point, and calculate fluctuations at finite μ_B.

The Wuppertal Budapest Houston [5] and the HotQCD [6] collaboration have published results for the QCD transition temperature at finite μ_B. The most recent value for the chiral transition temperature from the HotQCD is $T_c = 156.5 \pm 1.5$ MeV, obtained from chiral observables. Both collaborations found that the curvature of the phase diagram at $\mu_B = 0$ is extremely small (see Fig. 59.2 left).

After the publication of the QCD Equation of State at $\mu_B = 0$ [7–11], both collaborations proceeded to extend them to finite density. This is difficult due to the sign problem. One of the methods to circumvent the sign problem is to expand the thermodynamic observables as a Taylor series in powers of μ_B/T. The Taylor coefficients can be calculated in two ways, either by direct simulations (method chosen by the HotQCD collaboration), or simulations at imaginary chemical potentials (choice of the WBH collaboration). Besides, a finite μ_B chemical potential implies a choice for the strangeness and electric charge chemical potentials, μ_S and μ_Q respectively. The two possibilities that were considered are $\mu_S = \mu_Q = 0$ or $\mu_S(T, \mu_B)$ and $\mu_Q(T, \mu_B)$ such that the average strangeness density $\langle \rho_S \rangle = 0$ and the average electric charge density $\langle \rho_Q \rangle = 0.4 \langle \rho_B \rangle$. Both collaborations published results for the Taylor coefficients up to sixth order, in the case of $\mu_S = \mu_Q = 0$ [12, 13] and in the case of strangeness neutrality [13–15]. All results for the EoS are consistent between the two collaborations. More recently, a lattice-based Taylor expansion for the equation of state containing all conserved charges has been developed within the BEST collaboration [16, 17].

Fluctuations of conserved charges can be used to study criticality, as they are expected to diverge with powers of the correlation length near the critical point. Recent results by the HotQCD collaboration are showing the baryon number variance and the disconnected chiral susceptibility, extrapolated to finite μ_B along the crossover line. Both are expected to diverge at the critical point, but they do not show signs of criticality up to μ_B=250 MeV [18]. Both collaborations have expanded the higher order fluctuations as functions of μ_B/T, finding a behavior that is compatible with the experimental results [12, 19] (see Fig. 59.2 right). Both collaborations also have results for correlators between conserved charges [20, 21].

The lattice QCD Equation of state developed by the lattice effort of the BEST collaboration has been modified by introducing a critical point in the 3D Ising model universality class [22], in order to test its effects on thermodynamics and eventually dynamical observables. The hydrodynamics working group within BEST has already

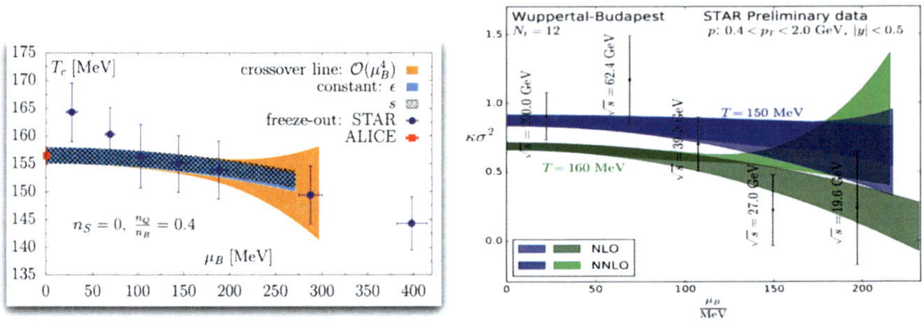

Fig. 59.2 Left: QCD transition line from [6]. Right: extrapolation of higher order fluctuations to finite μ_B [12]

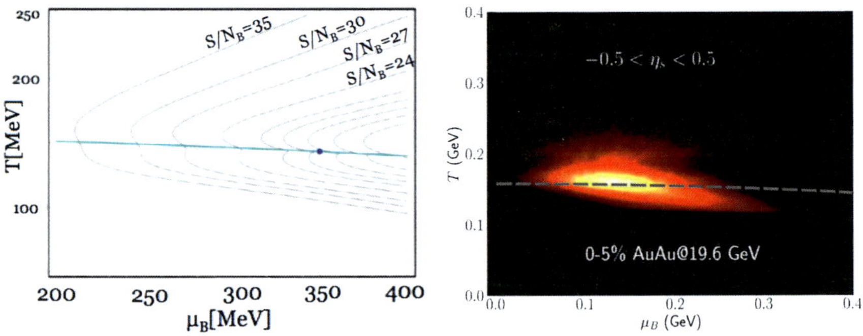

Fig. 59.3 Left: effect of the critical point on the isentropic trajectories in the QCD phase diagram [22]. Right: fireball trajectory of a central gold-gold collision at 19.6 GeV [25]

started testing it, and the predictions will be compared to the experimental data as they become available. The community also benefits from this achievement, since the code to generate the EoS is open source [23]. It was assumed that the lattice QCD Taylor expansion coefficients are the sum of a regular and a singular contributions. The latter are obtained by mapping the Ising model Equation of State onto the QCD one. The left panel of Fig. 59.3 shows the effect of the critical point on the isentropic trajectories in the QCD phase diagram.

59.3 Other Working Groups

Initial conditions for all three conserved charges have been developed, which take into account the fact that the nuclei at the lowest collision energies are not so Lorentz contracted and so there can be multiple scatterings between them [24–27]. The hydrodynamics simulations have been extended to include the propagation of net-baryon current including dissipative effects [28]. This has allowed the authors of [25] to extract the fireball trajectory of a central gold-gold collision at 19.6 GeV, shown in the right panel of Fig. 59.3.

One of the goals of BEST is to develop a quantitative understanding of fluctuations near the CP, which includes a modification of hydro to couple with a slow mode. This formalism has been developed [29–32]. In [29] it was shown that the dynamics of fluctuations (more precisely, the Wigner transform of the two-point correlation function of the fluctuations) and the influence of the fluctuations on the bulk hydrodynamic evolution can be studied together, self-consistently, by solving deterministic equations. This formalism is being implemented numerically within BEST [33, 34].

The chiral magnetic effect has been the goal of an active research program within the collaboration [35–38]. One of the goals is to model initial conditions not only for the conserved charges, but also for axial charges. In a series of papers [39–41] it has been demonstrated that glasma provides the appropriate methodology for addressing this question. Another goal is to develop anomalous magneto hydrodynamics [42–44]. A first step in this direction is the development of a code [45], which calculates electromagnetic fields from the spectators and participant nucleons in relativistic heavy-ion collisions. Finally, we want to quantify the experimental signal of the chiral magnetic effect. One first example is the prediction for CME-induced charge asymmetry of azimuthal correlations in 200 GeV Au+Au collisions, which happens to be in good agreement with the STAR H-correlation measurements [45].

I will finally touch on the topic of particlization and hadron dynamics. It was found out that the standard Cooper-Frye procedure is not suitable to study fluctuations because the Poisson sampling adds unphysical fluctuations and washes away correlations [46, 47]. A new micro-canonical sampling method has been proposed [48], which conserves all the charges as well as energy and momentum. This method has been tested in a toy model and it was shown that the obtained fluctuations are in agreement with the ones which were computed analytically, contrary to the Cooper-Frye ones.

For other important developments within the collaboration, which are not at the core of the BEST project but have important implications for the success of the BES program as a whole, see [49–58].

Acknowledgments This material is based upon work supported by the National Science Foundation under grant no. PHY-1654219 and by the U.S. Department of Energy, Office of Science, Office of Nuclear Physics, within the framework of the Beam Energy Scan Topical (BEST) Collaboration. The author acknowledges useful discussions with her colleagues and collaborators from BEST.

References

1. Y. Aoki et al., Nature **443**, 675–678 (2006)
2. H.T. Ding, F. Karsch, S. Mukherjee, Int. J. Mod. Phys. E **24**(10), 1530007 (2015)
3. C. Ratti, Rept. Prog. Phys. **81**(8), 084301 (2018)
4. A. Bzdak et al., http://arxiv.org/abs/1906.00936arXiv:1906.00936 [nucl-th]

5. R. Bellwied et al., Phys. Lett. B **751**, 559 (2015)
6. A. Bazavov et al., HotQCD collaboration. Phys. Lett. B **795**, 15 (2019)
7. S. Borsanyi et al., JHEP **1011**, 077 (2010)
8. S. Borsanyi et al., Phys. Lett. B **730**, 99 (2014)
9. A. Bazavov et al., HotQCD collaboration. Phys. Rev. D **90**, 094503 (2014)
10. S. Borsanyi et al., Nature **539**(7627), 69 (2016)
11. A. Bazavov, P. Petreczky, J.H. Weber, Phys. Rev. D **97**(1), 014510 (2018)
12. S. Borsanyi et al., JHEP **1810**, 205 (2018)
13. A. Bazavov et al., Phys. Rev. D **95**(5), 054504 (2017)
14. J. Guenther et al., Nucl. Phys. A **967**, 720 (2017)
15. J. Guenther et al., EPJ Web Conf. **137**, 07008 (2017)
16. A. Monnai, B. Schenke, C. Shen, Phys. Rev. C **100**(2), 024907 (2019)
17. J. Noronha-Hostler, P. Parotto, C. Ratti, J.M. Stafford. http://arxiv.org/abs/1902.06723arXiv:1902.06723 [hep-ph]
18. P. Steinbrecher [HotQCD Collaboration], Nucl. Phys. A **982**, 847 (2019)
19. A. Bazavov et al. [HotQCD Collaboration], Phys. Rev. D **96**(7), 074510 (2017)
20. A. Bazavov et al., HotQCD collaboration. Phys. Rev. D **86**, 034509 (2012)
21. R. Bellwied et al.,. http://arxiv.org/abs/1910.14592arXiv:1910.14592 [hep-lat]
22. P. Parotto et al.,. http://arxiv.org/abs/1805.05249arXiv:1805.05249 [hep-ph]
23. The code on which the work is based can be downloaded at the following link. https://www.bnl.gov/physics/best/resources.php
24. C. Shen, B. Schenke, Phys. Rev. C **97**(2), 024907 (2018)
25. C. Shen, B. Schenke, Nucl. Phys. A **982**, 411 (2019)
26. C. Shen et al., Nucl. Phys. A **967**, 796 (2017)
27. L. Du, U. Heinz, G. Vujanovic, Nucl. Phys. A **982**, 407 (2019)
28. G.S. Denicol et al., Phys. Rev. C **98**(3), 034916 (2018)
29. M. Stephanov, Y. Yin, Phys. Rev. D **98**(3), 036006 (2018)
30. M. Martinez, T. Schäfer, Phys. Rev. C **99**(5), 054902 (2019)
31. Y. Akamatsu et al., Phys. Rev. C **100**(4), 044901 (2019)
32. M. Nahrgang et al., Phys. Rev. D **99**(11), 116015 (2019)
33. K. Rajagopal et al.,. http://arxiv.org/abs/1908.08539arXiv:1908.08539 [hep-ph]
34. L. Du and U. Heinz, http://arxiv.org/abs/1906.11181arXiv:1906.11181 [nucl-th]
35. Y. Jiang et al., Chin. Phys. C **42**(1), 011001 (2018)
36. S. Shi et al., Annals Phys. **394**, 50 (2018)
37. U. Guersoy et al., Phys. Rev. C **98**(5), 055201 (2018)
38. D.E. Kharzeev et al., Prog. Part. Nucl. Phys. **88**, 1 (2016)
39. M. Mace et al., Phys. Rev. D **95**(3), 036023 (2017)
40. M. Mace, S. Schlichting, R. Venugopalan, Phys. Rev. D **93**(7), 074036 (2016)
41. N. Müller, S. Schlichting, S. Sharma, Phys. Rev. Lett. **117**(14), 142301 (2016)
42. Y. Hirono, D.E. Kharzeev, Y. Yin, Nucl. Phys. A **967**, 840 (2017)
43. K. Hattori, Y. Hirono, H.U. Yee, Y. Yin, Phys. Rev. D **100**(6), 065023 (2019)
44. D.E. Kharzeev, M.A. Stephanov, H.U. Yee, Phys. Rev. D **95**(5), 051901 (2017)
45. S. Shi, H. Zhang, D. Hou, J. Liao, Nucl. Phys. A **982**, 539 (2019)
46. J. Steinheimer, V. Koch, Phys. Rev. C **96**(3), 034907 (2017)
47. J. Weil et al., Phys. Rev. C **94**(5), 054905 (2016)
48. D. Oliinychenko, V. Koch, Phys. Rev. Lett. **123**(18), 182302 (2019)
49. J. Brewer, S. Mukherjee, K. Rajagopal, Y. Yin, Phys. Rev. C **98**(6), 061901 (2018)
50. S. Mukherjee, R. Venugopalan, Y. Yin, Phys. Rev. Lett. **117**(22), 222301 (2016)
51. A. Bzdak, V. Koch, V. Skokov, Eur. Phys. J. C **77**(5), 288 (2017)
52. B. Ling, M.A. Stephanov, Phys. Rev. C **93**(3), 034915 (2016)
53. A. Bzdak et al., Phys. Rev. C **98**(5), 054901 (2018)
54. A. Bzdak, V. Koch, http://arxiv.org/abs/1811.04456arXiv:1811.04456 [nucl-th]
55. K.J. Sun et al., Phys. Lett. B **781**, 499 (2018)
56. D. Oliinychenko et al., Phys. Rev. C **99**(4), 044907 (2019)
57. P. Alba et al., Phys. Rev. C **98**(3), 034909 (2018)
58. R. Bellwied et al., Phys. Rev. C **99**(3), 034912 (2019)

Chapter 60
Overview of Experimental Critical Point Search

Tobiasz Czopowicz

Abstract The existence and location of the QCD critical point is an object of vivid experimental and theoretical studies. Rich and beautiful data recorded by experiments at SPS and RHIC allow for a systematic search for the critical point—the search for a non-monotonic dependence of various correlation and fluctuation observables on collision energy and size of colliding nuclei.

60.1 Critical Point Search Strategies

A sketch of the most popular phase diagram of strongly-interacting matter is shown in Fig. 60.1. At low temperatures and baryon chemical potential, the system consists of quarks and gluons confined inside hadrons. At higher temperature and/or baryon chemical potential, quarks and gluons may act like quasi-free particles, forming a different state of matter—the Quark-Gluon Plasma. Between the two phases, a first-order transition is expected at high μ. Critical point (CP) is a hypothetical end point of this first-order phase transition line that has properties of a second-order phase transition [1, 2].

It is commonly expected that the QCD critical point should lead to an anomaly in fluctuations in a narrow domain of the phase diagram. However predictions on the CP existence, its location and what and how should fluctuate are model dependent [3].

The experimental search for the critical point requires a two-dimensional scan in freeze-out parameters (T, μ) by changing collision parameters controlled in laboratory, i.e. energy and size of the colliding nuclei (or collision centrality).

T. Czopowicz (✉)
Jan Kochanowski University, Kielce, Poland
e-mail: Tobiasz.Czopowicz@cern.ch

Warsaw University of Technology, Warsaw, Poland

D. Elia et al. (eds.), *The XVIII International Conference on Strangeness in Quark Matter (SQM 2019)*, Springer Proceedings in Physics 250,
https://doi.org/10.1007/978-3-030-53448-6_60

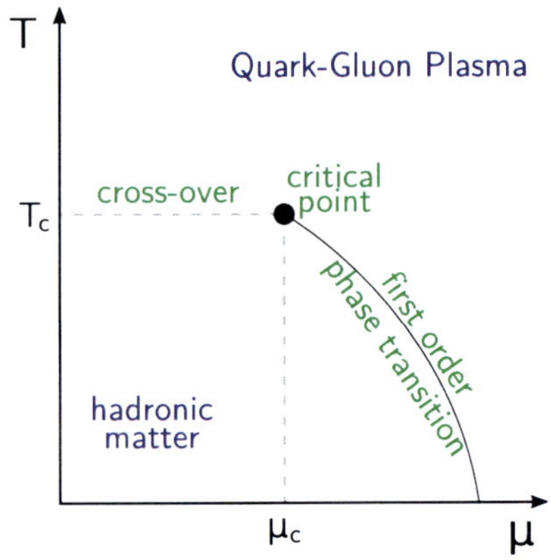

Fig. 60.1 A sketch of the phase diagram of strongly-interacting matter

60.2 Experimental Measures

60.2.1 Extensive Quantities

An extensive quantity is a quantity that is proportional to the number of Wounded Nucleons (W) in the Wounded Nucleon Model [4] (WNM) or to the volume (V) in the Ideal Boltzmann Grand Canonical Ensemble (IB-GCE). The most popular are particle number (multiplicity) distribution $P(N)$ cumulants:

$$
\begin{aligned}
\kappa_1 &= \langle N \rangle, \\
\kappa_2 &= \left\langle (\delta N)^2 \right\rangle = \sigma^2, \\
\kappa_3 &= \left\langle (\delta N)^3 \right\rangle = S\sigma^3, \\
\kappa_4 &= \left\langle (\delta N)^4 \right\rangle - 3\left\langle (\delta N)^2 \right\rangle^2 = \kappa\sigma^4.
\end{aligned}
$$

60.2.2 Intensive Quantities

Ratio of any two extensive quantities is independent of W (WNM) or V (IB-GCE) for an event sample with fixed W (or V)—it is an intensive quantity. For example:

$$\langle A \rangle \,/\, \langle B \rangle = W \cdot \langle a \rangle \,/\, W \cdot \langle b \rangle = \langle a \rangle \,/\, \langle b \rangle\,,$$

where A and B are any extensive event quantities, i.e. $\langle A \rangle \sim W$, $\langle B \rangle \sim W$ and $\langle a \rangle = \langle A \rangle$ and $\langle b \rangle = \langle B \rangle$ for $W = 1$. Popular examples are:

$$
\begin{aligned}
\kappa_2/\kappa_1 &= \omega[N] = \tfrac{\sigma^2[N]}{\langle N \rangle} = \tfrac{W \cdot \sigma^2[n]}{W \cdot \langle n \rangle} = \omega[n] \text{ (scaled variance)}, \\
\kappa_3/\kappa_2 &= S\sigma, \\
\kappa_4/\kappa_2 &= \kappa\sigma^2.
\end{aligned}
$$

60.2.3 Strongly Intensive Quantities

For an event sample with varying W (or V), cumulants are not extensive quantities any more. For example:

$$\kappa_2 = \sigma^2[N] = \sigma^2[n]\,\langle W\rangle + \langle n\rangle^2\,\sigma^2[W].$$

However, having two extensive event quantities, one can construct quantities that are independent of the fluctuations of W (or V). Popular examples include [5, 6]:

$\langle K\rangle\,/\,\langle\pi\rangle$,
$\Delta[N, P_T] = (\omega[N]\,\langle P_T\rangle - \omega[P_T]\,\langle N\rangle)/c$,
$\Sigma[N, P_T] = (\omega[N]\,\langle P_T\rangle + \omega[B]\,\langle N\rangle - 2(\langle N P_T\rangle - \langle P_T\rangle\,\langle N\rangle)/c$,
where $P_T = \sum_{i=1}^{N} p_{T,i}$ and C is any extensive quantity (e.g. $\langle N\rangle$).

60.2.4 Short-Range Correlations

Quantum statistics leads to short-range correlations in momentum space, which are sensitive to particle correlations in configuration space (e.g. of CP origin).

Popular measures include momentum difference in Longitudinal Comoving System (LCMS), **q**, that is decomposed into three components: q_{long}—denoting momentum difference along the beam, q_{out}—parallel to the pair transverse-momentum vector ($\mathbf{k}_t = (\mathbf{p}_{T,1} + \mathbf{p}_{T,2})/2$) and q_{side}—perpendicular to q_{out} and q_{long}. The two-particle correlation function $C(q)$ is often approximated by a three-dimensional Gauss function:

$$C(\mathbf{q}) \cong 1 + \lambda \cdot \exp\left(-R^2_{long}q^2_{long} - R^2_{out}q^2_{out} - R^2_{side}q^2_{side}\right),$$

where λ describes the correlation strength and R_{out}, R_{side}, R_{long} denote Gaussian HBT radii.

A more parametrization of the correlation function is possible via introducing Lévy-shaped source (1-D) [7]:

$$C(q) \cong 1 + \lambda \cdot e^{(-qR)^{\alpha}},$$

where $q = |p_1 - p_2|_{LCMS}$, λ describes correlation length, R determines the length of homogenity and Lévy exponent α determines source shape:
$\alpha = 2$: Gaussian, predicted from a simple hydro, $\alpha < 2$: anomalous diffusion, generalized central limit theorem, $\alpha = 0.5$: conjectured value at the critical point.

60.2.5 Fluctuations as a Function of Momentum Bin Size

When a system crosses the second-order phase transition, it becomes scale invariant, which leads to power-law form of correlation function. The second factorial moment is calculated as a function of the momentum cell size (or bin number M):

$$F_2(M) \equiv \left\langle \frac{1}{M} \sum_{i=1}^{M} n_i(n_i - 1) \right\rangle \Big/ \left\langle \frac{1}{M} \sum_{i=1}^{M} n_i \right\rangle,$$

where n_i is particle multiplicity in cell i.

At the second-order phase transition the system is a simple fractal and the factorial moment exhibits a power-law dependence on M [8–11]:

$$F_2(M) \sim (M)^{\varphi_2}.$$

In case the system freezes-out in the vicinity of the critical point, $\varphi_2 = 5/6$.

To cancel the $F_2(M)$ dependence on the single-particle inclusive momentum distribution, one needs a uniform distribution of particles in bins or subtraction of the $F_2(M)$ values for mixed events:

$$\Delta F_2(M) = F_2^{data}(M) - F_2^{mixed}(M).$$

60.2.6 Light Nuclei Production

Based on coalescence model, particle ratios of light nuclei are sensitive to the nucleon density fluctuations at kinetic freeze-out and thus to CP. In the vicinity of the critical point or the first-order phase transition, density fluctuation becomes larger [12, 13].

Nucleon density fluctuation can be expressed by proton, triton and deuteron yields as:

$$\Delta n = \frac{\langle (\delta n)^2 \rangle}{\langle n \rangle} \approx \frac{1}{2\sqrt{3}} \frac{N_p \cdot N_t}{N_d^2} - 1.$$

60.3 Experimental Results

60.3.1 Multiplicity Fluctuations

Results on energy dependence of multiplicity fluctuations by NA61/SHINE [14] quantified by the scaled variance are presented in Fig. 60.2 (*left*). No prominent structures that could be related to the critical point are observed.

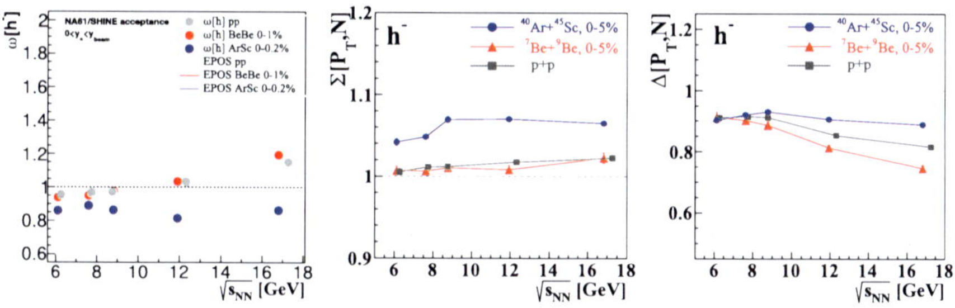

Fig. 60.2 Results on multiplicity [14] (*left*) and multiplicity-transverse momentum [15] (*center, right*) fluctuations for all negatively charged particles recorded by NA61/SHINE

60.3.2 Multiplicity-Transverse Momentum Fluctuations

Results on energy dependence of multiplicity-transverse momentum fluctuations by NA61/SHINE [15] expressed in Δ and Σ strongly intensive quantities are presented in Fig. 60.2 (*center, right*). No prominent structures that could be attributed to the critical point are observed.

60.3.3 Net-Proton Fluctuations

Figure 60.3 (*left*) presents energy dependence of fourth-order net-proton fluctuation in 5% most central Au+Au collisions recorded by STAR [16]. The observed non-monotonic dependence is consistent with theoretical predictions [17] and might suggest a critical point around $\sqrt{s_{NN}} \approx 7$ GeV.

60.3.4 Net-Kaon and Net-Charge Fluctuations

The STAR Collaboration has also studied net-kaon and net-charge distributions in central Au+Au collisions [18, 19]. However, the results, presented in Fig. 60.3 (*right*), show no (within errors) energy dependence.

60.3.5 Short-Range Correlations

Finite-Size Scaling Figure 60.4 presents compilations of Au+Au ($\sqrt{s_{NN}} = 7.7$–200 GeV) data from STAR [21] and Pb+Pb ($\sqrt{s_{NN}} = 2.76$ TeV) data from ALICE [22]. The Gaussian emission source radii ($R^2_{out} - R^2_{side}$) [20] show clear non-monotonic

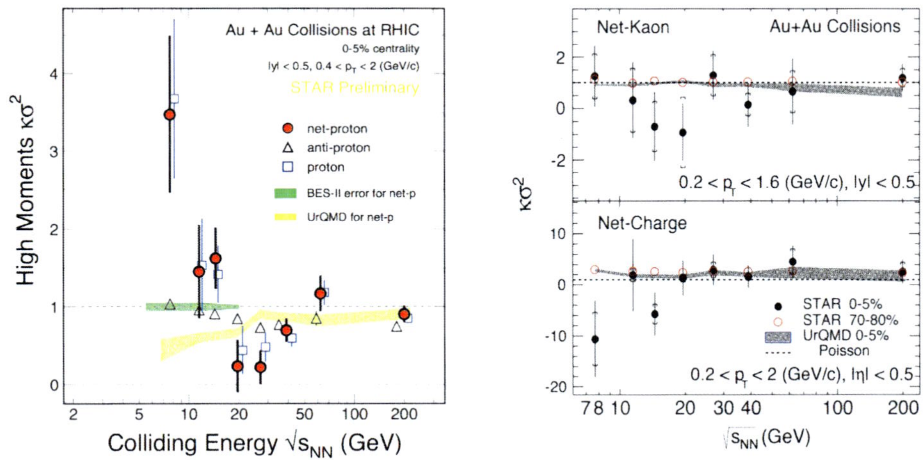

Fig. 60.3 Results on $\kappa\sigma^2$ of net-proton [16] (*left*) as well as net-kaon and net-charge [18, 19] (*right*) distributions measured by STAR

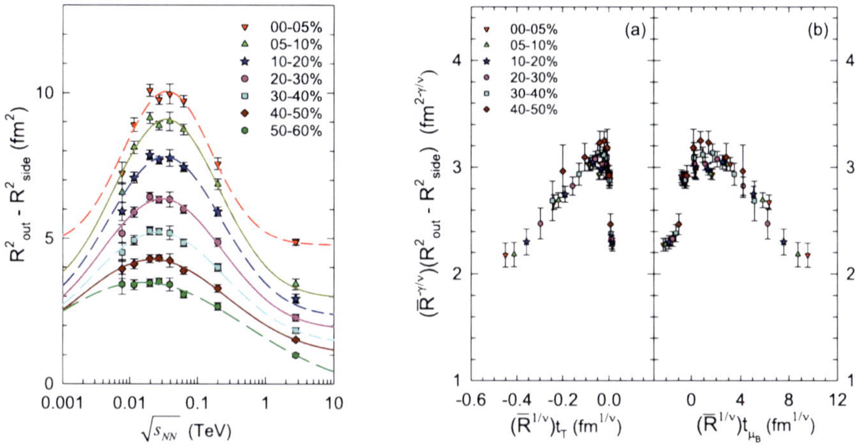

Fig. 60.4 Compilations of Au+Au ($\sqrt{s_{NN}}$ = 7.7–200 GeV, STAR [21]) and Pb+Pb ($\sqrt{s_{NN}}$ = 2.76 TeV, ALICE [22]) data: energy dependence of Gaussian emission source radii [20] (*left*) and one of the result for initial Finite-Size Scaling analysis [20] (*right*)

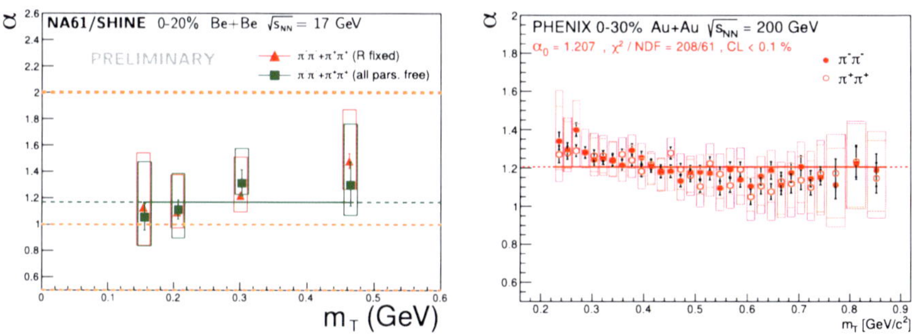

Fig. 60.5 Transverse mass dependence of the Lévy exponent α for 20% most central Be+Be collisions at 17 GeV by NA61/SHINE [23] (*left*) and for 30% Au+Au at 200 GeV by PHENIX [24] (*right*)

energy dependence with a maximum at $\sqrt{s_{NN}} \approx 47.5$ GeV. The initial Finite-Size Scaling analysis [20] suggests the critical point position: $T = 165$ MeV and $\mu = 95$ MeV.

Transverse-Mass Dependence of Lévy Exponent Transverse-mass dependence of Lévy exponent α have been studied both at SPS and RHIC. Figure 60.5 presents the results for Be+Be at 17 GeV by NA61/SHINE [23] and for Au+Au at 200 GeV by PHENIX [24]. Both studies revealed similar results, i.e. $\alpha \approx 1.2$, a value significantly above the CP prediction.

60.3.6 Fluctuations as a Function of Momentum Bin Size

NA49 and NA61/SHINE have studied the second factorial moment, ΔF_2, for mid-rapidity protons at 17 GeV.

Although in central Be+Be, C+C, Ar+Sc and Pb+Pb no signal has been observed, a deviation of ΔF_2 from zero seems apparent in central Si+Si and mid-central Ar+Sc as shown in Fig. 60.6.

60.3.7 Light Nuclei Production

The nucleon density fluctuations, Δn, for central Pb+Pb by NA49 [27] and central Au+Au by STAR [28, 29] show a non-monotonic dependence on collision energy with a peak for $\sqrt{s_{NN}} \approx 20$ GeV [30] as presented in Fig. 60.7.

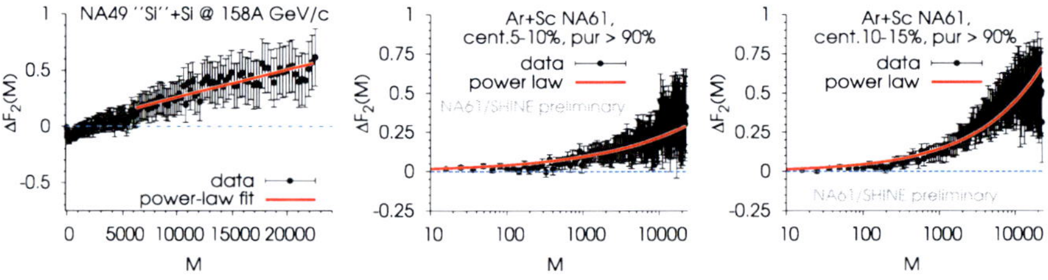

Fig. 60.6 Second factorial moment, ΔF_2, for mid-rapidity protons at 17 GeV in Si+Si by NA49 [26] (*left*) and in 5–10% and 10–15% Ar+Sc by NA61/SINE [25] (*center, right*)

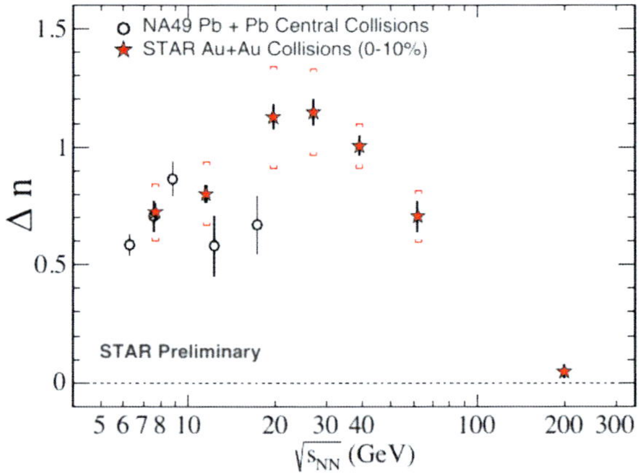

Fig. 60.7 Nucleon density fluctuation, Δn, for central Pb+Pb [27] and Au+Au [28, 29] collisions

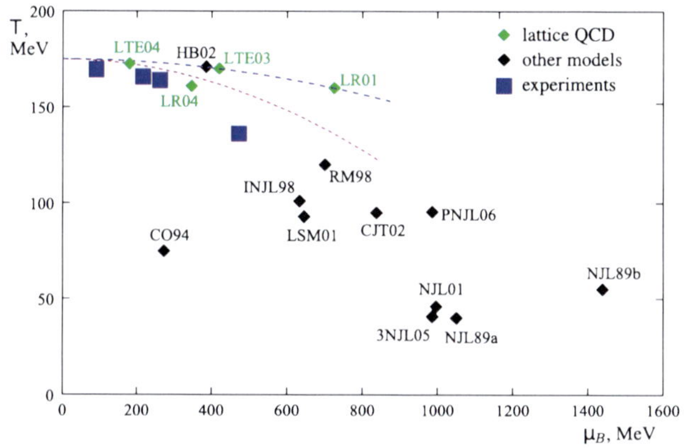

Fig. 60.8 Compilation of theoretical predictions [3] and experimental hints on the critical point location

60.4 Summary

The experimental search for the critical point is ongoing. There are four indications of anomalies in fluctuations in heavy-ion collisions at different collision energies ($\sqrt{s_{NN}} \approx 7, 17, 20, 47$ GeV). Interpreting them as due to CP allows one to estimate four hypothetical CP locations depicted in Fig. 60.8.

Fortunately, there are high-quality, beautiful new data coming soon both from SPS (NA61/SHINE) and RHIC (STAR Beam energy Scan II).

References

1. M. Asakawa, K. Yazaki, Nucl. Phys. A **504**, 668 (1989)
2. A. Barducci, R. Casalbuoni, S. De Curtis, R. Gatto, G. Pettini, Phys. Lett. B **231**, 463 (1989)
3. M.A. Stephanov, PoS LAT **2006**, 024 (2006)
4. A. Bialas, M. Bleszynski, W. Czyz, Nucl. Phys. B **111**, 461 (1976)
5. M.I. Gorenstein, M. Gazdzicki, Phys. Rev. C **84**, 014904 (2011)
6. M. Gazdzicki, M.I. Gorenstein, M. Mackowiak-Pawlowska, Phys. Rev. C **88**(2), 024907 (2013)
7. T. Csorgo, S. Hegyi, T. Novak, W.A. Zajc, A.I.P. Conf. Proc. **828**(1), 525 (2006)
8. J. Wosiek, Acta Phys. Polon. B **19**, 863 (1988)
9. A. Bialas, R.C. Hwa, Phys. Lett. B **253**, 436 (1991)
10. A. Bialas, R.B. Peschanski, Nucl. Phys. B **273**, 703 (1986)
11. N.G. Antoniou, F.K. Diakonos, A.S. Kapoyannis, K.S. Kousouris, Phys. Rev. Lett. **97**, 032002 (2006)
12. K.J. Sun, L.W. Chen, C.M. Ko, Z. Xu, Phys. Lett. B **774**, 103 (2017)
13. E. Shuryak, J. Torres-Rincon, Nucl. Phys. A **982**, 831 (2019)
14. M. Gazdzicki [NA61/SHINE Collaboration], PoS CPOD **2017**, 012 (2018)
15. E. Andronov [NA61/SHINE Collaboration], Acta Phys. Polon. Supp. **10** 449 (2017)
16. L. Adamczyk et al., STAR collaboration. Phys. Rev. Lett. **112**, 032302 (2014)
17. M.A. Stephanov, Phys. Rev. Lett. **107**, 052301 (2011)

18. L. Adamczyk et al., STAR collaboration. Phys. Lett. B **785**, 551 (2018)
19. L. Adamczyk et al., STAR collaboration. Phys. Rev. Lett. **113**, 092301 (2014)
20. R.A. Lacey, Phys. Rev. Lett. **114**(14), 142301 (2015)
21. L. Adamczyk et al. [STAR Collaboration], Phys. Rev. C **92**(1), 014904 (2015)
22. K. Aamodt et al., ALICE collaboration. Phys. Lett. B **696**, 328 (2011)
23. B. Porfy [NA61/SHINE Collaboration], Acta Phys. Polon. Supp. **12** 451 (2019)
24. A. Adare et al. [PHENIX Collaboration], Phys. Rev. C **97**(6), 064911 (2018)
25. N. Davis et al., NA61/SHINE collaboration. PoS CPOD **2017**, 054 (2018)
26. T. Anticic et al. [NA49 Collaboration], Eur. Phys. J. C **75**(12), 587 (2015)
27. T. Anticic et al. [NA49 Collaboration], Phys. Rev. C **94**(4), 044906 (2016)
28. L. Adamczyk et al. [STAR Collaboration], Phys. Rev. Lett. **121**(3), 032301 (2018)
29. N. Yu [STAR Collaboration], Nucl. Phys. A **967**, 788 (2017)
30. D. Zhang [STAR Collaboration], https://indico.cern.ch/event/656452/contributions/2859773/

Chapter 61
Patterns and Partners Within the QCD Phase Diagram Including Strangeness

Angel Gómez Nicola, Jacobo Ruiz de Elvira, and Andrea Vioque-Rodríguez

Abstract We review the current situation of the pattern of chiral symmetry restoration. In particular, we analyze partner degeneration for $O(4)$ and $U(1)_A$ symmetries within the context of Ward Identities and Effective Theories. The application of Ward Identities to the thermal scaling of screening masses is also discussed. We present relevant observables for which an Effective Theory description in terms of Chiral Perturbation Theory and its unitarized extension are compatible with lattice data even around the transition region. We pay special attention to the role of strangeness in this context.

61.1 Introduction

Over recent years, we are progressively reaching a deeper understanding of the QCD phase diagram and its main properties. Combined efforts from experiment, lattice simulations and phenomenology are allowing to access regions of the (T, μ_B) plane increasingly richer in baryon density. In particular, beam energy scans [1] would reveal whether a critical point exists and the behaviour of QCD matter around it. This is actually one of the main objectives of the current program of hot and dense QCD matter in lattice and heavy-ion collisions [2, 3].

In this context, a significative advance has been to realize that the phase boundary lies close to the chemical freeze-out for physical conditions of net baryon number B, electric charge Q and strangeness S, accesible to experimental heavy-ion exper-

A. Gómez Nicola (✉) · A. Vioque-Rodríguez
Departamento de Física Teórica and IPARCOS. Univ. Complutense., 28040 Madrid, Spain
e-mail: gomez@ucm.es

A. Vioque-Rodríguez
e-mail: elvira@itp.unibe.ch

J. Ruiz de Elvira
Albert Einstein Center for Fundamental Physics, Institute for Theoretical Physics, University of Bern, Sidlerstrasse 5, 3012 Bern, Switzerland
e-mail: avioque@ucm.es

D. Elia et al. (eds.), *The XVIII International Conference on Strangeness in Quark Matter (SQM 2019)*, Springer Proceedings in Physics 250,
https://doi.org/10.1007/978-3-030-53448-6_61

iments. Thus, using hadron statistical models [4], which have been very successful in the past for this purpose, one can fit hadron yields from ALICE data. The result of such fits are points on the freeze-out (T, μ_B) curve which turn out to overlap with the critical line obtained from lattice collaborations where μ_B is treated within Taylor expansions to avoid the so-called sign problem [5]. In addition, the study of fluctuations of those very same conserved changes opens up interesting possibilities. A particularly interesting analysis in this context regarding strangeness is the study of crossed susceptibilities performed in lattice works [6]. This is relevant because a combination of BS and QS crossed susceptibilities provides a relation between chemical potentials $\mu_{B,S,Q}$. Such relation can also be tested at freeze-out with experimental hadron yields fits or with theoretical models such as the Hadron Resonance Gas (HRG).

The $\mu_B = 0$ regime is in principle much better understood. Regarding the transition, the most analyzed signals have been the inflection point of the (subtracted) light quark condensate $\langle \bar{q}q \rangle_l = \langle \bar{u}u + \bar{d}d \rangle$ and the peak of the scalar (or chiral) susceptibility χ_S. Both reveal a crossover-like transition in the physical case ($N_f = 2 + 1$ light flavors and physical quark masses) at $T_c \simeq 156$ MeV [5, 7] which in the chiral limit reduces to $T_c^0 \simeq 132$ MeV [8] and becomes a true phase transition, most likely of second order, for two massless flavours [9].

An open problem in this context is to determine not only the order but the universality class (pattern) of the chiral phase transition. This depends crucially on whether the $U(1)_A$ anomalous symmetry is sufficiently restored at T_c [9–11], which may even affect the properties of the possible critical point at $\mu_B \neq 0$ [12]. A second-order $O(4) \equiv SU(2) \times SU(2)$ transition would be preferred in a scenario with $U(1)_A$ breaking at T_c, while a second-order $U(2) \times U(2)$ one would correspond to a $U(1)_A$ restored situation. The latter may even degenerate into a first order transition for strong enough $U(1)_A$ restoration [11].

A useful perspective to explore this problem is the analysis of partners, i.e., hadronic states which should become degenerate under those symmetries. Consider for instance the pseudoscalar and scalar nonets $\pi^a = i\bar{\psi}_l\gamma_5\tau^a\psi_l$, $\delta^a = \bar{\psi}_l\tau^a\psi_l$ for isospin $I = 1$, $\eta_l = i\bar{\psi}_l\gamma_5\psi_l$, $\eta_s = i\bar{s}\gamma_5 s$, $\sigma_l = \bar{\psi}_l\psi_l$, $\sigma_s = \bar{s}s$ for $I = 0$, $K^a = i\bar{\psi}\gamma_5\lambda^a\psi$, $\kappa^a = i\bar{\psi}\lambda^a\psi$ $(a = 4, 5, 6, 7)$ for $I = 1/2$. Here, ψ_l is the light quark doublet and those states correspond respectively to the quantum numbers of the pion, $a_0(980)$, light and strange component of the η/η', light and strange components of the $f_0(500)/f_0(980)$, kaon and $K(800)$ (or κ). For the isospin $I = 0, 1$ sector, chiral and $U(1)_A$ transformations connect the bilinears

$$\pi^a \overset{SU_A(2)}{\longleftrightarrow} \sigma, \quad \delta^a \overset{SU_A(2)}{\longleftrightarrow} \eta_l, \tag{61.1}$$

$$\pi^a \overset{U(1)_A}{\longleftrightarrow} \delta^a, \quad \sigma \overset{U(1)_A}{\longleftrightarrow} \eta_l, \tag{61.2}$$

which are the partners that have been studied in recent lattice and theoretical works on this subject. The lattice results are not fully conclusive. On the one hand, for $N_f = 2 + 1$ flavors and physical quark masses, the analysis of [13] shows degeneracy of $U(1)_A$ partners well above the $O(4)$ ones. On the other hand, $N_f = 2$ works [14–17] point to $U(1)_A$ restoration at T_c in the chiral limit, while for massive quarks in those works the strength of $U(1)_A$ breaking increases with the volume [11].

61.2 Ward Identities

We have recently analyzed the chiral pattern commented above, exploiting Ward Identities derived formally from the QCD generating functional [18, 19]. In particular, the following identity connects susceptibilities (two-point correlators at $p = 0$) in the pseudoscalar η_l, π and crossed $\eta_l\eta_s$ channels with the topological susceptibility of the anomaly operator $A(x) = \frac{3g^2}{16\pi^2}\mathrm{Tr}_c G_{\mu\nu}\tilde{G}^{\mu\nu}$:

$$\chi_P^{ls}(T) = -2\frac{\hat{m}}{m_s}\chi_{5,disc}(T) = -\frac{2}{\hat{m}m_s}\chi_{top}(T), \tag{61.3}$$

where $\chi_{5,disc}(T) = \frac{1}{4}\left[\chi_P^{\pi}(T) - \chi_P^{ll}(T)\right]$ and $\hat{m} = m_u = m_d$. Now, one can choose a $SU(2)_A$ transformation so that

$$\eta_l(x) \xrightarrow{SU_A(2)} -\delta^b(x) \Rightarrow \chi_P^{ls} \xrightarrow{SU_A(2)} 0, \tag{61.4}$$

since η_s is invariant under $SU(2)_A$ transformations and the $\delta\eta_s$ correlator vanishes by parity. Therefore, from (61.3), the conclusion is that for exact chiral restoration, where δ and η_l should degenerate, $\chi_{5,disc}$ should vanish as well. Thus, $\pi^a - \eta$ degenerate and the $O(4) \times U(1)_A$ pattern is realized. This should be then the scenario in the chiral limit for two massless flavours at T_c, consistently with the lattice results in [11, 14–17]. For $N_f = 2 + 1$ flavours and physical masses, the strangeness contribution and the large uncertainties for $\delta - \eta_l$ degeneration [13] might explain a stronger $U(1)_A$ breaking, consistently also with the chiral limit analysis of that collaboration [5].

An interesting application of WI in this context is related to the temperature dependence of lattice spatial screening masses M_i [19, 20] for different i channels. Assuming a scaling $M_i(T)/M_i(0) \sim [\chi_i(T)/\chi_i(0)]^{-1/2}$, the WI allow to connect $M_i(T)$ with suitably subtracted quark condensates, well under control in lattice simulations. This assumption implies that the zero momentum propagator given by the susceptibilities χ_i dominates the thermal dependence. One can actually test such scaling laws directly for lattice collaborations providing data on both screening masses and quark condensates for the same lattice setup. Such test has been performed in [19] for the π, K, $\bar{s}s$ and κ channels, which according to the WI scale as the inverse square root of $\langle\bar{q}q\rangle_l$, $\langle\bar{q}q\rangle_l + 2\langle\bar{s}s\rangle$, $\langle\bar{s}s\rangle$ and $\langle\bar{q}q\rangle_l - 2\langle\bar{s}s\rangle$ respectively. The agree-

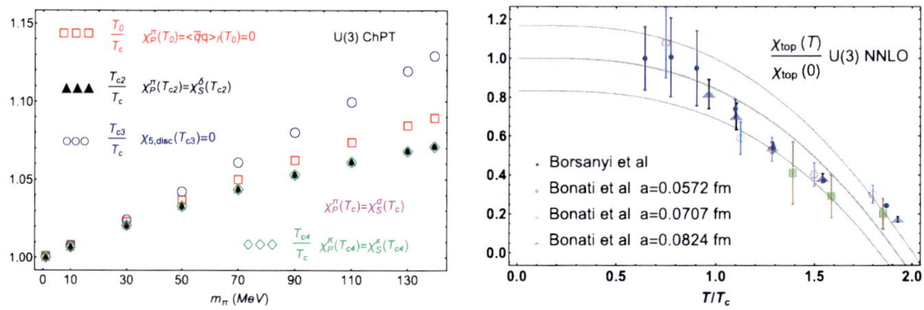

Fig. 61.1 Left: Evolution towards the chiral limit of the different $O(4)$ and $U(1)_A$ restoration temperatures within $U(3)$ ChPT. Right: Temperature dependence of the topological susceptibility calculated within the U(3) formalism compared to lattice data from [21] and [22] with $T_c = 155$ MeV

ment is quite good, with only two fit parameters related to the definition of subtracted condensates. It explains also the qualitative behaviour of the $M_i(T)$ around T_c, from the expected one of the quark condensates involved. Thus, for instance, the rapid growth of $M_\pi(T)$ would be explained by the inverse dependence $\left[\langle\bar{q}q\rangle_l(T)\right]^{-1/2}$ while M_K and $M_{\bar{s}s}$ are softened by the $\langle\bar{s}s\rangle(T)$ component.

61.3 Effective Theories

Hadronic effective approaches like the HRG or ChPT (for the lightest states) are needed to provide a physically meaningful description below the transition. In connection with our previous discussion, it is worth mentioning that recent analysis within $U(3)$ ChPT (where N_c^{-1} is included in the standard chiral power counting) have allowed on the one hand to verify the previously mentioned WI [20] and on the other hand to confirm the pattern of $O(4) \times U(1)_A$ restoration in the chiral limit [19]. The latter is showed in Fig. 61.1 where pseudocritical temperatures associated to the degeneracy of different $O(4)$ and $O(4) \times U(1)_A$ partners, converge as the pion mass vanishes.

The $U(3)$ ChPT framework allows also to obtain a quite accurate description of the topological susceptibility χ_{top} and its thermal dependence [23]. The leading order yields

$$\chi_{top}^{U(3),LO} = \Sigma \frac{M_0^2 \bar{m}}{M_0^2 + 6B_0\bar{m}} \tag{61.5}$$

with $\Sigma = B_0 F^2$ the single-flavor quark condensate in the chiral limit, $B_0 = M_{0\pi^\pm}^2/(m_u + m_d)$, with $M_{0\pi^\pm}$ the tree-level mass of the charged pions, F the pion decay constant in the chiral limit, M_0 the anomalous part of the η' mass and $\bar{m}^{-1} = \sum_{i=u,d,s} m_i^{-1}$. Expression (61.5) reproduces the known results for two and three light flavours in the limit $M_0 \to \infty$ [24] as well as the quenched gluodynamics limit

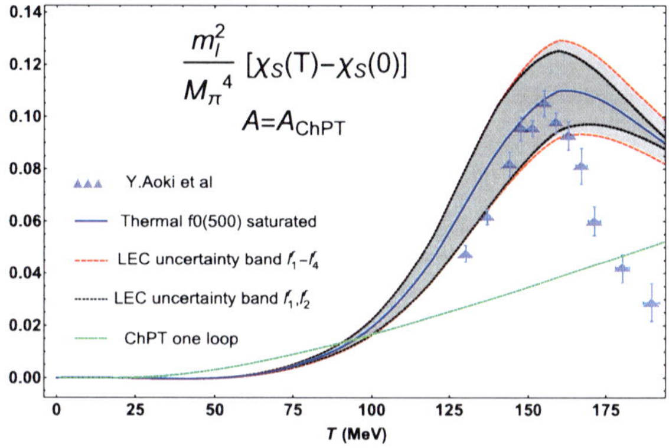

Fig. 61.2 Scalar susceptibility saturated by the unitarized thermal $f_0(500)$ pole, according to [31]. A normalization factor A has been chosen to match the perturbative ChPT result at $T = 0$ and the uncertainty bands given by the low-energy constants (LEC) is shown. Lattice points are taken from [7]

for $m_i \to \infty$ [25, 26]. The NLO corrections can be found in [27], while the NLO and NNLO $U(3)$ results at $T = 0$ are given in [23], including the fourth-order cumulant of the topological charge. The contribution of η' loops and $\eta - \eta'$ mixing corrections provided by the $U(3)$ formalism are of the same order as the K, η $SU(3)$ ones and are compatible with the lattice results in [21, 28]. In addition, the large-N_c behaviour of both quantities arising naturally within this formalism agrees also with lattice analysis [28].

The temperature evolution of the topological susceptibility within the $U(3)$ ChPT analysis, showed in Fig. 61.1, is consistent with lattice data, even far beyond the applicability range of the theory. Although $\chi_{top}(T)$ scales perturbatively as $\langle \bar{q}q \rangle_l (T)$ (actually both quantities are proportional at LO), deviations from this behaviour are expected around the transition. Actually, from the WI in (61.3) and the WI $\chi_P^\pi = -\langle \bar{q}q \rangle_l / \hat{m}$, an additional contribution proportional to $\chi_P^{ll}(T)$ is present, consistently with the existence of a sizable gap between chiral and $U(1)_A$ restoration.

Finally, we remark that combining the standard ChPT expansion with unitarization arguments, one can generate thermal resonances, which show up as second-sheet Riemann poles of meson scattering amplitudes at finite temperature [29]. The case of the thermal $f_0(500)$ is particularly important in the present context since it saturates the scalar susceptibility, giving rise to a peak around the crossover transition compatible with lattice data, as shown in Fig. 61.2, even more accurately than the HRG description [30, 31].

61.4 Conclusions

Despite the recent advances in the understanding of the QCD phase diagram, there are still many relevant open problems such as the nature of the transition, the description of matter rich in baryon density and the critical point. We have showed that the use of theoretical tools such as Ward Identities and Effective Theories allow us to make strong claims about the pattern of the transition. It points towards $O(4) \times$

$U(1)_A$ restoration in the limit of two massless flavours, from the analysis of partner degeneration. Related observables accurately described within this framework are screening masses, the topological charge distribution and the scalar susceptibility through thermal unitarity.

Acknowledgments Work partially supported by research contract FPA2016-75654-C2-2-P (spanish "Ministerio de Economía y Competitividad") and the Swiss National Science Foundation, project No. PZ00P2_174228. This work has also received funding from the European Union Horizon 2020 research and innovation programme under grant agreement No 824093. A. V-R acknowledges support from a fellowship of the UCM predoctoral program.

References

1. L. Adamczyk et al. [STAR Collaboration], Phys. Rev. C **96**(4), 044904 (2017)
2. C. Ratti, Rept. Prog. Phys. **81**(8), 084301 (2018)
3. A. Bazavov et al. [USQCD Collaboration], arXiv:1904.09951 [hep-lat]
4. A. Andronic, P. Braun-Munzinger, K. Redlich, J. Stachel, Nature **561**(7723), 321 (2018)
5. A. Bazavov et al., HotQCD collaboration. Phys. Lett. B **795**, 15 (2019)
6. A. Bazavov et al., Phys. Rev. Lett. **113**(7), 072001 (2014)
7. Y. Aoki, S. Borsanyi, S. Durr, Z. Fodor, S.D. Katz, S. Krieg, K.K. Szabo, JHEP **0906**, 088 (2009)
8. H.T. Ding et al., Phys. Rev. Lett. **123**(6), 062002 (2019)
9. R.D. Pisarski, F. Wilczek, Phys. Rev. D **29**, 338 (1984)
10. A. Pelissetto, E. Vicari, Phys. Rev. D **88**(10), 105018 (2013)
11. B.B. Brandt et al.,. http://arxiv.org/abs/1904.02384arXiv:1904.02384 [hep-lat]
12. M. Mitter, B.J. Schaefer, Phys. Rev. D **89**(5), 054027 (2014)
13. M.I. Buchoff et al., Phys. Rev. D **89**, 054514 (2014)
14. S. Aoki, H. Fukaya, Y. Taniguchi, Phys. Rev. D **86**, 114512 (2012)
15. G. Cossu et al., Phys. Rev. D **87**, no. 11, 114514 (2013) Erratum: [Phys. Rev. D **88**, no. 1, 019901 (2013)]
16. A. Tomiya, G. Cossu, S. Aoki, H. Fukaya, S. Hashimoto, T. Kaneko, J. Noaki, Phys. Rev. D **96**(3), 034509 (2017)
17. B.B. Brandt et al., JHEP **1612**, 158 (2016)
18. A. Gómez Nicola. J. Ruiz de Elvira, Phys. Rev. D **97**(7), 074016 (2018)
19. A. Gómez Nicola, J. Ruiz De Elvira, Phys. Rev. D **98**(1), 014020 (2018)
20. A. Gómez Nicola, J. Ruiz de Elvira, JHEP **1603**, 186 (2016)
21. C. Bonati et al., JHEP **1603**, 155 (2016)
22. S. Borsanyi *et al.*, Nature **539**(7627), 69 (2016); [arXiv:1606.07494 [hep-lat]] for additional information
23. A. Gómez Nicola, J. Ruiz De Elvira, A. Vioque-Rodríguez, arXiv:1907.11734 [hep-ph]
24. H. Leutwyler, A.V. Smilga, Phys. Rev. D **46**, 5607 (1992)
25. E. Witten, Nucl. Phys. B **156**, 269 (1979)
26. G. Veneziano, Nucl. Phys. B **159**, 213 (1979)
27. Y.Y. Mao et al., TWQCD Collaboration. Phys. Rev. D **80**, 034502 (2009)
28. C. Bonati, M. D'Elia, P. Rossi, E. Vicari, Phys. Rev. D **94**(8), 085017 (2016)
29. A. Dobado, A. Gomez Nicola, F.J. Llanes-Estrada, J.R. Pelaez, Phys. Rev. C **66**, 055201 (2002)
30. A. Gómez Nicola, J. Ruiz de Elvira, R. Torres Andres, Phys. Rev. D **88**, 076007 (2013)
31. S. Ferreres-Solé, A. Gómez Nicola, A. Vioque-Rodríguez, Phys. Rev. D **99**(3), 036018 (2019)

Part VIII
Strangeness in Astrophysics

Chapter 62
Production of Hypernuclei and Properties of Hyper-Nuclear Matter

Alexander Botvina, Marcus Bleicher, and Nihal Buyukcizmeci

Abstract The relativistic nucleus-nucleus collisions can produce hypernuclei and low-temperature hyper-matter as a result of hyperon capture by nuclear residues and free nucleons. We use the transport, coalescence and statistical models to describe the whole process, and point at the important advantages of such reactions: A broad variety of formed hypernuclei in masses and isospin allows for investigating properties of exotic hypernuclei, as well as the hypermatter both at high and low temperatures. The abundant production of multi-strange nuclei that can give an access to multi-hyperon systems and strange nuclear matter. The de-excitation of hot hyper-cluster will allow for the hyperon correlation studies. There is a saturation of the hypernuclei production at high energies, therefore, the optimal way to pursue this experimental study is to use the accelerator facilities of intermediate energies.

Embedding hyperons in the nuclear matter allows to explore the many-body aspects of the strong three-flavor interaction (i.e., including u, d, and s quarks) at low energies. Heavy hypernuclei open also opportunities to study the hyperon interactions and properties of strange matter, that is important for nuclear astrophysics (e.g., in neutron stars). Many various hyperons (Λ, Σ, Ξ, Ω) are produced in relativistic nuclear collisions. These hyperons can be captured by the produced baryons as well as by the projectile/target nuclear residues with the formation of hypernuclei. In such deep-inelastic reactions leading to fragmentation processes one can form hypernuclei of all sizes and isospin content, that gives advantages over traditional experimental hypernuclear methods. There are many experimental collaborations (STAR at RHIC,

A. Botvina (✉) · M. Bleicher
ITP and FIAS, J.W.Goethe University, Frankfurt am Main, Germany
e-mail: A.Botvina@gsi.de

A. Botvina
INR Russian Academy of Sciences, Moscow, Russia

M. Bleicher
GSI Helmholtz Center for Heavy Ion Research, Darmstadt, Germany

John-von-Neumann Institute for Computing (NIC), FZ Julich, Julich, Germany

N. Buyukcizmeci
Department of Physics, Selcuk University, 42079 Kampus, Konya, Turkey

D. Elia et al. (eds.), *The XVIII International Conference on Strangeness in Quark Matter (SQM 2019)*, Springer Proceedings in Physics 250,
https://doi.org/10.1007/978-3-030-53448-6_62

ALICE at LHC, PANDA, CBM, HypHI, Super-FRS, R3B at GSI/FAIR, BM@N, MPD at NICA) which plan to investigate hypernuclei and their properties in reactions induced by relativistic ions. The isospin space, particle unstable states, multi-strange nuclei and their precise lifetime can be explored in these fragmentation reactions. It was theoretically demonstrated with numerous models [1–8] that in nuclear collisions one can produce all kind of hypernuclei including multi-strange and exotic ones. Moreover the properties of the hypernuclei (e.g., the hyperon binding energies) can be extracted directly from their yields [9]. There also exist experimental confirmations of such processes leading to light hypernuclei [10–12] and fission hypernuclei [13, 14].

62.1 Formation of Hypernuclei in Relativistic Ion Collisions

Figure 62.1 we show our theoretical predictions on the fragment and Λ-hyperfragment production for the gold–gold and carbon–carbon relativistic collisions. The dynamical reaction stage is calculated with the Ultrarelativistic Quantum Molecular Dynamics (UrQMD) model, which is followed by the coalescence of baryons (CB) model [7]. The suggested mechanisms are the following: Nucleons of the projectile and target interact with binary collisions leading to the production of new particle

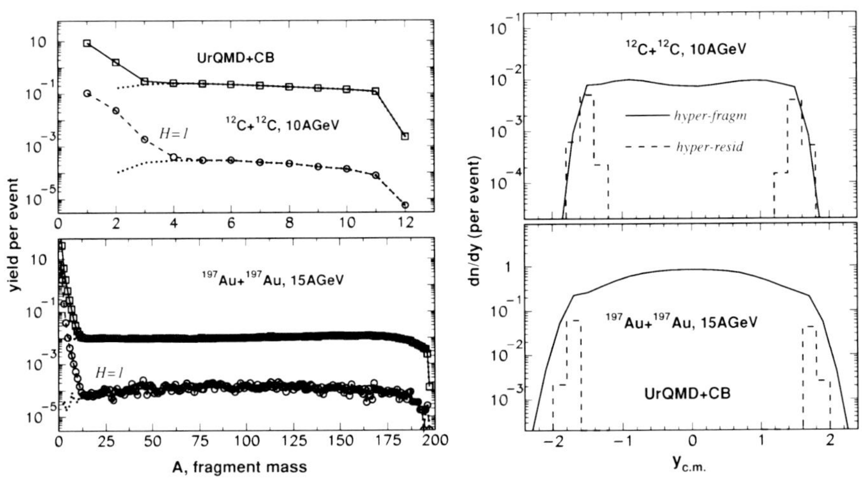

Fig. 62.1 The UrQMD plus CB model calculations of Au + Au and C + C ion collisions integrated over all impact parameters. The laboratory energies, projectile and targets are noted in panels. Left panels: Yields (per one inelastic event) of normal fragments (solid lines with squares) and hyperfragments with one captured Λ (notation H = 1, dashed lines with circles) versus their mass number (A). The dotted lines present the corresponding fragments originated from the projectile/target spectator residues. Right panels: Rapidity distributions (in the center of mass system, $y_{c.m.}$) of all produced hyper-fragments (solid lines) and, separately, the hyper-residues (dashed lines) [7]

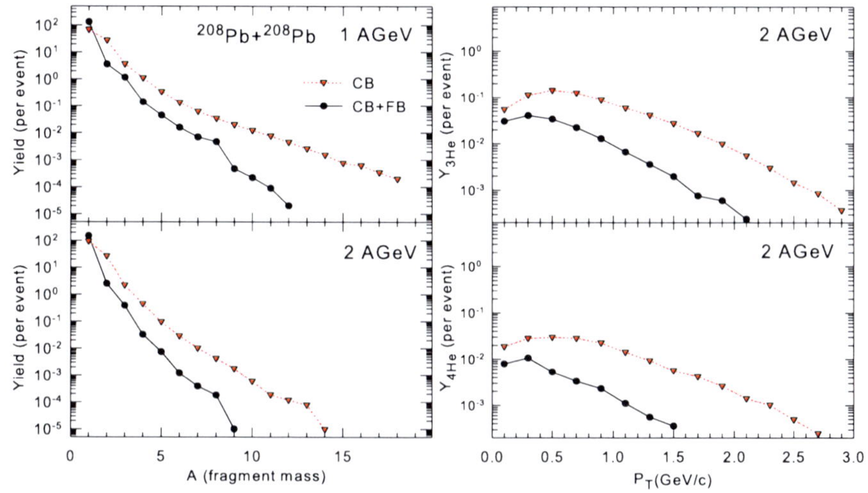

Fig. 62.2 Fragments formed from participant nucleons in the midrapidity region in lead on lead collisions at projectile energies of 1 and 2 GeV per nucleon. The dynamical stage and participant nucleons were calculated with DCM. Left panels: mass yields of excited fragments as obtained with CB model, and ones after de-excitation of these fragments via Fermi-break-up (FB) process. Right panels: transverse momenta distributions of ^{3}He (top) and ^{4}He (bottom) fragments, before and after the de-excitation of fragments

including resonances. As a result of primary and secondary interactions between the particles and other nucleons many hyperons can be created. These hyperons can be captured by spectator residues consisting of non-interacting (spectator) nucleons, therefore, big pieces of hypermatter around normal nuclear density are produced. Hyperons can also be captured by other free baryons which are stochastically occurred in their vicinity at the dynamical stage. Such a condensation-like process is responsible for light clusters. In these cases we expect mostly the production of excited fragments, since the baryon capture in the ground states has a very low probability according to the reaction phase space. As was established in studies of normal nuclear matter the excitation energy of big spectator residues may reach several MeV per nucleon [8, 15]. We obtain very broad mass distributions of both normal fragments and hyper-fragments, see Fig. 62.1. There are a lot of large fragments corresponding to the spectator residues. They concentrate in the regions of the projectile/target rapidities. The coalescence is mainly responsible for the lightest fragments and gives a contribution at all rapidities. The yields of coalescence fragments fall exponentially with A. We believe light hyper-fragments can be used for studying hyperon and nucleon interactions. Whereas moderately excited big hyper-fragments are suitable for the investigation of hyper-matter properties. Important features of hypernuclei production in relativistic ion collisions related to the projectile/target hyper-residues were discussed in [2–9].

62.2 De-Excitation of Coalescence Fragments

In this proceedings we concern the de-excitation process of light coalescence fragments produced abundantly in the midrapidity zone. Namely this zone is under study with the modern detectors [11, 12]. We show in Fig. 62.2 that at intermediate projectile energies around 1 GeV per nucleon primary coalescence clusters including tens of nucleons can be formed, and they undergo the following de-excitation. Here all dynamical participant baryons were generated within the Dubna Cascade Model (DCM) [8], while for the secondary de-excitation of clusters the statistical Fermi-break-up (FB) model was adopted, which was generalized for hypernuclei [5]. It is expected that the fragments become smaller after de-excitation (left panels). In addition, their isospin content changes and this can be seen in particle correlations. Important unstable states can be investigated in this way. We have also found an interesting isotope effect: In central collisions of big nuclei at low energies (less than 400 MeV per nucleon) the yield of ^{4}He nuclei can be larger than ^{3}He ones as a result of this de-excitation. (Note, that within a pure coalescence picture the yields of small clusters are always larger than big ones.) However, if we increase the projectile energy the yields of ^{3}He become again larger than ^{4}He ones, since the primary hot fragments will have smaller sizes. The kinetic energies of fragments change after de-excitation too. As seen from the right panels for helium particles the momentum distributions become steeper and maximums are shifted toward low energies. This is a consequence of that large primary fragments are formed by coalescence from slow baryons. We believe the extension of the coalescence towards hot clusters and their de-excitation is the qualitatively new development and it is consistent with the reaction physics.

A. Botvina acknowledges the support of BMBF (Germany). N.B. acknowledges the Turkish Scientific and Technological Research Council of Turkey (TUBITAK) support under Project No. 118F111. M.B. and N.B. acknowledges that the work has been performed in the framework of COST Action CA15213 THOR.

References

1. Z. Rudy, W. Cassing et al., Z. Phys. A **351**, 217 (1995)
2. A.S. Botvina, J. Pochodzalla, Phys. Rev. C **76**, 024909 (2007)
3. Th Gaitanos, H. Lenske, U. Mosel, Phys. Lett. B **675**, 297 (2009)
4. A.S. Botvina et al., Phys. Rev. C **84**, 064904 (2011)
5. A.S. Botvina, K.K. Gudima, J. Pochodzalla, Phys. Rev. C **88**, 054605 (2013)
6. N. Buyukcizmeci et al., Phys. Rev. C **88**, 014611 (2013)
7. A.S. Botvina et al., Phys. Lett. B **742**, 7 (2015)
8. A.S. Botvina et al., Phys. Rev. C **95**, 014902 (2017)
9. N. Buyukcizmeci et al., Phys. Rev. C **98**, 064603 (2018)
10. T.R. Saito et al., HypHI collaboration. Nucl. Phys. A **881**, 218 (2012)
11. The STAR collaboration, Science **328**, 58 (2010)

12. B. Dönigus et al., ALICE collaboration. Nucl. Phys. A **904–905**, 547c (2013)
13. T.A. Armstrong et al., Phys. Rev. C **47**, 1957 (1993)
14. H. Ohm et al., Phys. Rev. C **55**, 3062 (1997)
15. R. Ogul et al., Phys. Rev. C **83**, 024608 (2011)

Chapter 63
Recent AMADEUS Studies of Low-Energy K$^-$—Nucleus/Nuclei Interactions

Magdalena Skurzok, Massimiliano Bazzi, Mario Alexandru Bragadireanu, Damir Bosnar, Michael Cargnelli, Catalina Curceanu, Luca De Paolis, Raffaele Del Grande, Laura Fabbietti, Carlo Fiorini, Carlo Guaraldo, Mihai Iliescu, Paolo Levi Sandri, Johann Marton, Marco Miliucci, Pawel Moskal, Kristian Piscicchia, Angels Ramos, Alessandro Scordo, Michał Silarski, Diana Laura Sirghi, Florin Sirghi, Antonio Spallone, Oton Vazquez Doce, Eberhard Widmann, Sławomir Wycech, and Johann Zmeskal

Abstract We briefly report the recent results obtained by the AMADEUS collaboration on experimental studies of the K$^-$ low-energy interactions with light nuclei and outline the future perspectives.

M. Skurzok (✉) · M. Bazzi · C. Curceanu · L. De Paolis · R. Del Grande · C. Guaraldo · M. Iliescu · P. L. Sandri · M. Miliucci · K. Piscicchia · A. Scordo · D. Laura Sirghi · F. Sirghi · A. Spallone
INFN Laboratori Nazionali di Frascati, Frascati, Rome, Italy
e-mail: magdalena.skurzok@lnf.infn.it

M. Skurzok · P. Moskal · M. Silarski
Institute of Physics Jagiellonian University, Cracow, Poland

M. Alexandru Bragadireanu · D. Laura Sirghi · F. Sirghi
Horia Hulubei National Institute of Physics and Nuclear Engineering, Magurele, Romania

D. Bosnar
Department of Physics Faculty of Science, University of Zagreb, Zagreb, Croatia

M. Cargnelli · J. Marton · E. Widmann · J. Zmeskal
Stefan-Meyer-Institut für subatomare Physik, Vienna, Austria

L. De Paolis · R. Del Grande · M. Miliucci
Università degli Studi di Roma Tor Vergata, Rome, Italy

L. Fabbietti · O. V. Doce
Excellence Cluster Origin and Structure of the Universe, Garching, Germany

Physik Department E12, Technische Universität München, Garching, Germany

C. Fiorini
Politecnico di Milano Dip. di Elettronica Informazione e Bioingegneria, Milan, Italy

INFN Sezione di Milano, Milan, Italy

D. Elia et al. (eds.), *The XVIII International Conference on Strangeness in Quark Matter (SQM 2019)*, Springer Proceedings in Physics 250,
https://doi.org/10.1007/978-3-030-53448-6_63

63.1 Introduction

The low-energy QCD describing the strong interaction remains still poorly known in the strangeness sector due to the lack of fundamental experimental results. One of the key-issues is the investigation of the low-energy interaction between K^- mesons and nucleons/nuclei reflected by the $\Lambda(1405)$ resonance properties and possible kaonic bound state formation in the isospin $I = 0$ channel [1, 2]. Recently, two different theoretical approaches considered this issue. The phenomenological potential models [3–7] consider the $I = 0$ $\Lambda(1405)$ as a pure $\bar{K}N$ bound state and thus predicting the existence of deeply bound kaonic nuclear states. The chiral models [8–12] predict the $\Lambda(1405)$ as a superposition of two states, which results in a much less attractive K^-N and leads to the prediction of only slightly bound kaonic nuclear states. Therefore, to clarify this issue, experimental data are needed.

The AMADEUS collaboration performed measurements which set new experimental constraints to the K^-N strong interaction in the non-perturbative QCD exploiting the low-energy K^- hadronic interactions with light nuclei (e.g. H, ^{4}He, ^{9}Be and ^{12}C) [13]. The excellent quality low-momentum kaon beam ($p_K \sim 127$ MeV/c) delivered by the DAΦNE electron-positron collider [14] and the KLOE detector [15] as an active target were used to explore both stopped and in-flight K^- nuclear captures.

A complete characterization of the K^- two-, three- and four-nucleon absorptions (2NA, 3NA and 4NA) was performed for the first time in the Λp and $\Sigma^0 p$ final states studying the low-energy K^- captures on a solid ^{12}C target [16, 17]. Moreover, the possible contribution of a K^-pp bound state to the measured Λp spectrum was investigated. A summary of the analysis [16] is shown in Sect. 2.

The experimental investigation of the non-resonant hyperon-pion production in $I = 1$ channel ($K^-n \rightarrow \Lambda\pi^-$) [18] was carried out to provide important informations on the $\Lambda(1405)$ resonance structure. Sec. 3 presents a summary of the obtained results.

K. Piscicchia
Museo Storico della Fisica e Centro Studi e Ricerche Enrico Fermi, Rome, Italy

A. Ramos
Departament de Fisica Quantica i Astrofisica and Institut de Ciencies del Cosmos, Universitat de Barcelona, Marti i Franques 1, Barcelona, Spain

S. Wycech
National Centre for Nuclear Research, Warsaw, Poland

The described analyses have been performed for the data sample of 1.74 fb^{-1}, collected by the KLOE collaboration [15] during the 2004/2005 data taking period.

63.2 Characterisation of the K$^-$ Multi-nucleon Absorptions in Λp and Σ^0p Final States; Search for the K$^-$pp Bound State

The AMADEUS collaboration performed studies of $\Lambda(\Sigma^0)$p channels in K$^-$ absorption on ^{12}C [16] which allowed to extract for the first time two, three and four nucleon absorption branching ratios (BRs) and cross sections for low-momentum kaons in the investigated channels. The Λp direct production in 2NA-QF is expected to be phase space favoured with respect to the corresponding Σ^0p final state and the ratio between the final state phase spaces for the two processes is $\mathcal{R}' \simeq 1.22$. From the BRs we measure:

$$\mathcal{R} = \frac{BR(K^-pp \rightarrow \Lambda p)}{BR(K^-pp \rightarrow \Sigma^0 p)} = 0.7 \pm 0.2(stat.)^{+0.2}_{-0.3}(syst.). \tag{63.1}$$

The dominance of the Σ^0p channel is then evidence of the important dynamical effects involved in the measured processes.

The reconstruction of $\Lambda(\Sigma^0)$p channels allows also for the search of a signal corresponding to eventual intermediate formation of a K$^-$pp nuclear cluster. The performed analysis shows that the K$^-$ multi-nucleon absorption processes are sufficient to describe the Λp spectrum (right panel of Fig. 63.1). The contribution of

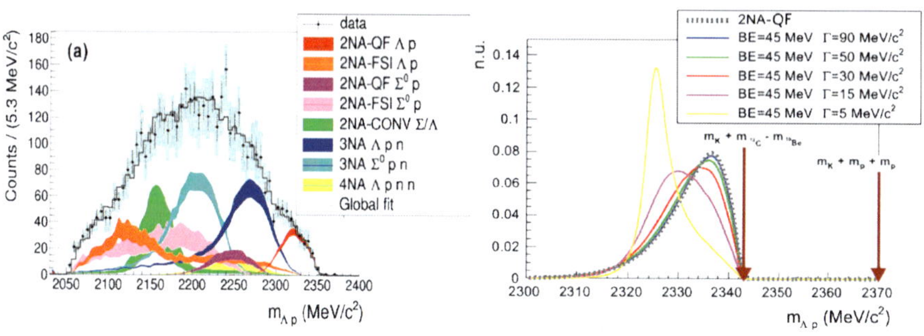

Fig. 63.1 (left) Λp invariant mass distribution for the K$^-$ absorption on ^{12}C listed in the legend. Black points represent the data, black error bars correspond to the statistical errors, cyan error bars correspond to the systematic errors. The gray line distributions represent the global fitting functions. (right) Calculated Λp invariant mass distributions for the process $K^- {}^{12}C \rightarrow K^-pp + {}^{10}Be \rightarrow \Lambda p + {}^{10}Be$, for a bound state with BE = 45 MeV and = 5, 15, 30, 50 and 90 MeV/c^2 (yellow, magenta, red, green and blue curves respectively). The gray curve is the shape of the 2NA-QF. The areas of the distributions are normalised to unity. These figures are adapted from [16]

the possible K^-pp bound state completely overlaps with 2NA-QF, except for small values of the bound state width (less than 15 MeV/c^2), as it is shown in the left panel of Fig. 63.1.

63.3 Investigation of Non-resonant $A_{K^-n \rightarrow \Lambda\pi^-}$ Transition Amplitude Below the $\bar{K}N$ Threshold

In order to investigate the $\Lambda(1405)$ resonance properties, produced through the K^- induced reaction in light nuclear targets, it is essential to take into account two biases. One bias is related to the invariant mass threshold due to the absorbing nucleon binding energy (for K^- capture at rest on ^4He and on ^{12}C the $\Sigma\pi$ invariant mass threshold is about 1412 MeV, and 1416 MeV, respectively). The $\bar{K}N$ sub-threshold region associated to the $\Lambda(1405)$ high-mass predicted pole (about 1420 MeV), can be explored by measurement of K^-N absorption in-flight. The in-flight contribution (for kaons with mean momentum of 100 MeV/c) shifts by about 10 MeV the $\Sigma\pi$ invariant mass threshold.

The second bias is the contribution of non-resonant $K^-N \rightarrow Y\pi$ reaction. The non-resonant transition amplitude modulus $|A_{K^-n \rightarrow \Lambda\pi^-}|$ was extracted for the first time in the $K^-n \rightarrow \Lambda\pi^-$ process, considering K^-n single nucleon absorptions in ^4He [18]. For this purpose experimentally extracted $\Lambda\pi^-$ invariant mass, momentum and angular distributions were simultaneously fitted using dedicated MC simulations that include non-resonant processes, resonant processes and the primary production of a Σ followed by the $\Sigma N \rightarrow \Lambda N'$ conversion process. The simulations of non-resonant/resonant processes were performed based on the results of phenomenological calculations presented in [19]. The data analysis gave $|A_{K^-n \rightarrow \Lambda\pi^-}|(33\pm6)$ MeV = 0.334 ± 0.018 (stat.) $^{+0.034}_{-0.058}$ (syst.) fm. This measurement allows to test and constrain recent calculations for S-wave $K^-n \rightarrow \Lambda\pi^-$ transition amplitude.

63.4 Conclusion

The interactions of low-momentum kaons K^- with nucleons/nuclei in light nuclear targets are investigated by AMADEUS with the aim to provide new experimental constraints to the K^-N strong interaction in the non-perturbative regime of the QCD in the strangeness sector. BRs and cross sections for the two-, three- and four-nucleon absorptions in the Λp and $\Sigma^0 p$ final states were determined by the studies of low-energy K^- interactions in a solid carbon target. The experimental investigation of the non-resonant $K^-N \rightarrow Y\pi$ production was performed for the first time for K^-n single nucleon absorption in ^4He. This result is substantial for the determination of the $I = 1$ background biasing the $\Lambda(1405)$ spectrum.

Acknowledgment We acknowledge the KLOE/KLOE-2 Collaboration for their support and for having provided us the data and the tools to perform the analysis presented in this paper. We acknowledge the CENTRO FERMI - Museo Storico della Fisica e Centro Studi e Ricerche "Enrico Fermi", for the project PAMQ. Part of this work was supported by the Austrian Science Fund (FWF): [P24756-N20]; Austrian Federal Ministry of Science and Research BMBWK 650962/0001VI/2/2009; the Croatian Science Foundation, under project 8570; Minstero degli Affari Esteri e della Cooperazione Internazionale, Direzione Generale per la Promozione del Sistema Paese (MAECI), StrangeMatter project; Polish National Science Center through grant No. UMO-2016/21/D/ST2/01155; EU STRONG-2020 (grant agreement 824093).

References

1. S. Wycech, Nucl. Phys. A **450**, 399c (1986)
2. Y. Akaishi, T. Yamazaki, Phys. Lett. B **535**, 70 (2002)
3. Y. Akaishi, T. Yamazaki, Phys. Rev. C **65**, 044005 (2002)
4. Y. Ikeda, T. Sato, Phys. Rev. C **76**, 035203 (2007)
5. S. Wycech, A.M. Green, Phys. Rev. C **79**, 014001 (2009)
6. J. Revai, N.V. Shevchenko, Phys. Rev. C **90**, 034004 (2014)
7. S. Maeda, Y. Akaishi, T. Yamazaki, Proc. Jpn. Acad. B **89**, 418 (2013)
8. A. Dote, T. Hyodo, W. Weise, Phys. Rev. C **79**, 014003 (2009)
9. N. Barnea, A. Gal, E.Z. Liverts, Phys. Lett. B **712**, 132 (2012)
10. Y. Ikeda, H. Kamano, T. Sato, Prog. Theor. Phys. **124**, 533 (2010)
11. P. Bicudo, Phys. Rev. D **76**, 031502 (2007)
12. M. Bayar, E. Oset, Nucl. Phys. A **914**, 349 (2013)
13. M. Skurzok et al., Recent Progress in Few-Body Physics FB22 2018. Springer Proc. Phys. **238**, (2018)
14. A. Gallo et al., Conf. Proc. **C060626**, 604 (2006)
15. F. Bossi et al., Riv. Nuovo Cim. **31**, 531 (2008)
16. Del Grande, R. et al., arXiv:1809.07212 (2018)
17. O. Vazques Doce et al., Phys. Lett. B **758**, 134 (2016)
18. K. Piscicchia et al., Phys. Lett. B **782**, 339 (2018)
19. K. Piscicchia, S. Wycech, C. Curceanu, Nucl. Phys. A **954**, 75 (2016)

Chapter 64
Hawking Radiation from the Relics of the Cosmic Quark Hadron Phase Transition

Bikash Sinha

Abstract It is entirely plausible that during the primordial quark- hadron transition, microseconds after the Big Bang, the universe may experience supercooling accompanied by mini inflation leading to a first order phase transition from quarks to hadrons. The relics, in the form of quark nuggets expected to consist of Strange Quark Matter, with a baryon number beyond a critical value will survive. It is conjectured that color confinement turns the physical vacuum to an event horizon for quarks and gluons. The horizon can be crossed only by quantum tunnelling. The process just mentioned is the QCD counterpart of Hawking radiation from gravitational black holes. Thus, when the Hawking temperature of the quark nuggets gets turned off, tunnelling will stop and the nuggets will survive forever. The baryon number and the mass of these nuggets are derived using this theoretical format. The results agree well with the prediction using other phenomenological models. Further, the variation of Hawking temperature as a function of baryon number and mass of the nugget mimicks chiral phase transition, somewhat similar to the QCD phase transition just described. Finally the strange quark nuggets may well be the candidates of baryonic dark matter.

Microseconds after the Big Bang the universe consisted of quarks, leptons and photons. As the universe expanded and cooled to a temperature around (150–200) MeV the chirally symmetric quark gluon plasma made a phase transition to the hadron phase of broken chiral symmetry. For very small baryon number $\eta = n_B/s \sim 10^{-9}$; ($n_B$, the net baryon density and s the entropy density), the wisdom of lattice suggests a rapid "cross- over" from the universe of quarks to the universe of hadrons. A first-order QCD phase transition seems unlikely and any imprint of the time before the phase transition gets erased out. However, lattice calculations for an expanding universe with massless bare quarks is not entirely satisfactory. Instead, chiral models of QCD have been used extensively [1–4].

It is conventionally assumed that the baryon asymmetry $\eta = (n_B - n_{\bar{B}})\gamma$ at that primordial epoch of phase transition is the same as that of today's universe $\eta \sim 10^{-10}$. There are however reasonably straightforward arguments [1–7] that η at that epoch

B. Sinha (✉)
Variable Energy Cyclotron Centre, 1/AF-Bidhannagar, Kolkata 700064, India
e-mail: bikash@vecc.gov.in

D. Elia et al. (eds.), *The XVIII International Conference on Strangeness in Quark Matter (SQM 2019)*, Springer Proceedings in Physics 250,
https://doi.org/10.1007/978-3-030-53448-6_64

is much higher and indeed of the order of $\eta \sim \mathcal{O}(1)$ unity. It can be shown however, that after the phase transition η goes back to 10^{-10}, as it is today. The consequence of such a possibility is discussed in the following.

Witten [7] and others [4–6, 8, 9] have argued that a first order phase transition is plausible with a "small" supercooling. In a recent private communication Witten [5] further asserted that if $\eta = n_B/\gamma$ remains $\sim 10^{-10}$ at the point of q-h phase transition, as it is in the current universe, then supercooling is not plausible. However, he also points out [5] that if the baryon to photon ratio is not small during the QCD phase transition and becomes small because of some phenomena at later times, then supercooling is plausible in principle, eventually leading to a first order phase transition from quark and gluons to hadrons. In the following, we demonstrate that this is entirely possible.

However, the physical process guiding the value of $\eta \sim \mathcal{O}(1)$ before the phase transition, to come down to a value of $\eta \sim 10^{-10}$ at CMB temperature later should emerge naturally.

This is what is achieved through a "little inflation" of about 7 e-folding occurring at a lower temperature, which may be identified with the QCD first order phase transition [1, 3]. Such an inflation naturally dilutes the baryon photon ratio to the observed range. Comparing this "little inflation" with the more standard Guth's inflationary model [10], one finds that the patterns of entropy variation in the two cases are very different. In the standard inflationary model [10] the entropy is conserved during exponential expansion, and increases, due to reheating when bubbles collide, at the end of the transition. However in Guth's scenario, supercooling is there though very large; In the little inflation scenario for the case of quark-hadron phase transition, the entropy is constantly increasing during the quark-hadron phase transition.

Bhattacharya et al. [11] however, have demonstrated, using Flux Tube model that quark nuggets of baryon number $\sim 10^{43}$, or larger will survive even upto now. These nuggets will consist of Strange Quarks (SQN) [7] and could be a viable candidates of cold dark matter [5, 7, 12], could even be candidates of primordial black hole [13].

We shall now discuss the evolution of these SQNs as the universe expands with time.

The SQNs remain in kinematic equilibrium due to radiation pressure (photons and neutrinos) acting on them with their velocity extremely non-relativistic. Effectively, the SQNs are almost static. It has been argued [12] that when the temperature of the universe is around couple of MeV, gravity dominates over radiation and the SQNs grow in size after collapsing under gravitational attraction, due to other SQN's; it can not grow however beyond a certain mass [12]. For baryon number 10^{42} the mass of the collapsed SQN is $0.24M_{\odot}$ but for baryon number 10^{46} it goes down to $0.0001M_{\odot}$ [12]. These inputs are crucial for the construction of the model developed in the following.

Color confinement in QCD does not allow colored species to exist in the physical vacuum and thus in some sense color confinement is similar to gravitational confinement provided by black holes. Thus, while 'G' ensures gravitational attraction and eventually to black holes, 'B', the Bag pressure ensures confinement of quarks leading to colorless physical vacuum (white holes) in QCD.

It has been argued [14] that quantum tunnelling through a color event horizon is the QCD counterpart of Hawking radiation from the gravitational black holes.

Let us consider the SQNs, after the gravitational collapse as QCD white holes, with neutrons tunnelling out from the event horizon. The natural length scale for SQNs, heuristically, can be argued as [8, 9] $L_B = M_N/M_\odot(B^{1/4})^{-1}$, M_N being the mass of the SQN and B the Bag Pressure which keeps the color confined. So, the entropy for QCD white hole becomes

$$S_{QN} = (A/4L_B^2) = \pi R^2 B^{1/2} (M_N/M_\odot)^{-2} \tag{64.1}$$

The celebrated Hawking entropy can be written as $S = \pi R^2/G$; for QCD [14], $G \equiv 1/2\sigma$; $\sigma = 0.16GeV^2$, σ being the string tension.

The confinement radius R of a SQN is equivalent to the "strong" Schwartzchild radius equivalent of the gravitational black hole.

$$S = 2\pi R^2\sigma = \pi R^2 B^{1/2} (M_N/M_\odot)^{-2} \tag{64.2}$$

thus $B^{1/2} (M_N/M_\odot)^{-2} = 2\sigma$, it is noted that in the final expression, "strong" schwartzchild radius gets crossed out as it should. For $B^{1/4} = 150MeV$; $M_N/M_\odot = 0.265$ and for $B^{1/4} = 200$ MeV $M_N/M_\odot = 0.35$; the result agrees closely with the results obtained by Banerjee et al. [12], although derived from a very different route of arguments.

Using the conjectured analogy between black hole thermodynamics and the thermodynamics of confined charged quarks, the SQNs, we represent black hole mass, charge and gravitational constant [14] by the mass M_N of the SQNs, baryon number B_N and the string tension σ

$$\{M, Q, G\}_{BH} \longleftrightarrow \{M_N, B_N, 1/2\sigma\}_{QN}$$

such that

$$T_{QN}(B_N, M_N) = T_{QN}(B_N, M_N = 0) \times \left[\frac{4\sqrt{1 - \frac{M_N^2}{2\sigma B_N^2}}}{\left(1 + \sqrt{1 - \frac{M_N^2}{2\sigma B_N^2}}\right)^2} \right] \tag{64.3}$$

Figure 64.1, $T_{SN}(B_N, M_N)/T_{SN}(B_N, M_N = 0)$ as a function of $(M_N^2/2\sigma B_N^2)$ is shown [14]; and $T_{QN}(B_N, M_N = 0)$ can be identified with QCD phase transition temperature ~150 MeV [14]. The mass of M_N being zero seems to imply complete restoration of chiral symmetry, similarly when $M_N = \sqrt{2\sigma} B_N \approx B_N$ all the baryons (quarks) have a total mass approximately the same as Baryon number, chiral symmetry is exactly and completely broken. For $M_N \sim 10^{44}$ (GeV) [12] $B_N = (1/\sqrt{2\sigma} M_N) \approx 6 \times 10^{43}$; Hawking radiation stops and no more tunneling out to the physical vacuum takes place. The SQNs survive the evolution of the uni-

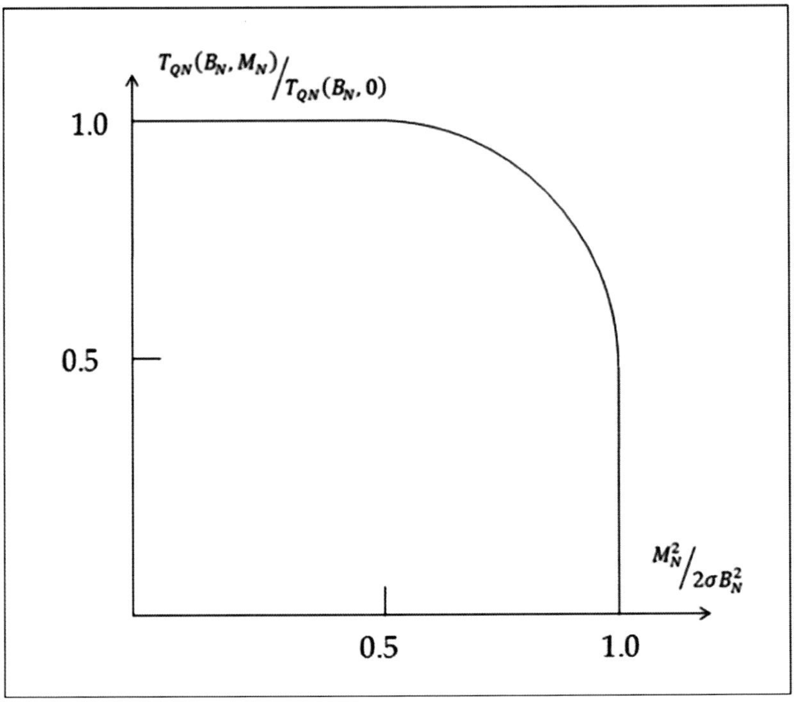

Fig. 64.1 Hawking temperature as a function of mass M_N and the baryon number B_N of the nuggets; σ is the string tension

verse. The surviving nugget with baryon number $\sim 6 \times 10^{43}$ agrees closely the value obtained by Bhattacharya et al. [11], although obtained from a flux tube model.

The phase transition is not instantaneous but happens over a period of time [15]; for a critical temperature of transition 100 MeV, the time it takes for the phase transition $\approx t_{ch} \sim 144\mu sec$ [15] and for a critical temperature 150 MeV $t_{ch} \equiv 64\mu sec$; during this time all the quarks inside the nugget, acquires mass, by breaking chiral symmetry.

For M_N less than $\approx 10^{44}$ (GeV), equivalent of Hawking radiation persists and neutrons tunnel out of the event horizon. After tunnelling of neutrons (proton tunnelling will be inhibited by coulomb repulsion) through the event horizon the temperature will rise and the mass of the evaporating SQN's will decrease; just like Hawking temp, $T = 1/8\pi GM$. Those SQN's with $M_N < 10^{44}C$ will eventually disappear leaving large entropy pockets in the universe associated with baryon inhomogeneity arising from the evaporated neutrons from the quarks nuggets [15] With expansion of the universe and the neutrino inflation [15], the inhomogeneity will be washed out, preserving the Big Bang nucleosynthesis.

One has attempted so far utilizing the possible equivalence of gravitational black hole and QCD white hole to drive the mass of QCD white hole which survives the evolution of the universe uptil now. Using the same conceptual framework we have derived the baryon number at which evaporation or equivalently tunneling out of the QCD white hole stops. Stopping of Hawking radiation from white holes indicate the survivability of the nuggets. The results agree closely with previously obtained results [11, 12] using somewhat different theoretical framework. The results obtained using black hole analogy is quite straightforward but based on fundamental physical laws, such as thermodynamics and Hawking radiation from black hole. The results obtained using black (white) hole thermodynamics lends credence to the framework used and the results obtained.

Where do these SQNs go? Originally, it was proposed that these SQNs are the MACHOs, the Massive Astrophysical Compact Objects [12], detected in the direction of the Large Magellanic Cloud (LMC) of mass range $(0.15–0.95)M_{\odot}$, with a probable mass of $0.5M_{\odot}$, and the total number being $N_{macho} \approx 10^{23-24}$. These MACHOs are candidate of baryonic dark matter [12].

Acknowledgments The author wishes to thank E. Witten for his perceptive comments on super-cooling. He is grateful to Dima Kharzeev, Larry Mc Larren and Roger Penrose for their suggestion. Comments from Pijus Bhattacharya, Debasish Mazumdar and Sibaji Raha are gratefully acknowledged. Finally the help rendered by Chiranjib Barman is appreciated.

References

1. T. Boeckel, J. Schaffner-Bielich, Phys. Rev. Lett. **105**(4), 041301 (2010)
2. M. Stephanov, K. Rajagopal, E. Shuryak, Phys. Rev. Lett. **81**(22), 4816 (1998)
3. T. Boeckel, J. Schaffner-Bielich, Phys. Rev. D **85**(10), 103506 (2012)
4. N. Borghini, W. Cottingham, R.V. Mau, J. Phys. G: Nucl. Part. Phys. **26**(6), 771 (2000)
5. E. Witten, Private Communication (2014)
6. B. Sinha, Int. J. Mod. Phys. A **29**(23), 1430024 (2014)
7. E. Witten, Phys. Rev. D **30**, 272 (1984). https://doi.org/10.1103/PhysRevD.30.272; https://link.aps.org/doi/10.1103/PhysRevD.30.272
8. B. Sinha, in *Journal of Physics: Conference Series*, vol. 668 (IOP Publishing, 2016), vol. 668, p. 012028
9. B. Sinha, Nucl. Phys. A **982**, 235 (2019)
10. A.H. Guth, Phys. Rev. D **23**(2), 347 (1981)
11. P. Bhattacharjee, J.e. Alam, B. Sinha, S. Raha, Phys. Rev. D **48**(10), 4630 (1993)
12. S. Banerjee, A. Bhattacharyya, S.K. Ghosh, S. Raha, B. Sinha, H. Toki, Mon. Not. R. Astron. Soc. **340**(1), 284 (2003)
13. K. Jedamzik, Phys. Rev. D **55**(10), R5871 (1997)
14. P. Castorina, D. Kharzeev, H. Satz, Eur. Phys. J. C **52**(1), 187 (2007)
15. A. Bhattacharyya, J.e. Alam, S. Sarkar, P. Roy, B. Sinha, S. Raha, P. Bhattacharjee, Phys. Rev. D **61**(8), 083509 (2000)

Chapter 65
Hyperon Interaction with Dense Nuclear Matter and Link to Neutron Stars

Laura Tolos

Abstract The theoretical status of the hyperon-nucleon and hyperon-hyperon interactions is reviewed, paying a special attention to chiral effective field theories. Results on hyperons in dense matter are presented and the consequences for the properties of neutron stars are analyzed.

65.1 Introduction

Over the past decades the properties of hyperons in dense matter have been object of high interest, in connection with on-going and future experiments on hypernuclei as well as the possible existence of hyperons in neutron stars.

In order to determine the properties of hyperons in a dense medium, it is of crucial importance to have a deep understanding of the underlying bare hyperon-nucleon (YN) and hyperon-hyperon (YY) interactions, and the modifications induced in a dense medium.

In this contribution, the theoretical status of the YN and YY interactions is reviewed, concentrating the attention on the recent developments within chiral effective field theory (χEFT). The properties of hyperons in dense nuclear matter within χEFT are studied, whereas the presence of hyperons in neutron stars and the consequences for their structure are analyzed.

L. Tolos (✉)
Institut für Theoretische Physik, University of Frankfurt,
Max-von-Laue-Str. 1, 60438 Frankfurt am Main, Germany
e-mail: tolos@th.physik.uni-frankfurt

Frankfurt Institute for Advanced Studies, University of Frankfurt,
Ruth-Moufang-Str. 1, 60438 Frankfurt am Main, Germany

Institute of Space Sciences (ICE, CSIC), Campus UAB,
Carrer de Can Magrans, 08193 Barcelona, Spain

Institut d' Estudis Espacials de Catalunya (IEEC), 08034 Barcelona, Spain

D. Elia et al. (eds.), *The XVIII International Conference on Strangeness in Quark Matter (SQM 2019)*, Springer Proceedings in Physics 250,
https://doi.org/10.1007/978-3-030-53448-6_65

65.2 Theoretical Approaches to YN and YY Interactions

Despite the scarce experimental information on YN and YY interactions, there has been an important effort from the theoretical point of view to describe the YN and YY interactions. Those include meson-exchange models, χEFT approaches, calculations on the lattice, low-momentum models and quark model approaches.

Meson-exchange models rely on the fact that the interaction between two baryons is given by the exchange of mesons. In particular, the determination of the YN and YY interactions is based on the nucleon-nucleon (NN) meson-exchange model under the assumption of $SU(3)_{\text{flavor}}$ symmetry. Among those models, one should mention the Jülich [1, 2] and Nijmegen potentials (see [3] and references therein).

The YN and YY *models based on* χ*EFT* have been also built recently by constructing a systematic approach that respects chiral symmetry. More precisely, the Jülich-Bonn-Munich group has constructed the YN and YY interactions within χEFT starting from their previous approach for the NN interaction [4–6]. In the next section we discuss this approach in more detail.

Another way of constructing the YN and YY interactions is based on solving QCD on the lattice. Within *lattice QCD*, Monte Carlo methods are used to solve the QCD path integral over the quark and gluon fields at each point of a four-dimensional space-time grid. The effort in this direction has been lead by the HALQCD (see contribution to these proceedings) and the NPLQCD [7] collaborations.

Other approaches include *low-momentum interactions* and *quark model potentials*. The former aims at obtaining a universal effective low-momentum potential for YN and YY using renormalization-group techniques [8], whereas the latter builds the YN and YY interactions within constituent quark models [9].

I refer the reader to the recent review of [10] and references therein.

65.3 Hyperons in Dense Matter Within χEFT

The χEFT has been used to described NN interaction to a high precision. However, only during the past decade the YN interaction has been object of analysis within the χEFT. The YN interaction has been obtained within the SU(3) χEFT deriving the different orders in the chiral expansion and improving calculations in a systematic way by going to higher orders in the Weinberg power counting. At leading order (LO) in the power counting, the YN potential consists of single pseudoscalar-meson exchanges and non-derivative four-baryon contact terms [4], whereas the next-to-leading order (NLO) accounts for two pseudoscalar-meson exchanges and contact interactions with two derivatives [5, 11].

The solution of a regularized Lippmann-Schwinger equation using, as kernel, the YN LO and NLO contributions allows for the calculation of scattering observables. In [5] the NLO13 interaction was determined by fixing the baryon-baryon-meson couplings constants using the available standard YN data points and the SU(3) symmetry. This symmetry has also helped to derive the various low-energy constants, although this symmetry is broken by using the hadron physical masses. As a result, the available ΛN and ΣN data is described consistently. However, the simultaneous determination of NN and YN interactions with contact terms completely respecting SU(3) symmetry was not possible [5].

The properties of Λ and Σ in dense matter have been later on analyzed using the LO and NLO YN interactions [12]. Within the Brueckner-Hartree-Fock framework, the single-particle potentials of the Λ and Σ hyperons in nuclear matter have been obtained. The Λ single-particle potential has been found to be in good qualitative agreement with the empirical values extracted from hypernuclear data, whereas the Σ-nuclear potential has been determined to be repulsive. These results have been improved in the subsequent analysis of [13]. Whereas the Σ-nuclear potential becomes moderately repulsive, the Λ one is repulsive starting at two-to-three times saturation density.

The effect of the three-body forces has also been studied for the Λ-nuclear interaction in dense matter [6]. The inclusion of three-body forces is needed in order to obtain, for example, the nuclear saturation in non-relativistic approaches, such as the Brueckner-Hartree-Fock. The Λ single-particle potential turns out to be more repulsive when three-body forces are considered [6].

More recently, there has been a reanalysis of the work in [5] regarding the ΛN and ΣN interactions. In this recent work [11] the number of constants is reduced by inferring some of them from the NN sector using SU(3) symmetry, leading to the NLO19 interaction. Whereas NLO13 of [5] and NLO19 of [11] yield equivalent results for ΛN and ΣN scattering observables, the in-medium Λ and Σ properties in matter are affected by the different choice. This is due to the fact that the strength of $\Lambda N \rightarrow \Sigma N$ transition potential changes. The Λ single-particle potential from the new NLO19 is much more attractive than the NLO13 one, whereas the NLO19 interaction provides slightly more repulsion for the Σ single-particle potential.

With regards to ΞN interaction and the Ξ-hyperon in nuclear matter, the calculation within χEFT up to NLO [14] shows that the $\Lambda\Lambda$ s-wave scattering length and upper bounds of the $\Xi^- p$ cross sections are compatible with the obtained ΞN interaction. The Ξ single-particle potential in nuclear matter ranges between -3 to -5 MeV, similar to other Brueckner-Hartree-Fock calculations, whereas the reported experimental value is larger.

65.4 Hyperons in Neutron Stars

The knowledge of the properties of hyperons in dense matter is of crucial importance in the context of neutron stars. Neutron stars are the most compact known objects without event horizons and, therefore, serve as a unique laboratory for dense matter physics [15]. Their bulk features, such as mass and radius, strongly depend on the properties of matter in their interior and, hence, on the equation of state (EoS). The most precise measurements of masses are located around the Hulse-Taylor pulsar. Accurate values of approximately $2M_\odot$ have been determined recently [16–18]. As for radii, precise determinations do not yet exist, being the simultaneous determination of both mass and radius the focus of the ongoing NICER (Neutron star Interior Composition ExploreR) [19] and the future eXTP (enhanced X-ray Timing and Polarimetry) missions [20].

The composition of the neutron star interior is determined by demanding equilibrium against weak interaction processes, the so-called β-stability. Traditionally the interior of neutron stars was modelled by a uniform fluid of neutron rich matter in β-equilibrium, where the main ingredients are neutron, protons and electrons. However, other degrees of freedom are expected, such as hyperons, because of the fast increase of the nucleon chemical potential with density as well as the fact that the density in the center of neutron stars is very high.

The presence of hyperons affects the EoS of the neutron star interior. The EoS and, hence, the pressure becomes softer with respect to the case when only neutrons and protons are present. The addition of one particle specie opens a set of new available low-energy states that can be filled, hence lowering the total energy of the system. Neutron stars are in hydrostatic equilibrium, that is, there exists a balance between the gravitational force and the internal pressure. Thus, the less pressure, the less mass a neutron star can sustain.

As indicated before, hyperons are energetically favoured in neutron stars. However, their presence induces a softening of the EoS. A softening of the EoS could then lead to masses for neutron stars below the $2M_\odot$ observations. In the literature this effect is usually referred as *the hyperon puzzle*. Over the years, several solutions have been put forward so as to have hyperonic $2M_\odot$ neutron stars. One possible solution is based on stiff YN and YY interactions. In this manner, a stiff EoS will be produced that overcomes the softening induced by hyperons, thus leading to $2M_\odot$ neutron stars. Another solution relies on the stiffening induced by hyperonic three-body forces. Nevertheless, there is not a general consensus whether models with hyperonic three-body forces will allow for $2M_\odot$ neutron stars. Also other solutions have been considered, such as those assuming new degrees of freedom (Δ baryons or condensates), or the presence of a transition to quark matter below the hyperon onset with strong quark interactions to still reach $2M_\odot$.

I refer the reader to the recent reviews of [10, 21] and references therein.

Acknowledgments This research is supported by the Heisenberg Programme of the Deutsche Forschungsgemeinschaft under the Project Nr. 383452331 (Heisenberg Programme) and Nr. 411563442; the Spanish Ministerio de Economía, Industria y Competitividad under contract FPA2016-81114-P; the EU STRONG-2020 project under the program H2020-INFRAIA-2018-1, grant agreement no. 824093; and the PHAROS COST Action CA16214.

References

1. B. Holzenkamp, K. Holinde, J. Speth, A meson exchange model for the hyperon nucleon interaction. Nucl. Phys. A **500**, 485–528 (1989)
2. J. Haidenbauer, U.-G. Meißner, The Julich hyperon-nucleon model revisited. Phys. Rev. **C72**, 044005 (2005)
3. T.A. Rijken, M.M. Nagels, Y. Yamamoto, Baryon-baryon interactions: nijmegen extended-soft-core models. Prog. Theor. Phys. Suppl. **185**, 14–71 (2010)
4. H. Polinder, J. Haidenbauer, U.-G. Meißner, Hyperon-nucleon interactions: a chiral effective field theory approach. Nucl. Phys. **A779**, 244–266 (2006)
5. J. Haidenbauer, S. Petschauer, N. Kaiser, U.-G. Meißner, A. Nogga, W. Weise, Hyperon-nucleon interaction at next-to-leading order in chiral effective field theory. Nucl. Phys. **A915**, 24–58 (2013)
6. J. Haidenbauer, U.-G. Meißner, N. Kaiser, W. Weise, Lambda-nuclear interactions and hyperon puzzle in neutron stars. Eur. Phys. J. **A53**, 121 (2017)
7. https://www.ub.edu/nplqcd/
8. B.J. Schaefer, M. Wagner, J. Wambach, T.T.S. Kuo, G.E. Brown, Low-momentum hyperon-nucleon interactions. Phys. Rev. C **73**, 011001 (2006)
9. Y. Fujiwara, Y. Suzuki, C. Nakamoto, Baryon-baryon interactions in the SU(6) quark model and their applications to light nuclear systems. Prog. Part. Nucl. Phys. **58**, 439–520 (2007)
10. L. Tolos, L. Fabbietti, Strangeness in nuclei and neutron stars. Prog. Part. Nucl. Phys. **112**, 103770 (2020)
11. J. Haidenbauer, U.-G. Meißner, A. Nogga, Hyperon-nucleon interaction within chiral effective field theory revisited, arXiv:1906.11681 [nucl-th]
12. J. Haidenbauer, U.-G. Meißner, A study of hyperons in nuclear matter based on chiral effective field theory. Nucl. Phys. **A936**, 29–44 (2015)
13. S. Petschauer, J. Haidenbauer, N. Kaiser, U.-G. Meißner, W. Weise, Hyperons in nuclear matter from SU(3) chiral effective field theory. Eur. Phys. J. **A52**, 15 (2016)
14. J. Haidenbauer, U.-G. Meißner, In-medium properties of a ΞN interaction derived from chiral effective field theory. Eur. Phys. J. **A55**, 23 (2019)
15. A.W. Watts et al., Colloquium : Measuring the neutron star equation of state using x-ray timing. Rev. Mod. Phys. **88**, 021001 (2016)
16. P. Demorest, T. Pennucci, S. Ransom, M. Roberts, J. Hessels, Shapiro delay measurement of a two solar mass neutron star. Nature **467**, 1081–1083 (2010)
17. J. Antoniadis et al., A massive pulsar in a compact relativistic binary. Science **340**, 6131 (2013)
18. H.T. Cromartie et al., Relativistic Shapiro delay measurements of an extremely massive millisecond pulsar. Nat. Astron. (2019). https://doi.org/10.1038/s41550-019-0880-2
19. https://www.nasa.gov/nicer
20. A.L. Watts et al., Dense matter with eXTP. Sci. China Phys. Mech. Astron. **62**, 29503 (2019)
21. I. Vidana, Hyperons: the strange ingredients of the nuclear equation of state. Proc. Roy. Soc. Lond. **A474**, 0145 (2018)

Part IX
Hydrodynamics, Chirality and Vorticity

Chapter 66
Spin Alignment Measurements of Vector Mesons in ALICE at the LHC

Sourav Kundu

Abstract Spin alignment of K^{*0} and ϕ vector mesons produced in non-central heavy-ion collisions could occur due to the presence of large initial angular momentum. This phenomenon leads to a non-uniform angular distribution of the decay daughters of vector mesons with respect to the quantization axis in the rest frame of vector mesons. We present the recent spin alignment measurements of K^{*0} and ϕ vector mesons at mid-rapidity ($|y| < 0.5$) in Pb–Pb collisions at $\sqrt{s_{NN}} = 2.76$ TeV with respect to the production plane and the second-order event plane. The measured value of the second diagonal spin density matrix element (ρ_{00}) is lower than 1/3 for both the vector mesons at low transverse momentum (p_T) and consistent with 1/3 (a value implying no spin alignment) at high p_T in mid-central Pb-Pb collisions. The results are also compared with the spin alignment of vector mesons in pp collisions at $\sqrt{s} = 13$ TeV and spin alignment of K^0_S (spin zero hadron) in Pb–Pb collisions at $\sqrt{s_{NN}} = 2.76$ TeV. In both cases the extracted ρ_{00} values are consistent with 1/3 within the uncertainty.

66.1 Introduction

A large initial angular momentum is expected to be created in non-central heavy-ion collisions. Its magnitude is predicted to be $\sim 10^5\hbar$ [1]. In presence of large initial angular momentum quarks could be polarized due to the spin-orbital interaction of quantum chromodynamics and quark polarization may lead to the polarization of vector mesons (spin = 1) [2]. Polarization/spin alignment of vector mesons is also sensitive to the different hadronization mechanisms [3]. Spin alignment of vector mesons can be studied by measuring the angular distributions of the decay daughters of vector mesons with respect to the quantization axis. The quantization axis can be

Sourav Kundu for the ALICE collaboration.

S. Kundu (✉)
National Institute of Science Education and Research, HBNI, Jatni 752050, India
e-mail: sourav.kundu@cern.ch

D. Elia et al. (eds.), *The XVIII International Conference on Strangeness in Quark Matter (SQM 2019)*, Springer Proceedings in Physics 250,
https://doi.org/10.1007/978-3-030-53448-6_66

taken either as the normal to the production plane or as the normal to the reaction plane. The production plane is defined by the momentum of the vector meson and the beam axis, whereas the reaction plane is subtended by the impact parameter and the beam axis. In the experiment the event plane is used as a proxy for the reaction plane as the impact parameter direction is not directly measured. Therefore the results with respect to the event plane are further corrected for the event-plane resolution. The angular distribution is expressed as [4],

$$\frac{\mathrm{d}N}{\mathrm{d}\cos\theta^*} \propto [1-\rho_{00}+\cos^2\theta^*(3\rho_{00}-1)]. \tag{66.1}$$

θ^* is the angle between the decay daughter of vector meson and the quantization axis in the rest frame of the vector meson. N_0 is a normalization constant. The second diagonal element of the spin density matrix is ρ_{00} which corresponds to the probability of finding a vector meson in spin state 0 out of the 3 spin states −1, 0 and 1. All spin states are equally probable in the absence of spin alignment, which makes the angular distribution flat. However, any deviation of ρ_{00} from 1/3 leads to a non-uniform angular distribution. This is the experimental signature of the spin alignment of vector mesons.

In this work we present the p_T and centrality dependence of ρ_{00} for K^{*0} and ϕ vector mesons in Pb–Pb collisions at $\sqrt{s_{NN}}$ = 2.76 TeV. The results for vector mesons in pp collisions at $\sqrt{s}$ = 13 TeV and for K^0_S mesons (spin zero) in Pb–Pb collisions at $\sqrt{s_{NN}}$ = 2.76 TeV are also shown, for which we do not expect any spin alignment.

66.2 Analysis Details

Spin alignment measurements of vector mesons are carried out by analyzing 14 million events in Pb–Pb collisions at $\sqrt{s_{NN}}$ = 2.76 TeV and 43 million events in pp collisions at $\sqrt{s}$ = 13 TeV. Measurements are performed at midrapidity ($|y| < 0.5$) in different p_T regions and centrality classes. The V0 detector [5, 6] is used for trigger logic, centrality determination and 2nd order event plane estimation. K^{*0} and ϕ vector mesons are reconstructed in each event from the invariant-mass distribution of oppositely charged $K\pi$ and KK pairs, respectively, as discussed in [6, 7]. The invariant-mass distribution of unlike charged pairs contains a large combinatorial background along with the signal. The combinatorial background is estimated by using the mixed-event technique [6]. K^0_S mesons are reconstructed from the invariant-mass distribution of oppositely charged $\pi\pi$ pairs, which are selected by using their distinctive V-shaped decay topology [8]. Charged π and K mesons are identified from the specific energy loss measured in the Time Projection Chamber (TPC) and the velocity measured by the Time-Of-Flight (TOF) detector. K^{*0}, ϕ and K^0_S signals are extracted for various p_T and $\cos\theta^*$ bins in different centrality classes. The raw yields are corrected for detector acceptance × efficiency, determined by using a dedicated Monte Carlo simulation of the ALICE detector response. Figure 66.1 shows

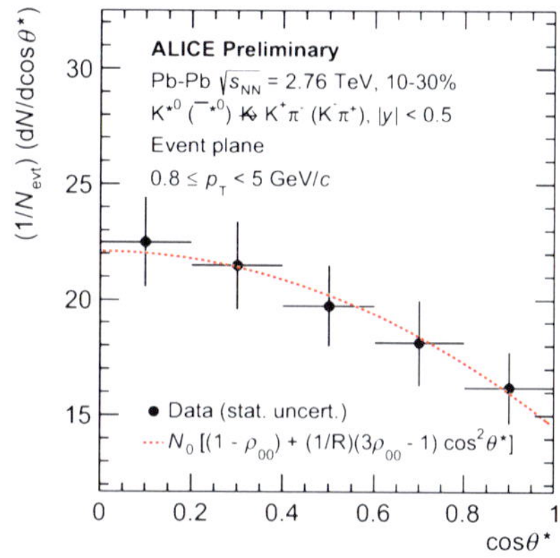

Fig. 66.1 $dN/d\cos\theta^*$ distribution at mid-rapidity in 10–30% Pb–Pb collisions at $\sqrt{s_{NN}} = 2.76$ TeV using the event plane

corrected $\cos\theta^*$ distribution for K^{*0} at mid-rapidity in 10–30% Pb–Pb collisions at $\sqrt{s_{NN}} = 2.76$ TeV for $0.8 \le p_T < 5.0$ GeV/c using the event plane. The efficiency and acceptance corrected $\cos\theta^*$ distributions [7] are fitted with the function defined in (66.1) to extract ρ_{00} values for each studied p_T range and centrality class.

66.3 Results

Figure 66.2 shows the extracted ρ_{00} values for K^{*0} as a function of p_T in 10–50% Pb–Pb collisions at $\sqrt{s_{NN}} = 2.76$ TeV with respect to both the production plane and the event plane. The ρ_{00} values for K^{*0} in Pb–Pb collisions show a deviation from 1/3 at $p_T < 2$ GeV/c for both the production and the event-plane measurements. For $p_T > 2$ GeV/c the extracted ρ_{00} values are consistent with 1/3. Measurements for vector mesons in heavy-ion collisions are compared with the results from pp collisions at $\sqrt{s} = 13$ TeV where the measured ρ_{00} values are consistent with 1/3 throughout the studied p_T range. The extracted ρ_{00} values for K^0_S in 20–40% Pb–Pb collisions are also consistent with 1/3. ρ_{00} versus $\langle N_{part} \rangle$ for K^{*0} and ϕ mesons in Pb–Pb collisions at $\sqrt{s_{NN}} = 2.76$ TeV are shown in Fig. 66.3. A clear centrality dependence is observed for the extracted ρ_{00} values of vector mesons in Pb–Pb collisions with the maximum deviation from 1/3 occurring in mid-central collisions. In peripheral and central collisions the measurements are consistent or close to 1/3 within uncertainty. This centrality dependence is similar to the dependency of the angular momentum on the impact parameter [1].

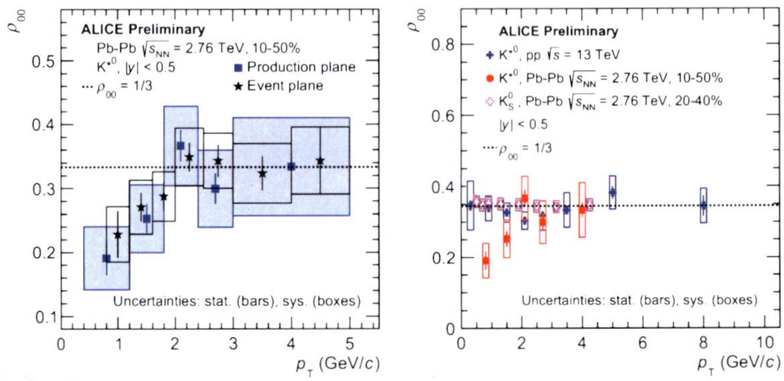

Fig. 66.2 Left panel: ρ_{00} as a function of p_T for K^{*0} with respect to both the production plane and the event plane in 10–50% Pb–Pb collisions at $\sqrt{s_{NN}}$ = 2.76 TeV. Right panel: ρ_{00} versus p_T for K^{*0} in 10–50% Pb–Pb collisions at $\sqrt{s_{NN}}$ = 2.76 TeV, along with the results for K^{*0} in pp collisions at $\sqrt{s}$ = 13 TeV and for K^0_S in 20–40% Pb–Pb collisions at $\sqrt{s_{NN}}$ = 2.76 TeV. Measurements are done at midrapidity with respect to the production plane

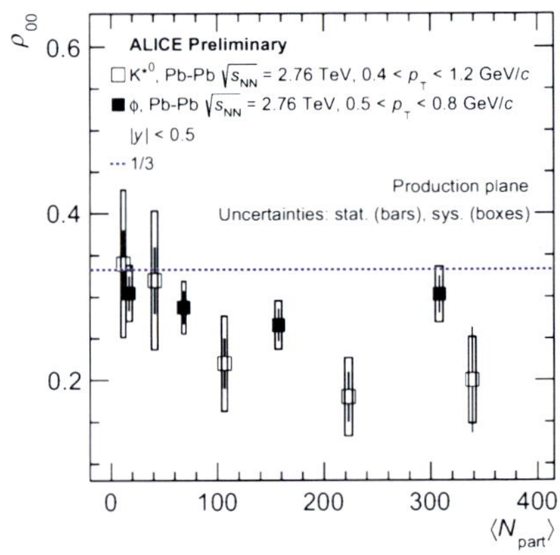

Fig. 66.3 ρ_{00} versus $\langle N_{part} \rangle$ for K^{*0} and ϕ mesons in 10–50% Pb–Pb collisions at $\sqrt{s_{NN}}$ = 2.76 TeV with respect to the production plane

66.4 Summary

We have reported the first observation of spin alignment of vector mesons in high-energy heavy-ion collisions with the ALICE detector at the LHC. The extracted ρ_{00} values for both K^{*0} and ϕ mesons deviate from 1/3 at low p_T. The deviation from 1/3 is largest in mid-central collisions which is expected from the presence of a large initial angular momentum in mid-central collisions [1].

References

1. F. Becattini, F. Piccinini, J. Rizzo, Phys. Rev. C **77**, 024906 (2008)
2. Z.T. Liang, X.N. Wang, Phys. Rev. Lett. **94**, 102301 (2005)
3. Z.T. Liang, X.N. Wang, Phys. Lett. B **629**, 20 (2005)
4. K. Schilling, P. Seyboth, G.E. Wolf, Nucl. Phys. B **15**, 397 (1970)
5. B. Abelev et al., ALICE Collaboration. Int. J. Mod. Phys. A **29**, 1430044 (2014)
6. J. Adam et al., ALICE Collaboration. Phys. Rev. C **95**, 064606 (2017)
7. B. Mohanty [ALICE Collaboration], EPJ. Web. Conf. 171, 16008 (2018)
8. B. Abelev et al., ALICE Collaboration. Phys. Rev. Lett. **111**, 222301 (2013)

Chapter 67
Vorticity and $\Lambda/\bar{\Lambda}$ Polarization in Heavy-Ion Collisions at FAIR and NICA Energies

Oleksandr Vitiuk, Larisa Bravina, Evgeny Zabrodin, Aleksandr Sorin, and Oleg Teryaev

Abstract Vorticity and polarization of Λ and $\bar{\Lambda}$ were calculated within the UrQMD model in non-central Au+Au collisions at energies of FAIR and NICA. To determine temperature and chemical potentials of the areas from where $\Lambda/\bar{\Lambda}$ were emitted, the fit to statistical model of ideal hadron gas was employed. Then the thermal vorticities of the areas and polarization of both hyperons were calculated. It is found that the polarization of $\Lambda/\bar{\Lambda}$ increases with decreasing energy of the collisions. The stronger $\bar{\Lambda}$ polarization compared to that of Λ is explained by different space-time distributions and different freeze-out conditions of both hyperons.

67.1 Introduction

The properties of deconfined partonic matter, the quark-gluon plasma (QGP), have been explored in heavy-ion collisions at the Relativistic Heavy Ion Collider (RHIC) and the Large Hadron Collider (LHC). One of the important characteristics related to collective behavior of the produced substance is the vorticity of nuclear matter. In non-central heavy-ion collisions huge orbital angular momenta of order $10^3 - 10^5\hbar$ are generated. Thus, the matter should exhibit rotational motion to conserve the initial angular momentum carried by two colliding nuclei. The direction of the angular

O. Vitiuk · L. Bravina · E. Zabrodin (✉)
Department of Physics, University of Oslo, Oslo, Norway
e-mail: zabrodin.vitiuk@fys.uio.no

O. Vitiuk
e-mail: oleksandr.vitiuk@fys.uio.no

O. Vitiuk
Taras Shevchenko National University of Kyiv, Kyiv, Ukraine

E. Zabrodin
Skobeltzyn Institute for Nuclear Physics, Moscow State University, Moscow, Russia

A. Sorin · O. Teryaev
Joint Institute for Nuclear Research, Dubna, Russia

D. Elia et al. (eds.), *The XVIII International Conference on Strangeness in Quark Matter (SQM 2019)*, Springer Proceedings in Physics 250,
https://doi.org/10.1007/978-3-030-53448-6_67

momentum is orthogonal to the reaction plane defined by incoming beam and the impact parameter vector. Several theoretical models suggest that the large angular momentum carried by two colliding nuclei [1–3] can be transferred to the created system. As a consequence, the spin of particles composing the system might be globally polarized along the direction of the system angular momentum, due to spin-orbit coupling. Hyperons are natural candidates to explore this phenomenon since in the parity violating weak decays of the hyperons the momentum vector of the decayed baryon is highly correlated with the hyperon spin. Recently STAR measured the global polarization of and in non-central Au+Au collisions in the Beam Energy Scan (BES) program [4]. From the measured polarization, the fluid vorticity of the strongly coupled QGP and the magnitude of the magnetic field created in non-central heavy-ion collisions are extracted. Quite astonishing is the observed difference in polarizations of Λ and $\bar{\Lambda}$ at intermediate and low energies. Therefore, it is worth to study the inherent correlation between the vorticity and global polarization in detail.

67.2 Results

We studied Au+Au collisions within the UrQMD model [5, 6] at $7.7 \leq \sqrt{s} \leq 19.6$ GeV corresponding to energies of BES, FAIR and NICA. At each energy one million collisions with the impact parameter $b = 6$ fm, corresponding to centrality $\sigma/\sigma_{geo} \approx 20\%$, were generated. The freeze-out conditions of both hyperons are studied first. Since there is no sharp freeze-out of hadrons in transport models [7, 8], we listed the mean values of the freeze-out times in Table 67.1. One can see that $\bar{\Lambda}$ are emitted, in average, about 1 fm/c earlier than Λ. But the fireball expands quickly, and its temperature can drop noticeably within this time. To check the temperatures of the areas from where the hyperons were emitted, the whole space was subdivided into small cubic cell with volume $V = 1$ fm^3. After calculation of the total energy density ε, net baryon density $\rho_{\rm B}$, and net strange density $\rho_{\rm S}$ for each cell in its local rest frame, the procedure described in [9, 10] was employed to find the temperature and the chemical potentials in each cell. Figure 67.1a shows that the temperature is not uniformly distributed within the whole volume. It drops from the inner to outer zones, and the spectators are the hottest areas of the fireball. Thus, the temperatures of emitted hyperons depend also on the location of their freeze-out areas.

Under assumption of local thermal equilibrium in the system, its thermal vorticity tensor reads

Table 67.1 Mean freeze-out time of Λ and $\bar{\Lambda}$ hyperons in UrQMD calculations of Au+Au collisions with $b = 6$ fm at $\sqrt{s} = 7.7 - 19.6$ GeV

$\sqrt{s}$ (GeV)	7.7	11.5	14.5	19.6
$\langle t_{\Lambda}^{FO} \rangle$ (fm/c)	21.3009	21.9568	23.066	24.3462
$\langle t_{\bar{\Lambda}}^{FO} \rangle$ (fm/c)	19.7806	21.0302	21.959	23.1288

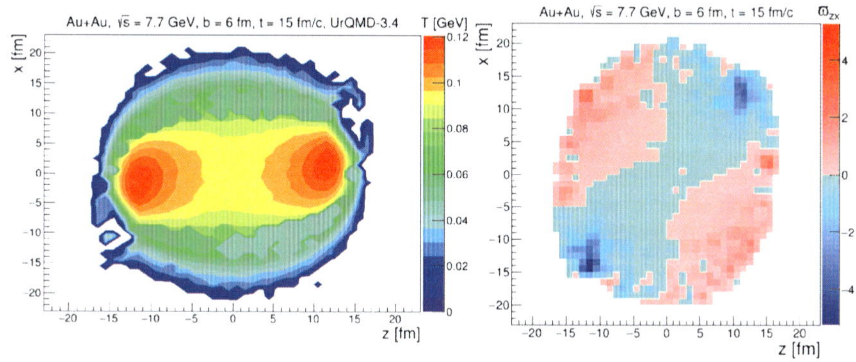

Fig. 67.1 Proper temperature T (left) and thermal vorticity component ϖ_{zx} (right) in the reaction plane of Au+Au collisions with $b = 6$ fm at $\sqrt{s} = 7.7$ GeV at $t = 15$ fm/c

$$\varpi_{\mu\nu} = \frac{1}{2}\left(\partial_\nu \beta_\mu - \partial_\mu \beta_\nu\right) \qquad (67.1)$$

Here $\beta^\mu = u^\mu/T$ is the inverse-temperature four-velocity, T is a proper temperature, and u^μ is hydrodynamic four-velocity, respectively. The reaction-plane component ϖ_{zx} displayed in Fig. 67.1b is the most important for calculation of $\Lambda/\bar{\Lambda}$ polarization because it is parallel to angular momentum of the system. As shown in [11], Λ are located both in the center of the fireball and in the spectator regions, whereas $\bar{\Lambda}$ are distributed mainly in the baryon-less areas. Because of the larger vorticity at early times, earlier frozen hyperons possess larger polarization, see Fig. 67.2a. A bit shorter freeze-out time of $\bar{\Lambda}$ together with its different distribution in space result to stronger $\bar{\Lambda}$ polarization at intermediate energies, as depicted in Fig. 67.2b, in

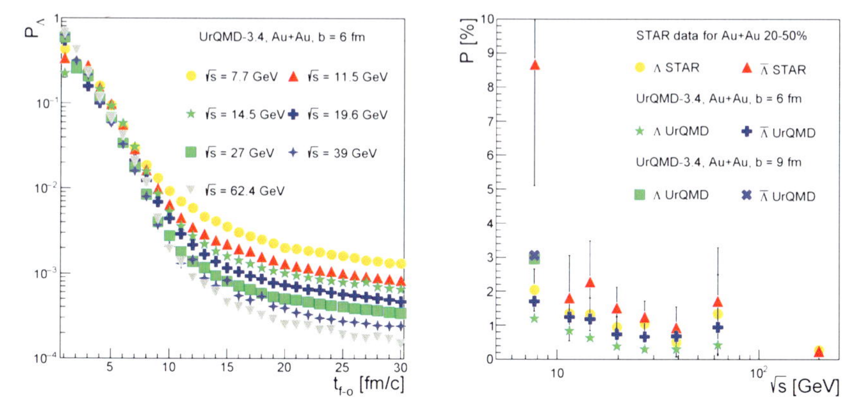

Fig. 67.2 Dependence of Λ polarization on the emission time (left), and (right) global polarization of $\Lambda/\bar{\Lambda}$ as function of $\sqrt{s}$ compared to the data

good agreement with the STAR experimental data. Here we plot also calculations of $\Lambda/\bar{\Lambda}$ polarization in Au+Au collisions at $\sqrt{s} = 7.7$ GeV with $b = 9$ fm, roughly corresponding to centrality 50%. To match the data on $\bar{\Lambda}$ at this energy one can, e.g., slightly increase the yield of early frozen $\bar{\Lambda}$ hyperons.

67.3 Conclusions

Method for calculation of thermal vorticity field and then Λ and $\bar{\Lambda}$ global polarization in transport model is developed. We found that Λ and $\bar{\Lambda}$ have different freeze-out conditions both in space and in time. Therefore, polarization of $\bar{\Lambda}$ is stronger than that of Λ at intermediate energies. The increase of the hyperon polarization with decreasing bombarding energy is reproduced. Strong $\bar{\Lambda}$ polarization at $\sqrt{s} = 7.7$ GeV seen in experiment can be reproduced by moderate rise of early frozen $\bar{\Lambda}$ in the transport model.

Acknowledgments The work of L.B. and E.Z. was supported by Russian Foundation for Basic Research (RFBR) under Grants No. 18-02-40084 and No. 18-02-40085, and by the Norwegian Research Council (NFR) under grant No. 255253/F50. O.V. acknowledges the financial support of the Norwegian Centre for International Cooperation in Education (SIU) under Grant CPEA-LT-2016/10094. Computer calculations were made at Abel (UiO, Oslo) and Govorun (JINR, Dubna) computer cluster facilities.

References

1. Z. Liang, X.N. Wang, Globally polarized quark-gluon plasma in non-central A+A collisions. Phys. Rev. Lett. **94**, 102301 (2005). https://doi.org/10.1103/PhysRevLett.94.102301
2. F. Becattini, F. Piccinini, J. Rizzo, Angular momentum conservation in heavy ion collisions at very high energy. Phys. Rev. **C77**, 024906 (2008). https://doi.org/10.1103/PhysRevC.77.024906
3. A. Sorin, O. Teryaev, Axial anomaly and energy dependence of hyperon polarization in heavy-ion collisions. Phys. Rev. **C95**, 011902(R) (2017). https://doi.org/10.1103/PhysRevC.95.011902
4. STAR Collaboration: Global Λ hyperon polarization in nuclear collisions: evidence for the most vortical fluid. Nature **548**, 62 (2017). https://doi.org/10.1038/nature23004
5. S.A. Bass et al.: Microscopic models for ultrarelativistic heavy ion collisions. Prog. Part. Nucl. Phys. **41**, 255 (1998). https://doi.org/10.1016/S0146-6410(98)00058-1
6. M. Bleicher et al., Relativistic hadron-hadron collisions in the ultra-relativistic quantum molecular dynamics model. J. Phys. **G25**, 1859 (1999). https://doi.org/10.1088/0954-3899/25/9/308
7. L.V. Bravina et al., Microscopic study of freezeout in relativistic heavy ion collisions at SPS energies. Phys. Rev. **C60**, 044905 (1999). https://doi.org/10.1103/PhysRevC.60.044905
8. L. Bravina et al., The spatial sub-separation of strangeness from anti-strangeness in heavy-ion collisions at energies of FAIR and NICA. *These Proceedings*
9. L.V. Bravina et al., Local equilibrium in heavy ion collisions: Microscopic model versus statistical model analysis. Phys. Rev. **C60**, 024904 (1999). https://doi.org/10.1103/PhysRevC.60.024904

10. L.V. Bravina et al., Microscopic models and effective equation of state in nuclear collisions in the vicinity of $E_{lab} = 30A$ GeV at the GSI Facility for Antiproton and Ion Research (FAIR) and beyond. Phys. Rev. **C78**, 014907 (2008). https://doi.org/10.1103/PhysRevC.78.014907
11. O. Vitiuk, L. Bravina, E. Zabrodin, Different space-time freeze-out picture - an explanation of different Λ and $\bar{\Lambda}$ polarization? eprint: arXiv:1910.06292 [hep-ph]

Chapter 68
Λ Polarization in Au+Au Collisions at $\sqrt{s}_{NN} = 2.4$ GeV Measured with HADES

Frédéric Julian Kornas

Abstract In April 2012, a large statistic sample of Au+Au collisions has been recorded by the High Acceptance Di-Electron Spectrometer (HADES). We describe the strategy to measure a possible global polarization through weakly decaying particles. The Λ polarization extracted so far is at most at a few percent following the trend of the STAR measurements. However, a significant background correlation is observed and requires detailed investigations.

68.1 Introduction

The system created in peripheral heavy-ion collision might obtain a large orbital momentum from the colliding nuclei. This may result in a global spin polarization of the particles produced in such a collision [1–4]. The polarization measurement has been performed for the strange hyperons Λ ($\bar{\Lambda}$. Due to the parity violation in the weak decay, the proton (antiproton) is preferably emitted in the spin direction of the mother particle. This turns the spin measurement into a momentum measurement which makes it experimentally accessible.

A non-zero polarization for Λ and $\bar{\Lambda}$ has been observed by the STAR collaboration at the RHIC beam energy scan (BES) [5]. In Au+Au collisions at center-of-mass energies ranging from $\sqrt{s}_{NN} = 7.7 - 39$GeV, it has been found the polarization to increase up to a few percent with decreasing beam energy. With a large data sample, the analysis was also performed at .., resulting in a non-zero polarization [6]. Recently, the ALICE collaboration measured the Λ and $\bar{\Lambda}$ polarization in Pb+Pb collisions at $\sqrt{s}_{NN} = 2.76$TeV and $\sqrt{s}_{NN} = 5.02$TeV [7]. They found both to be consistent with zero.

Frédéric Julian Kornas for the HADES Collaboration.

F. J. Kornas (✉)
Technische Universität Darmstadt, Karolinenplatz 5, 64289 Darmstadt, Germany
e-mail: f.kornas@gsi.de

D. Elia et al. (eds.), *The XVIII International Conference on Strangeness in Quark Matter (SQM 2019)*, Springer Proceedings in Physics 250,
https://doi.org/10.1007/978-3-030-53448-6_68

In summary, the polarization trends towards zero for high energies, while it increases towards lower beam energies. Hence, a measurement at lower energies might be helpful to learn about initial conditions and dynamics in the baryon dominated regime.

68.2 Data Analysis

The data presented in this paper has been recorded in April 2012 by the High Acceptance Di-Electron Spectrometer (HADES). A detailed description of the detector can be found in [8]. During this beamtime, seven billion events have been recorded. The Λ hyperon can be detected by reconstructing its weak decay into charged particles, namely $\Lambda \rightarrow \pi^- + p$. Out of all reconstructed tracks, a first set of protons and pions was pre-selected by a loose mass cut. Due to the flight length of the Λ of a few centimeters, the identification is mainly performed by its decay topology. If both, pion and proton originate from a Λ, the distance of closest approach of the two daughters should be small, while their combined vertex is more likely to be a few centimeters away from the global event vertex. Both daughter tracks are not pointing to the event vertex, while the combined mother track should. This set of topological parameters is a good tool, to remove most of the combinatorial background, while keeping most of the Λ s. Applying topological cuts allows one to reconstruct Λ signal, however the statistics and signal-to-background ratio is not satisfactory to perform the Λ polarization analysis.

For this purpose, an artificial neural network is used in the analysis. It allows to substantially suppress the combinatorial background while increasing the reconstruction efficiency significantly. Being among the best setups, the Multi-Layer Perceptron (MLP) included in the TMVA package is used [9]. As input parameters, the above described topological parameters are used together with the particles masses and the momentum of the Λ. The network has been trained on reconstructed particles from Monte-Carlo simulations for the signal, while the background generated from data is described by the mixed-event method. To account for the efficiency loss of close pairs in the same-event data, a 15° opening angle cut was necessary [10]. To avoid training on primary particles, loose cuts on the topological parameters are necessary before training. For each set of parameters, the network returns one single discriminant, which can be interpreted as the probability of one pair to originate from a Λ decay. It is chosen to maximize the significance and an overall number of $2.1 \cdot 10^5$ Λ s are available for the analysis of 10-40% centrality class as can be seen in Fig. 68.1.

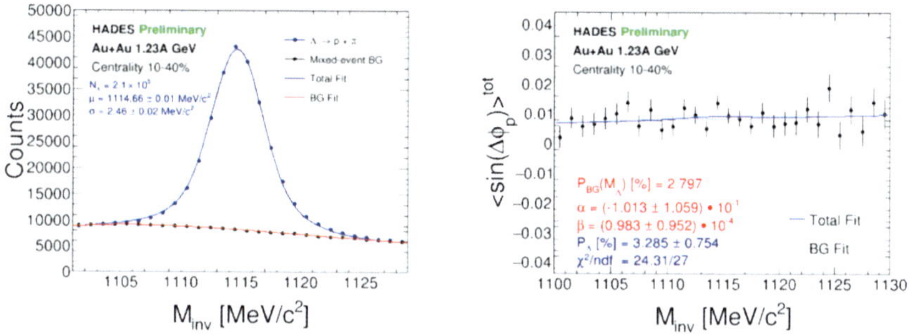

Fig. 68.1 Left panel: Blue: Invariant mass distribution of reconstructed $p\pi^-$-pairs. Black: Mixed-event background. The background has been fitted with a Landau function while for the signal a double Gaussian is used. Right panel: Extraction of the polarization with the Invariant Mass Method. The background shape is estimated by a linear function $\alpha + \beta M_{inv}$

68.3 Polarization Measurement of the Λ

As pointed out in the introduction, in the decay $\Lambda \rightarrow p + \pi^-$ the proton will mainly be emitted in the direction of the spin of the Λ. If the Λ has a polarization P_Λ, this means that in the rest frame of the Λ, the proton angular distribution will be modified to

$$\frac{dN}{d\cos\theta^*} = \frac{1}{4\pi}\left(1 + \alpha_\Lambda P_\Lambda \cos\theta^*\right), \qquad (68.1)$$

where θ^* is the angle between the polarization vector and the proton direction and $\alpha_\Lambda = 0.75 \pm 0.01$ is the decay parameter [11]. Note that in comparison to previous measurements [5, 6, 9] the value is about 15% higher (old value: $\alpha_\Lambda = 0.642 \pm 0.013$). The polarization vector is expected to be along the global angular momentum which is perpendicular to the reaction plane, spanned by the beam axis and the impact parameter of the two colliding nuclei. The true reaction plane (RP) can be estimated by the event plane (EP), reconstructed from the spectator hits collected far behind the target. Dividing all hits randomly in two subsamples and performing the reconstruction separately results in an estimate for the EP resolution R_{EP} [12]. Then (68.1) can be transformed to the azimuthal angles [13] resulting in

$$P_\Lambda = \frac{8}{\pi\alpha_\Lambda}\frac{\sin\left(\Psi_{EP} - \phi_p^*\right)}{R_{EP}}, \qquad (68.2)$$

where ϕ_p^* is the azimuthal angle of the proton in the Λ rest frame, and ... denote the average over all particles in all events. Thus, we can calculate a possible polarization from our Λ sample.

Two different methods have been used to perform polarization analysis [14]. For the Invariant Mass Method (IMM), the second factor in (68.2) is calculate as a function of the invariant mass M_{inv}. From the number distribution, for each M_{inv} the exact signal and background fraction is known, thus on can get the Λ polarization but fitting

$$\sin\left(\Psi_{EP} - \phi_p^*\right)_{tot} = \frac{S}{S+B}\sin\left(\Psi_{EP} - \phi_p^*\right)_{SG} + \frac{B}{S+B}\sin\left(\Psi_{EP} - \phi_p^*\right)_{BG}. \tag{68.3}$$

However, one has to assume the shape of the background correlation (as shown in the right panel of Fig. 68.1).

The second method is the Event Plane Method (EPM). Here the distribution of $\Psi_{EP} - \phi_p^*$ is divided into certain number of bins. For each bin, one measures the invariant mass distribution to get the net amount of Λ s in each bin (compare to Fig. 68.1). This second decomposition needs a lot of státistics, but therefore all background correlations are removed properly. Then the distribution of $N_\Lambda\left(\Psi_{EP} - \phi_p^*\right)$ is decomposed into the Fourier components, while the first sinus-term corresponds to the polarization (simply by multiplication with the prefactors according to (68.2)).

Both methods have been applied and are consistent with each other. However, a significant background correlation is observed and currently under investigation.

68.4 Summary and Outlook

The data collected in April 2012 with Au+Au collisions allows to reconstruct a high statistic data sample from the decay channel $\Lambda \rightarrow p + \pi^-$. Making use of an artificial neural network results in a significant increase in the efficiency of Λ reconstruction. Using this data sample, the first measurement of the Λ polarization at collision energies of $\sqrt{s_{NN}} = 2.4$GeV has been performed with two different methods, which are consistent with each other. A significant background correlation is observed which is currently under investigation. The event generator which allows to simulate background with realistic hadron multiplicities and kinematics as well as including Λ polarization signal is being developed. In addition, v_1 and v_2 of the Λ is measured and will be investigated further.

Acknowledgments The author gratefully acknowledges the support by VH-NG-823 and STIFTUNG GIERSCH.

References

1. Z.-T. Liang, X.-N. Wang, Globally polarized quark-gluon plasma in non-central A+A collisions. Phys. Rev. Lett. **94**, 102301 (2005)
2. S.A. Voloshin, Polarized secondary particles in unpolarized high energy hadron-hadron collisions? arXiv:nucl-th/0410089 (2005)
3. Z.-T. Liang, X.-N. Wang, Spin alignment of vector mesons in non-central A+A collisions. Phys. Lett. B **629**, 20–26 (2005)
4. J.-H. Gao et al., Global quark polarization in non-central A+A collisions. Phys. Rev. C **77**, 044902 (2008)
5. Adamczyk et al., STAR Collaboration, "Global Λ hyperon polarization in nuclear collisions: evidence for the most vortical fluid". Nature **548**, 62–65 (2017)
6. J. Adam et al., STAR Collaboration, "Global polarization of Λ hyperons in Au+Au collisions at $\sqrt{s_{NN}} = 200$GeV", Phys. Rev. **C98**, 014910 (2018)
7. S. Acharya et al., ALICE Collaboration, "Global polarization of Λ and $\bar{\Lambda}$ hyperons in Pb-Pb collisions at the LHC", arXiv:1909.01281 [nucl-ex]
8. G. Agakishiev et al., The high-acceptance dielectron spectrometer HADES. EPJA **41**, 243 (2007)
9. A. Hoecker et al., TMVA – Toolkit for Multivariate Data Analysis. PoS ACAT, 040 (2007)
10. T. Scheib, Λ and K^0_S Production in Au+Au Collisions at 1.23A GeV. PhD Thesis Goethe-University Frankfurt, (2017)
11. M. Ablikim et al., BESIII Collaboration, "Polarization and Entanglement in Baryon-Antibaryon Pair Production in Electron-Positron Annihilation". arXiv:1808.08917 (2018)
12. A.M. Poskanzer, S.A. Voloshin, Methods for analyzing anisotropic flow in relativistic nuclear collisions. Phys. Rev. C **58**, 1671 (1998)
13. B.I. Abelev et al., STAR Collaboration, "Global polarization measurement in Au+Au collisions". Phys. Rev. C **76**, 024915 (2007)
14. L. Adamczyk et al., Elliptic flow of identified hadrons in Au+Au collisions at $\sqrt{S_{NN}} = 7.7 - 62.4$GeV. Phys. Rev. **C88**, 014902 (2013)

Chapter 69
Relativistic Dissipative Hydrodynamics: Effective Fugacity Quasiparticle Description

Samapan Bhadury, Manu Kurian, Vinod Chandra, and Amaresh Jaiswal

Abstract The relativistic first order hydrodynamic evolution equation for the shear stress tensor, the bulk viscous pressure and the charge current have been investigated by employing a effective covariant kinetic theory. We have studied the mean field contribution to the dissipative evolution equation of the quark-gluon plasma at non-vanishing baryon chemical potential and finite quark mass. As an implication, dependence of hot QCD equation of state and the viscous corrections on the temperature evolution and pressure anisotropy have been investigated for one-dimensional boost invariant Bjorken expansion.

69.1 Introduction

The high energetic heavy-ion collisions in Relativistic Heavy Ion Collider and Large Hadron Collider have experimentally verified the existence of strongly coupled QCD matter: Quark-Gluon Plasma (QGP). The relativistic dissipative hydrodynamics serves as an efficient theoretical approach to describe the space-time evolution of the created QGP. In this contribution, we discuss the first order dissipative evolution equation of the QGP with finite baryon chemical potential and quark mass within a recently proposed effective covariant kinetic theory [1]. The effective fugacity quasiparticle model (EQPM) [2] have been utilized in the analysis to encode the hot QCD equation of state (EoS) in terms of temperature dependent fugacity parameter. The relativistic Boltzmann equation with EQPM mean field term has been solved within relaxation time approximation (RTA) [3] by employing the iterative Chapman-Enskog like expansion. We study the mean field correction to the dissipative quan-

S. Bhadury (✉) · A. Jaiswal
School of Physical Sciences, National Institute of Science Education and Research, HBNI, Jatni 752050, India
e-mail: samapan.bhadury@niser.ac.in

M. Kurian · V. Chandra
Indian Institute of Technology Gandhinagar, Gandhinagar 382355, Gujarat, India

D. Elia et al. (eds.), *The XVIII International Conference on Strangeness in Quark Matter (SQM 2019)*, Springer Proceedings in Physics 250,
https://doi.org/10.1007/978-3-030-53448-6_69

tities at finite baryon chemical potential. We investigate the viscous corrections to the temperature evolution and pressure anisotropy for the boost invariant Bjorken expansion.

69.2 Effective Covariant Kinetic Theory

The relativistic Boltzmann equation quantifies the change of momentum distribution of the particles in the system and takes the following form within RTA for the particle species k [1],

$$\tilde{p}_k^{\mu}\,\partial_{\mu} f_k(x,\tilde{p}_k) + F_k^{\mu}\,(u\cdot\tilde{p}_k)\,\partial_{\mu}^{(p)} f_k = -\,(u\cdot\tilde{p}_k)\,\frac{\delta f_k}{\tau_R}, \tag{69.1}$$

where τ_R is the thermal relaxation time and u_{μ} is the fluid velocity. The term $F_k^{\mu} = -\partial_{\nu}(\delta\omega_k u^{\nu} u^{\mu})$ describes the mean field force term that can be realized from the conservation laws [1]. The covariant form of EQPM distribution function for quarks, antiquarks and gluons at non-zero baryon chemical potential μ_q can be written as,

$$f_q^0 = \frac{z_q \exp[-\beta(u\cdot p_q - \mu_q)]}{1 + z_q \exp[-\beta(u\cdot p_q - \mu_q)]}, \tag{69.2}$$

$$f_{\bar{q}}^0 = \frac{z_{\bar{q}} \exp[-\beta(u\cdot p_{\bar{q}} + \mu_q)]}{1 + z_{\bar{q}} \exp[-\beta(u\cdot p_{\bar{q}} + \mu_q)]}, \tag{69.3}$$

$$f_g^0 = \frac{z_g \exp[-\beta\, u\cdot p_g]}{1 - z_g \exp[-\beta\, u\cdot p_g]}, \tag{69.4}$$

where z_q and z_g are the temperature dependent effective fugacity parameter for quarks and gluons, respectively. Note that the fugacity parameter z_k encodes the thermal medium effects and is same for quarks and antiquarks, i.e. $z_q = z_{\bar{q}}$, in the EQPM description of the QGP. The physical picture of the effective fugacity can be understood from the dispersion relation that relates the dressed (quasiparticle) four-momenta $\tilde{p}_k^{\mu}$ and the bare particle four-momenta p_k^{μ} as,

$$\tilde{p}_k{}^{\mu} = p_k^{\mu} + \delta\omega_k\, u^{\mu}, \qquad \delta\omega_k = T^2\,\partial_T\, \ln(z_k), \tag{69.5}$$

in which modified zeroth component of the four-momenta defined as $\tilde{p}_k{}^0 \equiv \omega_k = E_k + \delta\omega_k$. We solve the relativistic Boltzmann equation by employing an iterative Chapman-Enskog like expansion by assuming the system is near local equilibrium, i.e. $f_k = f_k^0 + \delta f_k$, where $\delta f_k / f_k^0 \ll 1$ and δf_k takes the following forms,

$$\delta f_q = \tau_R\left[\tilde{p}_q^{\gamma}\partial_\gamma\beta + \frac{\tilde{p}_q^{\gamma}}{u\cdot\tilde{p}_q}\Big(\beta\,\tilde{p}_q^{\phi}\partial_\gamma u_\phi - \partial_\gamma\alpha\Big) - \beta\theta\,\delta\omega_q\right] f_q\tilde{f}_q, \tag{69.6}$$

$$\delta f_{\bar{q}} = \tau_R\left[\tilde{p}_{\bar{q}}^{\gamma}\partial_\gamma\beta + \frac{\tilde{p}_{\bar{q}}^{\gamma}}{u\cdot\tilde{p}_{\bar{q}}}\Big(\beta\,\tilde{p}_{\bar{q}}^{\phi}\partial_\gamma u_\phi + \partial_\gamma\alpha\Big) - \beta\theta\,\delta\omega_{\bar{q}}\right] f_{\bar{q}}\tilde{f}_{\bar{q}}, \tag{69.7}$$

$$\delta f_g = \tau_R\left(\tilde{p}_g^{\gamma}\partial_\gamma\beta + \frac{\beta\,\tilde{p}_g^{\gamma}\,\tilde{p}_g^{\phi}}{u\cdot\tilde{p}_g}\partial_\gamma u_\phi - \beta\theta\,\delta\omega_g\right) f_g\tilde{f}_g. \tag{69.8}$$

Here, $\alpha = \beta\mu_q$ and $\theta \equiv \partial_\mu u^\mu$ is the expansion scalar. Note that we assume that the thermal relaxation time τ_R is independent of particle four-momenta.

69.3 Dissipative Evolution Equation

The macroscopic definition of shear stress tensor in terms of the non-equilibrium part of the distribution function δf_k within EQPM have the following forms,

$$\pi^{\mu\nu} = \sum_k g_k\,\Delta^{\mu\nu}_{\alpha\beta}\int d\tilde{P}_k\;\tilde{p}_k^{\alpha}\,\tilde{p}_k^{\beta}\,\delta f_k + \sum_k g_k\,\delta\omega_k\,\Delta^{\mu\nu}_{\alpha\beta}\int d\tilde{P}_k\;\tilde{p}_k^{\alpha}\,\tilde{p}_k^{\beta}\,\frac{1}{E_k}\delta f_k, \tag{69.9}$$

where g_k is the degeneracy factor and $d\tilde{P}_k \equiv \frac{d^3|\tilde{\mathbf{p}}_k|}{(2\pi)^3\omega_k}$ is the momentum integral factor. We define the projection operator $\Delta^{\mu\nu} \equiv g^{\mu\nu} - u^\mu u^\nu$ and a four-index tensor $\Delta^{\mu\nu}_{\alpha\beta} \equiv \frac{1}{2}(\Delta^{\mu}_{\alpha}\Delta^{\nu}_{\beta} + \Delta^{\mu}_{\beta}\Delta^{\nu}_{\alpha}) - \frac{1}{3}\Delta^{\mu\nu}\Delta_{\alpha\beta}$ which is a traceless symmetric projection operator orthogonal to the fluid velocity. Similarly, the bulk viscous pressure Π and the particle diffusion current n^μ can be defined respectively as,

$$\Pi = -\frac{1}{3}\sum_k g_k\,\Delta_{\alpha\beta}\int d\tilde{P}_k\;\tilde{p}_k^{\alpha}\,\tilde{p}_k^{\beta}\,\delta f_k - \frac{1}{3}\sum_k g_k\,\delta\omega_k\,\Delta_{\alpha\beta}\int d\tilde{P}_k\;\tilde{p}_k^{\alpha}\,\tilde{p}_k^{\beta}\,\frac{1}{E_k}\delta f_k, \tag{69.10}$$

$$n^\mu = g_q\Delta^{\mu}_{\alpha}\int d\tilde{P}_q\;\tilde{p}_q^{\alpha}\,(\delta f_q - \delta f_{\bar{q}}) - \delta\omega_q g_q\Delta^{\mu}_{\alpha}\int d\tilde{P}_q\;\tilde{p}_q^{\alpha}\,\frac{1}{E_q}(\delta f_q - \delta f_{\bar{q}}). \tag{69.11}$$

Substituting δf_k from (69.6)–(69.8) and keeping terms of first order gradients, we obtain the Naiver-Stokes like equation as follows,

$$\pi^{\mu\nu} = 2\,\tau_R\,\beta_\pi\,\sigma^{\mu\nu}, \qquad \Pi = -\tau_R\,\beta_\Pi\,\theta, \qquad n^\mu = \tau_R\,\beta_n\,\nabla^\mu\alpha, \tag{69.12}$$

with $\sigma^{\mu\nu} \equiv \Delta^{\mu\nu}_{\alpha\beta}\nabla^\alpha u^\beta$. The dissipative coefficients β_π, β_Π and β_n can be expressed in terms of thermodynamic integrals for massive and massless case and are described in detail in the [4]. We employ Bjorken's prescription to model the dissipative hydrodynamical evolution of the QGP [5]. In terms of Milne coordinates,(τ, x, y, η_s), where $\tau = \sqrt{t^2 - z^2}$, $\eta_s = \tanh^{-1}(z/t)$, $u^\mu = (1, 0, 0, 0)$ and $g^{\mu\nu} = (1, -1, -1, -1/\tau^2)$, the energy density ε evolution in the boost-invariant longitudinal expansion takes the form,

$$\frac{d\varepsilon}{d\tau} = -\left(\frac{\varepsilon + P}{\tau}\right) + \left(\frac{\zeta + 4\eta/3}{\tau^2}\right), \tag{69.13}$$

with $\theta = 1/\tau$, $\Pi = -\zeta/\tau$, $\pi^{\mu\nu}\sigma_{\mu\nu} = \Phi/\tau$, $\Phi = 4\eta/3\tau$.

69.4 Results and Discussions

The temperature behavior of the ratio of the coefficient of the bulk viscous tensor to that of the shear tensor (β_Π/β_π) at $\mu_q = 0.1$ GeV is depicted in Fig. 69.1. The ratio becomes $\beta_\Pi/\beta_\pi = \zeta/\eta$ within RTA, where ζ and η are the bulk and shear viscosities of the QGP medium. The ratio has a decreasing trend with the increase in temperature. The quark mass correction and mean field corrections are seen to have visible effects in the low temperature regime near to the transition temperature. In Fig. 69.2, we plotted the proper time evolution of temperature and pressure anisotropy in ideal and first order hydrodynamics. We considered the initial condition for shear and bulk viscous part as $\Phi = 4\eta/3\tau_0$ and $\Pi = -\zeta/\tau_0$ with the thermal relaxation time $\tau_R = 0.25$ fm. We observe that the viscous effects leads to slower cooling of the medium compared to ideal hydrodynamic evolution.

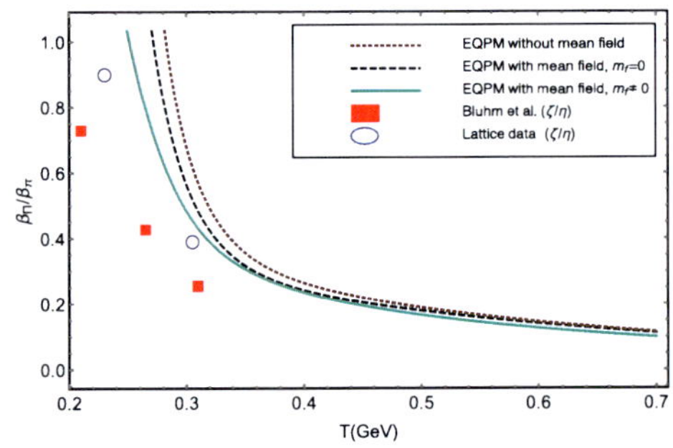

Fig. 69.1 Evolution of (β_Π/β_π) with temperature and comparison with results in [6–8]

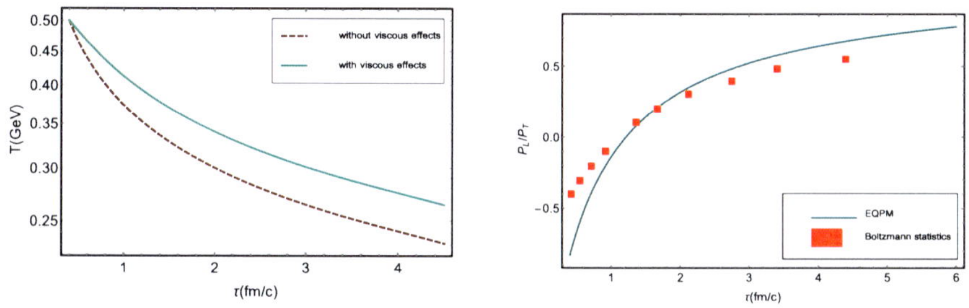

Fig. 69.2 The proper time temperature evolution (left panel) and behavior of pressure anisotropy (right panel) with initial temperature $T_0 = 500$ MeV at initial proper time $\tau_0 = 0.4$ fm. The evolution of P_L/P_T is compared with the result of [9]

Acknowledgments V. C. would like to acknowledge SERB for the Early Career Research Award (ECRA/2016), and DST, Govt. of India for INSPIRE-Faculty Fellowship (IFA-13/PH-55). A. J. is supported in part by the DST-INSPIRE faculty award under Grant No. DST/INSPIRE/04/2017/ 000038.

References

1. S. Mitra, V. Chandra, Phys. Rev. D **97**(3), 034032 (2018)
2. V. Chandra, V. Ravishankar, Phys. Rev. D **84**, 074013 (2011)
3. J.L. Anderson, H.R. Witting Physica **74**, 466 (1974)
4. S. Bhadury, M. Kurian, V. Chandra, A. Jaiswal,. arXiv:1902.05285 [hep-ph]
5. J.D. Bjorken, Phys. Rev. D **27**, 140 (1983)
6. M. Bluhm, B. Kampfer, K. Redlich, Phys. Lett. B **709**, 77 (2012)
7. H.B. Meyer, Phys. Rev. D **76**, 101701 (2007)
8. H.B. Meyer, Phys. Rev. Lett. **100**, 162001 (2008)
9. A. Jaiswal, Phys. Rev. C **88**, 021903 (2013)

Chapter 70
Transport Coefficients of Hot and Dense Matter

Olga Soloveva, Pierre Moreau, Lucia Oliva, Taesoo Song, Wolfgang Cassing, and Elena Bratkovskaya

Abstract We present calculations for the shear viscosity of the hot and dense quark-gluon plasma (QGP) using the partonic scattering cross sections as a function of temperature T and baryon chemical potential μ_B from the dynamical quasiparticle model (DQPM) that is matched to reproduce the equation of state of the partonic system above the deconfinement temperature T_c from lattice QCD. To this aim we calculate the collisional widths for the partonic degrees of freedom at finite T and μ_B in the time-like sector and conclude that the quasiparticle limit holds sufficiently well. Furthermore, the ratio of shear viscosity η over entropy density s, i.e. η/s, is evaluated using these collisional widths and are compared to lQCD calculations for $\mu_B = 0$ as well. We find that the ratio η/s is in agreement with the results of calculations within the original DQPM on the basis of the Kubo formalism. Furthermore, there is only a very modest change of η/s with the baryon chemical μ_B as a function of the scaled temperature $T/T_c(\mu_B)$.

70.1 Introduction

Transport coefficients of the hot and dense QGP are important ingridients for a fundamental description of medium properties. An exploration of temperature T and baryon chemical potential μ_B dependences of transport coefficients will provide useful information for the hydrodynamical simulations of heavy-ion collisions (HICs). It has been found that the QGP, produced in the central regions of HICs at the Relativistic Heavy Ion Collider (RHIC), is a strongly interacting system. The experimental

O. Soloveva (✉) · P. Moreau · L. Oliva · E. Bratkovskaya
Institute for Theoretical Physics, Johann Wolfgang Goethe-Universität, Frankfurt am Main, Germany
e-mail: soloveva@fias.uni-frankfurt.de

L. Oliva · T. Song · E. Bratkovskaya
GSI Helmholtzzentrum für Schwerionenforschung GmbH, Darmstadt, Germany

W. Cassing
Institut für Theoretische Physik, Universität, Gießen, Germany

D. Elia et al. (eds.), *The XVIII International Conference on Strangeness in Quark Matter (SQM 2019)*, Springer Proceedings in Physics 250,
https://doi.org/10.1007/978-3-030-53448-6_70

data for elliptic flow can be well reproduced by hydrodynamical simulations with a small value of the shear viscosity over entropy density [1, 2]. Calculations in the vicinity to the deconfinement phase transition T_c on the basis of perturbative QCD are notoriously difficult since the coupling rises in this region. In the strong interacting regime the methods of lattice gauge theories can be applied. Results from the lattice QCD calculations available only for $\mu_B = 0$, so far the presence of a non-zero chemical potential μ_B might affect the values of transport coefficients substantially. In light of these difficulties the evaluations of transport coefficients for a non-zero μ_B are necessary. One can evaluate transport coefficients for a non-zero μ_B within the effective approaches, which are found to match well the lQCD equation of state. We will present calculations for the shear viscosity within the DQPM for moderate values of $\mu_B \leq 450$ MeV, where we assume that in this region the transition between the hadronic and the quark-gluon plasma phase is a smooth cross-over.

70.2 Dynamical Quasiparticle Model (DQPM)

We describe the QGP in equilibrium on the base of the dynamical quasiparticle model (DQPM), which is based on partonic propagators with sizable imaginary parts of the selfenergies incorporated [3]. Whereas the real part of the self-energies can be attributed to a dynamically generated mass (squared) the imaginary parts contain the information about the interaction rates in the system. Furthermore, the imaginary parts of the propagators define the spectral functions of the degrees of freedom which might show quasiparticle peaks. A further advantage of a propagator-based approach is that one can formulate a consistent thermodynamics [4] as well as a causal theory for non-equilibrium configurations on the basis of the Kadanoff-Baym equations. In order to explore the transport properties of a partonic system we have calculated the interaction rates using the partonic differential cross sections as a function of T and μ_B, which are evaluated for the leading tree-level diagrams using the DQPM (cf. Appendices of [5]). In the on-shell case (energies of the particles are taken to be $E^2 = \mathbf{p}^2 + M^2$ with M being the pole mass) the interaction rate Γ_i^{on} is obtained as follows:

$$\Gamma_i^{on}(p_i, T, \mu_B) = \frac{1}{2E_i} \sum_{j=q,\bar{q},g} \int \frac{d^3 p_j}{(2\pi)^3 2E_j} d_j \; f_j(E_j, T, \mu_B) \int \frac{d^3 p_3}{(2\pi)^3 2E_3} \tag{70.1}$$

$$\times \int \frac{d^3 p_4}{(2\pi)^3 2E_4} (1 \pm f_3)(1 \pm f_4) |\bar{\mathcal{M}}|^2 (p_i, p_j, p_3, p_4) \, (2\pi)^4 \delta^{(4)} \left(p_i + p_j - p_3 - p_4\right)$$

$$= \sum_{j=q,\bar{q},g} \int \frac{d^3 p_j}{(2\pi)^3} d_j \; f_j \; v_{rel} \int d\sigma_{ij \to 34}^{on} \, (1 \pm f_3)(1 \pm f_4),$$

where d_j is the degeneracy factor for spin and color (for quarks $d_q = 2 \times N_c$ and for gluons $d_g = 2 \times (N_c^2 - 1)$), and with the shorthand notation $f_j = f_j(E_j, T, \mu_B)$ for the Fermi and Bose distribution functions. In (70.1) and in all the following sections, the notation $\sum_{j=q,\bar{q},g}$ includes the contribution from all possible partons which in our case are the gluons and the (anti-)quarks of three different flavors (u, d, s).

70.3 Shear Viscosity of the Hot and Dense QGP

The starting point to evaluate viscosity coefficients for the partonic matter is the Kubo formalism [6] which was also used to calculate the viscosities within the PHSD in a box with periodic boundary conditions (cf. [7]). We focus here on the calculation of the shear viscosity η based on [8] which reads:

$$\eta^{\text{Kubo}}(T, \mu_B) = -\int \frac{d^4p}{(2\pi)^4}\, p_x^2 p_y^2 \sum_{i=q,\bar{q},g} d_i\, \frac{\partial f_i(\omega)}{\partial \omega}\, \rho_i(\omega, \mathbf{p})^2, \tag{70.2}$$

where the notation $f_i(\omega) = f_i(\omega, T, \mu_B)$ is used again for the distribution functions, and ρ_i denotes the spectral function of the partons, while d_i stand for the degeneracy factors. We note that the derivative of the distribution function accounts for the Pauli-blocking (−) and Bose-enhancement (+) factors. Following [9], we can evaluate the integral over $\omega = p_0$ in (70.2) by using the residue theorem. When keeping only the leading order contribution in the width $\gamma(T, \mu_B)$ from the residue—evaluated at the poles of the spectral function $\omega_i = \pm\tilde{E}(\mathbf{p}) \pm i\gamma$—we finally obtain:

$$\eta^{\text{RTA}}(T, \mu_B) = \frac{1}{15T} \int \frac{d^3p}{(2\pi)^3} \sum_{i=q,\bar{q},g} \left(\frac{\mathbf{p}^4}{E_i^2\, \Gamma_i(\mathbf{p}_i, T, \mu_B)}\, d_i\, ((1 \pm f_i(E_i)) f_i(E_i)) \right), \tag{70.3}$$

which corresponds to the expression derived in the relaxation-time approximation (RTA) [10] by identifying the interaction rate Γ with 2γ as expected from transport theory in the quasiparticle limit [11]. We recall that γ is the width parameter in the parton propagator. The interaction rate $\Gamma_i(\mathbf{p}_i, T, \mu_B)$ (inverse relaxation time) here is calculated microscopically using the differential cross sections for parton scattering as described in Sect. 70.2.

The actual results are displayed in Fig. 70.1 for the ratio of shear viscosity to entropy density η/s as a function of the scaled temperature T/T_c for $\mu_B = 0$ in comparison to those from lattice QCD [13]. The solid green line (η^{Kubo}/s) shows the result from the original DQPM in the Kubo formalism while the dashed green line ($\eta_{2\gamma}^{\text{RTA}}/s$) shows the same result in the relaxation-time approximation (70.3) by replacing Γ_i by $2\gamma_i$. The solid red line ($\eta_{\Gamma^{\text{on}}}^{\text{RTA}}/s$) results from (70.3) using the interaction rate Γ^{on} calculated by the microscopic differential cross sections in the on-shell limit. We find that the ratios η/s do not differ very much and have a similar behavior as a function of temperature. The approximation (70.3) of the shear viscosity

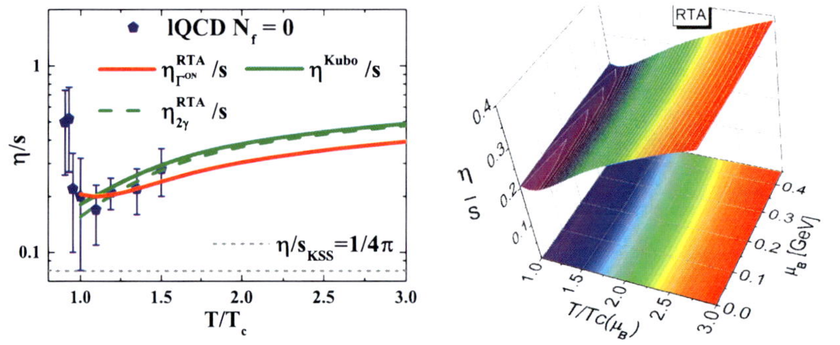

Fig. 70.1 Left: The ratio of shear viscosity to entropy density as a function of the scaled temperature T/T_c for $\mu_B = 0$ from (70.2–70.3). The solid green line (η^{Kubo}/s) shows the results from the original DQPM in the Kubo formalism while the dashed green line ($\eta^{\text{RTA}}_{2\gamma}/s$) shows the same result in the relaxation-time approximation (70.3). The solid red line ($\eta^{\text{RTA}}_{\Gamma^{\text{on}}}/s$) results from (70.3) using the interaction rate Γ^{on} calculated by the microscopic differential cross sections in the on-shell limit. The dashed gray line demonstrates the Kovtun-Son-Starinets bound [12] $(\eta/s)_{\text{KSS}} = 1/(4\pi)$. The symbols show lQCD data for pure SU(3) gauge theory taken from [13] (pentagons). Right: The ratio of shear viscosity to entropy density as a function of the scaled temperature T/T_c from the DQPM for μ_B in the range [0, 0.45] GeV from (70.3)

is found to be very close to the one from the Kubo formalism (70.2) indicating that the quasiparticle limit ($\gamma \ll M$) holds in the DQPM.

We note in passing that there is no strong variation with μ_B for fixed $T/T_c(\mu_B)$, however, the ratio increases slightly with μ_B in the on-shell limit while it slightly drops with μ_B in the Kubo formalism for the DQPM [5]. Accordingly, there is some model uncertainty when extracting the shear viscosity in the different approximations.

Acknowledgments The authors acknowledge inspiring discussions with J. Aichelin, H. Berrehrah, Claudia Ratti, E. Seifert, A. Palmese and T. Steinert. Furthermore, P.M., L.O. and E.B. acknowledge support by the DFG grant CRC-TR 211 'Strong-interaction matter under extreme conditions' - Project number 315477589 - TRR 211. O.S. acknowledges support from HGS-HIRe for FAIR; L.O. and E.B. thank the COST Action THOR, CA15213. The computational resources have been provided by the LOEWE-Center for Scientific Computing.

References

1. P. Romatschke, U. Romatschke, Phys. Rev. Lett. **99**, 172301 (2007)
2. H. Song, U.W. Heinz, Phys. Rev. C **78**, 024902 (2008)
3. O. Linnyk, E.L. Bratkovskaya, W. Cassing, Prog. Part. Nucl. Phys. **87**, 50 (2016)
4. B. Vanderheyden, G. Baym, J. Stat. Phys. **93**, 843 (1998)
5. P. Moreau, O. Soloveva, L. Oliva, T. Song, W. Cassing, E. Bratkovskaya, Phys. Rev. C **100**, 014911 (2019)
6. R. Kubo, J. Phys. Soc. Jpn. **12**, 570 (1957)

7. V. Ozvenchuk, O. Linnyk, M.I. Gorenstein, E.L. Bratkovskaya, W. Cassing, Phys. Rev. C **87**, 064903 (2013)
8. G. Aarts, J.M. Martinez Resco, J. High Energy Phys. **04**, 053 (2002)
9. R. Lang, N. Kaiser, W. Weise, Eur. Phys. J. A **48**, 109 (2012)
10. C. Sasaki, K. Redlich, Phys. Rev. C **79**, 055207 (2009)
11. J.-P. Blaizot, E. Iancu, Nucl. Phys. B **557**, 183 (1999)
12. P.K. Kovtun, D.T. Son, A.O. Starinets, Phys. Rev. Lett. **94**, 111601 (2005)
13. N. Astrakhantsev, V. Braguta, A. Kotov, J. High Energy Phys. **04**, 101 (2017)

Chapter 71
Evolution of Fluctuations in the Initial State of Heavy-Ion Collisions from RHIC to LHC

Giuliano Giacalone, Francçois Gelis, Pablo Guerrero-Rodríguez, Matthew Luzum, Cyrille Marquet, and Jean-Yves Ollitrault

Abstract Fluctuations in the initial state of heavy-ion collisions are larger at RHIC energy than at LHC energy. This fact can be inferred from recent measurements of the fluctuations of the particle multiplicities and of elliptic flow performed at the two different energies. We show that an analytical description of the initial energy-density field and its fluctuations motivated by the color glass condensate (CGC) effective theory predicts and quantitatively captures the measured energy evolution of these observables. The crucial feature is that fluctuations in the CGC scale like the inverse of the saturation scale of the nuclei.

Data collected at the CERN Large Hadron Collider (LHC) and at the BNL Relativistic Heavy Ion Collider (RHIC) indicates that initial-state fluctuations in relativistic heavy-ion collisions are larger at RHIC. Two observables support this statement: The relative fluctuations of the charged-particle multiplicity, N_{ch}, in central collisions, and the relative fluctuations of elliptic flow, v_2.

Fluctuations of N_{ch} in nucleus-nucleus collisions probe the fluctuations of the initial state because they are to a good approximation equal to the relative fluctuations

G. Giacalone (✉) · F. Gelis · J.-Y. Ollitrault
Institut de physique théorique Université Paris Saclay CNRS,
91191 Gif-sur-Yvette, France
e-mail: giuliano.giacalone@ipht.fr

P. Guerrero-Rodríguez
Department of Physics, University of Jyväskylä, 35, 40014 Jyväskylä, Finland

M. Luzum
Instituto de Física Universidade de São Paulo, R. do Matão 1371,
São Paulo, SP 05508-090, Brazil

C. Marquet
CPHT, CNRS École Polytechnique, Institut Polytechnique de Paris,
Route de Saclay, 91128 Palaiseau, France

D. Elia et al. (eds.), *The XVIII International Conference on Strangeness in Quark Matter (SQM 2019)*, Springer Proceedings in Physics 250,
https://doi.org/10.1007/978-3-030-53448-6_71

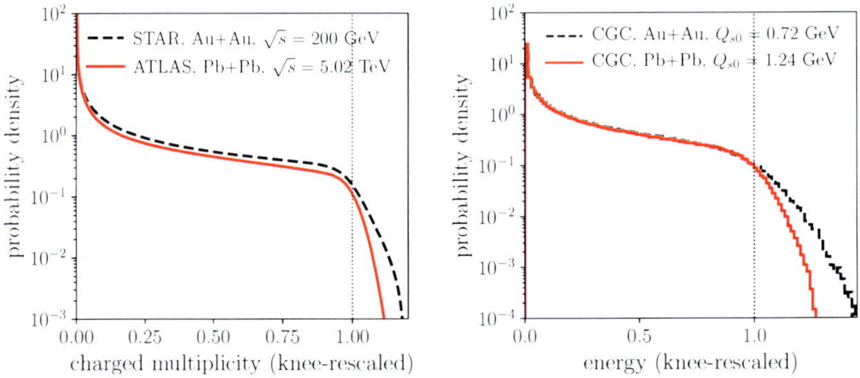

Fig. 71.1 Left: Rescaled histograms of the charged-particle multiplicity measured by the STAR Collaboration [1] and by the ATLAS Collaboration [2]. Right: Rescaled histogram of the total energy in the initial state of Pb+Pb and Au+Au collisions obtained in the CGC model (using the Monte Carlo implementation of [10])

of initial total entropy of the system. The left panel of Fig. 71.1 shows the distribution of charged multiplicity measured by the STAR Collaboration in Au+Au collisions at $\sqrt{s} = 200$ Gev [1], and by the ATLAS Collaboration in Pb+Pb collisions at $\sqrt{s} = 5.02$ TeV [2]. The histograms are here rescaled by their value at the *knee*, i.e., the mean value of $N_{\rm ch}$ at zero impact parameter, inferred with the Bayesian procedure of [3]. The fluctuations of $N_{\rm ch}$ around the knee quantify the width of the large-multiplicity tail, and this is 1.5 times larger in STAR data:

$$\left.\frac{\sigma[N_{\rm ch}](b=0)}{\langle N_{\rm ch}\rangle(b=0)}\right|_{\rm RHIC} = 0.065, \qquad \left.\frac{\sigma[N_{\rm ch}](b=0)}{\langle N_{\rm ch}\rangle(b=0)}\right|_{\rm LHC} = 0.044. \quad (71.1)$$

This implies that initial-state fluctuations are larger at RHIC.

Moving on to the relative fluctuations of v_2, they also serve as a probe of the initial state since elliptic flow is a measure of the initial eccentricity, ε_2 [4], of the system, a quantity which originates from the fluctuating geometry of the initial energy-density field. In general, at a given centrality we have $v_2 = \kappa\varepsilon_2$, where κ is a response coefficient. The relative fluctuation of v_2 can be quantified by the ratio of the first two cumulant of its distribution, $v_2\{4\}/v_2\{2\}$, which is simply equal to $\varepsilon_2\{4\}/\varepsilon_2\{2\}$ because the coefficient κ cancels in the ratio [5]. This allows us in particular to compare the relative v_2 fluctuations between RHIC and LHC without knowing how the coefficient κ evolves with energy. In absence of fluctuations, the ratio is equal to unity, while $\varepsilon_2\{4\}/\varepsilon_2\{2\} < 1$ for a fluctuating initial state. The deviation of this quantity from unity quantifies the amount of fluctuations in the system. I show in Fig. 71.2 the ratio $v_2\{4\}/v_2\{2\}$ measured at both RHIC and LHC. STAR data are lower than ATLAS data. The initial state fluctuates more at RHIC.

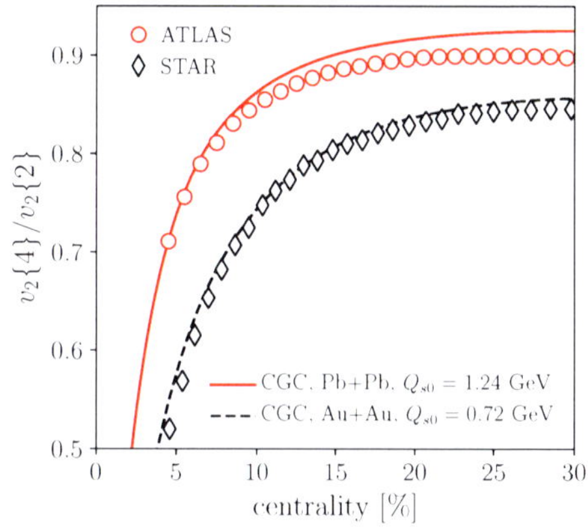

Fig. 71.2 Ratio of the first two cumulants of elliptic flow, $v_2\{4\}/v_2\{2\}$. Symbols are STAR data (diamonds) [1] and ATLAS data (circles) [2]. Lines are the results of the CGC model. Figure from [6]

Armed with this knowledge, we argue now that the energy evolution of the previous observables is captured by the model we introduced in [6], which treats the initial condition of heavy-ion collisions as the energy density field produced immediately after ($\tau = 0^+$) two sheets of Color Glass Condensate [7] cross each other. The statistics of energy-density fluctuations in this model was derived in [8]. The local average (1-point function) of energy density reads:

$$\langle \rho(\mathbf{s}) \rangle = \frac{4}{3g^2} Q_A^2(\mathbf{s}) Q_B^2(\mathbf{s}), \tag{71.2}$$

where $\mathbf{s}$ is a transverse coordinate, and $Q^2_{A/B}$ is the saturation scale (squared) of nucleus A/B, which we take proportional to the nuclear density integrated along the collision axis (*thickness* function), usually denoted by $T_{A/B}$, with a coefficient in front, Q^2_{s0}, which gives the value of the saturation scale at the center of the nucleus. Fluctuations of energy density are, at leading logarithmic accuracy, given instead by the following short-range connected 2-point function:

$$\langle \rho(\mathbf{s}_1)\rho(\mathbf{s}_2) \rangle - \langle \rho(\mathbf{s}_1) \rangle \langle \rho(\mathbf{s}_2) \rangle = \delta(\mathbf{r})\xi(\mathbf{s}),$$

with $\mathbf{s} = (\mathbf{s}_1 + \mathbf{s}_2)/2$ and $\mathbf{r} = \mathbf{s}_1 - \mathbf{s}_2$, and

$$\xi(\mathbf{s}) = \frac{16\pi}{9g^4} Q_A^2(\mathbf{s}) Q_B^2(\mathbf{s}) \left[Q_A^2(\mathbf{s}) \ln\left(\frac{Q_B^2(\mathbf{s})}{m^2} \right) + Q_B^2(\mathbf{s}) \ln\left(\frac{Q_A^2(\mathbf{s})}{m^2} \right) \right], \tag{71.3}$$

where m is an infrared scale which cuts off the correlation of two color sources in the transverse plane. We shall use $m = 0.14$ GeV, i.e., the pion mass. Let us study, then, the previous observables within this model.

The relative fluctuations of the total energy per unit longitudinal length, E, are equal to:

$$\frac{\sigma[E]}{\langle E\rangle} = \frac{\sqrt{\int_{\mathbf{s}} \xi(\mathbf{s})}}{\int_{\mathbf{s}} \langle \rho(\mathbf{s})\rangle} \propto \frac{1}{Q_{s0}}, \tag{71.4}$$

neglecting slowly-varying logarithms. Now, fits of the Bjorken-x evolution of structure functions measured at HERA indicate that the saturation scale evolves with collision energy as follows [9]:

$$\frac{Q_s[\text{LHC}]}{Q_s[\text{RHIC}]} = \left(\frac{\sqrt{s_{\text{LHC}}}}{\sqrt{s_{\text{RHIC}}}}\right)^{0.14} \approx 1.5. \tag{71.5}$$

This yields the same factor 1.5 obtained in the comparison between the relative fluctuations of multiplicity at RHIC and LHC in (71.1). Multiplicity fluctuations, hence, are quantitatively consistent with an $1/Q_s$ scaling, precisely as predicted by our model (71.4). To visualize how the fluctuations of E look like, we compute its distribution using the `magma` model [10], a Monte Carlo implementation of event-by-event profiles that fluctuate according to (71.2) and (71.3). The distributions of E for RHIC and LHC are shown in Fig. 71.1, on the right. Note that the different values of Q_{s0} used in the calculations were fitted from anisotropic flow data [6], and are consistent with the HERA scaling, (71.5). Computing (71.4), in central collisions we obtain:

$$\left.\frac{\sigma[E](b=0)}{\langle E\rangle(b=0)}\right|_{\text{RHIC}} = 0.134, \qquad \left.\frac{\sigma[E](b=0)}{\langle E\rangle(b=0)}\right|_{\text{LHC}} = 0.088. \tag{71.6}$$

These numbers are essentially larger by a factor 2 than those given in (71.1), as also suggested by the width of the tails of the histograms in Fig. 71.1. This is consistent with the fact that the distribution of E will receive an important correction from the pre-equilibrium dynamics of the system during the first fm/c of its evolution [11]. The relative fluctuations of entropy at equilibrium are in fact expected to be significantly smaller than those of the initial energy [12].

Finally, let us compute the fluctuations of elliptic flow. As anticipated, these originate from the fluctuations of the initial ε_2. Following Blaizot et al. [13], and neglecting slowly-varying logarithms, the rms eccentricity due to fluctuations can be written as:

$$\frac{\sqrt{\int_{\mathbf{s}} |\mathbf{s}|^4 \xi(\mathbf{s})}}{\int_{\mathbf{s}} |\mathbf{s}|^2 \langle \rho(\mathbf{s})\rangle} \propto \frac{1}{Q_{s0}}. \tag{71.7}$$

The saturation scale appears in the denominator. Interestingly, this implies that ε_2 is larger at RHIC than at LHC. Nevertheless, the measured v_2 is smaller at RHIC, because the response κ is strongly suppressed by the lower collision energy. Relative fluctuations, though, do not depend on κ and can be genuinely compared. The CGC, thus, naturally predicts that they are larger at RHIC, in agreement with the experimental data shown in Fig. 71.2. Results on $v_2\{4\}/v_2\{2\} = \varepsilon_2\{4\}/\varepsilon_2\{2\}$ in the CGC are reported as lines in Fig. 71.2, and describe quantitatively the data.

In summary, the energy evolution of fluctuations provides a powerful probe of the initial state of nucleus-nucleus collisions. Experiments indicate that initial-state fluctuations in heavy-ion collisions are larger at RHIC than at LHC. This feature is both qualitatively and quantitatively captured by the CGC-inspired model of [6], where fluctuations are inversely proportional to the saturation scale of the colliding nuclei.

Acknowledgments G.G. M.L. and J.-Y.O. acknowledge funding from USP-COFECUB (grant Uc Ph 160-16, 2015/13). The work of F.G. and C.M. was supported in part by the Agence Nationale de la Recherche under the project ANR-16-CE31-0019-02. M.L. acknowledges support from FAPESP projects 2016/24029-6, 2017/05685-2, 2018/24720-6, and project INCT-FNA Proc. No. 464898/2014-5.

References

1. L. Adamczyk et al. [STAR Collaboration], Phys. Rev. Lett. **115**, 22, 222301 (2015). https://doi.org/10.1103/PhysRevLett.115.222301 [arXiv:1505.07812 [nucl-ex]]
2. M. Aaboud et al. [ATLAS Collaboration], [arXiv:1904.04808 [nucl-ex]]
3. S.J. Das, G. Giacalone, P.A. Monard, J.Y. Ollitrault, Phys. Rev. C **97**, 1, 014905 (2018). https://doi.org/10.1103/PhysRevC.97.014905 [arXiv:1708.00081 [nucl-th]]
4. D. Teaney, L. Yan, Phys. Rev. C **83**, 064904 (2011). https://doi.org/10.1103/PhysRevC.83.064904. [arXiv:1010.1876 [nucl-th]]
5. G. Giacalone, J. Noronha-Hostler, J. Y. Ollitrault, Phys. Rev. C **95**, 5, 054910 (2017). https://doi.org/10.1103/PhysRevC.95.054910 [arXiv:1702.01730 [nucl-th]]
6. G. Giacalone, P. Guerrero-Rodríguez, M. Luzum, C. Marquet, J.Y. Ollitrault, Phys. Rev. C **100**, 2, 024905 (2019). https://doi.org/10.1103/PhysRevC.100.024905 [arXiv:1902.07168 [nucl-th]]
7. E. Iancu, R. Venugopalan, in Hwa, R.C. (ed.) et al., *Quark gluon plasma*, pp. 249–3363. https://doi.org/10.1142/9789812795533_0005 [hep-ph/0303204]
8. J.L. Albacete, P. Guerrero-Rodríguez, C. Marquet, JHEP **1901**, 073 (2019). https://doi.org/10.1007/JHEP01(2019)073. [arXiv:1808.00795 [hep-ph]]
9. J.L. Albacete, C. Marquet, Prog. Part. Nucl. Phys. **76**, 1 (2014). https://doi.org/10.1016/j.ppnp.2014.01.004. [arXiv:1401.4866 [hep-ph]]
10. F. Gelis, G. Giacalone, P. Guerrero-Rodríguez, C. Marquet, J.Y. Ollitrault, arXiv:1907.10948 [nucl-th]
11. S. Schlichting, D. Teaney, arXiv:1908.02113 [nucl-th]
12. G. Giacalone, A. Mazeliauskas, S. Schlichting, arXiv:1908.02866 [hep-ph]
13. J.P. Blaizot, W. Broniowski, J.Y. Ollitrault, Phys. Lett. B **738**, 166 (2014). https://doi.org/10.1016/j.physletb.2014.09.028. arXiv:1405.3572 [nucl-th]

Chapter 72
Jet-Fluid Interaction in the EPOS3-Jet Framework

Iurii Karpenko, Joerg Aichelin, Pol Bernard Gossiaux, Martin Rohrmoser, and Klaus Werner

Abstract EPOS3-Jet is an integrated framework for jet modeling in heavy ion collisions, where the initial hard (jet) partons are produced along with soft (medium) partons in the initial state EPOS approach. The jet partons then propagate in the hydrodynamically expanding medium. The energy and momentum lost by the jet partons is added to the hydrodynamic medium via the source terms. The full evolution proceeds in a concurrent mode, without separating hydrodynamic and jet parts. In this report we examine the medium recoil effects in Pb–Pb collisions at $\sqrt{s_{\rm NN}} = 2.76$ TeV LHC energy in the EPOS3-Jet framework.

72.1 Introduction

A consistent modeling of back reaction of the hydrodynamic medium on the jet evolution is important for understanding the substructure of jets produced in heavy ion collisions. The majority of existing models implement only one-way jet-hydro interaction by coupling jets to a fixed hydrodynamic expansion and not including the energy deposition in the medium itself. On the other hand, some recent studies, e.g. [1] conclude that the medium recoil effect has to be taken into account in order to reproduce the modification of the radial momentum distribution within a jet in AA

I. Karpenko (✉) · J. Aichelin · P. B. Gossiaux · K. Werner
SUBATECH, Université de Nantes, IMT Atlantique, IN2P3/CNRS,
4 rue Alfred Kastler, 44307 Nantes cedex 3, France
e-mail: yu.karpenko@gmail.com

I. Karpenko
FNSPE, České vysoké učení technické v Praze, Břehová 7,
11519 Prague 1, Czech Republic

M. Rohrmoser
Institute of Physics, Jan Kochanowski University, 25-406 Kielce, Poland

H. Niewodniczański Institute of Nuclear Physics PAN, 31-342 Cracow, Poland

D. Elia et al. (eds.), *The XVIII International Conference on Strangeness in Quark Matter (SQM 2019)*, Springer Proceedings in Physics 250,
https://doi.org/10.1007/978-3-030-53448-6_72

collisions as compared to the pp case. Therefore, in this report we examine the back reaction of the medium due to energy loss of jet partons in the EPOS3-Jet framework.

72.2 Model

The initial hard partons—seeds of the jets—are sampled from the initial state calculations in EPOS3 [2]. For the hydrodynamic expansion of the medium, the averaged initial conditions are taken. This makes it easier to visualize the effect of the jet energy loss on the medium.

Each initial hard parton leads to the development of a time-like parton cascade, due to collinear parton splitting caused by bremsstrahlung. The evolution of the parton cascade is performed with a Monte Carlo algorithm [3] representing the Dokshitzer-Gribov-Lipatov-Altarelli-Parisi (DGLAP) equation with leading order $q \to qg, g \to gg$ and $g \to q\bar{q}$ splitting functions. The evolution of each parton cascade proceeds from an initial virtuality scale $Q_\uparrow$, which we set to be equal to parton's transverse momentum $p_\perp$, down to a minimal virtuality scale of $Q_\downarrow = 0.6$ GeV.

The DGLAP evolution proceeds in momentum space. In order to couple the parton cascade to the medium, one has to make assumptions about its spacetime evolution. Therefore, we assume that in the global frame a parton has a mean life time (or the time before its next splitting occurs) of $\Delta t = E/Q^2$.

The parton shower has vacuum splitting functions. For the medium modifications of the parton shower, we consider at present two effects: (i) an effective increase of the virtuality the off-mass-shell partons $\frac{dQ^2}{dt} = \hat{q}_R(T)$, where t is time and T the temperature in the local fluid (medium) rest frame. This mimics the medium induced radiation at the stage of jet formation; (ii) collisional energy loss which is modelled via a Langevin-type longitudinal drag and random transverse kicks to each jet parton: $\Delta p_\parallel = -A(t, x)\Delta t, \quad \Delta p_\perp = n_\perp \sqrt{\hat{q}_C \Delta t}$. Different from the previously reported results [4], $A(t, x)$ and $\hat{q}_C$ are taken to be both temperature- and momentum-dependent according to [5], where they are evaluated with pQCD cross sections with running α_s.

Here we are interested in a qualitative study of the jet-medium interaction and do not aim to fit the experimental data. Therefore, the transport coefficients $A(t, x)$ and $\hat{q}_C$ are further multiplied by a correction factor (or so-called K-factor) $K = 0.166$ to approach the temperature-dependent jet transport coefficient from MARTINI model reported in [6]. This results in a parton-level $R_{AA} \approx 0.6$ at $p_\perp^{\text{parton}} = 10...30$ GeV.

Back reaction on the medium. The space-time evolution of the jet partons proceeds in the same time steps as the hydrodynamic evolution of the medium. During each time step, the energy and momentum loss of each jet parton in the computational frame of the fluid, is added to the corresponding fluid cells via an additional source term in the hydrodynamic equations: $\partial_\nu T^{\mu\nu} = J^\mu$. The source term J^μ is essentially a 4-vector of the lost energy and momentum in Milne coordinates, multiplied by a Gaussian smearing kernel with the width $R_g = 0.4$ fm/c in the transverse direction around the position of the jet parton.

72.3 Results and Conclusions

The simulations are performed for most central Pb–Pb collisions at $\sqrt{s_{NN}} = 2.76$ TeV collision energy. To see the effects of medium recoil we run the "semi-EbE" simulations, where the sampled ensembles of initial hard partons are taken from the initial state EPOS with a requirement that in each event there must be at least 1 hard parton with $p_\perp > 50$ GeV (so-called $p_\perp$ trigger), whereas the hydrodynamic evolution always starts from the smooth, event-averaged initial state. In other words, initial hard partons are sampled with Monte Carlo (EPOS) whereas the hydrodynamic initial state is not fluctuating.

In Fig. 72.1 we show one of the events where the back reaction to the medium is visually expressed. On the top panel of the figure, which represents the snapshots of the x component of the transverse flow velocity one can see an irregular spot which emerged already at $\tau = 1$ fm/c. This local flip of collective flow velocity is caused by the energy loss of an energetic jet parton with $p_\perp \approx 50$ GeV, which propagates through the medium in the $-x$ direction.

The bottom panel of Fig. 72.1 shows perturbations in the local rest frame energy density of the fluid, i.e. a difference between a hydrodynamic evolution with the additional source term enabled and an unperturbed hydrodynamic evolution. In the energy density perturbations, one can see more structures formed at $\tau = 1$ fm/c, which all expand subsequently by a hydrodynamic evolution. This indicates that there are some 8 energetic jet partons at $\tau = 1$ fm/c, which all lose their energy to the medium. One may recognize the "Mach cones" [7] here, though the grid resolution does not permit to see them very clearly.

Conclusions. Let us draw conclusions from Fig. 72.1. First, the *relative* scale of energy density perturbations is very small in most of the cases. The perturbations become significant only at the periphery of the system where the background energy density is not large—which also results in visible velocity perturbations (top panel of Fig. 72.1). Second, the hydrodynamic evolution smears out the early time perturbations, as one can see as well from the top panel of Fig. 72.1, and the expansion at late times becomes smooth again. Therefore we conclude that the present model for the jet-medium interaction with the assumption of instant thermalization of the energy-momentum loss by the jet partons makes it difficult to observe the medium recoil effects on the jet observables.

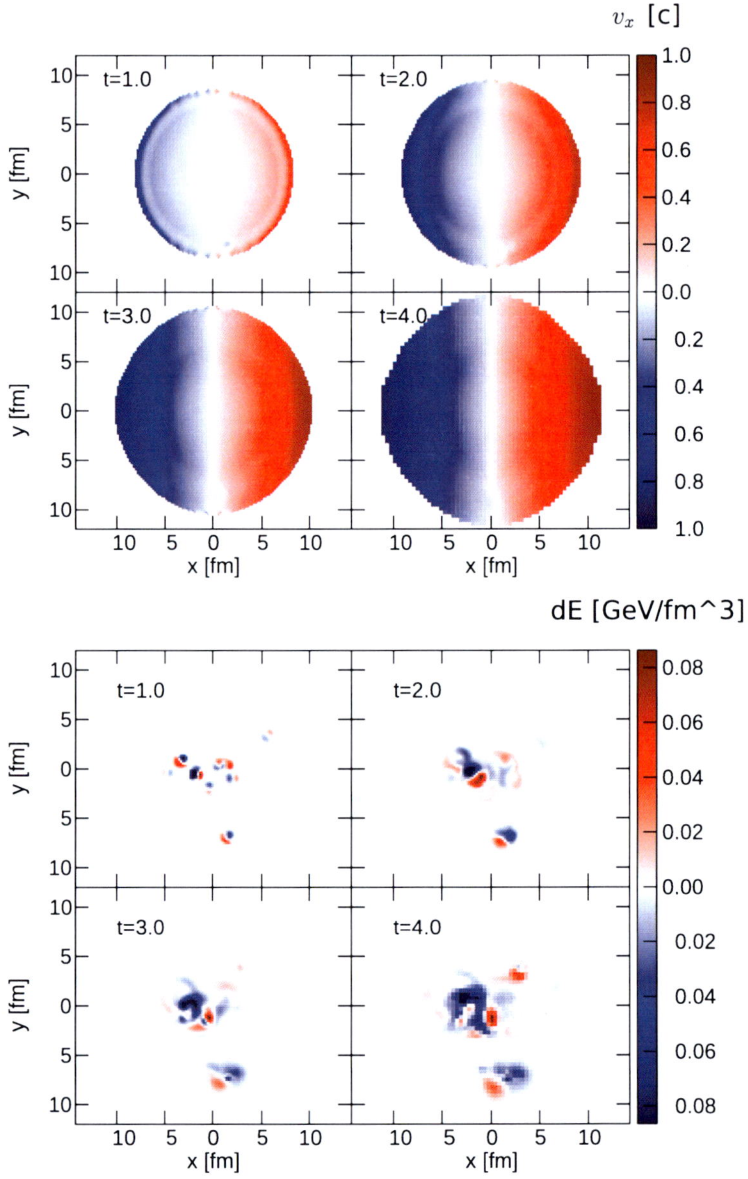

Fig. 72.1 x component of transverse flow velocity (top panel) and corresponding perturbations in the local rest frame energy density of the medium caused by the energy lost by the jets (bottom panel). A jet event displayed here is simulated for 0–5% central Pb-Pb collision at $\sqrt{s_{\mathrm{NN}}} = 2.76$ TeV

References

1. C. Park, S. Jeon, C. Gale, Jet modification with medium recoil in quark-gluon plasma. Nucl. Phys. A **982**, 643 (2019)
2. K. Werner, B. Guiot, I. Karpenko, T. Pierog, Analysing radial flow features in p-Pb and p-p collisions at several TeV by studying identified particle production in EPOS3. Phys. Rev. C **89**(6), 064903 (2014)
3. M. Rohrmoser, P.B. Gossiaux, T. Gousset, J. Aichelin, Discrimination of effective radiative and collisional in-medium energy-loss models by their effects on angular jet structure. Acta Phys. Polon. B **49**, 1325 (2018)
4. I. Karpenko, M. Rohrmoser, P. Gossiaux, J. Aichelin, K. Werner, Jet structure in integrated EPOS3-HQ approach. PoS HardProbes **2018**, 078 (2018)
5. P.B. Gossiaux, R. Bierkandt, J. Aichelin, Phys. Rev. C **79**, 044906 (2009)
6. K.M. Burke et al., Extracting the jet transport coefficient from jet quenching in high-energy heavy-ion collisions. Phys. Rev. C **90**(1), 014909 (2014)
7. B. Betz, M. Gyulassy, D.H. Rischke, H. Stocker, G. Torrieri, Jet propagation and Mach Cones in (3+1)d ideal hydrodynamics. J. Phys. G **35**, 104106 (2008)

Chapter 73
Radial Flow Induced by Inhomogeneous Magnetic Field in Heavy Ion Collisions

Mosen Haddadi Moghaddam, Behnam Azadegan, Ahmad F. Kord, and Wanda M. Alberico

Abstract In this paper, we study the effects of an in-homogeneous magnetic field on the quak gluon plasma dynamics, within the Magneto-Hydrodynamic framework. We have investigated the effect of an inhomogeneous external magnetic field on the transverse expansion of in-viscid fluid created in high energy nuclear collisions. Transverse velocity and energy density are modified by the presence of the magnetic field. This effects can also influence the transverse momentum spectrum for particles at the freeze-out surface. In this context we obtained an interesting comparison with data extracted from heavy-ion collisions.

73.1 Introduction

Collisions of two heavy nuclei at high energy produce a hot and dense fireball. Quarks and gluons could reach the deconfined state, called quark gluon plasma (QGP), in a very short time ($\sim$1 fm/c) after the initial hard parton collisions of nuclei. recently a wide range of studies has shown that relativistic heavy-ion collisions create huge magnetic field due to the relativistic motion of the colliding heavy ions carrying large positive electric charge (For more references and details refer to it [1–3]).

Recently, some efforts in numerical and analytical works have been made, based on the relativistic magneto-hydrodynamic (RMHD) setup, to describe high energy heavy ion collisions (See, for example, [4–11]). The aim of our work is to generalize the Bjorken model by considering an inhomogeneous external magnetic field acting on the medium. We show that the presence of the magnetic field leads to non-zero radial flow [9].

M. Haddadi Moghaddam (✉) · W. M. Alberico
Department of Physics, University of Turin and INFN, Turin, Via P. Giuria 1, 10125 Turin, Italy
e-mail: hadadi_65@yahoo.com

M. Haddadi Moghaddam · B. Azadegan · A. F. Kord
Department of Physics, Hakim Sabzevari University (HSU), P.O. Box 397 Sabzevar, Iran

D. Elia et al. (eds.), *The XVIII International Conference on Strangeness in Quark Matter (SQM 2019)*, Springer Proceedings in Physics 250,
https://doi.org/10.1007/978-3-030-53448-6_73

73.2 Relativistic Magneto-Hydrodynamic

We deal with the case of an ideal non-resistive plasma, with vanishing electric field in the local rest-frame ($e^\mu = 0$), which is embedded in an external magnetic field (b_μ) [12, 13]. The energy momentum conservation equations read:

$$d_\mu(T^{\mu\nu}_{pl} + T^{\mu\nu}_{em}) = 0, \tag{73.1}$$

where

$$T^{\mu\nu}_{pl} = (\epsilon + P)u^\mu u^\nu + P g^{\mu\nu} \tag{73.2}$$

$$T^{\mu\nu}_{em} = b^2 u^\mu u^\nu + \frac{1}{2} b^2 g^{\mu\nu} - b^\mu b^\nu. \tag{73.3}$$

In the above $g_{\mu\nu}$ is the metric tensor, ϵ and P are the energy density and pressure, respectively. Moreover d_μ is the covariant derivative. And the four velocity is defined as

$$u_\mu = \gamma(1, \mathbf{v}), \ \gamma = \frac{1}{\sqrt{1 - v^2}}$$

satisfying the condition $u^\mu u_\mu = -1$.

Canonically one takes projections of the equation $d_\mu(T^{\mu\nu}_{pl} + T^{\mu\nu}_{em}) = 0$ along the parallel and perpendicular directions to u_ν, which gives:

$$D(\epsilon + b^2/2) + (\epsilon + P + b^2)\Theta + u_\nu b^\mu (d_\mu b^\nu) = 0, \tag{73.4}$$

$$(\epsilon + P + b^2) D u^\alpha + \nabla^\alpha (P + \frac{1}{2} b^2) - d_\mu (b^\mu b^\alpha) - u^\alpha u_\nu d_\mu (b^\mu b^\nu) = 0. \tag{73.5}$$

Notice that α should be a spacelike index. Moreover

$$D = u^\mu d_\mu, \quad , \Theta = d_\mu u^\mu, \quad , \nabla^\alpha = \Delta^\alpha{}_\nu d^\nu. \tag{73.6}$$

73.2.1 *Induced Radial Flow in B-Field*

We consider the medium expands both radially and along the beam axis, the only nonzero components of $u_\mu = (u_\tau, u_\perp, 0, 0)$ are u_τ, which describes the boost-invariant longitudinal expansion [14], and $u_\perp$, which describes the transverse expansion. And we suppose the external magnetic field to be located in transverse plane as $b_\mu = (0, 0, b_\phi, 0)$ (Fig. 73.1).

We now seek the perturbation solution in the presence of a weak external magnetic field pointing along the ϕ direction in an inviscid fluid with infinite electrical conductivity:

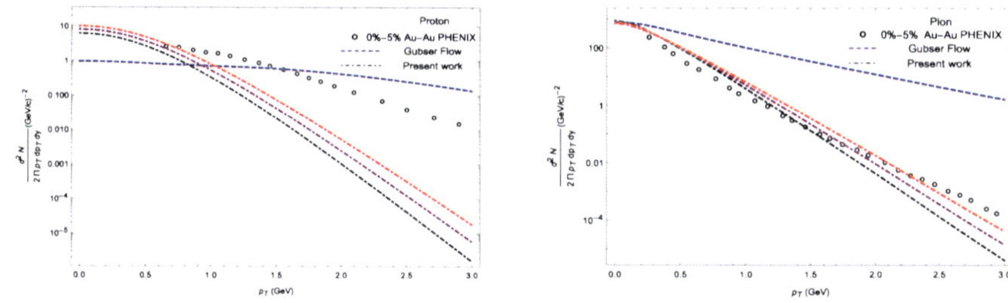

Fig. 73.1 Particles transverse momentum spectrum from central Au–Au collisions: black, purple and red lines correspond to a freeze out temperature of 140, 150 and 160 MeV, respectively. Circles: PHENIX data [15]

$$\begin{aligned} u_\mu &= (1, \lambda^2 u_\perp, 0, 0),\ b_\mu = (0, 0, \lambda b_\phi, 0),\ b^2 \equiv b^\mu b_\mu \\ \epsilon &= \epsilon_0(\tau) + \lambda^2 \epsilon_1(\tau, x_\perp),\ \epsilon_0(\tau) = \frac{\epsilon_c}{\tau^{4/3}} \end{aligned} \tag{73.7}$$

In such setup, the conservation equations (73.4), (73.5) reduce to the following partial differential equations

$$\begin{aligned} & u_\perp - \tau^2 \partial_\perp \left(\frac{u_\perp}{x_\perp}\right) - \tau^2 \partial_\perp^2 u_\perp - \tau \partial_\tau u_\perp + 3\tau^2 \partial_\tau^2 u_\perp \\ & - \frac{3\tau^{7/3}}{x_\perp \epsilon_c} b_\phi^2 - \frac{3\tau^{7/3}}{4\epsilon_c} \partial_\perp b_\phi^2 - \frac{9\tau^{10/3}}{4 x_\perp \epsilon_c} \partial_\tau b_\phi^2 - \frac{3\tau^{10/3}}{4\epsilon_c} \partial_\perp \partial_\tau b_\phi^2 = 0. \end{aligned} \tag{73.8}$$

For non-vanishing b_ϕ we assume a space-time profile of the magnetic field in central collisions in the form:

$$b_\phi^2(\tau, x_\perp) = B_c^2 \tau^{-1} \sqrt{\alpha} x_\perp e^{-\alpha x_\perp^2}. \tag{73.9}$$

We see that the magnitude of b_ϕ is zero at $x_\perp = 0$. Finally, one can find the solutions for transverse velocity $u_\perp$ and correspondingly modified energy density ϵ_1; for more details refer to [9].

73.3 Particle Transverse Momentum Spectrum

From the local equilibrium hadron distribution the transverse spectrum is calculated via the Cooper–Frye formula in the freeze out surface

$$\begin{aligned} S = E\frac{d^3N}{dp^3} &= \frac{g_i}{2\pi^2} \int_0^{x_f} x_\perp\, \tau_f(x_\perp)\, dx_\perp \left[m_T K_1\left(\frac{m_T u_\tau}{T_f}\right) I_0\left(\frac{m_T u_\perp}{T_f}\right) \right. \\ & \left. + p_T R_f K_0\left(\frac{m_T u_\tau}{T_f}\right) I_1\left(\frac{m_T u_\perp}{T_f}\right) \right] \end{aligned} \tag{73.10}$$

where $\tau_f(x_\perp)$ is the solution of the $T(\tau_f, x_\perp) = T_f$ and the degeneracy is $g_i = 2$ for both the pions and the protons. The above integral over $x_\perp$ on the freeze-out surface is evaluated numerically.

The spectrum (73.10) is illustrated in the following figures for three different values of the freeze out temperature (140, 150 and 160 MeV) and compared with experimental results obtained at PHENIX [15] in central collisions. Our proton spectrum appear to underestimate the experimental data, except at low p_T, but their behavior with $p_T$has the correct trend of a monotonically decrease. The pion spectrum, instead, appears in fair agreement with the experimental results, which are very close to the theoretical curves. This is an indication that hadrons with different masses have different sensitivities to the underlying hydrodynamic flow and to the electromagnetic fields. Indeed, the difference between the charge-dependent flow of light pions and heavy protons might arise because the former are more affected by the weak magnetic field than the heavy protons [16].

For comparison, we also show the results obtained by Gubser [17], which appear to be more flat and typically overestimate the experiment. We also notice that, for the proton case, the highest value of the freeze out temperature we employed (as suggested, e.g. in [18]) slightly brings (for protons) the calculation closer to the experimental data; however it also shows a kind of saturation phenomenon and points to the need of including other effects not considered in the present work.

References

1. K. Tuchin, Time and space dependence of the electromagnetic field in relativistic heavy-ion collisions. Phys. Rev. C **88**(2), 024911 (2013)
2. H. Li, X.-L. Sheng, Q. Wang, Electromagnetic fields with electric and chiral magnetic conductivities in heavy ion collisions. Phys. Rev. C **94**, 044903 (2016)
3. U. Gursoy, D. Kharzeev, K. Rajagopal, Magnetohydrodynamics, charged currents and directed flow in heavy ion collisions," Phys. Rev. C **89**(5), 054905 (2014)
4. Pu Shi, Di-Lun Yang, Transverse flow induced by inhomogeneous magnetic fields in the Bjorken expansion. Phys. Rev. D **93**, 054042 (2016)
5. V. Roy, S. Pu, L. Rezzolla, D. Rischke, Analytic Bjorken flow in one-dimensional relativistic magnetohydrodynamics. Phys. Lett. B **750** (2015)
6. S. Pu, V. Roy, L. Rezzolla, D.H. Rischke, Bjorken flow in one-dimensional relativistic magnetohydrodynamics with magnetization. Phys. Rev. D **93**, 074022 (2016)
7. G. Inghirami, L. Del Zanna, A. Beraudo, M. Haddadi Moghaddam, F. Becattini, M. Bleicher, " Numerical magneto-hydrodynamics for relativistic nuclear collisions. Eur. Phys. J. C **76**, 659 (2016)
8. M.H. Moghaddam, B. Azadegan, A.F. Kord, W.M. Alberico, Non-relativistic approximate numerical ideal-magneto-hydrodynamics of (1+1D) transverse flow in Bjorken scenario. Eur. Phys. J. C **78**, 255 (2018)
9. M.H. Moghaddam, B. Azadegan, A.F. Kord, W.M. Alberico, Transverse expansion of hot magnetized Bjorken flow in heavy ion collisions. Eur. Phys. J. C **79**, 619 (2019)
10. V. Roy, S. Pu, L. Rezzolla, D.H. Rischke, Effects of intense magnetic fields on reduced-MHD evolution in $\sqrt{s_{NN}} = 200$ GeV Au+Au collisions. Phys. Rev. C **96**, 054909 (2017)
11. A. Das, S.S. Dave, P.S. Saumia, A.M. Srivastava, Effects of magnetic field on the plasma evolution in relativistic heavy-ion collisions. Phys. Rev. C **96**, 034902 (2017)

12. J. Goedbloed, R. Keppens, S. Poedts, *Advanced Magnetohydrodynamics with Applications to Laboratory and Astrophysical Plasmas*. Cambridge University Press (2010)
13. A.M. Anile, *Relativistic Fluids and Magneto-Fluids*. Cambridge University Press (1989)
14. J.D. Bjorken, Phys. Rev. D **27**, 140 (1983)
15. K. Adcox et al., (PHENIX Collaboration), Formation of dense partonic matter in relativistic nucleus-nucleus collisions at RHIC: Experimental evaluation by the PHENIX collaboration. Nucl. Phys. A **757**, 184 (2005)
16. U. Gursoy, D. Kharzeev. E. Marcus, K. Rajagopal, C. Shen, Charge dependent flow induced by magnetic and electric field in heavy ion collisions. Phys. Rev. C **98**, 055201 (2018)
17. S.S. Gubser, Symmetry constraints on generalizations of Bjorken flow. Phys. Rev. D **82**, 085027 (2010)
18. C. Ratti, R. Bellwied, J. Noronha-Hostler, P. Parotto, I. Portillo Vazquez, J.M. Stafford, arXiv:1805.00088 [hep-ph]

Part X
Upgrades and New Experiments

Chapter 74
Upgrade of the NA61/SHINE Detector

Dariusz Tefelski

Abstract The NA61/SHINE detector is being upgraded during the Long Shutdown 2 period (2019–2021). In order to match the requirements for the upcoming open charm measurement program. The main goal of the upgrade is to accelerate the readout rate by a factor of 10 and increase acceptance in the high density tracks environment. The following elements of the detector are parts of the upgrade: Time Projection Chambers, Vertex Detector, Beam Position Detectors and Projectile Spectator Detector. Besides the detectors, a new Trigger and Data Acquisition system is being developed.

74.1 Introduction

The NA61/SHINE (SPS Heavy Ion and Neutrino Experiment) is a large acceptance, fixed target hadron spectrometer. It is located at the H2 beamline of the SPS synchrotron in CERN North Area. NA61/SHINE was approved by the CERN Research Board in 2007 and started taking physics data in 2009. It is a direct descendant of NA49 experiment [1]. The NA61/SHINE experiment facility is described in details in [2]. Design of the present data acquisition system is discussed in [3]. The main motivation for the experiment was to study the properties of the onset of deconfinement, search for the critical point of strongly interacting matter and precise reference measurements for neutrino physics (neutrino oscillation experiments) and for improving simulations of cosmic-ray air showers.

The physics program of NA61/SHINE for the period 2022–2024, after the Long Shutdown 2 (LS2), includes measurements of the open charm hadron production in Pb+Pb collisions for heavy ion physics, measurements of nuclear fragmentation cross-sections for cosmic ray physics and measurements of hadron production

---

Dariusz Tefelski for the NA61/SHINE Collaboration.

D. Tefelski (✉)
Faculty of Physics, Warsaw University of Technology,Warszawa, Poland
e-mail: dariusz.tefelski@cern.ch

D. Elia et al. (eds.), *The XVIII International Conference on Strangeness in Quark Matter (SQM 2019)*, Springer Proceedings in Physics 250,
https://doi.org/10.1007/978-3-030-53448-6_74

induced by proton, kaon and pion beams for neutrino-physics. A detailed description of this physics program is given in [4]. The goal of the upgrade of the NA61/SHINE experiment during the LS2 is to increase the trigger rate to 1 kHz, where the slowest detectors are Time Projection Chambers (TPC). Moreover, the acceptance and efficiency of Vertex Detector (VD) [5], and the radiation tolerance of the Projectile Spectator Detector (PSD) will be improved. Old and deteriorated detectors e.g. the trigger detectors, the Beam Position Detectors (BPD), the Time of Flight detectors (ToF) and outdated data acquisition systems based on CAMAC and FASTBUS buses will be replaced. On top of this new Trigger and Data Acquisition system (TDAQ) is developed.

74.2 Upgrade of the Subsystems

A sketch of the NA61/SHINE experiment including the location of the detector systems under upgrade is shown in Fig. 74.1. The experiment relies on eight TPCs, which will be upgraded with the readout electronics previously used in the ALICE-TPC. The Vertex TPC 1 and 2 are located inside superconducting magnets in a rather confined space. Which requires careful mechanical development and also proper routing and shielding of sensitive front end cards (FEC), adapters and cabling. As the ALICE FECs are bigger than the previously used boards, they do not match straight forwardly the mechanical constraints. The adapters and additional cables designed to overcome this point were considered to inject interference and noise. To counteract this phenomenon, additional shielding was developed and tested under real beam conditions before LS2. The data were analyzed and the feasibility of track reconstruction was confirmed. The tests demonstrates both, the need for the shielding and the success of the shielding measures implemented. Besides the cable adapters, an adapted water cooling system is required.

VD is of particular importance for charm measurement program. So far, the so-called Small Acceptance Vertex Detector (SAVD) was used in the NA61/SHINE experiment. This detector was based on MIMOSA-26 pixel sensors and a control and the TRBv3 readout boards as used in the prototype of the CBM Micro Vertex Detector (MVD). The TRBv3 boards were used for readout. Due to the limited data bandwidth of the 10 years old MIMOSA-26 sensors, this solution cannot be operated at a 1 kHz trigger. Therefore it was decided to use the state-of-the-art ALPIDE sensors which provide a higher rate capability and a substantially lower dark rate. For reading out the ALPIDE sensors, MOSAIC boards were chosen, as tests in similar configuration to NA61/SHINE were already performed in the ALICE Collaboration. A setup with one stave was also successfully tested in the NA61/SHINE environment.

The new BPD and trigger detectors (start detectors) will be realized from scintillating fibres (SciFi) and diamond based sensors. The precise configuration is still under debate. The readout of the detectors will be done with DRS4 boards (designed at the University of Geneva), which act as versatile fast waveform recorders (at 1 GS/s). Signals from start detectors will pass through constant fraction discriminators

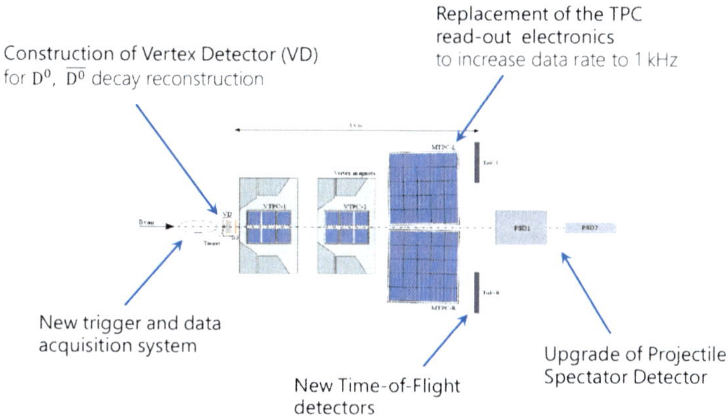

Fig. 74.1 Planned upgrades in the NA61/SHINE experiment. Beam is coming from the left side, first to trigger and beam position detectors, then to Vertex Detector and fixed target inside, then tracks of charged particles created in collisions are bent in magnetic field and registered by TPCs and ToF downstream. Finally, projectile spectators hit PSD

to specialized FPGAs with a custom configuration suited for the NA61/SHINE. This trigger system will also include a busy logic for all incorporated detectors. The heart of the trigger system is CAEN V2495 FPGA VME module which will cooperate with CAEN V812 CFD and CAEN V1290 MHTDC modules. Details are provided in the related Technical Design Report [6].

The ToF subsystem will base on Multigap Resistive Plate Chamber (MRPC) technology. This device is designed by the MPD/BM@N Collaboration at the Joint Institute for Nuclear Research (JINR) and will need dedicated closed loop gas system as a greenhouse gas (Freon) will be used. The readout electronics will incorporate DRS4 boards. The new MRPC detectors were tested on beam at NA61/SHINE experiments in place of old ToF-left wall and satisfactory time resolution was achieved (47.5 ps).

The Projectile Spectator Detector was already divided in two separate parts: Main PSD with a through-hole and the Forward PSD. This design will have much better radiation tolerance and signal dynamic range. Readout will be performed also with DRS4 boards.

The most important part of the upgrade consists in extending the data bandwidth of the experiment by means of designing a new TDAQ. The increased data rate forces to distribute the data recording over a multiple computers. Moreover, we conclude that it will be impossible to store all raw data from detectors. Therefore, various data reduction algorithms have to be applied online. Both, the data recording and reduction will be carried out by a novel server farm composed from 160 computers with 12 cores each. Block diagram of new TDAQ system is presented in Fig. 74.2.

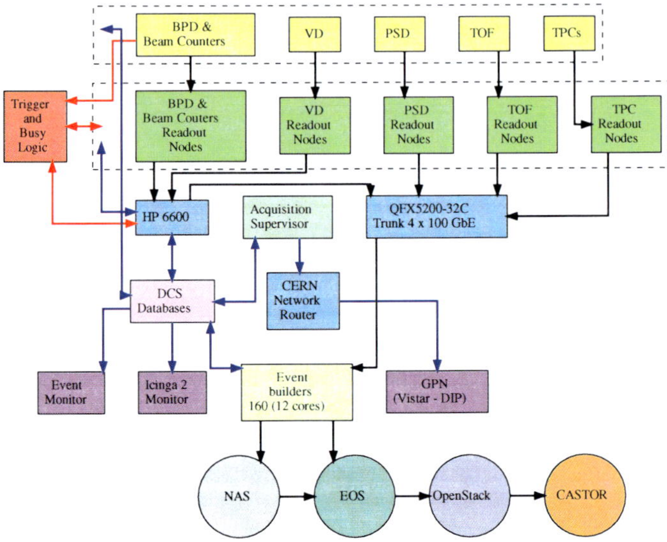

Fig. 74.2 Block diagram of TDAQ system. Red lines depict trigger and busy, blue lines denote Detector Control System and Monitoring network, black lines show data flow from detectors up to CERN data storage system. Blue boxes represent key elements of the network infrastructure

74.3 Conclusions

Tests which were performed on H2 line in experimental hall of the NA61/SHINE experiment showed that ALICE electronics work well with NA61/SHINE Time Projection Chambers. Further electronic matching and software preparation are ongoing. New Vertex Detector will be build using ALPIDE sensors which will help to achieve 1 kHz readout rate and will improve signal to noise ratio. Tests performed with DRS4 boards showed that these are versatile waveform recording boards suitable for data taking with detectors such as Projectile Spectator Detector, Trigger Detectors and Time-of-Flight detectors. The work on TDAQ system design is ongoing.

Acknowledgments The NA61/SHINE upgrade profits from synergies and the kind support of ALICE, CBM and MPD/BM@N. Thanks to them for the excellent collaboration and help. This work was supported by the Ministry of Science and Higher Education, Poland (DIR/WK/2016/2017/10-1).

References

1. S. Afanasiev et al., [NA49 Collaboration], The NA49 large acceptance hadron detector. Nucl. Instrum. Meth. A **430**, 210–244 (1999)
2. N. Abgrall et al., [NA61/SHINE Collaboration], NA61/SHINE facility at the CERN SPS: beams and detector system. JIST **9**, P06005 (2014)

3. A. László et al., Design and performance of the data acquisition system for the NA61/SHINE experiment at CERN. Nucl. Instrum. Meth. A **798**, 1–11 (2015)
4. A. Aduszkiewicz et al., [NA61/SHINE Collaboration], Addendum to the NA61/SHINE Proposal, SPSC-P-330-ADD-10 (2018)
5. M. Deveaux et al., [NA61/SHINE Collaboration], The small acceptance vertex detector for NA61/SHINE. EPJ Web. Conf. **171**, 21001 (2018)
6. P. Podlaski, Technical design report for the upgrade of the Trigger System of the NA61/SHINE experiment, [inner NA61/SHINE document] (2019)

Chapter 75
Status and Performance of the Detector Upgrades for STAR in the BES-II and Beyond

Florian Seck

Abstract The STAR collaboration has installed three detector upgrades for the beam energy scan phase II program at RHIC to aid the exploration of the QCD phase diagram. After BES-II the forward upgrade of STAR will enable novel measurements in pp, pA and AA collisions motivated by cold QCD and heavy-ion physics. Results from the commissioning of the new detectors and their performance during the first year of running in BES-II will be discussed together with the general progress of BES-II and the status of the forward upgrade.

75.1 Introduction

One of the driving forces for the construction of the Relativistic Heavy Ion Collider (RHIC) was the aim to study the QCD phase-diagram over a wide range of temperatures and baryon densities. The beam energy scan phase I program (BES-I) conducted between 2010 and 2014 revealed interesting hints in several observables that are considered to be sensitive to the turn-off of QGP signatures and a first-order transition in the QCD phase diagram at collision energies below $\sqrt{s_{\mathrm{NN}}} = 20\,\mathrm{GeV}$ [1]. These observables warrant a closer investigation with a beam energy scan phase II program (BES-II) that covers seven center-of-mass energies ranging from 7.7 to 19.6 GeV in the years 2019–2021 as shown in Fig. 75.1. Multi-differential studies will profit from datasets that are typically a factor 20 larger compared to the statistics collected during BES-I. In addition the STAR detector was upgraded with three new subsystems to enhance its experimental reach. Supplementing the collider program the STAR collaboration developed a fixed-target program (FXT) to measure heavy-ion collisions down to $\sqrt{s_{\mathrm{NN}}} = 3\,\mathrm{GeV}$. Collisions at 7.7 GeV can be measured in both setups and serve as an important cross check.

Florian Seck for the STAR collaboration.

F. Seck (✉)
Technische Universität Darmstadt, 65289 Darmstadt, Germany
e-mail: F.Seck@gsi.de

D. Elia et al. (eds.), *The XVIII International Conference on Strangeness in Quark Matter (SQM 2019)*, Springer Proceedings in Physics 250,
https://doi.org/10.1007/978-3-030-53448-6_75

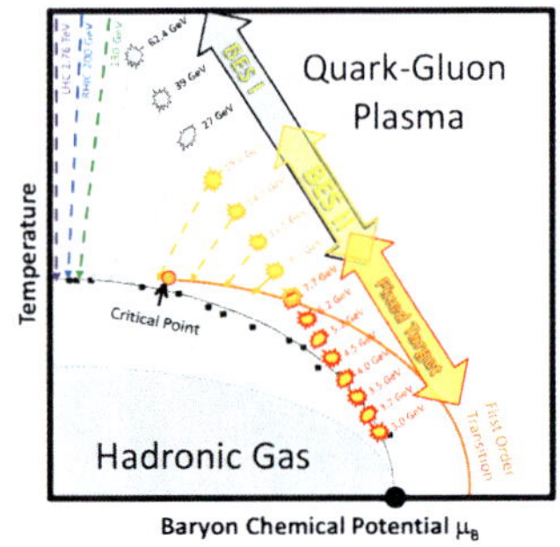

Fig. 75.1 Sketch of the QCD phase diagram with the BES-I and II programs highlighted

The RHIC program in the years 2021 and beyond will be devoted to address fundamental questions in both cold QCD and heavy-ion physics opening a portal towards the Electron-Ion Collider (EIC). RHIC has the unique capability to accelerate polarized protons and provide data in a kinematic regime in which nuclear modifications of the sea quark and gluon distributions are expected to be sizable and currently poorly constrained. The nuclear parton distribution functions (nPDF) for gluons and sea-quarks can be accessed with processes that experience no QCD final-state effects: the direct photon and Drell-Yan production. The forward upgrade at STAR will enable key measurements in pp, pA and AA collisions to improve for example our understanding of the spin of the proton and the longitudinal structure of the initial state in heavy-ion collisions. It consists of the Forward Tracking System (FTS) and the Forward Calorimeter System (FCS), both covering an acceptance of $2.5 < \eta < 4$ on the west side of STAR [2, 3].

75.2 BES-II

To allow for higher luminosity at the lowest BES-II collider mode energies the first RF linac-based electron cooler LEReC (Low Energy RHIC Electron Cooling) is under development. In 2019 electron cooling for two beams with a life time extension and a reduced transverse size was demonstrated. Further commissioning is ongoing to provide stable operation in 2020/21. Data at the first two collider energies ($\sqrt{s_{\mathrm{NN}}} =$ 19.6 and 14.6 GeV) were collected from February to July 2019 with a surplus of events compared to the original expectation. In addition one fixed-target dataset at $\sqrt{s_{\mathrm{NN}}} = 3.2\,\mathrm{GeV}$ was completed and small subsets of data were taken at several other fixed-target energies [4].

The Event Plane Detector [5] (EPD) was already fully installed for the data taking in 2018. With its higher granularity and larger acceptance compared to the Beam-Beam Counters (BBC), it improves significantly the event plane reso- lution and supplies means to define the centrality independent from the tracks measured in the TPC. In 2019 the EPD was also included into the minimum-bias trigger. An

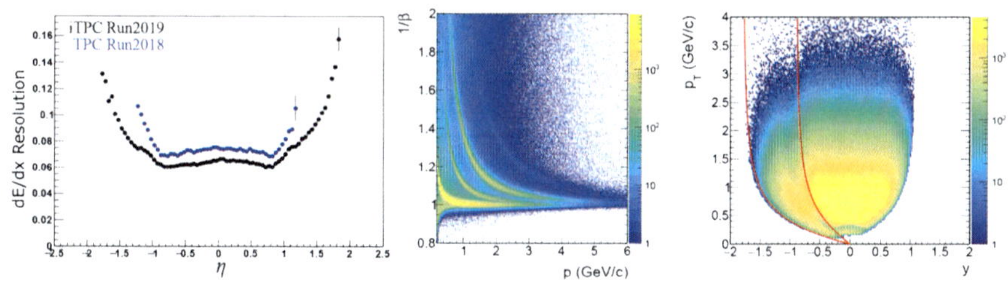

Fig. 75.2 Left: Comparison of dE/dx resolution in the TPC before and after the iTPC upgrade. Middle: $1/\beta$ measured with eTOF versus momentum (GeV/c). Right: Raw phase-space distribution of protons identified via m^2 cut in bTOF and eTOF in Au + Au collisions at $\sqrt{s_{\mathrm{NN}}} = 14.6\,\mathrm{GeV}$

asymmetry cut on the activity in the east and west side EPD detectors was developed to veto beampipe collisions at the lowest collider energies.

The upgrade of the inner Time Projection Chamber [6] (iTPC) increased the segmentation of the inner pad plane of the TPC and therefore the maximum number of TPC hits a track can have from 45 to 72. This leads to an extended rapidity coverage (from $|\eta| < 1$ to $|\eta| < 1.5$) and low p_{T} acceptance (from $p_{\mathrm{T}} > 125\,\mathrm{MeV}/c$ to $p_{\mathrm{T}} < 60\,\mathrm{MeV}/c$). The increased number of hits also improves the particle identification (PID) via the ionization energy loss dE/dx by 15–30% as shown in the left panel of Fig. 75.2.

The endcap Time Of Flight detector [7] (eTOF) installed on the east pole tip of STAR complements the particle identification capabilities at forward rapidities in collider mode allowing to study the rapidity dependence of several key observables from BES-I. It is also essential to provide PID in the fixed-target program at mid-rapidity. With online calibrations a time resolution better than 85 ps has been achieved in accordance with the design goals. The middle panel of Fig. 75.2 shows an eTOF PID plot with particle bands separated over a large momentum range. The right panel of Fig. 75.2 shows for Au + Au collisions at 14.6 GeV the raw phase-space distribution of protons identified via a squared mass m^2 cut in the barrel TOF (bTOF) and the eTOF. The acceptance in which the eTOF contributes is framed by the red lines $-1.8 < \eta < -0.9$ which is larger than the nominal eTOF acceptance of $-1.6 < \eta < -1.1$ due to a wide selection of event vertexes along the beam direction.

75.3 Forward Upgrade

The forward tracking system (FTS) provides precision tracking with a position resolution better than $100\,\mu\mathrm{m}$, charge discrimination, photon suppression and p_{T} measurements at forward rapidity. It consists of 3 layers of silicon mini-strip discs located 140–190 cm along the beam line from the center of the TPC and four layers of double-sided small-strip Thin Gap Chambers (sTGC) 270–360 cm away from the nominal interaction point inside the magnet pole tip opening (see Fig. 75.3 left panel). The

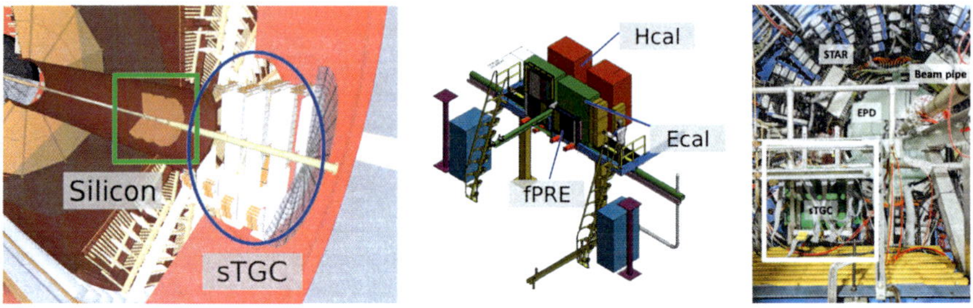

Fig. 75.3 Left: Schematic view of the FTS. Middle: Schematic view of the FCS. Right: sTGC prototype installed at STAR in 2019

detector design for the silicon tracker builds on the STAR experience with the Intermediate Silicon Tracker installed between 2014 and 2016. A first prototype of a silicon module was assembled and tested in the lab. A $30 \times 30\,\text{cm}^2$ sTGC prototype (see Fig. 75.3 right panel) was tested with cosmic rays, integrated into the STAR DAQ and successfully commissioned with beam during the last weeks of RHIC operation in 2019. A full-size prototype ($60 \times 60\,\text{cm}^2$) will be installed at STAR for the upcoming run in 2020. A new tracking algorithm for the forward region in a non-uniform magnetic field is under development.

The forward calorimeter system (FCS) consists of a preshower scintillator hodoscope (fPRE) as well as electromagnetic and hadronic calorimeters (Ecal, Hcal) shown in the middle panel of Fig. 75.3. The Ecal reuses a lead-scintillator sandwich calorimeter from PHENIX. The Hcal will be the first hadronic calori-meter at STAR and utilize $10 \times 10\,\text{cm}^2$ towers out of iron-scintillator sandwich. The readout for all three sub-detectors will be based on SiPMs. Large-scale prototypes were tested at Fermilab in April 2019 demonstrating that the requirements on the energy resolution can be met. At the end of the 2019 beam operation at RHIC, prototypes of all FCS components were installed at STAR for commissioning with 200 GeV Au + Au collisions.

75.4 Summary

STAR has extended its experimental reach for the BES-II program with three detector upgrades (EPD, iTPC, eTOF) that will not only provide high quality data during the BES-II but also enable unique opportunities for mid-rapidity physics at top RHIC energy. Datasets at the two highest collider energies (19.6 and 14.6 GeV) and several fixed-target energies have been collected in 2019. The forward upgrade at STAR is targeted at addressing fundamental questions in QCD paving the way towards the EIC. Prototypes of the new tracking and calorimeter systems have been tested at STAR and will be ready for the post-BES-II era in autumn 2021 and beyond.

References

1. STAR Collaboration: STAR Note 598 (2014), https://drupal.star.bnl.gov/STAR/starnotes/public/sn0598
2. STAR Collaboration: STAR Note 648 (2016), https://drupal.star.bnl.gov/STAR/starnotes/public/sn0648
3. STAR Collaboration: STAR Note 669 (2017), https://drupal.star.bnl.gov/STAR/starnotes/public/sn0669
4. STAR Collaboration: STAR Note 721 (2019), https://drupal.star.bnl.gov/STAR/starnotes/public/sn0721
5. STAR Collaboration: STAR Note 666 (2016), https://drupal.star.bnl.gov/STAR/starnotes/public/sn0666
6. STAR Collaboration: STAR Note 644 (2015), https://drupal.star.bnl.gov/STAR/starnotes/public/sn0644
7. STAR Collaboration: STAR Note 665 (2016), https://drupal.star.bnl.gov/STAR/starnotes/public/sn0665

Chapter 76
ALICE Inner Tracking System Upgrade: Construction and Commissioning

Domenico Colella

Abstract ALICE (A Large Ion Collider Experiment) is the CERN LHC experiment optimized for the study of the strongly interacting matter produced in heavy-ion collisions and devoted to the characterization of the Quark–Gluon Plasma. To achieve the physics program for LHC Run 3 and 4, ALICE is undergoing a major upgrade of the experimental apparatus during the ongoing second long LHC shutdown. A key element of the ALICE upgrade is the substitution of the Inner Tracking System (ITS) with a completely new silicon-based detector whose features will allow the reconstruction of rare physics channels which could not be accessed before with the ITS layout used during LHC Run 1 and 2. The enabling technology for such performance boost is the adoption of custom-designed MAPS (Monolithic Active Pixel Sensors) as detecting element. In this proceedings, the status of the construction and commissioning of the ITS upgrade will be detailed. The completion of the modules construction will be achieved soon and, in the meantime, the commissioning in laboratory is proceeding using the components already integrated in the detector.

76.1 ALICE Inner Tracking Upgrade

The main objective of the ALICE experimental program for LHC Run 3 and 4 is a detailed exploration of the Quark–Gluon Plasma properties via high precision measurements of rare probes in pp, p–Pb and Pb–Pb collisions [1]. In order to be able to measure a large sample of short lived systems such as heavy-flavor hadrons, quarkonia, and low mass dileptons over a wide range of transverse momenta, it is necessary to enhance the tracking and readout rate capabilities with respect to the ones of the detector that worked during the LHC Run 1 and 2. During the LHC Long

---

Domenico Colella for the ALICE Collaboration.

D. Colella (✉)
Istituto Nazionale di Fisica Nucleare, via E. Orabona 4, Bari 70125, Italy
e-mail: domenico.colella@cern.ch

D. Elia et al. (eds.), *The XVIII International Conference on Strangeness in Quark Matter (SQM 2019)*, Springer Proceedings in Physics 250,
https://doi.org/10.1007/978-3-030-53448-6_76

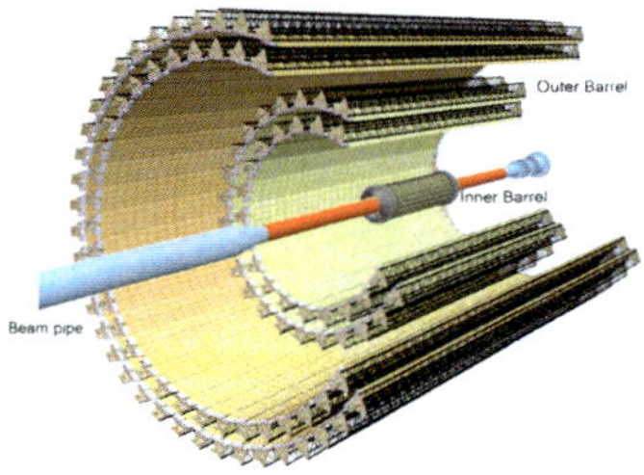

BARREL	LAYERS	RADIUS (cm)	LENGHT (cm)	NUM. STAVES	NUM. CHIPS
Inner	Layer 0	22.4	271	12	108
Inner	Layer 1	30.1	271	16	144
Inner	Layer 2	37.8	271	20	180
Outer	Layer 3	194.4	843	24	2688
Outer	Layer 4	243.9	843	30	3360
Outer	Layer 5	342.3	1475	42	8232
Outer	Layer 6	391.8	1475	48	9408

Fig. 76.1 (Left) Schematic layout of the upgraded ITS. (Right) Summary table containing the detector layers geometric characteristics

Shutdown 2 in 2019–2020, several sub-detectors will be upgraded and the Inner Tracking System (ITS) will be replaced.

The key improvements of the ITS upgrade are [2]: a completely new layout envisaging seven cylindrical layers (three innermost layers referred as Inner Barrel, IB, and four outermost layers referred as Outer Barrel, OB) all equipped with ALPIDE Monolithic Active Pixel Sensors (MAPS) chips of about $30 \times 30\ \mu m^2$ pixel size, a large reduction of the material budget ($\sim$0.3% X_0 for the IB and $\sim$1% X_0 for the OB), and a reduced distance of the innermost layer to the interaction point. These design characteristics will allow a strong improvement of the tracking efficiency at low transverse momenta as well as the impact parameter resolution [3]. A schematic view of the detector layout and a table reporting geometrical details for each layer are shown in Fig. 76.1.

The very low value for the material budget has been obtained adopting monolithic silicon pixel chips, integrating the sensor and the readout electronics functionalities in the same substrate, thinned to 50 μm in the IB (100 μm in the OB). The pixel chips are mounted on low-mass polyimide Flexible Printed Circuits (FPC, for powering and data stream), glued on a carbon fiber structure (space frame), equipped with water cooling pipes integrated in the structure.

The chips (9 for the IB and 14 for the OB) glued and wire-bonded to the FPC constitute the Hybrid Integrated Circuit (HIC). The HIC(s) (1 for the IB, 8 or 14 for the two innermost and the two outermost layers of the OB respectively) glued to the space frame constitute the Stave. More details on the ALPIDE chip characteristics and R&D can be found in [4].

76.2 Component Production Status

The electrical and functional tests of the ALPIDE chips have been performed at CERN (CH) for the 50 μm thinned chips and at Yonsei and Pusan/Inha (KR) for the 100 μm thinned chips, between September 2017 and May 2019. A total of almost 4×10^4 chips have been tested with a detector-grade quality yield of 63.7%.

The full production of HICs and Staves for the IB has been carried out at CERN and concluded at mid 2019. A total of 95 Staves, enough to build two copies of the three inner barrel layers, have been assembled with a yield of 73%. The OB HICs have been produced in five sites (Bari (IT), Liverpool (UK), Pusan/Inha, Strasbourg (FR) and Wuhan(CN)), while the FPCs have been tested in Trieste (IT). It took 80 weeks to assembly the needed 2500 HICs; this quantity includes the HICs needed to cover the whole detector acceptance plus spares and assumes an overall production yield of 74% (convolution of 82% for the HIC and 90% for the Stave). The HIC production has been completed and a yield of 84% has been achieved. The OB Staves have been assembled in five sites (Berkeley (US), Daresbury (UK), Frascati (IT), NIKHEF (NL) and Torino (IT)). Enough Staves to cover the full OB acceptance have been assembled and qualified as detector grade. The Stave assembly yield is close to 90% and production of few more spare Staves will be completed by October 2019.

Full readout logic is implemented in the ALPIDE chip that sends the digitized and zero-suppressed hit data to the off-detector electronics. A total of 192 FPGA-based readout units [5] (CERN, Bergen (NO), NIKHEF) control and monitor the sensors and their power supply modules, receive the trigger and detector control information, and deliver the sensor data to the counting room. A total of 142 boards (Berkeley), able to provide analog and digital 1.8 V supplies to each HIC plus a negative voltage output for the reverse bias, are needed to power the full detector. Production and qualification of the full set of readout and power boards are completed.

Detailed description of the mechanical support structure layout can be found in [2]. All the components have been produced and verified (Berkeley, CERN, Padova (IT)). A dry insertion test of a dummy version of IB half-detector has been successfully carried out. Cables placing in the supporting structures is ongoing and required smart solutions to fit all in the best way.

76.3 Layer Assembly and Commissioning

A large clean room has been built at CERN to allow the full detector assembly and the on-surface commissioning activities, before the installation in the ALICE cavern in May 2020. Here the same backend system that will be used in the experiment is available, including powering system, cooling system, full readout chain. Integration of the Staves in the layer structure and connection to the services is done once the Staves are shipped to CERN, in parallel with the commissioning on the already installed part of the system.

Detector Control System (DCS) and Data Acquisition System (DAQ) softwares are available and running on machines housed in a control room adjacent to the clean room. Starting from June 2019, acquisition of the first data using the Staves in the first available half-layer, IB-HL0 (innermost layer of the IB), began. Threshold, DAC tuning and noise occupancy scans are periodically performed to monitor the performance of the detector. One example of threshold map obtained after the

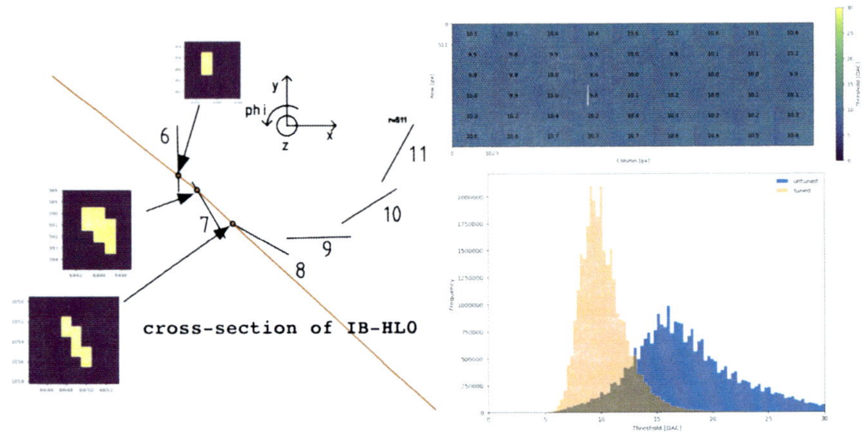

Fig. 76.2 (Left) First cosmic ray identified in the IB-HL0, including cluster shape of the track hit in the three crossed Staves. (Right) Map of the threshold value for all the pixels in the six Staves included in the IB-HL0 after tuning of the DAC parameters

tuning of the relevant DAC parameters and the effect of this tuning on the threshold distribution is visible in the right part of Fig. 76.2 for the IB-HL0. Still in the same Figure, the representation of the first reconstructed track of a cosmic ray is shown. As can be seen from the detector sketch, the track reconstruction is possible thanks to the partial overlap of adjacent Staves. It is also interesting to notice how the shape of the pixel cluster connected to the hit changes in the Staves sequentially crossed: in the first Stave (number 6), where the particle crossing is almost perpendicular, the cluster is small; in the second and third Staves (numbers 7 and 8), where the angle between the track and the chip plane decreases, the cluster becomes bigger and starts to resemble a piece of track.

From July 2019 the detector is kept powered and running; continuous monitoring is provided by three shift crews alternating all along the 24 h. Software development and hardware integration proceed in parallel as well as preparation of all the infrastructures in the experimental area for the final installation.

References

1. B. Abelev et al., (ALICE Collaboration). J. Phys. G **41**, 087001
2. B. Abelev et al., (ALICE Collaboration). J. Phys. G **41**, 087002
3. P. Camerini (On behalf of ALICE Collaboration). PoS **(EPS-HEP 2017)**, 487 (2017)
4. G. Aglieri Rinella, (On behalf of ALICE Collaboration). J. NIM A **845**, 583–587 (2017)
5. J. Schambach et al., https://doi.org/10.1109/NSSMIC.2018.8824419

Chapter 77
LHCb Fixed Target Results and Prospects

Luciano Libero Pappalardo

Abstract LHCb has the unique capability to study collisions of the LHC beams on fixed targets. Internal gas targets of He, Ne and Ar have been used so far to collect samples corresponding to integrated luminosities up to 0.1 pb^{-1}. An upgraded target, allowing a wider choice of target gas species and to increase the gas density by up to two orders of magnitude, is going to be installed for the LHC Run 3. Results and prospects on open and hidden charm productions are presented, which can provide crucial constraints on cold nuclear-matter effects and nPDF at large x. These measurements, together with production of $\bar{p}$ and other light hadrons, are of great interest to cosmic-ray physics as well.

77.1 The LHCb Fixed-Target Systems SMOG and SMOG2

Fixed-target pA collisions with a proton beam at the TeV scale allow to cover a broad and diversified physics program at unique kinematic conditions. In particular, due to the substantial boost of the reaction products in the laboratory frame, fixed-target collisions measured with a forward spectrometer, such as LHCb, allow to access the backward center-of-mass rapidity region ($-3 < y^* < 0$), corresponding to the poorly explored high x-Bjorken regime. Furthermore, the use of a gaseous target has the advantage of allowing for a broad variety of nuclear targets, thus providing novel probes for the study of the nucleon and nuclear structure, and for measurements of great interest ranging from heavy-ion physics to cosmic rays physics and Dark Matter (DM) search.

Among the main LHC experiments, LHCb is the only detector that can already run both in collider and fixed-target mode. The LHCb fixed-target system, called SMOG (System for Measuring the Overlap with Gas), was originally conceived for

---

Luciano Libero Pappalardo for the LHCb Collaboration.

L. L. Pappalardo (✉)
Universitá di Ferrara and INFN Sezione di Ferrara, 44122 Ferrara, Italy
e-mail: pappalardo@fe.infn.it

D. Elia et al. (eds.), *The XVIII International Conference on Strangeness in Quark Matter (SQM 2019)*, Springer Proceedings in Physics 250,
https://doi.org/10.1007/978-3-030-53448-6_77

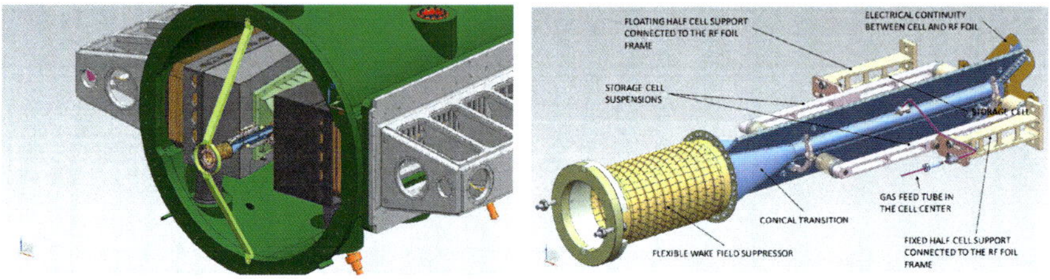

Fig. 77.1 Left: SMOG2 storage cell mounted inside LHCb, in front of the VELO detector. Right: details of the storage cell

precise luminosity determinations through the beam-gas imaging technique [1]. The SMOG system allows to inject a low flow rate of noble gas into the vacuum vessel of the LHCb VErtex LOcator (VELO) detector. A temporary local pressure bump of about 10^{-7} mbar is obtained in the LHCb beam-pipe section, which is about two orders of magnitude higher than the nominal LHC vacuum pressure and one order of magnitude lower than the LHC vacuum interlock limit. The resulting beam-gas collision rate allows for a precise determination of the beam density profiles. As an additional important feature, SMOG gives the unique opportunity to operate LHCb in fixed-target mode, and to study proton-nucleus and nucleus-nucleus collisions on various target types and at different center-of-mass energies. Several dedicated runs have already been performed since 2015 using He, Ar, or Ne targets with proton and Pb beams, and first interesting physics results have recently been published [2, 3]. The system is now being upgraded (SMOG2) with the implementation of a storage cell for the target gas [4, 5] to be installed at the upstream edge of the VELO, coaxial with the LHC beam. This new setup has the major advantage of increasing by up to two orders of magnitude the target areal density (and thus the luminosity), by injecting the same amount of gas of SMOG. Other important advantages are: the implementation of a more sophisticated gas-feed system with multiple injection lines, which will allow for a precise (at few percent-level) determination of the target density and for the injection of additional gas species (H, D, N, O, Kr and Xe); a well defined interaction region (limited by the cell length, 20 cm); the possibility to run in parallel (i.e. simultaneously) with the collider mode, thanks to the significantly displaced interaction regions for the two configurations. The final design of the storage cell and its arrangement inside the VELO vessel are shown in Fig. 77.1. Possible scenarios for the integrated luminosities to be collected with SMOG2 using different gas targets during the 3 years of LHC Run 3 data-taking are reported in [6].

77.2 Physics Results with SMOG

The LHCb Collaboration has recently reported first measurements of J/Ψ and D^0 production using fixed-target p–He and p–Ar collisions with SMOG, at $\sqrt{s_{NN}} = 87$ GeV and $\sqrt{s_{NN}} = 110$ GeV, respectively [3]. Events with J/Ψ and D^0 candidates

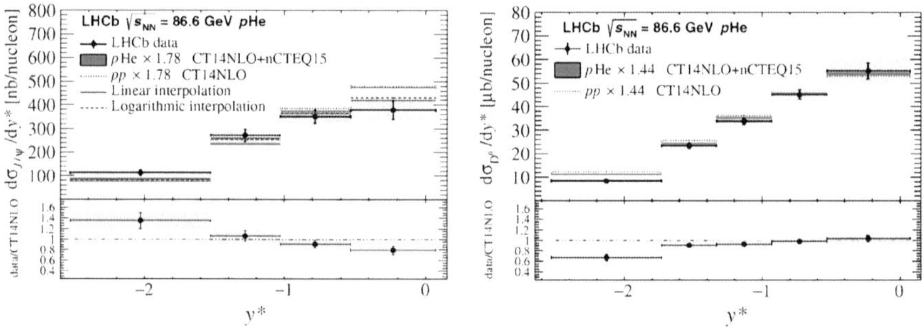

Fig. 77.2 Differential cross section for prompt production of J/Ψ (Left) and D^0 (Right) in a He target as a function of the center-of-mass rapidity compared with HELAC-ONIA predictions based on different parametrizations of nPDFs (details in [3])

have been selected by requiring a reconstructed primary vertex in the fiducial region -200 mm $< z_{PV} <$ 200 mm, where z_{PV} denotes the primary vertex position along the beam direction. While J/Ψ candidates are obtained by reconstructing $\mu^+\mu^-$ pairs originating from the Interaction Point (IP), D^0 candidates are reconstructed through the $K\pi$ decay products, requiring both tracks to be originated from a common displaced vertex. Prompt J/Ψ and D^0 signal yields are then obtained through an unbinned Maximum Likelihood fit of the invariant mass distribution of the respective decay products. The measured differential cross section for prompt production of J/Ψ and D^0 on a He target as a function of the center-of-mass rapidity y^* are shown in Fig. 77.2. These measurements are relevant for the study of cold nuclear-matter effects and allow to constrain the nPDFs parametrizations in the high-x region. The comparison between data and HELAC-ONIA predictions in the most negative y^* bin, corresponding to the Bjorken-x region $0.17 < x < 0.37$, does not reveal clear effects of intrinsic charm contributions, expected to be substantial at high x [7].

The SMOG system has also been exploited for direct measurements of antiproton production in p–He collisions at $\sqrt{s_{NN}} = 110$ GeV [2]. These measurements are of great relevance in the context of DM search. An excess of antiprotons over current predictions based on spallation of primary cosmic rays on interstellar medium (H and He) has been recently observed by the space-borne PAMELA and AMS-02 experiments. However, present predictions for the $\bar{p}/p$ flux ratio from the known production sources are limited by large uncertainties on $\bar{p}$ production cross sections, especially from He. At LHCb antiprotons are clearly identified by the RICH detectors in the wide momentum range $12 < p < 100$ GeV/c. Figure 77.3 shows, as an example, the measured differential production cross section as a function of the antiproton momenta in the transverse-momentum range $1.2 < p_T < 2.8$ GeV/c. Noteworthy, the experimental uncertainties are much smaller than the theoretical predictions spread.

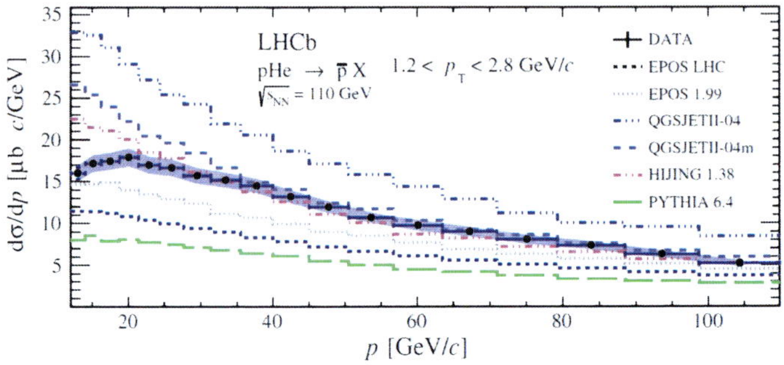

Fig. 77.3 Antiproton production cross section per He nucleus as a function of momentum in the $1.2 < p_T < 2.8$ GeV/c range. The data points are compared with predictions from theoretical models

77.3 Future Prospects

The implementation of the upgraded SMOG2 system will allow to continue and extend these studies with a significantly increased statistical precision and with the possibility to use additional target species, ranging from H up to Kr and Xe. Relevant achievements, in the field of cold-nuclear matter effects and heavy ion physics, include the possibility to measure prompt beauty production in pA collisions and to study the sequential charmonia production and QGP-related flow observables and correlations in PbA collisions. Interesting studies can also be performed in the field of cosmic rays physics. Measurements of $\bar{p}$ production from anti-hyperon decays (e.g. $\bar{\Lambda}$, $\bar{\Sigma}$) will be possible, and the production of light anti-nuclei ($\bar{d}$, $\bar{He}$) is also foreseen. Furthermore, thanks to the possibility to use also H and D targets, one can measure the ratios $\sigma(pHe \rightarrow \bar{p}X)/\sigma(pH \rightarrow \bar{p}X)$ and $\sigma(pD \rightarrow \bar{p}X)/\sigma(pH \rightarrow \bar{p}X)$. Finally, SMOG2 operated with H and D targets offers unique conditions to probe quark and gluon PDFs in nucleons and nuclei, especially at high-x and moderately-high Q^2, where present experimental data are largely insufficient to constraint the theoretical distributions.

References

1. C. Barschel, Precision luminosity measurements at LHCb with beam-gas imaging. CERN-THESIS-2013-301, RWTH Aachen University (2014)
2. LHCb collaboration, R. Aaij et al., Measurement of antiproton production in pHe collisions at $\sqrt{s_{NN}} = 110$ GeV. Phys. Rev. Lett. 121, 222001 (2018). arXiv:1808.06127
3. LHCb collaboration, R. Aaij et al., First measurement of charm production in its fixed-target configuration at the LHC. Phys. Rev. Lett. 122, 132002 (2019). arXiv:1810.07907v2
4. V. Carassiti et al., SMOG2 Technical Proposal. CERN-PBC-Notes-2018-007 (2018), http://cds.cern.ch/record/2651269

5. LHCb collaboration SMO,G2 Technical Design Report (2018), http://cds.cern.ch/record/2673690/files/LHCB-TDR-020.pdf
6. PBC Report, Physics opportunities with the fixed target program of the LHCb experiment using an unpolarized gas target. LHCb-PUB-2018-015 (2018)
7. J. Pumplin, H.L. Lai, W.K. Tung, The Charm Parton content of the nucleon. Phys. Rew. D **75**, 054029 (2007). arXiv:hep-ph/0701220

Chapter 78
Perspectives on Strangeness Physics with the CBM Experiment at FAIR

Iouri Vassiliev

Abstract The main goal of the CBM experiment at FAIR is to study the behavior of nuclear matter at very high baryonic density. This includes the exploration of the high density equation of state (EoS), search for the transition to a deconfined and chirally restored phase, critical endpoint. The promising diagnostic probes for this new states are the enhanced production of multi-strange (anti-)particles. The CBM detector is designed to measure such rare diagnostic probes multi-differentially with unprecedented precision and statistics. Important key observables are the production of hypernuclei. The discovery and investigation of new (doubly strange-)hypernuclei and of hyper-matter will shed light on the hyperon-nucleon and hyperon-hyperon interactions.

78.1 Introduction

Heavy-ion collision experiments allow to create and investigate extreme states of strongly interacting matter in the wide range of collision energies and colliding systems. In nature, such extreme states are predicted to exist in the cores of neutron stars or inside the colliding neutron stars. In the Compressed Baryonic Matter (CBM) experiment at the Facility for Antiproton and Ion Research (FAIR) ultra-high net-baryon densities and moderate temperatures will be reached in heavy-ion collisions. The structures in the QCD phase diagram like the critical endpoint followed by a first order phase transition at large baryon chemical potential or even new phases, such as quarkyonic matter, are predicted by different models [1, 2].

Experimental discovery of the structures in the QCD phase diagram provide an unique information about properties of strongly interacting matter at extreme states, with fundamental consequences for our understanding on the structure of neutron

Iouri Vassiliev for the CBM Collaboration.

I. Vassiliev (✉)
GSI Helmholtz Center for Heavy Ion Research GmbH, Darmstadt, Germany
e-mail: i.vassiliev@gsi.de

D. Elia et al. (eds.), *The XVIII International Conference on Strangeness in Quark Matter (SQM 2019)*, Springer Proceedings in Physics 250,
https://doi.org/10.1007/978-3-030-53448-6_78

stars, chiral symmetry restoration, and the origin of hadron masses. The CBM experiment will measure very rare probes including multi-strange (anti-)hyperons and hypernuclei [3] at interaction rates of up to 10 MHz.

78.2 CBM Detector

The CBM detector is a forvard spectrometer designed for heavy-ion collisions at SIS100 at FAIR. The SIS100 synchrotron will provide heavy-ion (up to Au) beam energies from 2 to 11 AGeV. The expected collision rate is in the range of 10^4–10^7 collisions per second. The Micro-Vertex Detector (MVD) comprises four stations which are placed between 4 and 20 cm downstream the target and inside vacuum. It will host highly-granular, next-generation Monolithic Active Pixel Sensors, with a spatial precision of 5 μm, a time resolution of 5 μs, and a peak rate capability of 700 kHz/mm^2. Eight double-sided Silicon Tracking Stations (STS) used as a main tracker. MVD and STS detectors are located inside a magnetic dipole field. Electrons are identified with the Ring Imaging Cherenkov (RICH) detector and the Transition Radiation Detector (TRD), hadrons are identified by the Time-of-Flight (TOF) detector. The CBM detector allows to replace RICH and the first TRD layers with sofisticated muon detector (MUCH) in order to provide muons identification. The setup is completed by the Projectile Spectator Detector (PSD) for centrality and reaction plane determination. All sub-systems are designed to provide rate capability and precision to the CBM physics program with rare probes. The high-speed data acquisition system will collect time-stamped subdetector signals.

78.3 First-Level Event Selection

The First-Level Event Selection (FLES) package of the CBM experiment is created to reconstruct online the full event topology including tracks of charged particles, short-lived multi-strange particles and hypernuclei, low mass vector mesons and resonances in about 200 channels. Tracks of the charged particles are reconstructed by the Cellular Automaton (CA) track finder [5] using STS or MVD plus STS hits. The Kalman filter (KF) based track fit is used for precise estimation of the track parameters and parameters errors. Covariance matrix contains essential information about tracking and detector performance. The KF particle finder [6] is used in order to find and reconstruct parameters of short-lived particles by combining the already found tracks of long-lived charged particles. The FLES package is platform and operating system independent.

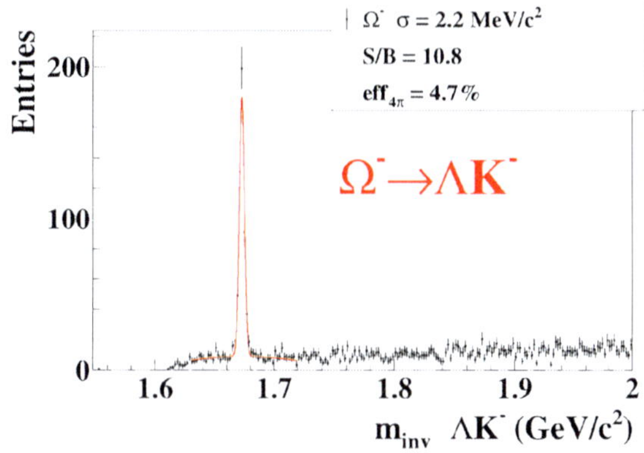

Fig. 78.1 The reconstructed invariant-mass distribution of ΛK^- pairs in $5 \cdot 10^6$ central Au + Au PHSD collisions at 10 AGeV. The red line indicates the signal plus background fit by a polynomial plus Gaussian function

78.4 Multi-strange Particles

The strange (anti-)particle production has been proposed as one of the most sensitive observables that could spot out the creation of a Quark-Gluon-Plasma (QGP) during the early stages of a heavy-ion collisions. Droplets of QGP allow to interact $s\bar{s}$-quarks and create more multi-strange (anti-)baryons. Moreover, yield of multi-strange hyperons at FAIR energy is sensitive to the effect of the chiral symmetry restoration [7]. To study the performance of multi-strange hyperon and hypernuclei reconstruction, several sets of $5 \cdot 10^6$ central Au + Au Parton-Hadron-String Dynamics (PHSD) [8] events at the FAIR energy range have been simulated. High statistics allows to calculate directly the reconstruction efficiency even for rarely produced $\bar{\Omega}^+$ particles, avoiding signal embedding into simulated events. The TOF detector identifies hadrons i.e. pions, kaons, protons and fragments, i.e. d, t, ^{3}He and ^{4}He in the angular range covered by the STS detector. A typical reconstructed ΛK^- invariant-mass spectrum is shown in Fig. 78.1. The red line indicates the signal and background fit by a polynomial plus Gaussian function. The Ω^- reconstruction efficiency results in about 4.7% for central PHSD events. A good signal to background ratio of about 10.8 is observed due to the TOF K^- and proton particles identification procedure and sophisticated $\Omega^- \rightarrow \Lambda K^-$ decay reconstruction with the KF Particle package.

78.5 Hypernuclei

Singly and doubly strange hypernuclei will be produced in heavy-ion collisions with the maximum yield in the region of SIS100 energies [9, 10]. Beam energies available at SIS100 appear to be especially well suited for generating signals of the phase transition, and offer an opportunity to address fundamental scientific questions: how far can we extend the chart of nuclei towards the third (strange) dimension by pro-

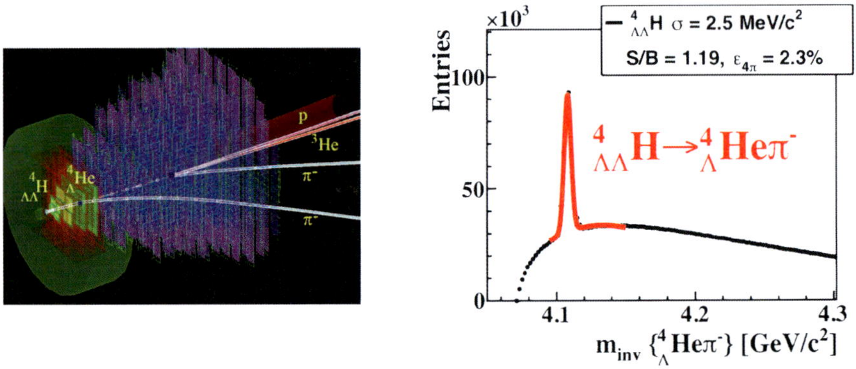

Fig. 78.2 A reconstructed $^4_{\Lambda\Lambda}$H→ π^- $^4_\Lambda$He → $2\pi^- p^3$He decay topology (left). The reconstructed invariant-mass spectrum of $^4_\Lambda$Heπ^- (right) in 10^{12} central Au + Au collisions at 10 *A*GeV. The red line indicates the signal plus background fit by a polynomial plus Gaussian function

ducing singly and doubly strange hypernuclei? Does strange matter exist in the form of heavy multi-strange objects? The current KF Particle package includes searching for the bound state of $\{\Lambda n\}_b$ and $\{\Lambda nn\}_b$, singly strange $^3_\Lambda$H, $^4_\Lambda$H, $^4_\Lambda$He, $^5_\Lambda$He, doubly strange hypernuclei-$^4_{\Lambda\Lambda}$H, $^5_{\Lambda\Lambda}$H and $^6_{\Lambda\Lambda}$He and corresponding anti-particles. A reconstructed $^4_{\Lambda\Lambda}$H → π^- $^4_\Lambda$He → $2\pi^- p^3$He decay is shown in Fig. 78.2(left). $^4_{\Lambda\Lambda}$H hypernuclei are created at the primary vertex and decay to π^- $^4_\Lambda$He several centimeters downstream of the target, $^4_\Lambda$He decay in the STS detector to $\pi^- p^3$He. Presence of the ^{3}He fragment in a 3-prong detached vertex is a unique signature of such decay. The reconstructed invariant-mass spectrum of $\pi^- p^3$He is shown in Fig. 78.2(right). Combinatorial background events is due to misidentified high-momentum hadrons in the sample of ^{3}He selected particles.

Assuming the $^4_{\Lambda\Lambda}$H production multiplicity for central Au + Au events at 10 *A*GeV is about $1 \cdot 10^{-4}$ [9, 10], a reaction rate of 10^6 central events/s, a branching ratio of 10% for two sequential weak decays, and an efficiency of 2.3%, one would expect to measure within one week about 14k of $^4_{\Lambda\Lambda}$H.

78.6 Conclusions

The main goal of the CBM research program is to explore the QCD phase diagram in the region of neutron star core densities. This includes study of the EoS of nuclear matter and search for new hypothetic phases. The start version of the CBM detector will be able to measure rare diagnostic probes, such as multi-strange (anti-)hyperons, hypernuclei and lepton pairs with unprecedented precision and statistics. In order to achieve the required precision for multi-differential observables, the measurements will be performed at reaction rates between 10 and 10 MHz.

References

1. J. Luecker et al., arXiv:1308.4509v1 [hep-ph]
2. L. McLerran, R.D. Pisarski, Nucl. Phys. A **796**, 83 (2007)
3. I. Vassiliev et al., JPS Conf. Proc. **17**, 092001 (2017). https://doi.org/10.7566/JPSCP.17.092001
4. C. Pinkenburg et al., Phys. Rev. Lett. **83**, 1295 (1999)
5. I. Kisel, Nucl. Instrum. Meth. **A566**, 85 (2006)
6. I. Kisel et al., IEEE Trans. Nucl. Sci. **60**(5), 3703 (2013)
7. W. Cassing, A. Palmese, P. Moreau, E.L. Bratkovskaya, Phys. Rev. **C93** (2016)
8. W. Cassing, E.L. Bratkovskaya, Nucl. Phys. A **831**, 215 (2009)
9. A. Andronic et al., Phys. Lett. B **697**, 203 (2011)
10. H. Stoecker et al., Nucl. Phys. A **827**, 1077 (2009)

Part XI
Future Experiments, Facilities and Physics Perspectives

Chapter 79
Status and Prospects at NICA

Vladimir Kekelidze, Vadim Kolesnikov, Viktor Matveev, and Alexander Sorin

Abstract The new accelerator complex NICA (Nuclotron-based Ion Collider fAcility) is under construction at JINR (Dubna, Russia). The main physics goal of the NICA project is the experimental exploration of the properties of nuclear matter under extreme condition. The MPD experiment is foreseen for study heavy-ion collisions at the NICA collider. The current status of the NICA project is presented.

79.1 Introduction

Relativistic nucleus-nucleus collisions provide the unique possibility for experimental investigation of dense nuclear matter in the laboratory. The structure of the phase diagram of strongly interacting matter in the region of high baryon density has gained a great attention in recent years. Since the existing theoretical methods are limited in describing fundamental characteristics of nuclear matter at finite baryochemical potential, including confinement and chiral symmetry breaking, new experimental data in this domain are of great importance. Due to these reasons, a new project has been launched at JINR aimed in building of a new accelerator facility for heavy ions and conducting comprehensive studies of nuclear matter in the range of collision energies where the net-baryon density has the maximum [1]. The NICA scientific program will address a variety of physics phenomena, including study of the nature of phase transitions in dense hadronic matter, in-medium modification of hadron properties, features of hyperon-nucleon interaction and critical phenomena in the QCD matter at high baryon density [2]. In particular, an energy scan will be performed in the energy range $4 < \sqrt{s_{NN}} < 11$ GeV with the intent to investigate the nuclear matter Equation of State (EoS), to study the excitation function of the production of hadrons and (hyper)nuclei, to investigate the energy and system

V. Kekelidze (✉) · V. Kolesnikov · V. Matveev · A. Sorin
Joint Institute for Nuclear Research, 141980 Dubna, Russia
e-mail: kekelidze@jinr.ru

A. Sorin
National Research Nuclear University (MEPhI), 115409 Moscow, Russia

D. Elia et al. (eds.), *The XVIII International Conference on Strangeness in Quark Matter (SQM 2019)*, Springer Proceedings in Physics 250,
https://doi.org/10.1007/978-3-030-53448-6_79

size dependence for event-by-event fluctuations and correlations, and to address the modifications of dilepton invariant mass spectra in dense medium [3]. For a proper normalization of the measurements from heavy-ion collisions, a study of elementary reactions is foreseen. The energy range at the NICA collider will be extended to the lower energy region with the Nuclotron accelerator extracting beams focusing on the investigation of the reaction dynamics, in-medium hadron properties, production of multistrange hyperons at the threshold, and search for hyper-nuclei [4]. Production of strangeness is of particular interest. Strange hadrons are valuable probes in the study of reaction dynamics and particle production mechanisms. Moreover, it was predicted that if the deconfinement phase transition has happened in a heavy-ion collision, enhanced production of strange hadrons (relative to the yields from elementary pp reactions) is expected [5]. The enhancement of strangeness increases with the strangeness content exhibiting non-trivial energy and centrality dependence. In order to better understand the nature and dynamics of phase transformations in hadronic matter, new precise experimental data on the production of hyperons are requested, including results for the yields, rapidity distributions, transverse momentum spectra, and azimuthal anisotropy. Strongly interacting matter, created in a heavy-ion collision at finite impact parameter, has a non-zero vorticity due to a large global orbital angular momentum in the system. The vorticity field may cause the global polarization characterized by the polarization of the secondary particles along the direction of this orbital momentum. Experimental data from the RHIC accelerator and theoretical calculations indicate that the polarization degree of produced hyperons is growing towards NICA energies [6]. Dense nuclear matter favors formation of exotic nuclei with strange degrees of freedom—hypernuclei. Hypernuclear spectroscopy is a unique experimental method to probe the properties of the hyperon-nucleon interactions. Since the energy range of the NICA covers the maximum-density region for net-baryons, the yields of nuclear clusters with strangeness are predicted to be enhanced considerably [7].

79.2 The NICA Accelerator Complex

Figure 79.1 (left panel) shows the NICA accelerator complex including an injector, a booster, the Nuclotron accelerator, and a collider. The NICA injector, which comprises a heavy-ion source and linear accelerator for heavy ions, should provide beams of ions (up to gold nuclei) with a design intensity of $2 \cdot 10^9$ particles per cycle. The booster synchrotron with its magnetic ring of 211 m circumference placed inside of the window of the existing Dubna Synchrophasotron yoke should accelerate ions up to 600 MeV/nucleon energy. The superconducting synchrotron Nuclotron, which is in operation since 1993 and recently was essentially upgraded, should provide p, d, and heavy-ion beams with the maximum kinetic energy per nucleon of 5.8 GeV for $A/Z = 0.5$ specie and 4.5 GeV for lead nuclei. The collider is composed of two rings, of about 503 m circumference each, separated vertically. The magnetic structure of the collider is based on double-aperture superconductive (SC) magnets, which pro-

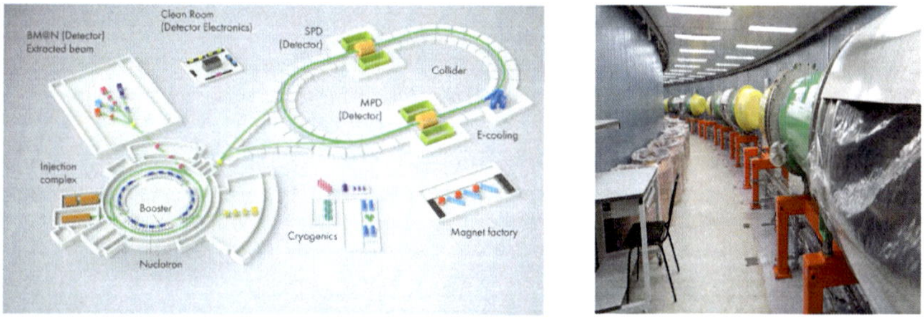

Fig. 79.1 Left: A schematic view of the NICA accelerator complex. Right: Booster magnets in the tunnel

vide the maximum field up to 1.8 T. Given to the modern electron and stochastic cooling systems at NICA, the achieved collider luminosity for gold ions will be of the order of 10^{27} cm^{-2}s^{-1}. NICA is also the accelerator of polarized protons and deuterons. It should provide beams of polarized protons up to center-of-mass energy of 27 GeV with the luminosity of 10^{32} cm^{-2}s^{-1}.

At present, all magnets for the booster has being constructed and tested, commissioning of the booster is ongoing (see Fig. 79.1, right panel). The Nuclotron accelerator is now routinely used to provide beam for fixed target experiments. Serial fabrication of the magnets for the NICA collider and construction of the new NICA infrastructure, including buildings, transfer lines, service and supply systems, is progressing according to plan.

79.3 The MPD Detector at NICA

The MultiPurpose Detector (MPD) at the NICA collider is a large acceptance solenoidal detector designed to detect hadrons, electrons and gammas [8]. It consists of several subsystems (see Fig. 79.2) located inside a superconducting solenoid generating a magnetic field of up to 0.6 T. In order to ensure the good tracking resolution, the magnetic field inhomogeneity inside the MPD tracking volume should be less than 10^{-3}.

The main tracker in the MPD detector is a cylindrical Time-Projection Chamber (TPC) 3.4 m long and 2.8 m in diameter. The TPC covers the pseudorapidity range $|\eta| < 2$. It provides more than 95% efficiency in track reconstruction with a relative momentum resolution $\Delta p_t/p_t$ below 3% within $0.2 < p_t < 2$ GeV/c. In addition to that, the TPC will enable particle identification via the specific energy loss (*dE/dx*) measurement with a precision better than 8%. At forward pseudorapidities tracking in TPC will be supplemented by a multi-layer end cap tracker (ECT) located just after the TPC end plates. The Inner Tracker (IT), which will be made of several layers

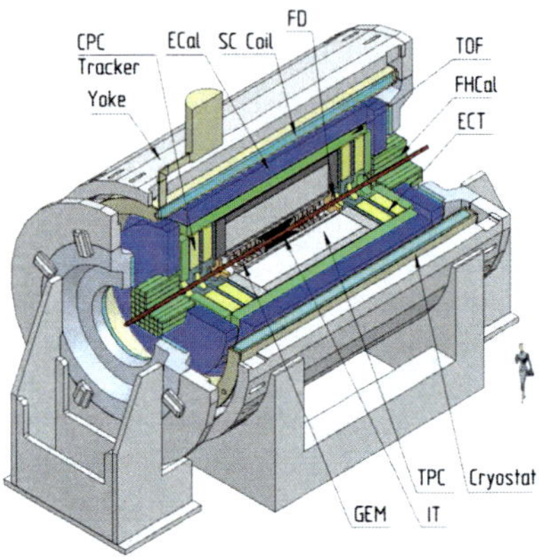

Fig. 79.2 A view of the MPD detector at the NICA collider

of double-sided silicon microstrip detectors. Its primary function will be to provide precision tracking points allowing determination of the position of secondary decay vertices displaced from the primary vertex. The MPD Time-of-Flight (TOF) system, which is situated just behind the TPC, covers a 25 m^2 area over the pseudorapidity range $|\eta| < 1.4$. The purpose of the TOF system, based on the multigap resistive plate chamber (MRPC) technology, is to extent the TPC particle identification capability into the intermediate momentum region. With the overall time resolution of the system of 80 ps, π/K separation up to 1.3 GeV/c and K/p separation up to 2 GeV/c will be real. The Electromagnetic Calorimeter (ECAL) is designed to measure the spatial position and energy of electrons and photons. It comprises of about 38000 of lead-scintillator sampling towers of 18 radiation length thickness, which are organized in a projective geometry to ensure uniform detector performance of the measured value of the energy deposit over a wide pseudorapidity range. Arrays of quartz counters (FD) should define the minimum-bias trigger and provide fast timing to the experiment. Two arms of hadron calorimeters (FHCAL) covering the peudorapidity range $2.5 < |\eta| < 4$ will detect the energy deposit of forward going particles providing data for the offline centrality estimate and event plane analysis. A more detailed description of the detector components can be found elsewhere [9].

The MPD Collaboration is now in the process of industrial production of detector elements. For example, the yoke for the unique MPD superconductive magnet has been delivered to Dubna recently (see Fig. 79.3, left panel), the magnet solenoid has been fabricated and now is under tests (see Fig. 79.3, right panel).

79.4 Results of Feasibility Study at NICA

In order to get realistic estimates for the performance of the MPD experimental setup, a dedicated ROOT-based software framework—MPDRoot has been developed [10]. The MPRoot system comprises interfaces to various event generators and provides

Fig. 79.3 Left: The yoke of the MPD magnet. Right: The MPD magnet solenoid

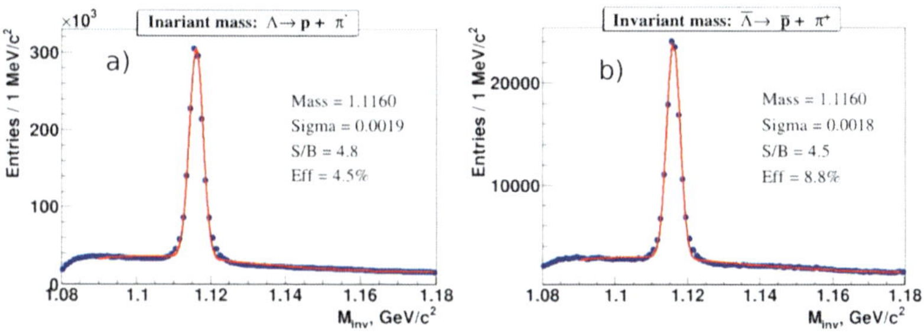

Fig. 79.4 Invariant mass spectra for: Λ (**a**) and $\overline{\Lambda}$ (**b**)

propagation of particles through the detector material based on the GEANT package. A collection of special class libraries describes relevant responses of all MPD subsystems and provides multiple reconstruction algorithms. In the following, several selected results of multiple Monte Carlo studies are outlined, more feasibility study results for MPD can be found elsewhere [11].

The MPD performance for hyperons and hypernuclei measurements was studied for the combination of TPC and TOF detectors within the midrapidity region ($|\eta| < 1.2$). A sample of 2×10^6 minimum bias Au+Au collisions at $\sqrt{s_{NN}} = 9$ GeV from the PHSD event generator [12] was used as an input for the study of hyperon production, and the event generator DCM-QGSM [13], implementing a coalescence-based algorithm for (hyper)nuclei formation, was used for simulation of hypernuclei. The analysis procedure includes particle propagation, detector response simulation, track reconstruction, primary and secondary vertex finding, and particle identification. Hyperons and hypernuclei were reconstructed using the secondary vertex finding technique with an optimized set of topological and track quality cuts in order to reduce the combinatorial background. This allowed invariant mass spectra of Λ-hyperons and $\bar{\Lambda}$ in the whole p_t-range to be plotted as in Fig. 79.4. The estimated efficiencies and signal-to-background (S/B) ratios are also indicated in the figure.

Figure 79.5a shows the invariant mass distribution of (^{3}He, π^-) pairs after subtraction of the combinatorial background. The results for reconstruction of hyperhelium-4 are shown in Fig. 79.5b. Given to the results achieved in MPD feasibility study, we

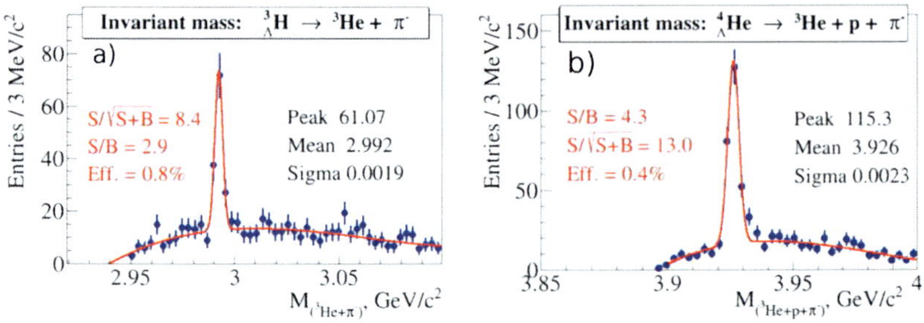

Fig. 79.5 **a** Invariant mass spectra of (^{3}He, π^-) pairs; and **b** invariant mass spectra for (^{3}He, p, π^-) candidates

can estimated that with a typical event rate of 7 kHz for the design NICA luminosity a sample of about 10^5 hypertritons can be collected in a week of data taking.

79.5 Summary

Realization of the NICA project at JINR will provide new opportunities for the experimental exploration of nuclear matter of the highest baryonic density. The construction of the accelerator complex and experimental setups at NICA is going close to the schedule enabling the start-up configurations of the collider and detectors be ready in 2022.

References

1. V.D. Kekelidze, R. Lednicky, V.A. Matveev, I.N. Meshkov, A.S. Sorin, G.V. Trubnikov, Eur. Phys. J. A **52**, 211 (2016)
2. A.S. Sorin, V.D. Kekelidze, A.D. Kovalenko et al., Nucl. Phys. A **855**, 510–513
3. V. Golovatyuk, V. Kekelidze, V. Kolesnikov, O. Rogachevsky, A. Sorin, Eur. Phys. J. A **52**, 212 (2016)
4. Mikhail Kapishin for the BM@N Collaboration. Eur. Phys. J. A **52**, 213 (2016)
5. J. Rafelski, B. Müller, Phys. Rev. Lett. **48**, 1066 (1982)
6. M.I. Baznat, K.K. Gudima, A.S. Sorin, O.V. Teryaev, Phys. Rev. C **93**, 031902 (2016)
7. J. Steinheimer, K. Gudima, A. Botvina et al., Phys. Lett. B **714**, 85 (2012)
8. V. Golovatyuk, V. Kekelidze, V. Kolesnikov, O. Rogachevsky, A. Sorin, Nucl. Phys. A **982**, 963–966 (2019)
9. K.U. Abraamyan et al., Nucl. Instrum. Meth. A **628**, 99 (2011)
10. http://mpd.jinr.ru
11. Vadim Kolesnikov for the MPD Collaboration, Universe **4**(12), 145 (2018)
12. W. Ehehalt, W. Cassing, Nucl. Phys. A **602**, 449 (1996)
13. N.S. Amelin et al., Sov. J. Nucl. Phys. **52**, 272 (1990)

Chapter 80
Prospects for Electron-Ion Colliders

Silvia Dalla Torre

Abstract Several projects of electron-ion colliders are being proposed worldwide involving a broad scientific community. The overall panorama is presented, with emphasis on the projects with shorter timescale and more mature technology, namely the EIC in USA and the LHeC proposed at CERN. In particular, the first level of the decisional process for EIC is expected within 2019.

80.1 The Worldwide Panorama

A compilation of the proposals for electron-ion colliders around the world from [1] is presented in Table 80.1. The number of the proposed projects and their worldwide distribution test of the wide scientific interest: it is estimated that the involved community includes 2000 scientists [1].

All the projects foresee, beyond electron scattering off proton, scattering off a variety of nuclei. Electron and light nuclei polarization is a major tool characterizing some of the projects. A large variety of center of mass energy E^{cm} and luminosity L can be observed, corresponding more to complementary aspects of the projects than to competition. The most promising new proposals couple high energy and high luminosity. Therefore, they offer the possibility of precision measurements and extended exploration in large phase-space domains.

The double handle of high E^{cm} and high L was not available till now. In fact, so far, all the experiments dedicated to electron-hadron scattering have been fixed target ones, a part the experiments at HERA, DESY. In fix target experiments, electron beams can provide high luminosity, even if at modest E^{cm}, as in the past experiments at SLAC and the ongoing program at Jlab. Higher energies are available at the unique high energy μ beam at CERN SPS, where high L is more difficult. Nevertheless, key steps in the exploration of Deep Inelastic Scattering (DIS) physics have been

S. Dalla Torre (✉)
INFN, Sezione di Trieste, Trieste, Italy
e-mail: Silvia.DallaTorre@ts.infn.it
URL: https://www.users.ts.infn.it/dallator/SilviaDALLATORRE/

D. Elia et al. (eds.), *The XVIII International Conference on Strangeness in Quark Matter (SQM 2019)*, Springer Proceedings in Physics 250,
https://doi.org/10.1007/978-3-030-53448-6_80

Table 80.1 Overview of proposed electron-hadron colliders. (data from [1])

Facility	Years	E^{cm} (GeV)	Luminosity (10^{33}cm^{-2}s^{-1})	Ions	Polarization
EIC in US	>2028	20–100 → 140	2–30	p → U	e, p, d, ^{3}He, Li
EicC in China	>2028	16–34	1 → 100	p → Pb	e, p, light nuclei
(HE-)LHeC at CERN	>2030	200–1300 (1800)	10	Depends on LHC	e possible
PEPIC at CERN	>2025	530 → 1400	<10^{-1}	Depends on LHC	e possible
VHEeP at CERN	>2030	1000–9000	10^{-5}–10^{-4}	Depends on LHC	e possible
FCC-eh at CERN	>2044	3500	15	Depends on FCC-hh	e possible

performed with the experiments EMC, BCDMS, NMC, SMC and COMPASS, the last one still in operation.

HERA, the only e-p collider so far, has offered extremely high energy even if with modest *L*. Also the HERMES experiment, running at the HERA electron beam and making use of an internal gas jet target, was limited in *L*.

The proposed collider projects have very different maturity level and some of them are very far in time. In particular, FCC-eh at CERN, included in the recent CERN FCC Conceptual Design Report [2] cannot become operational in less than 25 years, while PEPIC and VIEeP [3] are based on a novel electron acceleration technology, namely proton-driven plasma wakefield acceleration, currently been developed. They promise extremely high E^{cm}, even if with low *L*.

EIC [4], EicC [5] and LHeC [6], discussed in the following sections, are projects based on better consolidate technologies with perspectives for physics in a time term of about ten years.

80.2 The Large Hadron Electron Collider (LHeC) at CERN

In this proposal, one of the high energy LHC ion beams collides with electrons at maximum nominal energy of 60 GeV; 40 GeV and 50 GeV options are also been considered. The electrons are accelerated by an Energy Recovery Linac (ERL), a technology under development thanks to the project PERLE based in Orsay. The same scheme can later be proposed for the HE-LHC and the FCC. The corresponding E_{cm} is 1.3 TeV (LHC), 1.8 TeV (He-LHC) and 3.5 TeV (FCC). The LHeC expected ep *L* is 10^{34} cm^{-2}s^{-1}, 10^{31} cm^{-2}s^{-1} for eD and 3×10^{32} cm^{-2}s^{-1} for ePb.

The LHeC scientific case is multi-folded. The capability for high resolution exploration of the nucleon and nucleus structure in an extremely wide phase space domain will contribute to progress in QCD understanding. Present limitations in the knowledge of the gluon structure function at small and at large x^{Bjorken} will be overcome thanks to high energy and high L, respectively. LHeC has a mission in empowering the LHC program, where the uncertainties in Parton Distribution Functions (PDF) limit the precision in measuring the Higgs properties and the search of Physics Beyond the Standard Model (BSM). It can also act as low-cost Higgs factory, where the Higgs boson is produced via the WW or ZZ channel, with production rates lower, but comparable to those of dedicated e^+e^- factories [7]. The clean pattern of lepton scattering events offers novel opportunities in search of exotic particles.

80.3 The Electron-Ion Collider (EIC) in USA

80.3.1 The Electron-Ion Collider Accelerator Facility

The facility specifications are characterized by a wide coverage in E_{cm}, high luminosity, a large variety of nuclear beams and the polarization of the beams of both electron and light nuclei. More in detail: E_{cm} in the range 20–140 GeV; nuclear beams from H to Pb/U; L up to 10^{34} cm^{-2}s^{-1} for ep collisions; highly polarized (80%) e beams; highly polarized (70%) p beams; interaction regions specifically designed to ensure full 4-π coverage.

Two options are proposed [8] in two different sites: JLEIC to be constructed at Jefferson Lababoratory (Jlab) and eRHIC to be constructed at Brookhaven National Labaratory (BNL). Both designs benefit from existing nuclear physics infrastructure and are based on the same accelerator principles: electron storage rings with frequent injection of fresh polarized beams and hadron storage rings with strong cooling or alternatively frequent injections.

JLEIC profits of the existing CEBAF facility providing 12 GeV polarized electron, while all the other facility components are new. Electrons from CEBAF are injected in an eight-shaped storage ring. The h complex consists in a source, acceleration stages and an eight-shaped collider ring. The eight-figure ring design ensures polarization preservation, polarization management and spin-flip. High L is via frequent e and ion injection, full-size ion injection and multistage magnetized electron cooling of the ion beam. Operation at 140 GeV is possible changing the magnet technology in order to use 12 T bending dipoles.

eRHIC makes use of the whole existing RHIC complex, with minimal modifications: the storage ring, injectors, ion sources, tunnel and experimental halls. The electron complex is new and it can provide polarized electrons with energy up to 18 GeV, injected on energy in a novel storage ring added in the existing RHIC tunnel. In particular, the design of the e rapid cycling synchrotron is innovative: all systematic depolarizing resonances in e acceleration are suppressed up to energies higher than 18 GeV. High luminosity can profit of the technologically challenging strong h cooling. A valid alternative is the use of the second RHIC superconducting ring, for on-energy injections into the collider ring, replacing the hadron bunches after 1 h of storage.

A crossing angle of the two beams is necessary to avoid parasitic collisions due to short bunch spacing, make space for machine elements, improve detection and reduce detector background. The angle is 50 mrad at JLEIC and 25 mrad at eRHIC.

80.3.2 Detectors for Physics at the Electron-Ion Collider

The physics opportunities at EIC are intimately connected to the overall design of the experiments and to the performance of the required detectors. From the experimental point of view, the broad physics EIC program can be accomplished by the study of (i) inclusive reactions, (ii) semi-inclusive and (iii) exclusive processes. Correspondingly, the main requirements for the experimental apparatus follow, adding more and more complexity while moving from reactions (i) to (iii):

- Precise scattered electron identification and extremely fine resolution in the measurement of its angle and energy, where the electromagnetic calorimeter system must also provide highly efficient trigger capabilities for inclusive reactions (i) and all the other experimental channels; the pseudorapidity η interval to be cover is $|\eta| <3.5$ and electron-hadron discrimination at the 10^{-4} level results in different requirements for the energy resolution in different phase space regions: from 2 to $3\%/\sqrt(E)$ to $7\%/\sqrt(E)$; very low material budget in the rear direction of the order of 5% X_0 is also required;
- More is needed to access the semi-inclusive processes (ii): excellent hadron identification over a wide momentum and rapidity range, from below 1 GeV/c up to 50 GeV/c, full 2π acceptance for tracking and momentum analysis and excellent vertex resolution of 10–20 μm by a low-mass vertex detector;
- Exclusive reactions (iii) impose the capability to accurately reconstruct all particles in the event, the complete hermeticity of the setup with very forward detectors as Roman pots, and large acceptance zero-degree calorimetry to effectively detect neutrons from the breakup of nuclei;
- The entire experimental program will require the precise control of the luminosity;
- The polarized beams impose the use of electron and light nucleus polarimeters.

Four detector concepts are being discussed, all including the basic ingredients mentioned above. **TOPSIDE** is almost fully based on silicon detectors. **ePHENIX** is a detector, primarely designed for eRHIC, where the sPHENIX setup is completed with hermeticity and particle identification. sPHENIX TPC performs central tracking. In **BEAST**, also designed for eRHIC, the central tracking is again by a TPC in a strong magnetic field of 3 T. The dedicated **JLEIC** detector uses a set of gaseous trackers around the central vertex detector.

80.3.3 The Physics Case for the Electron-Ion Collider

EIC mission is understanding QCD with particular attention paid to the non-perturbative effects. The goal is to answer key questions as the complete 3-D description of nucleons and nuclei, the understanding of their properties, including the mass, from their parton composition, the role of the colour in the interaction between coloured partons and uncoulored particles and the formation of coulorless confined states. The following examples test of the many EIC potentialities.

A fundamental tool for the exploration depicted above is polarization. The open question concerning the nature the nucleon spin can only be attacked with high precision and high statistic measurements to explore the elusive contribution of the orbital angular momentum of the partons. The polarized function distributions depending on the transverse polarization are very marginally known and need EIC for a systematic exploration.

The study of the evolution of the parton see in the nuclei needs high statistics, variety of nuclei and variable E_{cm}. The intriguing question of the gluon saturation at small values of x^{Bjorken} can be attacked with delicate exclusive measurements in diffractive channels.

EIC can largely contribute to increase the precision in the knowledge of the nuclear PDFs, where, once more, the variety of the available nuclei represents a powerful handle.

At the highest luminosities, it will be possible to perform test of the electroweek fundamental symmetry, opening a window on possible effects of new physics.

80.3.4 Realization Perspectives for the Electron-Ion Collider

The EIC-dedicated white book, published in 2012 and revised in 2014 [4], has set the basis for one of the main recommendation of The 2015 Long Range Plan for Nuclear Science [9]: EIC is the main priority for new facility construction. The project has been scrutinized by the National Accademy of Science-Engineering-Medicine: a very positive judgement has been express in the Accademy report [10], formally opening the way to the project approval. The USA Department Of Energy (DOE) has included EIC in its budget justification for 2020 [11], planning the mission approval in 2019.

This extremely positive perspectives are complemented by the wide international community supporting EIC. The EIC User Group (EICUG) [12] has almost 900 members, extremely active: EICUG annual meetings, the annual dedicated conference Physics Opportunities at an ElecTron-Ion Collider(POETIC) [13], the working groups and more.

The mature decision-making process, the wide and committed community as well as the advanced status of the accelerator designs indicate that the project is mature and very near to approval.

80.4 The Electron-Ion Collider (EicC) in China

A High-Intensity Heavy Ion Accelerator Facility (HIAF) is presently under construction in Huizhou, Guandong province, for interdisciplinary studies. An electron-ion collider can be formed adding an adequate electron complex and an ion acceleration and interaction ring. In a first version (EicC-I), the beam energies are 3.5 GeV for e and 20 GeV for p, with polarization of, respectively, 80% and 70% and L of $(1\text{–}5)\times 10^{33}$ cm^{-2}s^{-1}. p, D and ^{3}He beams are foreseen. A following phase (EicC-II) can offer higher energies: 10 GeV for e and 60–200 GeV for p and higher L: 5×10^{34} cm^{-2}s^{-1}.

A white book about the EicC is in preparation to be submitted within 2019.

80.5 Conclusions

Perspectives for the Electron-Ion Colliders have been presented by facts, the main ones recalled in the following. A large and active community is at work, as also tested by the several proposals all over the world. The physics reach is broad, mainly centered in exploring and understanding QCD, while these facilities have also a mission in empowering high energy hadron colliders and they have discovery capability in the BSM sector. The different projects, covering different phase space regions, and offering different research handles, as, for instance, polarization, are intrinsically complementary.

There are concrete perspectives for USA EIC approval in 2019.

A future where key issues in physics are accessed by Electron-Ion Colliders is concretely in front of us.

References

1. A. Caldwell, R. Ent, A. Levy, P. Newman, F. Olness, The "DIS and Related Subjects" Strategy Document: Fundamental Science from Lepton-Hadron Scattering, Contribution to the Update of the European Strategy on Particle Physics (2018), https://arxiv.org/abs/1812.08110
2. Future Circular Collider Conceptual Design Report (2019), https://fcc-cdr.web.cern.ch/
3. A. Caldwell, M. Wing, VHEeP: a very high energy electron-proton collider. Eur. Phys. J. C **76**(8), 463 (2016)
4. A. Accardi et al., Electron Ion Collider: the next QCD frontier. Eur. Phys. J. A **52**(9), 268 (2016)
5. X. Chen, A plan for electron Ion Collider in China (2018), arXiv:1809.00448
6. J. L. Abelleira Fernandez et al., A Large Hadron Electron Collider at CERN: Report on the Physics and Design Concepts for Machine and Detector, arXiv:1206.2913
7. M. Cepeda, Higgs precision measurements at future colliders, talk at the EPPSU Open Symposium (Granada, Spain, 2019)
8. F. Willeke, EIC Accelerator Overview, talk at the EIUG annual meeting, 22-26 2019, Paris, France (2019)

9. A. Aprahamian et al., Reaching for the horizon: the 2015 long range plan for nuclear science, INSPIRE-1398831 (2015)
10. The National Academy of Science-Engineering-Medicine, Division on Engineering and Physical Sciences, Board on Physics and Astronomy, Committee on U.S.-Based Electron-Ion Collider Science Assessment, An assessment of US-Based Electron-Ion Collider Science (2018), https://www.nap.edu/catalog/25171/an-assessment-of-us-based-electron-ion-collider-science
11. The Department of Energy, FY 2020 Budget Justification, https://www.energy.gov/cfo/downloads/fy-2020-budget-justification
12. http://www.eicug.org/
13. The 9th POETIC Conference, Lawrence Berkeley National Laboratory, Berkeley, USA, 16–21 (2019), https://conferences.lbl.gov/event/196/overview

Part XII
Summary and Closing

Chapter 81
Theory Summary at Strangeness in Quark Matter 2019

Jacquelyn Noronha-Hostler

Abstract This is the theory summary of Strangeness in Quark Matter 2019 conference. Results include the state-of-the-art updates to the Quantum Chromodynamics (QCD) phase diagram with contributions both from heavy-ion collisions and nuclear astrophysics, studies on the QCD freeze-out lines, and several aspects regarding small systems including collectivity, heavy flavor dynamics, strangeness, and hard probes.

Keywords Heavy-ion Collisions · Quantum chromodynamics · Nuclear astrophysics

81.1 Introduction

One of the crucial signatures for the discovery of the Quark Gluon Plasma (QGP) was the measurement of strangeness enhancement due to the ease of producing more strangeness particles from gluon interactions or an annihilation of a light quark anti-quark pair. Since that time the study of strangeness has evolved significantly. For instance, now that it is understood that the Quantum Chromodynamic (QCD) phase transition is a cross-over [1], contrasting observables of light versus strange hadrons can provide insight into properties of this transition. Connected to the same QCD phase diagram but in the baryon rich regime, the interactions of strange hadrons are a necessary input to the QCD equation of state and can put constraints on the mass radius relationship of neutron stars.

The natural next step is then studying the charm quark, which may not be thermalized with the rest of the QGP and can provide orthogonal information about its properties. In fact, charm quarks appear to be a particularly interesting probe in small systems [2] and may provide key information to determine the limits of size of the

J. Noronha-Hostler (✉)
University of Illinois at Urbana-Champaign, Urbana, IL 61801, USA
e-mail: jnorhoss@illinois.edu

D. Elia et al. (eds.), *The XVIII International Conference on Strangeness in Quark Matter (SQM 2019)*, Springer Proceedings in Physics 250,
https://doi.org/10.1007/978-3-030-53448-6_81

QGP. However, one should caution that it is important to use realistic medium in theoretical descriptions, otherwise the results may be misleading.

In this theory summary of Strangeness in Quark Matter 2019, I provide an overview of the latest breakthroughs in relativistic heavy-ion collisions and nuclear astrophysics that give insight into strange and charm quarks.

81.2 QCD Phase Diagram: From Heavy-Ion Collisions to Neutron Stars

Over the last few years significant advancements have been made to the QCD phase diagram. The left side of Fig. 81.1 demonstrates the most recent reconstruction of the QCD equation of state based on Lattice QCD results for an equation of state with three conserved charges: baryon number, strangeness, and electric charge where one finds that the assumption of strangeness neutrality pushes experiments to larger baryon chemical potentials [3]. This implies that relativistic heavy ion collisions may possibly be closer to nuclear astrophysics than originally expected. In nuclear astrophysics it is possible to reach quite high temperatures in neutron star mergers (see Fig. 81.1 on the right) and gravitational waves may provide hints if deconfined matter is at the core of neutron stars [4].

At $\mu_B = 0$ the phase transition is a cross-over [1] and it is anticipated that at larger baryon densities a critical point may be discovered followed by a first-order phase transition line that may be reached in neutron star mergers or proto-neutron stars. One of the primary signals of such a critical point is the kurtosis of net-proton fluctuations [5, 6]. However, caveats exist once one considers finite size effects, centrality binning, and detector efficiencies [7, 8].

One of the crucial questions in nuclear astrophysics is: what is the state of matter at the core of a neutron star? Is it deconfined matter, just protons/neutrons, or strange

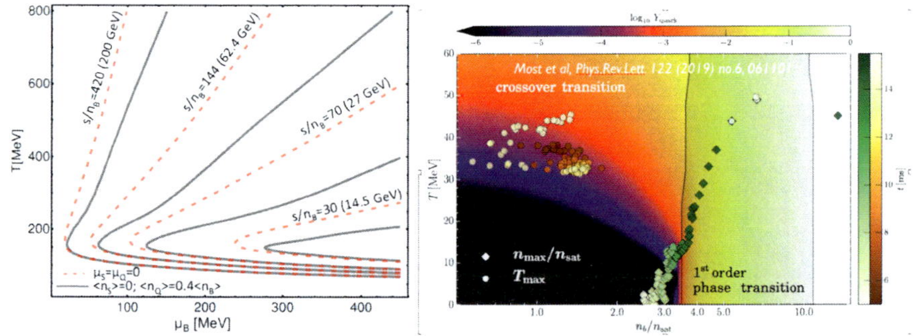

Fig. 81.1 QCD phase diagram from Lattice QCD (left) with BSQ conserved charges from [3]. On the right is the estimated range in the QCD that neutron star mergers may reach using ideal relativistic hydrodynamic calculations coupled to GR [4]

baryons (and their non-trivial interactions)? A possible signal for deconfiment would be the measurement of mass twins, which are stars that have the same mass but vastly different radii [9–11]. Additionally, it is important to properly understand repulsive versus attractive hyperon interactions since the addition of hyperons can affect the mass radius relationship [12–14].

Unlike in most studies of phase transitions in fields like condensed matter, in nuclear physics the system may be far from equilibrium and transport coefficients play a significant role in the search for a critical point/first order phase transition. Shear and bulk viscosity need to be calculated at finite baryon densities [15]. In heavy-ions where BSQ conserved charges are relevant, this also then leads to three diffusion transport coefficients for each conserved charges, which have been thus far been calculated in kinetic theory or non-conformal holographic models [16–21]. Unlike in heavy-ions, in neutron star mergers the transport coefficients stem from weak interactions [22, 23], which must eventually be incorporated into relativistic hydrodynamic calculations coupled to general relativity [24] (specifically bulk viscosity).

81.2.1 Freeze-Out

While each event in heavy-ion collisions (or single neutron star merger[1]) passes through the phase diagram in a unique manner depending on its initial conditions (expanding and cooling over time), one can measure the point of chemical freeze-out using identified particle yields [25, 26] and fluctuations [6, 27–29].

There is a tension between yields of light and strange particles in hadronic yield comparisons with thermal fits [26]. Fluctuations of conserved charges [30, 31] demonstrate a preference for a flavor hierarchy i.e. strange hadrons freezing out at a higher temperature than light hadrons. Additionally, transport models reveal a similar conclusion [32]. To explore this further, STAR has measured cross-correlations between conserved charges using certain identified hadrons [33]. However, it may be that there are alternative proxies for BSQ conserved charges that would be a better for direct comparisons to Lattice QCD [34].

Following chemical freeze-out, one expects that kinetic freeze-out is achieved (although at very low beam energies the two appear to converge to the same temperature [35]). Recently, HADES used virtual photons to estimate temperatures reached at very low beam energies [36] and found that the temperatures may be somewhat higher than originally expected.

One challenge to the thermalization picture comes from the recently measured light nuclei that appear to follow yields calculated from thermal fits. Questions remain in terms of interpreting these results and their implications [37, 38].

[1] I am currently unaware of an equivalent measurement to freeze-out in neutron star mergers but hypothesize that nuclei abundances would be a potential candidate.

81.2.2 Hadron Spectrum

Understanding both the total number of possible hadrons and their interactions are fundamental to both the understanding of the hadron gas phase in heavy ion collisions and the composition of the core of neutron stars. Recently, partial pressures were used to constrain the particle spectrum [39] and it was found that even resonances with the most experimental uncertainty are needed to reproduce Lattice QCD results. However, the inclusion of in-medium effects of the HRG could also reproduce these partial pressures [40]. Additionally, significant progress has been made in understanding hyperon interactions in Lattice QCD [41], which may have wide reaching effects.

81.3 Heavy Flavor and Hard Probes

Because of the large mass of the heavy quarks, charm quarks should likely have much longer thermalization times [42]. To understand the degree of thermalization of the charm quarks, the effect of charm conservation was studied [43] and predicts a very large differences in the D_s/D_0 ratio, if charm is conserved.

Since the very first event-by-event heavy flavor [44] and hard probe calculations [45] many new observables have been proposed that correlate the soft and hard/heavy sectors [46, 47]. For instance, it was first suggested in [48] that the event plane angle of higher order harmonics or hard probes would be less and less correlated with the soft event plane angle. This was further confirmed in [46, 47, 49]. Then in [50] a new correlation function was proposed in order to study the interplay with the soft and heavy flavor sectors, which also found that the higher harmonics are less correlated with the soft sector.

One important caveat in most heavy flavor studies is that the hydrodynamic background can play a significant role if not tuned properly to the soft sector. In [51] it was found that multiple heavy flavor models may appear to all simultaneously match both R_{AA} and v_2, however, once identical backgrounds were taken for all models wide variations were seen comparing to the same observables. In [49] two different choices in initial conditions were compared to experimental data and $v_2\{4\}/v_2\{2\}(p_T)$ of D mesons appeared to be the best choice to distinguish between the two initial condition models.

Heavy flavor studies are sensitive to coalescence and fragmentation [52]. Additionally, understanding the origin of the heavy flavor transport coefficients in a strongly coupled [53] versus weakly coupled approach is important to understanding the properties of the QGP. While the soft gluon approximation is well motivated, it was found that its effect on suppression is negligible [54]. Finally, taking the ratio of the R_{AA} in different collision sizes, may give insight into the path length dependence [55].

81.4 Small Systems

One of the newest frontiers of high-energy nuclear physics is the understanding on the limits of the smallest droplet of the Quark Gluon Plasma. Relativistic hydrodynamic models reproduce collective flow observables reasonably well, however, other signatures of the Quark Gluon Plasma are not as well understood. Even for collective flow, significant questions remain about the nature of the initial conditions [56] and the approach to hydrodynamics [57–60]. Alternative approaches are also being explored using PYTHIA+URQMD [61] and fluctuations derived from QCD interactions [62].

One fundamental question in small systems is if quarks of different flavors have sufficient time to reach thermalization. Recently, the ALICE collaboration published a paper [63] where they found an enhancement of strangeness in small systems that could not be explained by existing models. Using effective kinetic theory, in [64] they estimated that the minimum multiplicity to live long enough to reach thermalization would be $dN/d\eta \gtrsim 100$. However, if one considers a core-corona model, it appears that there is still a significant contribution from the core down to $dN/d\eta \gtrsim 10$ [65, 66].

It was originally thought that the chiral magnetic effect (CME) should only appear in large systems, however, in [67] it was found to have a significant signal in pPb collisions. Interestingly, enough it appears that when one fully incorporates electromagnetic fields in PHSD that one can obtain a splitting of charge in pPb [68]. Further developments in magnetohydrodynamics have also been made [69] that will also be relevant to future CME studies.

While collective flow and strangeness enhancement have been measured in small systems, the suppression of hard probes and heavy flavor has not (i.e. $R_{pPb} \sim 1$). In [70] an intermediate system size scan for D mesons was proposed to see the progression of $R_{AA} \rightarrow 1$ with shrinking system size and to make centrality comparisons of v_2, which has a non-trivial relationship with system size due to the increase in eccentricity with decreasing system size. This is an especially interest proposal considering that D mesons appear to be sensitive to out-of-equilibrium dynamics [71]. For the effect of quarkonium in small systems see [72].

81.5 Outlook

The study of strange and charm quarks has branched off into many new and unexpected directions that probe the fundamental theory of strongly interactions. For instance, D mesons may be used to further study far-from-equilibrium hydrodynamic behavior in small systems because of the unique information they can provide in contrast to light flow observables. For the QCD equation of state, one expects that new collaborations will spring up between heavy-ion physicists and nuclear astrophysicists who are willing to work together to better map out the QCD phase diagram at finite baryon densities. This could lead to full BSQ hydrodynamic calculations in heavy-ion collisions and the possibility of viscous fluid calculations coupled to GR in neutron star mergers. Further developments into magnetohydrodynamics and vorticity are also needed to better understand effects especially important at low beam energies. Of course, much needed context for these theoretical calculations will be provided by the Beam Energy Scan II and NICER data that are expected to appear soon.

Acknowledgments J.N.H. Acknowledges the support of the Alfred P. Sloan Foundation and support from the US-DOE Nuclear Science Grant No. DE-SC0019175.

References

1. Y. Aoki, G. Endrodi, Z. Fodor, S.D. Katz, K.K. Szabo, Nature **443**, 675 (2006). hep-lat/0611014
2. CMS, A.M. Sirunyan et al., Phys. Rev. Lett. **121**, 082301 (2018), 804.09767
3. J. Noronha-Hostler, P. Parotto, C. Ratti, J. M. Stafford **1902**, 06723 (2019)
4. E.R. Most et al., Phys. Rev. Lett. **122**, 061101 (2019), 1807.03684
5. M.A. Stephanov, Phys. Rev. Lett. **107**, 052301 (2011), 1104.1627
6. R. Critelli et al., Phys. Rev. **D96**, 096026 (2017), 1706.00455
7. S. Sombun et al., J. Phys. **G45**, 025101 (2018), 1709.00879
8. M. Agah Nouhou et al., PoS **CORFU2018**, 179 (2019), 1906.02647
9. M.G. Alford, S. Han, M. Prakash, Phys. Rev. **D88**, 083013 (2013), 1302.4732
10. S. Benic, D. Blaschke, D.E. Alvarez-Castillo, T. Fischer, S. Typel, Astron. Astrophys. **577**, A40 (2015), 1411.2856
11. G. Montana, L. Tolos, M. Hanauske, L. Rezzolla, Phys. Rev. **D99**, 103009 (2019), 1811.10929
12. D. Chatterjee, I. Vidaña, Eur. Phys. J. **A52**, 29 (2016), 1510.06306
13. I. Vidaña, Proc. Roy. Soc. Lond. **A474**, 0145 (2018), 1803.00504
14. P. Ribes, A. Ramos, L. Tolos, C. Gonzalez-Boquera, M. Centelles, Astrophys. J. **883**, 168 (2019), 1907.08583
15. P. Moreau et al., Phys. Rev. **C100**, 014911 (2019), 1903.10257
16. S. Bhadury, M. Kurian, V. Chandra, A. Jaiswal **1902**, 05285 (2019)
17. R. Rougemont, J. Noronha, J. Noronha-Hostler, Phys. Rev. Lett. **115**, 202301 (2015), 1507.06972
18. R. Rougemont, R. Critelli, J. Noronha-Hostler, J. Noronha, C. Ratti, Phys. Rev. **D96**, 014032 (2017), 1704.05558
19. M. Greif, J. A. Fotakis, G. S. Denicol, C. Greiner, Phys. Rev. Lett. **120**, 242301 (2018), 1711.08680
20. G. S. Denicol et al., Phys. Rev. **C98**, 034916 (2018), 1804.10557
21. M. Martinez, T. Schäfer, V. Skokov, (2019), 1906.11306
22. M. Alford, A. Harutyunyan, A. Sedrakian **1907**, 04192 (2019)
23. M.G. Alford, L. Bovard, M. Hanauske, L. Rezzolla, K. Schwenzer, Phys. Rev. Lett. **120**, 041101 (2018), 1707.09475
24. F.S. Bemfica, M.M. Disconzi, J. Noronha, Phys. Rev. Lett. **122**, 221602 (2019), 1901.06701
25. ALICE, F. Bellini, Nucl. Phys. **A982**, 427 (2019), 1808.05823
26. ALICE, F. Bellini, Proceeding this conference
27. S. Borsanyi et al., Phys. Rev. Lett. **113**, 052301 (2014), 1403.4576
28. P. Alba et al., Phys. Lett. **B738**, 305 (2014), 1403.4903
29. J. Noronha-Hostler et al., (2016), 1607.02527
30. R. Bellwied et al., Phys. Rev. **C99**, 034912 (2019), 1805.00088

31. M. Bluhm, M. Nahrgang, Eur. Phys. J. **C79**, 155 (2019), 1806.04499
32. K.A. Bugaev et al., (2018), 1812.02509
33. STAR, J. Adam et al., Phys. Rev. **C100**, 014902 (2019), 1903.05370
34. R. Bellwied et al., (2019), 1910.14592
35. STAR, L. Adamczyk et al., (2017), 1701.07065
36. J. Adamczewski-Musch et al., Nat. Phys. **15**, 1040 (2019)
37. D. Oliinychenko, L.-G. Pang, H. Elfner, V. Koch, Phys. Rev. **C99**, 044907 (2019), 1809.03071
38. Y. Cai, T. D. Cohen, B. A. Gelman, Y. Yamauchi, Phys. Rev. **C100**, 024911 (2019), 1905.02753
39. P. Alba et al., Phys. Rev. **D96**, 034517 (2017), 1702.01113
40. G. Aarts, C. Allton, D. De Boni, B. Jäger, Phys. Rev. **D99**, 074503 (2019), 1812.07393
41. E. Hiyama et al., (2019), 1910.02864
42. G.D. Moore, D. Teaney, Phys. Rev. C **71**, 064904 (2005). hep-ph/0412346
43. J. Zhao, S. Shi, N. Xu, P. Zhuang, EPJ Web Conf. **202**, 06004 (2019)
44. M. Nahrgang, J. Aichelin, S. Bass, P. B. Gossiaux, K. Werner, Phys. Rev. **C91**, 014904 (2015), 1410.5396
45. J. Noronha-Hostler, B. Betz, J. Noronha, M. Gyulassy, Phys. Rev. Lett. **116**, 252301 (2016), 1602.03788
46. B. Betz et al., Phys. Rev. **C95**, 044901 (2017), 1609.05171
47. C.A.G. Prado et al., Phys. Rev. **C96**, 064903 (2017), 1611.02965
48. J. Jia, Phys. Rev. **C87**, 061901 (2013), 1203.3265
49. R. Katz, C.A.G. Prado, J. Noronha-Hostler, J. Noronha, A.A.P. Suaide, (2019), 1906.10768
50. S. Plumari, G. Coci, S.K. Das, V. Minissale, V. Greco, Nucl. Phys. **A982**, 655 (2019), 1901.07815
51. S. Cao et al., Phys. Rev. **C99**, 054907 (2019), 1809.07894
52. V. Minissale, S. Plumari, G. Coci, S.K. Das, V. Greco, MDPI Proc. **10**, 9 (2019)
53. R. Hambrock, W.A. Horowitz, PoS HardProbes2018, 047 (2018)
54. B. Blagojevic, M. Djordjevic, M. Djordjevic, MDPI Proc. **10**, 13 (2019)
55. M. Djordjevic, D. Zigic, M. Djordjevic, J. Auvinen, Phys. Rev. **C99**, 061902 (2019), 1805.04030
56. A. Bzdak, B. Schenke, P. Tribedy, R. Venugopalan, Phys. Rev. **C87**, 064906 (2013), 1304.3403
57. M. Strickland, J. Noronha, G. Denicol, Phys. Rev. **D97**, 036020 (2018), 1709.06644
58. A. Kurkela, A. Mazeliauskas, J.-F. Paquet, S. Schlichting, D. Teaney, Phys. Rev. Lett. **122**, 122302 (2019), 1805.01604
59. A. Kurkela, U.A. Wiedemann, B. Wu **1907**, 08101 (2019)
60. U.W. Heinz, J.S. Moreland, J. Phys. Conf. Ser. **1271**, 012018 (2019), 1904.06592
61. C. Bierlich, G. Gustafson, L. Lönnblad, Phys. Lett. **B779**, 58 (2018), 1710.09725
62. G. Giacalone, P. Guerrero-Rodríguez, M. Luzum, C. Marquet, J.-Y. Ollitrault, Phys. Rev. **C100**, 024905 (2019), 1902.07168
63. ALICE, J. Adam et al., Nat. Phys. **13**, 535 (2017), 1606.07424
64. A. Kurkela, A. Mazeliauskas, Phys. Rev. Lett. **122**, 142301 (2019), 1811.03040
65. Y. Kanakubo, Y. Tachibana, T. Hirano **1910**, 10556 (2019)
66. Y. Kanakubo, M. Okai, Y. Tachibana, T. Hirano, PTEP **2018**, 121D01 (2018), 1806.10329
67. CMS, V. Khachatryan et al., Phys. Rev. Lett. **118**, 122301 (2017), 1610.00263
68. L. Oliva, P. Moreau, V. Voronyuk, E. Bratkovskaya **1909**, 06770 (2019)
69. G. Inghirami et al., J. Phys. Conf. Ser. **1024**, 012043 (2018)
70. R. Katz, C.A.G. Prado, J. Noronha-Hostler, A.A.P. Suaide, (2019), 1907.03308
71. Y. Xu et al., Phys. Rev. **C96**, 024902 (2017), 1703.09178
72. E. Ferreiro, Proceeding this conference

Chapter 82
Strangeness in Quark Matter 2019: Experimental Highlights

David Dobrigkeit Chinellato

Abstract This document summarizes selected experimental highlights of the Strangeness in Quark Matter 2019 conference grouped into a few broad categories.

82.1 Precision Studies of Light-Flavour Particle Production Mechanisms

A significant number of measurements of identified particle yields has already been performed over the past several years at the LHC, which has led to unprecedented understanding of the statistical properties of hadronization. Still, some details, such as the anomalous proton-to-pion ratio, remain illusive and have prompted further studies. One recent approach in which a statistical hadronization model is employed considering also resonant and non-resonant πN and $\pi\pi$N interactions succeeded in describing identified particle yields within uncertainties [1], while other phenomenological efforts incorporating a sequential freezeout also have been shown to succeed [2]. It is clear that further input must also come from experiment in order to shed further light on particle production mechanisms.

Many innovative techniques are being explored to provide further information, with specific emphasis on the transition from the low-multiplicity regime of pp collisions to the high-multiplicity environment of nucleus–nucleus collisions, where hadrochemistry still exhibits significant dynamics, as seen in Fig. 82.1. Charged-particle spectra measurements by ALICE in pp collisions selected according to a transverse activity classifier R_T show that this technique allows for jet-like structures to be disentangled from the underlying event (UE) [4]. Similarly, two-particle correlation techniques applied to ϕ-meson yield measurements enabled the ALICE collaboration to measure ϕ yields in association with jets and the UE, proving that the increase of the relative ϕ meson production is predominantly associated to a change in the UE [5], where most ϕ are produced. In all studies shown at the conference,

D. Dobrigkeit Chinellato (✉)
Universidade Estadual de Campinas, São Paulo, Brazil
e-mail: daviddc@ifi.unicamp.br

D. Elia et al. (eds.), *The XVIII International Conference on Strangeness in Quark Matter (SQM 2019)*, Springer Proceedings in Physics 250,
https://doi.org/10.1007/978-3-030-53448-6_82

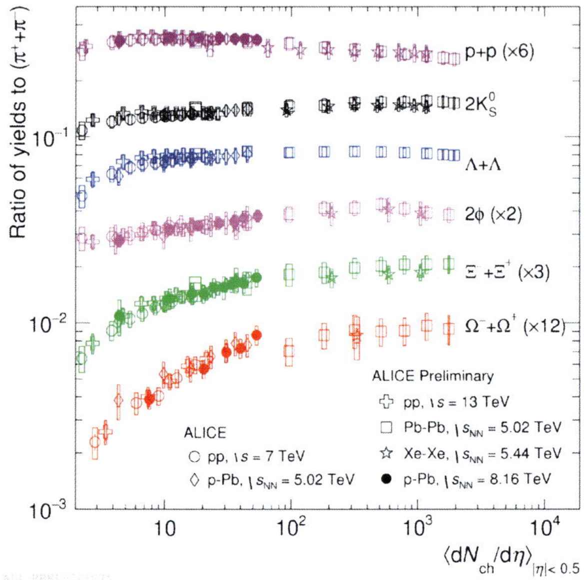

Fig. 82.1 Identified particle ratios as a function of charged-particle multiplicity density in pp, p–Pb, Xe–Xe and Pb–Pb collisions measured by the ALICE collaboration. Figure from [3]. Copyright CERN, reused with permission

it is clear that the focus now lies in pinpointing the phase space in which certain hadrochemistry dynamics occurs, with even more results to be expected in the near future.

Further measurements from the RHIC Beam Energy Scan (BES) as well as the NA61/SHINE experiment indicate substantial changes in strangeness-related observables at a beam energy of around 8–10 GeV/c [6, 7]. While these are not fully conclusive regarding the nature of a potential phase transition at those energies, they certainly add to the experimental information available now to theorists. Among other observables, the study of fluctuations of conserved charges may also provide further constraints and were discussed as well [8, 9], with indications that effects from strangeness and baryon number conservation are now within experimental reach.

82.2 Bulk Matter, Soft Physics and Collectivity

The study of the onset of collective behaviour when transitioning from elementary collisions, such as e^+e^- collisions, to nucleus–nucleus collisions continues to be an intensely debated topic. The CMS collaboration reports that the ratio of elliptic flow coefficients of charged particles $v_2\{4\}/v_2\{2\}$ is approximately equal to $v_3\{4\}/v_3\{2\}$ in p–Pb collisions at $\sqrt{s_{NN}}$ = 5.02 TeV, contrary to what is observed in Pb–Pb

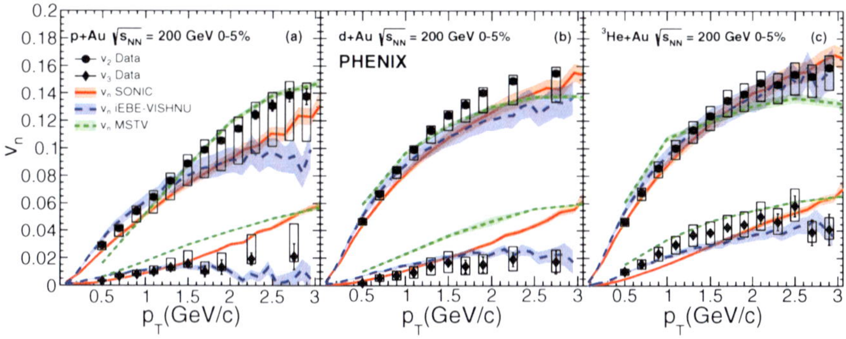

Fig. 82.2 Elliptic flow coefficients v_2 and v_3 in p–Au, d–Au and ^{3}He–Au measured by the PHENIX collaboration compared with hydrodynamics-based models. Figure from [12]

collisions, indicating that these flow coefficients may have their origins in fluctuations in small systems [10]. A similar observation of different fluctuation patterns for low multiplicity events with $N_{ch}^{|\eta|<0.8}$ below ~100 is also seen by the ALICE collaboration in a symmetric cumulants analysis performed systematically in pp, p–Pb, Xe–Xe and Pb–Pb [11].

In a complementary study of significant impact, the PHENIX collaboration systematically studied elliptic and triangular flow coefficients in p–Au, d–Au and ^{3}He–Au collisions [12], finding that elliptic and triangular flow follow the same ordering as the initial state eccentricities ϵ_2 and ϵ_3, as seen in Fig. 82.2. This observation supports the hypothesis that the observed flow is related to the initial condition geometry. Further studies will be needed to disentangle precisely the role of geometry and the aforementioned fluctuation-like behaviour measured by the CMS and ALICE collaborations.

82.3 Constraining Nuclei Production Mechanisms

New measurements on nuclei production in Pb–Pb collisions at $\sqrt{s_{NN}}$ = 5.02 TeV reported by the ALICE Collaboration include the elliptic and triangular flow coefficients of deuterons and ^{3}He [13]. Coupled with spectra measurements, these provide further constraints on production models, with a coalescence approach describing the data well provided phase space distributions from a hybrid model are used. The HADES collaboration also made systematic measurements of v_2 of protons, deuterons and tritons, where a particularly interesting result is that the v_2 of protons as a function of beam energy is observed to be negative (see Fig. 82.3), i.e. elliptic flow is out-of-plane, for $0.1 < E_{beam}/A(\text{GeV}) < 5.0$ [14]. It is clear that further theoretical work is needed to fully account for this observation.

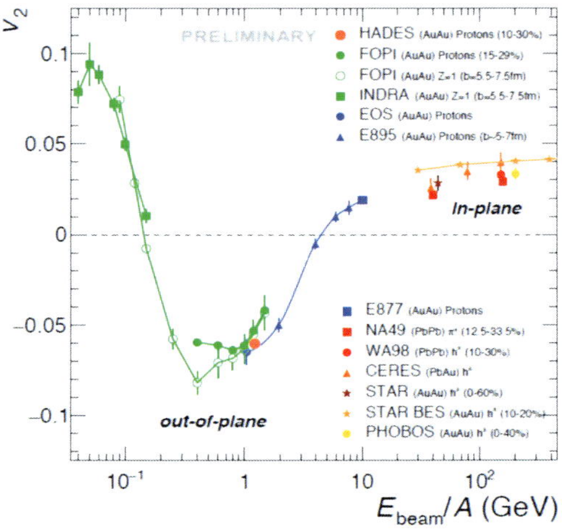

Fig. 82.3 Elliptic flow coefficient v_2 of protons and inclusive charged hadrons as a function of beam energy. Figure from [14]

82.4 Heavy-Flavour Production and Its Relation to the Bulk

In the heavy-flavour (HF) sector, a significant amount of measurements across systems exists for open HF, such as D and B mesons. New results from both the ALICE [15] and CMS [16] collaborations quantify the nuclear modification factor with unprecedented precision not only for D mesons but also for non-prompt $D \leftarrow b$, with data suggesting a hierarchy at high p_{T} in which $R_{\mathrm{AA}}(\pi) < R_{\mathrm{AA}}(D) < R_{\mathrm{AA}}(B)$, as expected from transport models. New data collected in 2018 also allowed for improved measurements of the open HF v_2, so that the combined information from R_{AA} and flow can now be used for characterising the interaction of heavy flavour quarks with the medium.

Further light can be shed on HF jet fragmentation by comparing D-tagged jets and their structure in various collision systems. TheALICE collaboration reports on such measurements in pp [17], while the CMS collaboration compares a similar measurement in pp with a Pb–Pb analysis and noted that the radial distributions of D^0 mesons seems broader in Pb–Pb than the one observed in pp [18]. In a related study, the ALICE collaboration reports that, by selecting 30–50% central Pb–Pb events with a small (larg) flow vector q_2 for charged particles, the v_2 observed for D mesons also decreases (increases), providing crucial information on the correlation between heavy and light-flavour hadron flow [19].

Another subject of noted interest is the p_{T}-differential baryon-to-meson ratio Λ_c/D^0. With new Pb-Pb data samples collected in 2018, measurements by the ALICE [20] and CMS [21] collaborations now suggest dynamics that is reminiscent of the baryon anomaly observed in the p/π and Λ/K^0_S ratios when going from pp to central Pb-Pb collisions, as seen in Fig. 82.4. While these observations are already exceptionally precise, even further improvements are needed to constrain model predictions well enough for a conclusion to be drawn. It is noteworthy that a

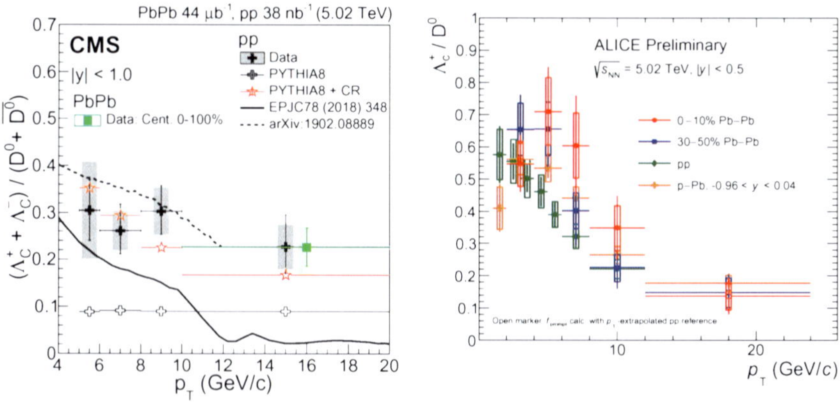

Fig. 82.4 Transverse momentum-differential Λ_c/D^0 ratio for various collision systems measured by the CMS (left) and the ALICE (right) collaborations. Figures from [21] (CMS) and [20] (ALICE). Copyright CERN, reused with permission

multiplicity-differential Λ_c/D^0 analysis is yet to be done in pp and p–Pb collisions and would provide crucial information.

Finally, the CMS collaboration also reports on the R_{AA} of the more exotic $\Upsilon(1S)$, $\Upsilon(2S)$ and $\Upsilon(3S)$ quarkonia in Pb-Pb collisions at $\sqrt{s_{NN}} = 5.02$ TeV [22], with the strongest suppression being observed for the higher mass states, consistent with model calculations that include suppression and regeneration. Such findings are complementary to the $\Upsilon(1S)$ elliptic flow measurement from the ALICE collaboration [23], which show flow consistent with zero for this quarkonium species, albeit with large uncertainties. It is clear that future measurements with increased precision and increased p_T coverage will be needed to fully characterise the energy loss and elliptic flow of Υ quarkonia in nucleus-nucleus collisions.

82.5 Hadronic Interactions, Astrophysics and More

The unprecedented amount of data collected at the LHC coupled to the tracking and particle identification capabilities of the ALICE experiment also allow for hadronic interactions between the outgoing particles to be studied in great detail, opening up an entirely new field of study that is not strictly related to high-energy heavy-ion physics. The ALICE collaboration has studied momentum correlation functions $C(k)$ that provide direct information about the interaction potential for a variety of particle species pairs [24]. The latest result in such efforts provides further insight on p-Ξ and even p-Ω interactions, reproduced in Fig. 82.5, with indications for an attractive potential for both of these pairs once comparisons with models are considered. These measurements are relevant not only to QCD studies but even to cosmology, since the

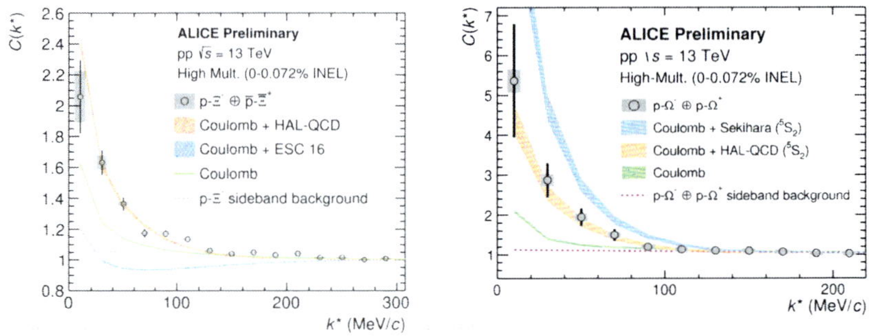

Fig. 82.5 Momentum correlation functions $C(k)$ for p-Ξ and p-Ω pairs measured by the ALICE collaboration compared with expectations from various models. Figure from [24]. Copyright CERN, reused with permission

nature of the interaction between protons and hyperons has a significant impact on the equation of state of matter contained in neutron stars. Future studies including the LHC Run 3 data will lead to even more precise information about hadronic interactions.

82.6 In Conclusion: The Future

Among the common themes discussed in the conference, a few highlights clearly emerge, and it is clear that, in general, ever more precise information on hadronization and production mechanisms is being attained. In light flavour studies, the interplay between soft and hard physics is being progressively disentangled, with the promise that in the near future the exact role of jets and the underlying event will be better quantified in future measurements. A similar discussion is evolving in heavy-flavour production studies, where the exact role of recombination as well as fragmentation in the formation of hadrons such as the Lambda$_c$ remains to be ascertained in more detail. Likewise, nuclei production mechanisms are also not a matter of consensus and a complete picture involving coalescence or thermal production is still not fully established. In addition, future, even higher-precision studies on fluctuations of conserved charges may provide more details on the phase transition and, at sub-LHC energies, the critical point. Finally, though a substantial amount of measurements regarding the onset of elliptic flow-like phenomena in small collision systems already exists, a precise, quantitative interpretation of these phenomena and their potential relation with initial geometry still remains a field of intense debate. All of these topics are expected to be explored in the coming years by the community at large, including new facilities such as BM@N and the upgraded HADES experiment.

References

1. A. Andronic et al. Phys. Lett. B **792** (2019) 304–309. https://doi.org/10.1016/j.physletb.2019.03.052
2. R. Bellwied, EPJ Web of Conferences. **171**, 02006 (2018). https://doi.org/10.1051/epjconf/201817102006
3. ALICE Collaboration, These proceedings/Willsher (2019)
4. ALICE Collaboration, These proceedings/Zaccolo (2019)
5. ALICE Collaboration, These proceedings/Blair (2019)
6. STAR Collaboration (2019). https://arxiv.org/abs/1906.03732
7. NA61/SHINE Collaboration, These proceedings/Podlaski (2019)
8. ALICE Collaboration, These proceedings/Arslandok (2019)
9. ALICE Collaboration, These proceedings/Bellwied (2019)
10. CMS Collaboration (2019). https://arxiv.org/abs/1904.11519
11. ALICE Collaboration. Phys. Rev. Lett. **123**(14) (2019). https://doi.org/10.1103/physrevlett.123.142301
12. PHENIX Collaboration. Nat. Phys. **15**(3), 214–220 (2018)
13. ALICE Collaboration, These proceedings/Caliva (2019)
14. HADES Collaboration, These proceedings/Chlad (2019)
15. ALICE Collaboration, These proceedings/Terrevoli (2019)
16. CMS Collaboration. Phys. Rev. Lett. **123**(2) (2019)
17. ALICE Collaboration, These proceedings/Mohanty (2019)
18. CMS Collaboration, These proceedings/X. Wang (2019)
19. ALICE Collaboration, These proceedings/Jaelani (2019)
20. ALICE Collaboration, These proceedings/Zampolli (2019)
21. CMS Collaboration (2019). https://arxiv.org/abs/1906.03322
22. CMS Collaboration. Phys. Lett. B **790**, 270–293 (2019). https://doi.org/10.1016/j.physletb.2019.01.006
23. ALICE Collaboration. Phys. Rev. Lett. **123**(19) (2019). https://doi.org/10.1103/physrevlett.123.192301
24. ALICE Collaboration, These proceedings/Fabbietti (2019)

Collaboration Membership Lists

ALICE Collaboration

S. Acharya[141], D. Adamová[94], S.P. Adhya[141], A. Adler[73], J. Adolfsson[80], M.M. Aggarwal[99], G. Aglieri Rinella[33], M. Agnello[30], N. Agrawal[10,48,53], Z. Ahammed[141], S. Ahmad[16], S.U. Ahn[75], A. Akindinov[91], M. Al-Turany[106], S.N. Alam[141], D.S.D. Albuquerque[122], D. Aleksandrov[87], B. Alessandro[58], H.M. Alfanda[6], R. Alfaro Molina[70], B. Ali[16], Y. Ali[14], A. Alici[26,10,53], A. Alkin[2], J. Alme[21], T. Alt[67], L. Altenkamper[21], I. Altsybeev[112], M.N. Anaam[6], C. Andrei[47], D. Andreou[33], H.A. Andrews[110], A. Andronic[144], M. Angeletti[33], V. Anguelov[103], C. Anson[15], T. Antičić[107], F. Antinori[56], P. Antonioli[53], R. Anwar[125], N. Apadula[79], L. Aphecetche[114], H. Appelshäuser[67], S. Arcelli[26], R. Arnaldi[58], M. Arratia[79], I.C. Arsene[20], M. Arslandok[103], A. Augustinus[33], R. Averbeck[106], S. Aziz[77], M.D. Azmi[16], A. Badalà[55], Y.W. Baek[40], S. Bagnasco[58], X. Bai[106], R. Bailhache[67], R. Bala[100], A. Baldisseri[137], M. Ball[42], S. Balouza[104], R.C. Baral[85], R. Barbera[27], L. Barioglio[25], G.G. Barnaföldi[145], L.S. Barnby[93], V. Barret[134], P. Bartalini[6], K. Barth[33], E. Bartsch[67], F. Baruffaldi[28], N. Bastid[134], S. Basu[143], G. Batigne[114], B. Batyunya[74], D. Bauri[48], J.L. Bazo Alba[111], I.G. Bearden[88], C. Bedda[62], N.K. Behera[60], I. Belikov[136], A.D.C. Bell Hechavarria[144], F. Bellini[33], R. Bellwied[125], V. Belyaev[92], G. Bencedi[145], S. Beole[25], A. Bercuci[47], Y. Berdnikov[97], D. Berenyi[145], R.A. Bertens[130], D. Berzano[58], M.G. Besoiu[66], L. Betev[33], A. Bhasin[100], I.R. Bhat[100], M.A. Bhat[3], H. Bhatt[48], B. Bhattacharjee[41], A. Bianchi[25], L. Bianchi[25], N. Bianchi[51], J. Bielčík[36], J. Bielčíková[94], A. Bilandzic[104,117], G. Biro[145], R. Biswas[3], S. Biswas[3], J.T. Blair[119], D. Blau[87], C. Blume[67], G. Boca[139], F. Bock[33,95], A. Bogdanov[92], L. Boldizsár[145], A. Bolozdynya[92], M. Bombara[37], G. Bonomi[140], H. Borel[137], A. Borissov[144,92], M. Borri[127], H. Bossi[146], E. Botta[25], L. Bratrud[67], P. Braun-Munzinger[106], M. Bregant[121], T.A. Broker[67], M. Broz[36], E.J. Brucken[43], E. Bruna[58], G.E. Bruno[32,105], M.D. Buckland[127], D. Budnikov[108], H. Buesching[67], S. Bufalino[30], O. Bugnon[114], P. Buhler[113], P. Buncic[33], Z. Buthelezi[71], J.B. Butt[14],

D. Elia et al. (eds.), *The XVIII International Conference on Strangeness in Quark Matter (SQM 2019)*, Springer Proceedings in Physics 250,
https://doi.org/10.1007/978-3-030-53448-6

J.T. Buxton[96], S.A. Bysiak[118], D. Caffarri[89], A. Caliva[106], E. Calvo Villar[111], R.S. Camacho[44], P. Camerini[24], A.A. Capon[113], F. Carnesecchi[26], R. Caron[137], J. Castillo Castellanos[137], A.J. Castro[130], E.A.R. Casula[54], F. Catalano[30], C. Ceballos Sanchez[52], P. Chakraborty[48], S. Chandra[141], W. Chang[6], S. Chapeland[33], M. Chartier[127], S. Chattopadhyay[141], S. Chattopadhyay[109], A. Chauvin[23], C. Cheshkov[135], B. Cheynis[135], V. Chibante Barroso[33], D.D. Chinellato[122], S. Cho[60], P. Chochula[33], T. Chowdhury[134], P. Christakoglou[89], C.H. Christensen[88], P. Christiansen[80], T. Chujo[133], C. Cicalo[54], L. Cifarelli[10,26], F. Cindolo[53], J. Cleymans[124], F. Colamaria[52], D. Colella[52], A. Collu[79], M. Colocci[26], M. Concas[II,58], G. Conesa Balbastre[78], Z. Conesa del Valle[77], G. Contin[127,59], J.G. Contreras[36], T.M. Cormier[95], Y. Corrales Morales[58,25], P. Cortese[31], M.R. Cosentino[123], F. Costa[33], S. Costanza[139], J. Crkovská[77], P. Crochet[134], E. Cuautle[68], P. Cui[6], L. Cunqueiro[95], D. Dabrowski[142], T. Dahms[104,117], A. Dainese[56], F.P.A. Damas[137,114], S. Dani[64], M.C. Danisch[103], A. Danu[66], D. Das[109], I. Das[109], P. Das[3], S. Das[3], A. Dash[85], S. Dash[48], A. Dashi[104], S. De[49,85], A. De Caro[29], G. de Cataldo[52], C. de Conti[121], J. de Cuveland[38], A. De Falco[23], D. De Gruttola[10], N. De Marco[58], S. De Pasquale[29], R.D. De Souza[122], S. Deb[49], H.F. Degenhardt[121], K.R. Deja[142], A. Deloff[84], S. Delsanto[25,131], D. Devetak[106], P. Dhankher[48], D. Di Bari[32], A. Di Mauro[33], R.A. Diaz[8], T. Dietel[124], P. Dillenseger[67], Y. Ding[6], R. Divià[33], D.U. Dixit[19], Ø. Djuvsland[21], U. Dmitrieva[61], A. Dobrin[33,66], B. Dönigus[67], O. Dordic[20], A.K. Dubey[141], A. Dubla[106], S. Dudi[99], M. Dukhishyam[85], P. Dupieux[134], R.J. Ehlers[146], D. Elia[52], H. Engel[73], E. Epple[146], B. Erazmus[114], F. Erhardt[98], A. Erokhin[112], M.R. Ersdal[21], B. Espagnon[77], G. Eulisse[33], J. Eum[17], D. Evans[110], S. Evdokimov[90], L. Fabbietti[104,117], M. Faggin[28], J. Faivre[78], F. Fan[6], A. Fantoni[51], M. Fasel[95], P. Fecchio[30], A. Feliciello[58], G. Feofilov[112], A. Fernández Téllez[44], A. Ferrero[137], A. Ferretti[25], A. Festanti[33], V.J.G. Feuillard[103], J. Figiel[118], S. Filchagin[108], D. Finogeev[61], F.M. Fionda[21], G. Fiorenza[52], F. Flor[125], S. Foertsch[71], P. Foka[106], S. Fokin[87], E. Fragiacomo[59], U. Frankenfeld[106], G.G. Fronze[25], U. Fuchs[33], C. Furget[78], A. Furs[61], M. Fusco Girard[29], J.J. Gaardhøje[88], M. Gagliardi[25], A.M. Gago[111], A. Gal[136], C.D. Galvan[120], P. Ganoti[83], C. Garabatos[106], E. Garcia-Solis[11], K. Garg[27], C. Gargiulo[33], A. Garibli[86], K. Garner[144], P. Gasik[104,117], E.F. Gauger[119], M.B. Gay Ducati[69], M. Germain[114], J. Ghosh[109], P. Ghosh[141], S.K. Ghosh[3], P. Gianotti[51], P. Giubellino[106,58], P. Giubilato[28], P. Glässel[103], D.M. Goméz Coral[70], A. Gomez Ramirez[73], V. Gonzalez[106], P. González-Zamora[44], S. Gorbunov[38], L. Görlich[118], S. Gotovac[34], V. Grabski[70], L.K. Graczykowski[142], K.L. Graham[110], L. Greiner[79], A. Grelli[62], C. Grigoras[33], V. Grigoriev[92], A. Grigoryan[1], S. Grigoryan[74], O.S. Groettvik[21], J.M. Gronefeld[106], F. Grosa[30], J.F. Grosse-Oetringhaus[33], R. Grosso[106], R. Guernane[78], B. Guerzoni[26], M. Guittiere[114], K. Gulbrandsen[88], T. Gunji[132], A. Gupta[100], R. Gupta[100], I.B. Guzman[44], R. Haake[146], M.K. Habib[106], C. Hadjidakis[77], H. Hamagaki[81], G. Hamar[145], M. Hamid[6], R. Hannigan[119], M.R. Haque[62], A. Harlenderova[106], J.W. Harris[146], A. Harton[11], J.A. Hasenbichler[33], H. Hassan[78], D. Hatzifotiadou[53,10], P. Hauer[42], S. Hayashi[132], S.T. Heckel[67], E. Hellbär[67], H. Helstrup[35], A. Herghelegiu[47], E.G. Hernandez[44], G. Herrera Corral[9], F. Herrmann[144], K.F. Hetland[35], T.E. Hilden[43],

H. Hillemanns[33], C. Hills[127], B. Hippolyte[136], B. Hohlweger[104], D. Horak[36], S. Hornung[106], R. Hosokawa[133], P. Hristov[33], C. Huang[77], C. Hughes[130], P. Huhn[67], T.J. Humanic[96], H. Hushnud[109], L.A. Husova[144], N. Hussain[41], S.A. Hussain[14], T. Hussain[16], D. Hutter[38], D.S. Hwang[18], J.P. Iddon[127,33], R. Ilkaev[108], M. Inaba[133], M. Ippolitov[87], M.S. Islam[109], M. Ivanov[106], V. Ivanov[97], V. Izucheev[90], B. Jacak[79], N. Jacazio[53,26], P.M. Jacobs[79], M.B. Jadhav[48], S. Jadlovska[116], J. Jadlovsky[116], S. Jaelani[62], C. Jahnke[121], M.J. Jakubowska[142], M.A. Janik[142], M. Jercic[98], O. Jevons[110], R.T. Jimenez Bustamante[106], M. Jin[125], F. Jonas[144,95], P.G. Jones[110], A. Jusko[110], P. Kalinak[63], A. Kalweit[33], J.H. Kang[147], V. Kaplin[92], S. Kar[6], A. Karasu Uysal[76], O. Karavichev[61], T. Karavicheva[61], P. Karczmarczyk[33], E. Karpechev[61], U. Kebschull[73], R. Keidel[46], M. Keil[33], B. Ketzer[42], Z. Khabanova[89], A.M. Khan[6], S. Khan[16], S.A. Khan[141], A. Khanzadeev[97], Y. Kharlov[90], A. Khatun[16], A. Khuntia[118,49], B. Kileng[35], B. Kim[60], B. Kim[133], D. Kim[147], D.J. Kim[126], E.J. Kim[72], H. Kim[147], J. Kim[147], J.S. Kim[40], J. Kim[103], J. Kim[147], J. Kim[72], M. Kim[103], S. Kim[18], T. Kim[147], T. Kim[147], S. Kirsch[38], I. Kisel[38], S. Kiselev[91], A. Kisiel[142], J.L. Klay[5], C. Klein[67], J. Klein[58], S. Klein[79], C. Klein-Bösing[144], M. Kleiner[67], S. Klewin[103], A. Kluge[33], M.L. Knichel[33], A.G. Knospe[125], C. Kobdaj[115], M.K. Köhler[103], T. Kollegger[106], A. Kondratyev[74], N. Kondratyeva[92], E. Kondratyuk[90], P.J. Konopka[33], L. Koska[116], O. Kovalenko[84], V. Kovalenko[112], M. Kowalski[118], I. Králik[63], A. Kravčáková[37], L. Kreis[106], M. Krivda[110,63], F. Krizek[94], K. Krizkova Gajdosova[36], M. Krüger[67], E. Kryshen[97], M. Krzewicki[38], A.M. Kubera[96], V. Kučera[60], C. Kuhn[136], P.G. Kuijer[89], L. Kumar[99], S. Kumar[48], S. Kundu[85], P. Kurashvili[84], A. Kurepin[61], A.B. Kurepin[61], A. Kuryakin[108], S. Kushpil[94], J. Kvapil[110], M.J. Kweon[60], J.Y. Kwon[60], Y. Kwon[147], S.L. La Pointe[38], P. La Rocca[27], Y.S. Lai[79], R. Langoy[129], K. Lapidus[146,33], A. Lardeux[20], P. Larionov[51], E. Laudi[33], R. Lavicka[36], T. Lazareva[112], R. Lea[24], L. Leardini[103], S. Lee[147], F. Lehas[89], S. Lehner[113], J. Lehrbach[38], R.C. Lemmon[93], I. León Monzón[120], E.D. Lesser[19], M. Lettrich[33], P. Lévai[145], X. Li[12], X.L. Li[6], J. Lien[129], R. Lietava[110], B. Lim[17], V. Lindenstruth[38], S.W. Lindsay[127], C. Lippmann[106], M.A. Lisa[96], V. Litichevskyi[43], A. Liu[19], S. Liu[96], W.J. Llope[143], I.M. Lofnes[21], V. Loginov[92], C. Loizides[95], P. Loncar[34], X. Lopez[134], E. López Torres[8], J.R. Luhder[144], M. Lunardon[28], G. Luparello[59], M. Lupi[73], A. Maevskaya[61], M. Mager[33], S.M. Mahmood[20], T. Mahmoud[42], A. Maire[136], R.D. Majka[146], M. Malaev[97], Q.W. Malik[20], L. Malinina[III, 74], D. Mal'Kevich[91], P. Malzacher[106], A. Mamonov[108], G. Mandaglio[55], V. Manko[87], F. Manso[134], V. Manzari[52], Y. Mao[6], M. Marchisone[135], J. Mareš[65], G.V. Margagliotti[24], A. Margotti[53], J. Margutti[62], A. Marín[106], C. Markert[119], M. Marquard[67], N.A. Martin[103], P. Martinengo[33], J.L. Martinez[125], M.I. Martínez[44], G. Martínez García[114], M. Martinez Pedreira[33], S. Masciocchi[106], M. Masera[25], A. Masoni[54], L. Massacrier[77], E. Masson[114], A. Mastroserio[138,52], A.M. Mathis[104,117], O. Matonoha[80], P.F.T. Matuoka[121], A. Matyja[118], C. Mayer[118], M. Mazzilli[52], M.A. Mazzoni[57], A.F. Mechler[67], F. Meddi[22], Y. Melikyan[92], A. Menchaca-Rocha[70], C. Mengke[6], E. Meninno[29], M. Meres[13], S. Mhlanga[124], Y. Miake[133], L. Micheletti[25], M.M. Mieskolainen[43], D.L. Mihaylov[104], K. Mikhaylov[74,91], A. Mischke[I, 62], A.N. Mishra[68], D. Miśkowiec[106], C.M. Mitu[66], A. Modak[3], N. Mohammadi[33], A.P. Mohanty[62],

B. Mohanty[85], M. Mohisin Khan[IV, 16], M. Mondal[141], M.M. Mondal[64], C. Mordasini[104], D.A. Moreira De Godoy[144], L.A.P. Moreno[44], S. Moretto[28], A. Morreale[114], A. Morsch[33], T. Mrnjavac[33], V. Muccifora[51], E. Mudnic[34], D. Mühlheim[144], S. Muhuri[141], J.D. Mulligan[79], M.G. Munhoz[121], K. Münning[42], R.H. Munzer[67], H. Murakami[132], S. Murray[71], L. Musa[33], J. Musinsky[63], C.J. Myers[125], J.W. Myrcha[142], B. Naik[48], R. Nair[84], B.K. Nandi[48], R. Nania[53,10], E. Nappi[52], M.U. Naru[14], A.F. Nassirpour[80], H. Natal da Luz[121], C. Nattrass[130], R. Nayak[48], T.K. Nayak[141,85], S. Nazarenko[108], A. Neagu[20], R.A. Negrao De Oliveira[67], L. Nellen[68], S.V. Nesbo[35], G. Neskovic[38], D. Nesterov[112], B.S. Nielsen[88], S. Nikolaev[87], S. Nikulin[87], V. Nikulin[97], F. Noferini[53,10], P. Nomokonov[74], G. Nooren[62], J. Norman[78], N. Novitzky[133], P. Nowakowski[142], A. Nyanin[87], J. Nystrand[21], M. Ogino[81], A. Ohlson[103,80], J. Oleniacz[142], A.C. Oliveira Da Silva[121], M.H. Oliver[146], C. Oppedisano[58], R. Orava[43], A. Ortiz Velasquez[68], A. Oskarsson[80], J. Otwinowski[118], K. Oyama[81], Y. Pachmayer[103], V. Pacik[88], D. Pagano[140], G. Paić[68], P. Palni[6], J. Pan[143], A.K. Pandey[48], S. Panebianco[137], V. Papikyan[1], P. Pareek[49], J. Park[60], J.E. Parkkila[126], S. Parmar[99], A. Passfeld[144], S.P. Pathak[125], R.N. Patra[141], B. Paul[58,23], H. Pei[6], T. Peitzmann[62], X. Peng[6], L.G. Pereira[69], H. Pereira Da Costa[137], D. Peresunko[87], G.M. Perez[8], E. Perez Lezama[67], V. Peskov[67], Y. Pestov[4], V. Petráček[36], M. Petrovici[47], R.P. Pezzi[69], S. Piano[59], M. Pikna[13], P. Pillot[114], L.O.D.L. Pimentel[88], O. Pinazza[53,33], L. Pinsky[125], C. Pinto[27], S. Pisano[10,51], D.B. Piyarathna[125], M. Płoskoń[79], M. Planinic[98], F. Pliquett[67], J. Pluta[142], S. Pochybova[I, 145], M.G. Poghosyan[95], B. Polichtchouk[90], N. Poljak[98], W. Poonsawat[115], A. Pop[47], H. Poppenborg[144], S. Porteboeuf-Houssais[134], V. Pozdniakov[74], S.K. Prasad[3], R. Preghenella[53], F. Prino[58], C.A. Pruneau[143], I. Pshenichnov[61], M. Puccio[25,33], V. Punin[108], K. Puranapanda[141], J. Putschke[143], R.E. Quishpe[125], S. Ragoni[110], S. Raha[3], S. Rajput[100], J. Rak[126], A. Rakotozafindrabe[137], L. Ramello[31], F. Rami[136], R. Raniwala[101], S. Raniwala[101], S.S. Räsänen[43], R. Rath[49], V. Ratza[42], I. Ravasenga[30], K.F. Read[95,130], K. Redlich[V, 84], A. Rehman[21], P. Reichelt[67], F. Reidt[33], X. Ren[6], R. Renfordt[67], A. Reshetin[61], J.-P. Revol[10], K. Reygers[103], V. Riabov[97], T. Richert[80,88], M. Richter[20], P. Riedler[33], W. Riegler[33], F. Riggi[27], C. Ristea[66], S.P. Rode[49], M. Rodríguez Cahuantzi[44], K. Røed[20], R. Rogalev[90], E. Rogochaya[74], D. Rohr[33], D. Röhrich[21], P.S. Rokita[142], F. Ronchetti[51], E.D. Rosas[68], K. Roslon[142], P. Rosnet[134], A. Rossi[28,56], A. Rotondi[139], F. Roukoutakis[83], A. Roy[49], P. Roy[109], O.V. Rueda[80], R. Rui[24], B. Rumyantsev[74], A. Rustamov[86], E. Ryabinkin[87], Y. Ryabov[97], A. Rybicki[118], H. Rytkonen[126], O.A.M. Saarimaki[43], S. Sadhu[141], S. Sadovsky[90], K. Šafařík[33,36], S.K. Saha[141], B. Sahoo[48], P. Sahoo[48,49], R. Sahoo[49], S. Sahoo[64], P.K. Sahu[64], J. Saini[141], S. Sakai[133], S. Sambyal[100], V. Samsonov[97,92], A. Sandoval[70], A. Sarkar[71], D. Sarkar[143], N. Sarkar[141], P. Sarma[41], V.M. Sarti[104], M.H.P. Sas[62], E. Scapparone[53], B. Schaefer[95], J. Schambach[119], H.S. Scheid[67], C. Schiaua[47], R. Schicker[103], A. Schmah[103], C. Schmidt[106], H.R. Schmidt[102], M.O. Schmidt[103], M. Schmidt[102], N.V. Schmidt[67,95], A.R. Schmier[130], J. Schukraft[33,88], Y. Schutz[136,33], K. Schwarz[106], K. Schweda[106], G. Scioli[26], E. Scomparin[58], M. Šefčík[37], J.E. Seger[15], Y. Sekiguchi[132], D. Sekihata[132,45], I. Selyuzhenkov[106,92], S. Senyukov[136], D. Serebryakov[61], E. Serradilla[70], P. Sett[48], A. Sevcenco[66], A. Shabanov[61], A. Shabetai[114],

R. Shahoyan[33], W. Shaikh[109], A. Shangaraev[90], A. Sharma[99], A. Sharma[100], H. Sharma[118], M. Sharma[100], N. Sharma[99], A.I. Sheikh[141], K. Shigaki[45], M. Shimomura[82], S. Shirinkin[91], Q. Shou[39], Y. Sibiriak[87], S. Siddhanta[54], T. Siemiarczuk[84], D. Silvermyr[80], C. Silvestre[78], G. Simatovic[89], G. Simonetti[33,104], R. Singh[85], R. Singh[100], V.K. Singh[141], V. Singhal[141], T. Sinha[109], B. Sitar[13], M. Sitta[31], T.B. Skaali[20], M. Slupecki[126], N. Smirnov[146], R.J.M. Snellings[62], T.W. Snellman[126], J. Sochan[116], C. Soncco[111], J. Song[60,125], A. Songmoolnak[115], F. Soramel[28], S. Sorensen[130], I. Sputowska[118], J. Stachel[103], I. Stan[66], P. Stankus[95], P.J. Steffanic[130], E. Stenlund[80], D. Stocco[114], M.M. Storetvedt[35], L.D. Stritto[29], P. Strmen[13], A.A.P. Suaide[121], T. Sugitate[45], C. Suire[77], M. Suleymanov[14], M. Suljic[33], R. Sultanov[91], M. Šumbera[94], S. Sumowidagdo[50], K. Suzuki[113], S. Swain[64], A. Szabo[13], I. Szarka[13], U. Tabassam[14], G. Taillepied[134], J. Takahashi[122], G.J. Tambave[21], S. Tang[6,134], M. Tarhini[114], M.G. Tarzila[47], A. Tauro[33], G. Tejeda Muñoz[44], A. Telesca[33], C. Terrevoli[28,125], D. Thakur[49], S. Thakur[141], D. Thomas[119], F. Thoresen[88], R. Tieulent[135], A. Tikhonov[61], A.R. Timmins[125], A. Toia[67], N. Topilskaya[61], M. Toppi[51], F. Torales-Acosta[19], S.R. Torres[120], A. Trifiro[55], S. Tripathy[49], T. Tripathy[48], S. Trogolo[28,25], G. Trombetta[32], L. Tropp[37], V. Trubnikov[2], W.H. Trzaska[126], T.P. Trzcinski[142], B.A. Trzeciak[62], T. Tsuji[132], A. Tumkin[108], R. Turrisi[56], T.S. Tveter[20], K. Ullaland[21], E.N. Umaka[125], A. Uras[135], G.L. Usai[23], A. Utrobicic[98], M. Vala[116,37], N. Valle[139], S. Vallero[58], N. van der Kolk[62], L.V.R. van Doremalen[62], M. van Leeuwen[62], P. Vande Vyvre[33], D. Varga[145], Z. Varga[145], M. Varga-Kofarago[145], A. Vargas[44], R. Varma[48], M. Vasileiou[83], A. Vasiliev[87], O. Vázquez Doce[104,117], V. Vechernin[112], A.M. Veen[62], E. Vercellin[25], S. Vergara Limón[44], L. Vermunt[62], R. Vernet[7], R. Vértesi[145], M.G.D.L.C. Vicencio[9], L. Vickovic[34], Z. Vilakazi[131], O. Villalobos Baillie[110], A. Villatoro Tello[44], G. Vino[52], A. Vinogradov[87], T. Virgili[29], V. Vislavicius[88], A. Vodopyanov[74], B. Volkel[33], M.A. Völkl[102], K. Voloshin[91], S.A. Voloshin[143], G. Volpe[32], B. von Haller[33], I. Vorobyev[104], D. Voscek[116], J. Vrláková[37], B. Wagner[21], M. Weber[113], S.G. Weber[144,106], A. Wegrzynek[33], D.F. Weiser[103], S.C. Wenzel[33], J.P. Wessels[144], E. Widmann[113], J. Wiechula[67], J. Wikne[20], G. Wilk[84], J. Wilkinson[53,10], G.A. Willems[33], E. Willsher[110], B. Windelband[103], W.E. Witt[130], Y. Wu[128], R. Xu[6], S. Yalcin[76], K. Yamakawa[45], S. Yang[21], S. Yano[137], Z. Yin[6], H. Yokoyama[133,62], I.-K. Yoo[17], J.H. Yoon[60], S. Yuan[21], A. Yuncu[103], V. Yurchenko[2], V. Zaccolo[58,24], A. Zaman[14], C. Zampolli[33], H.J.C. Zanoli[62,121], N. Zardoshti[33], A. Zarochentsev[112], P. Závada[65], N. Zaviyalov[108], H. Zbroszczyk[142], M. Zhalov[97], X. Zhang[6], Z. Zhang[6], V. Zherebchevskii[112], N. Zhigareva[91], D. Zhou[6], Y. Zhou[88], Z. Zhou[21], J. Zhu[6], Y. Zhu[6], A. Zichichi[10,26], M.B. Zimmermann[33], G. Zinovjev[2], N. Zurlo[140]

Affiliation Notes

[I] Deceased

[II] Also at: Dipartimento DET del Politecnico di Torino, Turin, Italy

[III] Also at: M.V. Lomonosov Moscow State University, D.V. Skobeltsyn Institute of Nuclear, Physics, Moscow, Russia

[IV] Also at: Department of Applied Physics, Aligarh Muslim University, Aligarh, India
[V] Also at: Institute of Theoretical Physics, University of Wroclaw, Poland

Collaboration Institutes

[1] A.I. Alikhanyan National Science Laboratory (Yerevan Physics Institute) Foundation, Yerevan, Armenia
[2] Bogolyubov Institute for Theoretical Physics, National Academy of Sciences of Ukraine, Kiev, Ukraine
[3] Bose Institute, Department of Physics and Centre for Astroparticle Physics and Space Science (CAPSS), Kolkata, India
[4] Budker Institute for Nuclear Physics, Novosibirsk, Russia
[5] California Polytechnic State University, San Luis Obispo, California, United States
[6] Central China Normal University, Wuhan, China
[7] Centre de Calcul de l'IN2P3, Villeurbanne, Lyon, France
[8] Centro de Aplicaciones Tecnológicas y Desarrollo Nuclear (CEADEN), Havana, Cuba
[9] Centro de Investigación y de Estudios Avanzados (CINVESTAV), Mexico City and Mérida, Mexico
[10] Centro Fermi - Museo Storico della Fisica e Centro Studi e Ricerche "Enrico Fermi', Rome, Italy
[11] Chicago State University, Chicago, Illinois, United States
[12] China Institute of Atomic Energy, Beijing, China
[13] Comenius University Bratislava, Faculty of Mathematics, Physics and Informatics, Bratislava, Slovakia
[14] COMSATS University Islamabad, Islamabad, Pakistan
[15] Creighton University, Omaha, Nebraska, United States
[16] Department of Physics, Aligarh Muslim University, Aligarh, India
[17] Department of Physics, Pusan National University, Pusan, Republic of Korea
[18] Department of Physics, Sejong University, Seoul, Republic of Korea
[19] Department of Physics, University of California, Berkeley, California, United States
[20] Department of Physics, University of Oslo, Oslo, Norway
[21] Department of Physics and Technology, University of Bergen, Bergen, Norway
[22] Dipartimento di Fisica dell'Università 'La Sapienza' and Sezione INFN, Rome, Italy
[23] Dipartimento di Fisica dell'Università and Sezione INFN, Cagliari, Italy
[24] Dipartimento di Fisica dell'Università and Sezione INFN, Trieste, Italy
[25] Dipartimento di Fisica dell'Università and Sezione INFN, Turin, Italy
[26] Dipartimento di Fisica e Astronomia dell'Università and Sezione INFN, Bologna, Italy
[27] Dipartimento di Fisica e Astronomia dell'Università and Sezione INFN, Catania, Italy

[28] Dipartimento di Fisica e Astronomia dell'Università and Sezione INFN, Padova, Italy
[29] Dipartimento di Fisica 'E.R. Caianiello' dell'Università and Gruppo Collegato INFN, Salerno, Italy
[30] Dipartimento DISAT del Politecnico and Sezione INFN, Turin, Italy
[31] Dipartimento di Scienze e Innovazione Tecnologica dell'Università del Piemonte Orientale and INFN Sezione di Torino, Alessandria, Italy
[32] Dipartimento Interateneo di Fisica 'M. Merlin' and Sezione INFN, Bari, Italy
[33] European Organization for Nuclear Research (CERN), Geneva, Switzerland
[34] Faculty of Electrical Engineering, Mechanical Engineering and Naval Architecture, University of Split, Split, Croatia
[35] Faculty of Engineering and Science, Western Norway University of Applied Sciences, Bergen, Norway
[36] Faculty of Nuclear Sciences and Physical Engineering, Czech Technical University in Prague, Prague, Czech Republic
[37] Faculty of Science, P.J. Šafárik University, Košice, Slovakia
[38] Frankfurt Institute for Advanced Studies, Johann Wolfgang Goethe-Universität Frankfurt, Frankfurt, Germany
[39] Fudan University, Shanghai, China
[40] Gangneung-Wonju National University, Gangneung, Republic of Korea
[41] Gauhati University, Department of Physics, Guwahati, India
[42] Helmholtz-Institut für Strahlen- und Kernphysik, Rheinische Friedrich-Wilhelms-Universität Bonn, Bonn, Germany
[43] Helsinki Institute of Physics (HIP), Helsinki, Finland
[44] High Energy Physics Group, Universidad Autónoma de Puebla, Puebla, Mexico
[45] Hiroshima University, Hiroshima, Japan
[46] Hochschule Worms, Zentrum für Technologietransfer und Telekommunikation (ZTT), Worms, Germany
[47] Horia Hulubei National Institute of Physics and Nuclear Engineering, Bucharest, Romania
[48] Indian Institute of Technology Bombay (IIT), Mumbai, India
[49] Indian Institute of Technology Indore, Indore, India
[50] Indonesian Institute of Sciences, Jakarta, Indonesia
[51] INFN, Laboratori Nazionali di Frascati, Frascati, Italy
[52] INFN, Sezione di Bari, Bari, Italy
[53] INFN, Sezione di Bologna, Bologna, Italy
[54] INFN, Sezione di Cagliari, Cagliari, Italy
[55] INFN, Sezione di Catania, Catania, Italy
[56] INFN, Sezione di Padova, Padova, Italy
[57] INFN, Sezione di Roma, Rome, Italy
[58] INFN, Sezione di Torino, Turin, Italy
[59] INFN, Sezione di Trieste, Trieste, Italy
[60] Inha University, Incheon, Republic of Korea
[61] Institute for Nuclear Research, Academy of Sciences, Moscow, Russia

[62] Institute for Subatomic Physics, Utrecht University/Nikhef, Utrecht, Netherlands
[63] Institute of Experimental Physics, Slovak Academy of Sciences, Košice, Slovakia
[64] Institute of Physics, Homi Bhabha National Institute, Bhubaneswar, India
[65] Institute of Physics of the Czech Academy of Sciences, Prague, Czech Republic
[66] Institute of Space Science (ISS), Bucharest, Romania
[67] Institut für Kernphysik, Johann Wolfgang Goethe-Universität Frankfurt, Frankfurt, Germany
[68] Instituto de Ciencias Nucleares, Universidad Nacional Autónoma de México, Mexico City, Mexico
[69] Instituto de Física, Universidade Federal do Rio Grande do Sul (UFRGS), Porto Alegre, Brazil
[70] Instituto de Física, Universidad Nacional Autónoma de México, Mexico City, Mexico
[71] iThemba LABS, National Research Foundation, Somerset West, South Africa
[72] Jeonbuk National University, Jeonju, Republic of Korea
[73] Johann-Wolfgang-Goethe Universität Frankfurt Institut für Informatik, Fachbereich Informatik und Mathematik, Frankfurt, Germany
[74] Joint Institute for Nuclear Research (JINR), Dubna, Russia
[75] Korea Institute of Science and Technology Information, Daejeon, Republic of Korea
[76] KTO Karatay University, Konya, Turkey
[77] Laboratoire de Physique des 2 Infinis, IrÃ¨ne Joliot-Curie, Orsay, France
[78] Laboratoire de Physique Subatomique et de Cosmologie, Université Grenoble-Alpes, CNRS-IN2P3, Grenoble, France
[79] Lawrence Berkeley National Laboratory, Berkeley, California, United States
[80] Lund University Department of Physics, Division of Particle Physics, Lund, Sweden
[81] Nagasaki Institute of Applied Science, Nagasaki, Japan
[82] Nara Women's University (NWU), Nara, Japan
[83] National and Kapodistrian University of Athens, School of Science, Department of Physics , Athens, Greece
[84] National Centre for Nuclear Research, Warsaw, Poland
[85] National Institute of Science Education and Research, Homi Bhabha National Institute, Jatni, India
[86] National Nuclear Research Center, Baku, Azerbaijan
[87] National Research Centre Kurchatov Institute, Moscow, Russia
[88] Niels Bohr Institute, University of Copenhagen, Copenhagen, Denmark
[89] Nikhef, National institute for subatomic physics, Amsterdam, Netherlands
[90] NRC Kurchatov Institute IHEP, Protvino, Russia
[91] NRC Â«Kurchatov InstituteÂ» - ITEP, Moscow, Russia
[92] NRNU Moscow Engineering Physics Institute, Moscow, Russia
[93] Nuclear Physics Group, STFC Daresbury Laboratory, Daresbury, United Kingdom

[94] Nuclear Physics Institute of the Czech Academy of Sciences, Řež u Prahy, Czech Republic
[95] Oak Ridge National Laboratory, Oak Ridge, Tennessee, United States
[96] Ohio State University, Columbus, Ohio, United States
[97] Petersburg Nuclear Physics Institute, Gatchina, Russia
[98] Physics department, Faculty of science, University of Zagreb, Zagreb, Croatia
[99] Physics Department, Panjab University, Chandigarh, India
[100] Physics Department, University of Jammu, Jammu, India
[101] Physics Department, University of Rajasthan, Jaipur, India
[102] Physikalisches Institut, Eberhard-Karls-Universität Tübingen, Tübingen, Germany
[103] Physikalisches Institut, Ruprecht-Karls-Universität Heidelberg, Heidelberg, Germany
[104] Physik Department, Technische Universität München, Munich, Germany
[105] Politecnico di Bari and Sezione INFN, Bari, Italy
[106] Research Division and ExtreMe Matter Institute EMMI, GSI Helmholtzzentrum für Schwerionenforschung GmbH, Darmstadt, Germany
[107] Rudjer Bošković Institute, Zagreb, Croatia
[108] Russian Federal Nuclear Center (VNIIEF), Sarov, Russia
[109] Saha Institute of Nuclear Physics, Homi Bhabha National Institute, Kolkata, India
[110] School of Physics and Astronomy, University of Birmingham, Birmingham, United Kingdom
[111] Sección Física, Departamento de Ciencias, Pontificia Universidad Católica del Perú, Lima, Peru
[112] St. Petersburg State University, St. Petersburg, Russia
[113] Stefan Meyer Institut für Subatomare Physik (SMI), Vienna, Austria
[114] SUBATECH, IMT Atlantique, Université de Nantes, CNRS-IN2P3, Nantes, France
[115] Suranaree University of Technology, Nakhon Ratchasima, Thailand
[116] Technical University of Košice, Košice, Slovakia
[117] Technische Universität München, Excellence Cluster 'Universe', Munich, Germany
[118] The Henryk Niewodniczanski Institute of Nuclear Physics, Polish Academy of Sciences, Cracow, Poland
[119] The University of Texas at Austin, Austin, Texas, United States
[120] Universidad Autónoma de Sinaloa, Culiacán, Mexico
[121] Universidade de São Paulo (USP), São Paulo, Brazil
[122] Universidade Estadual de Campinas (UNICAMP), Campinas, Brazil
[123] Universidade Federal do ABC, Santo Andre, Brazil
[124] University of Cape Town, Cape Town, South Africa
[125] University of Houston, Houston, Texas, United States
[126] University of Jyväskylä, Jyväskylä, Finland
[127] University of Liverpool, Liverpool, United Kingdom
[128] University of Science and Technology of China, Hefei, China

[129] University of South-Eastern Norway, Tonsberg, Norway
[130] University of Tennessee, Knoxville, Tennessee, United States
[131] University of the Witwatersrand, Johannesburg, South Africa
[132] University of Tokyo, Tokyo, Japan
[133] University of Tsukuba, Tsukuba, Japan
[134] Université Clermont Auvergne, CNRS/IN2P3, LPC, Clermont-Ferrand, France
[135] Université de Lyon, Université Lyon 1, CNRS/IN2P3, IPN-Lyon, Villeurbanne, Lyon, France
[136] Université de Strasbourg, CNRS, IPHC UMR 7178, F-67000 Strasbourg, France, Strasbourg, France
[137] Université Paris-Saclay Centre d'Etudes de Saclay (CEA), IRFU, Départment de Physique Nucléaire (DPhN), Saclay, France
[138] Università degli Studi di Foggia, Foggia, Italy
[139] Università degli Studi di Pavia and Sezione INFN, Pavia, Italy
[140] Università di Brescia and Sezione INFN, Brescia, Italy
[141] Variable Energy Cyclotron Centre, Homi Bhabha National Institute, Kolkata, India
[142] Warsaw University of Technology, Warsaw, Poland
[143] Wayne State University, Detroit, Michigan, United States
[144] Westfälische Wilhelms-Universität Münster, Institut für Kernphysik, Münster, Germany
[145] Wigner Research Centre for Physics, Budapest, Hungary
[146] Yale University, New Haven, Connecticut, United States
[147] Yonsei University, Seoul, Republic of Korea

ATLAS Collaboration

G. Aad[102], B. Abbott[129], D.C. Abbott[103], A. Abed Abud[71a,71b], K. Abeling[53], D.K. Abhayasinghe[94], S.H. Abidi[167], O.S. AbouZeid[40], N.L. Abraham[156], H. Abramowicz[161], H. Abreu[160], Y. Abulaiti[6], B.S. Acharya[67a,67b,o], B. Achkar[53], S. Adachi[163], L. Adam[100], C. Adam Bourdarios[5], L. Adamczyk[84a], L. Adamek[167], J. Adelman[121], M. Adersberger[114], A. Adiguzel[12c], S. Adorni[54], T. Adye[144], A.A. Affolder[146], Y. Afik[160], C. Agapopoulou[65], M.N. Agaras[38], A. Aggarwal[119], C. Agheorghiesei[27c], J.A. Aguilar-Saavedra[140f,140a,aj], F. Ahmadov[80], W.S. Ahmed[104], X. Ai[18], G. Aielli[74a,74b], S. Akatsuka[86], T.P.A. Åkesson[97], E. Akilli[54], A.V. Akimov[111], K. Al Khoury[65], G.L. Alberghi[23b,23a], J. Albert[176], M.J. Alconada Verzini[161], S. Alderweireldt[36], M. Aleksa[36], I.N. Aleksandrov[80], C. Alexa[27b], D. Alexandre[19], T. Alexopoulos[10], A. Alfonsi[120], F. Alfonsi[23b,23a], M. Alhroob[129], B. Ali[142], M. Aliev[155], G. Alimonti[69a], J. Alison[37], S.P. Alkire[148], C. Allaire[65], B.M.M. Allbrooke[156], B.W. Allen[132], P.P. Allport[21], A. Aloisio[70a,70b], A. Alonso[40], F. Alonso[89], C. Alpigiani[148], A.A. Alshehri[57], M. Alvarez Estevez[99], D. Álvarez Piqueras[174], M.G. Alviggi[70a,70b], Y. Amaral Coutinho[81b], A. Ambler[104], L. Ambroz[135], C. Amelung[26], D. Amidei[106], S.P. Amor Dos Santos[140a], S. Amoroso[46], C.S. Amrouche[54], F. An[79], C. Anastopoulos[149], N. Andari[145], T. Andeen[11], C.F. Anders[61b], J.K. Anders[20], A. Andreazza[69a,69b], V. Andrei[61a], C.R. Anelli[176], S. Angelidakis[38], A. Angerami[39], A.V. Anisenkov[122b,122a], A. Annovi[72a], C. Antel[61a], M.T. Anthony[149], E. Antipov[130], M. Antonelli[51], D.J.A. Antrim[171], F. Anulli[73a], M. Aoki[82], J.A. Aparisi Pozo[174], L. Aperio Bella[15a], G. Arabidze[107], J.P. Araque[140a], V. Araujo Ferraz[81b], R. Araujo Pereira[81b], C. Arcangeletti[51], A.T.H. Arce[49], F.A. Arduh[89], J-F. Arguin[110], S. Argyropoulos[78], J.-H. Arling[46], A.J. Armbruster[36], A. Armstrong[171], O. Arnaez[167], H. Arnold[120], Z.P. Arrubarrena Tame[114], G. Artoni[135], S. Artz[100], S. Asai[163], N. Asbah[59], E.M. Asimakopoulou[172], L. Asquith[156], J. Assahsah[35d], K. Assamagan[29], R. Astalos[28a], R.J. Atkin[33a], M. Atkinson[173], N.B. Atlay[19], H. Atmani[65], K. Augsten[142], G. Avolio[36], R. Avramidou[60a], M.K. Ayoub[15a], A.M. Azoulay[168b], G. Azuelos[110,ax], H. Bachacou[145], K. Bachas[68a,68b], M. Backes[135], F. Backman[45a,45b], P. Bagnaia[73a,73b], M. Bahmani[85], H. Bahrasemani[152], A.J. Bailey[174], V.R. Bailey[173], J.T. Baines[144], M. Bajic[40], C. Bakalis[10], O.K. Baker[183], P.J. Bakker[120], D. Bakshi Gupta[8], S. Balaji[157], E.M. Baldin[122b,122a], P. Balek[180], F. Balli[145], W.K. Balunas[135], J. Balz[100], E. Banas[85], A. Bandyopadhyay[24], Sw. Banerjee[181,j], A.A.E. Bannoura[182], L. Barak[161], W.M. Barbe[38], E.L. Barberio[105], D. Barberis[55b,55a], M. Barbero[102], G. Barbour[95], T. Barillari[115], M-S. Barisits[36], J. Barkeloo[132], T. Barklow[153], R. Barnea[160], S.L. Barnes[60c], B.M. Barnett[144], R.M. Barnett[18], Z. Barnovska-Blenessy[60a], A. Baroncelli[60a], G. Barone[29], A.J. Barr[135], L. Barranco Navarro[45a,45b], F. Barreiro[99], J. Barreiro Guimarães da Costa[15a], S. Barsov[138], R. Bartoldus[153], G. Bartolini[102], A.E. Barton[90], P. Bartos[28a], A. Basalaev[46], A. Bassalat[65,aq], M.J. Basso[167], R.L. Bates[57], S. Batlamous[35e], J.R. Batley[32], B. Batool[151], M. Battaglia[146], M. Bauce[73a,73b], F. Bauer[145], K.T. Bauer[171], H.S. Bawa[31,m], J.B. Beacham[49], T. Beau[136], P.H. Beauchemin[170], F. Becherer[52], P. Bechtle[24], H.C. Beck[53], H.P. Beck[20,s], K. Becker[52], M. Becker[100],

C. Becot[46], A. Beddall[12d], A.J. Beddall[12a], V.A. Bednyakov[80], M. Bedognetti[120], C.P. Bee[155], T.A. Beermann[182], M. Begalli[81b], M. Begel[29], A. Behera[155], J.K. Behr[46], F. Beisiegel[24], A.S. Bell[95], G. Bella[161], L. Bellagamba[23b], A. Bellerive[34], P. Bellos[9], K. Beloborodov[122b,122a], K. Belotskiy[112], N.L. Belyaev[112], D. Benchekroun[35a], N. Benekos[10], Y. Benhammou[161], D.P. Benjamin[6], M. Benoit[54], J.R. Bensinger[26], S. Bentvelsen[120], L. Beresford[135], M. Beretta[51], D. Berge[46], E. Bergeaas Kuutmann[172], N. Berger[5], B. Bergmann[142], L.J. Bergsten[26], J. Beringer[18], S. Berlendis[7], G. Bernardi[136], C. Bernius[153], F.U. Bernlochner[24], T. Berry[94], P. Berta[100], C. Bertella[15a], I.A. Bertram[90], O. Bessidskaia Bylund[182], N. Besson[145], A. Bethani[101], S. Bethke[115], A. Betti[42], A.J. Bevan[93], J. Beyer[115], D.S. Bhattacharya[177], P. Bhattarai[26], R. Bi[139], R.M. Bianchi[139], O. Biebel[114], D. Biedermann[19], R. Bielski[36], K. Bierwagen[100], N.V. Biesuz[72a,72b], M. Biglietti[75a], T.R.V. Billoud[110], M. Bindi[53], A. Bingul[12d], C. Bini[73a,73b], S. Biondi[23b,23a], M. Birman[180], T. Bisanz[53], J.P. Biswal[161], D. Biswas[181,j], A. Bitadze[101], C. Bittrich[48], K. Bjørke[134], K.M. Black[25], T. Blazek[28a], I. Bloch[46], C. Blocker[26], A. Blue[57], U. Blumenschein[93], G.J. Bobbink[120], V.S. Bobrovnikov[122b,122a], S.S. Bocchetta[97], A. Bocci[49], D. Boerner[46], D. Bogavac[14], A.G. Bogdanchikov[122b,122a], C. Bohm[45a], V. Boisvert[94], P. Bokan[53,172], T. Bold[84a], A.S. Boldyrev[113], A.E. Bolz[61b], M. Bomben[136], M. Bona[93], J.S. Bonilla[132], M. Boonekamp[145], C.D. Booth[94], H.M. Borecka-Bielska[91], A. Borisov[123], G. Borissov[90], J. Bortfeldt[36], D. Bortoletto[135], D. Boscherini[23b], M. Bosman[14], J.D. Bossio Sola[104], K. Bouaouda[35a], J. Boudreau[139], E.V. Bouhova-Thacker[90], D. Boumediene[38], S.K. Boutle[57], A. Boveia[127], J. Boyd[36], D. Boye[33b,ar], I.R. Boyko[80], A.J. Bozson[94], J. Bracinik[21], N. Brahimi[102], G. Brandt[182], O. Brandt[32], F. Braren[46], B. Brau[103], J.E. Brau[132], W.D. Breaden Madden[57], K. Brendlinger[46], L. Brenner[46], R. Brenner[172], S. Bressler[180], B. Brickwedde[100], D.L. Briglin[21], D. Britton[57], D. Britzger[115], I. Brock[24], R. Brock[107], G. Brooijmans[39], W.K. Brooks[147c], E. Brost[121], J.H Broughton[21], P.A. Bruckman de Renstrom[85], D. Bruncko[28b], A. Bruni[23b], G. Bruni[23b], L.S. Bruni[120], S. Bruno[74a,74b], M. Bruschi[23b], N. Bruscino[139], P. Bryant[37], L. Bryngemark[97], T. Buanes[17], Q. Buat[36], P. Buchholz[151], A.G. Buckley[57], I.A. Budagov[80], M.K. Bugge[134], F. Bührer[52], O. Bulekov[112], T.J. Burch[121], S. Burdin[91], C.D. Burgard[120], A.M. Burger[130], B. Burghgrave[8], J.T.P. Burr[46], C.D. Burton[11], J.C. Burzynski[103], V. Büscher[100], E. Buschmann[53], P.J. Bussey[57], J.M. Butler[25], C.M. Buttar[57], J.M. Butterworth[95], P. Butti[36], W. Buttinger[36], C.J. Buxo Vazquez[107], A. Buzatu[158], A.R. Buzykaev[122b,122a], G. Cabras[23b,23a], S. Cabrera Urbán[174], D. Caforio[56], H. Cai[173], V.M.M. Cairo[153], O. Cakir[4a], N. Calace[36], P. Calafiura[18], A. Calandri[102], G. Calderini[136], P. Calfayan[66], G. Callea[57], L.P. Caloba[81b], A. Caltabiano[74a,74b], S. Calvente Lopez[99], D. Calvet[38], S. Calvet[38], T.P. Calvet[155], M. Calvetti[72a,72b], R. Camacho Toro[136], S. Camarda[36], D. Camarero Munoz[99], P. Camarri[74a,74b], D. Cameron[134], R. Caminal Armadans[103], C. Camincher[36], S. Campana[36], M. Campanelli[95], A. Camplani[40], A. Campoverde[151], V. Canale[70a,70b], A. Canesse[104], M. Cano Bret[60c], J. Cantero[130], T. Cao[161], Y. Cao[173], M.D.M. Capeans Garrido[36], M. Capua[41b,41a], R. Cardarelli[74a], F. Cardillo[149], G. Carducci[41b,41a], I. Carli[143], T. Carli[36], G. Carlino[70a], B.T. Carlson[139], L. Carminati[69a,69b], R.M.D. Carney[45a,45b], S. Caron[119],

E. Carquin[147c], S. Carrá[46], J.W.S. Carter[167], M.P. Casado[14,e], A.F. Casha[167], D.W. Casper[171], R. Castelijn[120], F.L. Castillo[174], V. Castillo Gimenez[174], N.F. Castro[140a,140e], A. Catinaccio[36], J.R. Catmore[134], A. Cattai[36], J. Caudron[24], V. Cavaliere[29], E. Cavallaro[14], V. Cavasinni[72a,72b], E. Celebi[12b], L. Cerda Alberich[174], K. Cerny[131], A.S. Cerqueira[81a], A. Cerri[156], L. Cerrito[74a,74b], F. Cerutti[18], A. Cervelli[23b,23a], S.A. Cetin[12b], Z. Chadi[35a], D. Chakraborty[121], S.K. Chan[59], W.S. Chan[120], W.Y. Chan[91], J.D. Chapman[32], B. Chargeishvili[159b], D.G. Charlton[21], T.P. Charman[93], C.C. Chau[34], S. Che[127], S. Chekanov[6], S.V. Chekulaev[168a], G.A. Chelkov[80,aw], M.A. Chelstowska[36], B. Chen[79], C. Chen[60a], C.H. Chen[79], H. Chen[29], J. Chen[60a], J. Chen[39], S. Chen[137], S.J. Chen[15c], X. Chen[15b,av], Y. Chen[83], Y-H. Chen[46], H.C. Cheng[63a], H.J. Cheng[15a], A. Cheplakov[80], E. Cheremushkina[123], R. Cherkaoui El Moursli[35e], E. Cheu[7], K. Cheung[64], T.J.A. Chevalérias[145], L. Chevalier[145], V. Chiarella[51], G. Chiarelli[72a], G. Chiodini[68a], A.S. Chisholm[21], A. Chitan[27b], I. Chiu[163], Y.H. Chiu[176], M.V. Chizhov[80], K. Choi[66], A.R. Chomont[73a,73b], S. Chouridou[162], Y.S. Chow[120], M.C. Chu[63a], X. Chu[15a,15d], J. Chudoba[141], A.J. Chuinard[104], J.J. Chwastowski[85], L. Chytka[131], D. Cieri[115], K.M. Ciesla[85], D. Cinca[47], V. Cindro[92], I.A. Cioară[27b], A. Ciocio[18], F. Cirotto[70a,70b], Z.H. Citron[180,k], M. Citterio[69a], D.A. Ciubotaru[27b], B.M. Ciungu[167], A. Clark[54], M.R. Clark[39], P.J. Clark[50], C. Clement[45a,45b], Y. Coadou[102], M. Cobal[67a,67c], A. Coccaro[55b], J. Cochran[79], H. Cohen[161], A.E.C. Coimbra[36], L. Colasurdo[119], B. Cole[39], A.P. Colijn[120], J. Collot[58], P. Conde Muiño[140a,f], E. Coniavitis[52], S.H. Connell[33b], I.A. Connelly[57], S. Constantinescu[27b], F. Conventi[70a,ay], A.M. Cooper-Sarkar[135], F. Cormier[175], K.J.R. Cormier[167], L.D. Corpe[95], M. Corradi[73a,73b], E.E. Corrigan[97], F. Corriveau[104,af], A. Cortes-Gonzalez[36], M.J. Costa[174], F. Costanza[5], D. Costanzo[149], G. Cowan[94], J.W. Cowley[32], J. Crane[101], K. Cranmer[125], S.J. Crawley[57], R.A. Creager[137], S. Crépé-Renaudin[58], F. Crescioli[136], M. Cristinziani[24], V. Croft[120], G. Crosetti[41b,41a], A. Cueto[5], T. Cuhadar Donszelmann[149], A.R. Cukierman[153], W.R. Cunningham[57], S. Czekierda[85], P. Czodrowski[36], M.J. Da Cunha Sargedas De Sousa[60b], J.V. Da Fonseca Pinto[81b], C. Da Via[101], W. Dabrowski[84a], T. Dado[28a], S. Dahbi[35e], T. Dai[106], C. Dallapiccola[103], M. Dam[40], G. D'amen[29], V. D'Amico[75a,75b], J. Damp[100], J.R. Dandoy[137], M.F. Daneri[30], N.P. Dang[181,j], N.S. Dann[101], M. Danninger[175], V. Dao[36], G. Darbo[55b], O. Dartsi[5], A. Dattagupta[132], T. Daubney[46], S. D'Auria[69a,69b], W. Davey[24], C. David[46], T. Davidek[143], D.R. Davis[49], I. Dawson[149], K. De[8], R. De Asmundis[70a], M. De Beurs[120], S. De Castro[23b,23a], S. De Cecco[73a,73b], N. De Groot[119], P. de Jong[120], H. De la Torre[107], A. De Maria[15c], D. De Pedis[73a], A. De Salvo[73a], U. De Sanctis[74a,74b], M. De Santis[74a,74b], A. De Santo[156], K. De Vasconcelos Corga[102], J.B. De Vivie De Regie[65], C. Debenedetti[146], D.V. Dedovich[80], A.M. Deiana[42], M. Del Gaudio[41b,41a], J. Del Peso[99], Y. Delabat Diaz[46], D. Delgove[65], F. Deliot[145,r], C.M. Delitzsch[7], M. Della Pietra[70a,70b], D. Della Volpe[54], A. Dell'Acqua[36], L. Dell'Asta[74a,74b], M. Delmastro[5], C. Delporte[65], P.A. Delsart[58], D.A. DeMarco[167], S. Demers[183], M. Demichev[80], G. Demontigny[110], S.P. Denisov[123], D. Denysiuk[120], L. D'Eramo[136], D. Derendarz[85], J.E. Derkaoui[35d], F. Derue[136], P. Dervan[91], K. Desch[24], C. Deterre[46], K. Dette[167], C. Deutsch[24], M.R. Devesa[30], P.O. Deviveiros[36], A. Dewhurst[144],

F.A. Di Bello[54], A. Di Ciaccio[74a,74b], L. Di Ciaccio[5], W.K. Di Clemente[137], C. Di Donato[70a,70b], A. Di Girolamo[36], G. Di Gregorio[72a,72b], B. Di Micco[75a,75b], R. Di Nardo[103], K.F. Di Petrillo[59], R. Di Sipio[167], D. Di Valentino[34], C. Diaconu[102], F.A. Dias[40], T. Dias Do Vale[140a], M.A. Diaz[147a], J. Dickinson[18], E.B. Diehl[106], J. Dietrich[19], S. Díez Cornell[46], A. Dimitrievska[18], W. Ding[15b], J. Dingfelder[24], F. Dittus[36], F. Djama[102], T. Djobava[159b], J.I. Djuvsland[17], M.A.B. Do Vale[81c], M. Dobre[27b], D. Dodsworth[26], C. Doglioni[97], J. Dolejsi[143], Z. Dolezal[143], M. Donadelli[81d], B. Dong[60c], J. Donini[38], A. D'onofrio[93], M. D'Onofrio[91], J. Dopke[144], A. Doria[70a], M.T. Dova[89], A.T. Doyle[57], E. Drechsler[152], E. Dreyer[152], T. Dreyer[53], A.S. Drobac[170], D. Du[60b], Y. Duan[60b], F. Dubinin[111], M. Dubovsky[28a], A. Dubreuil[54], E. Duchovni[180], G. Duckeck[114], A. Ducourthial[136], O.A. Ducu[110], D. Duda[115], A. Dudarev[36], A.C. Dudder[100], E.M. Duffield[18], L. Duflot[65], M. Dührssen[36], C. Dülsen[182], M. Dumancic[180], A.E. Dumitriu[27b], A.K. Duncan[57], M. Dunford[61a], A. Duperrin[102], H. Duran Yildiz[4a], M. Düren[56], A. Durglishvili[159b], D. Duschinger[48], B. Dutta[46], D. Duvnjak[1], G.I. Dyckes[137], M. Dyndal[36], S. Dysch[101], B.S. Dziedzic[85], K.M. Ecker[115], R.C. Edgar[106], M.G. Eggleston[49], T. Eifert[36], G. Eigen[17], K. Einsweiler[18], T. Ekelof[172], H. El Jarrari[35e], M. El Kacimi[35c], R. El Kosseifi[102], V. Ellajosyula[172], M. Ellert[172], F. Ellinghaus[182], A.A. Elliot[93], N. Ellis[36], J. Elmsheuser[29], M. Elsing[36], D. Emeliyanov[144], A. Emerman[39], Y. Enari[163], M.B. Epland[49], J. Erdmann[47], A. Ereditato[20], M. Errenst[36], M. Escalier[65], C. Escobar[174], O. Estrada Pastor[174], E. Etzion[161], H. Evans[66], A. Ezhilov[138], F. Fabbri[57], L. Fabbri[23b,23a], V. Fabiani[119], G. Facini[95], R.M. Faisca Rodrigues Pereira[140a], R.M. Fakhrutdinov[123], S. Falciano[73a], P.J. Falke[5], S. Falke[5], J. Faltova[143], Y. Fang[15a], Y. Fang[15a], G. Fanourakis[44], M. Fanti[69a,69b], M. Faraj[67a,67c,u], A. Farbin[8], A. Farilla[75a], E.M. Farina[71a,71b], T. Farooque[107], S. Farrell[18], S.M. Farrington[50], P. Farthouat[36], F. Fassi[35e], P. Fassnacht[36], D. Fassouliotis[9], M. Faucci Giannelli[50], W.J. Fawcett[32], L. Fayard[65], O.L. Fedin[138,p], W. Fedorko[175], A. Fehr[20], M. Feickert[42], L. Feligioni[102], A. Fell[149], C. Feng[60b], E.J. Feng[36], M. Feng[49], M.J. Fenton[57], A.B. Fenyuk[123], J. Ferrando[46], A. Ferrante[173], A. Ferrari[172], P. Ferrari[120], R. Ferrari[71a], D.E. Ferreira de Lima[61b], A. Ferrer[174], D. Ferrere[54], C. Ferretti[106], F. Fiedler[100], A. Filipčič[92], F. Filthaut[119], K.D. Finelli[25], M.C.N. Fiolhais[140a,140c,a], L. Fiorini[174], F. Fischer[114], W.C. Fisher[107], I. Fleck[151], P. Fleischmann[106], R.R.M. Fletcher[137], T. Flick[182], B.M. Flierl[114], L. Flores[137], L.R. Flores Castillo[63a], F.M. Follega[76a,76b], N. Fomin[17], J.H. Foo[167], G.T. Forcolin[76a,76b], A. Formica[145], F.A. Förster[14], A.C. Forti[101], A.G. Foster[21], M.G. Foti[135], D. Fournier[65], H. Fox[90], P. Francavilla[72a,72b], S. Francescato[73a,73b], M. Franchini[23b,23a], S. Franchino[61a], D. Francis[36], L. Franconi[20], M. Franklin[59], A.N. Fray[93], P.M. Freeman[21], B. Freund[110], W.S. Freund[81b], E.M. Freundlich[47], D.C. Frizzell[129], D. Froidevaux[36], J.A. Frost[135], C. Fukunaga[164], E. Fullana Torregrosa[174], E. Fumagalli[55b,55a], T. Fusayasu[116], J. Fuster[174], A. Gabrielli[23b,23a], A. Gabrielli[18], G.P. Gach[84a], S. Gadatsch[54], P. Gadow[115], G. Gagliardi[55b,55a], L.G. Gagnon[110], C. Galea[27b], B. Galhardo[140a], G.E. Gallardo[135], E.J. Gallas[135], B.J. Gallop[144], G. Galster[40], R. Gamboa Goni[93], K.K. Gan[127], S. Ganguly[180], J. Gao[60a], Y. Gao[50], Y.S. Gao[31,m], C. García[174], J.E. García Navarro[174], J.A. García Pascual[15a], C. Garcia-Argos[52],

M. Garcia-Sciveres[18], R.W. Gardner[37], N. Garelli[153], S. Gargiulo[52], V. Garonne[134], A. Gaudiello[55b,55a], G. Gaudio[71a], I.L. Gavrilenko[111], A. Gavrilyuk[124], C. Gay[175], G. Gaycken[46], E.N. Gazis[10], A.A. Geanta[27b], C.M. Gee[146], C.N.P. Gee[144], J. Geisen[53], M. Geisen[100], C. Gemme[55b], M.H. Genest[58], C. Geng[106], S. Gentile[73a,73b], S. George[94], T. Geralis[44], L.O. Gerlach[53], P. Gessinger-Befurt[100], G. Gessner[47], S. Ghasemi[151], M. Ghasemi Bostanabad[176], A. Ghosh[65], A. Ghosh[78], B. Giacobbe[23b], S. Giagu[73a,73b], N. Giangiacomi[23b,23a], P. Giannetti[72a], A. Giannini[70a,70b], G. Giannini[14], S.M. Gibson[94], M. Gignac[146], D. Gillberg[34], G. Gilles[182], D.M. Gingrich[3,ax], M.P. Giordani[67a,67c], F.M. Giorgi[23b], P.F. Giraud[145], G. Giugliarelli[67a,67c], D. Giugni[69a], F. Giuli[74a,74b], S. Gkaitatzis[162], I. Gkialas[9,h], E.L. Gkougkousis[14], P. Gkountoumis[10], L.K. Gladilin[113], C. Glasman[99], J. Glatzer[14], P.C.F. Glaysher[46], A. Glazov[46], G.R. Gledhill[132], M. Goblirsch-Kolb[26], D. Godin[110], S. Goldfarb[105], T. Golling[54], D. Golubkov[123], A. Gomes[140a,140b], R. Goncalves Gama[53], R. Gonçalo[140a], G. Gonella[52], L. Gonella[21], A. Gongadze[80], F. Gonnella[21], J.L. Gonski[59], S. González de la Hoz[174], S. Gonzalez-Sevilla[54], G.R. Gonzalvo Rodriguez[174], L. Goossens[36], P.A. Gorbounov[124], H.A. Gordon[29], B. Gorini[36], E. Gorini[68a,68b], A. Gorišek[92], A.T. Goshaw[49], M.I. Gostkin[80], C.A. Gottardo[119], M. Gouighri[35b], D. Goujdami[35c], A.G. Goussiou[148], N. Govender[33b], C. Goy[5], E. Gozani[160], I. Grabowska-Bold[84a], E.C. Graham[91], J. Gramling[171], E. Gramstad[134], S. Grancagnolo[19], M. Grandi[156], V. Gratchev[138], P.M. Gravila[27f], F.G. Gravili[68a,68b], C. Gray[57], H.M. Gray[18], C. Grefe[24], K. Gregersen[97], I.M. Gregor[46], P. Grenier[153], K. Grevtsov[46], C. Grieco[14], N.A. Grieser[129], A.A. Grillo[146], K. Grimm[31,l], S. Grinstein[14,aa], J.-F. Grivaz[65], S. Groh[100], E. Gross[180], J. Grosse-Knetter[53], Z.J. Grout[95], C. Grud[106], A. Grummer[118], L. Guan[106], W. Guan[181], J. Guenther[36], A. Guerguichon[65], J.G.R. Guerrero Rojas[174], F. Guescini[115], D. Guest[171], R. Gugel[52], T. Guillemin[5], S. Guindon[36], U. Gul[57], J. Guo[60c], W. Guo[106], Y. Guo[60a,t], Z. Guo[102], R. Gupta[46], S. Gurbuz[12c], G. Gustavino[129], M. Guth[52], P. Gutierrez[129], C. Gutschow[95], C. Guyot[145], C. Gwenlan[135], C.B. Gwilliam[91], A. Haas[125], C. Haber[18], H.K. Hadavand[8], N. Haddad[35e], A. Hadef[60a], S. Hageböck[36], M. Haleem[177], J. Haley[130], G. Halladjian[107], G.D. Hallewell[102], K. Hamacher[182], P. Hamal[131], K. Hamano[176], H. Hamdaoui[35e], G.N. Hamity[149], K. Han[60a,z], L. Han[60a], S. Han[15a], Y.F. Han[167], K. Hanagaki[82,x], M. Hance[146], D.M. Handl[114], B. Haney[137], R. Hankache[136], E. Hansen[97], J.B. Hansen[40], J.D. Hansen[40], M.C. Hansen[24], P.H. Hansen[40], E.C. Hanson[101], K. Hara[169], T. Harenberg[182], S. Harkusha[108], P.F. Harrison[178], N.M. Hartmann[114], Y. Hasegawa[150], A. Hasib[50], S. Hassani[145], S. Haug[20], R. Hauser[107], L.B. Havener[39], M. Havranek[142], C.M. Hawkes[21], R.J. Hawkings[36], D. Hayden[107], C. Hayes[155], R.L. Hayes[175], C.P. Hays[135], J.M. Hays[93], H.S. Hayward[91], S.J. Haywood[144], F. He[60a], M.P. Heath[50], V. Hedberg[97], L. Heelan[8], S. Heer[24], K.K. Heidegger[52], W.D. Heidorn[79], J. Heilman[34], S. Heim[46], T. Heim[18], B. Heinemann[46,as], J.J. Heinrich[132], L. Heinrich[36], C. Heinz[56], J. Hejbal[141], L. Helary[61b], A. Held[175], S. Hellesund[134], C.M. Helling[146], S. Hellman[45a,45b], C. Helsens[36], R.C.W. Henderson[90], Y. Heng[181], S. Henkelmann[175], A.M. Henriques Correia[36], G.H. Herbert[19], H. Herde[26], V. Herget[177], Y. Hernández Jiménez[33d], H. Herr[100], M.G. Herrmann[114],

T. Herrmann[48], G. Herten[52], R. Hertenberger[114], L. Hervas[36], T.C. Herwig[137], G.G. Hesketh[95], N.P. Hessey[168a], A. Higashida[163], S. Higashino[82], E. Higón-Rodriguez[174], K. Hildebrand[37], E. Hill[176], J.C. Hill[32], K.K. Hill[29], K.H. Hiller[46], S.J. Hillier[21], M. Hils[48], I. Hinchliffe[18], F. Hinterkeuser[24], M. Hirose[133], S. Hirose[52], D. Hirschbuehl[182], B. Hiti[92], O. Hladik[141], D.R. Hlaluku[33d], X. Hoad[50], J. Hobbs[155], N. Hod[180], M.C. Hodgkinson[149], A. Hoecker[36], F. Hoenig[114], D. Hohn[52], D. Hohov[65], T.R. Holmes[37], M. Holzbock[114], L.B.A.H. Hommels[32], S. Honda[169], T.M. Hong[139], J.C. Honig[52], A. Hönle[115], B.H. Hooberman[173], W.H. Hopkins[6], Y. Horii[117], P. Horn[48], L.A. Horyn[37], S. Hou[158], A. Hoummada[35a], J. Howarth[101], J. Hoya[89], M. Hrabovsky[131], J. Hrdinka[77], I. Hristova[19], J. Hrivnac[65], A. Hrynevich[109], T. Hryn'ova[5], P.J. Hsu[64], S.-C. Hsu[148], Q. Hu[29], S. Hu[60c], Y.F. Hu[15a,15d], D.P. Huang[95], Y. Huang[60a], Y. Huang[15a], Z. Hubacek[142], F. Hubaut[102], M. Huebner[24], F. Huegging[24], T.B. Huffman[135], M. Huhtinen[36], R.F.H. Hunter[34], P. Huo[155], A.M. Hupe[34], N. Huseynov[80,ah], J. Huston[107], J. Huth[59], R. Hyneman[106], S. Hyrych[28a], G. Iacobucci[54], G. Iakovidis[29], I. Ibragimov[151], L. Iconomidou-Fayard[65], Z. Idrissi[35e], P. Iengo[36], R. Ignazzi[40], O. Igonkina[120,ac,*], R. Iguchi[163], T. Iizawa[54], Y. Ikegami[82], M. Ikeno[82], D. Iliadis[162], N. Ilic[119,167,af], F. Iltzsche[48], G. Introzzi[71a,71b], M. Iodice[75a], K. Iordanidou[168a], V. Ippolito[73a,73b], M.F. Isacson[172], M. Ishino[163], W. Islam[130], C. Issever[19,46], S. Istin[160], F. Ito[169], J.M. Iturbe Ponce[63a], R. Iuppa[76a,76b], A. Ivina[180], H. Iwasaki[82], J.M. Izen[43], V. Izzo[70a], P. Jacka[141], P. Jackson[1], R.M. Jacobs[24], B.P. Jaeger[152], V. Jain[2], G. Jäkel[182], K.B. Jakobi[100], K. Jakobs[52], T. Jakoubek[141], J. Jamieson[57], K.W. Janas[84a], R. Jansky[54], J. Janssen[24], M. Janus[53], P.A. Janus[84a], G. Jarlskog[97], N. Javadov[80,ah], T. Javůrek[36], M. Javurkova[52], F. Jeanneau[145], L. Jeanty[132], J. Jejelava[159a,ai], A. Jelinskas[178], P. Jenni[52,b], J. Jeong[46], N. Jeong[46], S. Jézéquel[5], H. Ji[181], J. Jia[155], H. Jiang[79], Y. Jiang[60a], Z. Jiang[153,q], S. Jiggins[52], F.A. Jimenez Morales[38], J. Jimenez Pena[115], S. Jin[15c], A. Jinaru[27b], O. Jinnouchi[165], H. Jivan[33d], P. Johansson[149], K.A. Johns[7], C.A. Johnson[66], R.W.L. Jones[90], S.D. Jones[156], S. Jones[7], T.J. Jones[91], J. Jongmanns[61a], P.M. Jorge[140a], J. Jovicevic[36], X. Ju[18], J.J. Junggeburth[115], A. Juste Rozas[14,aa], A. Kaczmarska[85], M. Kado[73a,73b], H. Kagan[127], M. Kagan[153], A. Kahn[39], C. Kahra[100], T. Kaji[179], E. Kajomovitz[160], C.W. Kalderon[97], A. Kaluza[100], A. Kamenshchikov[123], M. Kaneda[163], L. Kanjir[92], Y. Kano[117], V.A. Kantserov[112], J. Kanzaki[82], L.S. Kaplan[181], D. Kar[33d], K. Karava[135], M.J. Kareem[168b], S.N. Karpov[80], Z.M. Karpova[80], V. Kartvelishvili[90], A.N. Karyukhin[123], L. Kashif[181], R.D. Kass[127], A. Kastanas[45a,45b], C. Kato[60d,60c], J. Katzy[46], K. Kawade[150], K. Kawagoe[88], T. Kawaguchi[117], T. Kawamoto[163], G. Kawamura[53], E.F. Kay[176], V.F. Kazanin[122b,122a], R. Keeler[176], R. Kehoe[42], J.S. Keller[34], E. Kellermann[97], D. Kelsey[156], J.J. Kempster[21], J. Kendrick[21], K.E. Kennedy[39], O. Kepka[141], S. Kersten[182], B.P. Kerševan[92], S. Ketabchi Haghighat[167], M. Khader[173], F. Khalil-Zada[13], M. Khandoga[145], A. Khanov[130], A.G. Kharlamov[122b,122a], T. Kharlamova[122b,122a], E.E. Khoda[175], A. Khodinov[166], T.J. Khoo[54], E. Khramov[80], J. Khubua[159b], S. Kido[83], M. Kiehn[54], C.R. Kilby[94], Y.K. Kim[37], N. Kimura[95], O.M. Kind[19], B.T. King[91,*], D. Kirchmeier[48], J. Kirk[144], A.E. Kiryunin[115], T. Kishimoto[163], D.P. Kisliuk[167], V. Kitali[46], O. Kivernyk[5], T. Klapdor-Kleingrothaus[52], M. Klassen[61a], M.H. Klein[106], M. Klein[91], U. Klein[91],

K. Kleinknecht[100], P. Klimek[121], A. Klimentov[29], T. Klingl[24], T. Klioutchnikova[36], F.F. Klitzner[114], P. Kluit[120], S. Kluth[115], E. Kneringer[77], E.B.F.G. Knoops[102], A. Knue[52], D. Kobayashi[88], T. Kobayashi[163], M. Kobel[48], M. Kocian[153], P. Kodys[143], P.T. Koenig[24], T. Koffas[34], N.M. Köhler[36], T. Koi[153], M. Kolb[61b], I. Koletsou[5], T. Komarek[131], T. Kondo[82], N. Kondrashova[60c], K. Köneke[52], A.C. König[119], T. Kono[126], R. Konoplich[125,an], V. Konstantinides[95], N. Konstantinidis[95], B. Konya[97], R. Kopeliansky[66], S. Koperny[84a], K. Korcyl[85], K. Kordas[162], G. Koren[161], A. Korn[95], I. Korolkov[14], E.V. Korolkova[149], N. Korotkova[113], O. Kortner[115], S. Kortner[115], T. Kosek[143], V.V. Kostyukhin[166,166], A. Kotsokechagia[65], A. Kotwal[49], A. Koulouris[10], A. Kourkoumeli-Charalampidi[71a,71b], C. Kourkoumelis[9], E. Kourlitis[149], V. Kouskoura[29], A.B. Kowalewska[85], R. Kowalewski[176], C. Kozakai[163], W. Kozanecki[145], A.S. Kozhin[123], V.A. Kramarenko[113], G. Kramberger[92], D. Krasnopevtsev[60a], M.W. Krasny[136], A. Krasznahorkay[36], D. Krauss[115], J.A. Kremer[84a], J. Kretzschmar[91], P. Krieger[167], F. Krieter[114], A. Krishnan[61b], K. Krizka[18], K. Kroeninger[47], H. Kroha[115], J. Kroll[141], J. Kroll[137], K.S. Krowpman[107], J. Krstic[16], U. Kruchonak[80], H. Krüger[24], N. Krumnack[79], M.C. Kruse[49], J.A. Krzysiak[85], T. Kubota[105], O. Kuchinskaia[166], S. Kuday[4b], J.T. Kuechler[46], S. Kuehn[36], A. Kugel[61a], T. Kuhl[46], V. Kukhtin[80], R. Kukla[102], Y. Kulchitsky[108,ak], S. Kuleshov[147c], Y.P. Kulinich[173], M. Kuna[58], T. Kunigo[86], A. Kupco[141], T. Kupfer[47], O. Kuprash[52], H. Kurashige[83], L.L. Kurchaninov[168a], Y.A. Kurochkin[108], A. Kurova[112], M.G. Kurth[15a,15d], E.S. Kuwertz[36], M. Kuze[165], A.K. Kvam[148], J. Kvita[131], T. Kwan[104], A. La Rosa[115], L. La Rotonda[41b,41a], F. La Ruffa[41b,41a], C. Lacasta[174], F. Lacava[73a,73b], D.P.J. Lack[101], H. Lacker[19], D. Lacour[136], E. Ladygin[80], R. Lafaye[5], B. Laforge[136], T. Lagouri[33d], S. Lai[53], S. Lammers[66], W. Lampl[7], C. Lampoudis[162], E. Lançon[29], U. Landgraf[52], M.P.J. Landon[93], M.C. Lanfermann[54], V.S. Lang[46], J.C. Lange[53], R.J. Langenberg[36], A.J. Lankford[171], F. Lanni[29], K. Lantzsch[24], A. Lanza[71a], A. Lapertosa[55b,55a], S. Laplace[136], J.F. Laporte[145], T. Lari[69a], F. Lasagni Manghi[23b,23a], M. Lassnig[36], T.S. Lau[63a], A. Laudrain[65], A. Laurier[34], M. Lavorgna[70a,70b], S.D. Lawlor[94], M. Lazzaroni[69a,69b], B. Le[105], E. Le Guirriec[102], M. LeBlanc[7], T. LeCompte[6], F. Ledroit-Guillon[58], A.C.A. Lee[95], C.A. Lee[29], G.R. Lee[17], L. Lee[59], S.C. Lee[158], S.J. Lee[34], S. Lee[79], B. Lefebvre[168a], H.P. Lefebvre[94], M. Lefebvre[176], F. Legger[114], C. Leggett[18], K. Lehmann[152], N. Lehmann[182], G. Lehmann Miotto[36], W.A. Leight[46], A. Leisos[162,y], M.A.L. Leite[81d], C.E. Leitgeb[114], R. Leitner[143], D. Lellouch[180,*], K.J.C. Leney[42], T. Lenz[24], B. Lenzi[36], R. Leone[7], S. Leone[72a], C. Leonidopoulos[50], A. Leopold[136], G. Lerner[156], C. Leroy[110], R. Les[167], C.G. Lester[32], M. Levchenko[138], J. Levêque[5], D. Levin[106], L.J. Levinson[180], D.J. Lewis[21], B. Li[15b], B. Li[106], C-Q. Li[60a], F. Li[60c], H. Li[60a], H. Li[60b], J. Li[60c], K. Li[153], L. Li[60c], M. Li[15a,15d], Q. Li[15a,15d], Q.Y. Li[60a], S. Li[60d,60c], X. Li[46], Y. Li[46], Z. Li[60b], Z. Liang[15a], B. Liberti[74a], A. Liblong[167], K. Lie[63c], S. Lim[29], C.Y. Lin[32], K. Lin[107], T.H. Lin[100], R.A. Linck[66], J.H. Lindon[21], A.L. Lionti[54], E. Lipeles[137], A. Lipniacka[17], M. Lisovyi[61b], T.M. Liss[173,au], A. Lister[175], J.D. Little[8], B. Liu[79], B.L. Liu[6], H.B. Liu[29], H. Liu[106], J.B. Liu[60a], J.K.K. Liu[135], K. Liu[136], M. Liu[60a], P. Liu[18], Y. Liu[15a,15d], Y.L. Liu[106], Y.W. Liu[60a], M. Livan[71a,71b], A. Lleres[58], J. Llorente Merino[152], S.L. Lloyd[93], C.Y. Lo[63b], F. Lo Sterzo[42], E.M. Lobodzinska[46], P. Loch[7], S. Loffredo[74a,74b],

T. Lohse[19], K. Lohwasser[149], M. Lokajicek[141], J.D. Long[173], R.E. Long[90], L. Longo[36], K.A. Looper[127], J.A. Lopez[147c], I. Lopez Paz[101], A. Lopez Solis[149], J. Lorenz[114], N. Lorenzo Martinez[5], M. Losada[22], P.J. Lösel[114], A. Lösle[52], X. Lou[46], X. Lou[15a], A. Lounis[65], J. Love[6], P.A. Love[90], J.J. Lozano Bahilo[174], M. Lu[60a], Y.J. Lu[64], H.J. Lubatti[148], C. Luci[73a,73b], A. Lucotte[58], C. Luedtke[52], F. Luehring[66], I. Luise[136], L. Luminari[73a], B. Lund-Jensen[154], M.S. Lutz[103], D. Lynn[29], H. Lyons[91], R. Lysak[141], E. Lytken[97], F. Lyu[15a], V. Lyubushkin[80], T. Lyubushkina[80], H. Ma[29], L.L. Ma[60b], Y. Ma[60b], G. Maccarrone[51], A. Macchiolo[115], C.M. Macdonald[149], J. Machado Miguens[137], D. Madaffari[174], R. Madar[38], W.F. Mader[48], N. Madysa[48], J. Maeda[83], T. Maeno[29], M. Maerker[48], A.S. Maevskiy[113], V. Magerl[52], N. Magini[79], D.J. Mahon[39], C. Maidantchik[81b], T. Maier[114], A. Maio[140a,140b,140d], K. Maj[84a], O. Majersky[28a], S. Majewski[132], Y. Makida[82], N. Makovec[65], B. Malaescu[136], Pa. Malecki[85], V.P. Maleev[138], F. Malek[58], U. Mallik[78], D. Malon[6], C. Malone[32], S. Maltezos[10], S. Malyukov[80], J. Mamuzic[174], G. Mancini[51], I. Mandić[92], L. Manhaes de Andrade Filho[81a], I.M. Maniatis[162], J. Manjarres Ramos[48], K.H. Mankinen[97], A. Mann[114], A. Manousos[77], B. Mansoulie[145], I. Manthos[162], S. Manzoni[120], A. Marantis[162], G. Marceca[30], L. Marchese[135], G. Marchiori[136], M. Marcisovsky[141], L. Marcoccia[74a,74b], C. Marcon[97], C.A. Marin Tobon[36], M. Marjanovic[129], Z. Marshall[18], M.U.F. Martensson[172], S. Marti-Garcia[174], C.B. Martin[127], T.A. Martin[178], V.J. Martin[50], B. Martin dit Latour[17], L. Martinelli[75a,75b], M. Martinez[14,aa], V.I. Martinez Outschoorn[103], S. Martin-Haugh[144], V.S. Martoiu[27b], A.C. Martyniuk[95], A. Marzin[36], S.R. Maschek[115], L. Masetti[100], T. Mashimo[163], R. Mashinistov[111], J. Masik[101], A.L. Maslennikov[122b,122a], L. Massa[74a,74b], P. Massarotti[70a,70b], P. Mastrandrea[72a,72b], A. Mastroberardino[41b,41a], T. Masubuchi[163], D. Matakias[10], A. Matic[114], P. Mättig[24], J. Maurer[27b], B. Maček[92], D.A. Maximov[122b,122a], R. Mazini[158], I. Maznas[162], S.M. Mazza[146], S.P. Mc Kee[106], T.G. McCarthy[115], W.P. McCormack[18], E.F. McDonald[105], J.A. Mcfayden[36], G. Mchedlidze[159b], M.A. McKay[42], K.D. McLean[176], S.J. McMahon[144], P.C. McNamara[105], C.J. McNicol[178], R.A. McPherson[176,af], J.E. Mdhluli[33d], Z.A. Meadows[103], S. Meehan[36], T. Megy[52], S. Mehlhase[114], A. Mehta[91], T. Meideck[58], B. Meirose[43], D. Melini[174], B.R. Mellado Garcia[33d], J.D. Mellenthin[53], M. Melo[28a], F. Meloni[46], A. Melzer[24], S.B. Menary[101], E.D. Mendes Gouveia[140a,140e], L. Meng[36], X.T. Meng[106], S. Menke[115], E. Meoni[41b,41a], S. Mergelmeyer[19], S.A.M. Merkt[139], C. Merlassino[20], P. Mermod[54], L. Merola[70a,70b], C. Meroni[69a], O. Meshkov[113,111], J.K.R. Meshreki[151], A. Messina[73a,73b], J. Metcalfe[6], A.S. Mete[171], C. Meyer[66], J. Meyer[160], J-P. Meyer[145], H. Meyer Zu Theenhausen[61a], F. Miano[156], M. Michetti[19], R.P. Middleton[144], L. Mijović[50], G. Mikenberg[180], M. Mikestikova[141], M. Mikuž[92], H. Mildner[149], M. Milesi[105], A. Milic[167], D.A. Millar[93], D.W. Miller[37], A. Milov[180], D.A. Milstead[45a,45b], R.A. Mina[153,q], A.A. Minaenko[123], M. Miñano Moya[174], I.A. Minashvili[159b], A.I. Mincer[125], B. Mindur[84a], M. Mineev[80], Y. Minegishi[163], L.M. Mir[14], A. Mirto[68a,68b], K.P. Mistry[137], T. Mitani[179], J. Mitrevski[114], V.A. Mitsou[174], M. Mittal[60c], O. Miu[167], A. Miucci[20], P.S. Miyagawa[149], A. Mizukami[82], J.U. Mjörnmark[97], T. Mkrtchyan[184], M. Mlynarikova[143], T. Moa[45a,45b], K. Mochizuki[110], P. Mogg[52], S. Mohapatra[39], R. Moles-Valls[24],

M.C. Mondragon[107], K. Mönig[46], J. Monk[40], E. Monnier[102], A. Montalbano[152], J. Montejo Berlingen[36], M. Montella[95], F. Monticelli[89], S. Monzani[69a], N. Morange[65], D. Moreno[22], M. Moreno Llácer[174], C. Moreno Martinez[14], P. Morettini[55b], M. Morgenstern[120], S. Morgenstern[48], D. Mori[152], M. Morii[59], M. Morinaga[179], V. Morisbak[134], A.K. Morley[36], G. Mornacchi[36], A.P. Morris[95], L. Morvaj[155], P. Moschovakos[36], B. Moser[120], M. Mosidze[159b], T. Moskalets[145], H.J. Moss[149], J. Moss[31,n], E.J.W. Moyse[103], S. Muanza[102], J. Mueller[139], R.S.P. Mueller[114], D. Muenstermann[90], G.A. Mullier[97], D.P. Mungo[69a,69b], J.L. Munoz Martinez[14], F.J. Munoz Sanchez[101], P. Murin[28b], W.J. Murray[178,144], A. Murrone[69a,69b], M. Muškinja[18], C. Mwewa[33a], A.G. Myagkov[123,ao], J. Myers[132], M. Myska[142], B.P. Nachman[18], O. Nackenhorst[47], A.Nag Nag[48], K. Nagai[135], K. Nagano[82], Y. Nagasaka[62], M. Nagel[52], J.L. Nagle[29], E. Nagy[102], A.M. Nairz[36], Y. Nakahama[117], K. Nakamura[82], T. Nakamura[163], I. Nakano[128], H. Nanjo[133], F. Napolitano[61a], R.F. Naranjo Garcia[46], R. Narayan[42], I. Naryshkin[138], T. Naumann[46], G. Navarro[22], P.Y. Nechaeva[111], F. Nechansky[46], T.J. Neep[21], A. Negri[71a,71b], M. Negrini[23b], C. Nellist[53], M.E. Nelson[135], S. Nemecek[141], P. Nemethy[125], M. Nessi[36,d], M.S. Neubauer[173], M. Neumann[182], P.R. Newman[21], Y.S. Ng[19], Y.W.Y. Ng[171], B. Ngair[35e], H.D.N. Nguyen[102], T. Nguyen Manh[110], E. Nibigira[38], R.B. Nickerson[135], R. Nicolaidou[145], D.S. Nielsen[40], J. Nielsen[146], N. Nikiforou[11], V. Nikolaenko[123,ao], I. Nikolic-Audit[136], K. Nikolopoulos[21], P. Nilsson[29], H.R. Nindhito[54], Y. Ninomiya[82], A. Nisati[73a], N. Nishu[60c], R. Nisius[115], I. Nitsche[47], T. Nitta[179], T. Nobe[163], Y. Noguchi[86], I. Nomidis[136], M.A. Nomura[29], M. Nordberg[36], N. Norjoharuddeen[135], T. Novak[92], O. Novgorodova[48], R. Novotny[142], L. Nozka[131], K. Ntekas[171], E. Nurse[95], F.G. Oakham[34,ax], H. Oberlack[115], J. Ocariz[136], A. Ochi[83], I. Ochoa[39], J.P. Ochoa-Ricoux[147a], K. O'Connor[26], S. Oda[88], S. Odaka[82], S. Oerdek[53], A. Ogrodnik[84a], A. Oh[101], S.H. Oh[49], C.C. Ohm[154], H. Oide[165], M.L. Ojeda[167], H. Okawa[169], Y. Okazaki[86], M.W. O'Keefe[91], Y. Okumura[163], T. Okuyama[82], A. Olariu[27b], L.F. Oleiro Seabra[140a], S.A. Olivares Pino[147a], D. Oliveira Damazio[29], J.L. Oliver[1], M.J.R. Olsson[171], A. Olszewski[85], J. Olszowska[85], D.C. O'Neil[152], A.P. O'neill[135], A. Onofre[140a,140e], P.U.E. Onyisi[11], H. Oppen[134], M.J. Oreglia[37], G.E. Orellana[89], D. Orestano[75a,75b], N. Orlando[14], R.S. Orr[167], V. O'Shea[57], R. Ospanov[60a], G. Otero y Garzon[30], H. Otono[88], P.S. Ott[61a], M. Ouchrif[35d], J. Ouellette[29], F. Ould-Saada[134], A. Ouraou[145], Q. Ouyang[15a], M. Owen[57], R.E. Owen[21], V.E. Ozcan[12c], N. Ozturk[8], J. Pacalt[131], H.A. Pacey[32], K. Pachal[49], A. Pacheco Pages[14], C. Padilla Aranda[14], S. Pagan Griso[18], M. Paganini[183], G. Palacino[66], S. Palazzo[50], S. Palestini[36], M. Palka[84b], D. Pallin[38], I. Panagoulias[10], C.E. Pandini[36], J.G. Panduro Vazquez[94], P. Pani[46], G. Panizzo[67a,67c], L. Paolozzi[54], C. Papadatos[110], K. Papageorgiou[9,h], S. Parajuli[43], A. Paramonov[6], D. Paredes Hernandez[63b], S.R. Paredes Saenz[135], B. Parida[166], T.H. Park[167], A.J. Parker[31], M.A. Parker[32], F. Parodi[55b,55a], E.W. Parrish[121], J.A. Parsons[39], U. Parzefall[52], L. Pascual Dominguez[136], V.R. Pascuzzi[167], J.M.P. Pasner[146], F. Pasquali[120], E. Pasqualucci[73a], S. Passaggio[55b], F. Pastore[94], P. Pasuwan[45a,45b], S. Pataraia[100], J.R. Pater[101], A. Pathak[181,j], T. Pauly[36], B. Pearson[115], M. Pedersen[134], L. Pedraza Diaz[119], R. Pedro[140a], T. Peiffer[53], S.V. Peleganchuk[122b,122a], O. Penc[141], H. Peng[60a], B.S. Peralva[81a],

M.M. Perego[65], A.P. Pereira Peixoto[140a], D.V. Perepelitsa[29], F. Peri[19], L. Perini[69a,69b], H. Pernegger[36], S. Perrella[70a,70b], K. Peters[46], R.F.Y. Peters[101], B.A. Petersen[36], T.C. Petersen[40], E. Petit[102], A. Petridis[1], C. Petridou[162], P. Petroff[65], M. Petrov[135], F. Petrucci[75a,75b], M. Pettee[183], N.E. Pettersson[103], K. Petukhova[143], A. Peyaud[145], R. Pezoa[147c], L. Pezzotti[71a,71b], T. Pham[105], F.H. Phillips[107], P.W. Phillips[144], M.W. Phipps[173], G. Piacquadio[155], E. Pianori[18], A. Picazio[103], R.H. Pickles[101], R. Piegaia[30], D. Pietreanu[27b], J.E. Pilcher[37], A.D. Pilkington[101], M. Pinamonti[74a,74b], J.L. Pinfold[3], M. Pitt[161], L. Pizzimento[74a,74b], M.-A. Pleier[29], V. Pleskot[143], E. Plotnikova[80], P. Podberezko[122b,122a], R. Poettgen[97], R. Poggi[54], L. Poggioli[65], I. Pogrebnyak[107], D. Pohl[24], I. Pokharel[53], G. Polesello[71a], A. Poley[18], A. Policicchio[73a,73b], R. Polifka[143], A. Polini[23b], C.S. Pollard[46], V. Polychronakos[29], D. Ponomarenko[112], L. Pontecorvo[36], S. Popa[27a], G.A. Popeneciu[27d], L. Portales[5], D.M. Portillo Quintero[58], S. Pospisil[142], K. Potamianos[46], I.N. Potrap[80], C.J. Potter[32], H. Potti[11], T. Poulsen[97], J. Poveda[36], T.D. Powell[149], G. Pownall[46], M.E. Pozo Astigarraga[36], P. Pralavorio[102], S. Prell[79], D. Price[101], M. Primavera[68a], S. Prince[104], M.L. Proffitt[148], N. Proklova[112], K. Prokofiev[63c], F. Prokoshin[80], S. Protopopescu[29], J. Proudfoot[6], M. Przybycien[84a], D. Pudzha[138], A. Puri[173], P. Puzo[65], J. Qian[106], Y. Qin[101], A. Quadt[53], M. Queitsch-Maitland[46], A. Qureshi[1], M. Racko[28a], P. Rados[105], F. Ragusa[69a,69b], G. Rahal[98], J.A. Raine[54], S. Rajagopalan[29], A. Ramirez Morales[93], K. Ran[15a,15d], T. Rashid[65], S. Raspopov[5], D.M. Rauch[46], F. Rauscher[114], S. Rave[100], B. Ravina[149], I. Ravinovich[180], J.H. Rawling[101], M. Raymond[36], A.L. Read[134], N.P. Readioff[58], M. Reale[68a,68b], D.M. Rebuzzi[71a,71b], A. Redelbach[177], G. Redlinger[29], K. Reeves[43], L. Rehnisch[19], J. Reichert[137], D. Reikher[161], A. Reiss[100], A. Rej[151], C. Rembser[36], M. Renda[27b], M. Rescigno[73a], S. Resconi[69a], E.D. Resseguie[137], S. Rettie[175], E. Reynolds[21], O.L. Rezanova[122b,122a], P. Reznicek[143], E. Ricci[76a,76b], R. Richter[115], S. Richter[46], E. Richter-Was[84b], O. Ricken[24], M. Ridel[136], P. Rieck[115], O. Rifki[46], M. Rijssenbeek[155], A. Rimoldi[71a,71b], M. Rimoldi[46], L. Rinaldi[23b], G. Ripellino[154], I. Riu[14], J.C. Rivera Vergara[176], F. Rizatdinova[130], E. Rizvi[93], C. Rizzi[36], R.T. Roberts[101], S.H. Robertson[104,af], M. Robin[46], D. Robinson[32], J.E.M. Robinson[46], C.M. Robles Gajardo[147c], A. Robson[57], A. Rocchi[74a,74b], E. Rocco[100], C. Roda[72a,72b], S. Rodriguez Bosca[174], A. Rodriguez Perez[14], D. Rodriguez Rodriguez[174], A.M. Rodríguez Vera[168b], S. Roe[36], O. Røhne[134], R. Röhrig[115], R.A. Rojas[147c], C.P.A. Roland[66], J. Roloff[29], A. Romaniouk[112], M. Romano[23b,23a], N. Rompotis[91], M. Ronzani[125], L. Roos[136], S. Rosati[73a], K. Rosbach[52], G. Rosin[103], B.J. Rosser[137], E. Rossi[46], E. Rossi[75a,75b], E. Rossi[70a,70b], L.P. Rossi[55b], L. Rossini[69a,69b], R. Rosten[14], M. Rotaru[27b], J. Rothberg[148], D. Rousseau[65], G. Rovelli[71a,71b], A. Roy[11], D. Roy[33d], A. Rozanov[102], Y. Rozen[160], X. Ruan[33d], F. Rühr[52], A. Ruiz-Martinez[174], A. Rummler[36], Z. Rurikova[52], N.A. Rusakovich[80], H.L. Russell[104], L. Rustige[38,47], J.P. Rutherfoord[7], E.M. Rüttinger[149], M. Rybar[39], G. Rybkin[65], E.B. Rye[134], A. Ryzhov[123], J.A. Sabater Iglesias[46], P. Sabatini[53], G. Sabato[120], S. Sacerdoti[65], H.F-W. Sadrozinski[146], R. Sadykov[80], F. Safai Tehrani[73a], B. Safarzadeh Samani[156], P. Saha[121], S. Saha[104], M. Sahinsoy[61a], A. Sahu[182], M. Saimpert[46], M. Saito[163], T. Saito[163], H. Sakamoto[163], A. Sakharov[125,an], D. Salamani[54], G. Salamanna[75a,75b], J.E. Salazar Loyola[147c], A. Salnikov[153],

J. Salt[174], D. Salvatore[41b,41a], F. Salvatore[156], A. Salvucci[63a,63b,63c], A. Salzburger[36], J. Samarati[36], D. Sammel[52], D. Sampsonidis[162], D. Sampsonidou[162], J. Sánchez[174], A. Sanchez Pineda[67a,36,67c], H. Sandaker[134], C.O. Sander[46], I.G. Sanderswood[90], M. Sandhoff[182], C. Sandoval[22], D.P.C. Sankey[144], M. Sannino[55b,55a], Y. Sano[117], A. Sansoni[51], C. Santoni[38], H. Santos[140a,140b], S.N. Santpur[18], A. Santra[174], A. Sapronov[80], J.G. Saraiva[140a,140d], O. Sasaki[82], K. Sato[169], F. Sauerburger[52], E. Sauvan[5], P. Savard[167,ax], N. Savic[115], R. Sawada[163], C. Sawyer[144], L. Sawyer[96,al], C. Sbarra[23b], A. Sbrizzi[23a], T. Scanlon[95], J. Schaarschmidt[148], P. Schacht[115], B.M. Schachtner[114], D. Schaefer[37], L. Schaefer[137], J. Schaeffer[100], S. Schaepe[36], U. Schäfer[100], A.C. Schaffer[65], D. Schaile[114], R.D. Schamberger[155], N. Scharmberg[101], V.A. Schegelsky[138], D. Scheirich[143], F. Schenck[19], M. Schernau[171], C. Schiavi[55b,55a], S. Schier[146], L.K. Schildgen[24], Z.M. Schillaci[26], E.J. Schioppa[36], M. Schioppa[41b,41a], K.E. Schleicher[52], S. Schlenker[36], K.R. Schmidt-Sommerfeld[115], K. Schmieden[36], C. Schmitt[100], S. Schmitt[46], S. Schmitz[100], J.C. Schmoeckel[46], U. Schnoor[52], L. Schoeffel[145], A. Schoening[61b], P.G. Scholer[52], E. Schopf[135], M. Schott[100], J.F.P. Schouwenberg[119], J. Schovancova[36], S. Schramm[54], F. Schroeder[182], A. Schulte[100], H-C. Schultz-Coulon[61a], M. Schumacher[52], B.A. Schumm[146], Ph. Schune[145], A. Schwartzman[153], T.A. Schwarz[106], Ph. Schwemling[145], R. Schwienhorst[107], A. Sciandra[146], G. Sciolla[26], M. Scodeggio[46], M. Scornajenghi[41b,41a], F. Scuri[72a], F. Scutti[105], L.M. Scyboz[115], C.D. Sebastiani[73a,73b], P. Seema[19], S.C. Seidel[118], A. Seiden[146], B.D. Seidlitz[29], T. Seiss[37], J.M. Seixas[81b], G. Sekhniaidze[70a], K. Sekhon[106], S.J. Sekula[42], N. Semprini-Cesari[23b,23a], S. Sen[49], C. Serfon[77], L. Serin[65], L. Serkin[67a,67b], M. Sessa[60a], H. Severini[129], T. Šfiligoj[92], F. Sforza[55b,55a], A. Sfyrla[54], E. Shabalina[53], J.D. Shahinian[146], N.W. Shaikh[45a,45b], D. Shaked Renous[180], L.Y. Shan[15a], J.T. Shank[25], M. Shapiro[18], A. Sharma[135], A.S. Sharma[1], P.B. Shatalov[124], K. Shaw[156], S.M. Shaw[101], A. Shcherbakova[138], M. Shehade[180], Y. Shen[129], A.D. Sherman[25], P. Sherwood[95], L. Shi[158,at], S. Shimizu[82], C.O. Shimmin[183], Y. Shimogama[179], M. Shimojima[116], I.P.J. Shipsey[135], S. Shirabe[88], M. Shiyakova[80,ad], J. Shlomi[180], A. Shmeleva[111], M.J. Shochet[37], J. Shojaii[105], D.R. Shope[129], S. Shrestha[127], E.M. Shrif[33d], E. Shulga[180], P. Sicho[141], A.M. Sickles[173], P.E. Sidebo[154], E. Sideras Haddad[33d], O. Sidiropoulou[36], A. Sidoti[23b,23a], F. Siegert[48], Dj. Sijacki[16], M.Jr. Silva[181], M.V. Silva Oliveira[81a], S.B. Silverstein[45a], S. Simion[65], E. Simioni[100], R. Simoniello[100], S. Simsek[12b], P. Sinervo[167], V. Sinetckii[113,111], N.B. Sinev[132], M. Sioli[23b,23a], I. Siral[106], S.Yu. Sivoklokov[113], J. Sjölin[45a,45b], E. Skorda[97], P. Skubic[129], M. Slawinska[85], K. Sliwa[170], R. Slovak[143], V. Smakhtin[180], B.H. Smart[144], J. Smiesko[28a], N. Smirnov[112], S.Yu. Smirnov[112], Y. Smirnov[112], L.N. Smirnova[113,v], O. Smirnova[97], J.W. Smith[53], M. Smizanska[90], K. Smolek[142], A. Smykiewicz[85], A.A. Snesarev[111], H.L. Snoek[120], I.M. Snyder[132], S. Snyder[29], R. Sobie[176,af], A. Soffer[161], A. Søgaard[50], F. Sohns[53], C.A. Solans Sanchez[36], E.Yu. Soldatov[112], U. Soldevila[174], A.A. Solodkov[123], A. Soloshenko[80], O.V. Solovyanov[123], V. Solovyev[138], P. Sommer[149], H. Son[170], W. Song[144], W.Y. Song[168b], A. Sopczak[142], F. Sopkova[28b], C.L. Sotiropoulou[72a,72b], S. Sottocornola[71a,71b], R. Soualah[67a,67c,g], A.M. Soukharev[122b,122a], D. South[46], S. Spagnolo[68a,68b],

M. Spalla[115], M. Spangenberg[178], F. Spanò[94], D. Sperlich[52], T.M. Spieker[61a], R. Spighi[23b], G. Spigo[36], M. Spina[156], D.P. Spiteri[57], M. Spousta[143], A. Stabile[69a,69b], B.L. Stamas[121], R. Stamen[61a], M. Stamenkovic[120], E. Stanecka[85], B. Stanislaus[135], M.M. Stanitzki[46], M. Stankaityte[135], B. Stapf[120], E.A. Starchenko[123], G.H. Stark[146], J. Stark[58], S.H. Stark[40], P. Staroba[141], P. Starovoitov[61a], S. Stärz[104], R. Staszewski[85], G. Stavropoulos[44], M. Stegler[46], P. Steinberg[29], A.L. Steinhebel[132], B. Stelzer[152], H.J. Stelzer[139], O. Stelzer-Chilton[168a], H. Stenzel[56], T.J. Stevenson[156], G.A. Stewart[36], M.C. Stockton[36], G. Stoicea[27b], M. Stolarski[140a], S. Stonjek[115], A. Straessner[48], J. Strandberg[154], S. Strandberg[45a,45b], M. Strauss[129], P. Strizenec[28b], R. Ströhmer[177], D.M. Strom[132], R. Stroynowski[42], A. Strubig[50], S.A. Stucci[29], B. Stugu[17], J. Stupak[129], N.A. Styles[46], D. Su[153], S. Suchek[61a], V.V. Sulin[111], M.J. Sullivan[91], D.M.S. Sultan[54], S. Sultansoy[4c], T. Sumida[86], S. Sun[106], X. Sun[3], K. Suruliz[156], C.J.E. Suster[157], M.R. Sutton[156], S. Suzuki[82], M. Svatos[141], M. Swiatlowski[37], S.P. Swift[2], T. Swirski[177], A. Sydorenko[100], I. Sykora[28a], M. Sykora[143], T. Sykora[143], D. Ta[100], K. Tackmann[46,ab], J. Taenzer[161], A. Taffard[171], R. Tafirout[168a], H. Takai[29], R. Takashima[87], K. Takeda[83], T. Takeshita[150], E.P. Takeva[50], Y. Takubo[82], M. Talby[102], A.A. Talyshev[122b,122a], N.M. Tamir[161], J. Tanaka[163], M. Tanaka[165], R. Tanaka[65], S. Tapia Araya[173], S. Tapprogge[100], A. Tarek Abouelfadl Mohamed[136], S. Tarem[160], K. Tariq[60b], G. Tarna[27b,c], G.F. Tartarelli[69a], P. Tas[143], M. Tasevsky[141], T. Tashiro[86], E. Tassi[41b,41a], A. Tavares Delgado[140a,140b], Y. Tayalati[35e], A.J. Taylor[50], G.N. Taylor[105], W. Taylor[168b], A.S. Tee[90], R. Teixeira De Lima[153], P. Teixeira-Dias[94], H. Ten Kate[36], J.J. Teoh[120], S. Terada[82], K. Terashi[163], J. Terron[99], S. Terzo[14], M. Testa[51], R.J. Teuscher[167,af], S.J. Thais[183], T. Theveneaux-Pelzer[46], F. Thiele[40], D.W. Thomas[94], J.O. Thomas[42], J.P. Thomas[21], P.D. Thompson[21], L.A. Thomsen[183], E. Thomson[137], E.J. Thorpe[93], R.E. Ticse Torres[53], V.O. Tikhomirov[111,ap], Yu.A. Tikhonov[122b,122a], S. Timoshenko[112], P. Tipton[183], S. Tisserant[102], K. Todome[23b,23a], S. Todorova-Nova[5], S. Todt[48], J. Tojo[88], S. Tokár[28a], K. Tokushuku[82], E. Tolley[127], K.G. Tomiwa[33d], M. Tomoto[117], L. Tompkins[153,q], B. Tong[59], P. Tornambe[103], E. Torrence[132], H. Torres[48], E. Torró Pastor[148], C. Tosciri[135], J. Toth[102,ae], D.R. Tovey[149], A. Traeet[17], C.J. Treado[125], T. Trefzger[177], F. Tresoldi[156], A. Tricoli[29], I.M. Trigger[168a], S. Trincaz-Duvoid[136], W. Trischuk[167], B. Trocmé[58], A. Trofymov[145], C. Troncon[69a], M. Trovatelli[176], F. Trovato[156], L. Truong[33b], M. Trzebinski[85], A. Trzupek[85], F. Tsai[46], J.C-L. Tseng[135], P.V. Tsiareshka[108,ak], A. Tsirigotis[162,y], V. Tsiskaridze[155], E.G. Tskhadadze[159a], M. Tsopoulou[162], I.I. Tsukerman[124], V. Tsulaia[18], S. Tsuno[82], D. Tsybychev[155], Y. Tu[63b], A. Tudorache[27b], V. Tudorache[27b], T.T. Tulbure[27a], A.N. Tuna[59], S. Turchikhin[80], D. Turgeman[180], I. Turk Cakir[4b,w], R.J. Turner[21], R.T. Turra[69a], P.M. Tuts[39], S. Tzamarias[162], E. Tzovara[100], G. Ucchielli[47], K. Uchida[163], I. Ueda[82], M. Ughetto[45a,45b], F. Ukegawa[169], G. Unal[36], A. Undrus[29], G. Unel[171], F.C. Ungaro[105], Y. Unno[82], K. Uno[163], J. Urban[28b], P. Urquijo[105], G. Usai[8], Z. Uysal[12d], V. Vacek[142], B. Vachon[104], K.O.H. Vadla[134], A. Vaidya[95], C. Valderanis[114], E. Valdes Santurio[45a,45b], M. Valente[54], S. Valentinetti[23b,23a], A. Valero[174], L. Valéry[46], R.A. Vallance[21], A. Vallier[36], J.A. Valls Ferrer[174], T.R. Van Daalen[14],

P. Van Gemmeren[6], I. Van Vulpen[120], M. Vanadia[74a,74b], W. Vandelli[36], E.R. Vandewall[130], A. Vaniachine[166], D. Vannicola[73a,73b], R. Vari[73a], E.W. Varnes[7], C. Varni[55b,55a], T. Varol[158], D. Varouchas[65], K.E. Varvell[157], M.E. Vasile[27b], G.A. Vasquez[176], J.G. Vasquez[183], F. Vazeille[38], D. Vazquez Furelos[14], T. Vazquez Schroeder[36], J. Veatch[53], V. Vecchio[75a,75b], M.J. Veen[120], L.M. Veloce[167], F. Veloso[140a,140c], S. Veneziano[73a], A. Ventura[68a,68b], N. Venturi[36], A. Verbytskyi[115], V. Vercesi[71a], M. Verducci[72a,72b], C.M. Vergel Infante[79], C. Vergis[24], W. Verkerke[120], A.T. Vermeulen[120], J.C. Vermeulen[120], M.C. Vetterli[152,ax], N. Viaux Maira[147c], M. Vicente Barreto Pinto[54], T. Vickey[149], O.E. Vickey Boeriu[149], G.H.A. Viehhauser[135], L. Vigani[61b], M. Villa[23b,23a], M. Villaplana Perez[69a,69b], E. Vilucchi[51], M.G. Vincter[34], G.S. Virdee[21], A. Vishwakarma[46], C. Vittori[23b,23a], I. Vivarelli[156], M. Vogel[182], P. Vokac[142], S.E. von Buddenbrock[33d], E. Von Toerne[24], V. Vorobel[143], K. Vorobev[112], M. Vos[174], J.H. Vossebeld[91], M. Vozak[101], N. Vranjes[16], M. Vranjes Milosavljevic[16], V. Vrba[142], M. Vreeswijk[120], R. Vuillermet[36], I. Vukotic[37], P. Wagner[24], W. Wagner[182], J. Wagner-Kuhr[114], S. Wahdan[182], H. Wahlberg[89], V.M. Walbrecht[115], J. Walder[90], R. Walker[114], S.D. Walker[94], W. Walkowiak[151], V. Wallangen[45a,45b], A.M. Wang[59], C. Wang[60c], C. Wang[60b], F. Wang[181], H. Wang[18], H. Wang[3], J. Wang[63a], J. Wang[157], J. Wang[61b], P. Wang[42], Q. Wang[129], R.-J. Wang[100], R. Wang[60a], R. Wang[6], S.M. Wang[158], W.T. Wang[60a], W. Wang[15c,ag], W.X. Wang[60a,ag], Y. Wang[60a,am], Z. Wang[60c], C. Wanotayaroj[46], A. Warburton[104], C.P. Ward[32], D.R. Wardrope[95], N. Warrack[57], A. Washbrook[50], A.T. Watson[21], M.F. Watson[21], G. Watts[148], B.M. Waugh[95], A.F. Webb[11], S. Webb[100], C. Weber[183], M.S. Weber[20], S.A. Weber[34], S.M. Weber[61a], A.R. Weidberg[135], J. Weingarten[47], M. Weirich[100], C. Weiser[52], P.S. Wells[36], T. Wenaus[29], T. Wengler[36], S. Wenig[36], N. Wermes[24], M.D. Werner[79], M. Wessels[61a], T.D. Weston[20], K. Whalen[132], N.L. Whallon[148], A.M. Wharton[90], A.S. White[106], A. White[8], M.J. White[1], D. Whiteson[171], B.W. Whitmore[90], W. Wiedenmann[181], M. Wielers[144], N. Wieseotte[100], C. Wiglesworth[40], L.A.M. Wiik-Fuchs[52], F. Wilk[101], H.G. Wilkens[36], L.J. Wilkins[94], H.H. Williams[137], S. Williams[32], C. Willis[107], S. Willocq[103], J.A. Wilson[21], I. Wingerter-Seez[5], E. Winkels[156], F. Winklmeier[132], O.J. Winston[156], B.T. Winter[52], M. Wittgen[153], M. Wobisch[96], A. Wolf[100], T.M.H. Wolf[120], R. Wolff[102], R.W. Wölker[135], J. Wollrath[52], M.W. Wolter[85], H. Wolters[140a,140c], V.W.S. Wong[175], N.L. Woods[146], S.D. Worm[21], B.K. Wosiek[85], K.W. Woźniak[85], K. Wraight[57], S.L. Wu[181], X. Wu[54], Y. Wu[60a], T.R. Wyatt[101], B.M. Wynne[50], S. Xella[40], Z. Xi[106], L. Xia[178], X. Xiao[106], I. Xiotidis[156], D. Xu[15a], H. Xu[60a,c], L. Xu[29], T. Xu[145], W. Xu[106], Z. Xu[60b], Z. Xu[153], B. Yabsley[157], S. Yacoob[33a], K. Yajima[133], D.P. Yallup[95], D. Yamaguchi[165], Y. Yamaguchi[165], A. Yamamoto[82], M. Yamatani[163], T. Yamazaki[163], Y. Yamazaki[83], Z. Yan[25], H.J. Yang[60c,60d], H.T. Yang[18], S. Yang[78], X. Yang[60b,58], Y. Yang[163], W-M. Yao[18], Y.C. Yap[46], Y. Yasu[82], E. Yatsenko[60c,60d], J. Ye[42], S. Ye[29], I. Yeletskikh[80], M.R. Yexley[90], E. Yigitbasi[25], K. Yorita[179], K. Yoshihara[137], C.J.S. Young[36], C. Young[153], J. Yu[79], R. Yuan[60b,i], X. Yue[61a], S.P.Y. Yuen[24], M. Zaazoua[35e], B. Zabinski[85], G. Zacharis[10], E. Zaffaroni[54], J. Zahreddine[136], A.M. Zaitsev[123,ao], T. Zakareishvili[159b], N. Zakharchuk[34], S. Zambito[59], D. Zanzi[36], D.R. Zaripovas[57], S.V. Zeißner[47], C. Zeitnitz[182], G. Zemaityte[135], J.C. Zeng[173], O. Zenin[123],

T. Ženiš[28a], D. Zerwas[65], M. Zgubič[135], B. Zhang[15c], D.F. Zhang[15b], G. Zhang[15b], H. Zhang[15c], J. Zhang[6], L. Zhang[15c], L. Zhang[60a], M. Zhang[173], R. Zhang[24], X. Zhang[60b], Y. Zhang[15a,15d], Z. Zhang[63a], Z. Zhang[65], P. Zhao[49], Y. Zhao[60b], Z. Zhao[60a], A. Zhemchugov[80], Z. Zheng[106], D. Zhong[173], B. Zhou[106], C. Zhou[181], M.S. Zhou[15a,15d], M. Zhou[155], N. Zhou[60c], Y. Zhou[7], C.G. Zhu[60b], C. Zhu[15a,15d], H.L. Zhu[60a], H. Zhu[15a], J. Zhu[106], Y. Zhu[60a], X. Zhuang[15a], K. Zhukov[111], V. Zhulanov[122b,122a], D. Zieminska[66], N.I. Zimine[80], S. Zimmermann[52], Z. Zinonos[115], M. Ziolkowski[151], L. Živković[16], G. Zobernig[181], A. Zoccoli[23b,23a], K. Zoch[53], T.G. Zorbas[149], R. Zou[37], L. Zwalinski[36].

[1]Department of Physics, University of Adelaide, Adelaide; Australia.
[2]Physics Department, SUNY Albany, Albany NY; United States of America.
[3]Department of Physics, University of Alberta, Edmonton AB; Canada.
[4] [(a)]Department of Physics, Ankara University, Ankara;[(b)]Istanbul Aydin University, Istanbul;[(c)]Division of Physics, TOBB University of Economics and Technology, Ankara; Turkey.
[5]LAPP, Université Grenoble Alpes, Université Savoie Mont Blanc, CNRS/IN2P3, Annecy; France.
[6]High Energy Physics Division, Argonne National Laboratory, Argonne IL; United States of America.
[7]Department of Physics, University of Arizona, Tucson AZ; United States of America.
[8]Department of Physics, University of Texas at Arlington, Arlington TX; United States of America.
[9]Physics Department, National and Kapodistrian University of Athens, Athens; Greece.
[10]Physics Department, National Technical University of Athens, Zografou; Greece.
[11]Department of Physics, University of Texas at Austin, Austin TX; United States of America.
[12] [(a)]Bahcesehir University, Faculty of Engineering and Natural Sciences, Istanbul;[(b)]Istanbul Bilgi University, Faculty of Engineering and Natural Sciences, Istanbul;[(c)]Department of Physics, Bogazici University, Istanbul;[(d)]Department of Physics Engineering, Gaziantep University, Gaziantep; Turkey.
[13]Institute of Physics, Azerbaijan Academy of Sciences, Baku; Azerbaijan.
[14]Institut de Física d'Altes Energies (IFAE), Barcelona Institute of Science and Technology, Barcelona; Spain.
[15] [(a)]Institute of High Energy Physics, Chinese Academy of Sciences, Beijing;[(b)]Physics Department, Tsinghua University, Beijing;[(c)]Department of Physics, Nanjing University, Nanjing;[(d)]University of Chinese Academy of Science (UCAS), Beijing; China.
[16]Institute of Physics, University of Belgrade, Belgrade; Serbia.
[17]Department for Physics and Technology, University of Bergen, Bergen; Norway.
[18]Physics Division, Lawrence Berkeley National Laboratory and University of California, Berkeley CA; United States of America.

[19]Institut für Physik, Humboldt Universität zu Berlin, Berlin; Germany.
[20]Albert Einstein Center for Fundamental Physics and Laboratory for High Energy Physics, University of Bern, Bern; Switzerland.
[21]School of Physics and Astronomy, University of Birmingham, Birmingham; United Kingdom.
[22]Facultad de Ciencias y Centro de Investigaciónes, Universidad Antonio Nariño, Bogota; Colombia.
[23] [(a)]INFN Bologna and Universita' di Bologna, Dipartimento di Fisica;[(b)]INFN Sezione di Bologna; Italy.
[24]Physikalisches Institut, Universität Bonn, Bonn; Germany.
[25]Department of Physics, Boston University, Boston MA; United States of America.
[26]Department of Physics, Brandeis University, Waltham MA; United States of America.
[27] [(a)]Transilvania University of Brasov, Brasov;[(b)]Horia Hulubei National Institute of Physics and Nuclear Engineering, Bucharest;[(c)]Department of Physics, Alexandru Ioan Cuza University of Iasi, Iasi;[(d)]National Institute for Research and Development of Isotopic and Molecular Technologies, Physics Department, Cluj-Napoca;[(e)]University Politehnica Bucharest, Bucharest;[(f)]West University in Timisoara, Timisoara; Romania.
[28] [(a)]Faculty of Mathematics, Physics and Informatics, Comenius University, Bratislava;[(b)]Department of Subnuclear Physics, Institute of Experimental Physics of the Slovak Academy of Sciences, Kosice; Slovak Republic.
[29]Physics Department, Brookhaven National Laboratory, Upton NY; United States of America.
[30]Departamento de Física, Universidad de Buenos Aires, Buenos Aires; Argentina.
[31]California State University, CA; United States of America.
[32]Cavendish Laboratory, University of Cambridge, Cambridge; United Kingdom.
[33] [(a)]Department of Physics, University of Cape Town, Cape Town;[(b)]Department of Mechanical Engineering Science, University of Johannesburg, Johannesburg;[(c)]University of South Africa, Department of Physics, Pretoria;[(d)]School of Physics, University of the Witwatersrand, Johannesburg; South Africa.
[34]Department of Physics, Carleton University, Ottawa ON; Canada.
[35] [(a)]Faculté des Sciences Ain Chock, Réseau Universitaire de Physique des Hautes Energies - Université Hassan II, Casablanca;[(b)]Faculté des Sciences, Université Ibn-Tofail, Kénitra;[(c)]Faculté des Sciences Semlalia, Université Cadi Ayyad, LPHEA-Marrakech;[(d)]Faculté des Sciences, Université Mohamed Premier and LPTPM, Oujda;[(e)]Faculté des sciences, Université Mohammed V, Rabat; Morocco.
[36]CERN, Geneva; Switzerland.
[37]Enrico Fermi Institute, University of Chicago, Chicago IL; United States of America.
[38]LPC, Université Clermont Auvergne, CNRS/IN2P3, Clermont-Ferrand; France.

[39]Nevis Laboratory, Columbia University, Irvington NY; United States of America.
[40]Niels Bohr Institute, University of Copenhagen, Copenhagen; Denmark.
[41] [(a)]Dipartimento di Fisica, Università della Calabria, Rende;[(b)]INFN Gruppo Collegato di Cosenza, Laboratori Nazionali di Frascati; Italy.
[42]Physics Department, Southern Methodist University, Dallas TX; United States of America.
[43]Physics Department, University of Texas at Dallas, Richardson TX; United States of America.
[44]National Centre for Scientific Research "Demokritos", Agia Paraskevi; Greece.
[45] [(a)]Department of Physics, Stockholm University;[(b)]Oskar Klein Centre, Stockholm; Sweden.
[46]Deutsches Elektronen-Synchrotron DESY, Hamburg and Zeuthen; Germany.
[47]Lehrstuhl für Experimentelle Physik IV, Technische Universität Dortmund, Dortmund; Germany.
[48]Institut für Kern- und Teilchenphysik, Technische Universität Dresden, Dresden; Germany.
[49]Department of Physics, Duke University, Durham NC; United States of America.
[50]SUPA - School of Physics and Astronomy, University of Edinburgh, Edinburgh; United Kingdom.
[51]INFN e Laboratori Nazionali di Frascati, Frascati; Italy.
[52]Physikalisches Institut, Albert-Ludwigs-Universität Freiburg, Freiburg; Germany.
[53]II. Physikalisches Institut, Georg-August-Universität Göttingen, Göttingen; Germany.
[54]Département de Physique Nucléaire et Corpusculaire, Université de Genève, Genève; Switzerland.
[55] [(a)]Dipartimento di Fisica, Università di Genova, Genova;[(b)]INFN Sezione di Genova; Italy.
[56]II. Physikalisches Institut, Justus-Liebig-Universität Giessen, Giessen; Germany.
[57]SUPA - School of Physics and Astronomy, University of Glasgow, Glasgow; United Kingdom.
[58]LPSC, Université Grenoble Alpes, CNRS/IN2P3, Grenoble INP, Grenoble; France.
[59]Laboratory for Particle Physics and Cosmology, Harvard University, Cambridge MA; United States of America.
[60] [(a)]Department of Modern Physics and State Key Laboratory of Particle Detection and Electronics, University of Science and Technology of China, Hefei;[(b)]Institute of Frontier and Interdisciplinary Science and Key Laboratory of Particle Physics and Particle Irradiation (MOE), Shandong University, Qingdao;[(c)]School of Physics and Astronomy, Shanghai Jiao Tong University, KLPPAC-MoE, SKLPPC, Shanghai;[(d)]Tsung-Dao Lee Institute, Shanghai; China.

[61] [(a)]Kirchhoff-Institut für Physik, Ruprecht-Karls-Universität Heidelberg, Heidelberg;[(b)]Physikalisches Institut, Ruprecht-Karls-Universität Heidelberg, Heidelberg; Germany.
[62]Faculty of Applied Information Science, Hiroshima Institute of Technology, Hiroshima; Japan.
[63] [(a)]Department of Physics, Chinese University of Hong Kong, Shatin, N.T., Hong Kong;[(b)]Department of Physics, University of Hong Kong, Hong Kong;[(c)]Department of Physics and Institute for Advanced Study, Hong Kong University of Science and Technology, Clear Water Bay, Kowloon, Hong Kong; China.
[64]Department of Physics, National Tsing Hua University, Hsinchu; Taiwan.
[65]IJCLab, Université Paris-Saclay, CNRS/IN2P3, 91405, Orsay; France.
[66]Department of Physics, Indiana University, Bloomington IN; United States of America.
[67] [(a)]INFN Gruppo Collegato di Udine, Sezione di Trieste, Udine;[(b)]ICTP, Trieste;[(c)]Dipartimento Politecnico di Ingegneria e Architettura, Università di Udine, Udine; Italy.
[68] [(a)]INFN Sezione di Lecce;[(b)]Dipartimento di Matematica e Fisica, Università del Salento, Lecce; Italy.
[69] [(a)]INFN Sezione di Milano;[(b)]Dipartimento di Fisica, Università di Milano, Milano; Italy.
[70] [(a)]INFN Sezione di Napoli;[(b)]Dipartimento di Fisica, Università di Napoli, Napoli; Italy.
[71] [(a)]INFN Sezione di Pavia;[(b)]Dipartimento di Fisica, Università di Pavia, Pavia; Italy.
[72] [(a)]INFN Sezione di Pisa;[(b)]Dipartimento di Fisica E. Fermi, Università di Pisa, Pisa; Italy.
[73] [(a)]INFN Sezione di Roma;[(b)]Dipartimento di Fisica, Sapienza Università di Roma, Roma; Italy.
[74] [(a)]INFN Sezione di Roma Tor Vergata;[(b)]Dipartimento di Fisica, Università di Roma Tor Vergata, Roma; Italy.
[75] [(a)]INFN Sezione di Roma Tre;[(b)]Dipartimento di Matematica e Fisica, Università Roma Tre, Roma; Italy.
[76] [(a)]INFN-TIFPA;[(b)]Università degli Studi di Trento, Trento; Italy.
[77]Institut für Astro- und Teilchenphysik, Leopold-Franzens-Universität, Innsbruck; Austria.
[78]University of Iowa, Iowa City IA; United States of America.
[79]Department of Physics and Astronomy, Iowa State University, Ames IA; United States of America.
[80]Joint Institute for Nuclear Research, Dubna; Russia.
[81] [(a)]Departamento de Engenharia Elétrica, Universidade Federal de Juiz de Fora (UFJF), Juiz de Fora;[(b)]Universidade Federal do Rio De Janeiro COPPE/EE/IF, Rio de Janeiro;[(c)]Universidade Federal de São João del Rei (UFSJ), São João del Rei;[(d)]Instituto de Física, Universidade de São Paulo, São Paulo; Brazil.
[82]KEK, High Energy Accelerator Research Organization, Tsukuba; Japan.

[83]Graduate School of Science, Kobe University, Kobe; Japan.
[84] [(a)]AGH University of Science and Technology, Faculty of Physics and Applied Computer Science, Krakow;[(b)]Marian Smoluchowski Institute of Physics, Jagiellonian University, Krakow; Poland.
[85]Institute of Nuclear Physics Polish Academy of Sciences, Krakow; Poland.
[86]Faculty of Science, Kyoto University, Kyoto; Japan.
[87]Kyoto University of Education, Kyoto; Japan.
[88]Research Center for Advanced Particle Physics and Department of Physics, Kyushu University, Fukuoka ; Japan.
[89]Instituto de Física La Plata, Universidad Nacional de La Plata and CONICET, La Plata; Argentina.
[90]Physics Department, Lancaster University, Lancaster; United Kingdom.
[91]Oliver Lodge Laboratory, University of Liverpool, Liverpool; United Kingdom.
[92]Department of Experimental Particle Physics, Jožef Stefan Institute and Department of Physics, University of Ljubljana, Ljubljana; Slovenia.
[93]School of Physics and Astronomy, Queen Mary University of London, London; United Kingdom.
[94]Department of Physics, Royal Holloway University of London, Egham; United Kingdom.
[95]Department of Physics and Astronomy, University College London, London; United Kingdom.
[96]Louisiana Tech University, Ruston LA; United States of America.
[97]Fysiska institutionen, Lunds universitet, Lund; Sweden.
[98]Centre de Calcul de l'Institut National de Physique Nucléaire et de Physique des Particules (IN2P3), Villeurbanne; France.
[99]Departamento de Física Teorica C-15 and CIAFF, Universidad Autónoma de Madrid, Madrid; Spain.
[100]Institut für Physik, Universität Mainz, Mainz; Germany.
[101]School of Physics and Astronomy, University of Manchester, Manchester; United Kingdom.
[102]CPPM, Aix-Marseille Université, CNRS/IN2P3, Marseille; France.
[103]Department of Physics, University of Massachusetts, Amherst MA; United States of America.
[104]Department of Physics, McGill University, Montreal QC; Canada.
[105]School of Physics, University of Melbourne, Victoria; Australia.
[106]Department of Physics, University of Michigan, Ann Arbor MI; United States of America.
[107]Department of Physics and Astronomy, Michigan State University, East Lansing MI; United States of America.
[108]B.I. Stepanov Institute of Physics, National Academy of Sciences of Belarus, Minsk; Belarus.
[109]Research Institute for Nuclear Problems of Byelorussian State University, Minsk; Belarus.
[110]Group of Particle Physics, University of Montreal, Montreal QC; Canada.

[111]P.N. Lebedev Physical Institute of the Russian Academy of Sciences, Moscow; Russia.
[112]National Research Nuclear University MEPhI, Moscow; Russia.
[113]D.V. Skobeltsyn Institute of Nuclear Physics, M.V. Lomonosov Moscow State University, Moscow; Russia.
[114]Fakultät für Physik, Ludwig-Maximilians-Universität München, München; Germany.
[115]Max-Planck-Institut für Physik (Werner-Heisenberg-Institut), München; Germany.
[116]Nagasaki Institute of Applied Science, Nagasaki; Japan.
[117]Graduate School of Science and Kobayashi-Maskawa Institute, Nagoya University, Nagoya; Japan.
[118]Department of Physics and Astronomy, University of New Mexico, Albuquerque NM; United States of America.
[119]Institute for Mathematics, Astrophysics and Particle Physics, Radboud University Nijmegen/Nikhef, Nijmegen; Netherlands.
[120]Nikhef National Institute for Subatomic Physics and University of Amsterdam, Amsterdam; Netherlands.
[121]Department of Physics, Northern Illinois University, DeKalb IL; United States of America.
[122] [(a)]Budker Institute of Nuclear Physics and NSU, SB RAS, Novosibirsk;[(b)]Novosibirsk State University Novosibirsk; Russia.
[123]Institute for High Energy Physics of the National Research Centre Kurchatov Institute, Protvino; Russia.
[124]Institute for Theoretical and Experimental Physics named by A.I. Alikhanov of National Research Centre "Kurchatov Institute", Moscow; Russia.
[125]Department of Physics, New York University, New York NY; United States of America.
[126]Ochanomizu University, Otsuka, Bunkyo-ku, Tokyo; Japan.
[127]Ohio State University, Columbus OH; United States of America.
[128]Faculty of Science, Okayama University, Okayama; Japan.
[129]Homer L. Dodge Department of Physics and Astronomy, University of Oklahoma, Norman OK; United States of America.
[130]Department of Physics, Oklahoma State University, Stillwater OK; United States of America.
[131]Palacký University, RCPTM, Joint Laboratory of Optics, Olomouc; Czech Republic.
[132]Center for High Energy Physics, University of Oregon, Eugene OR; United States of America.
[133]Graduate School of Science, Osaka University, Osaka; Japan.
[134]Department of Physics, University of Oslo, Oslo; Norway.
[135]Department of Physics, Oxford University, Oxford; United Kingdom.
[136]LPNHE, Sorbonne Université, Université de Paris, CNRS/IN2P3, Paris; France.

[137]Department of Physics, University of Pennsylvania, Philadelphia PA; United States of America.
[138]Konstantinov Nuclear Physics Institute of National Research Centre "Kurchatov Institute", PNPI, St. Petersburg; Russia.
[139]Department of Physics and Astronomy, University of Pittsburgh, Pittsburgh PA; United States of America.
[140] [(a)]Laboratório de Instrumentação e Física Experimental de Partículas - LIP, Lisboa;[(b)]Departamento de Física, Faculdade de Ciências, Universidade de Lisboa, Lisboa;[(c)]Departamento de Física, Universidade de Coimbra, Coimbra;[(d)]Centro de Física Nuclear da Universidade de Lisboa, Lisboa;[(e)]Departamento de Física, Universidade do Minho, Braga;[(f)]Departamento de FÃsica TeÃ³rica y del Cosmos, Universidad de Granada, Granada (Spain);[(g)]Dep Física and CEFITEC of Faculdade de Ciências e Tecnologia, Universidade Nova de Lisboa, Caparica;[(h)]Instituto Superior Técnico, Universidade de Lisboa, Lisboa; Portugal.
[141]Institute of Physics of the Czech Academy of Sciences, Prague; Czech Republic.
[142]Czech Technical University in Prague, Prague; Czech Republic.
[143]Charles University, Faculty of Mathematics and Physics, Prague; Czech Republic.
[144]Particle Physics Department, Rutherford Appleton Laboratory, Didcot; United Kingdom.
[145]IRFU, CEA, Université Paris-Saclay, Gif-sur-Yvette; France.
[146]Santa Cruz Institute for Particle Physics, University of California Santa Cruz, Santa Cruz CA; United States of America.
[147] [(a)]Departamento de Física, Pontificia Universidad Católica de Chile, Santiago;[(b)]Universidad Andres Bello, Department of Physics, Santiago;[(c)]Departamento de Física, Universidad Técnica Federico Santa María, Valparaíso; Chile.
[148]Department of Physics, University of Washington, Seattle WA; United States of America.
[149]Department of Physics and Astronomy, University of Sheffield, Sheffield; United Kingdom.
[150]Department of Physics, Shinshu University, Nagano; Japan.
[151]Department Physik, Universität Siegen, Siegen; Germany.
[152]Department of Physics, Simon Fraser University, Burnaby BC; Canada.
[153]SLAC National Accelerator Laboratory, Stanford CA; United States of America.
[154]Physics Department, Royal Institute of Technology, Stockholm; Sweden.
[155]Departments of Physics and Astronomy, Stony Brook University, Stony Brook NY; United States of America.
[156]Department of Physics and Astronomy, University of Sussex, Brighton; United Kingdom.
[157]School of Physics, University of Sydney, Sydney; Australia.
[158]Institute of Physics, Academia Sinica, Taipei; Taiwan.

[159] [(a)]E. Andronikashvili Institute of Physics, Iv. Javakhishvili Tbilisi State University, Tbilisi;[(b)]High Energy Physics Institute, Tbilisi State University, Tbilisi; Georgia.
[160]Department of Physics, Technion, Israel Institute of Technology, Haifa; Israel.
[161]Raymond and Beverly Sackler School of Physics and Astronomy, Tel Aviv University, Tel Aviv; Israel.
[162]Department of Physics, Aristotle University of Thessaloniki, Thessaloniki; Greece.
[163]International Center for Elementary Particle Physics and Department of Physics, University of Tokyo, Tokyo; Japan.
[164]Graduate School of Science and Technology, Tokyo Metropolitan University, Tokyo; Japan.
[165]Department of Physics, Tokyo Institute of Technology, Tokyo; Japan.
[166]Tomsk State University, Tomsk; Russia.
[167]Department of Physics, University of Toronto, Toronto ON; Canada.
[168] [(a)]TRIUMF, Vancouver BC;[(b)]Department of Physics and Astronomy, York University, Toronto ON; Canada.
[169]Division of Physics and Tomonaga Center for the History of the Universe, Faculty of Pure and Applied Sciences, University of Tsukuba, Tsukuba; Japan.
[170]Department of Physics and Astronomy, Tufts University, Medford MA; United States of America.
[171]Department of Physics and Astronomy, University of California Irvine, Irvine CA; United States of America.
[172]Department of Physics and Astronomy, University of Uppsala, Uppsala; Sweden.
[173]Department of Physics, University of Illinois, Urbana IL; United States of America.
[174]Instituto de Física Corpuscular (IFIC), Centro Mixto Universidad de Valencia - CSIC, Valencia; Spain.
[175]Department of Physics, University of British Columbia, Vancouver BC; Canada.
[176]Department of Physics and Astronomy, University of Victoria, Victoria BC; Canada.
[177]Fakultät für Physik und Astronomie, Julius-Maximilians-Universität Würzburg, Würzburg; Germany.
[178]Department of Physics, University of Warwick, Coventry; United Kingdom.
[179]Waseda University, Tokyo; Japan.
[180]Department of Particle Physics, Weizmann Institute of Science, Rehovot; Israel.
[181]Department of Physics, University of Wisconsin, Madison WI; United States of America.
[182]Fakultät für Mathematik und Naturwissenschaften, Fachgruppe Physik, Bergische Universität Wuppertal, Wuppertal; Germany.
[183]Department of Physics, Yale University, New Haven CT; United States of America.
[184]Yerevan Physics Institute, Yerevan; Armenia.

[a] Also at Borough of Manhattan Community College, City University of New York, New York NY; United States of America.
[b] Also at CERN, Geneva; Switzerland.
[c] Also at CPPM, Aix-Marseille Université, CNRS/IN2P3, Marseille; France.
[d] Also at Département de Physique Nucléaire et Corpusculaire, Université de Genève, Genève; Switzerland.
[e] Also at Departament de Fisica de la Universitat Autonoma de Barcelona, Barcelona; Spain.
[f] Also at Departamento de FÃsica, Instituto Superior TÃ©cnico, Universidade de Lisboa, Lisboa; Portugal.
[g] Also at Department of Applied Physics and Astronomy, University of Sharjah, Sharjah; United Arab Emirates.
[h] Also at Department of Financial and Management Engineering, University of the Aegean, Chios; Greece.
[i] Also at Department of Physics and Astronomy, Michigan State University, East Lansing MI; United States of America.
[j] Also at Department of Physics and Astronomy, University of Louisville, Louisville, KY; United States of America.
[k] Also at Department of Physics, Ben Gurion University of the Negev, Beer Sheva; Israel.
[l] Also at Department of Physics, California State University, East Bay; United States of America.
[m] Also at Department of Physics, California State University, Fresno; United States of America.
[n] Also at Department of Physics, California State University, Sacramento; United States of America.
[o] Also at Department of Physics, King's College London, London; United Kingdom.
[p] Also at Department of Physics, St. Petersburg State Polytechnical University, St. Petersburg; Russia.
[q] Also at Department of Physics, Stanford University, Stanford CA; United States of America.
[r] Also at Department of Physics, University of Adelaide, Adelaide; Australia.
[s] Also at Department of Physics, University of Fribourg, Fribourg; Switzerland.
[t] Also at Department of Physics, University of Michigan, Ann Arbor MI; United States of America.
[u] Also at Dipartimento di Matematica, Informatica e Fisica, Università di Udine, Udine; Italy.
[v] Also at Faculty of Physics, M.V. Lomonosov Moscow State University, Moscow; Russia.
[w] Also at Giresun University, Faculty of Engineering, Giresun; Turkey.
[x] Also at Graduate School of Science, Osaka University, Osaka; Japan.
[y] Also at Hellenic Open University, Patras; Greece.
[z] Also at IJCLab, Université Paris-Saclay, CNRS/IN2P3, 91405, Orsay; France.

[aa] Also at Institucio Catalana de Recerca i Estudis Avancats, ICREA, Barcelona; Spain.
[ab] Also at Institut für Experimentalphysik, Universität Hamburg, Hamburg; Germany.
[ac] Also at Institute for Mathematics, Astrophysics and Particle Physics, Radboud University Nijmegen/Nikhef, Nijmegen; Netherlands.
[ad] Also at Institute for Nuclear Research and Nuclear Energy (INRNE) of the Bulgarian Academy of Sciences, Sofia; Bulgaria.
[ae] Also at Institute for Particle and Nuclear Physics, Wigner Research Centre for Physics, Budapest; Hungary.
[af] Also at Institute of Particle Physics (IPP), Vancouver; Canada.
[ag] Also at Institute of Physics, Academia Sinica, Taipei; Taiwan.
[ah] Also at Institute of Physics, Azerbaijan Academy of Sciences, Baku; Azerbaijan.
[ai] Also at Institute of Theoretical Physics, Ilia State University, Tbilisi; Georgia.
[aj] Also at Instituto de Fisica Teorica, IFT-UAM/CSIC, Madrid; Spain.
[ak] Also at Joint Institute for Nuclear Research, Dubna; Russia.
[al] Also at Louisiana Tech University, Ruston LA; United States of America.
[am] Also at LPNHE, Sorbonne Université, Université de Paris, CNRS/IN2P3, Paris; France.
[an] Also at Manhattan College, New York NY; United States of America.
[ao] Also at Moscow Institute of Physics and Technology State University, Dolgoprudny; Russia.
[ap] Also at National Research Nuclear University MEPhI, Moscow; Russia.
[aq] Also at Physics Department, An-Najah National University, Nablus; Palestine.
[ar] Also at Physics Dept, University of South Africa, Pretoria; South Africa.
[as] Also at Physikalisches Institut, Albert-Ludwigs-Universität Freiburg, Freiburg; Germany.
[at] Also at School of Physics, Sun Yat-sen University, Guangzhou; China.
[au] Also at The City College of New York, New York NY; United States of America.
[av] Also at The Collaborative Innovation Center of Quantum Matter (CICQM), Beijing; China.
[aw] Also at Tomsk State University, Tomsk, and Moscow Institute of Physics and Technology State University, Dolgoprudny; Russia.
[ax] Also at TRIUMF, Vancouver BC; Canada.
[ay] Also at Universita di Napoli Parthenope, Napoli; Italy.
* Deceased

BM@N Collaboration

Kh. Abraamyan[1], S. Afanasiev[1], G. Agakishiev[1], E. P. Alekseev[3], V. Babkin[1], V. Balandin[1], N. Balashov[1], D. Baranov[1], N. Barbashina[4], P. Batyuk[1], M. Baznat[7], S. Bazylev[1], D. Bogoslovsky[1], A. Bolozdynya[4], V. Burtsev[1], M. Buryakov[1], S. Buzin[1], A. Cheremukhin[1], B. Dabrowska[1], D. Dementev[1], A. Dmitriev[1], A. Dryablov[1], P. Dulov[1], D. Egorov[1],, A. Fedunin[1], I. Filippov[1], I. Filozova[1], I. Gabdrakhmanov[1], A. Galavanov[4], O. Gavrischuk[1], K. Gertsenberger[1], V. Golovatyuk[1], M. Golubeva[2], F. Guber[2], Yu. Ivanova[1], A. Ivashkin[2], A. Izvestnyy[2], S. Kakurin[1], M. Kapishin[1], V. Karjavin[1], R. Kattabekov[1], V. Kekelidze[1], S. Khabarov[1], Yu. Kiryushin[1], V. Kolesnikov[1], A. Kolozhvari[1], Yu. Kopylov[1], L. Kovachev[5], A. Kovalenko[1], Yu. Kovalev[1], S. Kuklin[1], E. Kulish[1], A. Kuznetsov[1], E. Ladygin[1], N. Lashmanov[1], V. Lenivenko[1], E. Litvinenko[1], S. Lobastov[1], A. Makankin[1], A. Maksymchyuk[1], A. Malakhov[1], S. Merts[1], A. Morozov[1], S. Morozov[2], Yu. Murin[1], G. Musulmanbekov[1], R. Nagdasev[1], D. Nikitin[1], D. Oleynik[1], V. Palchik[1], A. Petrosyan[1], M. Patsyuk[1], Yu. Petukhov[1], S. Piyadin[1], V. Plotnikov[1], D. Podgainy[1], G. Pokatashkin[1], Yu. Potrebenikov[1], O. Rogachevsky[1], V. Rogov[1], I. Rufanov[1], P. Rukoyatkin[1], M. Rumyantsev[1], D. Sakulin[1], S. Sedykh[1], I. Selyuzhenkov[4], S. Sergeev[1], A. Sheremetev[1], A. Sheremeteva[1], A. Shipunov[1], M. Shitenkov[1], A. Shutov[1], V. Shutov[1], I. Slepnev[1], V. Slepnev[1], I. Slepov[1], A. Sorin[1], V. Sosnovtsev[4], V. Spaskov[1], A. Stavinskiy[3], E. Streletskaya[1], N. Sukhov[1], D. Suvarieva[1], A. Taranenko[4], N. Tarasov[1], O. Tarasov[1], A. Terletsky[1], O. Teryaev[1], V. Tcholakov[5], V. Tikhomirov[1], A. Timoshenko[1], N. Topilin[1], B. Topko[1], I. Tserruya[6], I. Tyapkin[1], V. Vasendina[1], A. Vishnevsky[1], A. Voronin[1], N. Voytishin[1], V. Yurevich[1], N. Zamiatin[1], M. Zavertiaev[1], N. Zhigareva[3], A. Zinchenko[1], E. Zubarev[1], M. Zuev[1]

1. Joint Institute for Nuclear Research (JINR), Dubna, Russia
2. Institute for Nuclear Research of the RAS (INR RAS), Moscow, Russia
3. Institute for Theoretical and Experimental Physics (ITEP), Moscow, Russia
4. National Research Nuclear University MEPhI, Moscow, Russia
5. Plovdiv University "Paisii Hilendarski", Plovdiv, Bulgaria
6. Weizmann Institute of Science, Rehovot, Israel
7. Institute of Applied Physics, ASM, Chisinau, Moldova

CMS Collaboration

Yerevan Physics Institute, Yerevan, Armenia
A.M. Sirunyan[†], A. Tumasyan
Institut für Hochenergiephysik, Wien, Austria
W. Adam, F. Ambrogi, T. Bergauer, J. Brandstetter, M. Dragicevic, J. Erö, A. Escalante Del Valle, M. Flechl, R. Frühwirth[1], M. Jeitler[1], N. Krammer, I. Krätschmer, D. Liko, T. Madlener, I. Mikulec, N. Rad, J. Schieck[1], R. Schöfbeck, M. Spanring, D. Spitzbart, W. Waltenberger, C.-E. Wulz[1], M. Zarucki
Institute for Nuclear Problems, Minsk, Belarus
V. Drugakov, V. Mossolov, J. Suarez Gonzalez
Universiteit Antwerpen, Antwerpen, Belgium
M.R. Darwish, E.A. De Wolf, D. Di Croce, X. Janssen, J. Lauwers, A. Lelek, M. Pieters, H. Rejeb Sfar, H. Van Haevermaet, P. Van Mechelen, S. Van Putte, N. Van Remortel
Vrije Universiteit Brussel, Brussel, Belgium
F. Blekman, E.S. Bols, S.S. Chhibra, J. D'Hondt, J. De Clercq, D. Lontkovskyi, S. Lowette, I. Marchesini, S. Moortgat, L. Moreels, Q. Python, K. Skovpen, S. Tavernier, W. Van Doninck, P. Van Mulders, I. Van Parijs
Université Libre de Bruxelles, Bruxelles, Belgium
D. Beghin, B. Bilin, H. Brun, B. Clerbaux, G. De Lentdecker, H. Delannoy, B. Dorney, L. Favart, A. Grebenyuk, A.K. Kalsi, J. Luetic, A. Popov, N. Postiau, E. Starling, L. Thomas, C. Vander Velde, P. Vanlaer, D. Vannerom, Q. Wang
Ghent University, Ghent, Belgium
T. Cornelis, D. Dobur, I. Khvastunov[2], C. Roskas, D. Trocino, M. Tytgat, W. Verbeke, B. Vermassen, M. Vit, N. Zaganidis
Université Catholique de Louvain, Louvain-la-Neuve, Belgium
O. Bondu, G. Bruno, C. Caputo, P. David, C. Delaere, M. Delcourt, A. Giammanco, V. Lemaitre, A. Magitteri, J. Prisciandaro, A. Saggio, M. Vidal Marono, P. Vischia, J. Zobec
Centro Brasileiro de Pesquisas Fisicas, Rio de Janeiro, Brazil
F.L. Alves, G.A. Alves, G. Correia Silva, C. Hensel, A. Moraes, P. Rebello Teles
Universidade do Estado do Rio de Janeiro, Rio de Janeiro, Brazil
E. Belchior Batista Das Chagas, W. Carvalho, J. Chinellato[3], E. Coelho, E.M. Da Costa, G.G. Da Silveira[4], D. De Jesus Damiao, C. De Oliveira Martins, S. Fonseca De Souza, L.M. Huertas Guativa, H. Malbouisson, J. Martins[5], D. Matos Figueiredo, M. Medina Jaime[6], M. Melo De Almeida, C. Mora Herrera, L. Mundim, H. Nogima, W.L. Prado Da Silva, L.J. Sanchez Rosas, A. Santoro, A. Sznajder, M. Thiel, E.J. Tonelli Manganote[3], F. Torres Da Silva De Araujo, A. Vilela Pereira
Universidade Estadual Paulista [a], Universidade Federal do ABC [b], São Paulo, Brazil
S. Ahuja[a], C.A. Bernardes[a], L. Calligaris[a], T.R. Fernandez Perez Tomei[a], E.M. Gregores[b], D.S. Lemos, P.G. Mercadante[b], S.F. Novaes[a], SandraS. Padula[a]

Institute for Nuclear Research and Nuclear Energy, Bulgarian Academy of Sciences, Sofia, Bulgaria
A. Aleksandrov, G. Antchev, R. Hadjiiska, P. Iaydjiev, A. Marinov, M. Misheva, M. Rodozov, M. Shopova, G. Sultanov
University of Sofia, Sofia, Bulgaria
M. Bonchev, A. Dimitrov, T. Ivanov, L. Litov, B. Pavlov, P. Petkov
Beihang University, Beijing, China
W. Fang[7], X. Gao[7], L. Yuan
Institute of High Energy Physics, Beijing, China
M. Ahmad, G.M. Chen, H.S. Chen, M. Chen, C.H. Jiang, D. Leggat, H. Liao, Z. Liu, S.M. Shaheen[8], A. Spiezia, J. Tao, E. Yazgan, H. Zhang, S. Zhang[8], J. Zhao
State Key Laboratory of Nuclear Physics and Technology, Peking University, Beijing, China
A. Agapitos, Y. Ban, G. Chen, A. Levin, J. Li, L. Li, Q. Li, Y. Mao, S.J. Qian, D. Wang
Tsinghua University, Beijing, China
Z. Hu, Y. Wang
Universidad de Los Andes, Bogota, Colombia
C. Avila, A. Cabrera, L.F. Chaparro Sierra, C. Florez, C.F. González Hernández, M.A. Segura Delgado
Universidad de Antioquia, Medellin, Colombia
J. Mejia Guisao, J.D. Ruiz Alvarez, C.A. Salazar González, N. Vanegas Arbelaez
University of Split, Faculty of Electrical Engineering, Mechanical Engineering and Naval Architecture, Split, Croatia
D. Giljanović, N. Godinovic, D. Lelas, I. Puljak, T. Sculac
University of Split, Faculty of Science, Split, Croatia
Z. Antunovic, M. Kovac
Institute Rudjer Boskovic, Zagreb, Croatia
V. Brigljevic, S. Ceci, D. Ferencek, K. Kadija, B. Mesic, M. Roguljic, A. Starodumov[9], T. Susa
University of Cyprus, Nicosia, Cyprus
M.W. Ather, A. Attikis, E. Erodotou, A. Ioannou, M. Kolosova, S. Konstantinou, G. Mavromanolakis, J. Mousa, C. Nicolaou, F. Ptochos, P.A. Razis, H. Rykaczewski, D. Tsiakkouri
Charles University, Prague, Czech Republic
M. Finger[10], M. Finger Jr.[10], A. Kveton, J. Tomsa
Escuela Politecnica Nacional, Quito, Ecuador
E. Ayala
Universidad San Francisco de Quito, Quito, Ecuador
E. Carrera Jarrin
Academy of Scientific Research and Technology of the Arab Republic of Egypt, Egyptian Network of High Energy Physics, Cairo, Egypt
H. Abdalla[11], A.A. Abdelalim[12,13]
National Institute of Chemical Physics and Biophysics, Tallinn, Estonia
S. Bhowmik, A. Carvalho Antunes De Oliveira, R.K. Dewanjee, K. Ehataht, M. Kadastik, M. Raidal, C. Veelken

Department of Physics, University of Helsinki, Helsinki, Finland
P. Eerola, L. Forthomme, H. Kirschenmann, K. Osterberg, M. Voutilainen
Helsinki Institute of Physics, Helsinki, Finland
F. Garcia, J. Havukainen, J.K. Heikkilä, T. Järvinen, V. Karimäki, R. Kinnunen, T. Lampén, K. Lassila-Perini, S. Laurila, S. Lehti, T. Lindén, P. Luukka, T. Mäenpää, H. Siikonen, E. Tuominen, J. Tuominiemi
Lappeenranta University of Technology, Lappeenranta, Finland
T. Tuuva
IRFU, CEA, Université Paris-Saclay, Gif-sur-Yvette, France
M. Besancon, F. Couderc, M. Dejardin, D. Denegri, B. Fabbro, J.L. Faure, F. Ferri, S. Ganjour, A. Givernaud, P. Gras, G. Hamel de Monchenault, P. Jarry, C. Leloup, E. Locci, J. Malcles, J. Rander, A. Rosowsky, M.Ö. Sahin, A. Savoy-Navarro[14], M. Titov
Laboratoire Leprince-Ringuet, CNRS/IN2P3, Ecole Polytechnique, Institut Polytechnique de Paris
C. Amendola, F. Beaudette, P. Busson, C. Charlot, B. Diab, G. Falmagne, R. Granier de Cassagnac, I. Kucher, A. Lobanov, C. Martin Perez, M. Nguyen, C. Ochando, P. Paganini, J. Rembser, R. Salerno, J.B. Sauvan, Y. Sirois, A. Zabi, A. Zghiche
Université de Strasbourg, CNRS, IPHC UMR 7178, Strasbourg, France
J.-L. Agram[15], J. Andrea, D. Bloch, G. Bourgatte, J.-M. Brom, E.C. Chabert, C. Collard, E. Conte[15], J.-C. Fontaine[15], D. Gelé, U. Goerlach, M. Jansová, A.-C. Le Bihan, N. Tonon, P. Van Hove
Centre de Calcul de l'Institut National de Physique Nucleaire et de Physique des Particules, CNRS/IN2P3, Villeurbanne, France
S. Gadrat
Université de Lyon, Université Claude Bernard Lyon 1, CNRS-IN2P3, Institut de Physique Nucléaire de Lyon, Villeurbanne, France
S. Beauceron, C. Bernet, G. Boudoul, C. Camen, N. Chanon, R. Chierici, D. Contardo, P. Depasse, H. El Mamouni, J. Fay, S. Gascon, M. Gouzevitch, B. Ille, Sa. Jain, F. Lagarde, I.B. Laktineh, H. Lattaud, M. Lethuillier, L. Mirabito, S. Perries, V. Sordini, G. Touquet, M. Vander Donckt, S. Viret
Georgian Technical University, Tbilisi, Georgia
A. Khvedelidze[10]
Tbilisi State University, Tbilisi, Georgia
Z. Tsamalaidze[10]
RWTH Aachen University, I. Physikalisches Institut, Aachen, Germany
C. Autermann, L. Feld, M.K. Kiesel, K. Klein, M. Lipinski, D. Meuser, A. Pauls, M. Preuten, M.P. Rauch, C. Schomakers, J. Schulz, M. Teroerde, B. Wittmer
RWTH Aachen University, III. Physikalisches Institut A, Aachen, Germany
A. Albert, M. Erdmann, S. Erdweg, T. Esch, B. Fischer, R. Fischer, S. Ghosh, T. Hebbeker, K. Hoepfner, H. Keller, L. Mastrolorenzo, M. Merschmeyer, A. Meyer, P. Millet, G. Mocellin, S. Mondal, S. Mukherjee, D. Noll, A. Novak, T. Pook, A. Pozdnyakov, T. Quast, M. Radziej, Y. Rath, H. Reithler, M. Rieger, J. Roemer, A. Schmidt, S.C. Schuler, A. Sharma, S. Thüer, S. Wiedenbeck

RWTH Aachen University, III. Physikalisches Institut B, Aachen, Germany
G. Flügge, W. Haj Ahmad[16], O. Hlushchenko, T. Kress, T. Müller, A. Nehrkorn, A. Nowack, C. Pistone, O. Pooth, D. Roy, H. Sert, A. Stahl[17]
Deutsches Elektronen-Synchrotron, Hamburg, Germany
M. Aldaya Martin, P. Asmuss, I. Babounikau, H. Bakhshiansohi, K. Beernaert, O. Behnke, U. Behrens, A. Bermúdez Martínez, D. Bertsche, A.A. Bin Anuar, K. Borras[18], V. Botta, A. Campbell, A. Cardini, P. Connor, S. Consuegra Rodríguez, C. Contreras-Campana, V. Danilov, A. De Wit, M.M. Defranchis, C. Diez Pardos, D. Domínguez Damiani, G. Eckerlin, D. Eckstein, T. Eichhorn, A. Elwood, E. Eren, E. Gallo[19], A. Geiser, J.M. Grados Luyando, A. Grohsjean, M. Guthoff, M. Haranko, A. Harb, A. Jafari, N.Z. Jomhari, H. Jung, A. Kasem[18], M. Kasemann, H. Kaveh, J. Keaveney, C. Kleinwort, J. Knolle, D. Krücker, W. Lange, T. Lenz, J. Leonard, J. Lidrych, K. Lipka, W. Lohmann[20], R. Mankel, I.-A. Melzer-Pellmann, A.B. Meyer, M. Meyer, M. Missiroli, G. Mittag, J. Mnich, A. Mussgiller, V. Myronenko, D. Pérez Adán, S.K. Pflitsch, D. Pitzl, A. Raspereza, A. Saibel, M. Savitskyi, V. Scheurer, P. Schütze, C. Schwanenberger, R. Shevchenko, A. Singh, H. Tholen, O. Turkot, A. Vagnerini, M. Van De Klundert, G.P. Van Onsem, R. Walsh, Y. Wen, K. Wichmann, C. Wissing, O. Zenaiev, R. Zlebcik
University of Hamburg, Hamburg, Germany
R. Aggleton, S. Bein, L. Benato, A. Benecke, V. Blobel, T. Dreyer, A. Ebrahimi, A. Fröhlich, C. Garbers, E. Garutti, D. Gonzalez, P. Gunnellini, J. Haller, A. Hinzmann, A. Karavdina, G. Kasieczka, R. Klanner, R. Kogler, N. Kovalchuk, S. Kurz, V. Kutzner, J. Lange, T. Lange, A. Malara, D. Marconi, J. Multhaup, M. Niedziela, C.E.N. Niemeyer, D. Nowatschin, A. Perieanu, A. Reimers, O. Rieger, C. Scharf, P. Schleper, S. Schumann, J. Schwandt, J. Sonneveld, H. Stadie, G. Steinbrück, F.M. Stober, M. Stöver, B. Vormwald, I. Zoi
Karlsruher Institut fuer Technologie, Karlsruhe, Germany
M. Akbiyik, C. Barth, M. Baselga, S. Baur, T. Berger, E. Butz, R. Caspart, T. Chwalek, W. De Boer, A. Dierlamm, K. El Morabit, N. Faltermann, M. Giffels, P. Goldenzweig, A. Gottmann, M.A. Harrendorf, F. Hartmann[17], U. Husemann, S. Kudella, S. Mitra, M.U. Mozer, Th. Müller, M. Musich, A. Nürnberg, G. Quast, K. Rabbertz, M. Schröder, I. Shvetsov, H.J. Simonis, R. Ulrich, M. Weber, C. Wöhrmann, R. Wolf
Institute of Nuclear and Particle Physics (INPP), NCSR Demokritos, Aghia Paraskevi, Greece
G. Anagnostou, P. Asenov, G. Daskalakis, T. Geralis, A. Kyriakis, D. Loukas, G. Paspalaki
National and Kapodistrian University of Athens, Athens, Greece
M. Diamantopoulou, G. Karathanasis, P. Kontaxakis, A. Panagiotou, I. Papavergou, N. Saoulidou, A. Stakia, K. Theofilatos, K. Vellidis
National Technical University of Athens, Athens, Greece
G. Bakas, K. Kousouris, I. Papakrivopoulos, G. Tsipolitis
University of Ioánnina, Ioánnina, Greece
I. Evangelou, C. Foudas, P. Gianneios, P. Katsoulis, P. Kokkas, S. Mallios, K. Manitara, N. Manthos, I. Papadopoulos, J. Strologas, F.A. Triantis, D. Tsitsonis

MTA-ELTE Lendület CMS Particle and Nuclear Physics Group, Eötvös Loránd University, Budapest, Hungary
M. Bartók[21], M. Csanad, P. Major, K. Mandal, A. Mehta, M.I. Nagy, G. Pasztor, O. Surányi, G.I. Veres
Wigner Research Centre for Physics, Budapest, Hungary
G. Bencze, C. Hajdu, D. Horvath[22], F. Sikler, T.Á. Vámi, V. Veszpremi, G. Vesztergombi[†]
Institute of Nuclear Research ATOMKI, Debrecen, Hungary
N. Beni, S. Czellar, J. Karancsi[21], A. Makovec, J. Molnar, Z. Szillasi
Institute of Physics, University of Debrecen, Debrecen, Hungary
P. Raics, D. Teyssier, Z.L. Trocsanyi, B. Ujvari
Eszterhazy Karoly University, Karoly Robert Campus, Gyongyos, Hungary
T. Csorgo, W.J. Metzger, F. Nemes, T. Novak
Indian Institute of Science (IISc), Bangalore, India
S. Choudhury, J.R. Komaragiri, P.C. Tiwari
National Institute of Science Education and Research, HBNI, Bhubaneswar, India
S. Bahinipati[24], C. Kar, G. Kole, P. Mal, V.K. Muraleedharan Nair Bindhu, A. Nayak[25], D.K. Sahoo[24], S.K. Swain
Panjab University, Chandigarh, India
S. Bansal, S.B. Beri, V. Bhatnagar, S. Chauhan, R. Chawla, N. Dhingra, R. Gupta, A. Kaur, M. Kaur, S. Kaur, P. Kumari, M. Lohan, M. Meena, K. Sandeep, S. Sharma, J.B. Singh, A.K. Virdi, G. Walia
University of Delhi, Delhi, India
A. Bhardwaj, B.C. Choudhary, R.B. Garg, M. Gola, S. Keshri, Ashok Kumar, S. Malhotra, M. Naimuddin, P. Priyanka, K. Ranjan, Aashaq Shah, R. Sharma
Saha Institute of Nuclear Physics, HBNI, Kolkata, India
R. Bhardwaj[26], M. Bharti[26], R. Bhattacharya, S. Bhattacharya, U. Bhawandeep[26], D. Bhowmik, S. Dey, S. Dutta, S. Ghosh, M. Maity[27], K. Mondal, S. Nandan, A. Purohit, P.K. Rout, A. Roy, G. Saha, S. Sarkar, T. Sarkar[27], M. Sharan, B. Singh[26], S. Thakur[26]
Indian Institute of Technology Madras, Madras, India
P.K. Behera, P. Kalbhor, A. Muhammad, P.R. Pujahari, A. Sharma, A.K. Sikdar
Bhabha Atomic Research Centre, Mumbai, India
R. Chudasama, D. Dutta, V. Jha, V. Kumar, D.K. Mishra, P.K. Netrakanti, L.M. Pant, P. Shukla
Tata Institute of Fundamental Research-A, Mumbai, India
T. Aziz, M.A. Bhat, S. Dugad, G.B. Mohanty, N. Sur, RavindraKumar Verma
Tata Institute of Fundamental Research-B, Mumbai, India
S. Banerjee, S. Bhattacharya, S. Chatterjee, P. Das, M. Guchait, S. Karmakar, S. Kumar, G. Majumder, K. Mazumdar, N. Sahoo, S. Sawant
Indian Institute of Science Education and Research (IISER), Pune, India
S. Chauhan, S. Dube, V. Hegde, A. Kapoor, K. Kothekar, S. Pandey, A. Rane, A. Rastogi, S. Sharma
Institute for Research in Fundamental Sciences (IPM), Tehran, Iran

S. Chenarani[28], E. Eskandari Tadavani, S.M. Etesami[28], M. Khakzad, M. Mohammadi Najafabadi, M. Naseri, F. Rezaei Hosseinabadi

University College Dublin, Dublin, Ireland

M. Felcini, M. Grunewald

INFN Sezione di Bari [a], Università di Bari [b], Politecnico di Bari [c], Bari, Italy

M. Abbrescia[a,b], C. Calabria[a,b], A. Colaleo[a], D. Creanza[a,c], L. Cristella[a,b], N. De Filippis[a,c], M. De Palma[a,b], A. Di Florio[a,b], L. Fiore[a], A. Gelmi[a,b], G. Iaselli[a,c], M. Ince[a,b], S. Lezki[a,b], G. Maggi[a,c], M. Maggi[a], G. Miniello[a,b], S. My[a,b], S. Nuzzo[a,b], A. Pompili[a,b], G. Pugliese[a,c], R. Radogna[a], A. Ranieri[a], G. Selvaggi[a,b], L. Silvestris[a], R. Venditti[a], P. Verwilligen[a]

INFN Sezione di Bologna [a], Università di Bologna [b], Bologna, Italy

G. Abbiendi[a], C. Battilana[a,b], D. Bonacorsi[a,b], L. Borgonovi[a,b], S. Braibant-Giacomelli[a,b], R. Campanini[a,b], P. Capiluppi[a,b], A. Castro[a,b], F.R. Cavallo[a], C. Ciocca[a], G. Codispoti[a,b], M. Cuffiani[a,b], G.M. Dallavalle[a], F. Fabbri[a], A. Fanfani[a,b], E. Fontanesi, P. Giacomelli[a], C. Grandi[a], L. Guiducci[a,b], F. Iemmi[a,b], S. Lo Meo[a,29], S. Marcellini[a], G. Masetti[a], F.L. Navarria[a,b], A. Perrotta[a], F. Primavera[a,b], A.M. Rossi[a,b], T. Rovelli[a,b], G.P. Siroli[a,b], N. Tosi[a]

INFN Sezione di Catania [a], Università di Catania [b], Catania, Italy

S. Albergo[a,b,30], S. Costa[a,b], A. Di Mattia[a], R. Potenza[a,b], A. Tricomi[a,b,30], C. Tuve[a,b]

INFN Sezione di Firenze [a], Università di Firenze [b], Firenze, Italy

G. Barbagli[a], R. Ceccarelli, K. Chatterjee[a,b], V. Ciulli[a,b], C. Civinini[a], R. D'Alessandro[a,b], E. Focardi[a,b], G. Latino, P. Lenzi[a,b], M. Meschini[a], S. Paoletti[a], G. Sguazzoni[a], D. Strom[a], L. Viliani[a]

INFN Laboratori Nazionali di Frascati, Frascati, Italy

L. Benussi, S. Bianco, D. Piccolo

INFN Sezione di Genova [a], Università di Genova [b], Genova, Italy

M. Bozzo[a,b], F. Ferro[a], R. Mulargia[a,b], E. Robutti[a], S. Tosi[a,b]

INFN Sezione di Milano-Bicocca [a], Università di Milano-Bicocca [b], Milano, Italy

A. Benaglia[a], A. Beschi[a,b], F. Brivio[a,b], V. Ciriolo[a,b,17], S. Di Guida[a,b,17], M.E. Dinardo[a,b], P. Dini[a], S. Fiorendi[a,b], S. Gennai[a], A. Ghezzi[a,b], P. Govoni[a,b], L. Guzzi[a,b], M. Malberti[a], S. Malvezzi[a], D. Menasce[a], F. Monti[a,b], L. Moroni[a], G. Ortona[a,b], M. Paganoni[a,b], D. Pedrini[a], S. Ragazzi[a,b], T. Tabarelli de Fatis[a,b], D. Zuolo[a,b]

INFN Sezione di Napoli [a], Università di Napoli 'Federico II' [b], Napoli, Italy, Università della Basilicata [c], *Potenza, Italy, Università G. Marconi* [d], **Roma, Italy**

S. Buontempo[a], N. Cavallo[a,c], A. De Iorio[a,b], A. Di Crescenzo[a,b], F. Fabozzi[a,c], F. Fienga[a], G. Galati[a], A.O.M. Iorio[a,b], L. Lista[a,b], S. Meola[a,d,17], P. Paolucci[a,17], B. Rossi[a], C. Sciacca[a,b], E. Voevodina[a,b]

INFN Sezione di Padova [a], Università di Padova [b], Padova, Italy, Università di Trento [c], Trento, Italy

P. Azzi[a], N. Bacchetta[a], A. Boletti[a,b], A. Bragagnolo, R. Carlin[a,b], P. Checchia[a], P. De Castro Manzano[a], T. Dorigo[a], U. Dosselli[a], F. Gasparini[a,b], U. Gasparini[a,b], A. Gozzelino[a], S.Y. Hoh, P. Lujan, M. Margoni[a,b], A.T. Meneguzzo[a,b], J. Pazzini[a]

[b], N. Pozzobon[a,b], M. Presilla[b], P. Ronchese[a,b], R. Rossin[a,b], F. Simonetto[a,b], A. Tiko, M. Tosi[a,b], M. Zanetti[a,b], P. Zotto[a,b], G. Zumerle[a,b]

INFN Sezione di Pavia [a], Università di Pavia [b], Pavia, Italy

A. Braghieri[a], P. Montagna[a,b], S.P. Ratti[a,b], V. Re[a], M. Ressegotti[a,b], C. Riccardi[a,b], P. Salvini[a], I. Vai[a,b], P. Vitulo[a,b]

INFN Sezione di Perugia [a], Università di Perugia [b], Perugia, Italy

M. Biasini[a,b], G.M. Bilei[a], C. Cecchi[a,b], D. Ciangottini[a,b], L. Fanò[a,b], P. Lariccia[a,b], R. Leonardi[a,b], E. Manoni[a], G. Mantovani[a,b], V. Mariani[a,b], M. Menichelli[a], A. Rossi[a,b], A. Santocchia[a,b], D. Spiga[a]

INFN Sezione di Pisa [a], Università di Pisa [b], Scuola Normale Superiore di Pisa [c], Pisa, Italy

K. Androsov[a], P. Azzurri[a], G. Bagliesi[a], V. Bertacchi[a,c], L. Bianchini[a], T. Boccali[a], R. Castaldi[a], M.A. Ciocci[a,b], R. Dell'Orso[a], G. Fedi[a], L. Giannini[a,c], A. Giassi[a], M.T. Grippo[a], F. Ligabue[a,c], E. Manca[a,c], G. Mandorli[a,c], A. Messineo[a,b], F. Palla[a], A. Rizzi[a,b], G. Rolandi[31], S. Roy Chowdhury, A. Scribano[a], P. Spagnolo[a], R. Tenchini[a], G. Tonelli[a,b], N. Turini, A. Venturi[a], P.G. Verdini[a]

INFN Sezione di Roma [a], Sapienza Università di Roma [b], Rome, Italy

F. Cavallari[a], M. Cipriani[a,b], D. Del Re[a,b], E. Di Marco[a,b], M. Diemoz[a], E. Longo[a,b], B. Marzocchi[a,b], P. Meridiani[a], G. Organtini[a,b], F. Pandolfi[a], R. Paramatti[a,b], C. Quaranta[a,b], S. Rahatlou[a,b], C. Rovelli[a], F. Santanastasio[a,b], L. Soffi[a,b]

INFN Sezione di Torino [a], Università di Torino [b], Torino, Italy, Università del Piemonte Orientale [c], Novara, Italy

N. Amapane[a,b], R. Arcidiacono[a,c], S. Argiro[a,b], M. Arneodo[a,c], N. Bartosik[a], R. Bellan[a,b], C. Biino[a], A. Cappati[a,b], N. Cartiglia[a], S. Cometti[a], M. Costa[a,b], R. Covarelli[a,b], N. Demaria[a], B. Kiani[a,b], C. Mariotti[a], S. Maselli[a], E. Migliore[a,b], V. Monaco[a,b], E. Monteil[a,b], M. Monteno[a], M.M. Obertino[a,b], L. Pacher[a,b], N. Pastrone[a], M. Pelliccioni[a], G.L. Pinna Angioni[a,b], A. Romero[a,b], M. Ruspa[a,c], R. Sacchi[a,b], R. Salvatico[a,b], V. Sola[a], A. Solano[a,b], D. Soldi[a,b], A. Staiano[a]

INFN Sezione di Trieste [a], Università di Trieste [b], Trieste, Italy

S. Belforte[a], V. Candelise[a,b], M. Casarsa[a], F. Cossutti[a], A. Da Rold[a,b], G. Della Ricca[a,b], F. Vazzoler[a,b], A. Zanetti[a]

Kyungpook National University, Daegu, Korea

B. Kim, D.H. Kim, G.N. Kim, M.S. Kim, J. Lee, S.W. Lee, C.S. Moon, Y.D. Oh, S.I. Pak, S. Sekmen, D.C. Son, Y.C. Yang

Chonnam National University, Institute for Universe and Elementary Particles, Kwangju, Korea

H. Kim, D.H. Moon, G. Oh

Hanyang University, Seoul, Korea

B. Francois, T.J. Kim, J. Park

Korea University, Seoul, Korea

S. Cho, S. Choi, Y. Go, D. Gyun, S. Ha, B. Hong, K. Lee, K.S. Lee, J. Lim, J. Park, S.K. Park, Y. Roh

Kyung Hee University, Department of Physics

J. Goh

Sejong University, Seoul, Korea
H.S. Kim
Seoul National University, Seoul, Korea
J. Almond, J.H. Bhyun, J. Choi, S. Jeon, J. Kim, J.S. Kim, H. Lee, K. Lee, S. Lee, K. Nam, M. Oh, S.B. Oh, B.C. Radburn-Smith, U.K. Yang, H.D. Yoo, I. Yoon, G.B. Yu
University of Seoul, Seoul, Korea
D. Jeon, H. Kim, J.H. Kim, J.S.H. Lee, I.C. Park, I. Watson
Sungkyunkwan University, Suwon, Korea
Y. Choi, C. Hwang, Y. Jeong, J. Lee, Y. Lee, I. Yu
Riga Technical University, Riga, Latvia
V. Veckalns[32]
Vilnius University, Vilnius, Lithuania
V. Dudenas, A. Juodagalvis, J. Vaitkus
National Centre for Particle Physics, Universiti Malaya, Kuala Lumpur, Malaysia
Z.A. Ibrahim, F. Mohamad Idris[33], W.A.T. Wan Abdullah, M.N. Yusli, Z. Zolkapli
Universidad de Sonora (UNISON), Hermosillo, Mexico
J.F. Benitez, A. Castaneda Hernandez, J.A. Murillo Quijada, L. Valencia Palomo
Centro de Investigacion y de Estudios Avanzados del IPN, Mexico City, Mexico
H. Castilla-Valdez, E. De La Cruz-Burelo, I. Heredia-De La Cruz[34], R. Lopez-Fernandez, A. Sanchez-Hernandez
Universidad Iberoamericana, Mexico City, Mexico
S. Carrillo Moreno, C. Oropeza Barrera, M. Ramirez-Garcia, F. Vazquez Valencia
Benemerita Universidad Autonoma de Puebla, Puebla, Mexico
J. Eysermans, I. Pedraza, H.A. Salazar Ibarguen, C. Uribe Estrada
Universidad Autónoma de San Luis Potosí, San Luis Potosí, Mexico
A. Morelos Pineda
University of Montenegro, Podgorica, Montenegro
N. Raicevic
University of Auckland, Auckland, New Zealand
D. Krofcheck
University of Canterbury, Christchurch, New Zealand
S. Bheesette, P.H. Butler
National Centre for Physics, Quaid-I-Azam University, Islamabad, Pakistan
A. Ahmad, M. Ahmad, Q. Hassan, H.R. Hoorani, W.A. Khan, M.A. Shah, M. Shoaib, M. Waqas
AGH University of Science and Technology Faculty of Computer Science, Electronics and Telecommunications, Krakow, Poland
V. Avati, L. Grzanka, M. Malawski
National Centre for Nuclear Research, Swierk, Poland
H. Bialkowska, M. Bluj, B. Boimska, M. Górski, M. Kazana, M. Szleper, P. Zalewski
Institute of Experimental Physics, Faculty of Physics, University of Warsaw, Warsaw, Poland

K. Bunkowski, A. Byszuk[35], K. Doroba, A. Kalinowski, M. Konecki, J. Krolikowski, M. Misiura, M. Olszewski, A. Pyskir, M. Walczak
Laboratório de Instrumentação e Física Experimental de Partículas, Lisboa, Portugal
M. Araujo, P. Bargassa, D. Bastos, A. Di Francesco, P. Faccioli, B. Galinhas, M. Gallinaro, J. Hollar, N. Leonardo, J. Seixas, K. Shchelina, G. Strong, O. Toldaiev, J. Varela
Joint Institute for Nuclear Research, Dubna, Russia
S. Afanasiev, P. Bunin, M. Gavrilenko, I. Golutvin, I. Gorbunov, A. Kamenev, V. Karjavine, A. Lanev, A. Malakhov, V. Matveev[36, 37], P. Moisenz, V. Palichik, V. Perelygin, M. Savina, S. Shmatov, S. Shulha, N. Skatchkov, V. Smirnov, N. Voytishin, A. Zarubin
Petersburg Nuclear Physics Institute, Gatchina (St. Petersburg), Russia
L. Chtchipounov, V. Golovtsov, Y. Ivanov, V. Kim[38], E. Kuznetsova[39], P. Levchenko, V. Murzin, V. Oreshkin, I. Smirnov, D. Sosnov, V. Sulimov, L. Uvarov, A. Vorobyev
Institute for Nuclear Research, Moscow, Russia
Yu. Andreev, A. Dermenev, S. Gninenko, N. Golubev, A. Karneyeu, M. Kirsanov, N. Krasnikov, A. Pashenkov, D. Tlisov, A. Toropin
Institute for Theoretical and Experimental Physics named by A.I. Alikhanov of NRC 'Kurchatov Institute', Moscow, Russia
V. Epshteyn, V. Gavrilov, N. Lychkovskaya, A. Nikitenko[40], V. Popov, I. Pozdnyakov, G. Safronov, A. Spiridonov, A. Stepennov, M. Toms, E. Vlasov, A. Zhokin
Moscow Institute of Physics and Technology, Moscow, Russia
T. Aushev
National Research Nuclear University 'Moscow Engineering Physics Institute' (MEPhI), Moscow, Russia
O. Bychkova, R. Chistov[41], M. Danilov[41], S. Polikarpov[41], E. Tarkovskii
P.N. Lebedev Physical Institute, Moscow, Russia
V. Andreev, M. Azarkin, I. Dremin, M. Kirakosyan, A. Terkulov
Skobeltsyn Institute of Nuclear Physics, Lomonosov Moscow State University, Moscow, Russia
A. Baskakov, A. Belyaev, E. Boos, V. Bunichev, M. Dubinin[42], L. Dudko, V. Klyukhin, N. Korneeva, I. Lokhtin, S. Obraztsov, M. Perfilov, V. Savrin, P. Volkov
Novosibirsk State University (NSU), Novosibirsk, Russia
A. Barnyakov[43], V. Blinov[43], T. Dimova[43], L. Kardapoltsev[43], Y. Skovpen[43]
Institute for High Energy Physics of National Research Centre 'Kurchatov Institute', Protvino, Russia
I. Azhgirey, I. Bayshev, S. Bitioukov, V. Kachanov, D. Konstantinov, P. Mandrik, V. Petrov, R. Ryutin, S. Slabospitskii, A. Sobol, S. Troshin, N. Tyurin, A. Uzunian, A. Volkov
National Research Tomsk Polytechnic University, Tomsk, Russia
A. Babaev, A. Iuzhakov, V. Okhotnikov
Tomsk State University, Tomsk, Russia
V. Borchsh, V. Ivanchenko, E. Tcherniaev

University of Belgrade: Faculty of Physics and VINCA Institute of Nuclear Sciences
P. Adzic[44], P. Cirkovic, D. Devetak, M. Dordevic, P. Milenovic, J. Milosevic, M. Stojanovic
Centro de Investigaciones Energéticas Medioambientales y Tecnológicas (CIEMAT), Madrid, Spain
M. Aguilar-Benitez, J. Alcaraz Maestre, A. Ãlvarez Fernández, I. Bachiller, M. Barrio Luna, J.A. Brochero Cifuentes, C.A. Carrillo Montoya, M. Cepeda, M. Cerrada, N. Colino, B. De La Cruz, A. Delgado Peris, C. Fernandez Bedoya, J.P. Fernández Ramos, J. Flix, M.C. Fouz, O. Gonzalez Lopez, S. Goy Lopez, J.M. Hernandez, M.I. Josa, D. Moran, Ã. Navarro Tobar, A. Pérez-Calero Yzquierdo, J. Puerta Pelayo, I. Redondo, L. Romero, S. Sánchez Navas, M.S. Soares, A. Triossi, C. Willmott
Universidad Autónoma de Madrid, Madrid, Spain
C. Albajar, J.F. de Trocóniz
Universidad de Oviedo, Instituto Universitario de Ciencias y Tecnologías Espaciales de Asturias (ICTEA), Oviedo, Spain
B. Alvarez Gonzalez, J. Cuevas, C. Erice, J. Fernandez Menendez, S. Folgueras, I. Gonzalez Caballero, J.R. González Fernández, E. Palencia Cortezon, V. Rodríguez Bouza, S. Sanchez Cruz
Instituto de Física de Cantabria (IFCA), CSIC-Universidad de Cantabria, Santander, Spain
I.J. Cabrillo, A. Calderon, B. Chazin Quero, J. Duarte Campderros, M. Fernandez, P.J. Fernández Manteca, A. García Alonso, G. Gomez, C. Martinez Rivero, P. Martinez Ruiz del Arbol, F. Matorras, J. Piedra Gomez, C. Prieels, T. Rodrigo, A. Ruiz-Jimeno, L. Russo[45], L. Scodellaro, N. Trevisani, I. Vila, J.M. Vizan Garcia
University of Colombo, Colombo, Sri Lanka
K. Malagalage
University of Ruhuna, Department of Physics, Matara, Sri Lanka
W.G.D. Dharmaratna, N. Wickramage
CERN, European Organization for Nuclear Research, Geneva, Switzerland
D. Abbaneo, B. Akgun, E. Auffray, G. Auzinger, J. Baechler, P. Baillon, A.H. Ball, D. Barney, J. Bendavid, M. Bianco, A. Bocci, E. Bossini, C. Botta, E. Brondolin, T. Camporesi, A. Caratelli, G. Cerminara, E. Chapon, G. Cucciati, D. d'Enterria, A. Dabrowski, N. Daci, V. Daponte, A. David, O. Davignon, A. De Roeck, N. Deelen, M. Deile, M. Dobson, M. Dünser, N. Dupont, A. Elliott-Peisert, F. Fallavollita[46], D. Fasanella, G. Franzoni, J. Fulcher, W. Funk, S. Giani, D. Gigi, A. Gilbert, K. Gill, F. Glege, M. Gruchala, M. Guilbaud, D. Gulhan, J. Hegeman, C. Heidegger, Y. Iiyama, V. Innocente, P. Janot, O. Karacheban[20], J. Kaspar, J. Kieseler, M. Krammer[1], C. Lange, P. Lecoq, C. Lourenço, L. Malgeri, M. Mannelli, A. Massironi, F. Meijers, J.A. Merlin, S. Mersi, E. Meschi, F. Moortgat, M. Mulders, J. Ngadiuba, S. Nourbakhsh, S. Orfanelli, L. Orsini, F. Pantaleo[17], L. Pape, E. Perez, M. Peruzzi, A. Petrilli, G. Petrucciani, A. Pfeiffer, M. Pierini, F.M. Pitters, D. Rabady, A. Racz, M. Rovere, H. Sakulin, C. Schäfer, C. Schwick, M. Selvaggi, A. Sharma, P. Silva, W. Snoeys, P. Sphicas[47], J. Steggemann, V.R. Tavolaro, D. Treille, A. Tsirou, A. Vartak, M. Verzetti, W.D. Zeuner

Paul Scherrer Institut, Villigen, Switzerland
L. Caminada[48], K. Deiters, W. Erdmann, R. Horisberger, Q. Ingram, H.C. Kaestli, D. Kotlinski, U. Langenegger, T. Rohe, S.A. Wiederkehr
ETH Zurich - Institute for Particle Physics and Astrophysics (IPA), Zurich, Switzerland
M. Backhaus, P. Berger, N. Chernyavskaya, G. Dissertori, M. Dittmar, M. Donegà, C. Dorfer, T.A. Gómez Espinosa, C. Grab, D. Hits, T. Klijnsma, W. Lustermann, R.A. Manzoni, M. Marionneau, M.T. Meinhard, F. Micheli, P. Musella, F. Nessi-Tedaldi, F. Pauss, G. Perrin, L. Perrozzi, S. Pigazzini, M. Reichmann, C. Reissel, T. Reitenspiess, D. Ruini, D.A. Sanz Becerra, M. Schönenberger, L. Shchutska, M.L. Vesterbacka Olsson, R. Wallny, D.H. Zhu
Universität Zürich, Zurich, Switzerland
T.K. Aarrestad, C. Amsler[49], D. Brzhechko, M.F. Canelli, A. De Cosa, R. Del Burgo, S. Donato, B. Kilminster, S. Leontsinis, V.M. Mikuni, I. Neutelings, G. Rauco, P. Robmann, D. Salerno, K. Schweiger, C. Seitz, Y. Takahashi, S. Wertz, A. Zucchetta
National Central University, Chung-Li, Taiwan
T.H. Doan, C.M. Kuo, W. Lin, S.S. Yu
National Taiwan University (NTU), Taipei, Taiwan
P. Chang, Y. Chao, K.F. Chen, P.H. Chen, W.-S. Hou, Y.y. Li, R.-S. Lu, E. Paganis, A. Psallidas, A. Steen
Chulalongkorn University, Faculty of Science, Department of Physics, Bangkok, Thailand
B. Asavapibhop, C. Asawatangtrakuldee, N. Srimanobhas, N. Suwonjandee, V. Wachirapusitanand
ukurova University, Physics Department, Science and Art Faculty, Adana, Turkey
A. Bat, F. Boran, S. Damarseckin[50], Z.S. Demiroglu, F. Dolek, C. Dozen, I. Dumanoglu, E. Eskut, G. Gokbulut, EmineGurpinar Guler[51], Y. Guler, I. Hos[52], C. Isik, E.E. Kangal[53], O. Kara, A. Kayis Topaksu, U. Kiminsu, M. Oglakci, G. Onengut, K. Ozdemir[54], A.E. Simsek, D. Sunar Cerci[55], B. Tali[55], U.G. Tok, S. Turkcapar, I.S. Zorbakir, C. Zorbilmez
Middle East Technical University, Physics Department, Ankara, Turkey
B. Isildak[56], G. Karapinar[57], M. Yalvac
Bogazici University, Istanbul, Turkey
I.O. Atakisi, E. Gülmez, O. Kaya[58], B. Kaynak, Ö. Özçelik, S. Ozkorucuklu[59], S. Tekten, E.A. Yetkin[60]
Istanbul Technical University, Istanbul, Turkey
A. Cakir, K. Cankocak, Y. Komurcu, S. Sen[61]
Institute for Scintillation Materials of National Academy of Science of Ukraine, Kharkov, Ukraine
B. Grynyov
National Scientific Center, Kharkov Institute of Physics and Technology, Kharkov, Ukraine
L. Levchuk
University of Bristol, Bristol, United Kingdom

F. Ball, E. Bhal, S. Bologna, J.J. Brooke, D. Burns, E. Clement, D. Cussans, H. Flacher, J. Goldstein, G.P. Heath, H.F. Heath, L. Kreczko, S. Paramesvaran, B. Penning, T. Sakuma, S. Seif El Nasr-Storey, D. Smith, V.J. Smith, J. Taylor, A. Titterton

Rutherford Appleton Laboratory, Didcot, United Kingdom

K.W. Bell, A. Belyaev[62], C. Brew, R.M. Brown, D. Cieri, D.J.A. Cockerill, J.A. Coughlan, K. Harder, S. Harper, J. Linacre, K. Manolopoulos, D.M. Newbold, E. Olaiya, D. Petyt, T. Reis, T. Schuh, C.H. Shepherd-Themistocleous, A. Thea, I.R. Tomalin, T. Williams, W.J. Womersley

Imperial College, London, United Kingdom

R. Bainbridge, P. Bloch, J. Borg, S. Breeze, O. Buchmuller, A. Bundock, GurpreetSingh CHAHAL[63], D. Colling, P. Dauncey, G. Davies, M. Della Negra, R. Di Maria, P. Everaerts, G. Hall, G. Iles, T. James, M. Komm, C. Laner, L. Lyons, A.-M. Magnan, S. Malik, A. Martelli, V. Milosevic, J. Nash[64], V. Palladino, M. Pesaresi, D.M. Raymond, A. Richards, A. Rose, E. Scott, C. Seez, A. Shtipliyski, M. Stoye, T. Strebler, S. Summers, A. Tapper, K. Uchida, T. Virdee[17], N. Wardle, D. Winterbottom, J. Wright, A.G. Zecchinelli, S.C. Zenz

Brunel University, Uxbridge, United Kingdom

J.E. Cole, P.R. Hobson, A. Khan, P. Kyberd, C.K. Mackay, A. Morton, I.D. Reid, L. Teodorescu, S. Zahid

Baylor University, Waco, USA

K. Call, J. Dittmann, K. Hatakeyama, C. Madrid, B. McMaster, N. Pastika, C. Smith

Catholic University of America, Washington, DC, USA

R. Bartek, A. Dominguez, R. Uniyal

The University of Alabama, Tuscaloosa, USA

A. Buccilli, S.I. Cooper, C. Henderson, P. Rumerio, C. West

Boston University, Boston, USA

D. Arcaro, T. Bose, Z. Demiragli, D. Gastler, S. Girgis, D. Pinna, C. Richardson, J. Rohlf, D. Sperka, I. Suarez, L. Sulak, D. Zou

Brown University, Providence, USA

G. Benelli, B. Burkle, X. Coubez, D. Cutts, Y.t. Duh, M. Hadley, J. Hakala, U. Heintz, J.M. Hogan[65], K.H.M. Kwok, E. Laird, G. Landsberg, J. Lee, Z. Mao, M. Narain, S. Sagir[66], R. Syarif, E. Usai, D. Yu

University of California, Davis, Davis, USA

R. Band, C. Brainerd, R. Breedon, M. Calderon De La Barca Sanchez, M. Chertok, J. Conway, R. Conway, P.T. Cox, R. Erbacher, C. Flores, G. Funk, F. Jensen, W. Ko, O. Kukral, R. Lander, M. Mulhearn, D. Pellett, J. Pilot, M. Shi, D. Stolp, D. Taylor, K. Tos, M. Tripathi, Z. Wang, F. Zhang

University of California, Los Angeles, USA

M. Bachtis, C. Bravo, R. Cousins, A. Dasgupta, A. Florent, J. Hauser, M. Ignatenko, N. Mccoll, W.A. Nash, S. Regnard, D. Saltzberg, C. Schnaible, B. Stone, V. Valuev

University of California, Riverside, Riverside, USA

K. Burt, R. Clare, J.W. Gary, S.M.A. Ghiasi Shirazi, G. Hanson, G. Karapostoli, E. Kennedy, O.R. Long, M. Olmedo Negrete, M.I. Paneva, W. Si, L. Wang, H. Wei, S. Wimpenny, B.R. Yates, Y. Zhang

University of California, San Diego, La Jolla, USA
J.G. Branson, P. Chang, S. Cittolin, M. Derdzinski, R. Gerosa, D. Gilbert, B. Hashemi, D. Klein, V. Krutelyov, J. Letts, M. Masciovecchio, S. May, S. Padhi, M. Pieri, V. Sharma, M. Tadel, F. Würthwein, A. Yagil, G. Zevi Della Porta
University of California, Santa Barbara - Department of Physics, Santa Barbara, USA
N. Amin, R. Bhandari, C. Campagnari, M. Citron, V. Dutta, M. Franco Sevilla, L. Gouskos, J. Incandela, B. Marsh, H. Mei, A. Ovcharova, H. Qu, J. Richman, U. Sarica, D. Stuart, S. Wang, J. Yoo
California Institute of Technology, Pasadena, USA
D. Anderson, A. Bornheim, O. Cerri, I. Dutta, J.M. Lawhorn, N. Lu, J. Mao, H.B. Newman, T.Q. Nguyen, J. Pata, M. Spiropulu, J.R. Vlimant, C. Wang, S. Xie, Z. Zhang, R.Y. Zhu
Carnegie Mellon University, Pittsburgh, USA
M.B. Andrews, T. Ferguson, T. Mudholkar, M. Paulini, M. Sun, I. Vorobiev, M. Weinberg
University of Colorado Boulder, Boulder, USA
J.P. Cumalat, W.T. Ford, A. Johnson, E. MacDonald, T. Mulholland, R. Patel, A. Perloff, K. Stenson, K.A. Ulmer, S.R. Wagner
Cornell University, Ithaca, USA
J. Alexander, J. Chaves, Y. Cheng, J. Chu, A. Datta, A. Frankenthal, K. Mcdermott, N. Mirman, J.R. Patterson, D. Quach, A. Rinkevicius[67], A. Ryd, S.M. Tan, Z. Tao, J. Thom, P. Wittich, M. Zientek
Fermi National Accelerator Laboratory, Batavia, USA
S. Abdullin, M. Albrow, M. Alyari, G. Apollinari, A. Apresyan, A. Apyan, S. Banerjee, L.A.T. Bauerdick, A. Beretvas, J. Berryhill, P.C. Bhat, K. Burkett, J.N. Butler, A. Canepa, G.B. Cerati, H.W.K. Cheung, F. Chlebana, M. Cremonesi, J. Duarte, V.D. Elvira, J. Freeman, Z. Gecse, E. Gottschalk, L. Gray, D. Green, S. Grünendahl, O. Gutsche, AllisonReinsvold Hall, J. Hanlon, R.M. Harris, S. Hasegawa, R. Heller, J. Hirschauer, B. Jayatilaka, S. Jindariani, M. Johnson, U. Joshi, B. Klima, M.J. Kortelainen, B. Kreis, S. Lammel, J. Lewis, D. Lincoln, R. Lipton, M. Liu, T. Liu, J. Lykken, K. Maeshima, J.M. Marraffino, D. Mason, P. McBride, P. Merkel, S. Mrenna, S. Nahn, V. O'Dell, V. Papadimitriou, K. Pedro, C. Pena, G. Rakness, F. Ravera, L. Ristori, B. Schneider, E. Sexton-Kennedy, N. Smith, A. Soha, W.J. Spalding, L. Spiegel, S. Stoynev, J. Strait, N. Strobbe, L. Taylor, S. Tkaczyk, N.V. Tran, L. Uplegger, E.W. Vaandering, C. Vernieri, M. Verzocchi, R. Vidal, M. Wang, H.A. Weber
University of Florida, Gainesville, USA
D. Acosta, P. Avery, P. Bortignon, D. Bourilkov, A. Brinkerhoff, L. Cadamuro, A. Carnes, V. Cherepanov, D. Curry, F. Errico, R.D. Field, S.V. Gleyzer, B.M. Joshi, M. Kim, J. Konigsberg, A. Korytov, K.H. Lo, P. Ma, K. Matchev, N. Menendez, G. Mitselmakher, D. Rosenzweig, K. Shi, J. Wang, S. Wang, X. Zuo
Florida International University, Miami, USA
Y.R. Joshi
Florida State University, Tallahassee, USA

T. Adams, A. Askew, S. Hagopian, V. Hagopian, K.F. Johnson, R. Khurana, T. Kolberg, G. Martinez, T. Perry, H. Prosper, C. Schiber, R. Yohay, J. Zhang
Florida Institute of Technology, Melbourne, USA
M.M. Baarmand, V. Bhopatkar, M. Hohlmann, D. Noonan, M. Rahmani, M. Saunders, F. Yumiceva
University of Illinois at Chicago (UIC), Chicago, USA
M.R. Adams, L. Apanasevich, D. Berry, R.R. Betts, R. Cavanaugh, X. Chen, S. Dittmer, O. Evdokimov, C.E. Gerber, D.A. Hangal, D.J. Hofman, K. Jung, C. Mills, T. Roy, M.B. Tonjes, N. Varelas, H. Wang, X. Wang, Z. Wu
The University of Iowa, Iowa City, USA
M. Alhusseini, B. Bilki[51], W. Clarida, K. Dilsiz[68], S. Durgut, R.P. Gandrajula, M. Haytmyradov, V. Khristenko, O.K. Köseyan, J.-P. Merlo, A. Mestvirishvili[69], A. Moeller, J. Nachtman, H. Ogul[70], Y. Onel, F. Ozok[71], A. Penzo, C. Snyder, E. Tiras, J. Wetzel
Johns Hopkins University, Baltimore, USA
B. Blumenfeld, A. Cocoros, N. Eminizer, D. Fehling, L. Feng, A.V. Gritsan, W.T. Hung, P. Maksimovic, J. Roskes, M. Swartz, M. Xiao
The University of Kansas, Lawrence, USA
C. Baldenegro Barrera, P. Baringer, A. Bean, S. Boren, J. Bowen, A. Bylinkin, T. Isidori, S. Khalil, J. King, G. Krintiras, A. Kropivnitskaya, C. Lindsey, D. Majumder, W. Mcbrayer, N. Minafra, M. Murray, C. Rogan, C. Royon, S. Sanders, E. Schmitz, J.D. Tapia Takaki, Q. Wang, J. Williams, G. Wilson
Kansas State University, Manhattan, USA
S. Duric, A. Ivanov, K. Kaadze, D. Kim, Y. Maravin, D.R. Mendis, T. Mitchell, A. Modak, A. Mohammadi
Lawrence Livermore National Laboratory, Livermore, USA
F. Rebassoo, D. Wright
University of Maryland, College Park, USA
A. Baden, O. Baron, A. Belloni, S.C. Eno, Y. Feng, N.J. Hadley, S. Jabeen, G.Y. Jeng, R.G. Kellogg, J. Kunkle, A.C. Mignerey, S. Nabili, F. Ricci-Tam, M. Seidel, Y.H. Shin, A. Skuja, S.C. Tonwar, K. Wong
Massachusetts Institute of Technology, Cambridge, USA
D. Abercrombie, B. Allen, A. Baty, R. Bi, S. Brandt, W. Busza, I.A. Cali, M. D'Alfonso, G. Gomez Ceballos, M. Goncharov, P. Harris, D. Hsu, M. Hu, M. Klute, D. Kovalskyi, Y.-J. Lee, P.D. Luckey, B. Maier, A.C. Marini, C. Mcginn, C. Mironov, S. Narayanan, X. Niu, C. Paus, D. Rankin, C. Roland, G. Roland, Z. Shi, G.S.F. Stephans, K. Sumorok, K. Tatar, D. Velicanu, J. Wang, T.W. Wang, B. Wyslouch
University of Minnesota, Minneapolis, USA
A.C. Benvenuti[†], R.M. Chatterjee, A. Evans, S. Guts, P. Hansen, J. Hiltbrand, Sh. Jain, S. Kalafut, Y. Kubota, Z. Lesko, J. Mans, R. Rusack, M.A. Wadud
University of Mississippi, Oxford, USA
J.G. Acosta, S. Oliveros
University of Nebraska-Lincoln, Lincoln, USA

K. Bloom, D.R. Claes, C. Fangmeier, L. Finco, F. Golf, R. Gonzalez Suarez, R. Kamalieddin, I. Kravchenko, J.E. Siado, G.R. Snow, B. Stieger

State University of New York at Buffalo, Buffalo, USA

G. Agarwal, C. Harrington, I. Iashvili, A. Kharchilava, C. Mclean, D. Nguyen, A. Parker, J. Pekkanen, S. Rappoccio, B. Roozbahani

Northeastern University, Boston, USA

G. Alverson, E. Barberis, C. Freer, Y. Haddad, A. Hortiangtham, G. Madigan, D.M. Morse, T. Orimoto, L. Skinnari, A. Tishelman-Charny, T. Wamorkar, B. Wang, A. Wisecarver, D. Wood

Northwestern University, Evanston, USA

S. Bhattacharya, J. Bueghly, T. Gunter, K.A. Hahn, N. Odell, M.H. Schmitt, K. Sung, M. Trovato, M. Velasco

University of Notre Dame, Notre Dame, USA

R. Bucci, N. Dev, R. Goldouzian, M. Hildreth, K. Hurtado Anampa, C. Jessop, D.J. Karmgard, K. Lannon, W. Li, N. Loukas, N. Marinelli, I. Mcalister, F. Meng, C. Mueller, Y. Musienko[36], M. Planer, R. Ruchti, P. Siddireddy, G. Smith, S. Taroni, M. Wayne, A. Wightman, M. Wolf, A. Woodard

The Ohio State University, Columbus, USA

J. Alimena, B. Bylsma, L.S. Durkin, S. Flowers, B. Francis, C. Hill, W. Ji, A. Lefeld, T.Y. Ling, B.L. Winer

Princeton University, Princeton, USA

S. Cooperstein, G. Dezoort, P. Elmer, J. Hardenbrook, N. Haubrich, S. Higginbotham, A. Kalogeropoulos, S. Kwan, D. Lange, M.T. Lucchini, J. Luo, D. Marlow, K. Mei, I. Ojalvo, J. Olsen, C. Palmer, P. Piroué, J. Salfeld-Nebgen, D. Stickland, C. Tully, Z. Wang

University of Puerto Rico, Mayaguez, USA

S. Malik, S. Norberg

Purdue University, West Lafayette, USA

A. Barker, V.E. Barnes, S. Das, L. Gutay, M. Jones, A.W. Jung, A. Khatiwada, B. Mahakud, D.H. Miller, G. Negro, N. Neumeister, C.C. Peng, S. Piperov, H. Qiu, J.F. Schulte, J. Sun, F. Wang, R. Xiao, W. Xie

Purdue University Northwest, Hammond, USA

T. Cheng, J. Dolen, N. Parashar

Rice University, Houston, USA

K.M. Ecklund, S. Freed, F.J.M. Geurts, M. Kilpatrick, Arun Kumar, W. Li, B.P. Padley, R. Redjimi, J. Roberts, J. Rorie, W. Shi, A.G. Stahl Leiton, Z. Tu, A. Zhang

University of Rochester, Rochester, USA

A. Bodek, P. de Barbaro, R. Demina, J.L. Dulemba, C. Fallon, T. Ferbel, M. Galanti, A. Garcia-Bellido, J. Han, O. Hindrichs, A. Khukhunaishvili, E. Ranken, P. Tan, R. Taus

Rutgers, The State University of New Jersey, Piscataway, USA

B. Chiarito, J.P. Chou, A. Gandrakota, Y. Gershtein, E. Halkiadakis, A. Hart, M. Heindl, E. Hughes, S. Kaplan, S. Kyriacou, I. Laflotte, A. Lath, R. Montalvo,

K. Nash, M. Osherson, H. Saka, S. Salur, S. Schnetzer, D. Sheffield, S. Somalwar, R. Stone, S. Thomas, P. Thomassen

University of Tennessee, Knoxville, USA

H. Acharya, A.G. Delannoy, J. Heideman, G. Riley, S. Spanier

Texas A&M University, College Station, USA

O. Bouhali[72], A. Celik, M. Dalchenko, M. De Mattia, A. Delgado, S. Dildick, R. Eusebi, J. Gilmore, T. Huang, T. Kamon[73], S. Luo, D. Marley, R. Mueller, D. Overton, L. Perniè, D. Rathjens, A. Safonov

Texas Tech University, Lubbock, USA

N. Akchurin, J. Damgov, F. De Guio, S. Kunori, K. Lamichhane, S.W. Lee, T. Mengke, S. Muthumuni, T. Peltola, S. Undleeb, I. Volobouev, Z. Wang, A. Whitbeck

Vanderbilt University, Nashville, USA

S. Greene, A. Gurrola, R. Janjam, W. Johns, C. Maguire, A. Melo, H. Ni, K. Padeken, F. Romeo, P. Sheldon, S. Tuo, J. Velkovska, M. Verweij

University of Virginia, Charlottesville, USA

M.W. Arenton, P. Barria, B. Cox, G. Cummings, R. Hirosky, M. Joyce, A. Ledovskoy, C. Neu, B. Tannenwald, Y. Wang, E. Wolfe, F. Xia

Wayne State University, Detroit, USA

R. Harr, P.E. Karchin, N. Poudyal, J. Sturdy, P. Thapa, S. Zaleski

University of Wisconsin - Madison, Madison, WI, USA

J. Buchanan, C. Caillol, D. Carlsmith, S. Dasu, I. De Bruyn, L. Dodd, F. Fiori, C. Galloni, B. Gomber[74], M. Herndon, A. Hervé, U. Hussain, P. Klabbers, A. Lanaro, A. Loeliger, K. Long, R. Loveless, J. Madhusudanan Sreekala, T. Ruggles, A. Savin, V. Sharma, W.H. Smith, D. Teague, S. Trembath-reichert, N. Woods

†: Deceased
1: Also at Vienna University of Technology, Vienna, Austria
2: Also at IRFU, CEA, Université Paris-Saclay, Gif-sur-Yvette, France
3: Also at Universidade Estadual de Campinas, Campinas, Brazil
4: Also at Federal University of Rio Grande do Sul, Porto Alegre, Brazil
5: Also at UFMS, Nova Andradina, Brazil
6: Also at Universidade Federal de Pelotas, Pelotas, Brazil
7: Also at Université Libre de Bruxelles, Bruxelles, Belgium
8: Also at University of Chinese Academy of Sciences, Beijing, China
9: Also at Institute for Theoretical and Experimental Physics named by A.I. Alikhanov of NRC 'Kurchatov Institute', Moscow, Russia
10: Also at Joint Institute for Nuclear Research, Dubna, Russia
11: Also at Cairo University, Cairo, Egypt
12: Also at Helwan University, Cairo, Egypt
13: Now at Zewail City of Science and Technology, Zewail, Egypt
14: Also at Purdue University, West Lafayette, USA
15: Also at Université de Haute Alsace, Mulhouse, France
16: Also at Erzincan Binali Yildirim University, Erzincan, Turkey

17: Also at CERN, European Organization for Nuclear Research, Geneva, Switzerland
18: Also at RWTH Aachen University, III. Physikalisches Institut A, Aachen, Germany
19: Also at University of Hamburg, Hamburg, Germany
20: Also at Brandenburg University of Technology, Cottbus, Germany
21: Also at Institute of Physics, University of Debrecen, Debrecen, Hungary, Debrecen, Hungary
22: Also at Institute of Nuclear Research ATOMKI, Debrecen, Hungary
23: Also at MTA-ELTE Lendület CMS Particle and Nuclear Physics Group, Eötvös Loránd University, Budapest, Hungary, Budapest, Hungary
24: Also at IIT Bhubaneswar, Bhubaneswar, India, Bhubaneswar, India
25: Also at Institute of Physics, Bhubaneswar, India
26: Also at Shoolini University, Solan, India
27: Also at University of Visva-Bharati, Santiniketan, India
28: Also at Isfahan University of Technology, Isfahan, Iran
29: Also at Italian National Agency for New Technologies, Energy and Sustainable Economic Development, Bologna, Italy
30: Also at Centro Siciliano di Fisica Nucleare e di Struttura Della Materia, Catania, Italy
31: Also at Scuola Normale e Sezione dell'INFN, Pisa, Italy
32: Also at Riga Technical University, Riga, Latvia, Riga, Latvia
33: Also at Malaysian Nuclear Agency, MOSTI, Kajang, Malaysia
34: Also at Consejo Nacional de Ciencia y Tecnología, Mexico City, Mexico
35: Also at Warsaw University of Technology, Institute of Electronic Systems, Warsaw, Poland
36: Also at Institute for Nuclear Research, Moscow, Russia
37: Now at National Research Nuclear University 'Moscow Engineering Physics Institute' (MEPhI), Moscow, Russia
38: Also at St. Petersburg State Polytechnical University, St. Petersburg, Russia
39: Also at University of Florida, Gainesville, USA
40: Also at Imperial College, London, United Kingdom
41: Also at P.N. Lebedev Physical Institute, Moscow, Russia
42: Also at California Institute of Technology, Pasadena, USA
43: Also at Budker Institute of Nuclear Physics, Novosibirsk, Russia
44: Also at Faculty of Physics, University of Belgrade, Belgrade, Serbia
45: Also at Università degli Studi di Siena, Siena, Italy
46: Also at INFN Sezione di Pavia [a], Università di Pavia [b], Pavia, Italy, Pavia, Italy
47: Also at National and Kapodistrian University of Athens, Athens, Greece
48: Also at Universität Zürich, Zurich, Switzerland
49: Also at Stefan Meyer Institute for Subatomic Physics, Vienna, Austria, Vienna, Austria
50: Also at Şırnak University, Sirnak, Turkey
51: Also at Beykent University, Istanbul, Turkey, Istanbul, Turkey

52: Also at Istanbul Aydin University, Istanbul, Turkey
53: Also at Mersin University, Mersin, Turkey
54: Also at Piri Reis University, Istanbul, Turkey
55: Also at Adiyaman University, Adiyaman, Turkey
56: Also at Ozyegin University, Istanbul, Turkey
57: Also at Izmir Institute of Technology, Izmir, Turkey
58: Also at Kafkas University, Kars, Turkey
59: Also at Istanbul University, Istanbul, Turkey
60: Also at Istanbul Bilgi University, Istanbul, Turkey
61: Also at Hacettepe University, Ankara, Turkey
62: Also at School of Physics and Astronomy, University of Southampton, Southampton, United Kingdom
63: Also at IPPP Durham University, Durham, United Kingdom
64: Also at Monash University, Faculty of Science, Clayton, Australia
65: Also at Bethel University, St. Paul, Minneapolis, USA, St. Paul, USA
66: Also at Karamanoğlu Mehmetbey University, Karaman, Turkey
67: Also at Vilnius University, Vilnius, Lithuania
68: Also at Bingol University, Bingol, Turkey
69: Also at Georgian Technical University, Tbilisi, Georgia
70: Also at Sinop University, Sinop, Turkey
71: Also at Mimar Sinan University, Istanbul, Istanbul, Turkey
72: Also at Texas A&M University at Qatar, Doha, Qatar
73: Also at Kyungpook National University, Daegu, Korea, Daegu, Korea
74: Also at University of Hyderabad, Hyderabad, India

HADES Collaboration

J. Adamczewski-Musch[4], B. Arnoldi-Meadows[8], A. Belounnas[16], A. Belyaev[7], A. Blanco[1], C. Blume[8], M. Böhmer[10], S Borisenko[13], L. Chlad[17], P. Chudoba[17], I. Ciepał[2], D. Dittert[5], J. Dreyer[6], M. Drr[11], W.A. Esmail[12],L. Fabbietti[10,9], O. Fateev[7], P. Fonte[1,a], J. Friese[10], I. Fröhlich[8], J. FÃ¶rtsch[19],T. Galatyuk[5,4], R. Gernhäuser[10], O. Golosov[15], M. Golubeva[13], R. Greifenhagen[6,b], F. Guber[13], M. Gumberidze[4], S. Harabasz[5,3], R. Hensch[8], C. Höhne[11,4], R. Holzmann[4], H. Huck[8], A. Ierusalimov[7], A. Ivashkin[13], B. Kämpfer[6,b], K-H. Kampert[19], B. Kardan[8], I. Koenig[4], W. Koenig[4], M. Kohls[8], G. Kornakov[5], F. Kornas[5], R. Kotte[6], A. Kozela[2], I. Kres[19], J. Kuboś[2], A. Kugler[17], P. Kulessa[12], V. Ladygin[7], R. Lalik[3], A. Lebedev[14], S. Lebedev[11,7], S. Linev[4], L. Lopes[1], M. Lorenz[8], G. Lykasov[7], T. Mahmoud[11], A. Malige[3], J. Markert[4], T. Matulewicz[18], S. Maurus[10], V. Metag[11], J. Michel[8], S. Morozov[13,15], C. Müntz[8], L. Naumann[6], K. Nowakowski[3], J.-H. Otto[11], V. Patel[19], C. Pauly[19], V. Pechenov[4], O. Pechenova[4], O. Petukhov[13], D. Pfeifer[19], K. Piasecki[18], J. Pietraszko[4], A. Prozorov[17], W. Przygoda[3], K. Pysz[2], B. Ramstein[16], N. Rathod[3], A. Reshetin[13], L. Ritman[12], P. Rodriguez-Ramos[17], A. Rost[5], P. Salabura[3], F. Scozzi[5,16], F. Seck[5], I. Selyuzhenkov[4,15], A. Shabanov[13], U. Singh[3], J. Smyrski[3], S. Spies[8], H. Ströbele[8], J. Stroth[8,4], J. Stumm[8], O. Svoboda[17], M. Szala[8], P. Tlusty[17], M. Traxler[4], C. Ungethm[5], O. Vazquez-Doce[10,9], V. Wagner[17], A.A. Weber[11], C. Wendisch[4], M.G. Wiebusch[8], P. Wintz[12], J. Wirth[10,9], A. Zhilin[14], P. Zumbruch[4]

LIP-Laboratório de Instrumentação e Física Experimental de Partículas , 3004-516 Coimbra, Portugal

Institute of Nuclear Physics, Polish Academy of Sciences, 31342 Kraków, Poland

Smoluchowski Institute of Physics, Jagiellonian University of Cracow, 30-059 Kraków, Poland

GSI Helmholtzzentrum für Schwerionenforschung GmbH, 64291 Darmstadt, Germany

Technische Universität Darmstadt, 64289 Darmstadt, Germany

Institut für Strahlenphysik, Helmholtz-Zentrum Dresden-Rossendorf, 01314 Dresden, Germany

Joint Institute of Nuclear Research, 141980 Dubna, Russia

Institut für Kernphysik, Goethe-Universität, 60438 Frankfurt, Germany

Excellence Cluster 'Origin and Structure of the Universe' , 85748 Garching, Germany

Physik Department E62, Technische Universität München, 85748 Garching, Germany

II.Physikalisches Institut, Justus Liebig Universität Giessen, 35392 Giessen, Germany

Forschungszentrum Juelich, 52428 Juelich, Germany

Institute for Nuclear Research, Russian Academy of Science, 117312 Moscow, Russia

Institute of Theoretical and Experimental Physics, 117218 Moscow, Russia

National Research Nuclear University MEPhI (Moscow Engineering Physics Institute), 115409 Moscow, Russia

Laboratoire de Physique des 2 infinis Irne Joliot-Curie, Universit Paris-Saclay, CNRS-IN2P3. , F-91405 Orsay , France
Nuclear Physics Institute, The Czech Academy of Sciences, 25068 Rez, Czech Republic
Uniwersytet Warszawski - Instytut Fizyki Doświadczalnej, 02-093 Warszawa, Poland
Bergische Universität Wuppertal, 42119 Wuppertal, Germany
Also at Coimbra Polytechnic - ISEC, Coimbra, Portugal
Also at Technische Universität Dresden, 01062 Dresden, Germany

LHCb Collaboration

R. Aaij[31], C. Abellán Beteta[49], T. Ackernley[59], B. Adeva[45], M. Adinolfi[53], H. Afsharnia[9], C.A. Aidala[80], S. Aiola[25], Z. Ajaltouni[9], S. Akar[66], J. Albrecht[14], F. Alessio[47], M. Alexander[58], A. Alfonso Albero[44], G. Alkhazov[37], P. Alvarez Cartelle[60], A.A. Alves Jr[45], S. Amato[2], Y. Amhis[11], L. An[21], L. Anderlini[21], G. Andreassi[48], M. Andreotti[20], F. Archilli[16], A. Artamonov[43], M. Artuso[67], K. Arzymatov[41], E. Aslanides[10], M. Atzeni[49], B. Audurier[11], S. Bachmann[16], J.J. Back[55], S. Baker[60], V. Balagura[11,b], W. Baldini[20], J. Baptista Leite[1], R.J. Barlow[61], S. Barsuk[11], W. Barter[60], M. Bartolini[23,47,h], F. Baryshnikov[77], J.M. Basels[13], G. Bassi[28], V. Batozskaya[35], B. Batsukh[67], A. Battig[14], A. Bay[48], M. Becker[14], F. Bedeschi[28], I. Bediaga[1], A. Beiter[67], V. Belavin[41], S. Belin[26], V. Bellee[48], K. Belous[43], I. Belyaev[38], G. Bencivenni[22], E. Ben-Haim[12], S. Benson[31], A. Berezhnoy[39], R. Bernet[49], D. Berninghoff[16], H.C. Bernstein[67], C. Bertella[47], E. Bertholet[12], A. Bertolin[27], C. Betancourt[49], F. Betti[19,e], M.O. Bettler[54], Ia. Bezshyiko[49], S. Bhasin[53], J. Bhom[33], M.S. Bieker[14], S. Bifani[52], P. Billoir[12], A. Bizzeti[21,t], M. Bjørn[62], M.P. Blago[47], T. Blake[55], F. Blanc[48], S. Blusk[67], D. Bobulska[58], V. Bocci[30], O. Boente Garcia[45], T. Boettcher[63], A. Boldyrev[78], A. Bondar[42,w], N. Bondar[37,47], S. Borghi[61], M. Borisyak[41], M. Borsato[16], J.T. Borsuk[33], T.J.V. Bowcock[59], A. Boyer[47], C. Bozzi[20], M.J. Bradley[60], S. Braun[65], A. Brea Rodriguez[45], M. Brodski[47], J. Brodzicka[33], A. Brossa Gonzalo[55], D. Brundu[26], E. Buchanan[53], A. Büchler-Germann[49], A. Buonaura[49], C. Burr[47], A. Bursche[26], A. Butkevich[40], J.S. Butter[31], J. Buytaert[47], W. Byczynski[47], S. Cadeddu[26], H. Cai[72], R. Calabrese[20,g], L. Calero Diaz[22], S. Cali[22], R. Calladine[52], M. Calvi[24,i], M. Calvo Gomez[44,l], P. Camargo Magalhaes[53], A. Camboni[44,l], P. Campana[22], D.H. Campora Perez[31], A.F. Campoverde Quezada[5], L. Capriotti[19,e], A. Carbone[19,e], G. Carboni[29], R. Cardinale[23,h], A. Cardini[26], I. Carli[6], P. Carniti[24,i], K. Carvalho Akiba[31], A. Casais Vidal[45], G. Casse[59], M. Cattaneo[47], G. Cavallero[47], S. Celani[48], R. Cenci[28,o], J. Cerasoli[10], M.G. Chapman[53], M. Charles[12], Ph. Charpentier[47], G. Chatzikonstantinidis[52], M. Chefdeville[8], V. Chekalina[41], C. Chen[3], S. Chen[26], A. Chernov[33], S.-G. Chitic[47], V. Chobanova[45], S. Cholak[48], M. Chrzaszcz[33], A. Chubykin[37], V. Chulikov[37], P. Ciambrone[22], M.F. Cicala[55], X. Cid Vidal[45], G. Ciezarek[47], F. Cindolo[19], P.E.L. Clarke[57], M. Clemencic[47], H.V. Cliff[54], J. Closier[47], J.L. Cobbledick[61], V. Coco[47], J.A.B. Coelho[11], J. Cogan[10], E. Cogneras[9], L. Cojocariu[36], P. Collins[47], T. Colombo[47], A. Contu[26], N. Cooke[52], G. Coombs[58], S. Coquereau[44], G. Corti[47], C.M. Costa Sobral[55], B. Couturier[47], D.C. Craik[63], J. Crkovska[66], A. Crocombe[55], M. Cruz Torres[1,z], R. Currie[57], C.L. Da Silva[66], E. Dall'Occo[14], J. Dalseno[45,53], C. D'Ambrosio[47], A. Danilina[38], P. d'Argent[47], A. Davis[61], O. De Aguiar Francisco[47], K. De Bruyn[47], S. De Capua[61], M. De Cian[48], J.M. De Miranda[1], L. De Paula[2], M. De Serio[18,d], P. De Simone[22], J.A. de Vries[31], C.T. Dean[66], W. Dean[80], D. Decamp[8], L. Del Buono[12], B. Delaney[54], H.-P. Dembinski[14], A. Dendek[34], V. Denysenko[49], D. Derkach[78], O. Deschamps[9], F. Desse[11], F. Dettori[26], B. Dey[7], A. Di Canto[47], P. Di Nezza[22], S. Didenko[77], H. Dijkstra[47], V. Dobishuk[51], F. Dordei[26], M. Dorigo[28,x], A.C. dos Reis[1], L. Douglas[58], A. Dovbnya[50], K. Dreimanis[59], M.W. Dudek[33], L. Dufour[47],

P. Durante[47], J.M. Durham[66], D. Dutta[61], M. Dziewiecki[16], A. Dziurda[33], A. Dzyuba[37], S. Easo[56], U. Egede[69], V. Egorychev[38], S. Eidelman[42,w], S. Eisenhardt[57], S. Ek-In[48], L. Eklund[58], S. Ely[67], A. Ene[36], E. Epple[66], S. Escher[13], J. Eschle[49], S. Esen[31], T. Evans[47], A. Falabella[19], J. Fan[3], Y. Fan[5], N. Farley[52], S. Farry[59], D. Fazzini[11], P. Fedin[38], M. Féo[47], P. Fernandez Declara[47], A. Fernandez Prieto[45], F. Ferrari[19,e], L. Ferreira Lopes[48], F. Ferreira Rodrigues[2], S. Ferreres Sole[31], M. Ferrillo[49], M. Ferro-Luzzi[47], S. Filippov[40], R.A. Fini[18], M. Fiorini[20,g], M. Firlej[34], K.M. Fischer[62], C. Fitzpatrick[61], T. Fiutowski[34], F. Fleuret[11,b], M. Fontana[47], F. Fontanelli[23,h], R. Forty[47], V. Franco Lima[59], M. Franco Sevilla[65], M. Frank[47], C. Frei[47], D.A. Friday[58], J. Fu[25,p], Q. Fuehring[14], W. Funk[47], E. Gabriel[57], T. Gaintseva[41], A. Gallas Torreira[45], D. Galli[19,e], S. Gallorini[27], S. Gambetta[57], Y. Gan[3], M. Gandelman[2], P. Gandini[25], Y. Gao[4], L.M. Garcia Martin[46], J. García Pardiñas[49], B. Garcia Plana[45], F.A. Garcia Rosales[11], L. Garrido[44], D. Gascon[44], C. Gaspar[47], D. Gerick[16], E. Gersabeck[61], M. Gersabeck[61], T. Gershon[55], D. Gerstel[10], Ph. Ghez[8], V. Gibson[54], A. Gioventù[45], P. Gironella Gironell[44], L. Giubega[36], C. Giugliano[20], K. Gizdov[57], V.V. Gligorov[12], C. Göbel[70], D. Golubkov[38], A. Golutvin[60,77], A. Gomes[1,a], P. Gorbounov[38], I.V. Gorelov[39], C. Gotti[24,i], E. Govorkova[31], J.P. Grabowski[16], R. Graciani Diaz[44], T. Grammatico[12], L.A. Granado Cardoso[47], E. Graugés[44], E. Graverini[48], G. Graziani[21], A. Grecu[36], R. Greim[31], P. Griffith[20], L. Grillo[61], L. Gruber[47], B.R. Gruberg Cazon[62], C. Gu[3], M. Guarise[20], E. Gushchin[40], A. Guth[13], Yu. Guz[43,47], T. Gys[47], P. A. Gnther[16], T. Hadavizadeh[62], G. Haefeli[48], C. Haen[47], S.C. Haines[54], P.M. Hamilton[65], Q. Han[7], X. Han[16], T.H. Hancock[62], S. Hansmann-Menzemer[16], N. Harnew[62], T. Harrison[59], R. Hart[31], C. Hasse[14], M. Hatch[47], J. He[5], M. Hecker[60], K. Heijhoff[31], K. Heinicke[14], A.M. Hennequin[47], K. Hennessy[59], L. Henry[25,46], J. Heuel[13], A. Hicheur[68], D. Hill[62], M. Hilton[61], P.H. Hopchev[48], J. Hu[16], J. Hu[71], W. Hu[7], W. Huang[5], W. Hulsbergen[31], T. Humair[60], R.J. Hunter[55], M. Hushchyn[78], D. Hutchcroft[59], D. Hynds[31], P. Ibis[14], M. Idzik[34], P. Ilten[52], A. Inglessi[37], K. Ivshin[37], R. Jacobsson[47], S. Jakobsen[47], E. Jans[31], B.K. Jashal[46], A. Jawahery[65], V. Jevtic[14], F. Jiang[3], M. John[62], D. Johnson[47], C.R. Jones[54], B. Jost[47], N. Jurik[62], S. Kandybei[50], M. Karacson[47], J.M. Kariuki[53], N. Kazeev[78], M. Kecke[16], F. Keizer[54,47], M. Kelsey[67], M. Kenzie[55], T. Ketel[32], B. Khanji[47], A. Kharisova[79], K.E. Kim[67], T. Kirn[13], V.S. Kirsebom[48], S. Klaver[22], K. Klimaszewski[35], S. Koliiev[51], A. Kondybayeva[77], A. Konoplyannikov[38], P. Kopciewicz[34], R. Kopecna[16], P. Koppenburg[31], M. Korolev[39], I. Kostiuk[31,51], O. Kot[51], S. Kotriakhova[37], L. Kravchuk[40], R.D. Krawczyk[47], M. Kreps[55], F. Kress[60], S. Kretzschmar[13], P. Krokovny[42,w], W. Krupa[34], W. Krzemien[35], W. Kucewicz[33,k], M. Kucharczyk[33], V. Kudryavtsev[42,w], H.S. Kuindersma[31], G.J. Kunde[66], T. Kvaratskheliya[38], D. Lacarrere[47], G. Lafferty[61], A. Lai[26], D. Lancierini[49], J.J. Lane[61], G. Lanfranchi[22], C. Langenbruch[13], O. Lantwin[49,77], T. Latham[55], F. Lazzari[28,u], R. Le Gac[10], S.H. Lee[80], R. Lefèvre[9], A. Leflat[39,47], O. Leroy[10], T. Lesiak[33], B. Leverington[16], H. Li[71], L. Li[62], X. Li[66], Y. Li[6], Z. Li[67], X. Liang[67], T. Lin[60], R. Lindner[47], V. Lisovskyi[14], G. Liu[71], X. Liu[3], D. Loh[55], A. Loi[26], J. Lomba Castro[45], I. Longstaff[58], J.H. Lopes[2], G. Loustau[49], G.H. Lovell[54], Y. Lu[6], D. Lucchesi[27,n], M. Lucio Martinez[31], Y. Luo[3], A. Lupato[27], E. Luppi[20,g],

O. Lupton[55], A. Lusiani[28,s], X. Lyu[5], S. Maccolini[19,e], F. Machefert[11], F. Maciuc[36], V. Macko[48], P. Mackowiak[14], S. Maddrell-Mander[53], L.R. Madhan Mohan[53], O. Maev[37], A. Maevskiy[78], D. Maisuzenko[37], M.W. Majewski[34], S. Malde[62], B. Malecki[47], A. Malinin[76], T. Maltsev[42,w], H. Malygina[16], G. Manca[26,f], G. Mancinelli[10], R. Manera Escalero[44], D. Manuzzi[19,e], D. Marangotto[25], J. Maratas[9,v], J.F. Marchand[8], U. Marconi[19], S. Mariani[21,21,47], C. Marin Benito[11], M. Marinangeli[48], P. Marino[48], J. Marks[16], P.J. Marshall[59], G. Martellotti[30], L. Martinazzoli[47], M. Martinelli[24,i], D. Martinez Santos[45], F. Martinez Vidal[46], A. Massafferri[1], M. Materok[13], R. Matev[47], A. Mathad[49], Z. Mathe[47], V. Matiunin[38], C.Matteuzzi[24], K.R. Mattioli[80], A. Mauri[49], E. Maurice[11,b], M. McCann[60], L. Mcconnell[17], A. McNab[61], R. McNulty[17], J.V. Mead[59], B. Meadows[64], C. Meaux[10], G. Meier[14], N. Meinert[74], D. Melnychuk[35], S. Meloni[24,i], M. Merk[31], A. Merli[25], L. Meyer Garcia[2], M. Mikhasenko[47], D.A. Milanes[73], E. Millard[55], M.-N. Minard[8], O. Mineev[38], L. Minzoni[20], S.E. Mitchell[57], B. Mitreska[61], D.S. Mitzel[47], A. Mödden[14], A. Mogini[12], R.D. Moise[60], T. Mombächer[14], I.A. Monroy[73], S. Monteil[9], M. Morandin[27], G. Morello[22], M.J. Morello[28,s], J. Moron[34], A.B. Morris[10], A.G. Morris[55], R. Mountain[67], H. Mu[3], F. Muheim[57], M. Mukherjee[7], M. Mulder[47], D. Müller[47], K. Müller[49], C.H. Murphy[62], D. Murray[61], P. Muzzetto[26], P. Naik[53], T. Nakada[48], R. Nandakumar[56], T. Nanut[48], I. Nasteva[2], M. Needham[57], I. Neri[20], N. Neri[25,p], S. Neubert[16], N. Neufeld[47], R. Newcombe[60], T.D. Nguyen[48], C. Nguyen-Mau[48,m], E.M. Niel[11], S. Nieswand[13], N. Nikitin[39], N.S. Nolte[47], C. Nunez[80], A. Oblakowska-Mucha[34], V. Obraztsov[43], S. Ogilvy[58], D.P. O'Hanlon[53], R. Oldeman[26,f], C.J.G. Onderwater[75], J. D. Osborn[80], A. Ossowska[33], J.M. Otalora Goicochea[2], T. Ovsiannikova[38], P. Owen[49], A. Oyanguren[46], P.R. Pais[48], T. Pajero[28,47,28,s], A. Palano[18], M. Palutan[22], G. Panshin[79], A. Papanestis[56], M. Pappagallo[57], L.L. Pappalardo[20], C. Pappenheimer[64], W. Parker[65], C. Parkes[61], C.J. Parkinson[45], G. Passaleva[21,47], A. Pastore[18], M. Patel[60], C. Patrignani[19,e], A. Pearce[47], A. Pellegrino[31], M. Pepe Altarelli[47], S. Perazzini[19], D. Pereima[38], P. Perret[9], K. Petridis[53], A. Petrolini[23,h], A. Petrov[76], S. Petrucci[57], M. Petruzzo[25,p], B. Pietrzyk[8], G. Pietrzyk[48], M. Pili[62], D. Pinci[30], J. Pinzino[47], F. Pisani[19], A. Piucci[16], V. Placinta[36], S. Playfer[57], J. Plews[52], M. Plo Casasus[45], F. Polci[12], M. Poli Lener[22], M. Poliakova[67], A. Poluektov[10], N. Polukhina[77,c], I. Polyakov[67], E. Polycarpo[2], G.J. Pomery[53], S. Ponce[47], A. Popov[43], D. Popov[52], S. Poslavskii[43], K. Prasanth[33], L. Promberger[47], C. Prouve[45], V. Pugatch[51], A. Puig Navarro[49], H. Pullen[62], G. Punzi[28,o], W. Qian[5], J. Qin[5], R. Quagliani[12], B. Quintana[8], N.V. Raab[17], R.I. Rabadan Trejo[10], B. Rachwal[34], J.H. Rademacker[53], M. Rama[28], M. Ramos Pernas[45], M.S. Rangel[2], F. Ratnikov[41,78], G. Raven[32], M. Reboud[8], F. Redi[48], F. Reiss[12], C. Remon Alepuz[46], Z. Ren[3], V. Renaudin[62], S. Ricciardi[56], D.S. Richards[56], S. Richards[53], K. Rinnert[59], P. Robbe[11], A. Robert[12], A.B. Rodrigues[48], E. Rodrigues[59], J.A. Rodriguez Lopez[73], M. Roehrken[47], A. Rollings[62], V. Romanovskiy[43], M. Romero Lamas[45], A. Romero Vidal[45], J.D. Roth[80], M. Rotondo[22], M.S. Rudolph[67], T. Ruf[47], J. Ruiz Vidal[46], A. Ryzhikov[78], J. Ryzka[34], J.J. Saborido Silva[45], N. Sagidova[37], N. Sahoo[55], B. Saitta[26,f], C. Sanchez Gras[31], C. Sanchez Mayordomo[46], R. Santacesaria[30], C. Santamarina Rios[45], M. Santimaria[22], E. Santovetti[29,j], G. Sarpis[61], M. Sarpis[16],

A. Sarti[30], C. Satriano[30,r], A. Satta[29], M. Saur[5], D. Savrina[38,39], L.G. Scantlebury Smead[62], S. Schael[13], M. Schellenberg[14], M. Schiller[58], H. Schindler[47], M. Schmelling[15], T. Schmelzer[14], B. Schmidt[47], O. Schneider[48], A. Schopper[47], H.F. Schreiner[64], M. Schubiger[31], S. Schulte[48], M.H. Schune[11], R. Schwemmer[47], B. Sciascia[22], A. Sciubba[22], S. Sellam[68], A. Semennikov[38], A. Sergi[52,47], N. Serra[49], J. Serrano[10], L. Sestini[27], A. Seuthe[14], P. Seyfert[47], D.M. Shangase[80], M. Shapkin[43], L. Shchutska[48], T. Shears[59], L. Shekhtman[42,w], V. Shevchenko[76], E. Shmanin[77], J.D. Shupperd[67], B.G. Siddi[20], R. Silva Coutinho[49], L. Silva de Oliveira[2], G. Simi[27,n], S. Simone[18,d], I. Skiba[20], N. Skidmore[16], T. Skwarnicki[67], M.W. Slater[52], J.G. Smeaton[54], A. Smetkina[38], E. Smith[13], I.T. Smith[57], M. Smith[60], A. Snoch[31], M. Soares[19], L. Soares Lavra[9], M.D. Sokoloff[64], F.J.P. Soler[58], B. Souza De Paula[2], B. Spaan[14], E. Spadaro Norella[25,p], P. Spradlin[58], F. Stagni[47], M. Stahl[64], S. Stahl[47], P. Stefko[48], O. Steinkamp[49,77], S. Stemmle[16], O. Stenyakin[43], M. Stepanova[37], H. Stevens[14], S. Stone[67], S. Stracka[28], M.E. Stramaglia[48], M. Straticiuc[36], S. Strokov[79], J. Sun[26], L. Sun[72], Y. Sun[65], P. Svihra[61], K. Swientek[34], A. Szabelski[35], T. Szumlak[34], M. Szymanski[47], S. Taneja[61], Z. Tang[3], T. Tekampe[14], F. Teubert[47], E. Thomas[47], K.A. Thomson[59], M.J. Tilley[60], V. Tisserand[9], S. T'Jampens[8], M. Tobin[6], S. Tolk[47], L. Tomassetti[20,g], D. Torres Machado[1], D.Y. Tou[12], E. Tournefier[8], M. Traill[58], M.T. Tran[48], E. Trifonova[77], C. Trippl[48], A. Tsaregorodtsev[10], G. Tuci[28,o], A. Tully[48], N. Tuning[31], A. Ukleja[35], A. Usachov[31], A. Ustyuzhanin[41,78], U. Uwer[16], A. Vagner[79], V. Vagnoni[19], A. Valassi[47], G. Valenti[19], M. van Beuzekom[31], H. Van Hecke[66], E. van Herwijnen[47], C.B. Van Hulse[17], M. van Veghel[75], R. Vazquez Gomez[44], P. Vazquez Regueiro[45], C. Vázquez Sierra[31], S. Vecchi[20], J.J. Velthuis[53], M. Veltri[21,q], A. Venkateswaran[67], M. Veronesi[31], M. Vesterinen[55], J.V. Viana Barbosa[47], D. Vieira[64], M. Vieites Diaz[48], H. Viemann[74], X. Vilasis-Cardona[44,l], G. Vitali[28], A. Vitkovskiy[31], A. Vollhardt[49], D. Vom Bruch[12], A. Vorobyev[37], V. Vorobyev[42,w], N. Voropaev[37], R. Waldi[74], J. Walsh[28], J. Wang[3], J. Wang[72], J. Wang[6], M. Wang[3], Y. Wang[7], Z. Wang[49], D.R. Ward[54], H.M. Wark[59], N.K. Watson[52], D. Websdale[60], A. Weiden[49], C. Weisser[63], B.D.C. Westhenry[53], D.J. White[61], M. Whitehead[53], D. Wiedner[14], G. Wilkinson[62], M. Wilkinson[67], I. Williams[54], M. Williams[63], M.R.J. Williams[61], T. Williams[52], F.F. Wilson[56], W. Wislicki[35], M. Witek[33], L. Witola[16], G. Wormser[11], S.A. Wotton[54], H. Wu[67], K. Wyllie[47], Z. Xiang[5], D. Xiao[7], Y. Xie[7], H. Xing[71], A. Xu[4], J. Xu[5], L. Xu[3], M. Xu[7], Q. Xu[5], Z. Xu[4], Z. Yang[3], Z. Yang[65], Y. Yao[67], L.E. Yeomans[59], H. Yin[7], J. Yu[7], X. Yuan[67], O. Yushchenko[43], K.A. Zarebski[52], M. Zavertyaev[15,c], M. Zdybal[33], M. Zeng[3], D. Zhang[7], L. Zhang[3], S. Zhang[4], W.C. Zhang[3,y], Y. Zhang[47], A. Zhelezov[16], Y. Zheng[5], X. Zhou[5], Y. Zhou[5], X. Zhu[3], V. Zhukov[13,39], J.B. Zonneveld[57], S. Zucchelli[19,e].

[1]*Centro Brasileiro de Pesquisas Físicas (CBPF), Rio de Janeiro, Brazil*
[2] *Universidade Federal do Rio de Janeiro (UFRJ), Rio de Janeiro, Brazil*
[3]*Center for High Energy Physics, Tsinghua University, Beijing, China*
[4]*School of Physics State Key Laboratory of Nuclear Physics and Technology, Peking University, Beijing, China*
[5]*University of Chinese Academy of Sciences, Beijing, China*

[6] *Institute Of High Energy Physics (IHEP), Beijing, China*
[7] *Institute of Particle Physics, Central China Normal University, Wuhan, Hubei, China*
[8] *Univ. Grenoble Alpes, Univ. Savoie Mont Blanc, CNRS, IN2P3-LAPP, Annecy, France*
[9] *Université Clermont Auvergne, CNRS/IN2P3, LPC, Clermont-Ferrand, France*
[10] *Aix Marseille Univ, CNRS/IN2P3, CPPM, Marseille, France*
[11] *Laboratoire de Physique des 2 Infinis Irène Joliot-Curie, IJCLab, Orsay, France*
[12] *LPNHE, Sorbonne Université, Paris Diderot Sorbonne Paris Cité, CNRS/IN2P3, Paris, France*
[13] *I. Physikalisches Institut, RWTH Aachen University, Aachen, Germany*
[14] *Fakultät Physik, Technische Universität Dortmund, Dortmund, Germany*
[15] *Max-Planck-Institut für Kernphysik (MPIK), Heidelberg, Germany*
[16] *Physikalisches Institut, Ruprecht-Karls-Universität Heidelberg, Heidelberg, Germany*
[17] *School of Physics, University College Dublin, Dublin, Ireland*
[18] *INFN Sezione di Bari, Bari, Italy*
[19] *INFN Sezione di Bologna, Bologna, Italy*
[20] *INFN Sezione di Ferrara, Ferrara, Italy*
[21] *INFN Sezione di Firenze, Firenze, Italy*
[22] *INFN Laboratori Nazionali di Frascati, Frascati, Italy*
[23] *INFN Sezione di Genova, Genova, Italy*
[24] *INFN Sezione di Milano-Bicocca, Milano, Italy*
[25] *INFN Sezione di Milano, Milano, Italy*
[26] *INFN Sezione di Cagliari, Monserrato, Italy*
[27] *INFN Sezione di Padova, Padova, Italy*
[28] *INFN Sezione di Pisa, Pisa, Italy*
[29] *INFN Sezione di Roma Tor Vergata, Roma, Italy*
[30] *INFN Sezione di Roma La Sapienza, Roma, Italy*
[31] *Nikhef National Institute for Subatomic Physics, Amsterdam, Netherlands*
[32] *Nikhef National Institute for Subatomic Physics and VU University Amsterdam, Amsterdam, Netherlands*
[33] *Henryk Niewodniczanski Institute of Nuclear Physics Polish Academy of Sciences, Kraków, Poland*
[34] *AGH - University of Science and Technology, Faculty of Physics and Applied Computer Science, Kraków, Poland*
[35][36] *National Center for Nuclear Research (NCBJ), Warsaw, Poland*
Horia Hulubei National Institute of Physics and Nuclear Engineering, Bucharest-Magurele, Romania
[37] *Petersburg Nuclear Physics Institute NRC Kurchatov Institute (PNPI NRC KI), Gatchina, Russia*
[38] *Institute of Theoretical and Experimental Physics NRC Kurchatov Institute (ITEP NRC KI), Moscow, Russia, Moscow, Russia*
[39] *Institute of Nuclear Physics, Moscow State University (SINP MSU), Moscow, Russia*

[40]*Institute for Nuclear Research of the Russian Academy of Sciences (INR RAS), Moscow, Russia*
[41]*Yandex School of Data Analysis, Moscow, Russia*
[42]*Budker Institute of Nuclear Physics (SB RAS), Novosibirsk, Russia*
[43]*Institute for High Energy Physics NRC Kurchatov Institute (IHEP NRC KI), Protvino, Russia, Protvino, Russia*
[44]*ICCUB, Universitat de Barcelona, Barcelona, Spain*
[45]*Instituto Galego de Física de Altas Enerxías (IGFAE), Universidade de Santiago de Compostela, Santiago de Compostela, Spain*
[46]*Instituto de Fisica Corpuscular, Centro Mixto Universidad de Valencia - CSIC, Valencia, Spain*
[47]*European Organization for Nuclear Research (CERN), Geneva, Switzerland*
[48]*Institute of Physics, Ecole Polytechnique Fédérale de Lausanne (EPFL), Lausanne, Switzerland*
[49]*Physik-Institut, Universität Zürich, Zürich, Switzerland*
[50]*NSC Kharkiv Institute of Physics and Technology (NSC KIPT), Kharkiv, Ukraine*
[51]*Institute for Nuclear Research of the National Academy of Sciences (KINR), Kyiv, Ukraine*
[52]*University of Birmingham, Birmingham, United Kingdom*
[53]*H.H. Wills Physics Laboratory, University of Bristol, Bristol, United Kingdom*
[54]*Cavendish Laboratory, University of Cambridge, Cambridge, United Kingdom*
[55]*Department of Physics, University of Warwick, Coventry, United Kingdom*
[56]*STFC Rutherford Appleton Laboratory, Didcot, United Kingdom*
[57]*School of Physics and Astronomy, University of Edinburgh, Edinburgh, United Kingdom*
[58]*School of Physics and Astronomy, University of Glasgow, Glasgow, United Kingdom*
[59]*Oliver Lodge Laboratory, University of Liverpool, Liverpool, United Kingdom*
[60]*Imperial College London, London, United Kingdom*
[61]*Department of Physics and Astronomy, University of Manchester, Manchester, United Kingdom*
[62]*Department of Physics, University of Oxford, Oxford, United Kingdom*
[63]*Massachusetts Institute of Technology, Cambridge, MA, United States*
[64]*University of Cincinnati, Cincinnati, OH, United States*
[65]*University of Maryland, College Park, MD, United States*
[66]*Los Alamos National Laboratory (LANL), Los Alamos, United States*
[67]*Syracuse University, Syracuse, NY, United States*
[68]*Laboratory of Mathematical and Subatomic Physics , Constantine, Algeria, associated to* [2]
[69]*School of Physics and Astronomy, Monash University, Melbourne, Australia, associated to*[55]
[70]*Pontifícia Universidade Católica do Rio de Janeiro (PUC-Rio), Rio de Janeiro, Brazil, associated to* [2]

[71] *Guangdong Provencial Key Laboratory of Nuclear Science, Institute of Quantum Matter, South China Normal University, Guangzhou, China, associated to* [3]
[72] *School of Physics and Technology, Wuhan University, Wuhan, China, associated to* [3]
[73] *Departamento de Fisica , Universidad Nacional de Colombia, Bogota, Colombia, associated to* [12]
[74] *Institut für Physik, Universität Rostock, Rostock, Germany, associated to* [16]
[75] *Van Swinderen Institute, University of Groningen, Groningen, Netherlands, associated to* [31]
[76] *National Research Centre Kurchatov Institute, Moscow, Russia, associated to* [38]
[77] *National University of Science and Technology "MISIS", Moscow, Russia, associated to* [38]
[78] *National Research University Higher School of Economics, Moscow, Russia, associated to* [41]
[79] *National Research Tomsk Polytechnic University, Tomsk, Russia, associated to* [38]
[80] *University of Michigan, Ann Arbor, United States, associated to* [67]
[a] *Universidade Federal do Triângulo Mineiro (UFTM), Uberaba-MG, Brazil*
[b] *Laboratoire Leprince-Ringuet, Palaiseau, France*
[c] *P.N. Lebedev Physical Institute, Russian Academy of Science (LPI RAS), Moscow, Russia*
[d] *Università di Bari, Bari, Italy*
[e] *Università di Bologna, Bologna, Italy*
[f] *Università di Cagliari, Cagliari, Italy*
[g] *Università di Ferrara, Ferrara, Italy*
[h] *Università di Genova, Genova, Italy*
[i] *Università di Milano Bicocca, Milano, Italy*
[j] *Università di Roma Tor Vergata, Roma, Italy*
[k] *AGH - University of Science and Technology, Faculty of Computer Science, Electronics and Telecommunications, Kraków, Poland*
[l] *DS4DS, La Salle, Universitat Ramon Llull, Barcelona, Spain*
[m] *Hanoi University of Science, Hanoi, Vietnam*
[n] *Università di Padova, Padova, Italy*
[o] *Università di Pisa, Pisa, Italy*
[p] *Università degli Studi di Milano, Milano, Italy*
[q] *Università di Urbino, Urbino, Italy*
[r] *Università della Basilicata, Potenza, Italy*
[s] *Scuola Normale Superiore, Pisa, Italy*
[t] *Università di Modena e Reggio Emilia, Modena, Italy*

[u] *Università di Siena, Siena, Italy*
[v] *MSU - Iligan Institute of Technology (MSU-IIT), Iligan, Philippines*
[w] *Novosibirsk State University, Novosibirsk, Russia*
[x] *INFN Sezione di Trieste, Trieste, Italy*
[y] *School of Physics and Information Technology, Shaanxi Normal University (SNNU), Xi'an, China*
[z] *Universidad Nacional Autonoma de Honduras, Tegucigalpa, Honduras*

NA61 Collaboration

A. Aduszkiewicz[15], E.V. Andronov[21], T. Antićić[3], V. Babkin[19], M. Baszczyk[13], S. Bhosale[10], A. Blondel[4], M. Bogomilov[2], A. Brandin[20], A. Bravar[23], W. Bryliński[17], J. Brzychczyk[12], M. Buryakov[19], O. Busygina[18], A. Bzdak[13], H. Cherif[6], M. Ćirković[22], M. Csanad [7], J. Cybowska[17], T. Czopowicz[9,17], A. Damyanova[23], N. Davis[10], M. Deliyergiyev[9], M. Deveaux[6], A. Dmitriev [19], W. Dominik[15], P. Dorosz[13], J. Dumarchez[4], R. Engel[5], G.A. Feofilov[21], L. Fields[24], Z. Fodor[7,16], A. Garibov[1], M. Gaździcki[6,9], O. Golosov[20], V. Golovatyuk [19], M. Golubeva[18], K. Grebieszkow[17], F. Guber[18], A. Haesler[23], S.N. Igolkin[21], S. Ilieva[2], A. Ivashkin[18], S.R. Johnson[25], K. Kadija[3], E. Kaptur[14], N. Kargin[20], E. Kashirin[20], M. Kiełbowicz[10], V.A. Kireyeu[19], V. Klochkov[6], V.I. Kolesnikov[19], D. Kolev[2], A. Korzenev[23], V.N. Kovalenko[21], S. Kowalski[14], M. Koziel[6], A. Krasnoperov[19], W. Kucewicz[13], M. Kuich[15], A. Kurepin[18], D. Larsen[12], A. László[7], T.V. Lazareva[21], M. Lewicki[16], K. Łojek[12], B. Łysakowski[14], V.V. Lyubushkin[19], M. Maćkowiak-Pawłowska[17], Z. Majka[12], B. Maksiak[11], A.I. Malakhov[19], A. Marcinek[10], A.D. Marino[25], K. Marton[7], H.-J. Mathes[5], T. Matulewicz[15], V. Matveev[19], G.L. Melkumov[19], A.O. Merzlaya[12], B. Messerly[26], Ł. Mik[13], S. Morozov[18,20], S. Mrówczyński[9], Y. Nagai[25], M. Naskręt[16], V. Ozvenchuk[10], V. Paolone[26], O. Petukhov[18], R. Płaneta[12], P. Podlaski[15], B.A. Popov[19,4], B. Porfy[7], M. Posiadała-Zezula[15], D.S. Prokhorova[21], D. Pszczel[11], S. Puławski[14], J. Puzović[22], M. Ravonel[23], R. Renfordt[6], D. Röhrich[8], E. Rondio[11], M. Roth[5], B.T. Rumberger[25], M. Rumyantsev[19], A. Rustamov[1,6], M. Rybczynski[9], A. Rybicki[10], A. Sadovsky[18], K. Schmidt[14], I. Selyuzhenkov[20], A.Yu. Seryakov[21], P. Seyboth[9], M. Słodkowski[17], P. Staszel[12], G. Stefanek[9], J. Stepaniak[11], M. Strikhanov[20], H. Ströbele[6], T. Šuša[3], A. Taranenko[20], A. Tefelska[17], D. Tefelski[17], V. Tereshchenko[19], A. Toia[6], R. Tsenov[2], L. Turko[16], R. Ulrich[5], M. Unger[5], F.F. Valiev[21], D. Veberič[5], V.V. Vechernin[21], A. Wickremasinghe[26,24], Z. Włodarczyk[9], O. Wyszyński[12], E.D. Zimmerman[25], and R. Zwaska[24]

[1] National Nuclear Research Center, Baku, Azerbaijan
[2] Faculty of Physics, University of Sofia, Sofia, Bulgaria
[3] Rudjer Bošković Institute, Zagreb, Croatia
[4] LPNHE, University of Paris VI and VII, Paris, France
[5] Karlsruhe Institute of Technology, Karlsruhe, Germany
[6] University of Frankfurt, Frankfurt, Germany
[7] Wigner Research Centre for Physics of the Hungarian Academy of Sciences, Budapest, Hungary
[8] University of Bergen, Bergen, Norway
[9] Jan Kochanowski University in Kielce, Poland
[10] Institute of Nuclear Physics, Polish Academy of Sciences, Cracow, Poland
[11] National Centre for Nuclear Research, Warsaw, Poland
[12] Jagiellonian University, Cracow, Poland
[13] AGH - University of Science and Technology, Cracow, Poland
[14] University of Silesia, Katowice, Poland
[15] University of Warsaw, Warsaw, Poland

[16] University of Wrocław, Wrocław, Poland
[17] Warsaw University of Technology, Warsaw, Poland
[18] Institute for Nuclear Research, Moscow, Russia
[19] Joint Institute for Nuclear Research, Dubna, Russia
[20] National Research Nuclear University (Moscow Engineering Physics Institute), Moscow, Russia
[21] St. Petersburg State University, St. Petersburg, Russia
[22] University of Belgrade, Belgrade, Serbia
[23] University of Geneva, Geneva, Switzerland
[24] Fermilab, Batavia, USA
[25] University of Colorado, Boulder, USA
[26] University of Pittsburgh, Pittsburgh, USA

PHENIX Collaboration

XXVIIIth International Conference on Ultrarelativistic Nucleus-Nucleus Collisions

(Quark Matter 2019)

[23]U.A. Acharya, [12]A. Adare, [42,47,48]C. Aidala, [70]N.N. Ajitanand**, [64,65]Y. Akiba*, [11]R. Akimoto, [56]H. Al-Bataineh, [70]J. Alexander, [26]M. Alfred, [56]H. Al-Ta'ani, [48]V. Andrieux, [13]A. Angerami, [35,38,64]K. Aoki, [31,71]N. Apadula, [11,64]Y. Aramaki, [38,64]H. Asano, [7]E.C. Aschenauer, [39,71]E.T. Atomssa, [71]R. Averbeck, [59]T.C. Awes, [48]C. Ayuso, [7]B. Azmoun, [27]V. Babintsev, [17]A. Bagoly, [6]M. Bai, [10]X. Bai, [21]G. Baksay, [21]L. Baksay, [47]N.S. Bandara, [71]B. Bannier, [8]K.N. Barish, [55]B. Bassalleck, [1]A.T. Basye, [5,8,65]S. Bathe, [62]V. Baublis, [7,49]C. Baumann, [64]S. Baumgart, [7]A. Bazilevsky, [8]M. Beaumier, [12]S. Beckman, [7]S. Belikov**, [12,48,57,76]R. Belmont, [71]R. Bennett, [67]A. Berdnikov, [67]Y. Berdnikov, [80]J.H. Bhom, [76]L. Bichon, [8]D. Black, [76]B. Blankenship, [37,54]D.S. Blau, [42]M. Boer, [55,56,80]J.S. Bok, [67]V. Borisov, [65,71]K. Boyle, [42]M.L. Brooks, [5,8]J. Bryslawskyj, [7]H. Buesching, [27]V. Bumazhnov, [7,65]G. Bunce, [23]C. Butler, [42,55]S. Butsyk, [13,31,71]S. Campbell, [71]V. Canoa Roman, [50]A. Caringi, [71]P. Castera, [71]R. Cervantes, [65,71]C.-H. Chen, [13]C.Y. Chi, [7]M. Chiu, [28,80]I.J. Choi, [33]J.B. Choi**, [69]S. Choi, [4]R.K. Choudhury, [44]P. Christiansen, [75]T. Chujo, [70]P. Chung, [8]O. Chvala, [59]V. Cianciolo, [71,78]Z. Citron, [13]B.A. Cole, [39]Z. Conesa del Valle, [23,65,71]M. Connors, [50,71]N. Cronin, [50]N. Crossette, [17]M. Csanád, [18,79]T. Csörgő, [71]T. Dahms, [38,64]S. Dairaku, [76]I. Danchev, [58]T.W. Danley, [22]K. Das, [47,55]A. Datta, [1]M.S. Daugherity, [7,16,71]G. David, [23]M.K. Dayananda, [55]K. DeBlasio, [71]K. Dehmelt, [27]A. Denisov, [7,65,71]A. Deshpande, [7]E.J. Desmond, [56]K.V. Dharmawardane, [68]O. Dietzsch, [31]L. Ding, [31,71]A. Dion, [46]P.B. Diss, [71]D. Dixit, [80]J.H. Do, [68]M. Donadelli, [46]L. D'Orazio, [39]O. Drapier, [71]A. Drees, [6]K.A. Drees, [78]M. Dumancic, [42,71]J.M. Durham, [27]A. Durum, [4]D. Dutta, [6,22]S. Edwards, [59]Y.V. Efremenko, [23]T. Elder, [12]F. Ellinghaus, [13]T. Engelmore, [59,64,66]A. Enokizono, [64,65]H. En'yo, [71]R. Esha, [75]S. Esumi, [7,8]K.O. Eyser, [50]B. Fadem, [71]W. Fan, [71]N. Feege, [55]D.E. Fields, [9]M. Finger, [9]M. Finger, Jr., [16]D. Firak, [48]D. Fitzgerald, [39]F. Fleuret, [37]S.L. Fokin, [78]Z. Fraenkel**, [58,71]J.E. Frantz, [7]A. Franz, [22]A.D. Frawley, [64]K. Fujiwara, [35,64]Y. Fukao, [75]Y. Fukuda, [52]T. Fusayasu, [1]K. Gainey, [71]C. Gal, [14]P. Gallus, [48]E.A. Gamez, [3,71]P. Garg, [73]A. Garishvili, [41,73]I. Garishvili, [71]H. Ge, [28]F. Giordano, [41]A. Glenn, [71]H. Gong, [70]X. Gong, [39]M. Gonin, [64,65]Y. Goto, [39]R. Granier de Cassagnac, [2,13]N. Grau, [76]S.V. Greene, [42]G. Grim, [28]M. Grosse Perdekamp, [70]Y. Gu, [11]T. Gunji, [42]L. Guo, [23]H. Guragain, [44]H.-Å. Gustafsson**, [53,64,65]T. Hachiya, [7]J.S. Haggerty, [19]K.I. Hahn, [11]H. Hamagaki, [73]J. Hamblen, [1]H.F. Hamilton, [61]R. Han, [19,36,64]S.Y. Han, [13,71]J. Hanks, [32]S. Hasegawa, [23]T.O.S. Haseler, [64,66]K. Hashimoto, [44]E. Haslum, [11]R. Hayano, [11]S. Hayashi, [23]X. He, [41]M. Heffner, [71]T.K. Hemmick, [8]T. Hester, [31]J.C. Hill, [12]K. Hill, [23]A. Hodges, [21]M. Hohlmann, [8]R.S. Hollis, [13]W. Holzmann, [25]K. Homma, [36]B. Hong, [25,75]T. Horaguchi, [11]Y. Hori, [73]D. Hornback, [25]T. Hoshino, [31]N. Hotvedt, [7,42]J. Huang, [76]S. Huang, [64,65]T. Ichihara, [64]R. Ichimiya, [35]H. Iinuma, [64,75]Y. Ikeda, [32,38,64]K. Imai, [64]Y. Imazu, [16]J. Imrek,

[75]M. Inaba, [8]A. Iordanova, [1]D. Isenhower, [64]M. Ishihara, [53]S. Ishimaru, [50]A. Isinhue, [76]M. Issah, [53]Y. Ito, [62]D. Ivanishchev, [25]Y. Iwanaga, [71]B.V. Jacak, [23]M. Javani, [51]S.J. Jeon, [23]M. Jezghani, [71]Z. Ji, [7,70]J. Jia, [42]X. Jiang, [13]J. Jin, [7,23]B.M. Johnson, [1]T. Jones, [36]E. Joo, [51]K.S. Joo, [71]V. Jorjadze, [60]D. Jouan, [1,28]D.S. Jumper, [11]F. Kajihara, [71]J. Kamin, [11,35,64]S. Kanda, [71]S. Kaneti, [24]B.H. Kang, [80]J.H. Kang, [24]J.S. Kang, [8]D. Kapukchyan, [42]J. Kapustinsky, [38,64]K. Karatsu, [71]S. Karthas, [64,66]M. Kasai, [47,65]D. Kawall, [64,66]M. Kawashima, [37]A.V. Kazantsev, [31]T. Kempel, [55]J.A. Key, [71]V. Khachatryan, [3]P.K. Khandai, [62]A. Khanzadeev, [42]A. Khatiwada, [75]K. Kihara, [25]K.M. Kijima, [77]J. Kikuchi, [19]A. Kim, [36]B.I. Kim, [8,36]C. Kim, [19]D.H. Kim, [34]D.J. Kim, [33]E.-J. Kim, [19]G.W. Kim, [80]H.-J. Kim, [80]H.J. Kim, [33]K.-B. Kim, [64,69]M. Kim, [36]M.H. Kim, [28]Y.-J. Kim, [24]Y.K. Kim, [50]B. Kimelman, [17]D. Kincses, [12]E. Kinney, [17]Á. Kiss, [7]E. Kistenev, [11]R. Kitamura, [22]J. Klatsky, [8]D. Kleinjan, [71]P. Kline, [12]T. Koblesky, [62]L. Kochenda, [17,79]M. Kofarago, [11,35]Y. Komatsu, [62]B. Komkov, [75]M. Konno, [28,65]J. Koster, [58]D. Kotchetkov, [62,67]D. Kotov, [14]A. Král, [13]A. Kravitz, [34]F. Krizek, [75]S. Kudo, [42]G.J. Kunde, [17]B. Kurgyis, [64,66]K. Kurita, [64,65]M. Kurosawa, [80]Y. Kwon, [56]G.S. Kyle, [70]R. Lacey, [13]Y.S. Lai, [31]J.G. Lajoie, [50]E.O. Lallow, [67]D. Larionova, [67]M. Larionova, [31]A. Lebedev, [24]B. Lee, [42]D.M. Lee, [33]G.H. Lee, [19,72]J. Lee, [36,42]K.B. Lee, [36]K.S. Lee, [80]S. Lee, [31,71]S.H. Lee, [33]S.R. Lee, [42]M.J. Leitch, [68]M.A.L. Leite, [28]M. Leitgab, [71]Y.H. Leung, [71]B. Lewis, [48]N.A. Lewis, [10]X. Li, [42]X. Li, [50]P. Lichtenwalner, [65]P. Liebing, [12,42,63,80]S.H. Lim, [12,41]L.A. Linden Levy, [14]T. Liška, [42]H. Liu, [61]L. D. Liu, [42]M.X. Liu, [28]V.-R. Loggins, [17,18]S. Lökös, [16]K. Lovasz, [76]B. Love, [7]D. Lynch, [76]C.F. Maguire, [16]T. Majoros, [6]Y.I. Makdisi, [78,81]M. Makek, [62]M. Malaev, [55]M.D. Malik, [71]A. Manion, [37]V.I. Manko, [7,13]E. Mannel, [61,64]Y. Mao, [66]H. Masuda, [75]H. Masui, [11,35]S. Masumoto, [13]F. Matathias, [12,42,71]M. McCumber, [42]P.L. McGaughey, [12,22,42]D. McGlinchey, [28]C. McKinney, [71]N. Means, [56]A. Meles, [8]M. Mendoza, [13,28]B. Meredith, [18]W.J. Metzger, [75]Y. Miake, [35]T. Mibe, [25]J. Midori, [46]A.C. Mignerey, [71]D.E. Mihalik, [64,75]K. Miki, [1]A.J. Miller, [7,78]A. Milov, [4]D.K. Mishra, [7]J.T. Mitchell, [67]Iu. Mitrankov, [35,64,65]G. Mitsuka, [64,74]Y. Miyachi, [64,74]S. Miyasaka, [64,75]S. Mizuno, [4]A.K. Mohanty, [70]S. Mohapatra, [28]P. Montuenga, [51]H.J. Moon, [36,80]T. Moon, [11]Y. Morino, [8]A. Morreale, [7]D.P. Morrison, [76]S.I. Morrow, [50]M. Moskowitz, [50]S. Motschwiller, [37]T.V. Moukhanova, [36,64]B. Mulilo, [38,64]T. Murakami, [64,66]J. Murata, [70]A. Mwai, [38]T. Nagae, [74]K. Nagai, [35,64]S. Nagamiya, [25,64]K. Nagashima, [66]T. Nagashima, [12]J.L. Nagle, [78]M. Naglis, [17,79]M.I. Nagy, [64,65]I. Nakagawa, [64,75]H. Nakagomi, [25]Y. Nakamiya, [38,64]K.R. Nakamura, [64]T. Nakamura, [64,74]K. Nakano, [19]S. Nam, [73]C. Nattrass, [50]A. Nederlof, [20]S. Nelson, [4]P.K. Netrakanti, [41]J. Newby, [71]M. Nguyen, [25,64]M. Nihashi, [75]T. Niida, [11,35]S. Nishimura, [53]R. Nishitani, [7,65]R. Nouicer, [18,79]T. Novák, [34,71,75]N. Novitzky, [14]R. Novotny, [11]A. Nukariya, [37]A.S. Nyanin, [23]C. Oakley, [25]H. Obayashi, [7]E. O'Brien, [11]S.X. Oda, [31]C.A. Ogilvie, [11]H. Oide, [75]M. Oka, [65]K. Okada, [64]Y. Onuki, [12]J.D. Orjuela Koop, [48]J.D. Osborn, [44]A. Oskarsson, [55]G.J. Ottino, [25,64]M. Ouchida, [11,35,75]K. Ozawa, [7]R. Pak, [29,71]V. Pantuev, [56]V. Papavassiliou, [24]B.H. Park, [19,72]I.H. Park, [69]J.S. Park, [64,69,71]S. Park, [36]S.K. Park, [36]W.J. Park, [56]S.F. Pate, [23]L. Patel, [31]M. Patel, [31]H. Pei, [28]J.-C. Peng,

[76]W. Peng, [15]H. Pereira, [7,12,13]D.V. Perepelitsa, [56]G.D.N. Perera, [37]D.Yu. Peressounko, [71]C.E. PerezLara, [31]J. Perry, [7,71]R. Petti, [7,28]M. Phipps, [7]C. Pinkenburg, [1]R. Pinson, [7]R.P. Pisani, [71]M. Proissl, [58]A. Pun, [7]M.L. Purschke, [1,23]H. Qu, [67]P.V. Radzevich, [34]J. Rak, [71]N. Ramasubramanian, [48]B.J. Ramson, [78]I. Ravinovich, [59,73]K.F. Read, [21]S. Rembeczki, [49]K. Reygers, [70]D. Reynolds, [54,62]V. Riabov, [62,67]Y. Riabov, [46]E. Richardson, [5]D. Richford, [28,31]T. Rinn, [58]N. Riveli, [76]D. Roach, [43]G. Roche**, [8]S.D. Rolnick, [31]M. Rosati, [12]C.A. Rosen, [44]S.S.E. Rosendahl, [5]Z. Rowan, [48]J.G. Rubin, [31]J. Runchey, [30]P. Ružička, [24]M.S. Ryu, [67]A.S. Safonov, [49,71]B. Sahlmueller, [35]N. Saito, [7]T. Sakaguchi, [64,74]K. Sakashita, [32]H. Sako, [54,62]V. Samsonov, [75]M. Sano, [11,77]S. Sano, [23]M. Sarsour, [75]K. Sato, [32,35]S. Sato, [75]T. Sato, [35]S. Sawada, [20]C.Y. Scarlett, [76]B. Schaefer, [73]B.K. Schmoll, [8]K. Sedgwick, [12,65]J. Seele, [28,64,65]R. Seidl, [11]Y. Sekiguchi, [23,31,73]A. Sen, [8]R. Seto, [4]P. Sett, [46]A. Sexton, [71,78]D. Sharma, [31]A. Shaver, [27]I. Shein, [64,74]T.-A. Shibata, [25]K. Shigaki, [31,53,75]M. Shimomura, [75]T. Shioya, [38,64]K. Shoji, [4]P. Shukla, [7,28]A. Sickles, [31,42]C.L. Silva, [44,59]D. Silvermyr, [15]C. Silvestre, [36]K.S. Sim, [3]B.K. Singh, [3]C.P. Singh, [3]V. Singh, [48]M.J. Skoby, [50]M. Skolnik, [9]M. Slunečka, [22]K.L. Smith, [42]M. Snowball, [50]S. Solano, [41]R.A. Soltz, [42]W.E. Sondheim, [73]S.P. Sorensen, [7]I.V. Sourikova, [59]P.W. Stankus, [7]P. Steinberg, [44]E. Stenlund, [47,56]M. Stepanov**, [79]A. Ster, [7]S.P. Stoll, [12]M.R. Stone, [25]T. Sugitate, [7]A. Sukhanov, [64]T. Sumita, [71]J. Sun, [23]X. Sun, [16]Z. Sun, [53]S. Suzuki, [23]S. Syed, [79]J. Sziklai, [68]E.M. Takagui, [11]A. Takahara, [53]A. Takeda, [64,65]A. Taketani, [75]R. Tanabe, [52]Y. Tanaka, [71]S. Taneja, [32,38,64,65,69]K. Tanida, [7]M.J. Tannenbaum, [3,76,78]S. Tarafdar, [54,70]A. Taranenko, [16]G. Tarnai, [56]E. Tennant, [71]H. Themann, [1]D. Thomas, [55]T.L. Thomas, [23,45]R. Tieulent, [31]A. Timilsina, [64,65,75]T. Todoroki, [65]M. Togawa, [71]A. Toia, [30]L. Tomášek, [14,30]M. Tomášek, [11,25]H. Torii, [1]C.L. Towell, [1]M. Towell, [1]R. Towell, [1]R.S. Towell, [78]I. Tserruya, [11,25]Y. Tsuchimoto, [11]T. Tsuji, [25]Y. Ueda, [16]B. Ujvari, [7]C. Vale, [76]H. Valle, [42]H.W. van Hecke, [17,79]M. Vargyas, [12]S. Vazquez-Carson, [13]E. Vazquez-Zambrano, [13,28]A. Veicht, [76]J. Velkovska, [79]R. Vértesi, [14]M. Virius, [31]B. Voas, [28]A. Vossen, [14,30]V. Vrba, [81]N. Vukman, [62]E. Vznuzdaev, [56,65]X.R. Wang, [5]Z. Wang, [25]D. Watanabe, [64,66,75]K. Watanabe, [64,65]Y. Watanabe, [11,35]Y.S. Watanabe, [31,56]F. Wei, [70]R. Wei, [49]J. Wessels, [31]S. Whitaker, [48]A.S. White, [7]S.N. White, [13]D. Winter, [28]S. Wolin, [23,42]C.P. Wong, [7]C.L. Woody, [1]R.M. Wright, [8]Y. Wu, [12,59]M. Wysocki, [58]B. Xia, [56]C. Xu, [76]Q. Xu, [23]L. Xue, [71]S. Yalcin, [11,64,65,71]Y.L. Yamaguchi, [75]H. Yamamoto, [25]K. Yamaura, [28]R. Yang, [27]A. Yanovich, [12]P. Yin, [23]J. Ying, [64,65]S. Yokkaichi, [36,65]J.H. Yoo, [69]I. Yoon, [42,61]Z. You, [59]G.R. Young, [40,55]I. Younus, [56,61]H. Yu, [37]I.E. Yushmanov, [13]W.A. Zajc, [6]A. Zelenski, [31]Y. Zhai, [67]S. Zharko, [10]S. Zhou, [8]L. Zou

[1]Abilene Christian University, Abilene, Texas 79699, USA
[2]Department of Physics, Augustana University, Sioux Falls, South Dakota 57197, USA
[3]Department of Physics, Banaras Hindu University, Varanasi 221005, India
[4]Bhabha Atomic Research Centre, Bombay 400 085, India
[5]Baruch College, City University of New York, New York, New York, 10010 USA

[6]Collider-Accelerator Department, Brookhaven National Laboratory, Upton, New York 11973-5000, USA
[7]Physics Department, Brookhaven National Laboratory, Upton, New York 11973-5000, USA
[8]University of California-Riverside, Riverside, California 92521, USA
[9]Charles University, Ovocný trh 5, Praha 1, 116 36, Prague, Czech Republic
[10]Science and Technology on Nuclear Data Laboratory, China Institute of Atomic Energy, Beijing 102413, People's Republic of China
[11]Center for Nuclear Study, Graduate School of Science, University of Tokyo, 7-3-1 Hongo, Bunkyo, Tokyo 113-0033, Japan
[12]University of Colorado, Boulder, Colorado 80309, USA
[13]Columbia University, New York, New York 10027 and Nevis Laboratories, Irvington, New York 10533, USA
[14]Czech Technical University, Zikova 4, 166 36 Prague 6, Czech Republic
[15]Dapnia, CEA Saclay, F-91191, Gif-sur-Yvette, France
[16]Debrecen University, H-4010 Debrecen, Egyetem tér 1, Hungary
[17]ELTE, Eötvös Loránd University, H-1117 Budapest, Pázmány P. s. 1/A, Hungary
[18]Eszterházy Károly University, Károly Róbert Campus, H-3200 Gyöngyös, Mátrai út 36, Hungary
[19]Ewha Womans University, Seoul 120-750, Korea
[20]Florida A&M University, Tallahassee, FL 32307, USA
[21]Florida Institute of Technology, Melbourne, Florida 32901, USA
[22]Florida State University, Tallahassee, Florida 32306, USA
[23]Georgia State University, Atlanta, Georgia 30303, USA
[24]Hanyang University, Seoul 133-792, Korea
[25]Hiroshima University, Kagamiyama, Higashi-Hiroshima 739-8526, Japan
[26]Department of Physics and Astronomy, Howard University, Washington, DC 20059, USA
[27]IHEP Protvino, State Research Center of Russian Federation, Institute for High Energy Physics, Protvino, 142281, Russia
[28]University of Illinois at Urbana-Champaign, Urbana, Illinois 61801, USA
[29]Institute for Nuclear Research of the Russian Academy of Sciences, prospekt 60-letiya Oktyabrya 7a, Moscow 117312, Russia
[30]Institute of Physics, Academy of Sciences of the Czech Republic, Na Slovance 2, 182 21 Prague 8, Czech Republic
[31]Iowa State University, Ames, Iowa 50011, USA
[32]Advanced Science Research Center, Japan Atomic Energy Agency, 2-4 Shirakata Shirane, Tokai-mura, Naka-gun, Ibaraki-ken 319-1195, Japan
[33]Jeonbuk National University, Jeonju, 54896, Korea
[34]Helsinki Institute of Physics and University of Jyväskylä, P.O.Box 35, FI-40014 Jyväskylä, Finland
[35]KEK, High Energy Accelerator Research Organization, Tsukuba, Ibaraki 305-0801, Japan
[36]Korea University, Seoul 02841, Korea
[37]National Research Center "Kurchatov Institute", Moscow, 123098 Russia

[38]Kyoto University, Kyoto 606-8502, Japan
[39]Laboratoire Leprince-Ringuet, Ecole Polytechnique, CNRS-IN2P3, Route de Saclay, F-91128, Palaiseau, France
[40]Physics Department, Lahore University of Management Sciences, Lahore 54792, Pakistan
[41]Lawrence Livermore National Laboratory, Livermore, California 94550, USA
[42]Los Alamos National Laboratory, Los Alamos, New Mexico 87545, USA
[43]LPC, Université Blaise Pascal, CNRS-IN2P3, Clermont-Fd, 63177 Aubiere Cedex, France
[44]Department of Physics, Lund University, Box 118, SE-221 00 Lund, Sweden
[45]IPNL, CNRS/IN2P3, Univ Lyon, UniversitÃ© Lyon 1, F-69622, Villeurbanne, France
[46]University of Maryland, College Park, Maryland 20742, USA
[47]Department of Physics, University of Massachusetts, Amherst, Massachusetts 01003-9337, USA
[48]Department of Physics, University of Michigan, Ann Arbor, Michigan 48109-1040, USA
[49]Institut für Kernphysik, University of Münster, D-48149 Münster, Germany
[50]Muhlenberg College, Allentown, Pennsylvania 18104-5586, USA
[51]Myongji University, Yongin, Kyonggido 449-728, Korea
[52]Nagasaki Institute of Applied Science, Nagasaki-shi, Nagasaki 851-0193, Japan
[53]Nara Women's University, Kita-uoya Nishi-machi Nara 630-8506, Japan
[54]National Research Nuclear University, MEPhI, Moscow Engineering Physics Institute, Moscow, 115409, Russia
[55]University of New Mexico, Albuquerque, New Mexico 87131, USA
[56]New Mexico State University, Las Cruces, New Mexico 88003, USA
[57]Physics and Astronomy Department, University of North Carolina at Greensboro, Greensboro, North Carolina 27412, USA
[58]Department of Physics and Astronomy, Ohio University, Athens, Ohio 45701, USA
[59]Oak Ridge National Laboratory, Oak Ridge, Tennessee 37831, USA
[60]IPN-Orsay, Univ. Paris-Sud, CNRS/IN2P3, Université Paris-Saclay, BP1, F-91406, Orsay, France
[61]Peking University, Beijing 100871, People's Republic of China
[62]PNPI, Petersburg Nuclear Physics Institute, Gatchina, Leningrad region, 188300, Russia
[63]Pusan National University, Pusan 46241, Korea
[64]RIKEN Nishina Center for Accelerator-Based Science, Wako, Saitama 351-0198, Japan
[65]RIKEN BNL Research Center, Brookhaven National Laboratory, Upton, New York 11973-5000, USA
[66]Physics Department, Rikkyo University, 3-34-1 Nishi-Ikebukuro, Toshima, Tokyo 171-8501, Japan
[67]Saint Petersburg State Polytechnic University, St. Petersburg, 195251 Russia

[68]Universidade de São Paulo, Instituto de Física, Caixa Postal 66318, São Paulo CEP05315-970, Brazil
[69]Department of Physics and Astronomy, Seoul National University, Seoul 151-742, Korea
[70]Chemistry Department, Stony Brook University, SUNY, Stony Brook, New York 11794-3400, USA
[71]Department of Physics and Astronomy, Stony Brook University, SUNY, Stony Brook, New York 11794-3800, USA
[72]Sungkyunkwan University, Suwon, 440-746, Korea
[73]University of Tennessee, Knoxville, Tennessee 37996, USA
[74]Department of Physics, Tokyo Institute of Technology, Oh-okayama, Meguro, Tokyo 152-8551, Japan
[75]Tomonaga Center for the History of the Universe, University of Tsukuba, Tsukuba, Ibaraki 305, Japan
[76]Vanderbilt University, Nashville, Tennessee 37235, USA
[77]Waseda University, Advanced Research Institute for Science and Engineering, 17 Kikui-cho, Shinjuku-ku, Tokyo 162-0044, Japan
[78]Weizmann Institute, Rehovot 76100, Israel
[79]Institute for Particle and Nuclear Physics, Wigner Research Centre for Physics, Hungarian Academy of Sciences (Wigner RCP, RMKI) H-1525 Budapest 114, POBox 49, Budapest, Hungary
[80]Yonsei University, IPAP, Seoul 120-749, Korea
[81]Department of Physics, Faculty of Science, University of Zagreb, Bijenička c. 32 HR-10002 Zagreb, Croatia
*Spokesperson
**Deceased

STAR Collaboration

J. Adam[6], L. Adamczyk[2], J. R. Adams[39], J. K. Adkins[30], G. Agakishiev[28], M. M. Aggarwal[40], Z. Ahammed[59], I. Alekseev[3,35], D. M. Anderson[53], A. Aparin[28], E. C. Aschenauer[6], M. U. Ashraf[11], F. G. Atetalla[29], A. Attri[40], G. S. Averichev[28], V. Bairathi[22], K. Barish[10], A. Behera[51], R. Bellwied[20], A. Bhasin[27], J. Bielcik[14], J. Bielcikova[38], L. C. Bland[6], I. G. Bordyuzhin[3], J. D. Brandenburg[48,6], A. V. Brandin[35], J. Butterworth[44], H. Caines[62], M. Calderón de la Barca Sánchez[8], D. Cebra[8], I. Chakaberia[29,6], P. Chaloupka[14], B. K. Chan[9], F-H. Chang[37], Z. Chang[6], N. Chankova-Bunzarova[28], A. Chatterjee[11], D. Chen[10], J. H. Chen[18], X. Chen[47], Z. Chen[48], J. Cheng[55], M. Cherney[13], M. Chevalier[10], S. Choudhury[18], W. Christie[6], H. J. Crawford[7], M. Csanád[16], M. Daugherity[1], T. G. Dedovich[28], I. M. Deppner[19], A. A. Derevschikov[42], L. Didenko[6], X. Dong[31], J. L. Drachenberg[1], J. C. Dunlop[6], T. Edmonds[43], N. Elsey[61], J. Engelage[7], G. Eppley[44], R. Esha[51], S. Esumi[56], O. Evdokimov[12], A. Ewigleben[32], O. Eyser[6], R. Fatemi[30], S. Fazio[6], P. Federic[38], J. Fedorisin[28], C. J. Feng[37], Y. Feng[43], P. Filip[28], E. Finch[50], Y. Fisyak[6], A. Francisco[62], L. Fulek[2], C. A. Gagliardi[53], T. Galatyuk[15], F. Geurts[44], A. Gibson[58], K. Gopal[23], D. Grosnick[58], W. Guryn[6], A. I. Hamad[29], A. Hamed[5], J. W. Harris[62], S. He[11], W. He[18], X. He[26], S. Heppelmann[8], S. Heppelmann[41], N. Herrmann[19], E. Hoffman[20], L. Holub[14], Y. Hong[31], S. Horvat[62], Y. Hu[18], H. Z. Huang[9], S. L. Huang[51], T. Huang[37], X. Huang[55], T. J. Humanic[39], P. Huo[51], G. Igo[9], D. Isenhower[1], W. W. Jacobs[25], C. Jena[23], A. Jentsch[6], Y. JI[47], J. Jia[6,51], K. Jiang[47], S. Jowzaee[61], X. Ju[47], E. G. Judd[7], S. Kabana[29], M. L. Kabir[10], S. Kagamaster[32], D. Kalinkin[25], K. Kang[55], D. Kapukchyan[10], K. Kauder[6], H. W. Ke[6], D. Keane[29], A. Kechechyan[28], M. Kelsey[31], Y. V. Khyzhniak[35], D. P. Kikoła [60], C. Kim[10], B. Kimelman[8], D. Kincses[16], T. A. Kinghorn[8], I. Kisel[17], A. Kiselev[6], A. Kisiel[60], M. Kocan[14], L. Kochenda[35], L. K. Kosarzewski[14], L. Kramarik[14], P. Kravtsov[35], K. Krueger[4], N. Kulathunga Mudiyanselage[20], L. Kumar[40], R. Kunnawalkam Elayavalli[61], J. H. Kwasizur[25], R. Lacey[51], S. Lan[11], J. M. Landgraf[6], J. Lauret[6], A. Lebedev[6], R. Lednicky[28], J. H. Lee[6], Y. H. Leung[31], C. Li[47], W. Li[44], W. Li[49], X. Li[47], Y. Li[55], Y. Liang[29], R. Licenik[38], T. Lin[53], Y. Lin[11], M. A. Lisa[39], F. Liu[11], H. Liu[25], P. Liu[51], P. Liu[49], T. Liu[62], X. Liu[39], Y. Liu[53], Z. Liu[47], T. Ljubicic[6], W. J. Llope[61], R. S. Longacre[6], N. S. Lukow[52], S. Luo[12], X. Luo[11], G. L. Ma[49], L. Ma[18], R. Ma[6], Y. G. Ma[49], N. Magdy[12], R. Majka[62], D. Mallick[36], S. Margetis[29], C. Markert[54], H. S. Matis[31], J. A. Mazer[45], N. G. Minaev[42], S. Mioduszewski[53], B. Mohanty[36], I. Mooney[61], Z. Moravcova[14], D. A. Morozov[42], M. Nagy[16], J. D. Nam[52], Md. Nasim[22], K. Nayak[11], D. Neff[9], J. M. Nelson[7], D. B. Nemes[62], M. Nie[48], G. Nigmatkulov[35], T. Niida[56], L. V. Nogach[42], T. Nonaka[11], G. Odyniec[31], A. Ogawa[6], S. Oh[62], V. A. Okorokov[35], B. S. Page[6], R. Pak[6], A. Pandav[36], Y. Panebratsev[28], B. Pawlik[2], D. Pawlowska[60], H. Pei[11], C. Perkins[7], L. Pinsky[20], R. L. Pintér[16], J. Pluta[60], J. Porter[31], M. Posik[52], N. K. Pruthi[40], M. Przybycien[2], J. Putschke[61], H. Qiu[26], A. Quintero[52], S. K. Radhakrishnan[29], S. Ramachandran[30], R. L. Ray[54], R. Reed[32], H. G. Ritter[31], J. B. Roberts[44], O. V. Rogachevskiy[28], J. L. Romero[8], L. Ruan[6], J. Rusnak[38], N. R. Sahoo[48], H. Sako[56], S. Salur[45], J. Sandweiss[62], S. Sato[56], W. B. Schmidke[6], N. Schmitz[33],

B. R. Schweid[51], F. Seck[15], J. Seger[13], M. Sergeeva[9], R. Seto[10], P. Seyboth[33], N. Shah[24], E. Shahaliev[28], P. V. Shanmuganathan[6], M. Shao[47], F. Shen[48], W. Q. Shen[49], S. S. Shi[11], Q. Y. Shou[49], E. P. Sichtermann[31], R. Sikora[2], M. Simko[38], J. Singh[40], S. Singha[26], N. Smirnov[62], W. Solyst[25], P. Sorensen[6], H. M. Spinka[4], B. Srivastava[43], T. D. S. Stanislaus[58], M. Stefaniak[60], D. J. Stewart[62], M. Strikhanov[35], B. Stringfellow[43], A. A. P. Suaide[46], M. Sumbera[38], B. Summa[41], X. M. Sun[11], Y. Sun[47], Y. Sun[21], B. Surrow[52], D. N. Svirida[3], P. Szymanski[60], A. H. Tang[6], Z. Tang[47], A. Taranenko[35], T. Tarnowsky[34], J. H. Thomas[31], A. R. Timmins[20], D. Tlusty[13], M. Tokarev[28], C. A. Tomkiel[32], S. Trentalange[9], R. E. Tribble[53], P. Tribedy[6], S. K. Tripathy[16], O. D. Tsai[9], Z. Tu[6], T. Ullrich[6], D. G. Underwood[4], I. Upsal[48,6], G. Van Buren[6], J. Vanek[38], A. N. Vasiliev[42], I. Vassiliev[17], F. Videbæk[6], S. Vokal[28], S. A. Voloshin[61], F. Wang[43], G. Wang[9], J. S. Wang[21], P. Wang[47], Y. Wang[11], Y. Wang[55], Z. Wang[48], J. C. Webb[6], P. C. Weidenkaff[19], L. Wen[9], G. D. Westfall[34], H. Wieman[31], S. W. Wissink[25], R. Witt[57], Y. Wu[10], Z. G. Xiao[55], G. Xie[31], W. Xie[43], H. Xu[21], N. Xu[31], Q. H. Xu[48], Y. F. Xu[49], Y. Xu[48], Z. Xu[6], Z. Xu[9], C. Yang[48], Q. Yang[48], S. Yang[6], Y. Yang[37], Z. Yang[11], Z. Ye[44], Z. Ye[12], L. Yi[48], K. Yip[6], H. Zbroszczyk[60], W. Zha[47], D. Zhang[11], S. Zhang[47], S. Zhang[49], X. P. Zhang[55], Y. Zhang[47], Y. Zhang[11], Z. J. Zhang[37], Z. Zhang[6], J. Zhao[43], C. Zhong[49], C. Zhou[49], X. Zhu[55], Z. Zhu[48], M. Zurek[31], M. Zyzak[17]

[1]Abilene Christian University, Abilene, Texas 79699
[2]AGH University of Science and Technology, FPACS, Cracow 30-059, Poland
[3]Alikhanov Institute for Theoretical and Experimental Physics NRC "Kurchatov Institute", Moscow 117218, Russia
[4]Argonne National Laboratory, Argonne, Illinois 60439
[5]American University of Cairo, New Cairo 11835, New Cairo, Egypt
[6]Brookhaven National Laboratory, Upton, New York 11973
[7]University of California, Berkeley, California 94720
[8]University of California, Davis, California 95616
[9]University of California, Los Angeles, California 90095
[10]University of California, Riverside, California 92521
[11]Central China Normal University, Wuhan, Hubei 430079
[12]University of Illinois at Chicago, Chicago, Illinois 60607
[13]Creighton University, Omaha, Nebraska 68178
[14]Czech Technical University in Prague, FNSPE, Prague 115 19, Czech Republic
[15]Technische Universität Darmstadt, Darmstadt 64289, Germany
[16]ELTE Eötvös Loránd University, Budapest, Hungary H-1117
[17]Frankfurt Institute for Advanced Studies FIAS, Frankfurt 60438, Germany
[18]Fudan University, Shanghai, 200433
[19]University of Heidelberg, Heidelberg 69120, Germany
[20]University of Houston, Houston, Texas 77204
[21]Huzhou University, Huzhou, Zhejiang 313000
[22]Indian Institute of Science Education and Research (IISER), Berhampur 760010 , India

[23]Indian Institute of Science Education and Research (IISER) Tirupati, Tirupati 517507, India
[24]Indian Institute Technology, Patna, Bihar 801106, India
[25]Indiana University, Bloomington, Indiana 47408
[26]Institute of Modern Physics, Chinese Academy of Sciences, Lanzhou, Gansu 730000
[27]University of Jammu, Jammu 180001, India
[28]Joint Institute for Nuclear Research, Dubna 141 980, Russia
[29]Kent State University, Kent, Ohio 44242
[30]University of Kentucky, Lexington, Kentucky 40506-0055
[31]Lawrence Berkeley National Laboratory, Berkeley, California 94720
[32]Lehigh University, Bethlehem, Pennsylvania 18015
[33]Max-Planck-Institut für Physik, Munich 80805, Germany
[34]Michigan State University, East Lansing, Michigan 48824
[35]National Research Nuclear University MEPhI, Moscow 115409, Russia
[36]National Institute of Science Education and Research, HBNI, Jatni 752050, India
[37]National Cheng Kung University, Tainan 70101
[38]Nuclear Physics Institute of the CAS, Rez 250 68, Czech Republic
[39]Ohio State University, Columbus, Ohio 43210
[40]Panjab University, Chandigarh 160014, India
[41]Pennsylvania State University, University Park, Pennsylvania 16802
[42]NRC "Kurchatov Institute", Institute of High Energy Physics, Protvino 142281, Russia
[43]Purdue University, West Lafayette, Indiana 47907
[44]Rice University, Houston, Texas 77251
[45]Rutgers University, Piscataway, New Jersey 08854
[46]Universidade de São Paulo, São Paulo, Brazil 05314-970
[47]University of Science and Technology of China, Hefei, Anhui 230026
[48]Shandong University, Qingdao, Shandong 266237
[49]Shanghai Institute of Applied Physics, Chinese Academy of Sciences, Shanghai 201800
[50]Southern Connecticut State University, New Haven, Connecticut 06515
[51]State University of New York, Stony Brook, New York 11794
[52]Temple University, Philadelphia, Pennsylvania 19122
[53]Texas A&M University, College Station, Texas 77843
[54]University of Texas, Austin, Texas 78712
[55]Tsinghua University, Beijing 100084
[56]University of Tsukuba, Tsukuba, Ibaraki 305-8571, Japan
[57]United States Naval Academy, Annapolis, Maryland 21402
[58]Valparaiso University, Valparaiso, Indiana 46383
[59]Variable Energy Cyclotron Centre, Kolkata 700064, India
[60]Warsaw University of Technology, Warsaw 00-661, Poland
[61]Wayne State University, Detroit, Michigan 48201
[62]Yale University, New Haven, Connecticut 06520

Acknowledgments

We thank the staff of the Collider-Accelerator and Physics Departments at Brookhaven National Laboratory and the staff of the other PHENIX participating institutions for their vital contributions. We acknowledge support from the Office of Nuclear Physics in the Office of Science of the Department of Energy, the National Science Foundation, Abilene Christian University Research Council, Research Foundation of SUNY, and Dean of the College of Arts and Sciences, Vanderbilt University (U.S.A), Ministry of Education, Culture, Sports, Science, and Technology and the Japan Society for the Promotion of Science (Japan), Conselho Nacional de Desenvolvimento Científico e Tecnológico and Fundação de Amparo à Pesquisa do Estado de São Paulo (Brazil), Natural Science Foundation of China (People's Republic of China), Croatian Science Foundation and Ministry of Science and Education (Croatia), Ministry of Education, Youth and Sports (Czech Republic), Centre National de la Recherche Scientifique, Commissariat à l'Énergie Atomique, and Institut National de Physique Nucléaire et de Physique des Particules (France), Bundesministerium für Bildung und Forschung, Deutscher Akademischer Austausch Dienst, and Alexander von Humboldt Stiftung (Germany), J. Bolyai Research Scholarship, EFOP, the New National Excellence Program (ÚNKP), NKFIH, and OTKA (Hungary), Department of Atomic Energy and Department of Science and Technology (India), Israel Science Foundation (Israel), Basic Science Research and SRC(CENuM) Programs through NRF funded by the Ministry of Education and the Ministry of Science and ICT (Korea). Physics Department, Lahore University of Management Sciences (Pakistan), Ministry of Education and Science, Russian Academy of Sciences, Federal Agency of Atomic Energy (Russia), VR and Wallenberg Foundation (Sweden), the U.S. Civilian Research and Development Foundation for the Independent States of the Former Soviet Union, the Hungarian American Enterprise Scholarship Fund, the US-Hungarian Fulbright Foundation, and the US-Israel Binational Science Foundation.

Author Index

D. Elia et al. (eds.), *The XVIII International Conference on Strangeness in Quark Matter (SQM 2019)*, Springer Proceedings in Physics 250,
https://doi.org/10.1007/978-3-030-53448-6

Subject Index

D. Elia et al. (eds.), *The XVIII International Conference on Strangeness in Quark Matter (SQM 2019)*, Springer Proceedings in Physics 250,
https://doi.org/10.1007/978-3-030-53448-6

MIX
Papier aus verantwortungsvollen Quellen
Paper from responsible sources
FSC® C105338

Printed by Books on Demand, Germany